Heribert Insam · Nuntavun Riddech · Susanne Klammer (Eds.)

Microbiology of Composting

Springer

Berlin
Heidelberg
New York
Barcelona
Hong Kong
London
Milan
Paris
Tokyo

H. Insam · N. Riddech · S. Klammer (Eds.)

Microbiology of Composting

With 230 Figures and 126 Tables

Springer

Univ. Doz. Dr. Heribert Insam
Mag. Susanne Klammer
Universität Innsbruck
Institut für Mikrobiologie
Technikerstraße 25
6020 Innsbruck
Austria

http://www.microbial-ecology.at

Nuntavun Riddech, M. Sc.
Khon Kaen University
Faculty of Science
Department of Microbiology
40002 Khon Kaen
Thailand

Gedruckt mit Unterstützung des Bundesministeriums für Wissenschaft und Verkehr in Wien

ISBN 3-540-67568-X Springer-Verlag Berlin Heidelberg New York

Die Deutsche Bibliothek - CIP-Einheitsaufnahme

Microbiology of composting : with 126 tables / Heribert Insam ... (ed.) -
Berlin ; Heidelberg ; New York ; Barcelona ; Hong Kong ; London ; Milan ;
Paris , Tokyo : Springer, 2002
ISBN 3-540-67568-X

Springer-Verlag Berlin Heidelberg New York
a member of BertelsmannSpringer Science+Business Media GmbH
http://www.springer.de

Cover design: Design & Production, Heidelberg
Cover photograph: Erwin Schreckensperger
Typesetting: Camera ready by authors
SPIN 10752845 31/3130 - 5 4 3 2 1 0 - Printed on acid-free paper

Preface

Historically, it was the scarcity of new materials that has been the driving factor for recycling of used materials. In the agricultural sector, it was the lack of external nutrient sources or problems with soil degradation that has forced people to care for a functioning cycling of organic matter, hitting two birds with one stone: nutrient supply and long-term soil sustainability.

Now, in many countries traditional recycling seems to be forgotten, or its economical value is disputed. Centralised dumps or incineration plants are often the simple answer to growing waste problems. Only in a few countries, many of them in Europe, where population density is high and the value of preserved nature, along with sustainable agriculture is increasingly demanded, environmentally sound recycling programmes became a question of political culture.

The approaches are very different and range from treatment of bulk wastes to source-separated collection of organic wastes. The latter yields the best quality substrate for any further treatment, but also needs a considerable level of understanding by the people. Also the approaches for treatment vary, both in their size and technology. Be it small or big units, or be it anaerobic digestion or composting, if precaution is taken for very few fundamentals like quality of input material, aeration and water content, the process will work without any further knowledge about it. For this reason, research to improve the process has long been neglected. In particular, little attention has been given to the organisms that actually do the job, the microorganisms.

This seems to be changing. Microbiologists and microbial ecologists have a variety of new tools at hand, and habitats that are so extremely variable in both time and space like compost heaps are increasingly attracting scientists. This is the chance to learn more about these processes and the microorganisms involved.

In this book, 51 research papers are compiled that give up-to-date insight into processes, the organisms involved, and different kinds of problems associated with compost production. The book is a compilation of the presentations given at the International Conference 'Microbiology of Composting' held in Innsbruck, Austria, from Oct. 18-20 (2000) which was attended by 300 scientists from 42 countries. Further papers, namely those related to compost application in agriculture and forestry, will be published in a special issue of the soil science journal GEODERMA.

The editors wish to thank the members of the Scientific Committee, H. Hoitink, USA, J. Garland, USA, M. Itävaara, Finland, M. DeBertoldi, Italy, C.A. Reddy, USA, S. Pillai, USA, C.C. Tebbe, Germany and C. Campbell, UK as well as numerous other reviewers for their help in editing this book. The editors also thank Johannes Feurle and particularly Brigitte Knapp for organisational and editorial assistance, as well as Springer Verlag, namely Anette Lindqvist and Christiane Glier, for the good cooperation during the production process. Further thanks are expressed to the University of Innsbruck, the Austrian Society for Soil Biology, the European Commission, the Federal Ministry for Forestry, Agriculture and Environment, the Federal Ministry of Science, the Province of Tyrol, the City of Innsbruck, Austrian Airlines Group, Thöni Industries, Linde Inc., Komptech Inc., TEG Environmental plc., who sponsored the meeting Microbiology of Composting. The production of this book was made possible by a grant of the Austrian Ministry of Science.

Heribert Insam, Nuntavun Riddech, Susanne Klammer,
Editors

Contents

Microbial Communities

How Resilient Are Microbial Communities to Temperature Changes During Composting? 3
J.N. Cooper, J.G. Anderson and C.D. Campbell

Survey of Fungal Diversity in Mushroom Compost Using Sequences of PCR-Amplified Genes Encoding 18S Ribosomal RNA 17
K. Ivors, D. Beyer, P. Wuest and S. Kang

Bacterial Community Structure During Yard Trimmings Composting 25
F.C. Michel Jr., T.J. Marsh and C.A. Reddy

Characterisation of Microbial Communities During Composting of Organic Wastes 43
N. Riddech, S. Klammer and H. Insam

Microbial Succession During Composting of Source-Separated Urban Organic Household Waste Under Different Initial Temperature Conditions 53
I. Sundh and S. Rönn

Bacterial Diversity in Livestock Manure Composts as Characterized by Terminal Restriction Fragment Length Polymorphisms (T-RLFP) of PCR-amplified 16S rRNA Gene Sequences 65
S. M. Tiquia and F.C. Michel Jr.

Amplified 16S Ribosomal DNA Restriction Analysis of Microbial Community Structure During Rapid Degradation of a Biopolymer, PHA, by Composting 83
K. Uchiyama, T. Suzuki, H. Tatsumi, H. Kanetake and S. Shioya

Comparative Investigation of Vermicompost Microbial Communities 99
N.V. Verkhovtseva, G.A. Osipov, T.N. Bolysheva, V.A. Kasatikov, N.V. Kuzmina, E.J. Antsiferova and A.S. Alexeeva

Processes and Controls

Heat Production During Thermophilic Decomposition of Municipal Wastes in the Dano-System Composting Plant 111
J.E. Dziejowski and J. Kazanowska

Composting of Different Horticultural Wastes: Effect of Fungal Inoculation 119
M.A. Elorrieta, M.J. López, F. Suárez-Estrella, M.C. Vargas-García and J. Moreno

Backyard Composting: General Considerations and a Case Study 133
P. Illmer

N-Dynamics During Composting – Overview and Experimental Results 143
I. Körner and R. Stegmann

Unsuitability of Anaerobic Compost from Solid Substrate Anaerobic Digestion as a Soil Amendment 155
H.M. Poggi-Varaldo, E. Gómez-Cisneros, R. Rodríguez-Vázquez, J. Trejo-Espino and N. Rinderknecht-Seijas

Pile Composting of Two-phase Centrifuged Olive Husks: Bioindicators of the Process 165
G. Ranalli, P. Principi, M. Zucchi, F. da Borso, L. Catalano and C. Sorlini

Organic Acids as a Decisive Limitation to Process Dynamics During Composting of Organic Matter 177
T. Reinhardt

Effects of Interrupted Air Supply on the Composting Process – Composition of Volatile Organic Acids 189
M. Robertsson

Reduction of Ammonia Emission and Waste Gas Volume by Composting at Low Oxygen Pressure 203
D.P. Rudrum, A.H.M. Veeken, V. de Wilde, W.H. Rulkens and B.V.M. Hamelers

Review of Compost Process-Control for Product Function 217
R.A.K. Szmidt

Using Agricultural Wastes for *Tricholoma crassum* (Berk.) Sacc. Production 231
N. Teaumroong, W. Sattayapisut, T. Teekachunhatean and N. Boonkerd

Microbial Transformation of Nitrogen During Composting 237
S.M. Tiquia

Effect of Additives on the Nitrification-Denitrification Activities During Composting of Chicken Manure 247
H. Yulipryanto, P. Morand, P. Robin, G. Tricot and C. Aubert

Biodegradability

Biodegradability Study on Films for Packaging Based on Isotactic Polypropylene Modified with Natural Terpene Resins 265
S. Cimmino, E. D'Alma, E. Ionata, F. La Cara and C. Silvestre

Isolation and Characterization of Thermophilic Microorganisms Able to Grow on Cellulose Acetate 273
F. Degli-Innocenti, G. Goglino, G. Bellia, M. Tosin, P. Monciardini and L. Cavaletti

PCB's Biotransformation by a White-Rot Fungus Under Composting and Liquid Culture Conditions 287
G. Ruiz-Aguilar, J.M. Fernández-Sánchez, R. Rodríguez-Vázquez, H.M. Poggi-Varaldo, F. Esparza-García and R. Vázquez-Duhalt

Tests on Composting of Degradable Polyethylene in Respect to the Quality of the End-Product Compost 299
B. Raninger, G. Steiner, D.M. Wiles and C.W.J. Hare

Microbial Degradation of Sulfonylurea Herbicides:Chlorsulfuron and Metsulfuron-Methyl 309
E. Zanardini, A. Arnoldi, G. Boschin, A. D'Agostina, M. Negri and C. Sorlini

Maturity Testing

Hydrogen Peroxide Effects on Composting and Its Use in Assessing the Degree of Maturity 323
C. Balis, V. Tassiopoulou and K. Lasaridi

Plant Performance in Relation to Oxygen Depletion, CO_2-Rate and Volatile Organic Acids in Container Media Composts of Varying Maturity 335
W.F. Brinton and E. Evans

Microbiological and Chemical Characterisation of Composts at Different Levels of Maturity, with Evaluation of Phytotoxicity and Enzymatic Activities 345
A.C. Cunha Queda, G. Vallini, M. Agnolucci, C.A. Coelho, L. Campos and R.B. de Sousa

Monitoring of a Composting Process: Thermal Stability of Raw Materials and Products 357
M.T. Dell´Abate and F. Titarelli

Compost Maturity – Problems Associated with Testing 373
M. Itävaara, O. Venelampi, M. Vikman and A. Kapanen

Use of CLPP for Evaluating the Role of Different Organic Materials in Composting 383
F. Pinzari, F. Tittarelli, A. Benedetti and H. Insam

Evaluation of Organic Matter Stability During the Composting Process of Agroindustrial Wastes 397
F. Tittarelli, A. Trinchera, F. Intrigliolo and A. Benedetti

Characterization of Organic Substances in Stabilized Composts of Rest Wastes 407
A. Zach

Application and Environmental Impact

Composting of *Posidonia oceanica* and Its Use in Agriculture 425
P. Castaldi and P. Melis

Practical Use of Quality Compost for Plant Health and Vitality Improvement 435
J.G. Fuchs

Environmental Impacts of Cattle Manure Composting 445
P.A. Gibbs, R.J. Parkinson, T.H. Misselbrook and S. Burchett

Agronomic Value and Environmental Impacts of Urban Composts Used in Agriculture 457
S. Houot, D. Clergeot, J. Michelin, C. Francou, S. Bourgeois, G. Caria and H. Ciesielski

Composting in the Framework of the EU Landfill Directive 473
M. Kranert, A. Behnsen, A. Schultheis and D. Steinbach

Occurence of *Aspergillus fumigatus* in a Compost Polluted with Heavy Metals 487
E. López Errasquin, B. Patiño, R.M. Fernández and C. Vázquez

Important Aspects of Biowaste Collection and Composting in Nigeria 495
K.T. Raheem, K.I. Hänninen and M. Odele

Plant and Human Pathogens

Use of Actinobacteria in Composting of Sheep Litter 505
S. Baccella, A.L. Botta, S. Manfroni, A. Trinchera, P. Imperiale, A. Benedeti, M. Del Gallo and A. Lepidi

Methods for Health Risk Assessment by *Clostridium botulinum* in Biocompost 517
H. Böhnel, B.-H. Briese and F. Gessler

The Fate of Plant Pathogens and Seeds During Backyard Composting of Vegetable, Fruit and Garden Waste 527
J. Ryckeboer, S. Cops and J. Coosemans

Survival of Phytopathogen Viruses During Semipilot-Scale Composting 539
F. Suárez-Estrella, M.J. López, M.A. Elorrieta, M.C. Vargas-García and J. Moreno

Air-Borne Emissions and Their Control

Composting Conditions Preventing the Development of Odorous Compounds 551
E. Binner, D. Grassinger and M. Humer

Odour Emissions During Yard Waste Composting: Effect of Turning Frequency 561
N. Defoer and H. Van Langenhove

Imission of Microorganisms from Composting Facilities 571
P. Kämpfer, C. Jureit, A. Albrecht and A. Neef

Molecular Identification of Airborne Microorganisms from Composting Facilities 585
A. Neef and P. Kämpfer

Bioaerosols and Public Health 595
S.D. Pillai

Passively Aerated Composting of Straw-Rich Organic Pig Manure 607
A. Veeken, V. de Wilde, G. Szanto and B. Hamelers

Index 623

Microbial Communities

How Resilient Are Microbial Communities to Temperature Changes During Composting?

J.N. Cooper[1,2,3], J.G. Anderson[3] and C.D. Campbell[1]

Abstract. The resilience of compost systems to perturbation is usually attributed to the highly active and diverse microbial population. Composting is characterized by distinct temperature changes that are associated with a succession of microbial communities adapted to the prevailing temperature. The temperature fluctuations are spatially and temporally variable. The transition between mesophilic and thermophilic communities can result in loss of degradation efficiency. We tested the resilience of microbial communities to temperature fluctuations in laboratory reactors containing tree bark and pulverized wood amended with pot ale liquor, a waste by-product of the whisky manufacturing process. The laboratory reactors were operated in fixed temperature mode at 50°C, 35°C, and 20°C, in three replicates of each treatment, from which COD, suspended solids, BOD and pH of the resultant leachate were monitored on a daily basis. After running at fixed temperatures for 45 days, the reactors, initially operated at 50, 35 and 20°C, were switched to 20, 50 and 35°C, respectively. A further temperature switch was made after 90 days so that, at the conclusion of the experiment after 135 days, each triplicate set of incubators had experienced a 45-day period at each of the three temperature settings.

Microbial community structure, measured using phospholipid fatty acid analysis, showed that communities at different temperatures were quite distinct. The communities conditioned at low temperature adapted to the higher temperature imposed and were indistinguishable from other thermophilic populations.

In contrast, the thermophilic communities that had been conditioned at high temperature and switched to low temperature evolved a mesophilic community that was distinct from the mesophilic communities previously conditioned at low temperature. The loss in process efficiency after the thermophilic phase may therefore be limited by the recovery of surviving populations and/or by slow recolonisation from low-temperature zones.

---

[1]The Macaulay Institute, Craigiebuckler, Aberdeen, AB15 8QH, UK
[2]Highland Distillers, Highland Park Distillery, Holm Road, Kirkwall, Orkney, UK
[3]University of Strathclyde, Royal College Building, Glasgow, UK
e-mail: c.campbell@mluri.sari.ac.uk; Telephone: +44 (0)1224 318611
Fax: +44 (0)1224 311556

H. Insam, N. Riddech, S. Klammer (Eds.)
Microbiology of Composting

Introduction

Ever since the early work of Waksman et al. (1939) on manure composting it has been recognised that temperature and its interaction with chemical changes are the primary driving forces in the succession of microbial communities during the process. When operated as a batch process the effect of the availability of C and energy sources is confounded with temperature and it is difficult to separate the two. Our overall view of the microbial ecology of composting has not changed significantly since this early research and only now that cultivation independent methods are being applied are we finding new organisms (Blanc et al. 1997) and being able to quantify shifts in microbial community structure (Frostegård et al. 1996; Hellmann et al. 1997; Herrmann and Shann 1997; Campbell and Cooper 1998; Carpenter-Boggs et al. 1998; Klamer and Bååth 1998; Lei and Van der Ghenst 2000). At the same time, composting is being used for an increasingly wider variety of different organic materials and it is also being used in novel ways e.g. to treat contaminated soils, liquid and air (Campbell and Cooper 1998).

One example of a novel application of composting is the TAPP (thermogenic, aerobic, plug-flow, percolation; UK Patent Application No GB 2322623A) process (Campbell and Cooper 1998) devised to treat a liquid waste, Pot ale, from the whisky industry. As the remnant liquor of the first distillation in whisky production, pot ale has a solid content of 2~3% that mainly consists of dead yeast cells sterilised during the 6 h of boiling required to drive off the alcohol from the fermented wort. Pot ale liquor cannot be discharged directly to waters because of its high biological oxygen demand (BOD), low pH and Cu content. The approximate 80 or more malt whisky distilleries operating in Scotland today produce a volume of pot ale which, in terms of sewage effluent loading, represents a population equivalent of about 3 million persons. Pot ale is usually boiled down to a heavy syrup which is then used directly or indirectly as animal feed material Castle and Watson 1982. Today, this process is still used to a large degree in the absence of any alternative method, but with increasing oil prices and decreasing livestock outlets, such techniques are no longer cost-effective. This left an important problem for the industry to address: how can it dispose of pot ale in a way that is both economically viable and environmentally acceptable?

The TAPP process was devised to treat pot ale liquor by percolating the liquor through compost stacks made from wood and bark wafers. In this process, a mixture of pulverised wood and bark, in the proportions of 3:1, provide a matrix in which the breakdown of Pot ale solubles, as well as insoluble dead yeast, takes place by percolating compost stacks with pot ale. BOD levels of 25, 000 parts per million (ppm) can be reduced to 50 ppm. Chemical oxygen demand (COD) levels falling from 50, 000 ppm to 2, 000 ppm, and suspended solids (SS) brought down from ca. 8000 ppm to less than 100 ppm (Campbell and Cooper 1998). At the same time, the acidity of the displaced leachate is effectively neutralised. Pot ale is a strongly buffered, moderately acidic (pH 3~4) protein-rich liquid with a total nitrogen content of ca. 2500 ppm. The TAPP process gives rise to a weakly buffered leachate with a pH value of 6 or more; nitrogen levels are reduced by

90~95% (predominantly by denitrification), phosphorus levels by 50%, and copper, which is present in pot ale at ca. 10 ppm, is reduced to less than 100 ppb (parts per billion) in the leachate runoff, having been largely sequestered in the phenolic polymers of the bark (Vaughan et al. 1984). The TAPP process therefore offers a solution to this waste problem by combining the use of two substantial waste materials, wood debris and distillery effluent, to produce, on the one hand, a clean leachate and, on the other, a high-quality compost called pot ale solids compost (PASCO).

In the TAPP system, the wood waste is arranged in windrows of up to 1.7 m in height, and not more than 4 m wide at the top. Ideally, the stack cross-section should be a square: a configuration nearly achievable once the wood waste has been wetted and the sidewalls reshaped to a near-vertical slope. Ventilation by passive airflow has been found to be adequate, providing the above dimensions are not exceeded. Pot ale feedstock application can be done manually using a hose and bowser for pilot-scale operations, but for full industrial-scale operations, however, involving treatment of tens of thousands of litres daily, it has been necessary to employ automated irrigation methods. The apron on which the stacks are constructed is designed to isolate and collect leachate. Temperatures can be depressed where unprotected stacks are subjected to heavy continual rainfall in cold conditions. In addition, applications of pot ale every 2 days has been observed to cause an initial cooling, after which core temperatures rise after 2 h to maximal levels before falling slowly back over the course of the next 2 days until further pot ale irrigation, and the cycle is then repeated. Variation in temperature in the full-scale stacks can therefore be both rapid and large, spatially and temporally. In the full-scale system the process efficiency, expressed as percentage removal of initial BOD, sometimes dropped to 80% and the question arose whether this was due entirely to these wide variations in temperature.

In this chapter we report on the consequences of imposing specific temperature regimes for extended periods of time on contained incubation systems and specifically we sought to determine what was the optimal operational temperature for TAPP and how resilient were the microbial populations and leachate quality to interchanging temperature over extended periods of time. Microbial community structure was measured by analysis of the compost phopholipid fatty acid composition, which is a technique that has proved useful for this purpose in the past (Frostegård et al. 1996; Hellmann et al. 1997; Herrmann and Shann 1997; Campbell and Cooper 1998; Carpenter-Boggs et al. 1998; Klamer and Bååth 1998; Lei and Van der Ghenst 2000).

Methods

Laboratory Reactors and Their Operation

Nine cylindrical PVC incubation vessels, 20 cm in diameter and 30 cm in height giving a total volume of about 9 litres each were used in these experiments. These were constructed specifically for simulating the TAPP process. Their design, temperature monitoring and control mechanisms were based on those of Campbell et al. (1990 a,b) and Sikora et al. (1983) except that additional feed input ports and leachate collection ports were built at the top and bottom respectively, of the reactor vessel. The reactors were operated individually in fixed temperature mode under thermostatic control.

The feedstock, or 'pot ale', was obtained from Highland Park distillery, Orkney, Scotland. Approximately 8 l of woodwaste comprising 25% pine bark and 75% pulverised pine wood were placed in each of the nine reactors. These were then subjected to irrigation from the top by pot ale at the rate of 100 ml day^{-1}. The woodwaste had been weathered for several weeks prior to use so that it was close to a water-saturated state (69% moisture) at the beginning of the experiment.

The reactors were aerated at the rate of 100 ml min^{-1} from an on-line compressor, the air from which was water-saturated and temperature-equilibrated prior to use. Flowmeters (Flowbits, CT Platon Ltd, Basingstoke, UK) were fitted on-line to each of the nine incubators. Internal temperatures of the reactors were continuously monitored, leachate was collected from each vessel daily, and leachate volumes from each were recorded and analysed. Bark and pulverised wood were supplied by Bannerman Transport, Ross-shire, UK.

Experimental design

In this experiment each of the temperature regimes was carried out in triplicate. The pattern for each set of reactors was therefore as follows:

Set 1: 50 °C day 1~45, 20 °C day 46~90, 35 °C day 91~135.
Set 2: 35 °C day 1~45, 50 °C day 46~90, 20 °C day 91~135.
Set 3: 20 °C day 1~45, 35 °C day 46~90, 50 °C day 91~135.

Leachate pH, BOD, COD and Suspended Solids

Biological oxygen demand (BOD) and pH measurements were made using ELE portable dissolved oxygen and pH meters (ELE International Ltd., Hemel Hempstead, UK). BOD levels were measured using air-saturated seeded controls and checked against internal standards. Tests were carried out using standard methods (American Standard Methods for the Examination of Water and Wastewater 1992) including checks for spurious enhancement of oxygen depletion due to nitrification, performed using the nitrification inhibitor 2-chloro-6-(trichoro methyl) pyridine (TCMP). Only a slight increase in BOD was seen over the 5-day period where TCMP was used. Chemical Oxygen Demand (COD) was determined using the HACH equipment (CAMLAB Ltd., Cambridge, UK) using standard methods (American Standard Methods for the Examination of Water and Wastewater 1992). A 2 ml sample of appropriate strength was added to the digestion solution and heated to 150 °C for 2 h. Cooled samples were measured at 600 nm in a Helios epsilon spectrophotometer (Norlab Instruments Ltd, Dyce,

Aberdeen, UK). Sample sets were in all cases measured in conjunction with blank and standard samples. Total Suspended Solids, dried at 103-105°C, was carried out using glass fibre filter discs grade 934AH (Whatman International Ltd, Maidstone, UK) using standard methods (American Standard Methods for the Examination of Water and Wastewater 1992).

Substrate Physicochemical Analysis

Moisture content was determined by oven drying at 105 °C for 24 h and pH was measured in suspension of 1 g compost in 10 ml of water using a pH electrode. The friability of the bark and wood was measured by means of a Friabilimeter that was used to apply a gentle crushing action on oven-dried samples of composted wood/bark. The amount of friable material falling through the rotating sieve of the Friabilimeter was expressed as a percentage of the total mass treated. This equipment (Chapon et al. 1980) was devised originally to assess the degree of modification of malted barley (Cooper 1986) and can be used as an indicator of structural changes. The standard method (Cooper 1986) was adapted such that composted material (20 g) was screened over a 1-cm sieve prior to testing and the running time was lowered from 8 min to 1 minute. All samples were dried overnight in a vented oven at 105 °C prior to friability testing. Percentage friability of dried composted samples from the incubators was tested at the end of the 135-day experiment and on control samples of the starting material.

Phospholipid Fatty Acid Analysis (PLFA)

Lipid extraction and PLFA (phospholipid fatty acid) analyses were performed (Frostegård et al. 1996) using the modified Bligh and Dyer method (Bligh and Dyer 1959). Briefly, 2.0 g (freeze-dried sample) was extracted with a chloroform-methanol-citrate butter mixture (1:2:0.8), and the phospholipids were separated from other lipids on a silicic acid column. The phospholipids were subjected to a mild-alkali methanolysis and the resulting fatty acid methyl esters were separated by gas chromatography. Individual fatty acids were designated in terms of total number of carbon atoms,: number of double bonds, followed by the position of the double bond from the methyl end of the molecule. The prefixes a and i indicate anteiso and iso branching, respectively, Me the presence of a methyl group, br indicates unknown branching and cy indicates a cyclopropane fatty acid. *Trans* and *cis* isomers are indicated by t and c, respectively. The PLFA 18:2ω 6,9 was taken to indicate predominantly fungal biomass (Frostegård et al. 1996). PLFA guilds were calculated by summing the component PLFAs for predominant Gram-positive PLFAs (i15:0, a15:0, i16:0, i17:0, and a17:0), predominant Gram-negative bacterial PLFAs (16:1ω5, 16:1ω7 t, 16:1ω9, cy17:0, 18:1ω5, 18:1ω7 and cy 19:0) and actinomycete PLFAs (16:0 10 Me, 17:0 10Me, 18:0 10Me). Samples were taken after 45, 90 and 135 days by bulking a number of small grab samples drawn from all parts of each of the nine reactors. These were immediately frozen and freeze-dried for PLFA analysis or analysed fresh for pH and moisture content.

Statistics

All ANOVA, regression and multivariate analyses were conducted using Genstat 5.3 (NAG Ltd., Oxford, UK). Treatment means and least significant differences at 5% ($LSD_{0.05}$) were calculated using a one way ANOVA. The PLFA data were expressed as mol% for multivariate analyses and as nmolg^{-1} and log-transformed for ANOVA. The data were analysed by canonical variate analysis (CVA), after first reducing the dimensionality by principal component analysis and by comparison of mean inter-group Mahalanobis distances with simulated confidence limits (SCL). SCL for ten groups (t = 0 and three treatments x three sample times) with three reps were 2.95 and 4.63 at the 95 and 99% confidence limit, respectively.

Results and Discussion

Temperature, BOD and pH Profiles

The temperature cycles were maintained close to their targets (Fig. 1a). The internal temperature of reactors set 1 and 2 reached their targets within 24 h and temperature switching at 45 and 90 days was also achieved within 24 h and was therefore rapid but not inconsistent with variations observed in the full-scale stacks.

Pot ale levels for COD and BOD were approximately 50, 000 ppm and 25, 000 ppm respectively. Leachate levels, running at ~1000 COD and ~500 BOD, indicate a removal efficiency of about 95% for both of these parameters. BOD was highly correlated with both suspended solids (r = 0.85, P<0.001) and COD (r = 0.90, P<0.001). Consequently, only BOD of the leachate is presented as an indication of the efficiency of the decomposition process (Fig. 1b). In the first few days of operation, BOD values fell rapidly in all treatments but this fall was significantly faster at 50 than at 35 and 20 °C. Thereafter, however, during the first 45 days the reactors at 35 °C achieved equally good BOD removal rates as those at 20 °C whilethe reactors at 50 °C were the least efficient with the highest BOD values (Fig. 1b). In order to test the effect of physical disruption on the process, all the reactors were mixed at day 25 to simulate the sampling procedure. This did not significantly affect the daily fluctations in BOD of any of the reactors.

Temperature switching at day 45 had a large and significant effect on the BOD of leachate from the reactors at 50 and 35 °C (Fig. 1b), which were changed to 20 and 50 °C, respectively (Fig. 1a). In particular, set 1 showed a dramatic increase in

BOD on day 46, which increased further on day 47 before recovering. Low levels of BOD were not recovered until day 64 . Reactor set 2 initially had higher BODs in the days immediately after the change in temperature and, although these BOD levels were lower than set 1, there appeared to be a longer recovery time with more erratic BOD removal rates than in either set 1 or 3. The efficiency of the reactor set 3, operated at 20 °C for days 0~45 was not, however, significantly affected by switching to 35 °C and remained low for the whole 45~90-day period of the second cycle.

The third temperature cycle again resulted in significant and large perturbations to BOD removal in reactor set 3, which had been increased to 50 from 35 °C, and then only gradually started to recover over the final 45 days. When set 1 was increased from 20 to 35 °C, in days 90~135, there was little disturbance and they had the lowest BOD values and were significantly more efficient than the other two sets. Reactor set 2, which was reduced from 50 to 20°C, showed dramatic effects of the temperature change with large increases in BOD that continued to fluctuate widely over days 90 to 115 but did eventually show some signs of recovery (Fig. 1b). Some of the erratic performance of set 2 in the final (90~135 day) period was due partly to poor aeration in one of the replicate reactors, which remained undetected for several days, giving rise to the anomalous peaks in both COD and BOD between days 114 and 123.

Over the whole 135-day period, the average BOD values were 547, 424 and 304 ppm ($LSD_{0.05}$=86) for reactor sets 1, 2 and 3 respectively. The superior performance of reactors operating at the two lower temperatures, which is particularly noticeable after the two changes of temperature at the 45- and the 90-day point, does not necessarily mean these are the best temperatures for this system. The fact that a 35 °C period in all cases follows on from a 20 °C period means that mesophilic microorganisms were in a better position to recover and take advantage of the rise in temperature, in a way that sets heated from 35 to 50 °C, and, to an even greater extent, sets cooled from 50 to 20 °C, were not.

The changes in temperature and BOD removal were mirrored by changes in leachate pH (Fig. 1c). Apart from the initial 14 days or so of the 135-day period, the leachate pH levels were significantly higher for the 50 °C incubations than for the 35 °C (Fig. 1c). This difference amounts to as much as 1 to 1.5 pH units for long periods of time and is less than 0.5 pH units only where temperature regimes cross over at the end of any given 45-day period. The pattern of leachate pH of sets held at 20 °C, was more complicated. In all three periods, these were seen to shadow the 50 °C sets, but then after about 3 weeks, they dropped to similar or even lower levels than those seen in the 35 °C sets. In all cases, the pH of the feedstock was at least 2 pH units lower than the final leachate so these changes were not due to variation in feedstock.

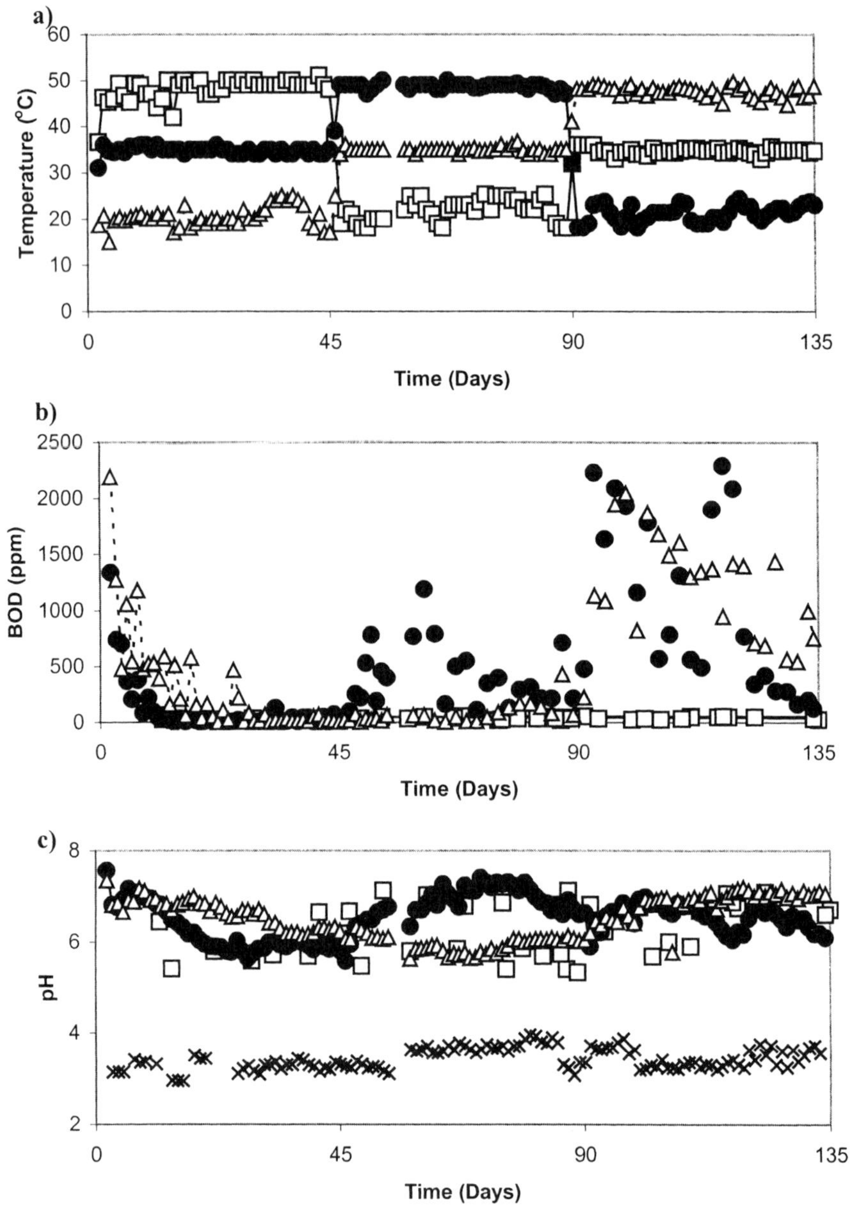

Fig 1. a) Mean internal temperature, **b)** mean BOD (ppm) and **c)** mean pH of leachate from reactor set 1 (□) held at 50, 20 and 35 °C for three consecutive 45-day periods; reactor set 2 (•) held at 35, 50 and 20 °C for three consecutive 45-day periods and reactor set 3 (Δ) °C held at 20, 35 and 50 °C for three consecutive 45-day periods. In **1c** (x) indicates the pH of the pot ale feedstock as used

Bark/Wood Compost

Samples obtained for the wood/bark matrix on the final 135 day showed that the final moisture contents were substantially higher than initially but were not significantly different for any treatment (Table 1). The pH of the composted material after 135 days showed little change except in set 1 where it was substantially higher. This set was running at 50 °C during the first 45-day period, the high pH recorded for the composted wood/bark at the end of this time agreed with the data for leachate pH profiles where elevated temperatures are associated with elevated pH levels (Fig. 1b). The pH of the matrix solids as against the leachate, shows that the leachate was less acidic than the matrix, but the difference was not particularly significant: overall average for leachate 6.43 ± 0.5 vs. matrix 6.07 ± 0.5 (±2 std deviations).

Table 1. Mean moisture content (%), pH in water, % dry weight loss and % friability of compost in reactors operated over different temperature cycles after 135 days

Reactor	%. Moisture	pH	% Dry Wt. loss	% Friability
Set 1	74.8	6.4	34.4	52.3
Set 2	75.1	5.6	20.0	39.3
Set 3	75.1	5.6	16.9	44.0
$LSD_{0.05}$	2.5	1.2	7.4	12.0

The % dry weight loss was significantly (P<0.05) more in reactor set 1 than in the other two reactors over the course of the whole 135 days (Table 1). The percent friability of uncomposted material was 32 (± 9) and this was increased in all treatments at the end of the 135 days (Table 1). Similarly to the dry weight loss there was weak evidence (P=0.08) of a significant effect on the percent friability of reactor set 1, which suggests that reactor set 1 was more friable and by inference more thoroughly composted than the other two sets. This suggests that a more rapid wood/bark degradation process was associated with the incubator set operating under the hot-cold-warm sequence (set 1), as against the other two: warm-hot-cold (set 2) and cold-warm-hot (set 3). While the reasons for this remain to be explored in further experimental work, it is clear that the sequence of temperature regimes employed can have a significant bearing on the outcome.

Phospholipid Fatty Acid Analysis (PLFA)

The total PLFAs in the samples taken at day 0, 45, 90 and 135 showed that after 45 days the biomass was significantly greater at 20 °C than at 35 or 50 °C (Fig. 2a). The range of values obtained for total PLFAs was much lower than that found on straw/manure compost at 5000 nmol/g (Klamer and Bååth 1998) but within the range found for other types of composted material structure (Hellmann et al. 1997; Herrmann and Shann 1997; Campbell and Cooper 1998; Carpenter-Boggs et al. 1998; Lei and Van der Ghenst 2000)]. Total PLFA in reactor set 1 remained constant over the first two temperature cycles, increasing significantly only in the third cycle when the temperature had been raised from 20 to 35 °C. Set 2 had a higher biomass in cycle 1 that remained constant in cycle 2 and then declined in the third cycle. Set 3 had the highest biomass coincident with the greatest decomposition (Fig. 1b), which declined when the temperature was held at 35 °C during the second cycle and remained constant during the third cycle. Similar changes were seen in the total Gram-negative PLFAs and fungal biomass PLFA (data not shown). Total PLFA and also the fungal PLFA 18:2ω6,9 did not decrease as found in other batch-operated compost studies (Hermann and Shann 1997) except when there was an increase in temperature (Fig. 2a). The actinomycete PLFA markers and Gram-positive PLFAs, however, both increased steadily over the 135-day period, (Fig. 2b, c). An increase in actinomycetes has also been previously observed in batch-operated sytems (Carpenter-Boggs et al. 1998; Klamer and Bååth 1998). In the first cycle actinomycete PLFAs were significantly higher at 50 than at 35 or 20 °C and remained higher even after the temperature had been lowered in the second cycle to 20 °C. The actinomycete PLFAs in reactor set 2 also continued to increase in the third cycle after the temperature had been lowered from 50 to 20 °C. This suggests that while the growth of actinomycetes was to some extent under temperature control, the continued increase, despite dramatic changes in temperature, was in part due to chemical changes and/or substrate availability. As the pot ale liquor was being fed continuously to the reactors, it might be argued that it was chemical changes in the wood/bark matrix that were responsible for this.

The multivariate analysis of the PLFA profiles of the different samples showed large and highly significant shifts in microbial populations due to temperature (Fig. 3a). After 45 days' composting the PLFA profiles were discriminated from the initial starting compost with increasing temperature.

Canonical variate analysis of the PLFA data showed the samples were clearly discriminated by their PLFA profile (Fig. 3a) with a mean Mahalanobis distance of 27.9, which was highly significant ($P < 0.001$). Canonical variate 1 (CV1) explained 75% of the variation and discriminated the samples according to temperature and time of sampling. CV 2 explained 14% of the variation in the data and discriminated the initial unamended compost from those in the reactors treated with pot ale liquor (Fig. 3a). In the first 45 days, reactor set 1 at 50 °C was most significantly different from the starting compost with low negative values on CV1 and low positive ordinate values on CV 2. Set 2 at 35 °C was more similar to set 3 operating at 20 °C than set 1. After the second temperature cycle, the PLFA profile of reactor set 1, which had been switched from 50 to 20 °C, changed again but did not return to a profile similar to either the initial material or the material in reactor set 3 that had been at 20 °C initially. Reactor set 2, which had been increased to 50 °C from 35 °C, had higher ordinate values on CV1 that decreased again after the third cycle at 20 °C. Reactor set 3, that started at 20 °C and was increased to 35 °C, also had a significant increase in ordinate values along CV1 but remained more closely similar to the samples incubated at 20 °C. Only in the third cycle did this microbial community PLFA pattern shift further to lower ordinate values on CV1 in the same direction as the samples previously incubated at high temperature.

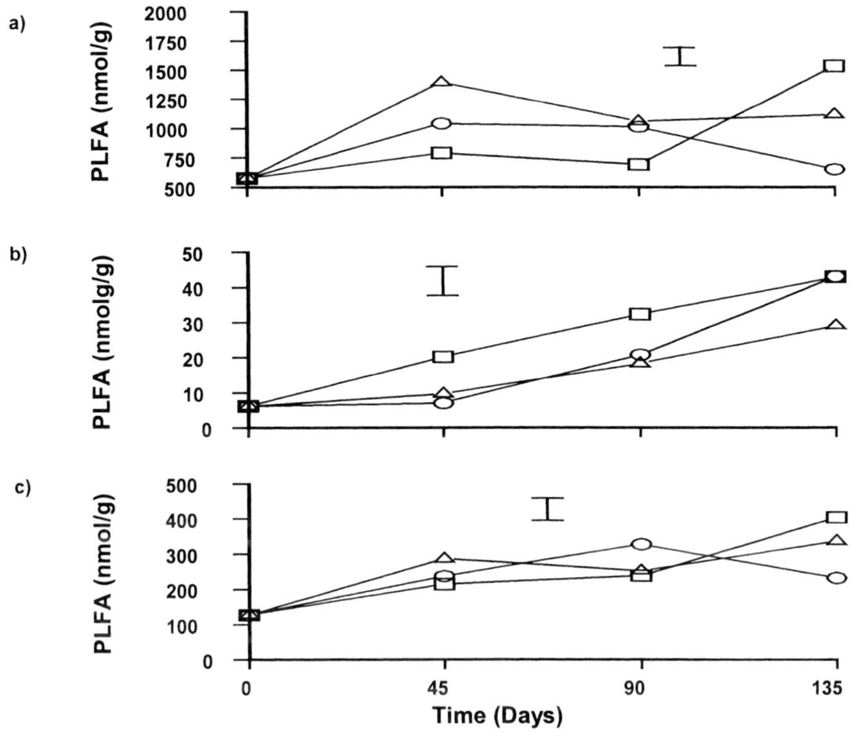

Fig 2. a) Total PLFA, **b)** actinomycete and **c)** PLFA Gram positive PLFA in PASCO sampled from reactor set 1 (□) held at 50, 20 and 35 °C for three consecutive 45-day periods; reactor set 2 (o) held at 35, 50 and 20 °C for three consecutive 45-day periods and reactor set 3 (Δ) °C held at 20, 35 and 50 °C for three consecutive 45-day periods

These results clearly show that temperature had a significant effect on community structure and that once it had been exposed to a high-temperature phase the structure was modified irreversibly. Similar responses to increasing temperature were observed for each reactor set in turn, primarily along CV1, and these were correlated with 19:1a, 16:0, 20:5,15:0i,15:0ai in the plot of the loadings from the CVA of this data (Fig. 3b). Branched PLFAs including 19:1a, 15:0i and 15:0ai are commonly associated with Gram-positive bacteria and were increasing over the period of the experiment (Fig. 2c). The PLFA 17:0cy and 16:1ω7t, 16:1ω7c, and 16:1ω9 had high positive ordinates on CV2 and may therefore have been due to the effect of pot ale addition and primarily associated with Gram negative bacteria.

Although PLFAs indicate viable cells, they do not necessarily distinguish between organisms that may be active at different temperatures and so the changes observed may indicate the presence of resting spores, for example, that are not necessarily involved in the decomposition process after the temperature has changed.

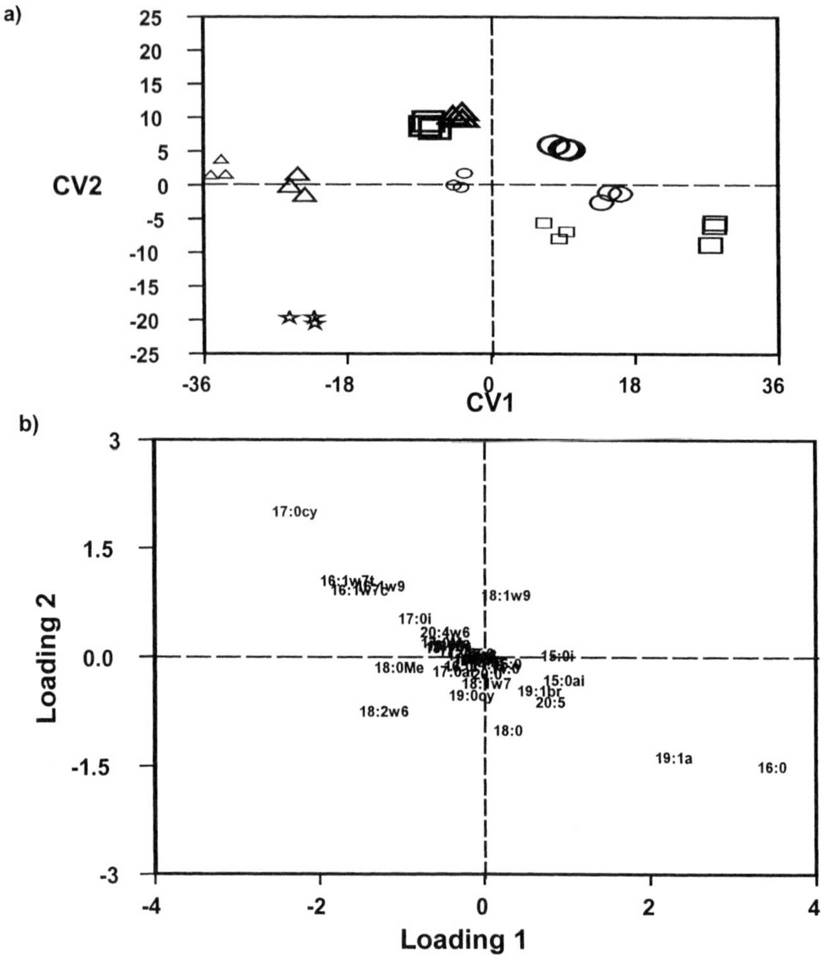

Fig 3. a) Ordination plot of canonical variates of initial wood/bark compost (☆) and wood/bark compost sampled from reactor set 1 (□) held at 50, 20 and 35 °C for three consecutive 45-day periods; reactor set 2 (o) held at 35, 50 and 20 °C for three consecutive 45-day periods and reactor set 3 (Δ) held at 20, 35 and 50 °C for three consecutive 45-day periods . *Small, medium and large symbols* refer to samples taken at 45, 90 or 135 days respectively. **b)** Loadings for individual PLFAs from CVA

Nevertheless, the changes in leachate pH and BOD patterns seen after each temperature changeover, strongly suggest that a specific microbial population rearrangement was initiated at these moments, and this has been confirmed by the PLFA analysis.

Taken together with the data on process efficiency, judged by leachate BOD, it appears that downshifts in temperature from 50 °C to lower temperatures result in greater disturbance than u shifts in temperature. There were very different microbial structures associated with each operating temperature. At higher temperatures the thermophilic population is probably less diverse (Sikora et al. 1983) and may therefore be more susceptible to perturbation. The decline in temperature after the initial self-heating phase of composting is usually accompanied by the death of thermophilic population and growth of a new mesophilic population on the necromass. The changes that occur during the usual batch-composting process are driven not only by temperature but also by the availability of C and energy sources that are depleted in the initial self-heating phase and therefore these factors are confounded. In our laboratory reactor system there was a continuous resupply of C and energy so that the changes observed are primarily driven by temperature alone and most of the changes observed in the PLFA patterns could be ordinated on CV 1 (Fig. 3a). A limitation in our approach to extrapolate to the full-scale system is that the reactors are essentially a closed system that do not afford the same opportunities for reinvasion from areas unaffected by temperature changes. Therefore, it is possible that the successional changes we have observed are primarily due to the survival of sporulating organisms.

Conclusions

It is difficult to be unequivocal about what is the optimal temperature for the operation of the TAPP process without paying due consideration to the operational circumstances, and the purpose for which the process is being used. For example, although the lower temperatures and especially 35 °C resulted in greater BOD removal, an additional objective of the process is to reduce the total volume of leachate collected by evaporation, and this can be promoted by operating at higher temperatures. What is clear is that severe fluctuations in temperature can have catastrophic effects on process efficiency, and in full-scale operations due regard must be given to this. That microbial populations undergo substantial alterations as a result of temperature changes is evident from the data, as is the difference in the ability of specific incubations to recover their capacity to degrade pot ale efficiently. Decomposition by thermophilic communities appeared to be less resilient than that by mesophilic communities when these changes are imposed on them. Therefore, it seems clear that in our application stable, moderate temperatures were more important than the temperature per se.

Acknowledgements. The work was funded by Highland Distillers Plc. Graham Gaskin and Gordon Ewan (MLURI) are thanked for help in the design and building of the laboratory reactor system.

References

American standard methods for the examination of water and wastewater. (1992) APHA,AWWA,WEF. (19th edition) Greenberg A E, Clesceri LS, Eaton AD (eds) American Public Health Association, Washington

Blanc M, Marilley L, Beffa T and Aragno M (1997) Rapid identification of heterotrophic, thermophilic, spore- forming bacteria isolated from hot composts. Inter J Syst Bacteriol 47: 1246-1248

Bligh EG, Dyer WJ (1959) A rapid method of total lipid extraction and purification. Can J Biochem Physio 37 : 911-917

Campbell CD Cooper JN (1998) Community succession of decomposition of microbial biomass during the composting of pot ale liquor. In: Bell CR, Brylinski M and P Johnson-Green (eds) Microbial biosystems: new frontiers. Proc 8th Int Symp on Microbial Ecology. Atlantic Canada Society for Microbial Ecology, Halifax, Canada.

Campbell CD, Darbyshire JF Anderson JG (1990a) The composting of tree bark in small reactors – self-heating experiments. Biol Wastes 31:145-161

Campbell CD, Darbyshire JF Anderson JG (1990b) The composting of tree bark in small reactors – adiabatic and fixed-temperature Experiments. Biol Wastes 31: 175-185

Carpenter-Boggs L, Kennedy AC, Reganold JP (1998) Use of phospholipid fatty acids and carbon source utilization patterns to track microbial community succession in developing compost. Appl Environ Microbiol 64:4062-4064

Castle ME Watson JN (1982) A mixture of malt distillers grains (draff) and pot ale syrup as a food for dairy-cows. Anim Prod 35:263-267

Chapon L, Eber H-L, Kretschmer K-F Kretschmer H (1980) The friability meter in the service of quality and efficiency in malting and brewing. Brauwissenschaft 33: 1-11

Cooper JN (1986) Analysis of non-friable residues remaining from Friability treatment of Scottish distilling malts. J Inst Brew 92:255-261

Frostegård A, Tunlid A, Bååth E, (1996) Changes in microbial community structure during long-term incubation in two soils experimentally contaminated with metals. Soil Biol Biochem 28: 55-63

Hellmann B, Zelles L, Palojarvi A, Bai QY (1997) Emission of climate-relevant trace gases and succession of microbial communities during open-windrow composting. Appl Environ Microbiol 63:1011-1018

Herrmann RF and Shann JF (1997) Microbial community changes during the composting of municipal solid waste. Microb Ecol 33:78-85

Klamer M, Bååth E (1998) Microbial community dynamics during composting of straw material studied using phospholipid fatty acid analysis. FEMS Microbiol Ecol. 27:9-20

Lei F and Van der Gheynst JS (2000) The effect of microbial inoculation and pH on microbial community structure changes during composting. Process Biochem 35: 923-929

Sikora LJ, Ramirez MA, Troeschel TA (1983) Laboratory composter for simulation studies. J Env. Qual 12 :219-224

Vaughan D, Wheatley RE, Ord BG (1984) Removal of ferrous iron from field drainage waters by conifer bark. J Soil Sci 35:149-153

Waksman SA, Cordon TC, Hulpoi N (1939) Influence of temperature upon the microbiological population and decomposition processes in composts of stable manure. Soil Sci 47: 83-113

Survey of Fungal Diversity in Mushroom Compost Using Sequences of PCR-Amplified Genes Encoding 18S Ribosomal RNA

K. Ivors, D. Beyer, P. Wuest and S. Kang[1]

Abstract. We have initiated a comprehensive survey of fungi in phase II commercial mushroom compost using sequences of genes encoding small subunit ribosomal RNA (rDNA). A specific segment of 18S rDNA from 26 individually cultured fungi, as well as from total DNA extracted from the compost sample, was amplified using polymerase chain reaction and sequenced. Taxonomic classification of fungal spp. identified by their rDNA sequence was performed by comparison with complete or partial rDNA sequences of previously identified fungi stored in public databases. Corresponding segments of the 18S rDNA of three species of thermophilic fungi, *Humicola grisea, H. insolens*, and *Torula thermophila,* were also amplified and sequenced to determine if any of the fungi identified in our compost sample are related to these previously characterized fungi associated with compost. The nature and complexity of the fungal community within mushroom compost is discussed.

Introduction

The aim of commercial mushroom composting is to produce a substrate which is optimal and selective for mushroom (*Agaricus bisporus* Lange) mycelial growth. This is largely accomplished by microbial activity. The preparation of mushroom compost is usually done in two stages. Phase I composting involves the initial breakdown of the raw ingredients. Phase I composting is followed by a controlled conditioning stage (phase II) in which the compost is pasteurized at 60 °C and allowed to cool, before mushroom spawn is incorporated. It is well known that thermophilic fungi play important roles in the composting process by decomposing cellulose, aiding thermogenesis, and contributing to the destruction of pathogens. A large number of thermophilic fungi have been isolated from commercial mushroom compost (Chang and Hudson 1967; Fergus 1978). Of the numerous fungal species associated with composting and the cultivation of *A. bisporus*, some may be pathogenic, causing serious diseases, while others are weed molds that have a negative influence on mushroom formation (Fermor et al. 1979). Many of these fungi are considered harmless, posing no threat to the mushroom, while others may benefit substrate preparation. More specifically, thermophilic fungi from the *Torula–Humicola* complex are considered to be

[1]Department of Plant Pathology, The Pennsylvania State University, University Park, PA 16802, USA; e mail: kli105@psu.edu

H. Insam, N. Riddech, S. Klammer (Eds.)
Microbiology of Composting

highly significant in compost preparation and nutrition of *A. bisporus* (Bels-Koning et al. 1962; Ross and Harris 1983; Straatsma et al. 1989). Although the importance of fungal activities during composting and mushroom cultivation has been well recognized, relatively limited information exists regarding the specific types and how they interact and influence various phases of production.

One problem with early investigations of mushroom compost microflora involves the methods available to identify and enumerate the microorganisms present. Culture-based methods are intrinsically biased, depending on growth conditions and media used. There are doubtless many fungi yet to be found in mushroom compost. The application of molecular-phylogenetic techniques to study microbial ecosystems has made it possible to identify microorganisms without cultivating them, which has also led to the discovery of many novel microorganisms (Hugenholtz and Pace 1996). These approaches are based primarily on the PCR amplification of DNA extracted from the substrate with universal primers targeted to genes encoding small subunit ribosomal RNA (rDNA). The objective of our research is to characterize the structure of fungal communities during phase II mushroom composting using the sequence of rDNA as a molecular tag. We characterized both cultivated and non-cultivated members of the fungal community.

Materials and Methods

Culture Conditions

Compost samples were taken 72 h after pasteurization during phase II composting from the Mushroom Test Demonstration Facility at the Pennsylvania State University. The compost formulation consisted of a standard wheat straw bedded-horse manure mixture. Compost was prepared by a conventional phase I and phase II process. Phase I composting was followed by a controlled conditioning stage (phase II) in which the compost was pasteurized at 60 °C for 2 hours on the second day and allowed to cool to 50 °C before taking the sample. Four 100-g compost samples were taken from various depths and locations of phase II compost trays, bulked, and mixed thoroughly to represent one composite sample.

Construction of rDNA Library Using Microbial DNA Extracted from Compost

Ground compost samples were subjected to DNA extraction using FastDNA Spin Kit for soil (BIO101, Vista, CA) according to the manufacturer's protocol except that 30 min of constant vortexing was used. Bulk DNA from the compost was used as a template in 50 µl of PCR mixture [10 mM Tris-HCl (pH 8.3), 50 mM

KCl, 1.5 mM $MgCl_2$, 0.001% gelatin, each dNTP at 0.25 mM, 0.2 μM of each primer, and 2.5 units of Taq polymerase] using primers NS1 and NS2 as described by White et al. (1990). This mixture was subjected to 30 cycles of amplification (one cycle of 1 min at 96 °C, 30 s at 55 °C, and 1 min at 72 °C followed by 29 cycles of 30 s at 94 °C, 20 s at 55 °C, and 1 min at 72 °C) followed by a 2-min extension at 72 °C. The rDNA fragment amplified from compost DNA was cloned into the pGEM-T vector (Promega, Madison, WI). Competent cells of *E. coli* were transformed with the ligated DNA and then plated on LB plates containing ampicillin and X-Gal (5-bromo-4-chloro-3-indolyl-b-D-galactopyranoside). White *E. coli* colonies, presumably carrying pGEM-T with an rDNA insert, were transferred to 96-well microtiter plates containing LB-ampicillin and cultured. The rDNA inserts from individual clones in the library were amplified with the SP6 and T7 primers and directly sequenced using the T7 primer and an automated ABI PRISM 377 DNA sequencer (PE Applied Biosystems, Foster City, CA). The sequences of rDNA were then compared with the sequences of previously identified microorganisms stored in the nonredundant nucleotide database available through the National Center for Biotechnology Information (NCBI; Bethesda, MD) using the BLAST algorithm (Altschul et al. 1997).

Isolation of Fungal Cultures from Compost

The compost sample was suspended in a sodium phosphate buffer, and diluted samples were inoculated onto Rose Bengal agar (amended with 50 mg streptomycin sulfate^{-1}), 1/3 potato dextrose agar (amended with 15 mg rifampicin and 15 mg penicillin G^{-1}), and compost extract agar (amended with 50 mg streptomycin sulfate^{-1}). One set of plates was incubated at 28 °C, and the other set at 40 °C, before colonies were picked and transferred to potato dextrose agar. Twenty-six fungal colonies were randomly isolated from these plates. Spore suspensions of each isolate were prepared, diluted with sterile water, and plated on potato dextrose agar to start single-spore cultures.

Fungal DNA Analyses

Type cultures of dominant thermophilic fungi known to be present during phase II mushroom composting, including: *Humicola grisea* var. *thermoidea* (CBS 622.91), *H. insolens* (CBS 619.91), *Torula thermophila* (CBS 671.88), and *Torula thermophila* (144.4a), were obtained (Straatsma et al. 1989). These isolates have already been morphologically characterized (Straatsma and Samson 1993). A modified CTAB fungal DNA extraction procedure (O'Donnell et al. 1997) was used to extract DNA from these type cultures, as well as from the 26 fungal colonies mentioned above. rDNA was amplified with NS1 and NS2 primers as described above and directly sequenced using the NS1 primer.

Results

Characterization of Fungal Clones

Seventy-three clones in the rDNA library were sequenced. The similarity of these sequences ranged from 91 to 99% with the rDNA sequences listed in the database (Table 1). Therefore all clones corresponded to previously unidentified strains of fungi belonging in two phyla (Ascomycota and Hyphochytriomycota) and three orders (Sordariales, Halosphaeriales, and Hyphochytriales). Sixty-six of the clones were highly homologous to *Aporothielavia leptoderma*, three were homologous to *Hyphochytrium catenoides*, two were homologous to *Chaetomium globosum*, and single clones were homologous to *Chaetomium elatum* and *Lulworthia fucicola*. Of the 66 *A. leptoderma*-related clones, 47 were identical in sequence while the remaining 19 had unique sequences. Of these unique sequences, 10 of them differed by only one nucleotide from that of the 47 identical clones, whereas the remaining 9 varied by a range of 2 to 20 nucleotides.

Table 1. Classification of distinct fungal clones from phase II compost based on their rDNA sequences

Clone	Size[a]	Closest relative[b]	Identity (%)	Accession no.[c]	Order
Group 1[d]	511	*Aporothielavia leptoderma*	99	AF096171	Sordariales
4NS06	511	*Aporothielavia leptoderma*	99	AF096171	Sordariales
4NS08	510	*Aporothielavia leptoderma*	99	AF096171	Sordariales
4NS10	511	*Aporothielavia leptoderma*	99	AF096171	Sordariales
4NS11	511	*Aporothielavia leptoderma*	99	AF096171	Sordariales
4NS18	511	*Aporothielavia leptoderma*	99	AF096171	Sordariales
4NS39	511	*Aporothielavia leptoderma*	99	AF096171	Sordariales
5NS29	511	*Aporothielavia leptoderma*	99	AF096171	Sordariales
5NS33	511	*Aporothielavia leptoderma*	99	AF096171	Sordariales
5NS37	511	*Aporothielavia leptoderma*	99	AF096171	Sordariales
6NS04	511	*Aporothielavia leptoderma*	99	AF096171	Sordariales
6NS16	511	*Aporothielavia leptoderma*	99	AF096171	Sordariales
8NS27	511	*Aporothielavia leptoderma*	99	AF096171	Sordariales
8NS29	511	*Aporothielavia leptoderma*	99	AF096171	Sordariales
8NS37	511	*Aporothielavia leptoderma*	99	AF096171	Sordariales
4NS01	511	*Aporothielavia leptoderma*	98	AF096171	Sordariales
7NS42	511	*Aporothielavia leptoderma*	98	AF096171	Sordariales
8NS35	510	*Aporothielavia leptoderma*	98	AF096171	Sordariales
5NS06	513	*Aporothielavia leptoderma*	94	AF096171	Sordariales
4NS04	503	*Chaetomium elatum*	96	M83257	Sordariales
7NS12	511	*Chaetomium globosum*	99	U20379	Sordariales
8NS25	511	*Chaetomium globosum*	99	U20379	Sordariales

Table 1. (cont.)

5NS47	518	*Hyphochytrium catenoides*	93	X80344	Hyphochytriales
8NS31	510	*Hyphochytrium catenoides*	92	X80344	Hyphochytriales
8NS34	516	*Hyphochytrium catenoides*	91	X80344	Hyphochytriales
4NS40	512	*Lulworthia fucicola*	94	AF050481	Halosphaeriales

[a] The number of nucleotides obtained by sequencing.

[b] Organism with rDNA sequences showing the highest homology to the clone based on BLAST search.

[c] NCBI database accession number of the closest relative.

[d] Clone group 1 consisted of 47 identical clones. All other individual listed clones were unique in sequence.

Characterization of Fungi Cultured from Compost

Twenty-six fungal cultures isolated from compost using three different media, as well as four previously typed strains of thermophilic compost fungi (*H. grisea* var. *thermoidea*, *H. insolens*, and *T. thermophila*), had identical NS1-NS2 sequences. These type cultures are now considered to form the *Torula-Humicola* complex (Austwick 1976) and the binomial *Scytalidium thermophilum* (Cooney & R. Emers.) Austwick is currently applied to these three strains (*S. thermophilum* ≡ *Humicola grisea* var. *thermoidea*, *H. insolens*, and *Torula thermophila*) (Straatsma and Samson 1993). The sequences of cultured and type fungi were also 100% identical to those of the 47 identical *A. leptoderma* clones from the rDNA library, therefore these clones are now identified as *S. thermophilum*. Until this study, the NS1-NS2 region of *S. thermophilum* had not been available in the NCBI database. Macroscopic colony morphology and color of these isolates and type strains were variable (Fig. 1).

Discussion

We have initiated a comprehensive survey of fungi in phase II mushroom compost using genes encoding rDNA. A specific segment of 18S rDNA from 26 individually cultured fungi, as well as from total DNA extracted from compost, was amplified using PCR and sequenced. Corresponding segments of four type specimens of thermophilic fungi previously isolated from mushroom compost were also amplified and sequenced to determine if any of the fungi identified in our compost sample were related to them.

Seventy-three fungal clone sequences corresponded to 26 unique strains representing 5 distinct species. Individual clones were highly homologous to *Aporothielavia leptoderma*, *Hyphochytrium catenoides*, *Chaetomium globosum*, *Chaetomium elatum*, and *Lulworthia fucicola*. Most of the fungal rDNA sequences (90%) were highly homologous to that of *A. leptoderma*. The NS1-NS2 sequences of the 47 identical *A. leptoderma* clones in this library were 100% identical to those of type cultures of *S. thermophilum*.

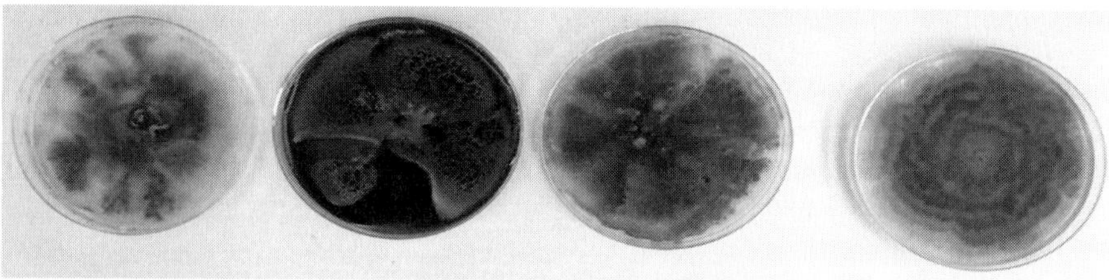

Fig. 1. Morphology and color of isolated compost fungi were highly variable on potato dextrose agar. The plate shown *on the far right* corresponds to a type culture of *S. thermophilum*. The *other plates* represent fungi isolated from compost in this study

All 26 isolated fungi, as well as the four type cultures (now considered *S. thermophilum*), also had identical NS1-NS2 sequences. Previously, Straatsma and Samson (1993) compared these four type strains and concluded that they represented one single variable species or a "morphologically indistinguishable species complex". In contrast to the high degree of complexity reported for bacterial communities in compost (Ivors et al. 2000), the complexity of the fungal community appears much simpler.

Among the 19 unique strains related to *A. leptoderma*, 10 of them differed by only one nucleotide, raising the possibility that some of them might have resulted from replication error by *Taq* polymerase, and thus the degree of diversity within these clones may not be as complex as our data suggest. It has been reported that DNA polymerases make errors during DNA synthesis, with the error rate depending on the DNA polymerase used and reaction conditions. Estimates of the error rate (mutations per base duplication) in PCR have been reported to be in the range of 2×10^{-4} with *Taq* polymerase (Keohavong and Thilly 1989). However, considering the size of the fungal rDNA fragments amplified (~600 bp), it seems unlikely that many of these unique rDNA fragments were artifacts due to replication error.

No 18S rDNA sequence diversity was detected among the fungal strains cultured from compost. In comparison to the results of our rDNA sequence survey, the cultivation method employed in this study did not allow a full description of the mycoflora of our phase II mushroom compost sample. This cultivation technique most likely favored the isolation of *S. thermophilum* and inhibited the detection of other fungal community members. In comparison, fungal clones from the rDNA-based library corresponded to two phyla and three different orders. To our knowledge, this is the first account of a hyphochytrid-related organism (*H. catenoides*) found within mushroom compost.

It is possible that the ability to use complex carbon sources and thrive at high temperatures are the two most important characteristics of successful fungal colonizers of mushroom compost. The high temperatures achieved in phase II composting have significant effects on compost selectivity. The increase in temperature up to 60 °C during peak heating increases the growth of fungi that define the selectivity of the substrate. Thermophilic fungi such as *S.*

thermophilum were the dominant community present during phase II mushroom composting.

Previous studies have employed other molecular tools to characterize microbial communities of mushroom composts. Koschinsky et al. (1998) adapted the single-strand conformational polymorphism technique (SSCP) to analyze the diversity of PCR products from community DNA, and found that both the fungal and bacterial communities consisted of several members, which changed in succession throughout the composting process. Another study also using PCR-SSCP profiles of total DNA extracted from compost identified 19 molecular isolates of bacteria and fungi. In comparison, the diversity of cultivated bacteria at the end of the composting process was rather low, with only 6 species identified from 290 isolates (Peters et al. 2000). The results of these studies, as well as our own findings, support the potential of molecular techniques to more accurately describe the microbial diversity of mushroom compost samples in comparison to cultivation-dependent methods. Optimization of compost quality can be directly linked to the composition and succession of the microbial populations during the composting process (Chanter and Spencer 1974; Derikx et al. 1990). These molecular tools can help monitor and characterize such communities during various phases of the composting process for determining and controlling quality of the substrate. We plan to use the knowledge and technology developed in this project to further enhance the understanding of how microbial changes influence odor emissions, the quality and yield of mushrooms, and mushroom diseases.

References

Altschul SF, Madden TL, Schäffer AA, Zhang J, Zhang Z, Miller W, Lipman DJ (1997) Gapped BLAST and PSI-BLAST: a new generation of protein database search programs. Nucleic Acids Res 25: 3389-3402

Austwick PKC (1976) Environmental aspects of *Mortierella wolfii* infection in cattle. N Z J Agric Res 19:25-33

Bels-Koning HC, Gerrits JPG, Vaandrager MH (1962) Some fungi appearing towards the end of composting. Mushroom Sci 5:165-169

Chang Y, Hudson HJ (1967) The fungi of wheat straw compost. I. Ecological studies. Trans Br Mycol Soc 50:649-666

Chanter DP, Spencer DM (1974) The importance of thermophilic bacteria in mushroom compost fermentation. Sci Hortic 2:249-256

Derikx PJL, Op Den Camp HJM, van der Drift C, Van Griensven LJLD, Vogels GD (1990) Biomass and biological activity during the production of compost used as a substrate in mushroom cultivation. Appl Environ Microbiol 56:3029-3034

Fergus CL (1978) The fungus flora of compost during mycelium colonization by the cultivated mushroom, *Agaricus bisporus*. Mycologia 70:636-644

Fermor TR, Smith JF, Spencer DM (1979) The microflora of experimental mushroom composts. J Hortic Sci 54:137-147

Hugenholtz P, Pace NR (1996) Identifying microbial diversity in the natural environment: a molecular phylogenetic approach. Trends Biotechnol 14:190-197

Ivors KL, Beyer DM, Wuest PJ, Kang S (2000) Survey of microbial diversity within mushroom substrate using molecular techniques. In: Van Griensven LJLD (ed) Mushroom science XV Science and cultivation of edible fungi. Balkema, Rotterdam, pp 401-407

Keohavong P, Thilly WG (1989) Fidelity of DNA polymerases in DNA amplification. Proc Natl Acad Sci USA 86: 9253-9257

Koschinsky S, Schwieger F, Peters S, Grabbe K, Tebbe CC (1998) Characterizing microbial communities of composts at the DNA level. Med Fac Landbouww Univ Gent 63/4b:1725-1732

O'Donnell K, Cigelnik E, Weber NS, Trappe JM (1997) Phylogenetic relationships among ascomycetous truffles and the true and false morels inferred from 18S and 28S ribosomal DNA sequence analysis. Mycologia 89:48-65

Peters S, Koschinsky S, Schwieger F, Tebbe CC (2000) Succession of microbial communities during hot composting as detected by PCR-single-strand-conformation-polymorphism-based genetic profiles of small-subunit rRNA genes. Appl Environ Microbiol 66:930-936

Ross RC, Harris PJ (1983) The significance of thermophilic fungi in mushroom compost preparation. Sci Hortic 20:61-70

Straatsma G, Gerrits JPG, Augustijn MPAM, Op den Camp HJM, Vogels GD, Van Griensven LJLD (1989) Population dynamics of *Scytalidium thermophilum* in mushroom compost and stimulatory effects on growth rate and yield. J Gen Microbiol 135:751-759

Straatsma G, Samson RA (1993) Taxonomy of *Scytalidium thermophilum*, an important thermophilic fungus in mushroom compost. Mycol Res 97: 321-328

White TJ, Bruns T, Lee S, Taylor J (1990) Amplification and direct sequencing of fungal ribosomal RNA genes for phylogenetics. In: Innis MA (ed) PCR protocols: a guide to methods and applications. Academic Press, San Diego, pp 315-322

Bacterial Community Structure During Yard Trimmings Composting

F.C. Michel, Jr[1], T. J. Marsh[2], and C.A. Reddy[2]

Abstract. A long-term objective of our group is to understand how various composting parameters affect microbial community structure in composts. In this study, we used terminal restriction fragment length polymorphisms (T-RFLP) of PCR-amplified 16S rRNA genes to analyze bacterial community structure during the composting of yard trimmings. Community DNA was isolated from samples collected on days 0, 8, 29, 64, and 136 from a compost windrow (consisting of leaves, grass, and brush in a 4:2:1 ratio) at a large-scale municipal facility. The DNA was PCR-amplified using fluorescently labeled primers targeted to bacterial domain 16S rRNA genes. The products were restriction-digested with *Hha1, Msp1,* and *Rsa1* to give fingerprints of the bacterial communities. Terminal restriction fragment (TRF) sizes obtained with the three digestions were compared to the three fragments determined by computer-simulated amplification and restriction digestions of complete 16S rRNA gene sequences. T-RFLP patterns indicated extensive bacterial diversity in all of the composts. A large percentage of the observed TRFs corresponded to sizes predicted for bacteria by computer-simulated digestion. Comparison of fragment sizes from three digestions to those predicted by computer-simulated digestions indicated a substantial shift from a bacterial community containing primarily Gram-negative α, β, and γ Proteobacteria (day 0) to communities containing many members of the Gram-positive *Bacillus-Clostridium* group (days 8, 29, and 64) and members of the CFB and Actinobacteria (days 29 and 64). Bacterial species identified on days 8, 29, and 64 included those previously isolated from thermophilic composts by cultivation such as *Bacillus* and *Pseudomonas* spp. as well as many not previously described in composts. Abundant TRFs corresponding to *E. coli* and other Gram-negative γ Proteobacteria, decreased dramatically after the first 8 days of composting. The day-136 composts contained a diverse group of bacteria including many fragment sizes consistent with known *Pantoea* and *Pseudomonas* biocontrol agents as well as *Xanthomonas* and *Bacillus* species. The greatest diversity of bacteria was observed in the stabilized day 64 and 136 composts where 115 and 111 TRFs corresponding to members of 7 and 6 different phylogenetic groups, respectively, were observed.

[1]Department of Food, Agricultural and Biological Engineering,
1680 Madison Ave., Ohio State University-OARDC, Wooster, Ohio, 44691, USA
email: michel.36@osu.edu, Tel: 330-263-3859, Fax: 330-263-3670
[2]NSF Center for Microbial Ecology and Department of Microbiology
Michigan State University, East Lansing, Michigan 48824, USA

H. Insam, N. Riddech, S. Klammer (Eds.)
Microbiology of Composting

Introduction

Composting is a complex, aerobic microbial process used to recycle organic materials such as yard trimmings, agricultural and industrial byproducts, and for the bioremediation of contaminated soils (Michel et al. 1996; Laine and Jorgensen 1997; Reddy and Michel 2000). Composting is a thermophilic process and temperatures within compost piles range from ambient levels to peak temperatures as high as 80 °C (Beffa et al. 1996b). Culture-based approaches have previously been used to study microorganisms active during composting (Finstein 1975; Strom 1985 a,b; Beffa et al. 1996a). However, only a small fraction of microbes present in environmental samples are typically culturable (Amman et al. 1995; Dunbar et al. 2001). As a result, important members of the composting microbial community may have been missed. Furthermore, very little is known about microbial community structure at different stages of composting. New molecular techniques such as density gradient gel electrophoresis (Muyzer et al. 1993), analysis of PCR-single-strand conformation polymorphisms (Tebbe and Schwieger 1998) and analysis of terminal restriction fragment length polymorphisms (T-RFLP) of PCR-amplified 16S rRNA genes (Liu et al. 1997) allow rapid profiling of microbial communities without the need for cultivation and can provide information about the specific phylogenetic groups present in a microbial community. In this study we used T-RFLP to analyze temporal changes in bacterial community structure during the composting of yard trimmings.

Materials and Methods

Windrow Composts

Compost samples were collected from a large-scale yard trimmings composting facility in Oakland County, Michigan, USA (Michel et al. 1996). Windrows were constructed using a 4:2:1 (v/v/v) mixture of leaves, grass and brush. The windrows were turned once every 4 weeks using a Scarab windrow turner. Multiple compost samples (1 kg) were collected from 61 cm and 122 cm (2 and 4 foot) depths and mixed on days 0, 8, 29, 64, and 136 of composting. Subsamples were homogenized using a Waring blender, and stored at –20 °C. Temperature, moisture and organic matter content, pH, and the rates of oxygen uptake, and organic matter conversion were determined as described previously (Michel et al. 1995, 1996) at the time of sampling.

DNA Extraction and Purification

Compost subsamples (0.75 g) were washed with 20 ml of 120 mM phosphate buffer, centrifuged (5000 *g*) and the pellet was treated with 20 ml (15 mg ml^{-1})

lysozyme (Sigma Biochemicals) for 30 min at 37 °C (Tsai and Olson 1991). DNA was then extracted as described by Zhou et al. (1996). Crude DNA preparations were purified by agarose (GTG SeaPlaque) gel electrophoresis followed by band excision, agarase digestion (Zhou et al. 1996), and ultrafiltration using Microcon 100 spin columns (LaMontagne et al., 2002). The purified DNA extract solutions were colorless, indicating that they were free of humic acid contamination. Agarose gel electrophoresis indicated that the aproximate DNA yield was 6 µg g^{-1} compost and that it had an MW greater than 20 kb.

16S rRNA Gene Amplification and Digestion

Ribosomal 16S rRNA genes were PCR amplified from purified DNA samples using universal eubacterial primers 8F hex (fluorescently labeled forward primer) and 1392R (reverse primer). The sequences (5'-3') of the forward and reverse primers were: AGAGTTTGATCCTGGCTCAG and ACGGGCGGTGTGTRC, respectively (Liu et al. 1997). Each 50 µl PCR reaction contained 50 ng template, 2.5 mM Mg^{2+}, 30 ng bovine serum albumin, 2.5 units *Taq* polymerase (Gibco, Bethesda, MD), 0.24 mM each dNTP, 25 pmol of primers, and 1 X PCR buffer. After heating the reaction mixtures at 94 °C for 9 min, the DNA was PCR–amplified by three steps as follows: denaturing (94 °C, 60 s), annealing (58 °C, 45 s), and primer extension (72 °C, 90s) for 30 cycles. The final primer extension step was held for 7 min.

T-RFLP Analysis

PCR products were purified with Wizard PCR preps (Promega, Madison, WI), subsampled, and digested with *HhaI* , *Rsa 1,* or *Msp* 1 (10 ul of the purified PCR product with 1 µl of restriction enzyme for 2–3 h at 37° C). The lengths of fluorescently labeled terminal restriction fragments (TRFs) were determined with an Applied Biosystems Instruments model 373A automated sequencer (Foster City, CA) as described by Liu et al. (1997). Fragment lengths were compared to fragment sizes determined by computer-simulated amplification (assuming a primer mismatch of up to 3 bp) and digestion of complete 16S rRNA gene sequences obtained from the Ribosomal Database (Maidak et al. 2000) using the PATSCAN program (http://www- unix.mcs.anl.gov/compbio/PatScan/HTML/ patscan.html). This simulation yielded a total of 1368 different species. A group of 15 *Pantoea* and *Pseudomonas* biocontrol agents that are not well represented on the RDP were added to this list. The fragment sizes for these strains were determined empirically in our laboratory using the PCR primers and restriction enzymes described above. These strains are marked with an asterisk in Table 3.

To compare observed fragment lengths with those determined by computer-simulated restriction digestions, a fragment length sizing error (E) was calculated using a polynomial expression (Eq.1)

$$E = 1.0\text{x}10^{-8}\ F^3 + 1.0\text{x}10^{-5}\ F^2 - 4.0\text{x}10^{-4}\ F + 1.1201 \qquad (1)$$

where F is the observed TR fragment length determined by electrophoresis.

This equation was used because sizing error on polyacrylamide sequencing gels increases with fragment size. Using this equation, the sizing error for a 100-bp fragment would be 1.2 bp and for a 300-bp fragment would be 2.2 bp. The magnitude of the sizing error determined using this equation approximates that reported by Liu et al. (1997) for a model community and observed by our group for known fragments of various sizes (data not shown).

The observed fragment sizes were compared to those predicted by computer simulation using a sorting program called FRAGSORT available from the authors. To estimate the abundance of each fragment size in the PCR product pool, TRF peak areas were normalized by dividing individual TRF peak areas by the sum of all TRF peak areas.

Cluster Analysis

T-RFLP community profiles were digitized to generate banding patterns that could be analyzed using GelCompar 4.0 software (Applied Maths, Kortrijk, Belgium). T-RFLP community profiles were clustered using the unweighted pair group method using average linkages (UPGMA) as described by Liu et al. (1997). Dendrograms were prepared using the GelCompar program.

Results and Discussion

Temporal Changes In Compost Bacterial Community T-RFLP Patterns

Samples were collected from a large-scale yard trimmings composting facility from day 0 to 136. Substantial changes in organic matter content, pH, oxygen uptake, and temperature were observed during composting (Table 1).

Temperatures in the thermophilic range (> 55 °C) were observed on days 8, 29, and 64. Approximately 64% of the volatile solids originally present in the feed stock were lost after 136 days of composting (Table 1). The final compost (day 136) was stable (O_2 uptake rate <0.1 mg O_2 gOM^{-1} h^{-1}) and the windrow temperature was in the mesophilic range.

Analysis of 16S rRNA gene T-RFLP patterns from the compost communities (e.g., Fig. 1) indicated extensive bacterial diversity. A total of 77, 78, 115, and 111 different *MspI*, *RsaI* and *HhaI* TRFs were observed on days 8, 29, 64, and 136 of composting, respectively. The total number of TRFs increased during composting, indicating an increase in bacterial diversity. The system presumably shifted from heterotrophy to oligotrophy during this period, as indicated by a shift from thermophilic to mesophilic temperatures (Table 1) and a decrease in compost O_2 uptake rate from 1.2 to less than 0.04 mg (O_2 g VS^{-1} h^{-1}).

Table 1. Characterization of composts taken from a yard trimmings compost windrow on days 0, 8, 29, 64, and 136. (Michel et al. 1996)

	Day				
Compost property[a]	0	8	29	64	136
Moisture % (g 100 g^{-1} wet)	65 ±1	60 ±4	59 ±1	50 ±1	56 ±2
Organic matter % (g 100g^{-1} dry)	72 ±5	69 ±6	62 ±2	57 ±6	48 ±2
pH	6.6 ±0.3	7.6 ±0.2	7.8 ±0.2	8.1 ±0.2	8.1±0.1
Volatile solids loss (g g^{-1} initial)	0%	16%	35%	49%	64%
O_2 uptake rate (mg O_2 gVS^{-1} h^{-1})	1.2	0.9	0.6	0.1	0.04
Mean temperature (°C)	32 ±2	59 ±8	55 ±1	59 ±2	42 ±5

[a]- Values are means ± one standard deviation of four composite samples (n=4) except temperature (n=6) and O_2 uptake rate (n=1).

UPGMA analysis of the combined TRF banding patterns after three restriction digestions (Liu et al. 1997) showed that the compost bacterial communities clustered into a thermophilic group consisting of days 8, 29, and 64 and a mesophilic group consisting of days 0 and 136 (Fig. 2).

Since each TRF represents at least one unique species, the high number of TRFs observed during thermophilic composting indicate greater bacterial species diversity than that reported previously using cultivation methods (Strom 1985a,b). These findings are in agreement with a recent study using primers targeting eubacterial 16S rRNA genes, 16S rRNA genes of actinomycetes, and fungal 18S rRNA genes that also found extensive bacterial diversity during composting and a a temporal increase in diversity concurrent with a decrease in composting temperature (Peters et al. 2000).

However, T-RFLP community profiles for days 0 and 136 were markedly different from each other (Jaccard Similarity coefficient (S_{ab}) < 30%), as well as from the thermophilic bacterial communities (Figure 2).

Comparison Between Individual TRFs And Those Predicted by Computer Simulation

A large percentage of TRFs corresponded to fragment sizes predicted for bacteria by computer-simulated digestions of complete 16S rRNA genes sequences (Table 2). However, the percentage of observed TRF sizes that corresponded to computer predicted TRF sizes varied with the restriction enzyme used.

For *MspI* 91–100% of the observed TRFs were found in the simulated digestion, whereas for *HhaI* only 39-77% of the TRFs corresponded to predicted fragment sizes (Table 2). The total TRF peak area percent corresponding to known fragments was lower for *HhaI* and *RsaI* digestions.

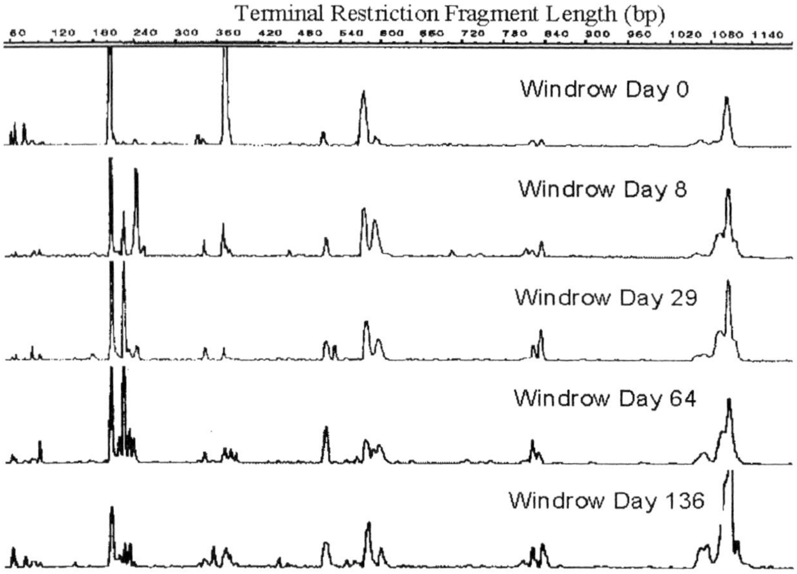

Fig. 1. Illustrative electropherograms of the 5' T-RFLP of HhaI digested 16S rRNA genes from a yard trimmings windrow. Samples were removed 0, 8, 29, 64 and 136 days after windrow formation. Electropherograms after *MspI* and *RsaI* digestions are not shown. The full-scale y-axis for each of the electropherograms is 900 fluorescence units.

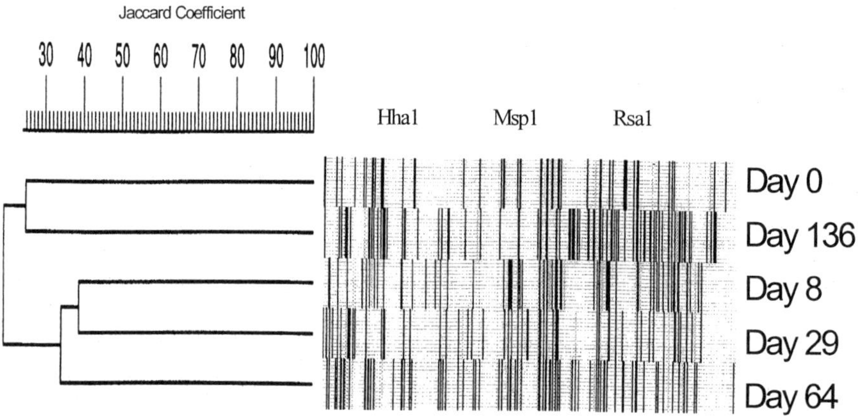

Fig. 2. Dendrogram of depicting the relatedness of microbial communities from a yard trimmings compost windrow. Each community was represented by a 5' T-RFLP banding pattern corresponding to *HhaI, MspI,* and *RsaI* digests. Samples were removed on days 0, 8, 29, 64, and 136 of composting

Table 2. Terminal restriction fragments after *Msp1*, *Rsa*, or *Hha*1 digestion of PCR-amplified compost DNA. RDP identifiable fragments are those that correspond to fragment sizes predicted based on a computer-simulated amplification and digestion of the ribosomal database (RDP)a.

	Day				
	0	8	29	64	136
Msp1					
MspI TRFs (total)	23	32	22	36	45
MspI TRFs (RDP identifiable)[b]	21	32	22	36	42
MspI TRFs (RDP identifiable, %)	91%	100%	100%	100%	93%
Peak area % of identifiable *MspI* TRFs[c]	95%	100%	100%	100%	98%
Rsa1					
RsaI TRFs (total)	15	22	27	38	31
RsaI TRFs (RDP identifiable)[b]	9	13	17	30	23
RsaI TRFs (RDP identifiable, %)	60%	59%	63%	79%	74%
Peak area % of identifiable *RsaI* TRFs[c]	21%	26%	26%	51%	28%
Hha1					
HhaI TRFs (total)	22	23	29	41	35
HhaI TRFs (RDP identifiable)[b]	17	9	12	27	18
HhaI TRFs (RDP identifiable, %)	77%	39%	41%	66%	51%
Peak area % of identifiable *HhaI* TRFs[c]	54%	9%	18%	47%	30%

a- Fragment sizing error (E) was estimated using the following formula:
$E=1.0x10^{-8} F^3 + 1.0x10^{-5} F^2 - 4.0x10^{-4} F + 1.1201$.

b- Identifiable TRFs are those that are within the fragment sizing error of fragment sizes predicted by a simulated amplification and restriction digestion of complete 16S rRNA gene sequences found on the RDP.

c- Peak area % of identifiable TRFs is the sum of the identifiable fragments normalized peak areas and is approximately equivalent to PCR product abundance.

For example, for *HhaI* TRFs only 9-54% of the normalized TRF peak area corresponded to predicted TRF sizes. This indicates that more of the less abundant TRFs corresponded to known fragment sizes. TRFs that did not correspond to those predicted by the computer simulated digestion are probably due to species that have not been completely sequenced or to the fact that only a limited supply of reference sequences is currently available in the ribosomal sequence database (Dunbar et al. 2001). The identification of bacteria based on observed TRF sizes (Table 3) was restricted to the fraction of fragments sizes corresponding to those predicted by computer simulated digestion.

TRF peak area corresponds to the amount of a fragment size amplified from community DNA. TRF peak area is not necessarily quantitatively related to the number of cells in a sample due to PCR biases, redundant fragment sizes, DNA extraction inefficiencies, and because some species have multiple copies of ribosomal genes (Liu et al. 1997; Klappenbach et al. 2000; Dunbar et al. 2001). However, comparison of the normalized peak area of a single TRF among different samples may indicate an increase or decrease in the presence of a corresponding ribotype. Specific TRF sizes that changed substantially during composting included a 373 bp *Hha*I fragment (Fig. 1), a 427-bp *RsaI* fragment and a 498-bp *MspI* fragment. This group of fragments accounted for the greatest peak areas in the day 0 T-RFLP profiles, but nearly disappeared after 8 days of composting (Fig. 1). These three fragments correspond to a variety of γ--Proteobacteria including *Escherichia coli* and various *Vibrio* and *Pantoea* species. The marked decrease in the peak area of these fragments indicates that members of this ribotype most likely declined substantially after 8 days of composting and remained low through days 29, 64, and 136. However, these TRF sizes did not completely disappear from the other compost samples, possibly indicating a potential for regrowth. Culturable plate counts of these organisms are known to decrease by 3 orders of magnitude after as few as 3 days of composting at temperatures above 55° C (Finstein 1975).

On days 8, 29, and 64, a 494-bp *MspI* TR fragment, 493-, 878-, and 885-bp *RsaI* TR fragments and a 207-bp *HhaI* TR fragment (Fig. 1) showed the greatest peak areas. The *HhaI* and *MspI* fragment sizes, both of which were also abundant in the feedstock, correspond to various *Pseudomonas* spp. Although this genus is not known to contain species capable of thermophilic growth, previous investigators have reported that thermophilic *Pseudomonas* spp. were isolated from hot composts (Strom 1985b; Peters et al. 2000). On day 8, the second largest *MspI* peak (150 bp), three large *RsaI* peaks (489, 477 and 479 bp), and the third largest *HhaI* peak (245 bp) (Fig. 1) correspond to nonthermophilic *Bacillus* species. Members of this genus have been shown to be a majority of the culturable isolates found during thermophilic solid waste composting (Strom 1985a,b; Blanc et al. 1997) and mushroom composting (Peters et al. 2000). Strom (1985a) found that many of the culturable isolates identified as mesophilic species using classical methods, were also capable of growth at temperatures as high as 65 °C. Blanc et al. (1997) further showed that phenotypic tests are unreliable for the classification of *Bacillus* species as thermophilic. The *MspI* TRF of 141 bp, *RsaI* TRFs of 461 and 463 bp, and a *HhaI* TRF of 579 bp found in various digests on days 8–64 correspond to a variety of other *Bacillus* species that include thermophilic species previously identified by Blanc et al. (1997) and Strom (1985b) as predominant culturable members of thermophilic compost bacterial communities.

Identification of Bacterial Species Based on Multiple TRF Sizes

The computer simulation predicted that the 16S rRNA from 1381 species would be amplified using the 8F primer and *MspI*, *RsaI,* and/or *HhaI* restriction sites. From these 1381, a list of putative bacterial species was assembled for which all three restriction fragments found in compost DNA were within the sizing error of three TRFs predicted by computer simulated digestion. A total of 123, 162, 164, 266, and 196 bacterial species corresponding to 23, 30, 26, 40, and 36 unique three-fragment combinations were identified on days 0, 8, 29, 64, and 136, respectively (data not shown). Species giving the same three-fragment combinations were grouped together and listed in order of greatest to least minimum normalized TRF peak area to reduce the complexity of the list and indicate changes in PCR product abundance (Table 3).

The identifications listed in Table 3 should be viewed as an overestimate of the community diversity. The assumed error in fragment sizing meant that a diverse group of bacterial species were identified. On the other hand, many TRF sizes found in the compost samples (e.g., *Msp1* fragment of 520 bp) were not found among restriction fragments determined by computer simulation (Table 2). Others did not correspond with two other TRF fragments to make a group of three corresponding to a known species, and therefore did not allow putative identification. These TRF sizes presumably represent unknown bacteria or 16S rRNA genes that have not been completely sequenced.

Table 3. Bacteria corresponding to three different observed terminal restriction fragment sizes in samples collected on days 0, 8, 29, 64 and 136 of composting. Identifications were based on TRFs after *Msp1*, *Rsa1*, and *Hha1* digestion. Minimum TRF area% is the normalized peak area percent of the TR fragment with the lowest normalized peak area of the three. A variable scale was used to determine if observed fragment sizes matched those predicted by computer--simulated restriction digestions.

Day 0 Species corresponding to three observed TRFs	Phylogenetic group	*Msp1* TRF size (bp)	*Rsa1* TRF size (bp)	*Hha1* TRF size (bp)	Min TRF Area (%)
Citrobacter freundii, Proteus mirabilis, Escherichia coli, Vibrio gazogenes, Salmonella enteritidis, V.ordalii, Hafnia alvei, V.ordalii	γ Proteobacteria	496	427	373	25.1
Vibrio species, Buchnera aphidicola	γ Proteobacteria	499	430	573	17.0
Vibrio aestuarianus, Pantoea ananas, Escherichia coli, P. agglomerens*, V. metschnikovii, V.aestuarianus*	γ Proteobacteria	497-500	428-429	374-375	16.9
Buchnera aphidicola	γ Proteobacteria	496-499	427-430	1098-1104	6.5
Methylobacillus flagellatus, Xylella fastidiosa	β Proteobacteria γ Proteobacteria	496	479	373	6.5
Pantoea herbicola pathovar gypso, Vibrio anguillarum, Escherichia coli, P.ananas*, V.damsela*	γ Proteobacteria	495-500	426	373	5.5
Pasteurella haemolytica, Actinobacillus capsulatus, Act.equuli, P.species strain Bisga, Act.suis, Act.hominis	γ Proteobacteria Actinobacteria	496	650	1099	1.6
Vibrio vulnificus, V.navarrensis, V.diazotrophicus,	γ Proteobacteria	496	649-650	373	1.6
Sphingomonas capsulata, S.paucimobilis, Erythrobacter longus	α Proteobacteria	150	422	82	1.2
Proteus vulgaris	γ Proteobacteria	495	426	212	0.8
Rhodobacter sphaeroides	α Proteobacteria	130	422	341	0.4
Beijerinckia indica	α Proteobacteria	150	422	341	0.4
Afipia felis	α Proteobacteria	440	423	342	0.4
Cyanophora paradoxa	Eukaryota	498	429	1052	0.4

Table 3. (cont.)

Day 8 Species corresponding to three observed TRFs	Phylogenetic group	*MspI* TRF size (bp)	*RsaI* TRF size (bp)	*HhaI* TRF size (bp)	Min TRF Area (%)
Bacillus pabuli	Bacill/Clostrid	150	489	244	4.3
Pasteurella species strain.Bisga, Actinobacillus capsulatus, P.haemolytica, Act.hominis, Act.suis, Act.equuli	γ Proteobacteria Actinobacteria	496	650	1099	3.6
Oceanospirillum commune	γ Proteobacteria	494	648	371	3.6
Bacillus macquariensis	Bacill/Clostrid	152	491	246	3.3
Bacillus fusiformis	Bacill/Clostrid	147	458	242	2.9
Streptococcus anginosus	Bacill/Clostrid	563	494	589	2.7
Bacillus alvei	Bacill/Clostrid	506	489	244	2.6
Clostridium herbivorans	Bacill/Clostrid	486	469	1091	2.3
Actinobacillus pleuropneum, Act.lignieresii	Actinobacteria	498	652	1101	1.9
Bacillus cereus, B.mycoides, B.thuringiensis, B.anthracis, B.medusa	Bacill/Clostrid	147	488	579	1.8
Alicyclobacillus acidoterrestris	Bacill/Clostrid	147	488	242	1.8
Vibrio vulnificus Vibrio navarrensis	γ Proteobacteria	496	650	373	1.5
Vibrio anguillarum Vibrio damsela	γ Proteobacteria	495	426	373	1.5
Methylobacillus flagellatus, Xylella fastidiosa	β Proteobacteria γ Proteobacteria	496	479	373	1.5
*Pantoea ananas**	γ Proteobacteria	498	426	373	1.5
*Pantoea herbicola pathovar gypso**	γ Proteobacteria	500	426	373	1.3
Bacillus polymyxa	Bacill/Clostrid	150	487	242	1.3
Lactobacillus thermophilus	Bacill/Clostrid	147	487	242	1.3
Mycoplasma pneumoniae Myc.genitalium	Bacill/Clostrid	545	475	226	1.0
Mycoplasma gallisepticum	Bacill/Clostrid	546	477	572	1.0
Listeria seeligeri L.ivanovii	Bacill/Clostrid	147	434	577	1.0
Buchnera aphidicola	γ Proteobacteria	501	432-433	1104-1107	1.0
Buchnera aphidicola	γ Proteobacteria	496	427	1098	0.9
Vibrio gazogenes, Escherichia coli	γ Proteobacteria	496-497	427	373	0.9
Vibrio aestuarianus	γ Proteobacteria	498	429	375	0.9
Vibrio species strain DSK1, Photobacterium angustum	γ Proteobacteria	498	429	572	0.9
Vibrio species	γ Proteobacteria	499	430	573	0.9
Salmonella enteritidis, Vibrio ordalii Escherichia coli	γ Proteobacteria	496	427	373	0.9
Buchnera aphidicola	γ Proteobacteria	499	430	1104	0.9
Vibrio ordalii, Escherichia coli, Proteus mirabilis, Citrobacter freundii, Hafnia alvei	γ Proteobacteria	496	427	373	0.9
Vibrio metschnikovii, V. aestuarianus	γ Proteobacteria	498	429	375	0.9
Vibrio marinus	γ Proteobacteria	498	429	572	0.9
Symbiont Acrt.P	unknown	498	429	574	0.9
Pantoea ananas, P. agglomerens**	γ Proteobacteria	500	428	374	0.9
Buchnera aphidicola	γ Proteobacteria	498	429	574	0.9
Spiroplasma species strain.TG-1.	Bacill/Clostrid	539	470	1080	0.9

Table 3. (cont.)

Mycoplasma species strain PG50	Bacill/Clostrid	540	471	832	0.9
Staphylococcus aerophilus, S.epidermidis.	Bacill/Clostrid	155	486	577	0.5
*Pseudomonas corrugata**	γ Proteobacteria	492	476	207	0.4
Pseudomonas flavescens	γ Proteobacteria	490	644	207	0.4
Bacillus alcalophilus	Bacill/Clostrid	165	485	239	0.4
Agromyces mediolanus	Actinobacteria	165	458	370	0.4

Day 29 Species corresponding to three observed TRFs	Phylogenetic group	*MspI* TRF size (bp)	*RsaI* TRF size (bp)	*HhaI* TRF size (bp)	Min TRF Area (%)
Flavobacterium balustinum, Fl.indologenes, Fl.indolthetic, Fl.gleum	CFB	202	93	225	3.7
Mycoplasma species strain PG50	Bacill/Clostrid	540	471	832	3.0
Haemophilus parasuis, H.paracuniculus, Pasteurella haemolytica, H. Parasuis	γ Proteobacteriau nknown	496-498	882-884	1099-1102	2.6
Buchnera aphidicola	γ Proteobacteria	496	427	1098	1.8
Escherichia coli	γ Proteobacteria	495	426	372	1.5
Klebsiella species	γ Proteobacteria	494	882	371	1.5
Listeria murrayi	Bacill/Clostrid	149	895	579	1.5
Legionella pneumophila	γ Proteobacteria	496	885	213	1.4
Proteus vulgaris	γ Proteobacteria	496	427	213	1.4
Buchnera aphidicola	γ Proteobacteria	501-502	432-433	1104-1107	1.0
Nitrosolobus multiformis	β Proteobacteria	461	459	207	0.8
Buchnera aphidicola	γ Proteobacteria	509-511	440	1116-1117	0.7
Spiroplasma species strain DU-1	Bacill/Clostrid	538	469	1052	0.6
Synechocystis species	Cyanobacteria	494	425	1048	0.6
Actinobacillus lignieresii, Act.pleuropneum, Pasteurella species strain Bisga, Act.hominis, Act.equuli, Act.capsulatus, Act.suis.	Actinobacteria, γ Proteobacteria	496-498	650-652	1099-1101	0.5
Oceanospirillum commune, Vibrio diazotrophicus	γ Proteobacteria	494-495	648-649	371-372	0.5
Pasteurella haemolytica	γ Proteobacteria	496	650	1099	0.5
Carnobacterium funditum	Bacill/Clostrid	564	901	590	0.5
Streptococcus anginosus	Bacill/Clostrid	563	494	589	0.5
Rhodospirillum centenum	α Proteobacteria	439	827	515	0.4
Escherichia coli	γ Proteobacteria	492	423	371	0.4
*Pseudomonas corrugata**	γ Proteobacteria	492	476	207	0.4
Carnobacterium divergens	Bacill/Clostrid	561	898	587	0.4
Vagococcus salmoninarum	Bacill/Clostrid	560	897	243	0.4
Globicatella sanguis	Bacill/Clostrid	556	894	582	0.4
Carnobacterium piscicola, Lactobacillus maltaromicus	Bacill/Clostrid	560	897	586	0.4
Clostridium herbivorans	Bacill/Clostrid	486	469	1091	0.3
Spiroplasma species strainTG-1.	Bacill/Clostrid	539	470	1080	0.3

Table 3. (cont.)

Day 64 Species corresponding to three observed TRFs	Phylogenetic group	*MspI* TRF size (bp)	*RsaI* TRF size (bp)	*HhaI* TRF size (bp)	Min TRF Area (%)
*Pseudomonas fluorescens**	γ Proteobacteria	495	888	207	16.1
Flavobacterium balustinum, Fl.indologenes, Fl.indolthetic, Fl.gleum	CFB	202	93	225	3.7
Haemophilus parasuis , H.paracuniculus, Pasteurella haemolytica H.parasuis, environmental isolate WHB462	γ Proteobacteria unknown	496-498	882-884	1099-1102	2.6
Buchnera aphidicola	γ Proteobacteria	496	427	1098	2.4
Flavobacterium odoratum	CFB	140	308	1085	1.8
Clostridium herbivorans	Bacill/Clostrid	486	469	1091	1.8
M.mlo.twb	unknown	486	469	600	1.8
Rhodospirillum centenum	α Proteobacteria	149	824	513	1.7
Rhodospirillum centenum	α Proteobacteria	439	827	515	1.7
Serratia marcescens, Yersinia enterocolitica	γ Proteobacteria	496	884	37 3	1.6
Escherichia coli, Vibrio anguillarum, Salmonella enteritidis, Citrobacter freundii, V.gazogenes str.PB 1, V.damsela, V.ordalii, Proteus mirabilis	γ Proteobacteria	495-496	426-497	372-373	1.6
Plesiomonas shigelloides, Erwinia carotovora	γ Proteobacteria	496	884	373	1.6
Listeria murrayi str.F-9.	Bacill/Clostrid	149	895	579	1.5
Ruminobacter amylophilus	β Proteobacteria	489	643	218	1.0
Pseudomonas flavescens	γ Proteobacteria	490	644	207	1.0
Vibrio vulnificus	γ Proteobacteria	496	650	373	1.0
Pasteurella haemolytica, Actinobacillus suis	γ Proteobacteria Actinobacteria	496	650	1099	1.0
Vibrio navarrensis, V. vulnificus, V. diazotrophicus	γ Proteobacteria	495-496	649-650	372-373	1.0
Pasteurella species strain Bisga, Actinobacillus capsulatus, P.haemolytica, Act.hominis, Act.suis, Act.equuli	γ Proteobacteria Actinobacteria	496	650	1099	1.0
*Pseudomonas cepacia**	γ Proteobacteria	143	887	207	0.8
Klebsiella species	γ Proteobacteria	494	882	371	0.8
Oceanospirillum commune	γ Proteobacteria	494	648	371	0.8
Anaerobranca horikoshii	Bacill/Clostrid	140	461	369	0.8
Weeksella virosa	CFB	140	93	371	0.8
Listeria seeligeri, L.ivanovii	Bacill/Clostrid	147	434	577	0.8
Bacillus cereus, B.mycoides, B.thuringiensis, B.anthracis, B.medusa	Bacill/Clostrid	147	488	579	0.8
Hirschia baltica	α Proteobacteria	403	422	513	0.7
Brochothrix thermosphacta.	Bacill/Clostrid	556	893	239	0.7
Lactococcus lactis, Streptococcus salivarius, Globicatella sanguis, L.garvieae, S.sanguis, S. bovis	Bacill/Clostrid	555-556	890-894	581-583	0.7
Bacillus fusiformis	Bacill/Clostrid	147	458	242	0.7
Alicyclobacillus acidoterrestris	Bacill/Clostrid	147	488	242	0.7
Buchnera aphidicola	γ Proteobacteria	502	433	1107	0.6

Table 3. (cont.)

Buchnera aphidicola	γ Proteobacteria	502	433	1107	0.6
Flavobacterium lutescens	CFB	490	878	564	0.6
Pseudomonas testosteroni	γ Proteobacteria	457	427	566	0.6
Bacillus brevis	Bacill/Clostrid	153	463	565	0.6
Buchnera aphidicola	γ Proteobacteria	501	432	1104	0.6
*Pantoea ananas (DC147)**	γ Proteobacteria	498	426	373	0.6
Vibrio species	γ Proteobacteria	499	430	573	0.6
Buchnera aphidicola	γ Proteobacteria	499	430	1104	0.6
*Pantoea herbicola pathovar gypso**	γ Proteobacteria	500	426	373	0.6
Haemophilus parasuis	γ Proteobacteria	498	884	1101	0.6
Buchnera aphidicola	γ Proteobacteria	498	429	574	0.6
Afipia felis	α Proteobacteria	440	423	342	0.6
Nitrosolobus multiformis	β Proteobacteria	461	459	207	0.5
Marinomonas vaga	γ Proteobacteria	459	647	567	0.5
Mycoplasma species strain PG50, M. capricolum species strain F38 M. putrefaciens, M. mycoides	Bacill/Clostrid	540-542	471-473	832-834	0.4
Spiroplasma citri str.R8A, Spiroplasma species strain DW-1.	Bacill/Clostrid	537-538	468-469	563-564	0.4
Spiroplasma species strain TG-1.	Bacill/Clostrid	539	470	1080	0.4
Clostridium sordellii	Bacill/Clostrid	457	117	1062	0.3
Mycoplasma orale str.CH19	Bacill/Clostrid	153	471	1072	0.3
Cyanophora paradoxa	Eukaryota	498	429	1052	0.3
Spiroplasma species strain DU-1.	Bacill/Clostrid	538	469	1052	0.3
Synechocystis species	Cyanobacteria	494	425	1048	0.3
Methylosinus species	α Proteobacteria	153	424	65	0.3
Lactobacillus thermophilus	Bacill/Clostrid	147	487	242	0.2
Proteus vulgaris	γ Proteobacteria	496	427	213	0.2
Legionella pneumophila	γ Proteobacteria	496	885	213	0.2
Methylococcus capsulatus	γ Proteobacteria	486	874	204	0.2

Day 136 Species corresponding to three observed TRFs	Phylogenetic group	*MspI* TRF size (bp)	*RsaI*TRF size (bp)	*HhaI* TRF size (bp)	Min TRF Area (%)
Haemophilus parasuis	γ Proteobacteria	498	884	1101	9.2
Pasteurella haemolytica H.parasuis H.paracuniculus environmental isolate WHB462,	γ Proteobacteria unknown	496	882-884	1099-1102	7.3
*Pseudomonas saccharophila**	γ Proteobacteria	494	887	208	4.6
Cyanophora paradoxa	Eukaryota	498	429	1052	2.9
Serratia marcescens, Plesiomonas shigelloides, Yersinia enterocolitica, P.shigelloides, Erwinia carotovora	γ Proteobacteria unknown	496	884	373	2.9
Escherichia coli, Citrobacter freundii Vibrio ordalii, Proteus mirabilis, Salmonella enteritidis, Hafnia alvei, V. gazogenes	γ Proteobacteria	496-497	427-428	373-374	2.4
Buchnera aphidicola	γ Proteobacteria	496	427	1098	2.4
Clostridium sticklandii	Bacill/Clostrid	445	428	1050	2.1
Klebsiella species	γ Proteobacteria	494	882	371	2.1

Table 3. (cont.)

Clostridium oroticum	Bacill/Clostrid	226	472	1087	2.0
Buchnera aphidicola, Symbiont *Acrt.P*	γ Proteobacteria unknown	498	429	574	2.0
Listeria murrayi str.F-9.	Bacill/Clostrid	149	895	579	2.0
Rhodospirillum centenum	α Proteobacteria	439	827	515	1.6
Rhodospirillum centenum	α Proteobacteria	149	824	513	1.6
*Pantoea ananas (DC 130)**	γ Proteobacteria	500	428	374	1.6
Oceanospirillum commune, Vibrio diazotrophicus	γ Proteobacteria	494-495	648-649	371-372	1.5
Mycoplasma putrefaciens, M.capricolum, M.species strain F38, M.mycoides	Bacill/Clostrid	541-542	472-473	833-834	1.1
Buchnera aphidicola	γ Proteobacteria	509-511	440	1116-1117	1.0
Wolbachia persica	α Proteobacteria	490	644	848	1.0
Spirillum volutans, Kinetoplastibacterium critidum	β Proteobacteria	489	472	206	1.0
*Pseudomonas chlororaphis**	γ Proteobacteria	492	891	206	1.0
Photobacterium leiognathi Vibrio splendidus, V. anguillarum	γ Proteobacteria	505-506	436-437	579-580	1.0
Vibrio damsela	γ Proteobacteria	506	894	580	1.0
Xanthomonas phasedi, X. campestris, X. oryzae	γ Proteobacteria	498	481	215	0.9
Methylobacillus flagellatus, Xylella fastidiosa	β Proteobacteria γ Proteobacteria	496	479	373	0.9
Escherichia coli	γ Proteobacteria	492	423	371	0.9
Escherichia coli, Vibrio damsela, V.anguillarum	γ Proteobacteria	495	426	372	0.9
Pantoea ananas (DC147), Pantoea herbicola pathovar gypso**	γ Proteobacteria	498-500	426	373	0.9
Spiroplasma species strain DU-1.	Bacill/Clostrid	538	469	1052	0.9
Spiroplasma species strain TG-1.	Bacill/Clostrid	539	470	1080	0.9
Synechococcus species	Cyanobacteria	491	422	340	0.8
Afipia felis	α Proteobacteria	440	423	342	0.8
Rhodopseudomonas species, Methylosinus species	α Proteobacteria	152	423-424	65	0.8
Mycoplasma pneumoniae, M.genitalium	Bacill/Clostrid	545	475	226	0.8
Oscillatoria williamsii	Cyanobacteria	538	423	552	0.7
Proteus vulgaris	γ Proteobacteria	496	427	213	0.7
Legionella pneumophila	γ Proteobacteria	496	885	213	0.7
Clostridium difficile	Bacill/Clostrid	457	118	1063	0.7
Rhodopseudomonas marina	α Proteobacteria	439	118	513	0.7
Magnetospirillum species	α Proteobacteria	439	826	82	0.5
Listeria grayi	Bacill/Clostrid	148	894	578	0.5

* biocontrol related species not in the RDP, TR-fragment sizes determined empirically.

The identified bacterial species from compost samples (Table 3) shifted from a community containing primarily Gram-negative γ–, β–, and α–Proteobacteria on day 0 (Table 3) to a community also containing many members of the Gram-positive Bacillus-Clostridium and Acintobacteria (days 8, 29, 64) and abundant TRFs corresponding to the Cytophaga-Flavobacterium-Bacteroides (CFB) group (days 29 and 64). No Bacillus-Clostridium or CFB group bacteria were found on day 0.

Many different γ-Proteobacteria were found in all five samples. However on day 0, the peak area % corresponding to this group was large (Table 3). For example the 497–500 Msp1 TRFs, 428–429 Rsa1 TRFs, and 374–375 Hha1 TRFs , corresponded to 17 to 32.1% of the normalized TRF peak area. These fragment sizes correspond to TRFs found for *E. coli* and *Vibrio* species. On day 8, the identified species with the greatest peak areas were members of the Bacillus--Clostridum group (Table 3). However, no members of this group were observed on day 0. This is consistent with previous reports showing thermophilic *Bacilli* to be predominant members of the culturable bacterial community during thermophilic composting (Strom 1985 a,b; Peters et al. 2000). The day-136 composts contained many fragment sizes consistent with known *Pantoea* and *Pseudomonas* biocontrol agents as well as *Xanthomonas,* and *Bacillus* species. This finding agrees with earlier reports on the presence of these species in plant disease suppressive mature composts (Hoitink and Boehm 1999).

The greatest diversity of bacteria was observed in the stabilized day 64 and 136 composts, where 115 and 111 TRFs corresponding to members of 7 and 6 different phylogenetic groups, respectively, were observed (Table 3). The least bacterial diversity was observed on days 8 and 29 (77 and 78 TRFs corresponding to identified species in 4 and 5 phylogenetic groups, respectively) when temperatures were in the thermophilic range (Tables 1 and 2). Similarly, Strom (1985a) found a fewer number of bacterial isolates on conventional bacteriological media from compost samples obtained during the thermophilic stage as compared to those found during the mesophilic stage.

Actinomycetes are commonly identified as one of the main groups of bacteria responsible for organic matter conversion during latter stages of composting. It was of interest that no groups of three TRFs corresponding to *Actinomycetes* were observed on days 64 and 136 (Table 3) even though *Actinomycetes* strains, including S*treptomyces* spp., were among those predicted by the computer simulation to be PCR amplified and digested under the conditions studied (data not shown). Many individual TRFs corresponding to *Actinomycetes* species (e.g., *Msp1* TRFs of 128, 159 or 189 bp; *Rsa1* TRFs of 450, 452, 506 or 607 bp; and *Hha1* TRFs of 172, 252, 351, or 466–468 bp) were found in the samples (data not shown). This indicates that this group was probably a relatively minor part of the overall bacterial community in the yard trimmings composts. This finding is not inconsistent with previous studies of *Actinomycetes* in composts. A survey of culture-based studies has shown that *Actinomycetes* comprise from 10% to less than 0.1% of the culturable bacteria in various composts (Epstein 1997). A recent analysis of phospholipid fatty acids extracted from straw/manure composts showed that fatty acids indicative of *Acintomycetes* were less than 1% of the total fatty acids extracted through 100 days of composting (Klamer and Baath 1998). On the other hand, many of the TRFs found in this study did not correspond to any species in the RDP (Table 2) and some of these fragments could be from *Actinomycetes* or other strains not yet sequenced and included in the RDP.

The species identifications (Table 3) were based on the fact that TRFs obtained from three different restriction digestions of PCR amplified compost DNA corresponded to the fragment sizes found after a computer-simulated amplification and digestion of complete 16S rRNA gene sequences. This analysis yielded a total of 1368 potential species. One cannot be certain without further analysis that all of the putative species identified in Table 3 are indeed present in the compost samples without further analysis. For example, *Buchnera* species identified on days 0, 8, and 136, are obligate symbionts and probably do not occur in composts. These organisms are closely related to the Enterobacteriaceae family, of which many of the identified species were members, however. Without cloning and completely sequencing the amplified 16S rRNA genes, the sequence similarity of the amplified ribotypes to known species is not known. Another factor is TRF size redundancy, where unsequenced microorganisms, or species from different phylogenetic groups, give similar TRF sizes (Dunbar et al. 2001). In spite of these caveats, these results provide evidence that a succession of microbial groups is present during composting and that a much more diverse group of microorganisms may be active than previously reported.

Conclusions

T-RFLP is a powerful tool for studying compost microbial communities. Analysis of compost community 16S rRNA gene T-RFLPs indicated that extensive bacterial species diversity exists during composting. Substantial shifts in bacterial communities were observed during composting from a community containing primarily Gram-negative α–, β–, and ϕ–Proteobacteria (day 0) to a community containing many members of the gram positive Bacillus-Clostridium group (days 8, 29, and 64) as well as CFB and Actinobacteria groups. Bacterial species identified in these compost samples included those previously isolated from thermophilic composts by cultivation such as *Bacillus* and *Pseudomonas* spp. as well as many not previously described in composts. Abundant TRFs corresponding to *E. coli* and other Gram-negative γ Proteobacteria decreased dramatically after 8 days of composting. A diverse group of potentially beneficial plant disease biocontrol agents were identified in day 136 samples. The greatest diversity of bacteria was observed in the stabilized day 64 and 136 composts, where 115 and 111 TRFs corresponding to members of 7 and 6 different phylogenetic groups, respectively, were observed.

References

Amman RI, Ludwig W, Schleifer KH (1995) Phylogenetic identification and in situ detection of microbial cells without cultivation. Microbiol Rev 59:143-169

Beffa T, Blanc M, Aragno M (1996a) Obligately and facultatively autotrophic, sulfur- and hydrogen-oxidizing thermophilic bacteria isolated from hot composts. Arch Microbiol 165:34-40

Beffa T, Blanc M, Lyon PF, Vogt G, Marchiani M, Lott-Fischer J and Aragno M (1996b) Isolation of Thermus strains from hot composts (60° to 80° C.). Appl. Environ. Microbiol. 62(5) 1723-1727.

Blanc M, Marilley L, Beffa T, Aragno M (1997) Rapid identification of heterotrophic, thermophilic, spore-forming bacteria isolated from hot composts. Int J Syst Bacteriol 47:1246-1248

Dunbar J, Ticknor LO, Kuske CR (2001) Phylogenetic specificity and reproducibility and new method for analysis of terminal restriction fragment profiles of 16S rRNA genes from bacterial communities. Appl Environ Microbiol 67:190-197

Epstein E (1997) The science of composting. Technomic Publishing Co, Lancaster, PA pgs 53-76

Finstein M (1975) The microbiology of municipal solid waste composting. Adv Appl Microbiol 19:113-151

Hoitink HAJ, Boehm MJ (1999) Biocontrol within the context of soil microbial communities: substrate-dependent phenomenon. Annu Rev Phytopathol 37:427-446

Klamer M, Baath E (1998) Microbial community dynamics during composting of straw material studied using phospholipid fatty acid analysis. FEMS Microbiol Ecol 27:9-20

Klappenbach J, Dunbar JM, Schmidt TM (2000) rRNA gene copy number predicts ecological strategies in bacteria. Appl Environ Microbiol 66:1328-1333

Laine MM, Jorgensen KS (1997) Effective and safe composting of chlorphenol--contaminated soil in pilot scale. Environ Sci Technol 31:371-378

LaMontagne ML, Michel, FC Jr. and Reddy CA. Evaluation of DNA extraction and purification methods for obtaining PCR amplifiable DNA from compost for community analysis. J Microbiol Meth (in press).

Liu WT, Marsh TL, Cheng H, Forney LJ (1997) Characterization of microbial diversity by determining terminal restriction fragment length polymorphisms of 16S ribosomal DNA. Appl Environ Microbiol 63:4516-4522

Maidak BL, Cole JR, Lilburn TG, Parker CT Jr, Saxman PR, Stredwick JM, Garrity GM, Li B, Olsen GJ, Pramanik S, Schmidt TM, Tiedje JM (2000) The RDP (ribosomal database project) continues. Nucleic Acids Res 28:173-174

Marsh TL, Saxman PR, Cole JR, Tiedje JM (2000) Terminal restriction fragment length polymorphism analysis program, a web based research tool for microbial community analysis. Appl Environ Microbiol 66:3615-3620

Michel FC Jr., Reddy CA, Forney LJ (1995) Microbial degradation and humification of 2,4 dichlorophenoxy acetic acid during the composting of yard trimmings. Appl Environ Microbiol 61:2566-2571

Michel FC Jr., Forney LJ, Huang AJF, Drew S, Czuprenski M, Lindeberg JD, Reddy CA (1996) Effects of turning frequency, leaves to grass mix ratio, and windrow vs. pile configurations on the composting of yard trimmings. Compost Sci Util 4(1):26-43

Muyzer GE, De Waal C, Uitterlinden AG (1993) Profiling of complex microbial populations by denaturing gradient gel electrophoresis analysis of polymerase chain reaction-amplified genes coding for 16S rRNA. Appl Environ Microbiol 59:695–700

Peters S., Koschinsky S, Schwieger F, Tebbe CC (2000) Succession of microbial communities during hot composting as detected by PCR–single-strand-conformation polymorphism-based genetic profiles of small-subunit rRNA genes. Appl Environ Microbiol 66:930–936

Reddy CA, Michel FC Jr (2000) Fate of xenobiotics during composting. Proceedings of the International Symposium of Microbial Ecology. Halifax NS August 20-25, 1998: pp 485-491

Strom PF (1985a) Effect of temperature on bacterial species diversity during thermophilic solid waste composting. Appl Environ Microbiol 50:899-905

Strom PF (1985b) Identification of thermophilic bacteria in solid-waste composting. Appl Environ Microbiol 50:906-913

Tebbe CC, Schwieger F (1998) A new approach to utilize PCR-single-strand-conformation polymorphism for 16S rRNA gene-based microbial community analysis. Appl Environ Microbiol 64: 4870-4876

Tsai YL, Olson BH (1991) Rapid method for direct extraction of DNA from soil and sediments Appl Environ Microbiol 57:1070-1074

Zhou J, Burns MA, Tiedje JM (1996) DNA recovery from soils of diverse composition. Appl Environ Microbiol 62:316-322

Characterisation of Microbial Communities During Composting of Organic Wastes

N. Riddech[1], S. Klammer[1] and H. Insam[1]

Abstract. Community level physiological profiles (CLPPs) tested with Biolog microplates were studied for a maturity sequence of compost from source-separate collected organic waste and prunings. The composts were 42 , 120 and 240 days old. Traditional maturity parameters like microbial biomass, pH and temperature significantly decreased with age. Principal component analysis and discriminant analysis showed distinctly different patterns of carbon source utilisation of the different composts. Five substrates were identified that significantly contributed to the difference in CLPPs.

Introduction

Composting is a term referring to the decomposition of organic materials by microorganisms under aerobic conditions, usually encompassing mesophilic and thermophilic phases. It is an environmentally sound way to reduce organic wastes and produce organic fertiliser or soil conditioner (Gajdos 1992). Although composting is a microbiological process, little is known about the microorganisms involved and their activities during the specific phases of the composting process. Several methods such as cultural and non-cultural techniques have been used for characterising the microorganisms during composting (Laine et al. 1997; Klamer and Bååth 1998; Tiquia and Tam 2000)

Community level physiological profiling is a simple method to characterise microbial communities from environmental samples (Garland and Mills 1991; Zak et al. 1994; Insam et al. 1996; Guckert et al. 1997; Ibekwe et al. 1998; Choi et al. 1999). Biolog microplates, such as the EcoPlate (Insam 1997) containing 31 different substrates as sole carbon sources (three replicates of each substrate in a 96-well microplate) can be directly inoculated with a mixed populations from sample extractions. The color formation from reduction of a tetrazolium dye in the Biolog microplates indicates the degradation of the specific C source. Changes in the patterns of carbon source utilisation assess the functional differences in the microbial communities. The aim of this study was to differentiate microbial communities during composting of source-separate collected organic waste by using CLPPs and to identify those substrates that contribute most to this differentiation.

[1]Institute of Microbiology, University of Innsbruck, Technikerstr. 25, 6020 Innsbruck, Austria

H. Insam, N. Riddech, S. Klammer (Eds.)
Microbiology of Composting

Material and Methods

Composting Site, Sampling and Sample Preparation

Samples were collected from the open-air windrow composting site at Völs, Austria. The substrates were source-separate collected organic waste, amended with suitable amounts of bulking agents (wood chips, prunings). The windrows were turned two times a month, watering was not necessary. The samples (1 kg) were taken at a depth of about 20 cm from three piles of different maturity. Each pile was sampled at three distinct locations. Samples were immediately sieved (2-mm mesh) for analysis.

The temperature at a depth of about 60 cm was recorded using a thermocouple (Sandberger, Austria). The moisture content was determined by drying at 105 °C for 24 h. For pH determination, 1 g of compost was diluted with 10 ml of 1 M KCl solution and measured with a pH meter (Sentron, 2001 pH) (Forster 1995).

Microbial Extraction

For the extraction, a modified procedure of Hopkins et al. (1991) (Insam et al., 2001) was used. Five g of sample was blended with 20 ml of 0.1% (w/v) sodium cholate solution, 8.5 g of cation exchange resin (Dowex 50 W x 8, 20-50 mesh, Sigma) and 30 glass beads. The suspension was shaken on a reciprocating shaker at 250 rpm for 2 h at 4 °C and centrifuged at 2200 rpm for 2 min. The supernatant was transferred to a sterile flask. The pellet was resuspended in 10 ml of Tris buffer (pH 7.4), shaken for 1 h at 4 °C, centrifuged (2200 rpm, 2 min) and the supernatant was added into the flask of step one.

Microbial Biomass

Microbial biomass (C_{mic}) was determined by substrate-induced respiration (SIR) (Anderson and Domsch 1978). Glucose (2%) was added to 20g of sample and the CO_2 release was measured by using a continuous flow infrared gas analyser at 22 °C (Heinemeyer et al. 1989)

Microplates and Inoculation

The Biolog EcoPlate contains 31 different C sources and a control well without a C source in three replications. The inoculum density in the compost extract was standardised by estimating the activity of microorganisms from the microbial biomass. The suspensions were diluted with sterile Ringer solution (Merck). Each well of the EcoPlate was inoculated with 125 µl of the diluted compost extract and incubated at 37 °C in the dark. The colour formation at 592 nm was measured every 8 h for 7 days by using an automatic plate reader (SLT SPECTRA, Grödig, Austria).

Data Analysis

The optical density of all wells was used for calculating the average well color development (AWCD). The AWCD was calculated at each reading time as follows: AWCD = Σ (R-C)/n, where R is color development in the well of each substrate, C is the optical density in the control well and n is the number of substrates on the Biolog EcoPlate (n=31) (Garland 1996). The value (R-C) for each well was transformed by dividing by the AWCD (Garland and Mills 1991). The resulting values were analysed by using principal component analysis (PCA) and discriminant analysis (DA). Pearson correlation coefficient were examined for all correlations. We used one–way analysis of variance (ANOVA) and the Bonferroni test (significance level 0.05) to determine which carbon substrates significantly differentiated between the various age of compost samples. Pearson correlation, ANOVA, principal component analysis (PCA) and discriminant analysis (DA) were computed with the SPSS program package (SPSS inc., Chicago, version 9.0,).

Results

Moisture Content, Temperature, pH and Microbial Biomass

The highest pH (8.3) and temperature (68 °C) were found at 42 d of composting. Both values decreased during the subsequent weeks. The moisture content declined only slightly from 41% to 36% (Fig.1). During the period of observation, the microbial biomass decreased from 2788 to 1946 and 770 μg C_{min} g^{-1} dry weight (Fig. 2). Microbial biomass, pH and temperature were significantly correlated ($p<0.01$) with the age of the composts, whereas no significant correlation between moisture content and age of compost was found.

Community Level Physiological Profiles

Principal component analysis, performed on the transformed dataset after 120 h of incubation, showed that the CLPPs of the three composts were distinctly different (Fig.3). PC1 mainly differentiated the young compost (42 d of composting) from the two old stages (120 and 240 d, respectively), explaining 23.6% of the variance. PC2 further separated the two older composts, explaining another 14% of the variance.

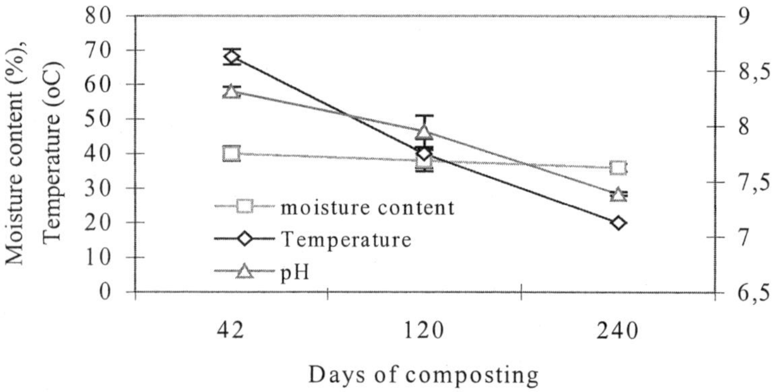

Fig. 1. Moisture content, temperature (°C) and pH after 42, 120 and 240 days of composting. Values reveal means ± standard deviation, $n = 3$

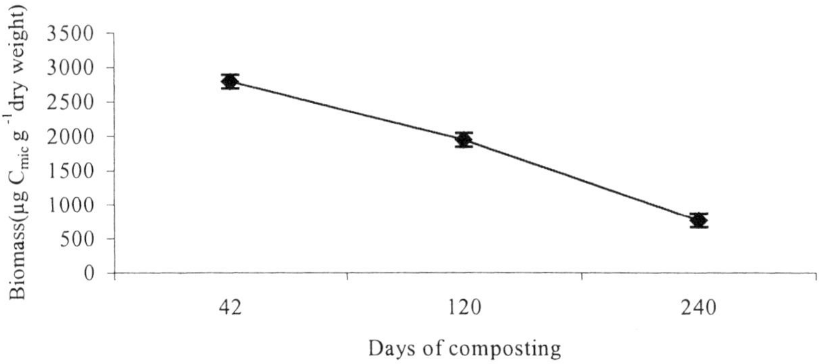

Fig. 2. Microbial biomass after 42, 120 and 240 days of composting. Values show mean ± standard deviation, $n = 3$

Carbon substrates that were most important for the separation of the different stages are shown in Table 1. PC1 indicated that the communities of the 42 d old composts utilised a number of carbohydrates (D-cellobiose, D-galactonic acid γ-lactone and D,L-α-glycerol phosphate), several carboxylic acids (D-glucosaminic acid, pyruvatic acid methyl ester, α-ketobutyric acid, D-malic acid), α-cyclodextrin, L-phenylalanine and phenyl ethylamine to a greater degree than the communities of 120 or 240 d of composting. On the other hand, microbial communities associated with 42 d old composts showed a poor utilisation of Tween 40, L-threonine and 4-hydroxybenzoic acid. PC 2 was positively correlated with glycogen, i-erythritol, two carboxylic acids (itaconic acid and γ-hydroxybutyric acid), L-serine and negatively correlated with Tween-80, two carbohydrates (D-mannitol and β-methyl-D-glucoside) and L-arginine utilisation.

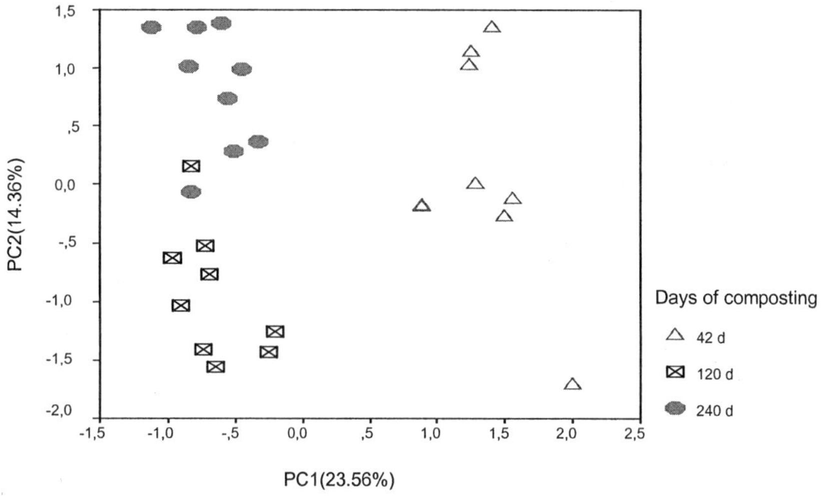

Fig.3. Principal component analysis of different compost maturity stages

Table 1. Correlation of carbon substrates with the PC1 and PC2 scores according to PCA (see Fig. 3). correlation significant at the $P < 0.01$ are given

Carbon substrates	PC1	PC2
Polymers: α-cyclodextrin	0.84	
Tween 40	-0.66	
Tween 80		-0.54
glycogen		0.49
Carbohydrates: D-cellobiose	0.64	
D-galactonic acid γ-lactone	0.46	
D,L-α-glycerol phosphate	0.45	
D-mannitol		-0.68
i-erythritol		0.61
β-methyl-D-glucoside		-0.46
Carboxylic acids: D-glucosaminic acid	0.79	
pyruvatic acid methyl ester	0.64	
α-ketobutyric acid		0.57
D-malic acid	0.53	
Itaconic acid		0.65
γ-hydroxybutyric acid		0.57
Amino acids: L-threonine	-0.90	
L-phenylalanine	0.77	
L-serine		0.55
L-arginine		-0.49
Amines: phenyl ethylamine		0.51
Phenolic compounds: 4-hydroxybenyoic acid	-0.67	

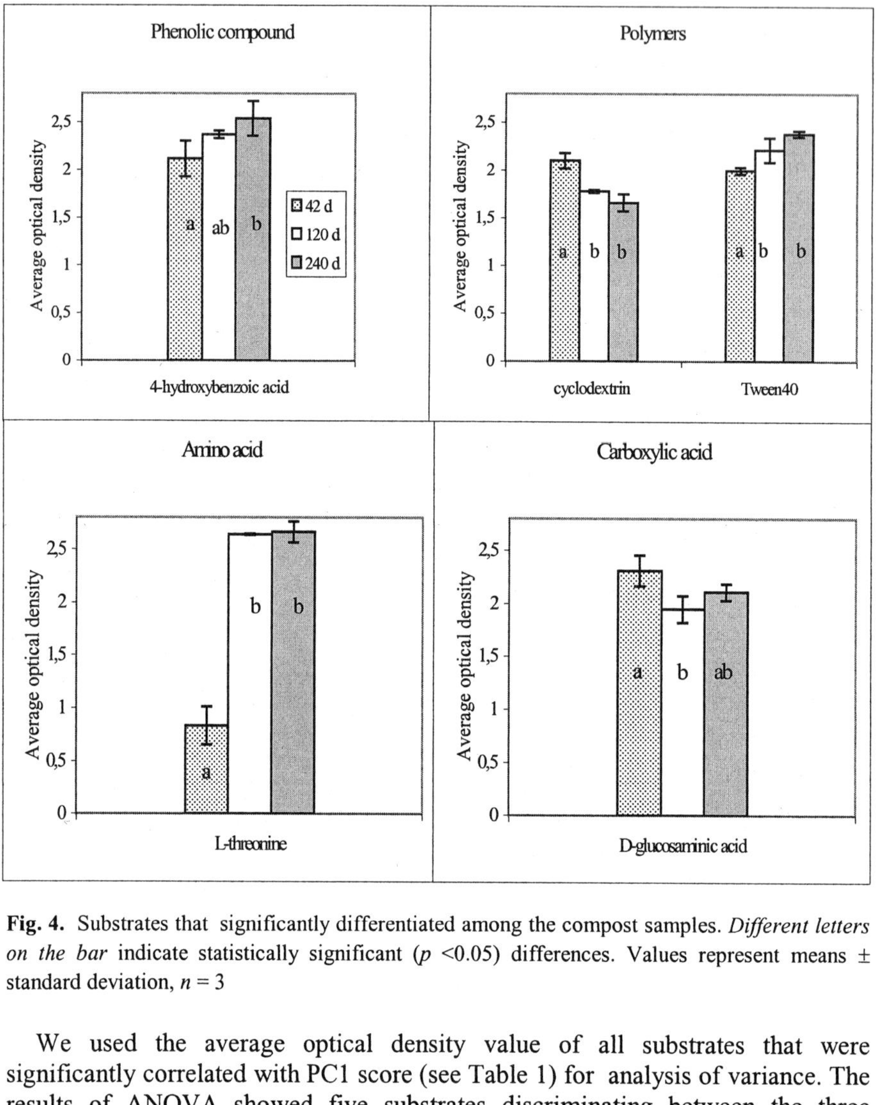

Fig. 4. Substrates that significantly differentiated among the compost samples. *Different letters on the bar* indicate statistically significant (p <0.05) differences. Values represent means ± standard deviation, $n = 3$

We used the average optical density value of all substrates that were significantly correlated with PC1 score (see Table 1) for analysis of variance. The results of ANOVA showed five substrates discriminating between the three compost samples (Fig.4). Two groups of the microbial communities were significantly classified by L-threonine, α-cyclodextrin and Tween 40. 4-hydroxybenzoic acid and D-glucosaminic acid significantly differentated only between the youngest and the oldest composts.

All the transformed data were used for discriminant analysis. In the case of DA, function 1 which explained 73.7% of the variance, differentiated between the young and the two older composts. Function 2 separated between the middle and the old compost, explaining 23.4% of variance. These results indicate differences in the functional properties of the three composts (Fig.5).

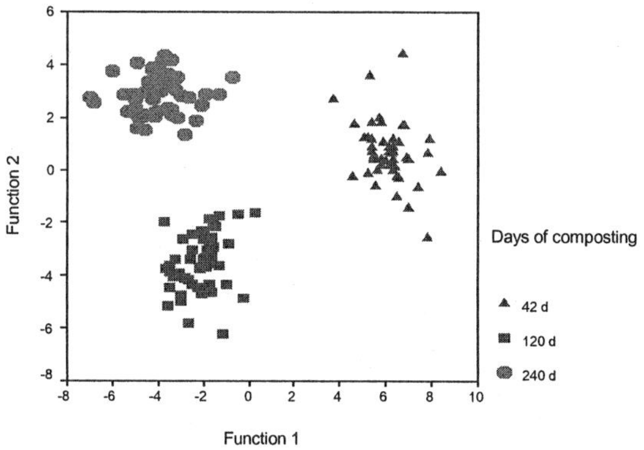

Fig. 5. Discriminant analysis after community level physiological profiles of composts of three different maturity stages

Substrate Classification

The carbon substrates on the Biolog EcoPlates were divided into six groups (polymers, carbohydrates, carboxylic acids, amino acids, amines and phenolic compounds) to determine the relative contribution during composting. We found that the average optical density of amines significantly increased ($p<0.05$) with compost age. In contrast, no significant correlation between carbohydrate to amino acid ratio, polymer, carbohydrate, carboxylic acid, amino acids, phenolic acid utilisation and the age of compost were found (data not shown).

Discussion

During the composting process, the pH dropped from 8.3 to 7.4, which may be due to the ammonification and mineralisation of organic matter by the activities of microorganisms as found by Wong et al. (2001). The decline in moisture content may be explained by microbial heat generation causing enhanced desiccation (Tiquia and Tam 2000). Physical characteristics such as temperature give a general idea of the decomposition stage reached (Bernal et al. 1998). In this experiment, the continuous decline in temperature indicated that the compost had gone through the thermophilic stage and approached maturity. The microbial biomass also decreased with composting age. Insam et al. (1996) reported that the microbial biomass during composting of manure significantly declined after depletion of the easily available substrates.

Sharma et al. (1998) reported that a decline in the carbohydrates-to-amino acid (CH/AA) ratio during incubation of litter-amended soil samples may be related to the change in available substrates, i.e. a shift from carbohydrate-rich plant tissue to protein-rich microbial mass. For this composting situation, we made a similar observation, with a decreasing utilisation of carbohydrates and an increasing use of amino acids during 42, 120 and 240 d of composting. However, no significant differences in the CH/AA ratio were found.

Community level physiological profiles were already earlier used for characterising microbial communities in compost (Insam et al. 1996; Laine et al. 1997; Carpenter-Boggs et al.1998). Carpenter-Boggs et al. (1998) found that the common plant sugars such as sucrose, galactose and fructose were utilised by the compost microbial communities during composting dairy waste. In the present investigation, the correlation between carbon substrate utilisation and the PC scores (Table 1) showed that the microbial communities in the 42 samples were clearly differentiated from the communities in 120- or 240 d samples. α-cyclodextrin, Tween 40, D-glucosaminic acid, L-threonine and 4-hydroxybenzoic revealed an ability to classify the microbial communities. We also found that the CLPPs of the different compost maturity stage were clearly separated by discriminant analysis. Lulu (2000) reported a shift in the CLPPs during composting of organic waste might be due to changes in polymers, carbohydrates, carboxylic acids and amino acids utilisation. Our results confirm that changes in the substrate utilisation profile are useful for characterising the microbial communities from compost of different age.

Acknowledgements. FWF project P13953-MOB supported this study. The authors thank Khon Kaen University for the KKU scholarship to N. Riddech. We would like to thank Meinhard Rudig for the compost samples from Völs.

References

Anderson JPE, Domsch KH (1978) A physiological method for quantitative measurement of microbial biomass in soils. Soil Biol Biochem 10:215-221

Bernal MP, Paredes C, Sanchez-Monedero MA, Cegarra J (1998) Maturity and stability parameters of composts prepared with a wide range of organic wastes. Biores Technol 63:91-99

Carpenter-Boggs L, Kennedy AC, Reganold JP (1998) Use of phospholipid fatty acids and carbon source utilisation patterns to track microbial community succession in developing compost. Appl Environ Microbiol 64:4062-4064

Choi K-H, Dobbs FC (1999) Comparison of two kinds of Biolog microplates (GN and ECO) in their ability to distinguish among aquatic microbial communities. J Microbiol Meth 36:203-213

Forster J (1995) Determination of soil pH. In: Alef K, Nannipieri P (eds) Methods in Applied Soil Microbiology and Biochemistry. Academic Press, Tokyo, p 55

Gajdos R (1992) The use of organic waste materials as organic fertilizers-recycling of plant nutrients. Acta Hortic 302:325-331.

Garland JL (1996) Patterns of potential C source utilisation by rhizosphere communities. Soil Biol Biochem 28: 223-230

Garland JL, Mills AL (1991) Classification and characterisation of heterotrophic microbial communities on the basis of patterns of community-level sole-carbon-source utilisation. Appl Environ Microbiol 57:2351-2359

Guckert JB, Carr GJ, Johnson TD, Hamm BG, Davidson DH, Kumagai Y (1996) Community analysis by Biolog: curve integration for statistical analysis of activated sludge microbial habitats. J Microbiol Meth 27:183-197

Heinemeyer O, Insam H, Kaiser EA, Walenzik G (1989) Soil microbial biomass measurements: an automated technique based on infra-red gas analysis. Plant Soil 116:191-195

Hopkins DW, MacNnaughton SJ, O'Donnell AG (1991) A dispersion and differential centrifugation technique for representatively sampling microorganisms from soil. Soil Biol Biochem 23:217-225

Ibekwe AM, Kennedy AC (1998) Phospholipid fatty acids profiles and carbon utilisation patterns for analysis of microbial community structure under field and greenhouse conditions. FEMS Microbiol Ecol 26:151-163

Insam H (1997) A new set of substrates proposed for community characterisation in environmental samples. In: Insam H, Rangger A (eds) Microbial communities: functional versus structural approaches. Springer, Berlin Heidelberg New York, pp 259-260

Insam H, Amor K, Renner M, Crepaz C (1996) Changes in functional abilities of the microbial community during composting of manure. Microb Ecol 31:77-87

Insam H, Feurle J, Rangger A (2001) Community level physiological profiles (Biolog substrate use tests) of environmental samples. In: Akkermans ADL, van Elsas JD, DeBruijn FJ (eds): Molecular Microbial Ecology Manual, 5th Supplement, Kluwer, Amsterdam

Klamer M, Bååth E (1998) Microbial community dynamics during composting of straw material studied using phospholipid fatty acid analysis. FEMS Microb Ecol 27:9-20

Laine M.M, Haario H, Jørgensen (1997) Microbial functional activity during composting of chlorophenol-contaminated sawmill soil. J Microbiol Meth 30:21-32

Lulu B (2000) Evaluation of soil organic matter pools of some Ethiopian agricultural soils and compost maturity: a microbial approach. Dissertation, University of Innsbruck, Innsbruck

Sharma S, Rangger A, Insam H (1998) Effects of decomposing maize litter on community level physiological profiles of soil bacteria. Microb Ecol 35:301-310

SPSS (1998) Statistical package of the social sciences. SPSS Inc, Chicago

Tiquia SM, Tam NFY (2000) Co-composting of spent pig litter and sludge with forced aeration. Biores Technol 72:1-7

Wong JWC, Mak KF, Chan NW, Lam A, Fang M, Zhou LX, Wu QT, Liao XD (2001) Co-compost of soybean residues and leaves in Hong Kong. Biores Technol 76:99-106

Zak JC, Willig MR, Moorhead DL, Wildman HG (1994) Functional diversity of microbial communities: a quantitative approach. Soil Biol Biochem 26:1101-1108

Microbial Succession During Composting of Source-Separated Urban Organic Household Waste Under Different Initial Temperature Conditions

I. Sundh and S. Rönn[1]

Abstract. The effects of initial temperature regime on microbial community succession during controlled composting of organic household waste in a laboratory reactor was determined by analysis of phospholipid fatty acids (PLFAs). Spontaneous self-heating of the substrate led to substantial microbial biomass increase (maximum PLFA concentration of 2000 nmol g^{-1} d.w.) and high CO_2 production in the thermophilic phase (regulated at 55 °C). In contrast, when the initial period of temperature increase was shortened by external heating, there was a negligible increase in biomass and only a small increase in CO_2 production. Thus, attempts to speed up the process initially by external heating are not advisable. The increase in PLFA concentration under self-heating conditions occurred mainly in iso- and anteiso-branched fatty acids (more than 100-fold increase, their maximum corresponding roughly to 10^{11} bacterial cells g^{-1} d.w.) from different types of thermophilic bacteria. One PLFA typical of actinomycetes (10Me18:0) had a low initial concentration, but started to increase during the thermophilic phase. The abundance of polyunsaturated PLFAs generally decreased during composting, indicating no growth of eukaryotes.

Introduction

A well-managed composting process has a comparatively small environmental impact and the stable product has a large potential as a nutrient additive and soil conditioner in farming, and as a horticultural growth substrate. Composting of the organic fractions of urban household waste is therefore expected to increase substantially in many countries, partly due to legislative changes (Barth and Kroeger 1998; de Bertoldi 1998).

Composting is a microbiological process and understanding its microbial ecology can be useful to optimize the process. Additionally, standardized microbiological analyses may potentially be used for judging the quality and maturity of the compost residue, a task that has proven difficult (Mathur et al. 1993; Hue and Liu 1995). Many studies employing plate counts on selective media have described the dynamics during composting of, e.g. aerobic and anaerobic bacteria, actinomycetes, mesophilic and thermophilic organisms, fungi

[1]Department of Microbiology, SLU, P.O. Box 7025, 750 07 Uppsala, Sweden

H. Insam, N. Riddech, S. Klammer (Eds.)
Microbiology of Composting

and nitrifying and denitrifying bacteria (Fermor et al. 1979; Davis et al. 1992; Tam 1995; Atkinson et al. 1996; Vuorinen and Saharinen 1997). Several reviews on microbiology of composting have been written (Finstein and Morris 1975; de Bertoldi et al. 1983; Miller 1993).

Although some organisms that are characteristic of different stages of composting are known, knowledge of the quantitative aspects of the microbial ecology of composting is still poor, due to the selectivity of the plate count method. The introduction of growth-independent, non-selective biomarker methods has been an important step forward, and a few studies have employed DNA/RNA-based methods (Koschinsky et al. 1998; Kowalchuk et al. 1999; Peters et al. 2000) and analysis of phospholipid fatty acids (PLFAs) (Herrmann and Shann 1997; Carpenter-Boggs et al. 1998; Klamer and Bååth 1998) in compost samples.

In order to study the effects of important environmental factors on composting of urban organic household waste, a 200-l indoor reactor has been built at SLU, allowing composting under tightly regulated temperature and oxygen concentrations (Smårs et al. 2001). Temperature regime has a drastic impact on the composting process and the first reactor runs investigated effects of different initial temperature conditions.

In this study, we compare the development of the microbial biomass and community structure in reactor runs where the substrate was either externally heated, quickly attaining thermophilic conditions, or allowed to heat spontaneously. The microbial analyses are set in perspective of overall changes of mineralization rates in the reactor. The microbial community was described by analysis of PLFAs (bacteria and eukaryotes) and intact diether lipids (archaea).

Materials and Methods

Substrate

The substrate consisted of the organic fraction of source-separated household waste from the city of Uppsala, Sweden, collected during February 1995. After collection, it was sorted, homogenized and stored frozen. The pH of the waste was 4.9 (±2%, =coefficient of variation) and the ash and C contents were 25% (±2%) and 37% (±4%) of the d.w., respectively. Contents of the major fractions of organic matter (% of the ash-free d.w.) were: cellulose 16 (±5%), hemicellulose 3.2 (±46%), lignin 10 (±7%), starch 13 (±10%), simple sugars 1.6 (±25%) and crude fat 15 (±9%). The N and P contents were 2.2% (±5%) and 0.4% (±20%) of the d.w., respectively. Further details about the composition of the waste are given by Eklind et al. (1997). As soon as possible, after some thawing, the substrate was mixed with wheat straw (0.3:1 straw to waste ratio on d.w. basis), homogenized with a mincer, and transferred to the reactor (Beck-Friis et al. 2001).

The Compost Reactor

The reactor is basically a stainless steel container of 200-l volume. The material in the reactor is kept uniform by circulating the compost air through the material with a fan, and by intermittent (every 24 h) mixing by rotation of the reactor. Oxygen supply, temperature and moisture are independently controlled and regulated. Temperature was registered automatically by thermistors in the reactor and the gas stream, and the CO_2 concentration in the outgoing gas was measured every 5 min (Smårs et al. 2001).

Experimental Reactor Runs

Replicate composting runs were made at different initial temperature regimes. In one type, the substrate mixed with straw was placed in the reactor and heated rather quickly (5–10 h) to 55 °C by external heating (=EH experiment). In the other type, the substrate was allowed to heat spontaneously to 55 °C (ca. 5 d) (=SH experiment). Thereafter, the temperature was kept at 55 °C, the oxygen concentration at 16% and the moisture content at approximately 65% for the rest of the experiments. The process parameters (e.g. the CO_2 production) were very similar within replicate runs (Beck-Friis et al. 2001; Smårs et al. 2001). Microbial communities were investigated in one run for each temperature treatment.

Samples for lipid analysis were taken once a day for approximately 2 weeks. Three replicate samples of 15-20 g were taken on each occasion, between thoroughly mixing the reactor contents. Samples were stored at -20 °C until analysis.

Lipid Analysis

Methods for the lipid extraction, preparation of methyl esters of the PLFAs, derivatization of diethers, and quantifications and identifications with GC and GC-MS analyses, respectively, were described earlier (Sundh et al. 1995; Virtue et al. 1996). However, in contrast to Virtue et al. (1996), we used helium as the carrier gas, and a HP-1MS column (length 30 m, i.d. 0.25 mm, film thickness 0.1 mm).

The fatty acids are named by the total number of carbon atoms, followed by a colon and the number of double bonds. Then ω and a number show the position of the double bond, relative to the aliphatic end of the chain. The prefixes i, a and 10Me refer to methyl branching at the iso and anteiso positions and at the 10th carbon from the carboxyl end, respectively. Cyclopropane fatty acids have the prefix cy. The PLFAs 18:2a and 18:2b are isomers where the double bond positions were uncertain.

Certain PLFAs were assigned to microbial groups based on previous studies (Lechevalier and Lechevalier 1988; Vestal and White 1989; White et al. 1996). Thus, we considered 10Me18:0 to be a signature of actinomycetes, all monounsaturates containing from 14 to 19 carbon (except 18:1ω9) as representing

Gram-negative bacteria, polyunsaturated (18:3 and 18:2) to be fungal markers, and iso- and anteiso-branched with 14–19 carbon + 15:0 and 17:0 as representing Gram-positive bacteria.

Concentrations of PLFAs were transformed to approximate cell numbers, assuming that bacteria contain 100 μmol PLFA g^{-1} d.w. (White et al. 1979). Conversion factors for cell volumes were taken from Bergey's Manual (1984) (using *Bacillus circulans* and *B. stearothermophilus* isolated from compost [Strom 1985b] as model bacteria) and the dry weight contents determined according to Norland et al. (1987).

Statistics

First, the variation in PLFA composition within the three replicate samples from each sampling occasion was determined for one occasion from experiment EH, and three occasions from experiment SH (beginning, middle and end of run). Since the variation among the replicate samples was much smaller (mean CV for single PLFAs was 18%) than the variation among the three sampling occasions of SH, subsequently only one replicate from each occasion was analyzed.

The overall patterns of variation in PLFA composition were investigated with principal component analysis (PCA), using the JMP software (SAS Institute Inc.). We used standardized principal components, where the PC scores are scaled to have unit variance instead of canonical variance, directly for concentrations in nmol g^{-1} d.w. General differences among samples were detected in two-dimensional plots of the first two PCs.

Results

In runs where the temperature was quickly raised to 55 °C by external heating (EH), the total mineralization (measured as CO_2 evolution) had a small peak after only 3 d, but declined quickly thereafter (Fig. 1a,b). Production of CO_2 was not measured after day 7 in this experiment since there were no signs of further activity in the compost. When the temperature was allowed to rise spontaneously, however, 4 d elapsed before the temperature reached 55 °C, and the mineralization rate peaked after 8 d (Fig. 1a,b). The total CO_2 production was much larger in SH than in EH. By day 14, the activity had decreased substantially in both runs and the experiments were terminated.

The total PLFA concentration at the start of experiments was approximately 400 nmol g^{-1} d.w. In EH, the concentration remained close to this level throughout the run (Fig. 1c). In SH, it parallelled the CO_2 production, with a maximum of 2000 nmol g^{-1} d.w. by day 10. This implies growth and a large microbial biomass increase in SH, but not in EH.

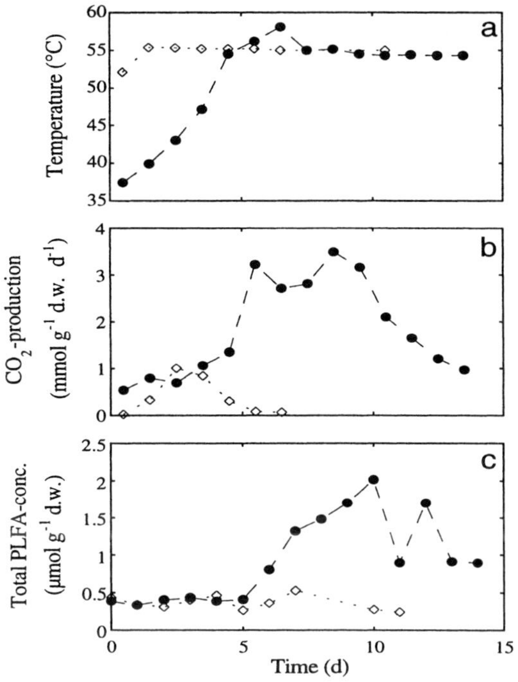

Fig. 1a-c. Temperature, CO2 production and total PLFA concentration during composting of organic household waste at different initial temperature conditions. Experiments EH: external heating (open squares); and SH: spontaneous heating (closed circles)

The PLFA composition at the start of the experiments (reflecting that in the frozen substrate), was dominated by four fatty acids, 16:0, 18:0, 18:2b and 18:1ω9.

The PCA demonstrated substantial, time-dependent variation in PLFA composition, and hence in the microbial community structure, in the self-heated process (Fig. 2a), but much less in the externally heated (data not shown). Similarly, the clustering of fatty acids into groups in the loading plot was stronger in SH (Fig. 2b).

In SH, one group of PLFAs (e.g. i16:0), had intermediate values in PC2 and high values in PC1 (Fig. 2b). These fatty acids had very low start concentrations but increased considerably, beginning after 5 days and peaking on day 10 (Fig. 3a). Another group of PLFAs (e.g. 18:2b), which had very low values in PC1, generally decreased in concentration over time in both runs (Fig. 3b). A few fatty acids were relatively high in both PCs (e.g. 16:0). Like i16:0, they increased substantially after day 5 in SH, but had relatively high concentrations from the start (Fig. 3c). Furthermore, they peaked in concentration earlier (day 7 instead of day 10). The PLFAs lowest in PC2 (e.g. 15:1) had low concentrations from the start and increased in SH, but not until day 8–9 (Fig. 3d).

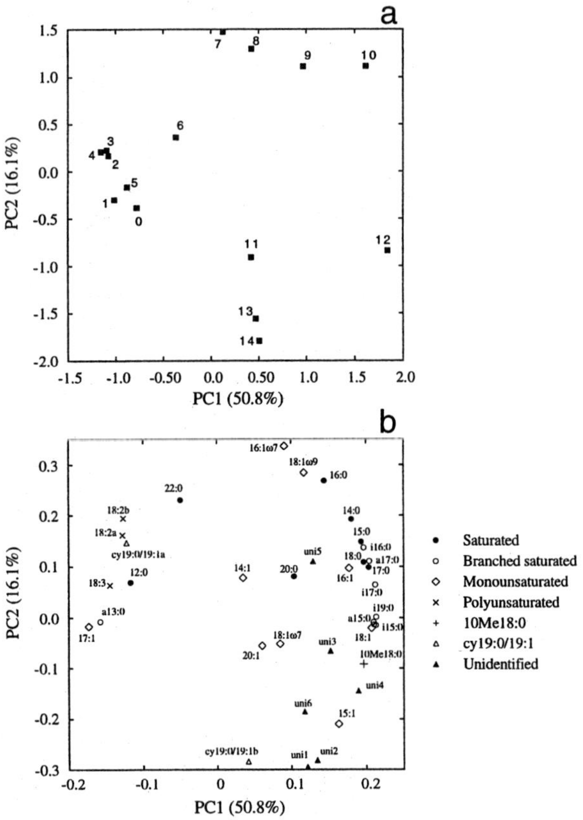

Fig. 2a-b. Principal component analysis of PLFA data from a reactor composting run. Experiment SH. **a**) Score plot for the different samples. *Numbers* denote day of sampling. **b**) Loading plot with all the individual fatty acids. (*uni*=unidentified PLFAs)

In the self-heated run, the sum of PLFAs from Gram-positive bacteria developed similarly to the concentration of i16:0 (Fig. 4a). The eukaryotic markers all belonged to the cluster with lowest PC1 and the sum of these fatty acids generally decreased over time (Fig. 4b). The sum of PLFAs from Gram-negative bacteria was small and changed very little, while the actinomycete marker (10Me18:0) occurred only after day 7.

Based on the maximum concentrations of PLFAs indicating Gram-positive bacteria, the maximum number of *Bacillus*-type cells was calculated to be 7.9 10^9 and 6.2 10^{10} cells g^{-1} d.w. during the externally and spontaneously heated compost runs, respectively.

With the experimental procedures used here, the diether lipids from archaea were in the region of the detection limit and therefore not quantified.

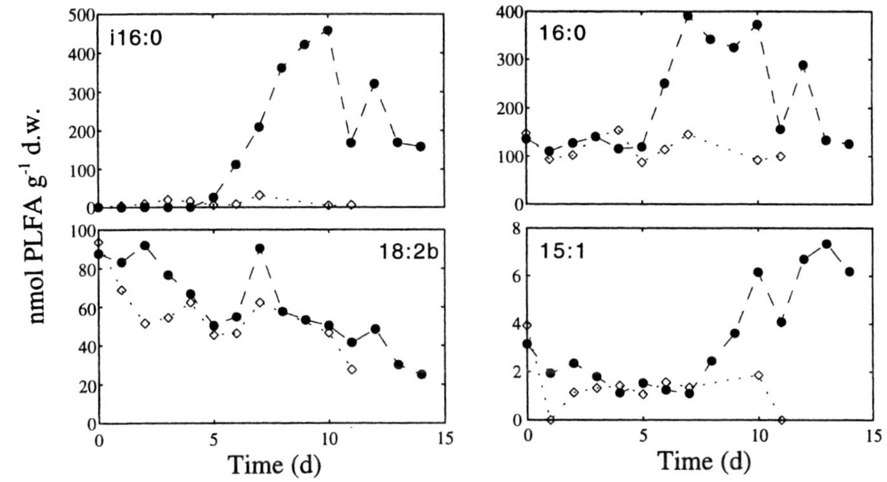

Fig. 3. Concentrations of typical individual PLFAs during composting of organic household waste. Experiments EH (*open squares*) and SH (*closed circles*)

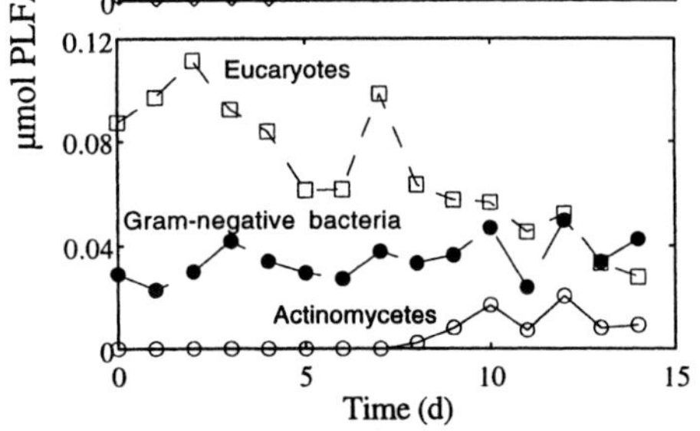

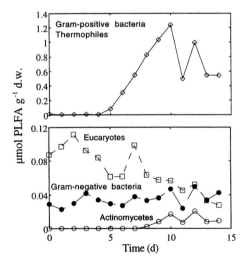

Fig. 4. Concentrations of PLFAs, typical of different groups of microorganisms, during reactor composting of household waste (experiment SH). Fatty acids from $Gram^+$ bacteria (*diamonds*), $Gram^-$ bacteria (*filled circles*), eukaryotes (*empty squares*) and actinomycetes (*empty circles*)

Discussion

The self-heated process was typical for thermophilic composting. During the first few days, temperature and mineralization rate increased progressively. During days 6–7 pH went up roughly from 5 to 9, simultaneously with a disappearance of volatile fatty acids (Beck-Friis et al. 2001). These conditions then persisted for the rest of the composting trial. In these runs, 60–65% of the C in the starting material was mineralized. In the externally heated process, however, there was only a small transient peak in CO_2 production and much less increase in pH and decrease in concentrations of volatile fatty acids (Beck-Friis et al. 2001). Under these conditions only 5–10% of the C was mineralized. In line with this there was a large increase in total PLFA concentration and hence in microbial biomass in SH, but only a very slight increase in EH (Fig. 1). These reactor composting runs were relatively short and probably did not reach the stage of a stable and mature residue. Still, they show that it is not recommended to try to speed up the composting process by external heating, at least not when the substrate has been frozen.

Although the analysis of microbial community dynamics demonstrated a lack of growth of thermophiles when the temperature was increased rather quickly to 55 °C by external heating, it offers no definite explanation for this. It is possible that some organisms, which were dependent on specific environmental conditions and necessary for the succession from meso- to thermophilic degradation, were excluded by the rapid temperature increase. As a consequence, some necessary step in the gradual degradation of the organic material may have been excluded, making further degradation difficult. In support of this assumption, a few PLFAs with comparatively low concentrations had a maximum during days 3–6 of SH (12:0, cy19:0/19:1a, 18.2a, 18:3). However, this group of PLFAs is not easily assigned to specific microorganisms.

It is not surprising that we found low concentrations of archaean ether lipids in the reactor. Although methanogens, which are the most likely archaea to appear in this system, may have been present in low numbers in anoxic microzones, our data suggest that they did not multiply under these composting conditions with a large oxygen supply.

The large increase in total PLFA concentration in the thermophilic phase of SH almost completely occurred in fatty acids typical of Gram-positive bacteria. This is in line with several other studies, which have shown that Gram-positives, and particularly *Bacillus*, are very common among isolates (Fermor et al. 1979; Strom 1985a; Blanc et al. 1997) and PCR-amplified 16S rRNA genes (Peters et al. 2000) recovered from thermophilic composts. Thus, it seems likely that the increase in PLFAs was due to growth of *Bacillus*. However, although iso- and anteiso-branched fatty acids are much less common in Gram-negative bacteria, they do in fact predominate in some. One example is the thermophilic taxus *Thermus*, which has a fatty acid composition similar to *Bacillus* (O'Leary and Wilkinson 1988; Wilkinson 1988; Nordström 1992). In fact, iso- and anteiso-branched fatty acids dominate in many thermophilic microorganisms, as an adaptation to high temperatures (Wilkinson 1988). One difference, however, seems to be a higher content of 16:0 in *Bacillus* than in *Thermus*. This PLFA also increased substantially, suggesting that *Bacillus* was more important than *Thermus*. On the other hand, Beffa et al. (1996) found that numbers of *Thermus* in thermophilic composts of urban organic waste were at least as high as those of *Bacillus*. Thus, the biomass increase in SH may have been due to growth of *Thermus*, *Bacillus*, a mixture of both, and perhaps also other thermophilic bacteria.

The PLFAs present at the beginning of the composting runs may have come from microorganisms in the frozen waste, but an alternative and very likely source is the biological material in the waste itself, of plant as well as animal origin. Indeed, the four fatty acids that dominated initially (16:0, 18:0, 18:2, 18:1ω9) are all common in many higher eukaryotes. Hence, maybe the decrease in polyunsaturated 18C-PLFAs in the reactor does not mainly reflect a decrease in fungal biomass, but rather the degradation of plant and animal remnants in the waste. Since the other three of the four fatty acids dominating initially are very common in many bacteria as well as in eukaryotes (16:0, 18:0, and 18:1ω9), it is not surprising that they increased in concentration after day 5 in the self-heated process (Fig. 3c).

There are few reports of the microbial community structure in composts, based on growth-independent methods. One research group has shown that SSCP (single-strand conformational polymorphism) profiles of 16S rRNA genes changed over the course of composting of plant material, demonstrating changes in the microbial community (Koschinsky et al. 1998; Peters et al. 2000). The shift to predominance of iso- and anteiso-branched PLFAs during the transition from meso- to thermophilic conditions in our reactor runs is perhaps a general pattern in thermophilic composting of various substrates (Herrmann and Shann 1997; Klamer and Bååth 1998). Additionally, the increase in actinomycete fatty acids (mainly 10Me18:0) after the initial thermophilic phase has also been seen in other studies (Hellmann et al. 1997; Herrmann and Shann 1997; Klamer and Bååth 1998) substantiating the view that actinomycetes are important for the degradation of polymers during the later, mesophilic stabilization phase of composting (Miller 1993).

Overall, our maximum total PLFA concentration (approximately 2000 nmol g^{-1} d.w.) was similar to the results of others. Thus, Hellmann et al. (1997) reported 1350 nmol g^{-1} d.w. and Klamer and Bååth (1998) 5000 nmol g^{-1} organic matter. This implies that the peak microbial biomass was similar among all these studies, and perhaps typical for thermophilic composting in general. Our maximum concentration of PLFAs from Gram-positive bacteria corresponded to almost 10^{11} ($6.2\ 10^{10}$) cells g^{-1} d.w. of *Bacillus*-type bacteria. This is higher than cell numbers obtained with the viable count method, of at most ca. 10^{10} cells g^{-1} d.w. in thermophilic compost (Strom 1985a; Beffa et al. 1996). One likely reason for higher cell numbers with the PLFA method is that, in contrast to the viable count method, it is not dependent on growth for detection.

In conclusion, analysis of signature phospholipid fatty acids was an excellent tool for following the overall microbial succession during composting of household waste in the reactor. It allowed a quantitative description of the development of the total microbial biomass as well as of different groups of eubacteria (including actinomycetes) and eukaryotes. However, the sensitivity needs to be increased for the quantification of diethers from archaea in this system. Analyses of lipid biomarkers can obviously be an excellent tool to assess differences in microbial community dynamics due to variations in composting conditions, such as substrate composition, temperature, moisture and O_2 supply.

Acknowledgements. We thank Barbro Beck-Friis (Department of Soil Sciences, SLU) and Sven Smårs (Department of Agricultural Engineering, SLU) for their careful sampling from the compost reactor. Thanks go also to Helena Carlsson for help with the GC-MS identifications of the PLFAs.

References

Atkinson CF, Jones DD, Gauthier JJ (1996) Putative anaerobic activity in aerated composts. J Ind Microbiol 16:182-188

Barth J, Kroeger B (1998) Composting progress in Europe. BioCycle April 1998:65-68

Beck-Friis B, Smårs S, Jönsson H, Kirchmann H (2001) Emissions of CO_2, NH_3 and N_2O from organic household waste in a compost reactor under different temperature regimes. J Agr Eng Res 78:423-430

Beffa T, Blanc M, Lyon P-F, Vogt G, Marchiani M, Lott Fischer J, Aragno M (1996) Isolation of *Thermus* strains from hot composts (60 to 80 °C). Appl Environ Microbiol 62:1723-1727

Bergey´s Manual of Systematic Bacteriology (1984) Williams & Wilkins, Baltimore

Blanc M, Marilley L, Beffa T, Aragno M (1997) Rapid identification of heterotrophic, thermophilic, spore-forming bacteria isolated from hot composts. Int J Syst Bacteriol 47:1246-1248

Carpenter-Boggs L, Kennedy AC, Reganold JP (1998) Use of phospholipid fatty acids and carbon source utilization patterns to track microbial community succession in developing compost. Appl Environ Microbiol 64:4062-4064

Davis CL, Hinch SA, Donkin CJ, Germishuizen PJ (1992) Changes in microbial population numbers during the composting of pine bark. Biores Technol 39:85-92

de Bertoldi M (1998) Composting in the European Union. BioCycle June 1998:74-75

de Bertoldi M, Vallini G, Pera A (1983) The biology of composting: a review. Waste Manage Res 1:157-176

Eklind Y, Beck-Friis B, Bengtsson S, Ejlertsson J, Kirchmann H, Mathisen B, Nordkvist E, Sonesson U, Svensson BH, Torstensson L (1997) Chemical characterization of source-separated organic household wastes. Swed J Agric Res 27:167-178

Fermor TR, Smith JF, Spencer DM (1979) The microflora of experimental mushroom composts. J Hortic Sci 54:137-147

Finstein MS, Morris ML (1975) Microbiology of solid waste composting. Adv Appl Microbiol 19:113-151

Hellmann B, Zelles L, Palojärvi A, Bai Q (1997) Emission of climate-relevant trace gases and succession of microbial communities during open-windrow composting. Appl Environ Microbiol 63:1011-1018

Herrmann RF, Shann JF (1997) Microbial community changes during the composting of municipal solid waste. Microb Ecol 33:78-85

Hue NV, Liu J (1995) Predicting compost stability. Compost Sci Util 3:8-15

Klamer M, Bååth E (1998) Microbial community dynamics during composting of straw material studied using phospholipid fatty acid analysis. FEMS Microbiol Ecol 27:9-20

Koschinsky S, Schwieger F, Peters S, Grabbe K, Tebbe CC (1998) Characterizing microbial communities of composts at the DNA level. Med Fac Landbouww Univ Gent 63/4:1725-1732

Kowalchuk GA, Naoumenko ZS, Derikx PJL, Felske A, Stephen JR, Arkhipchenko IA (1999) Molecular analysis of ammonia-oxidizing bacteria of the ß subdivision of the class *Proteobacteria* in compost and composted materials. Appl Environ Microbiol 65:396-403

Lechevalier H, Lechevalier MP (1988) Chemotaxonomic use of lipids – an overview. In: Ratledge C, Wilkinson SG (eds) Microbial lipids, vol 1. Academic Press, London, pp 869-902

Mathur SP, Owen G, Dinel H, Schnitzer M (1993) Determination of compost biomaturity. I. Literature review. Biol Agric Hortic 10:65-85

Miller FC (1993) Composting as a process based on the control of ecologically selective factors. In: Metting FB Jr (ed) Soil microbial ecology – applications in agricultural and environmental management. Marcel Dekker, New York, pp 515-544

Nordström KM (1992) Effect of growth phase on the fatty acid composition of *Thermus* spp. Arch Microbiol 158:452-455

Norland S, Heldal M, Tumyr O (1987) On the relation between dry matter and volume of bacteria. Microb Ecol 13:95-101

O'Leary WM, Wilkinson SG (1988) Gram-positive bacteria. In: Ratledge C, Wilkinson SG (eds) Microbial lipids, vol 1. Academic Press, London, pp 117-202

Peters S, Koschinsky S, Schwieger F, Tebbe CC (2000) Succession of microbial communities during hot composting as detected by PCR-single-strand-conformation polymorphism-based genetic profiles of small-subunit rRNA genes. Appl Environ Microbiol 66:930-936

Smårs S, Beck-Friis B, Jönsson H, Kirchmann H (2001) An advanced experimental composting reactor for systematic simulation studies. J Agr Eng Res 78:415-422

Strom PF (1985a) Effect of temperature on bacterial species diversity in thermophilic solid-waste composting. Appl Environ Microbiol 50:899-905

Strom PF (1985b) Identification of thermophilic bacteria in solid-waste composting. Appl Environ Microbiol 50:906-913

Sundh I, Borgå P, Nilsson M, Svensson BH (1995) Estimation of cell numbers of methanotrophic bacteria in boreal peatlands based on analysis of specific phospholipid fatty acids. FEMS Microbiol Ecol 18:103-112

Tam NFY (1995) Changes in microbiological properties during *in-situ* composting of pig manure. Environ Technol 16:445-456

Vestal JR, White DC (1989) Lipid analysis in microbial ecology. BioScience 39:535-541

Virtue P, Nichols PD, Boon PI (1996) Simultaneous estimation of microbial phospholipid fatty acids and diether lipids by capillary gas chromatography. J Microbiol Methods 25:177-185

Vuorinen AH, Saharinen MH (1997) Evolution of microbiological and chemical parameters during manure and straw co-composting in a drum composting system. Agric Ecosyst Environ 66:19-29

White DC, Bobbie RJ, Herron JS, King JD, Morrison SJ (1979) Biochemical measurements of microbial mass and activity from environmental samples. In: Costerton JW, Colwell RR (eds) Native aquatic bacteria: enumeration, activity, and ecology, ASTM STP 695. American Society for Testing and Materials, Philadelphia, PA, pp 69-81

White DC, Stair JO, Ringelberg DB (1996) Quantitative comparisons of *in situ* microbial biodiversity by signature biomarker analysis. J Ind Microbiol 17:185-196

Wilkinson SG (1988) Gram-negative bacteria. In: Ratledge C, Wilkinson SG (eds) Microbial lipids, vol 1. Academic Press, London, pp 299-488

Bacterial Diversity in Livestock Manure Composts as Characterized by Terminal Restriction Fragment Length Polymorphisms (T-RFLP) of PCR-amplified 16s rRNA Gene Sequences

S. M. Tiquia[1] and F. C. Michel Jr.[2]

Abstract. Composts contain a large and diverse community of microorganisms that play a central role in the decomposition of organic matter during the composting process. However, microbial communities active in composts have not been well described in the past. In the present study, the phylogenetic diversity of bacterial communities in livestock manure compost was determined based on terminal restriction fragment length polymorphisms (T-RFLP) of 16S rRNA genes. This technique uses a PCR in which one of the primers is fluorescently labeled. After amplification, the PCR product is then digested with restriction enzymes such as *Hha*I, *Msp*I, and *Rsa*I to generate T-RFLP fingerprints of bacterial communities. In the present study, a mixture of dairy and horse manure (dairy+horse manure; 1:1 ratio w/w) was composted in windrows and in-vessel to investigate compost bacterial diversity. The DNA was isolated from the feedstocks (day 0) and after 21 and 104 days of in-vessel and windrow composting, respectively, for T-RFLP analysis. A variety of techniques were then used to analyze T-RFLP data to gain insights about the structure of the bacterial community from these compost samples. Results of the T-RFLP analysis revealed high species diversity in the feedstocks sample. As many as 27 to 39 different terminal restriction fragments (T-RFs) were found in these samples, revealing high diversity in the livestock manure composts. After composting, an increase in the T-RFLP-based Shannon diversity index was observed in the in-vessel compost, while a decrease was found in the windrow compost. Differences in chemical properties were also observed in the windrow and in-vessel composts. The windrow compost had lower water, organic matter (OM) and C contents and higher C and OM loss than the in-vessel compost.

[1]Environmental Sciences Division, Oak Ridge National Laboratory, P.O. Box 2008, Oak Ridge, Tennessee 37831, USA
e mail: tiquias@ornl.gov/ tel: +1 865-5747302/ fax: +1 865-5768676
[2]Department of Food, Agricultural, and Biological Engineering, The Ohio State University, Ohio Agricultural Research and Development Center (OARDC), 1680 Madison Avenue, Wooster, OH 44691, USA

H. Insam, N. Riddech, S. Klammer (Eds.)
Microbiology of Composting

Introduction

Composting is the biological conversion of organic material into a stabilized end product that can be used as a soil amendment. The process involves dynamic changes in temperature, oxygen concentration, moisture content, and nutrient availability. The active component mediating the composting process is the resident microbial community (Beffa et al. 1996). Defining the diversity and structure of microbial communities of composts through their constituent populations has been of considerable interest to compost researchers in order to address basic ecological questions such as how similar are microbial communities in mature compost that were made from different feedstocks and using different composting methods. It has been known that composts typically contain very high numbers of microorganisms (10^{10}-10^{12} viable cells g^{-1}) (Beffa et al. 1996; Tiquia et al. 1996), the majority of which are bacteria (Epstein 1997). Previous studies of bacterial populations in composts have relied largely on culture-based methods to isolate microorganisms in composts and assess microbial diversity in compost (Strom 1985a; Fujio and Kume 1991; Blanc et al. 1996).

Recently, attempts have been made to characterize bacterial communities from various environmental samples by molecular techniques utilizing the polymerase chain reaction (PCR) (Massol-Deya et al., 1995; Liu et al. 1997). The initial step of bacterial community analysis begins with the extraction of total community DNA from environmental samples. The DNAs present in the community are then PCR-amplified using universal primers targeted to conserved regions of the gene common to all bacteria. Fingerprinting techniques such as denaturing gradient gel electrophoresis (DGGE) and temperature gradient gel electrophoresis (TGGE) (Heuer and Smalla 1997); single-strand-conformation polymorphism (SCCP) (Schwieger and Tebbe 1998); amplified rDNA restriction endonucleases analysis (ARDRA) (Massol-Deya et al. 1995), and terminal restriction fragment length polymorphism (T-RFLP) of the 16S rRNA genes (Liu et al. 1997) are then employed to understand the composition and diversity of the bacterial community. T-RFLP is a technique in which the PCR-amplified 16S rRNA genes are flourescently labeled on one end through the use of a single labeled primer during the PCR step (Liu et al. 1997). Therefore, only the sizes of the fluorescently labeled terminal restriction fragments (T-RF) are seen on polyacrylamide gels. The T-RFLP fingerprinting technique has been effectively used in the exploration of complex microbial environments and has provided a rapid means to assess community diversity in various environments (Liu et al. 1997; Kerkhof et al. 2000; Ludeman et al. 2000). However, this technique has not yet been employed in the characterization of bacterial diversity and community during composting of manure under different composting systems.

Several diversity indices that describe the species richness and evenness have been used to describe the assemblage of microbial populations within a community (Pielou 1969). A widely used measure of diversity is the Shannon index (Shannon-and Weaver 1949; Wiener 1948). This general diversity index is sensitive to both species richness and relative species abundace of the community.

In this study, the genetic diversity of the microbial community was assessed using T-RFLP based on the Shannon diversity index. The digested T-RFs separated by polyacrylamide electrophoresis were used to generate characteristic profile data for the estimation of diversity among different compost samples.

Principal components analysis (PCA) of T-RFLP patterns was also performed to detect statistical significance of changing composting conditions on microbial communities in the compost samples. PCA may provide a means to separate and group compost samples based on their complex T-RFLP patterns as this statistical analysis simultaneously considers many correlated variables, and then identifies the lowest number to accurately represent the structure of the data. In this study, the variables are the T-RFs or peak areas of a given T-RFLP.

This preliminary study was carried out to compare bacterial diversity of dairy+horse manure during the initial and final stage of composting in windrow and in-vessel composting systems using T-RFLP of the 16S rRNA genes. A variety of techniques were used to analyze T-RFLP data to gain insights about the structure of the bacterial communities in the dairy+horse manure compost.

Materials and Methods

Composting and Determination of Compost Physico-Chemical Properties

The manures used in this study were (1) non-separated dairy manure from a free stall barn, and (2) horse manure+wood shavings bedding. The manures were mixed homogeneously at a ratio of 1:1 (dairy manure:horse manure, wet w/w) using a mixer wagon. Table 1 shows the general properties of dairy manure, horse manure, and compost feedstock (dairy+horse manure) at the beginning and end of the composting process. The dairy+horse manure was then composted using two methods: in-vessel (for 21 days) and windrow (outdoor; for 104 days). For the in-vessel composting method, three vessels (205 l) were filled with dairy+horse manure compost. For the windrow composting, the dairy+horse manure was stacked at 2.7 m wide, 1.2 m high, and 12 m long. Temperatures in the vessels and windrows were recorded during composting. The day 0 (feedstock), in-vessel, and windrow composts were characterized for water content (105 °C for 24 h), pH, organic matter, C, total N, NH_4^+-N, and NO_3^-N.

DNA Extraction and PCR Amplification

The total community DNA from each replicate compost sample was extracted and purified using the UltraClean Soil DNA Isolation Kit (MoBio Laboratories, Inc., California, USA). A PCR inhibitor removal solution (UltraClean IRS solution;

MoBio Laboratories) was added to reduce humic acid contamination in compost and produce a PCR-quality DNA. Bacterial (16S rDNA) DNAs present in the community were PCR-amplified using the universal bacterial primers: 8F forward (5'-AGAGTTTGATCCTGGCTCAG-3') and 1406r (5'-ACGGGCGGTGTG TRC-3') reverse, with the 8F forward primer labeled with HEX (5-hexachloro-flourescein) (Operon, Inc., California, USA). Each PCR reaction mixture contained 50 ng DNA template, 2.5 mM $MgCl_2$, 2.5 units *Taq* polymerase [Boehringer Mannheim Biochemicals (BMB), Indiana, USA], 1X PCR reaction buffer, 0.2 mM PCR nucleotide mix (BMB, Indiana, USA), 0.5 μM DNA primers, and 0.6 μl bovine serum albumin (BMB, Indiana, USA) in a final volume of 50 μl. Reaction mixtures were heated at 94 °C for 9 min, and cycled 30 times through three steps: denaturing (94 °C; 60 s); annealing (58 °C; 45 s); and primer extension (72 °C; 90s); in a PTC-100 thermal cycler (MJ Research, Inc., Massachusetts, USA). To minimize PCR bias, amplicons from three PCR runs were combined and then purified, using a PCR purification kit (PCR Clean-up Kit; MoBio Laboratories, Inc, California, USA), and eluted in a final volume of 50 μl.

Positive Controls

Arthrobacter globiformis (a Gram-positive bacterium) and *Xanthomonas campestris* (a Gram-negative bacterium) were used as positive controls in this study. Both bacteria were obtained from the Department of Plant Pathology culture collection at The Ohio State University. Genomic DNA of each bacterium was isolated using the same kit (Soil DNA extraction kit, Mobiolab, California) used for extracting total DNAs from dairy+horse manure compost samples.

T-RFLP of 16 S rRNA Genes

Aliquots (10 μl) of amplified 16S rRNA genes obtained from the dairy+horse manure compost and the two pure cultures (*A. globiformis* and *X. campestris*) were digested separately with restriction endonucleases, *Hha*I (for 5 h), *Msp*I (for 3 h) and *Rsa*I (for 3 h) (BMB, Indiana, USA), to produce a mixture of variable length end-labeled 16S rRNA fragments. The labeled fragments were separated electrophoretically on a polyacrylamide gel (5.5%) in an ABI model 373 automated sequencer (Perkin Elmer, California, USA). Thereafter, the lengths of flourescently labeled terminal restriction fragments (T-RFs) were determined, by comparison with internal standards, and were analyzed using Genescan software (Perkin Elmer, California, USA) with a peak height detection of 50.

Similarity and Diversity of T-RFLP Patterns

The levels of similarity between T-RFLP fingerprints were determined using the T-RFLP profile matrix analysis program from the Ribosomal Database Project

(RDP) II web site (http://www.cme.edu/RDP/html). T-RFLP data generated from *Hha*I, *Msp*I, and *Rsa*I digestions from each compost sample were combined to produce a similarity matrix for the pattern fragments in the sample.

The T-RFLP-based Shannon diversity index and equitability index as per Atlas and Bartha (1997) was used as a measure of diversity, which was derived on Shannon and Weaver's formula based on the information theory (Wiener 1948; Shannon and Weaver 1949).

Principal Components Analysis of T-RFLP Patterns

Principal components analysis (PCA) of T-RFLPs was employed to cluster samples based on the presence or absence of T-RFs or relative abundance of the T-RFS (% peak area) from each T-RFLP pattern. For PCA based on presence or absence of T-RFs, the sample data were arranged based on Boolean character sets (1 or 0) (Hammer and Rudeanu 1968), which correspond to the absence and presence of a given T-RFLP pattern. Each T-RF size found from each T-RFLP fingerprint was scored 0 (if the T-RF size in the list was absent in a given T-RFLP fingerprint) or 1 (if the T-RF size in the list was present in the given T-RFLP fingerprint). For PCA based on the relative abundance of T-RFLP patterns, the peak area of each T-RF was standardized by calculating the percentage peak area from the total peak area of all T-RFs of each T-RFLP fingerprint. PCA was performed using SYSTAT statistical computing package (SYSTAT version 9.0) using the Boolean or % peak area datasets from T-RFLPs generated from *Hha*I, *Msp*I, and *Rsa*I digestions. PCA was also carried out from merged digest data by stacking the T-RFLP data derived from *Hha*I, *Msp*I, and *Rsa*I digestions of the same compost sample, to combine information from three different digests and provide accurate community characterization.

T-RFLP Analysis from 16s rRNA Gene Sequence Database

Phylogenetic groups of bacteria from each T-RFLP fingerprint were theoretically evaluated and matched with T-RF lengths of 16S rRNA gene sequences deposited at the RDP II web site. To access the database, the T-RFLP analysis program (TAP) from the RDP II web site was used. TAP is a web-based research tool that facilitates microbial community analysis using T-RFLPs of 16S rRNA genes. At present, TAP contains 16, 277 aligned 16S rRNA gene sequences (Maidak 2000). This program permits the user to perform simulated restriction digestions of the entire 16S sequence database and derive terminal restriction fragment sizes (T-RFs), measured in base pairs, from the 5' terminus of the user-specified primer to the 3' terminus of the restriction endonuclease target site (Marsh et al. 2000). Each sequence that is successfully recognized by the primer sequence is digested by the specified enzyme(s). The program displays each resulting T-RF size and enzyme with the organism's name.

Results and Discussion

Temperature Profiles and Chemical Properties

Temperatures of the in-vessel compost reached thermophilic levels (>55°C) by day 1 and then peaked at 70 °C ± 3 °C by day 2 (Fig. 1A). Vessel temperatures ranged between 56 and 65 °C until sampling. At the end of 21 days of in-vessel composting, compost temperatures were 54 °C ± 2 °C (Fig. 1A). In windrows, compost temperatures peaked at 73 °C. After 2 days, the temperature declined, due to windrow turning, but then rose to nearly 70 °C by day 6. Windrow temperatures fluctuated between 46 and 70 °C during the composting process and by the end of 109 days of composting, temperature in the windrow was 67 °C (Fig. 1B). Differences in chemical properties were observed in final composts (Table 1), probably due to differences in temperature profiles and duration of composting in the two trials (Fig. 1). Windrow compost had lower water content, organic matter (OM) and C concentrations but higher C and OM loss than the in-vessel compost (Table 1). This could be attributed to the fact that the windrow compost had been through a much longer (five times longer) composting process than the in-vessel compost. During composting, a major part of the process is the loss C, H, O, and N (Golueke 1972). As composting progressed, the amount of these elements further decreased (Table 1). Ammonium levels in the windrow compost samples were lower but yielded higher NO_3^-N, suggesting a much greater nitrification process than in the in-vessel compost.

Table 1. General properties of the dairy manure, horse manure, feedstock (day 0), and composted manure

Properties	Dairy manure[a]	Horse manure[b]	Feedstock (day 0)[b]	In-vessel compost[b]	Windrow compost[b]
pH	ND	8.2 ± 0.70	8.61 ± 0.02	7.77 ±0.02	8.04±0.04
Water content (%)	91	57 ± 2.10	73 ± 0.50	67 ± 1.10	30.0 ± 2.20
Dry matter content (%)	9	43 ± 2.13	26.6 ± 0.50	32.9 ± 1.10	70.0 ± 2.20
Ash content (%)	47.5	5.6 ± 1.10	15.5 ± 0.43	22.0 ± 2.97	42.31 ± 1.86
OM content (%)	52.5	94.4 ± 1.14	84.5 ± 0.40	78.0 ± 3.0	57.7 ± 1.90
Carbon (%)	25.5	46.9 ± 1.20	41.3 ± 2.60	41.1 ± 1.20	34.5 ± 0.70
Nitrogen (%)	4.9	1.0 ± 0.10	1.38 ± 0.02	2.11 ± 0.13	2.66 ± 0.30
NH_4^+-N (mg kg^{-1})	ND	ND	4258 ± 43.0	414 ± 12.0	109 ± 8.0
NO_3^-N (mg kg^{-1})	ND	ND	<50.0± 0.00	<50.0± 0.00	224.5 ± 35.1

Table 1. (cont.)

C:N ratio	5:1	47:1 ± 4.50	30:1 ± 1.50	19:1 ± 1.70	13:1 ± 0.30
C:OM ratio	0.49	0.50 ± 0.02	0.49 ± 0.03	0.53 ± 0.01	0.60 ± 0.01
Final temp. (°C)	-	-	-	54 ± 2.00	67
C loss (% of initial)	-	-	-	0.7 ± 0.03	16.5 ± 1.60
O.M. loss (%)	-	-	-	7.6 ± 0.04	31.7 ± 2.20

OM= organic matter; ND= not determined.
a Data obtained from Keener et al. (2000)
b Data obtained from this study. Mean and standard deviation of three replicates are shown.

Bacterial Community T-RFLPs from Dairy+horse Manure Compost

Figure 2A revealed diversity among three different samples. An increase in diversity, as indicated by the T-RFLP based Shannon diversity index, was observed in dairy+horse manure after 21 days of in-vessel composting, while a decrease in Shannon diversity (from 2.8 to 2.6) was found in the manure composted in windrows for 104 days (Table 2). The Shannon diversity value derived from the windrow compost was lower than that derived from the initial feedstock (day 0) (Table 2). The windrow compost had been subjected to prolonged high temperatures (Fig. 1B), which may have reduced the diversity of the biological community.

Microbial diversity has been shown to be markedly lower in habitats under conditions of stress or disturbance, and environmental fluctuations (Atlas and Bartha 1997). The Shannon diversity of the in-vessel compost was higher than the windrow compost due to the fact that the process was stopped at a lower temperature. The temperature of the in-vessel compost was 54 °C at the end of composting while that of the windrow compost was 67 °C (Fig. 1A and Table 1). Strom (1985b) showed that species diversity of culturable bacteria dropped markedly from 0.65 to 0.07 when compost temperature rose to 60 °C and above during a laboratory composting. In his study, the highest diversity was found when temperatures are between 50 and 57 °C.

It is interesting to note that in the present study, the decline in Shannon diversity index of the windrow compost bacterial communities also corresponded with a decrease the equitability index (a measure of dominance) (Table 2). Moreover, the in-vessel compost had a higher diversity index and showed an increase in the equitability index value, which indicated the presence of dominant T-RF peaks in this sample (Table 2). In these diversity calculations, both the number of peaks and peak areas were considered.

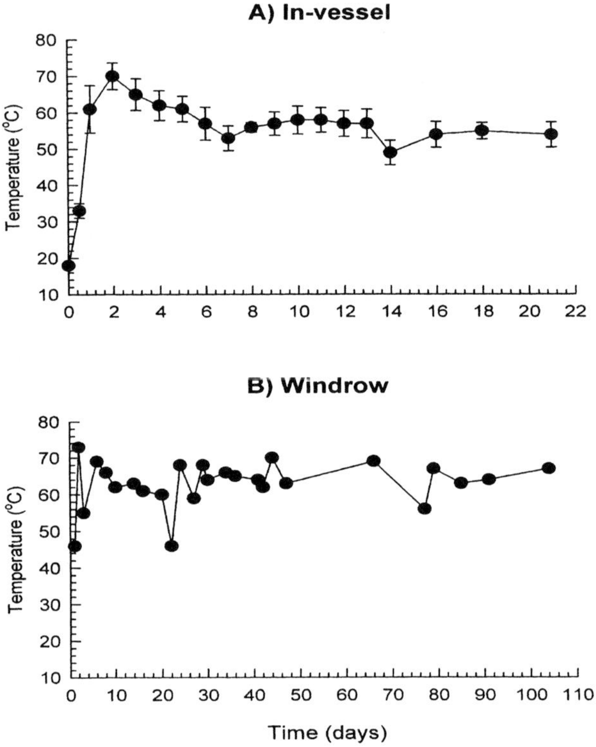

Fig. 1A,B. Temperature change during composting of dairy+ horse manure

However, there are ongoing debates as to whether peak areas should be included in PCR-based analyses, due to PCR amplification rates of difference templates (Suzuki and Giovanni 1996). Therefore the genetic diversity of the microbial community was determined, based on the number of T-RF fragments found in each T-RFLP profile. The diversity and overall similarities between samples were estimated by the number of bands found per sample. Therefore, the greater the number of fragments found, the greater the diversity. The number of T-RF peaks found in day 0 (feedstock) sample was about 27 to 39 (Table 2). This number increased after in-vessel composting to 31–44 peaks, and decreased after windrow composting (21–28). This pattern change in diversity was similar to the Shannon diversity and equitability index, where both the number of peaks and relative abundance of peaks were considered (Table 2). The T-RFLP similarity matrix (Table 2) showed that 59% (0.59) of the T-RF peaks in the in-vessel compost could be found in the initial feedstock (day 0 sample), while 44% (0.44) of the T-RF peaks in the windrow sample could be found in the day 0 sample (Table 2).

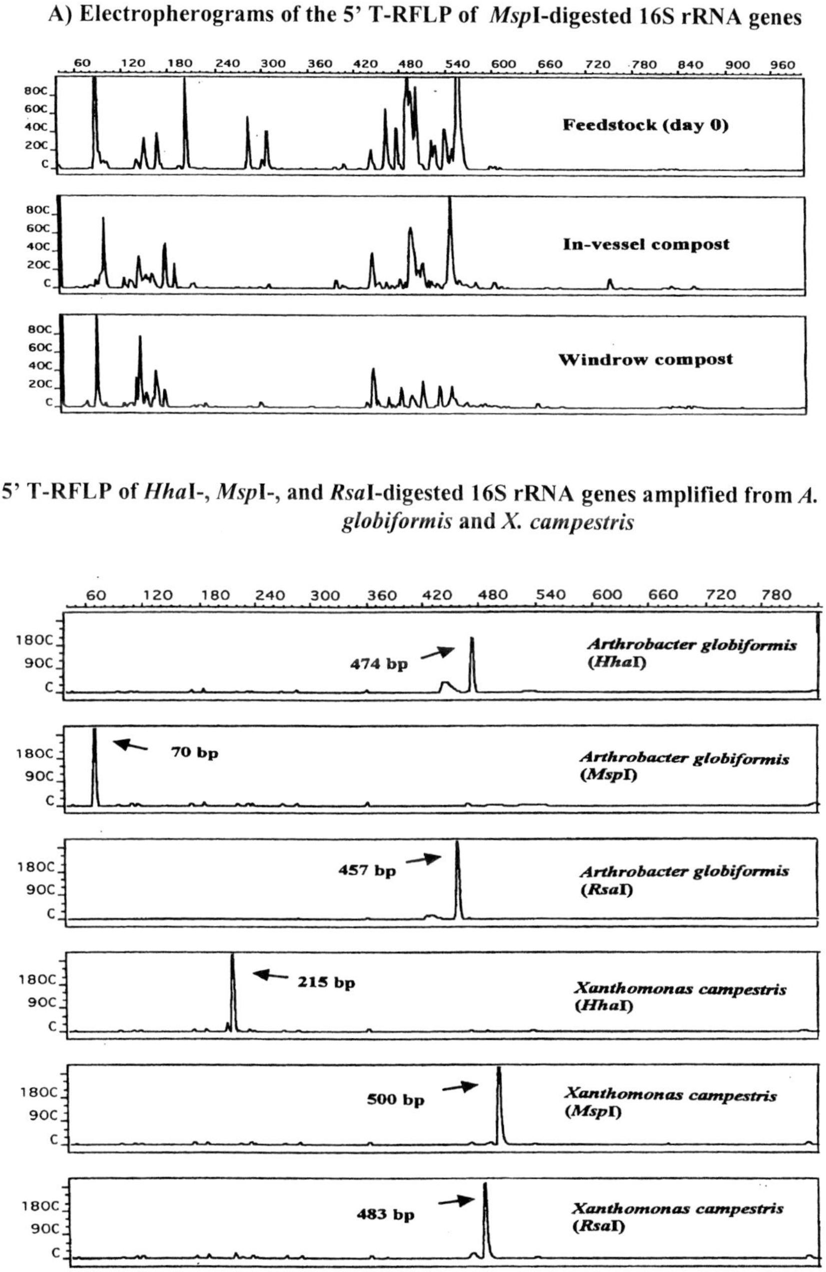

Fig. 2A,B. Electropherograms of the **A** 5' T-RFLP of *Msp*I-digested 16S rRNA genes amplified from dairy+horse manure composts and **B** 5' T-RFLP of *Hha*I-, *Msp*I-, and *Rsa*I-digested 16S rRNA genes amplified from *A. globiformis* and *X. campestris*

The final compost samples had a similarity of 61% (0.61) (Table 2), suggesting that the composted samples were more similar to each other than to the feedstock sample. Moreover, the feedstock samples were more similar to the in-vessel samples than the windrow samples.

Multivariate Analysis of T-RFLP Patterns

Figure 3A–D shows the PCA results based on the presence or absence (Boolean data) of T-RF peaks. T-RFLPs generated using *Hha*I, *Msp*I, and *Rsa*I digests and the combined digest data, which accounted for 89, 81, 80, and 70%, respectively of the total variance. The scree plot for these PCAs indicated that the variation in *Hha*1, *Msp*1, *Rsa*I and the combined digest data could be explained by principal components 1 (PC1) and PC2 (data not shown). The analysis arranged the three compost samples (feedstock, in-vessel compost, and windrow compost) differently depending on the restriction enzyme used to create the T-RFLP pattern (Fig. 3A–D). PCA of *Hha*I, *Msp*I, and *Rsa*I failed to produce significant pairing among samples (Fig. 3A–C), suggesting that the Boolean data (presence or absence of T-RFs) derived from T-RFLP patterns of the three compost samples were different from each other. Principal components 1 (PC1) of the *Hha*I T-RFLP, separated the windrow and in-vessel compost from the feedstock sample (Fig. 3A). Conversely, PCA of the *Rsa*I T-RFLP, separated the feedstock and windrow samples from the in-vessel compost (Fig. 3C). However, when the information from the three digests was combined, all three samples correlated with PC1, while no clear correlation could be found among samples with PC 2 (Fig. 3D).

The PCA based on % peak area from each T-RFLP patterns represented 73 (*Hha*I), 79 (*Msp*I), 74 (*Rsa*I), and 70% (combined digest data) of the total variance (Fig. 3F–H). The scree plot indicated that PC1 was responsible for the variation in *Hha*1, *Rsa*I, and the combined digest data, whereas PCI and PC2 are responsible for the observed variation in *Msp*1 (data not shown). PCA of *Hha*I and *Rsa*I separated the in-vessel and windrow composts from the feedstock (day 0) (Fig. 3E,G). The PCA of the combined data also separated the in-vessel and windrow composts from the feedstock (day 0) sample (Fig. 3H). Results of the PCA based on % peak areas of the *Hha*I, *Rsa*I T-RFLPs, and the combined digest data correlated with the similarity matrix data in Table 2. It appears that the in-vessel and windrow composts have slightly higher similarity than the feedstock (fresh manure) sample.

Principal components analysis has been used previously by Clement et al. (1998) to compare T-RFLP patterns from different fecal pellets, petroleum-hydrocarbon contaminated sands, and pristine sand samples. Clement et al. (1998) pointed out that the patterns derived from a single enzyme digest did not result in accurate community characterizations, and that accurate characterization reflecting the expected bacterial community biology, were achieved by combining T-RFLP data derived from different enzyme digestions. In their study, PCA of the combined data resulted in a clear separation of the two fecal samples from two petroleum-contaminated sands, and pristine sand.

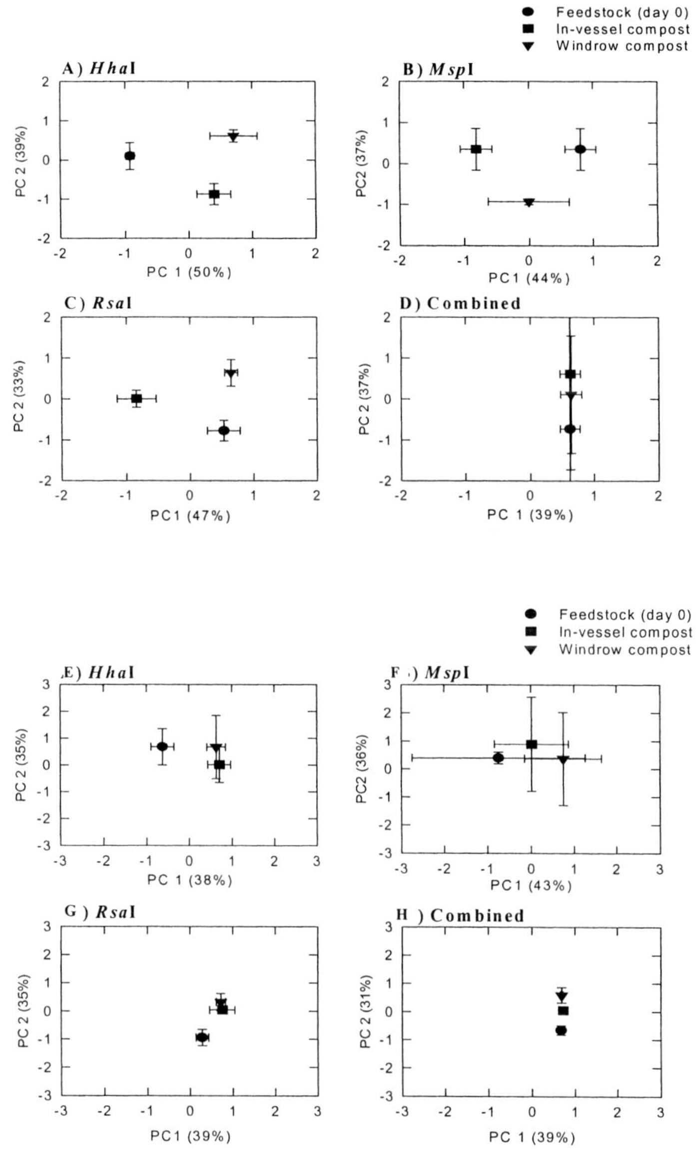

Fig. 3A–H. Principal components analyses of T-RFLP data sets based A–D on presence or absence of T-RFs and E–H % peak area from HhaI, MspI, RsaI, and D combined digests data. Error bars indicate standard error for each eigen vector element. Values in parentheses indicate variances of PCA derived from HhaI, MspI, RsaI, and combined digests data

Table 2. Number of terminal restriction fragments (T-RFs), diversity statistics and similarity matrix calculated from T-RFLP profiles of 16S rRNA genes amplified from dairy+horse manure composts

	No. of T-RFs			Shannon diversity index				Equitability index				Similarity matrix		
Sample	*Hha*I	*Msp*I	*Rsa*I	*Hha*I	*Msp*I	*Rsa*I	Average	*Hha*I	*Msp*I	*Rsa*I	Average	Day 0	Final (Bio-reactor)	Final (Windrow)
Feedstock (day 0)	29	39	27	2.7	3.1	2.6	2.8	0.81	0.84	0.79	0.81	1.00	0.59	0.44
In-vessel compost	31	44	34	3.1	3.2	2.9	3.1	0.89	0.86	0.82	0.85	0.59	1.00	0.61
Windrow compost	28	28	21	2.6	2.9	2.2	2.6	0.79	0.88	0.74	0.80	0.44	0.61	1.00

Calculations:

Shannon index ($\hat{H}$) = C/N (N log10 N- $\sum$ni log10 ni), where C = 2.3; N= sum of peak areas in a given T-RFLP; ni = area of T-RF i; and i = number of T-RFs of each T-RFLP pattern.

Equitability (J) = $\hat{H}$/Hmax, where $\hat{H}$= Shannon-diversity index, and Hmax = theoretical maximal Shannon-Weaver diversity index for the T-RFLP examined assuming that each peak represent only one member.

$H_{max} = C/N\,(N \log_{10} N - \sum n_i \log_{10} n_i)$, where $C = 2.3$, N= sum of peak areas in a given T-RFLP assuming that each peak has an area of 1, and n_i= area of T-RF i (1); i = number of T-RFs of each T-RFLP pattern.

Thus, the PCA results, correlated with the expected bacterial community biology of each pattern when the data from multiple sample enyme digestions were combined in their study. However, in the present study the separation among samples was not clear when the *Hha*I, *Msp*I, and *Rsa*I digest data were combined (Fig. 3D). There seems to be a complete overlap of variances for PC1 in spite of the combined variance for PC1 given as 39% in Table 2. Therefore, the PCA of the combined of digest data did not have much meaning in this study.

Fragment Length and Phylogeny

Since the current 16S rRNA gene sequence database in the TAP-TRFLP program of the RDP II can be distinguished based on T-RFLP analysis, an attempt has been made to identify phylogenetic groups or species from dairy+horse manure compost samples. Prior to this, efforts were also made to assure that the restriction digestions are specific. To do this, amplified products of two known microorganisms (*A. globiformis* and *X. campestris*) were subjected to T-RFLP analysis. The T-RFs of these two known organisms were then compared and matched with the aligned 16S rRNA gene sequences *A. globiformis and X. campestris* deposited in the TAP-TRFLP program. The observed sizes of the T-RF of *A. globiformis* were 474, 70, and 457 bp for *Hha*I, *Msp*I, and *Rsa*I, respectively (Figure 2B) while the predicted sizes found in the database were 472, 69, and 457 bp. For *X. campestris,* the observed T-RFs (*Hha*I=215 bp; *Msp*I= 500 bp; *Rsa*I=483 bp), which was very close to the predicted T-RFs (*Hha*I=215 bp; *Msp*I= 498 bp; *Rsa*I=481 bp) from the database. The accuracy for the three different digests was around 2± bp. This result therefore demonstrated that the digestions were specific and sizing of fragments was accurate for both of the two known organisms, and compost samples processed in the same way.

The ten T-RFs with the greatest peak area from *Hha*I, *Msp*I, and *Rsa*I digestions were determined based on % peak area percentage for each T-RFLP pattern and were matched with the aligned 16S rRNA gene sequences from the database. The fragments in the database were sorted first by the T-RF length from *Hha*I digestion, and then by T-RF lengths from *Msp*I and *Rsa*I digestions, respectively. Tables 4 to 6 present the putative organisms that correspond to T-RFs from the dairy+horse manure samples. It has been known that different organisms, even unrelated species, may produce the same T-RF length with a given restriction enzyme (Liu et al. 1997; Marsh et al. 2000). This information can be mitigated by the use of multiple enzymes so that organisms that produce identical T-RF lengths with one restriction enzyme may produce different T-RF length when digested with another enzyme. Although three different enzymes were used in the present study, the T-RFs produced from these digestions matched with more than one organism in the 16S rRNA sequence database. This made it difficult to identify a single organism that corresponds to the three different T-RF lengths derived from three different digestions. In most cases, the T-RFs from the dairy+horse manure samples could only be correctly matched with two digestions from the 16S rRNA gene sequence database (Table 3). Fragment sizes common to all compost samples corresponded to bacterial groups including *Bacillus, Mycoplasma,* and *Eubacterium.* These bacterial groups were resolved with at least two digestions. In some cases, T-RF sizes from all three digestions were out of range and could not be resolved from the database (Table 3). It could be possible that the specific phylogenetic groups/species that correspond to these T-RF sizes are unknown. These results, however, underscore the need to identify dominant T-RF peaks of specific importance by DNA sequencing in order to dissect complex compost microbial communities using T-RFLP technique.

The fact that the *in silico* digestions of the RDP indicate that a significant fraction of the TAP-TRFLP database can be distinguished on the basis of T-RF length does not imply that one can positively identify phylogenetic groups or species based on T-RF length. Therefore, one cannot be certain that all putative species identified in Tables 3 are indeed present in the samples. Marsh et al. (2000) pointed out that T-RF sizes greater than 600 bp are resolved poorly by gel systems. Hence, the T-RFLP profile may not be able to reveal or track all potentially resolvable populations of the community. To reveal the true identity of T-RF peaks, and confirm the putative organisms found in the compost samples, cloning and complete sequencing of the 16S rRNA genes obtained from the compost samples are necessary.

Table 3. Determination of putative of phylogenic groups based on 10 T-RF peaks with the greatest peak areas found in the day 0, in-vessel, and windrow composts.

Dominant T-RF	T-RF lengths (bp)			Putative phylogenetic groups/species
	*Hha*I	*Msp*I	*Rsa*I	
1	**583 (581-583)**	**557 (557-589)**	490	*Aerococcus viridans, Streptococcus bovis, S. salivarius, S. sanguis, Globicatella sanguis, Lactococcus garvieae,*
2	**567 (568)**	**491 (492)**	309	*Bordetella bronchiseptica, Bordetella parapertussis,*
	567	**491 (489)**	**309 (311)**	*Flexibacter tractuosus*
3	**226 (225)**	**495 (497)**	460	*Vibrio* sp.
	226 (227)	495	**460 (458)**	*Clostridium* sp.
4	**92 (90)**	**87 (87)**	486	*Capnocytophaga gingivalis, C. succinicans, C. aquatilis, C. flevensis, C. johnsonae, C. saccharophila, Flavobacterium aquatile*
	92	**87 (86)**	**486 (486-488)**	*Bacillus* sp.
	93 (90)	87	**486 (484)**	*Desulfovibrio* sp.
5	588	502	475	Unknown

Table 3. (cont.)

6	580	202	473	Unknown
7	239	**462 (458-461)**	**431 (429)**	*Clostridium* sp.
8	576	**489 (492)**	**93 (94)**	*Flavobacterium* sp.
9	572	**283 (283)**	**484 (485)**	*Eubacterium* sp.
10	**840 (841)**	561	**478 (476-478)**	*Mycoplasma* sp.
	(In-vessel compost)			
1	**234 (235)**	**545 (544)**	850	*Mycoplasma* sp.
2	215	**492 (490)**	**854 (857)**	*Eubacterium* sp.
3	**206 (207)**	497	**475 (474)**	*Nitrosolobus multiformis*
	206 (205)	**497 (496)**	475	*Oceanospirilum* sp.
4	352	**95 (95)**	**118 (117)**	*Flavobacterium* sp.
5	515	495	526	Unknown
6	217	**509 (508)**	**478 (476)**	*Bacillus* sp.
7	561	442	481	Unknown
8	893	174	836	Unknown
9	723	543	829	Unknown
10	552	140	842	Unknown
	(Windrow compost)			
1	188	**85 (86)**	**462 (461-464)**	*Bacillus brevis* , *Bacillus* sp.
2	**190 (189-192)**	**140 (141)**	426	*Eubacterium* sp.; *Clostridium innocuum*
3	385	**442 (442)**	**459 (460)**	*Eubacterium* sp.
4	293	**507 (508)**	**493 (490-492)**	*Bacillus* sp.
5	181	**161 (161)**	**477 (475-478)**	*Mycoplasma* sp.
	181 (180-183)	**161 (161-163)**	477	*Mycobacterium* sp., *Clostridium* sp.
6	**297 (294)**	162	**476 (478)**	*Flexibacter* sp.
	297	**162 (161)**	**476 (478)**	*Mycoplasma pulmonis*
7	287	545	79	Unknown
8	286	**493 (494)**	**186 (183)**	*Flexibacter* sp.
9	473	**148 (145)**	**118 (117-119)**	*Clostridium* sp.
10	423	530	727	Unknown

Values in bold are observed fragment sizes that correspond with sizes found in the RDP (TAP-TRFLP) database (RDP release 7.1). Non-bold values are observed fragment sizes that are not found in the TAP-TRFLP database. Values in parentheses are predicted values for the listed phylogenetic groups. Observed T-RF lengths are within ±2 bp of those T-RF lengths found in the TAP-TRFLP database.

Conclusions

Differences in chemical properties and bacterial community patterns were observed between in-vessel and windrow composts. An increase in diversity was observed in in-vessel compost after 21 days of composting, while a decrease in diversity was observed in the windrow compost after 109 days of composting. The in-vessel compost also showed an increase in equitability index, which indicated the presence of dominant T-RF peaks in this sample. Results of the PCA based on Boolean character sets and peak areas of the combined digest data correlated with the similarity matrix data. It appears that the in-vessel and windrow composts have similar bacterial community structure, which are different from the feedstock (fresh manure) sample in the present study. Although this study was rather preliminary, the research presented here begins to address the questions of critical factors controlling microbial diversity under different composting systems, and may provide important baseline information critical for the design and optimization of microbial-based composting systems in the future.

This study also demonstrated that distinctive community patterns from livestock composts could be rapidly generated using T-RFLP analysis. The T-RF peaks were useful in investigating the diversity of complex compost communities. The succession of peaks in combination of increasing and decreasing peak heights at using composting methods indicates the high potential of T-RFLP technique to monitor microbial communities and their variation qualitatively and quantitatively. The use of T-RFLP fingerprinting profile reduces the number of product that needs to be identified by DNA sequencing to those that are assumed to be of specific importance.

Acknowledgements. The authors thank T. Meulia of the Molecular and Cellular Imaging Center for the Genescan analysis, and W. A. Dick (Department of Natural Sciences) and H.A.J. Hoitink (Department of Plant Pathology) for providing laboratory space. The authors also thank the two anonymous reviewers for their improvements to the earlier draft of this chapter. This work was financially supported by funds from The Ohio State University, Agricultural Research and Development Center Seed Grant Program.

References

Atlas RM, Bartha R (1997) Microbial ecology: fundamentals and applications, 4th edn. Benjamin/Cummings, Menlo Park, California, 694 pp

Beffa T, Blanc M, Marilley L, Fischer JL, Lyon,PF, Aragno M (1996) Taxonomic and metabolic microbial diversity during composting. In: De Bertoldi M, Sequi P, Lemmes B, Papi T (eds) The science of composting. Part I. M. Chapman and Hall, London, pp 149–161

Blanc M, Beffa T, Aragno M (1996) Biodiversity of thermophilic bacteria isolated from hot compost piles. In: De Bertoldi M, Sequi P, Lemmes B, Papi T (eds) The science of composting. Part II. Chapman and Hall, London, pp 1087–1090

Clement BG, Kehl LE, DeBord KL, Kits CL (1998) Terminal restriction fragment patterns (TRFLPs), a rapid, PCR-based method for the comparison of complex bacterial communities. J Microbiol Methods 31: 135–142

Epstein E (1997) The Science of Composting. Technomic Publishing, Lancaster, Pennsylvania, 487 pp

Fujio Y, Kume SJ (1991) Isolation and identification of thermophilic bacteria from sewage sludge compost. J Ferment Bioeng 72: 334–337

Golueke CG (1972) Composting: a study of the process and its principles. Rodale Press, Emmaus, Pennsylvania, 110 pp

Hammer PL, Rudeanu S (1968) Boolean methods in operation research and related areas. Spinger, New York 329 pp

Heuer H, Smalla K (1997) Application of denaturing gradient gel electrophoresis and temperature gradient gel electrophoresis for studying soil microbial communities. In: Van Elsas JD, Trevors JT, Wellington EMH (eds) Modern Soil Microbiology. Marcel Dekker, New York, pp 352–373

Keener HM, Elwell DL, Reid GL, Michel FC Jr (2000) Composting non-separated dairy manure-theoretical limits and practical experience. Proc 8th Int Symp on Animal, Agricultural and Food Processing Wastes (ISAAFPW 2000). Des Moines, Iowa

Kerkhof L, Santoro M, Garland J (2000) Response of soybean rhizosphere communities to human hygiene water addition as determined by community level physiological profiling (CLPP) and terminal restriction fragment length polymorphism (TRFLP) analysis. FEMS Microbiol Lett 184: 95–101

Liu WT, Marsh T, Cheng H, Forney LJ (1997) Characterization of microbial community by determining terminal restriction fragment length polymorphisms of genes encoding 16S rRNA. Appl Environ Microbiol 63: 4516–4522

Ludemann H, Arth I, Liesack W (2000) Spatial changes in bacterial community structure along a vertical oxygen gradient in flooded paddy soil cores. Appl Environ Microbiol 66: 754–762

Maidak BL (2000) The Ribosomal Database Project II. Abstract presented at the Midwest Molecular Ecology 2000 Conference. July 16-18, 2000, Northern Illinois University, Illinois, p 16

Marsh TL, Saxman P, Cole J, Tiedje J (2000) Terminal restriction fragment length polymorphism analysis program, a web-based research tool for microbial community analysis. Appl Environ Microbiol 66: 3616–3620

Massol-Deya AA, Odelson DA, Hickey RF, Tiedjie JM (1995) Bacterial community fingerprinting of amplified 16S and 16-23S ribosomal gene sequences and restriction endonucleases analysis (ARDRA). Mol Microb Ecol Man 3:3:2: 1–8

Pielou EC (1969) An introduction to mathematical ecology. Wiley, New York

Schwieger F, Tebbe CC (1998) A new approach to utilize PCR-single-strand-conformation polymorphism for 16S rRNA gene based microbial community analysis. Appl Environ Microbiol 64: 4870–4876

Shannon CE, Weaver W (1949) The mathematical theory of communication. University of Illinois Press, Urbana, Illinois, 125 pp

Sparks DL (1996) Nitrogen-total. In: Methods of soil analysis. Part 3- Chemical methods. SSSA, Madison, Wisconsin

Strom PF (1985a) Identification of thermophilic bacteria in solid waste composting. Appl Environ Microbiol 50: 906–913

Strom PF (1985b) Effect of temperature on bacterial species diversity in thermophilic solid-waste composting. Appl Environ Microbiol 50: 899–905

Suzuki MT, Giovanni SJ (1996) Bias caused by template annealing in the amplication of mixtures of 16S rRNA genes by PCR. Appl Envrion Microbiol 62: 625–630

Tiquia SM, Tam NFY, Hodgkiss IJ (1996) Microbial activities during composting of spent pig-manure sawdust litter at different moisture contents. Biores Technol 55: 201–206

Wiener N (1948) Cybernetics or control and communication in the animal and the machine. The Massachusetts Institute of Technology Press, Cambridge, Massachusetts, 191 pp

Amplified 16S Ribosomal DNA Restriction Analysis of Microbial Community Structure During Rapid Degradation of a Biopolymer, PHA, by Composting

K. Uchiyama, T. Suzuki, H. Tatsumi, H. Kanetake and S. Shioya[1]*

Abstract. To investigate the changes in the microbial community structure during composting and to compare the differences in the structures between different composting processes, amplified ribosomal DNA restriction analysis (ARDRA) patterns of entire microbial communities were developed. Rapid degradation of polyhydroxyalkanoates (PHAs) in a model composting process was studied by changing the operation conditions and initial conditions as well as by addition of PHA-degrading microorganisms. *Azotobacter* sp. AZ34, which was isolated from our compost and could degrade PHA, was added at high concentrations at the beginning of composting. The differences in ARDRA patterns between composting with and without addition of the bacteria were investigated. There was a significant difference in microbial community structures between the two cases at the early stage, but ARDRA patterns at the middle and final stages appeared to be almost the same. No effect was observed at least in terms of PHA degradation rate compared, between AZ34 and the control, which was primed with the seed compost without addition of PHA-degrading bacteria. After autoclaving the raw materials of the compost, AZ34 was initially added and the seed compost was added at 72 hours and composting process continued. The effect of the PHA-degrading bacteria was observed, in this case. This experiment demonstrated that if the initial microbial population in the raw materials of the compost is significantly reduced, AZ34 can grow preferentially and promotes PHA degradation during the composting process. The same effect as that observed in this experiment was found by recycling fairly large amounts of the compost. This finding of the preferential growth of microorganisms in a microbial community will be utilized for the production of functional composts.

Introduction

In ecosystem processes, microorganisms have important functions in terms of nutrient recycling and decomposition for which many steps of the enzymatic

---

[1]Dept. of Biotechnology, Graduate School of Eng., Osaka Univ. Suita, Osaka 565-0871, Japan, e-mail: shioya@bio.eng.osaka-u.ac.jp; Tel/Fax: +81-6-6879-7444.
*to whom all correspondence should be addressed.

H. Insam, N. Riddech, S. Klammer (Eds.)
Microbiology of Composting

degradation of complex organic substrates and the release of nutrients and trace elements are required. That is, many microorganisms with their specific functions and the microbial community contribute to keeping the natural environment as it is. However, the lack of effective methods for analysis of microbial communities has limited the understanding of the microbial community structure. Traditional methods depended on the isolation of bacteria, which is time-consuming and misses a number of uncultivable bacterial species. Only a small proportion of microorganisms could be cultivated until now. Recently, molecular and biological, particularly taxonomic, knowledge on various microorganisms has been accumulated and many useful methods of describing the microbial community have been developed. Moreover, the recently developed direct extraction and purification methods of nucleic acids from an entire microbial community (Steffan et al. 1988; Tsai et al. 1991; Berthelet et al. 1996; Purdy et al. 1996; Zhou et al. 1996) provided the means of analyzing a microbial community because the isolation of microorganisms could be omitted.

The 16S ribosomal RNA genes can be used as a tool for studying the bacterial community because all bacteria have these genes that include both conserved and variable regions. The conserved regions can be used as primer-binding sites for polymerase chain reaction (PCR) and the variable regions can be used for the discrimination of individual bacteria based on the difference in DNA sequences. Many methods are applicable to distinguish the DNA sequences. The most simple and accurate method is DNA sequencing. However, this method requires cloning of individually amplified ribosomal DNA fragments into a certain cloning vector and detection of DNA sequences because a microbial community includes many bacteria. The sequencing method is actually superior for application in taxonomy. However, it is time-consuming and difficult to obtain information on the entire microbial community structure. As an alternative, DNA fragment analyses, for example, amplified ribosomal DNA restriction analysis (ARDRA) (Vaneechoutte et al. 1992, Massol-Deya et al. 1995, Blanc et al. 1997; Schmit et al. 1997), single-strand conformation polymorphism (SSCP) (Lee et al. 1996; Schwieger and Tebbe 1998; Peters et al. 2000), denaturing gradient gel electrophoresis (DGGE) (Fischer and Lerman 1979; Myers et al. 1985; Muyzer et al. 1993; Ferris et al. 1996; Duineveld et al. 1998), and temperature gradient gel electrophoresis (TGGE) (Riesner et al. 1989; Felske et al. 1998), can be applied to mixed DNA fragments and group them based on the difference in DNA sequences. It would be rather difficult to describe the microbial community in the individual level by DNA fragment analysis. However, these methods could easily provide profiles of the entire microbial community structure and have been successfully applied to microbial community analysis (Muyzer et al. 1993; Massol-Deya et al. 1995; Ferris et al. 1996; Lee et al. 1996; Blanc et al. 1997; Schmit et al. 1997; Duineveld et al. 1998; Felske et al. 1998; Schwieger and Tebbe 1998; Peters et al. 2000).

In this study, ARDRA, one type of DNA fragment analysis, was applied to the analyses of the microbial community in a composting process. Biodegradable polymers, such as polyhydroxyalcanoate (PHA), will eventually be degraded in the natural environment, however, they should be degraded in a shorter time when they are used as materials for plastic bags that are composted with garbage. Rapid degradation of PHA in a model composting process by changing the operation conditions and initial conditions as well as by addition of PHA-degrading microorganisms was studied. We compared the ARDRA patterns of an entire microbial community exhibiting rapid degradation activity of PHA with those of a microbial population in a normal composting process. The community-level analysis based on ARDRA patterns can provide valuable data for understanding the metabolic activity of PHA-degrading bacteria in composts.

Materials and Methods

Composting Experiment

A commercially available dog food (VITA-ONE: Nippon Pet Food Cooperation, Japan), which was used as the source of organic matter, was ground into powder and passed through a sieve of 1-mm mesh. The sieved dog food (400 g) was mixed with 400 g of sawdust in a 5-l styroform container and 800 g of water was added. In general, the seed compost was added at a ratio of 0.25% (w/w) as a starter material. The seed compost was obtained after 60 days of composting. Ventilation was provided through the bottom of the container at 0.02 m^3 kg^{-1} h^{-1}. The water content with respect to the total weight was maintained at 50–60% (w/w) throughout the composting experiment by adding water once a day. In this study, 50 films (10 x 50 mm, 20 mg) of poly(3-hydroxybutyric-co-3-hydroxy--valeric) acid (PHB/V), which is a PHA and has 95% of the 3-hydroxybutyric acid unit, were laid in the compost and three films were taken out at each sampling.

The temperature was measured at 1 h intervals using the THERMIC model 2100A (ETO DENKI, Tokyo, Japan). A 1-g compost sample was dried at 80 °C for 24 h. Then, water content with respect to the total weight was calculated. The percentage of PHB/V residual weight was obtained after three films of PHB/V were taken out from the compost, gently washed with water and dried. The organic carbon and total nitrogen contents of the compost were measured by the Tyurin (1931) and Kjeldahl (Kandeler 1996) methods, respectively. Then, the ratio of total carbon to nitrogen (C/N) was calculated. The number of bacteria in the compost was determined as follows. A 1-g compost sample was suspended in 9 ml of sterilized water. A series of ten times dilutions was prepared. Dilutions of 1 ml were transferred to petri dishes and mixed with melted YG agar [1 g l^{-1} yeast extract, 1 g l^{-1} glucose, 0.3 g l^{-1} K_2HPO_4, 0.2 g l^{-1} KH_2PO_4, 0.2 g l^{-1} $MgSO_4 \cdot 7H_2O$, 15 g l^{-1} agar (pH 6.8)]. They were incubated at 30 and 60 °C for 10 d. The colonies were counted and the number of bacteria per g of compost was determined. The PHA-degrading bacteria were also counted. In this case, mPY agar medium (Matavulj and Molitoris 1992) mixed with powder of PHB/V polymer (HV contents 5%, Biopol, USA) at a concentration of 3 g l^{-1} was used. After 10 d of incubation at 30 °C, colony-forming halos on the plate were counted as PHA-degrading bacteria.

We screened the PHA-degrading bacterium, *Azotobacter* sp. AZ34, from the matured compost initiated with dog food and PHA films, which can survive at the highest composting temperature of 60 °C and at a high pH of 9.0 (data not shown). By addition of *Azotobacter* AZ34, the composting process was also performed, in which case, the raw materials of composting were autoclaved (at 120 °C for 15') to eliminate other bacterial populations and used for the composting experiment.

Direct Extraction of DNA from Compost

The direct extraction method described by Tsai and Olson (1991) was modified and applied to the extraction of DNA from the entire microbial community in the compost. A 1-g sample was washed twice with 4 ml of 120 mM Na-phosphate buffer (pH 8.0). After centrifugation at 6,000 *g* for 10', the pellet was resuspended in 2 ml of lysis solution 1 [0.15 M NaCl, 0.1 M Na_2EDTA, 15 $mgml^{-1}$ lysozyme, (pH 8.0)] and incubated at 37 °C for 2 h. Then, 2 ml of lysis solution 2 [0.1 M NaCl, 0.5 M Tris-HCl, 10% (w/v) SDS, (pH 8.0)] was added. After being subjected to three cycles of cold (-80 °C ethanol bath)-thaw (65°C water bath) treatment, 4 ml of Tris-HCl (pH 8.0)-saturated phenol was added and mixed. The mixture was centrifuged at 12,000 *g* for 15'. Then, 3 ml of the upper aqueous phase was transferred to another tube and extracted with a phenol-chloroform-isoamyl alcohol (25:24:1) mixture. The extracted aqueous phase (2 ml) was precipitated with 2 ml of isopropanol by placing the mixture at –20 °C for more than 2 h. After centrifugation at 12,000 *g* for 15', the pellet was washed with 2 ml of 70% (v/v) ethanol. The pellet was vacuum-dried and resuspended in 200 μl of sterilized pure water. Finally, the crude sample was treated with RNase A (final concentration, 0.1 $mgml^{-1}$) at 37 °C for 2 h.

DNA Purification

The crude DNA samples were purified using polyvinylpolypyrrolidone (PVPP) (Sigma, USA) (Berthelet et al. 1996) and hydroxyapatite (HTP) (Bio-Gel HTP; BioRad , UK) spin column (Purdy et al. 1996). The DNA directly extracted from compost (100 μl) was diluted with 100 μl of 40 mM potassium phosphate buffer (pH 7.4). The acid-washed PVPP (0.5 ml) suspended in 20 mM buffer (pH 7.4) was transferred to a Wizard Minicolumn (Promega, USA) using a 1-ml syringe. The diluted DNA sample was applied to the PVPP spin column. After 1', the pellet was washed with 0.4 ml of ethanol and kept at –80 °C for 10'. After centrifugation at 12,000 *g* for 15', the pellet was washed with 0.4 ml of 70% ethanol at 4 °C. The pellet was dried and resuspended in 100 ml of sterilized pure water.

After PVPP treatment, 100 μl of DNA sample was applied to the HTP spin column which was prepared as follows: HTP dry powder was rehydrated with 120 mM sodium phosphate (pH 8.0) and the mixture was degassed. The Wizard

Minicolumn was supported with 100 μl of HTP. The DNA sample of 100 μl was applied to the HTP spin column after centrifugation at 12000 *g* for 2'. The column was washed three times with 200 μl of 120 mM potassium phosphate (pH 7.2) and the DNA was eluted twice into another tube using 200 μl of 300 mM potassium phosphate (pH 7.2). The eluted DNA was precipitated with 2 vol ethanol. After centrifugation at 12,000 *g* for 15', the pellet was washed with 70% (v/v) ethanol, and resuspended in 100 μl of sterilized pure water.

Amplification of 16S rDNA

A set of primers, pA(5'-AGAGTTGATCCTGGCTCAG-3') (position 8-27 in *E. coli* 16S rDNA) and pH(5'-AAGGAGGTGATCCAGCCGCA-3') (position 1521-1540) which are highly conserved fsequences (Massol-Deya et al. 1995), was used for the amplification of 16S rDNA genes. The PCR was performed in 50 μl mixture, containing 1.25 U of AmpliTaq DNA polymerase (Perkin Elmer), 1×PCR buffer (final conc., 1.5 mM $MgCl_2$), 200 μM each of dATP, dCTP, dGTP and dTTP, 1 μM each of primers pA and pH, and 500 ng of the PVPP-HTP-purified DNA. First, the mixture was incubated at 94 °C for 0.5', with annealing cycles at 50 °C for 0.75', and an extension at 72 °C for 1.5' Then, a 10' final extension was carried out and the reaction mixture was maintained at 4 °C.

Digestion and Electrophoresis

In this study, two restriction enzymes, *Alu*I and *Hap*II, were used for ARDRA. Reactions were carried out in a 20-μl volume which contained 10 μl of the PCR mixture and 5 U of the restriction enzymes. The mixture was incubated at 37 °C for 2 h or overnight. An ExcelGel DNA Analysis kit (Amersham Pharmacia Biotech, Sweden) and Multiphor II Electrophoresis system (Amersham Pharmacia Biotech, Sweden) were used for the separation of digested DNA fragments. After electrophoresis, the separated DNA fragments were detected using a DNA silver staining kit (Amersham Pharmacia Biotech, Sweden).

Results

Effect of Number of PCR Cycles on ARDRA Patterns

PCR is a useful technique in amplifying a specific DNA sequence and serves as the basis of ARDRA. However, in the case of DNA fingerprinting of an entire microbial community, the template DNA contained many target DNAs. Therefore, the number of cycles of PCR was important as in the quantitative PCR where two different target regions (target and internal standard) are usually amplified with the same set of primers. The effect of the number of cycles was, therefore, investigated. The amount of PCR-amplified fragments was quantified for three different microbial communities from the compost at every five cycles. After 15 cycles, the amplified fragments could be detected in agarose gel stained with ethidium bromide. The relationship between the number of cycles and the amount of products obtained is shown in Fig. 1A. It was found that the reaction reached a plateau after 25 cycles. Next, these amplified DNA fragments were digested with the restriction enzymes and the resulting ARDRA patterns were compared (Fig. 1B). ARDRA patterns at 15 and 20 cycles were clearly different from those at 25 and 30 cycles, respectively, for each different community. Some DNA fragments were distinctly detected and some fragments could not be detected after 25 cycles. Consequently, the diversity of ARDRA patterns after 25 cycles was lost but the patterns became stable. We employed 20 cycles of PCR amplification by taking into account the sensitivity of the diversity of the microbial communities.

Analysis of Composting Process based on ARDRA Patterns

The composting process requires metabolic activities of various microorganisms and the microbial community structure changes during composting processes. It is very important and useful for the understanding of the composting process to determine the microbial community structure. In this study, the relationship between a feature of the microbial community in the compost, that is, PHB/V degradation, and microbial community structure was investigated. A PHA-degrading bacterial strain, *Azotobacter* sp. AZ34, was added to the compost at the beginning of the composting reaction and the difference in ARDRA patterns between with (run 2, run 3) and without (run 1) addition of the bacteria was investigated. Figure 2 shows the time courses of temperature, pH, number of cells and percentage of residual weight of PHB/V during composting. It appeared that the addition of *Azotobacter* sp. AZ34 at a high concentration had some, but not significant, effect on the composting process in more than five trials. Here, the data of two example runs are shown.

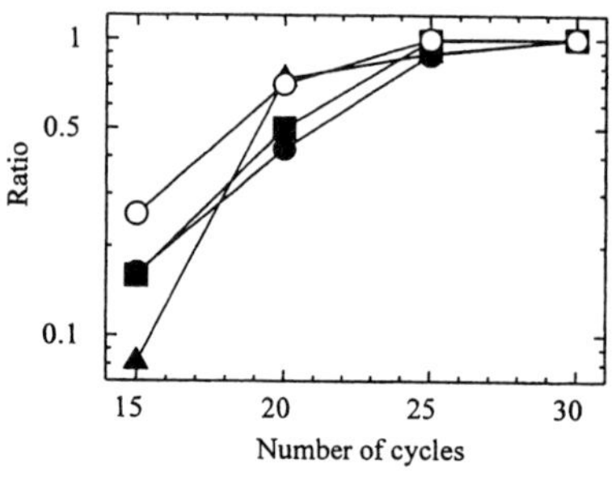

Fig. 1 A. Effect of the number of cycles of PCR amplification on ARDRA patterns of three different microbial communities using the set of primers (pA and pH). **A** Kinetics of PCR amplifications of three different microbial communities (●, ■ and ▲) and pure cultured bacteria (○) with the same set of primers. PCR products were detected and quantified in ethidium bromide-stained agarose gel (1%) before restriction enzyme digestion. The ratio of intensity at each cycle to that of 30 cycles was plotted with the logarithmic scale to the number of cycles.

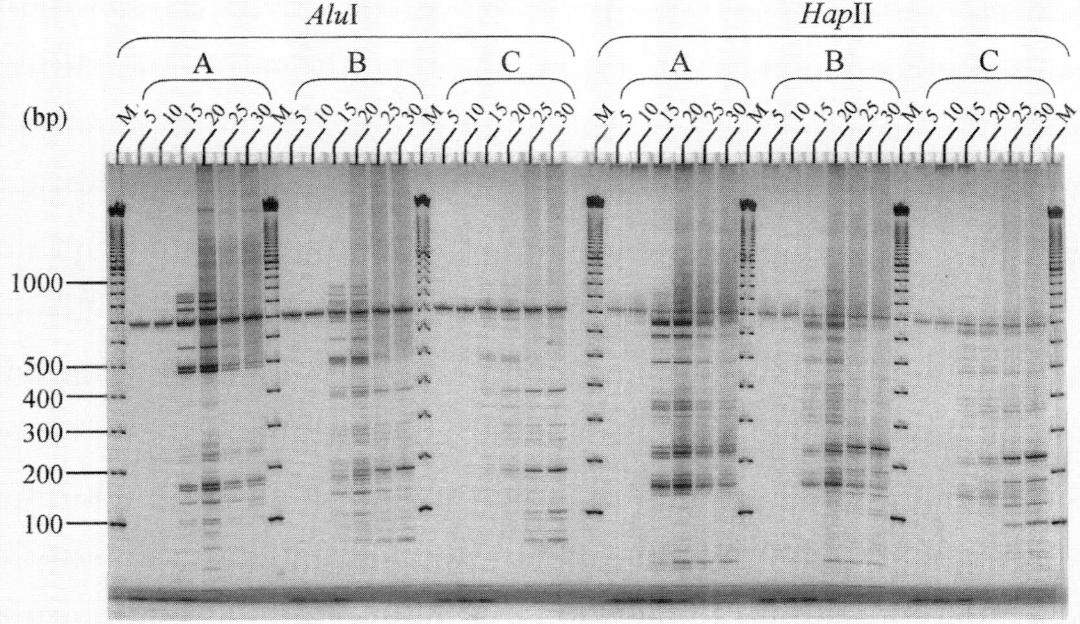

Fig1B. ARDRA patterns of three different microbial communities at every five cycles, digested with *Alu*I and *Hap*II, were obtained by silver-stained polyacrylamide gel electrophoresis. A, B and C correspond to ●, ■ and ▲, respectively.

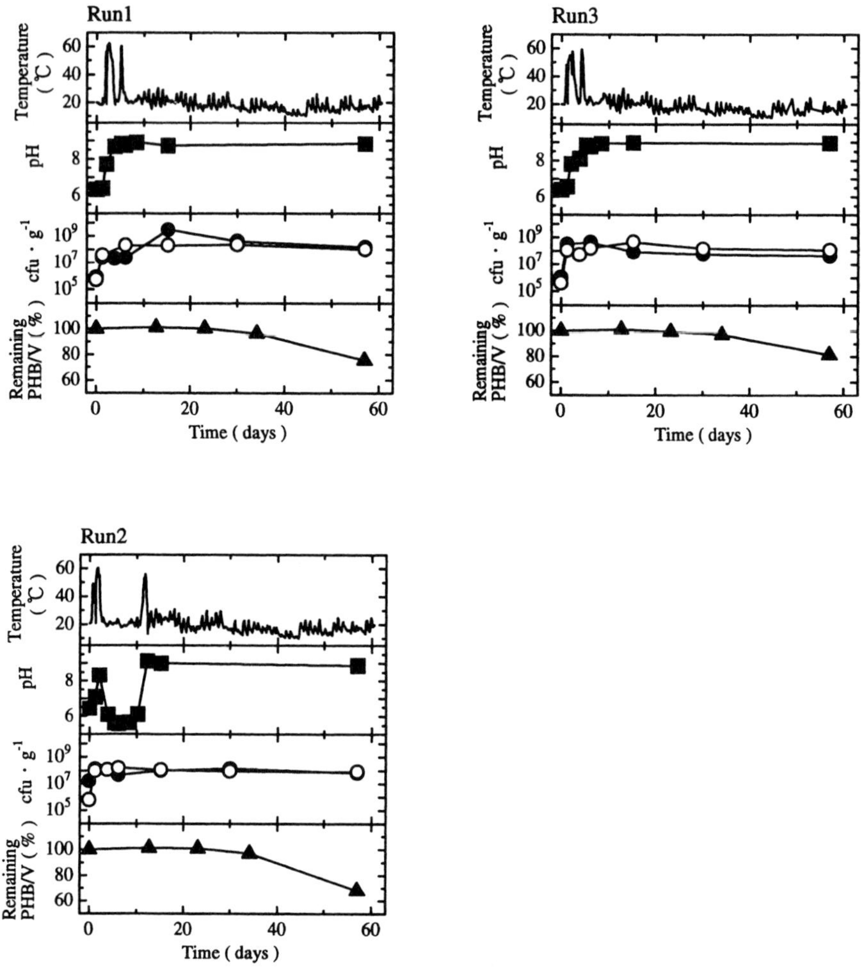

Fig. 2. Time courses of temperature, pH, cfu and percentage of remaining PHB/V during composting with and without addition of *Azotobacter sp.* AZ31. *Run 1*: *Azotobacter* sp. AZ34 was not added. *Runs 2 and 3 Azotobacter* sp. AZ34 of 10^5 cfu was added to 1 g of dry compost at the beginning of composting. ■, pH; ●, cfu of cells incubated in YG agar plate at 30 °C for 10 d; ○, cfu of cells incubated in YG agar plate at 60 °C for 10 d; ▲, % of remaining weight of PHB/V

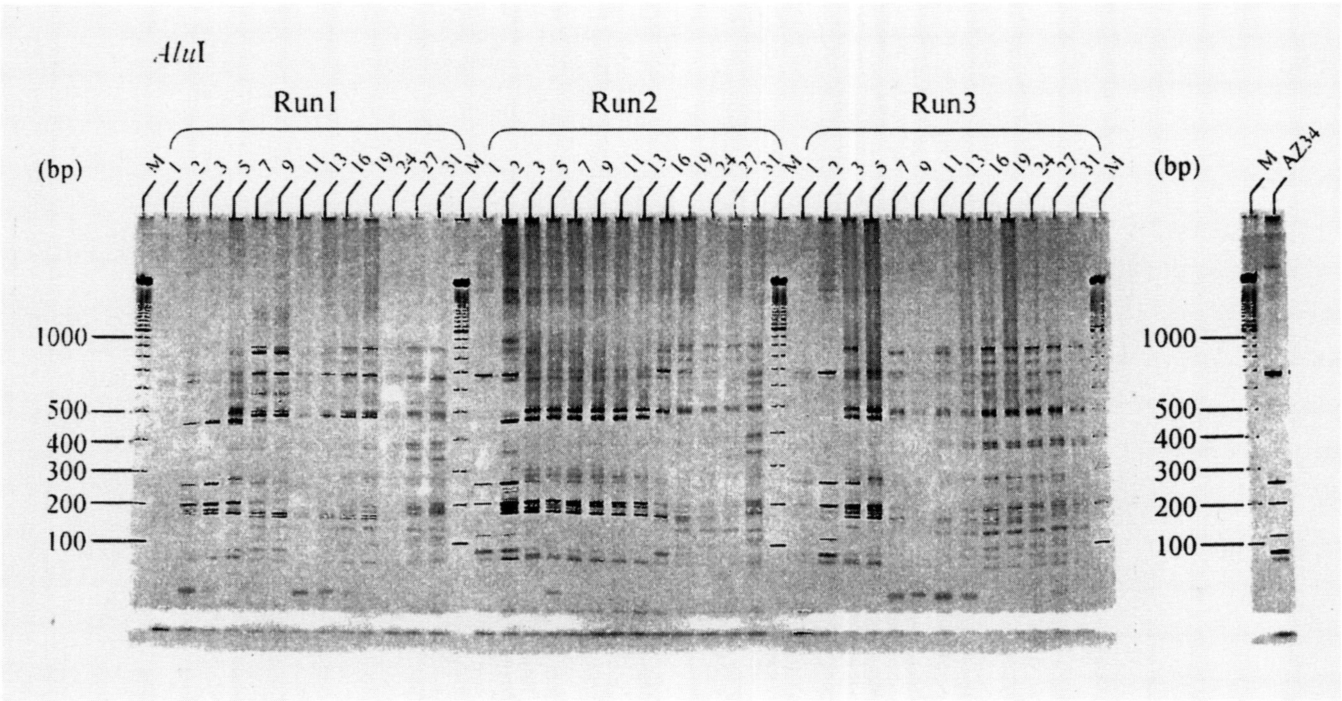

Fig. 3A. ARDRA patterns of the microbial community during composting shown in Fig. 2. *Runs 1, 2* and *3* are the ARDRA patterns for the microbial community during composting *runs 1, 2* and *3* shown in Fig. 2, respectively. **M** 100-bp DNA ladder. **Digest with *Alu*I**. Digest with *Hap*II is shown in Fig. 3B. ARDRA patterns for *Azotobacter* sp. AZ34 are also shown

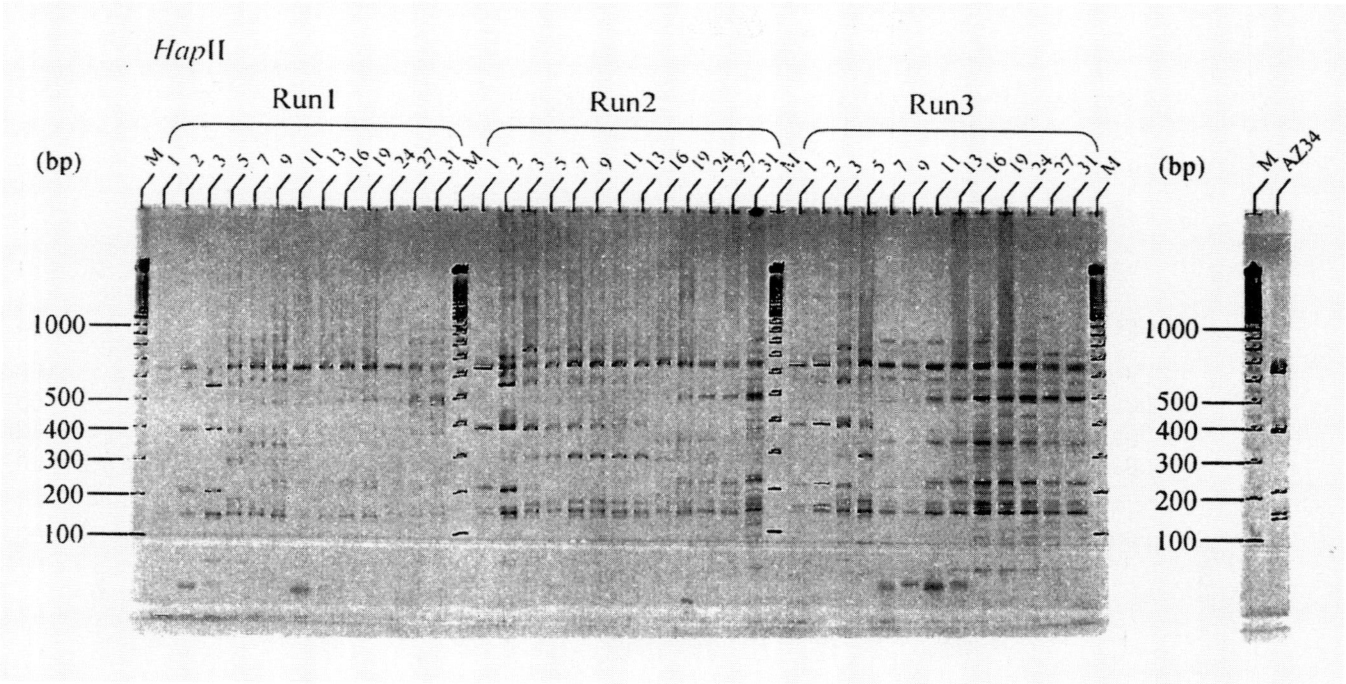

Fig. 3B. ARDRA patterns of the microbial community during composting shown in Fig. 2. *Runs 1, 2* and *3* are the ARDRA patterns for the microbial community during composting *runs 1, 2* and *3* shown in Fig. 2, respectively. **M** 100-bp DNA ladder. **Digest with *Hap*II**. Digest with *Alu*I is shown in Fig. 3A. ARDRA patterns for *Azotobacter* sp. AZ34 are also shown

Fig. 3 shows the ARDRA patterns during composting. The patterns indicate that the microbial community structures were clearly different and changed during composting. The difference in the microbial structure between run 1 and runs 2 and 3 was not clear from the data shown in Fig. 2, and it appeared that the addition of the bacterial strain did not affect the microbial community. However, ARDRA patterns for runs 2 and 3 were different from those of run 1 until day 3. ARDRA patterns for *Azotobacter* sp. AZ34 (Fig. 3) appeared distinctly at the beginning of the reaction. The community structure detected by ARDRA significantly changed in run 1 from day 1 to day 13 but became almost the same after day 16.

Rapid Degradation of PHAs by Addition of *Azotobacter* AZ34

The PHA-degrading bacterial strain, *Azotobacter* sp. AZ34, was simply added to the compost at the beginning of the composting reaction at 0.25% seed compost. As already mentioned, no effect was observed at least in terms of the PHA degradation rate compared with the control case, which was primed by seed compost without addition of the degrading bacteria. After autoclaving the raw materials of the compost, AZ34 was added and the experiment was started. After 72 h, the seed compost was added and the composting process continued. The result is shown in Fig. 4 and the effect of the addition of degrading bacteria was observed. In the figure, three cases, where one is with AZ34 and no autoclaving and the other two cases with AZ34 and with autoclaving, were compared. In these cases, the temperature and pH profiles were almost the same, which means that the same composting process proceeded. The ratio of PHA-degrading bacteria to the total bacterial population was initially high but decreased quickly after addition of seed compost and the start of the composting process, and finally became constant at approximately 0.1 to 0.001%. From this experiment, it is found that if the initial microbial population in the raw materials of compost is significantly reduced, AZ34 can grow preferentially and promotes PHB/V degradation through the composting process. The same effect as that observed in this experiment was found as shown later by recycling of fairly large amount of compost.

Fig. 5 shows ARDRA patterns obtained in the experiment, where raw materials of the compost were autoclaved and the AZ34 strain was added. After addition of seed compost, the ARDRA pattern started to change and after 16 days the fingerprinting pattern of this analysis had completely changed to a different one from AZ34. For example, from the ARDRA pattern obtained by *Alu*I digestion, the existence of the AZ34 strain in this population was not clear. Lanes A and B are ARDRA patterns of the PHA-degrading bacteria isolated from the matured compost after 30 days of this experiment.

Two strains were isolated, where one has the same ARDRA pattern (lane A) as that of AZ34, and thus is related to AZ34 while the other has a completely different ARDRA pattern (lane B). It is suggested that another PHA-degrading bacterium exists during the composting process.

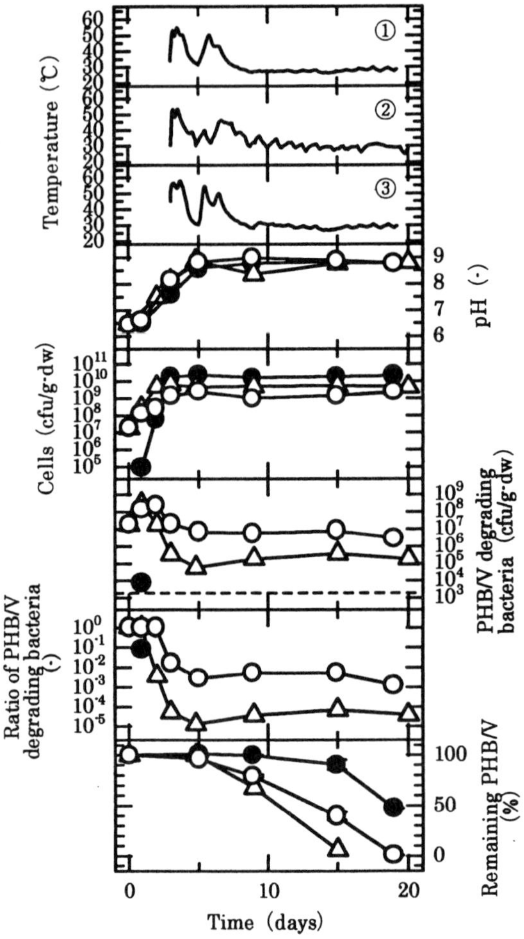

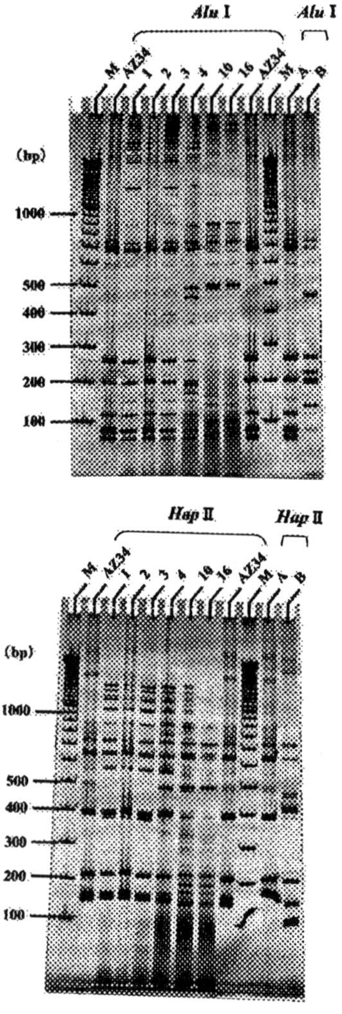

Fig. 4. Effect of addition of *Azotobacter* sp. AZ31 on PHB/V degradation during composting.
○ (1); Δ (2); with addition of AZ34 initially and 0.1% seed compost was added at 72 h.
• (3); control (without addition). Raw materials of the compost were autoclaved initially.

Fig. 5. ARDRA patterns of microbial community during composting with initial addition of AZ34, which corresponds to case ○ in Fig. 4. *Lanes A* and *B* are the ARDRA patterns of PHA-degrading microorganisms isolated from the matured compost after the PHA degradation experiment, the result of which is shown in Fig. 4.

From these experiments, the existence of the AZ34 strain in the final stage of composting could not be determined accurately based on ARDRA patterns. Thus, ARDRA patterns do not accurately reflect the activities of a microbial community because they can only provide information of the entire community structure. This is the limitation of this method.

Autoclaving of the raw materials of the compost increased the ratio of the AZ34 population, which resulted in preferential growth of AZ34 even after addition of seed compost. It is suggested that the initial preferential growth is most important for enhancing PHA degradation. Here, autoclaving is not practically available for actual composting plants. We found that the same effect was observed by recycling of fairly large amount of matured compost as seed compost. Fig. 6 shows the effect of recycling of the compost on the PHA degradation. If the recycle ratio was changed as 0.25 to 2.5, 12.5, and 25%, the remaining PHA contents after 20 d of composting decreased ranging from 99 to 38%. Thus, the same effect as that of AZ34 addition with autoclaving was obtained by recycling of a large amount of matured compost, e.g., 12.5% matured compost recycling attained a rapid degradation of PHA. This finding of preferential growth of specific microorganisms will be used for production of other functional composts.

Discussion

PCR has been applied to the analysis of microbial community structures based on the accumulated molecular biological information such as the ribosomal RNA sequence database. The advantage of PCR is its high sensitivity, that is, even a very small amount of DNA can be amplified and become detectable. Thus, we are able to understand the microbial community structure with the aid of PCR. However, in this study, it was found that the ARDRA pattern of the same microbial community, as well as the amount of PCR product, was dependent on the number of PCR cycles. The same phenomenon was observed in quantitative PCR, where two different targets were amplified in one reaction mixture (Wang et al. 1989; Chelly et al. 1990). In quantitative PCR, the quantitative relationship before PCR amplification could be kept within a certain number of cycles, that is, the exponential phase of amplification, due to the plateau effect (Chelly et al. 1990). The microbial community analysis was more complicated than the quantitative PCR because the DNA directly extracted from a microbial community included many genomic DNAs, and we could not accurately determine how many were included.

The result shown in Fig. 1B would not be caused by the differences in the length of the target DNA and PCR primers but mainly by the plateau effect, because the length of the PCR target region was almost the entire length of the 16S rDNA and only one set of primers was used. As shown in Fig. 1B, the diversity in ARDRA patterns decreased and some DNA bands were distinct at more than 25 cycles. Therefore, 20 cycles were used in this study because the sensitivity was also important. To promote PHB/V degradation, *Azotobacter* sp.

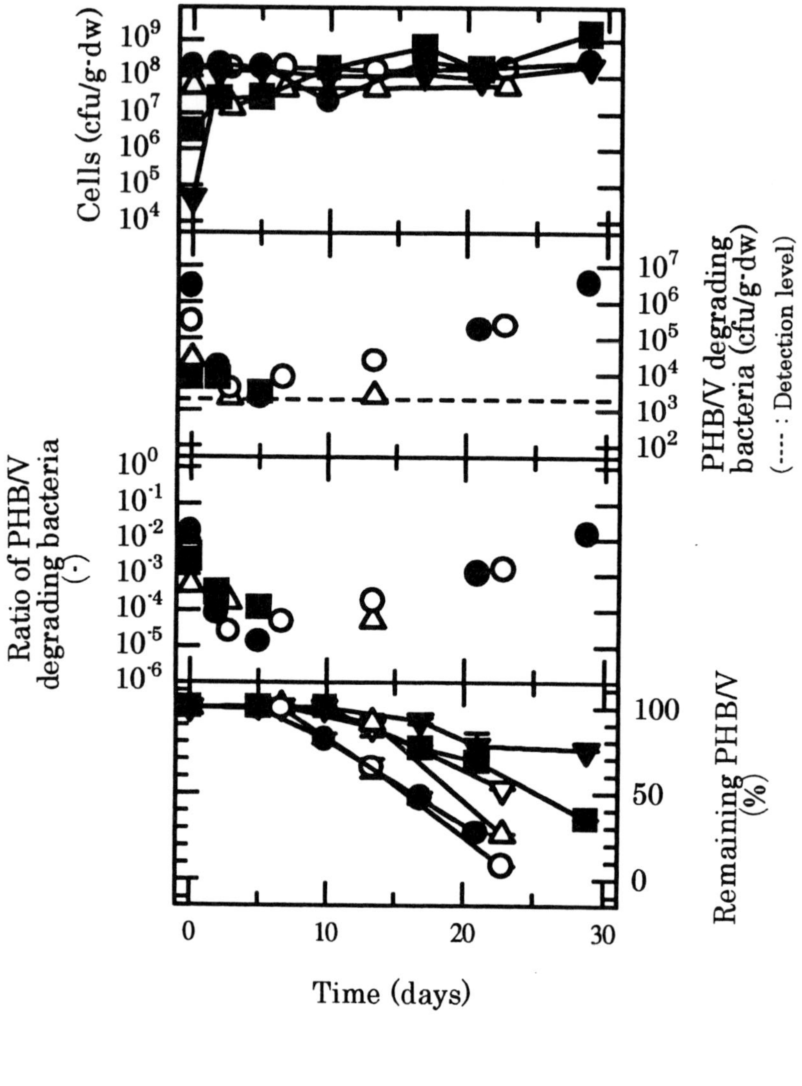

Fig. 6. Effect of recycling of the matured compost for PHA degradation. Recycling ratio (w/w%): ■; 0.25% of raw material, Δ; 2.5 %, ●; 12.5%, ○; 25%. After 20 days, 99, 72, 39 and 39% of the PHA films remained, compared to the control (∇, ▼: without AZ34 addition)

AZ34, which was isolated from the compost was added at the beginning of composting. The initial microbial structure was markedly changed by the addition of bacteria. However, PHB/V degradation was not promoted (Figs. 2, 3).

ARDRA is a potential tool for understanding the microbial ecosystem in the compost. However, ARDRA patterns do not accurately reflect the activities of a microbial community because they can only provide information of the entire community structure. The other methods, for example carbon utilization profiling (Garland et al. 1997; Konopka et al. 1998), phospholipid fatty profiling (Ibelwe and Kennedy 1998), and the method of in situ hybridization to mRNA (Poulsen et al. 1997), should be used in combination. Such a combination of community structure and metabolic activity analysis will expand the understanding of this highly active microbial ecosystem.

References

Berthelet M, Whyte LG, Greer CH (1996) Rapid, direct extraction of DNA from soils for PCR analysis using polyvinylpolypyrrolidone spin columns. FEMS Microbiol Lett 138: 17-22

Blanc M, Marilley L, Beffa T, Aragno M (1997) Rapid identification of heterotrophic, thermophilic, spore-forming bacteria isolated from hot composts. Int J Syst Bacteriol 47: 1246-1248

Chelly J, Montarras D, Pinset C, Berwald-Netter Y, Kaplan A (1990) Quantitative estimation of minor mRNAs by cDNA-polymerase chain reaction: application to dystrophin mRBA in cultured myogenic and brain cells. Eur J Biochem 187: 691-698

Duineveld B M, Rosado AS, Van Elsas JD, Van Veen JA (1998) Analysis of the dynamics of bacterial communities in the rhizosphere of the chrysanthemum via denaturing gradient gel electrophoresis and substrate utilization patterns. Appl Environ Microbiol 64: 4950-4957

Felske A, Akkermans ADL, De Vos WM (1998) Quantification of 16S rRNA in complex bacterial communities by multiple competitive reverse transcriptional-PCR in temperature gradient gel electrophoresis fingerprints. Appl Environ Microbiol 64: 4581-4587

Ferris MJ, Muyzer G, Ward DM (1996) Denaturing gradient gel electrophoresis profiles of 16S rRNA-defined populations inhabiting a hot spring microbial community. Appl Environ Microbiol 62: 340-346

Fischer SG, Lerman LS (1979) Length-independent separation of DNA restriction fragments in two-dimensional gel electrophoresis. Cell 16: 191-200

Garland JL (1997) Analysis and interpretation of community-level physiological profiles in microbial ecology. FEMS Microbiol Ecol 24: 289-300

Ibelwe AM, Kennedy AC (1998) Phospholipid fatty acid profiles and carbon utilization patterns for analysis of microbial community structure under field and greenhouse conditions. FEMS Microbiol Ecol 26: 151-163

Kandeler E (1996) Total Nitrogen. In: Schinner F, Öhlinger R, Kandeler E, Margesin R (eds.) Methods in soil biology, Springer, Berlin, Heidelberg, New York, pp. 403-406

Konopka A, Oliver L, Turco R F Jr (1998) The use of carbon substrate utilization patterns in environmental and ecological microbiology. Microb Ecol 35: 103-115

Lee DH, Zo YG, Kim SJ (1996) Nonradioactive method to study genetic profiles of natural bacterial communities by PCR-single-strand-conformation polymorphism. Appl Environ Microbiol 62: 3112-3120

Massol-Deya AA, Odelson DA, Hickey RF, Tiedje JM (1995) Bacterial community fingerprinting of amplified 16S and 16-23S ribosomal DNA gene sequences and restriction endonuclease analysis (ARDRA). In: Akkermans ADL, Van Elsas JD, de Bruijn FJ (eds) Molecular microbial ecology manual, Kluwer, Dordrecht, 3.3.2

Matavulj M, Molitoris H P (1992) FEMS Microbiol Rev 103: 323-332

Muyzer G, De Waal EC, Uitterlinden AG (1993) Profiling of complexed microbial populations by denaturing gradient gel electrophoresis analysis of polymerase chain reaction-amplified genes coding for 16S rRNA. Appl Environ Microbiol 59: 695-700

Myers R M, Fischer S G, Lerman L S, Maniatis T (1985) Nearly all single base substitutions in DNA fragments joined to a GC-clamp can be detected by denaturing gradient gel electrophoresis. Nucleic Acid Research 13: 3131-3144

Peters S, Koschinsky S, Schwieger F, Tebbe CC (2000) Succession of microbial communities during hot composting as detected by PCR-single-strand-conformation polymorphism-based genetic profiles of small-subunit rRNA genes. Appl Environ. Microbiol 66: 930-936

Poulsen LK, Dalton HM, Angles ML, Marshall KC, Molin S, Goodman AE (1997) Simultaneous determination of gene expression and bacterial identity in single cells in defined mixtures of pure cultures. Appl Environ Microbiol 63: 3698-3702

Purdy K J, Embley TM, Takii S, Nedwell DB (1996) Rapid extraction of DNA and rRNA from sediments by a novel hydroxyapatite spin-column method. Appl Environ Microbiol 62: 3905-3907

Riesner D, Steger G, Zimmat R, Owens RA, Wagenhofer M, Hillen W, Vollbach S, Henco K (1989) Temperature-gradient gel electrophoresis of nucleic acid: analysis of conformational transitions, sequence variations, and protein-nucleic acid interactions. Electrophoresis 10: 377-389

Schmit E, Leeflang P, Wernars K (1997) Detection of shifts in microbial community structure and diversity in soil caused by copper contamination using amplified ribosomal DNA restriction analysis. FEMS Microbiol Ecol 23: 249-261

Schwieger F, Tebbe CC (1998) A new approach to utilize PCR–single- strand-conformation polymorphism for 16S rRNA gene-based microbial community analysis. Appl Environ Microbiol 64: 4870-4876

Steffan RJ, Goksoyr J, Bej AK, Atlas RM (1988) Recovery of DNA from soils and sediments. Appl Environ Microbiol 54: 2908-2915

Tsai YL, Olson BH. (1991) Rapid method for direct extraction of DNA from soils and sediments. Appl Environ Microbiol 57:1070-1074

Tyurin IV (1931) A modification of a volumetric method of humus determination with chromic acid. Pochvovedenie 5-6: 36

Vaneechoutte M, Rossau R, De Vos P, Gillis M, Janssens D, Paepe N, De Rouck A, Fiers T, Claeys G, Kersters K (1992) Rapid identification of bacteria of Comamonadaceae with amplified ribosomal DNA-restriction analysis (ARDRA). FEMS Microbiol Lett 93: 227-234

Wang AM, Doyle MV, Mark DF (1989) Quantitation of mRNA by the polymerase chain reaction. Proc Natl Acad Sci USA 86: 9717-9721

Zhou J, Bruns MA, Tiedje JM (1996) DNA recovery from soils of diverse composition. Appl Environ Microbiol 62: 316-323

Comparative Investigation of Vermicompost Microbial Communities

N.V. Verkhovtseva[1], G.A. Osipov [2], T.N. Bolysheva[1], V.A. Kasatikov [3], N.V. Kuzmina[1], E.J. Antsiferova[1] and A.S. Alexeeva[1*]

Abstract. In the present study gas chromatography-mass spectrometry (GC-MS) methods were applied for investigation of worm compost (via *Eisenia fetida* culture). Sewage sludge (SS) and cattle manure (CM) were used as raw matter for compost preparation. Total percentage of carbon increased from 13% (for SS vermicompost) up to 50% (CM vermicompost), while the ratio of C:N from 22 – 42 up to 22 for SS and CM, respectively, has been shown. The contents of total nitrogen (N = 1–2 %), potassium (K_2O = 1.1–1.4%) and phosphorus (P_2O_5 = 0.6–1.5%) have been measured. The concentrations of heavy metals such as Cd, Cu, Cr, Ni, Pb, Zn in organic matter after digestion by earthworms decreased by two to six times. The greatest reduction was observed for Cr and Zn concentrations and the smallest for Ni. Lipid biomarker analysis revealed altogether 21 genera of microorganisms in vermicomposts community representing a number of trophical groups. The abundance of microorganisms from genera *Acetobacter, Sphingobacterium, Aeromonas, Vibrio* and *Streptomyces* increased remarkably after SS vermicomposting procedure. Quantitative comparison of microbial biodiversity showed, that three genera dominated in CM vermicompost: *Bdellovibrio* (and/or *Spirillum*) usually inhabitanting in wastewater (17%); *Bacteroides* (15%) and *Clostridium* (17%), which are common for rumen. The four following genera dominated in cenoses from SS composts: *Acetobacter* (20%), *Pseudomonas* (11 %), *Wolinella* (12%) and *Bacteroides* (10%), making totally 61%. Supposedly, prevalence of genera *Acetobacter* and *Wolinella* may improve nitrogen metabolism in compost matter. The SS community had two dominating genera: *Pseudomonas* (27%) and *Bacteroides* (37%). Acquired data demonstrates prevalence of microorganisms with high hydrolyzing activity. They are spore forming (*Bacillus*), mycelial (*Streptomyces, Nocardia,* representatives of the *Maduromycetes* group) and cellulolytic (*Cytophaga*) microorganisms. Their products serve as substrates for fermenting organisms (*Bacteroides, Clostridium, Vibrio, Wolinella*). Later, members of *Acetobacter, Pseudomonas, Sphingobacterium, Sphingomonas* use simple organic material formed after hydrolysis and fermentation in aerobic conditions.

---

[1]Department of Agrochemistry, Faculty of Soil Science, Moscow Lomonosov State University; 119899 Moscow, Russia; e-mail:ver@soil.msu.ru; Tel: +7-(095)-939-43-27
[2]Research group of Academician Ju.Isakov, Russian Academy of Medical Sciences, Moscow, Russia;
[3]Research Institute of Organic Fertilizers, Vladimir region, Russia

H. Insam, N. Riddech, S. Klammer (Eds.)
Microbiology of Composting

Introduction

Methods and technologies of worm compost preparation are of economical, environmental and biotechnological importance. Use of different organic waste types as ingredients for vermicomposts needs correct selection of specific microbiota to increase the value of the final product. The properties of worm composts are markedly affected by bacteria. Earthworms are also known as vectors for dispersal of soil microorganisms (Madsen and Alexander 1982) and bioreactors for certain kinds of microbes, providing changes in the soil bacterial community structure (Pedersen and Hendriksen 1993). So, the number of cultivable bacteria increased in earthworm casts, while fungal and protozoan populations decreased (Brown 1995). Besides, it was shown that the worm burrow microbial communities were markedly different from those of surrounding surface and subsurface sediments (Steward et al.1996). It is well known that Gram-negative bacteria are common inhabitants of the intestinal canal of soil animals, including earthworms. However, little is known about their taxonomic positions at genus or species level, although members of *Vibrio* sp. and *Aeromonas hydrophila* were reported to be frequent in the gut of the earthworms *Eisenia lucens* (Marialigeti 1979) and *E. foetida* (Toyota and Kimura 2000). Carbon source utilization studies, such as those done with Biolog plates, can distinguish microbial communities from different composts (Insam et al.1996) or soils on the basis of substrate utilization (Garland and Mills 1994) or metabolic potential. Chemical methods are becoming more popular in applied research in identification of microorganisms, as they are by far more rapid (Goodfellow and Minniken 1985). The two most widely used methods are gas chromatography (GC) and gas chromatography-mass spectrometry (GC-MS), which, when combined, provide extensive information on monomer chemical components of microbial cells and metabolites (Brondz and Olsen 1986). Specific cellular chemical components (or markers) are detected in situ from the background of other chemical constituents of the total biomass, indicating the presence of the relevant microbial genus or species in the community under study. Phospholipid fatty acids (PLFAs) metabolize rapidly, and thus, represent the current living community (White et al.1979). The PLFA approach was used for forest soil investigation (Pennanen et al.1996). Carbon source utilization tests and PLFA analysis were used to track microbial community succession in developing compost (Carpenter-Boggs et al. 1998). A new approach, allowing qualitative and quantitative analysis of microbial communities' genera-species composition by microbial markers and fitting FA profiles equations, has been developed during this study (Osipov and Turova 1997). The aim of the study was to compare microbial communities of the worm composts after preparation of sewage sludge and cattle manure vermicomposting (via *Eisenia foetida* culture) both qualitatively and quantitatively. The composition of microbial communities was investigated by gas chromatography-mass spectrometric (GC-MS) analysis of chemical signature markers as described previously (Osipov and Turova 1997) without the use of any conventional cultural techniques.

Materials and Methods

Vermicompost Preparation

These studies were curried out with two types of vermicomposts (VC). Sewage sludge (SS) from Trade Waste Purification Plant (Tractor Company, Vladimir city) with cattle manure (CM) addition (up to 20%) and separated solids of cattle manure were used as raw matter for worm compost preparation during 6–8 weeks. The height of the CM layer in boxes for vermicomposting process was no more than 60 cm. Cattle manure maintained within 2 weeks for passage of a stage of natural microbiological composting and heating of organic substrate prior to the addition of the worms ($50{\cdot}10^3$ worms m^{-3} organic matter). During SS and CM worm compost preparation, the heating of composts did not occur. The compost moisture content was 55–60%. The addition of water was made as necessary. Productivity of vermicomposting process was 0.6–0.8.

Chemical Analysis of Sewage Sludge and Vermicomposts

Measurements of organic matter content (1), nitrogen (2), phosphorus (3) and potassium (4) concentrations were carried out by standard agrochemical analysis: (1) measurement of percentage loss on ignition;(2,3,4) after digestion of samples in refluxing concentrated nitric acid and then (2) by Kjeldal's method, (3) with Denizhe colorimetric measurements. Total heavy metals (HM) and (4) concentrations in SS and VC were measured by flame atomic absorption spectrometry (HITACHI 18080, Japan, Tokyo).

Preparation of Samples for Gas Chromatography-Mass Spectrometry

Fatty acid extraction of microorganisms from SS and VC samples (15 mg) was carried out by the use of the whole biomass acid methanolysis in 0.4 ml HCl 1.2 N in methanol by heating to 80 ^{0}C for 1 h. Resulting fatty acid methyl esters were extracted twice with 0.4 ml hexane. The hexane fraction was dried, and the dry residue was silylated in 20 µl N,O-bis-trimethylsilyl-trifluoroacetamide (BSTFA) by heating at 80 ^{0}C for 15 min to obtain hydroxy-FA trimethylsylil derivatives. Measurements were performed on a Shimadzu QP-2000 system (Shimadzu Corporation, Kyoto, Japan) with a cross-linked methyl silicone capillary column (Ultra-1). The oven temperature was 2 min at 120 ^{0}C and then programmed to 320 ^{0}C at 5 ^{0}C min^{-1}; 1–2 µl of derivatised sample was injected in the gas chromatograph at 280 ^{0}C. The quadrupole mass spectrometer has a resolution of 0.5 mass units over the whole mass range of 2–1000 amu. Ionisation is performed by electrons at 70 eV. The sensitivity of GC-MS system is 1 ng of methyl stearate.

Fatty acids and other lipid components after separation in GC column were ionised by electron impact method and analysed in the selected ion monitoring (SIM) mode. The SIM program was arranged using the appropriate ions for more than 75 microbial markers distributed in five separate ion groups throughout the analytical range from decanoic acid to heaviest cholesterol metabolites. For instance, ion 87 was used for non hydroxy and ion 175 for hydroxy FA. Each substance was confirmed by another specific ion: M-15 ion for 3-OH acids and M-59 ions for 2-OH acids. Molecular ion and M-32 ion were used for confirmation of saturated and monounsaturated fatty acid. Other confirmations of markers were achieved by measuring the specific retention times and ratio of chromatographic peak areas for selected ions of single marker substance. A known quantity of *Methylococcus capsulatus* cells was examined in separate experiment for calibration along with CD_3-labelled tridecanoate.

Mathematical Processing and Calculation of Microbial Community Structure

Methodological and mathematical processing of quantitative determination of the species composition of microbial communities from GC-MS data was published earlier (Osipov and Turova 1997). A wide range of cellular components (lipids, sugars, proteins and volatile metabolites) form the basis of chemodifferentiation (Goodfellow and Minnikin 1985). Since microorganisms are differentiated by their trophic properties and habitat conditions and only a limited number of taxonomic units are expected in each particular case, all such units can be detected and identified by their chemical properties. Fatty acid composition of microorganisms contains ample components to form the basis for a comprehensive diagnostic system applicable to the strain level, when pure cultures are analysed. When natural samples containing microbes are subjected to acid methanolyses, free fatty acid methyl esters fraction are formed and extracted by hexane according to the common procedure. This fatty acid moiety is the superposition of inputs from single microorganisms of various species which inhabit the environment under study. The interference could be resolved by setting and solving the algebra equation simple proportion formula set. There is information on as many as 150 lipid components in our database. At the same time, single microorganism contains only 10 to 20 compounds from this list, i.e. the majority of compounds from this set are not included. The number of taxons in actual microbial communities studied ranges between 30 and 40, and the number of fatty acid can be detected in the total biomass of any particular community is 70 to 80. Selection of 30 to 40 equations can be made in such a way that the rank of the obtained subsystems is minimal. Finally, it should be taken into consideration that, within the community studied, many microbial constituents have the status of biomarkers, i.e. are specific to only one taxon of microorganisms. Such microorganisms can be enumerated from the concentration of their marker using a simple proportion formula.

Results and Discussion

Chemical Characteristics and Heavy Metals Concentration

Some chemical parameters of VC quality are given in Table 1. Mature SS and CM composts represent homogeneous structural substrate, consisting basically of worm casts. Both composts contain a substantial amount of nutrients such as C, N, P, K; pH values of vermicompost are also favourable for the growth of macro- and microorganisms. Both composts have a C:N ratio of 22–25%, which creates no threat to nitrogen mineralization (Paul and Clark 1989)

Table 1. Chemical characteristics of vermicomposts

Characteristic[a]	Sewage sludge	Sewage sludge vermicompost	Cattle manure	Cattle manure vermicompost
W,%	60.0	64.0	66.0	58.0
PH H_2O	6.9	6.5	7.2	6.8
Organic matter,%	26.7	27.3	74.0	51.0
C:N	24.0	25.0	26.0	22.0
N_{total}	1.1	1.1	2.9	2.3
P_2O_5 $_{total}$,%	2.0	2.0	1.1	1.3
K_2O $_{total}$,%	1.4	1.1	1.6	1.8

[a]Values are means of triplicates.

Concentrations of Cd, Cr, Cu, Ni, Pb, Zn in organic matter after earthworms destruction have decreased by two to six times. The greatest reduction was fixed for Cr and Zn concentrations and the smallest for Ni (Table 2). We suppose that a significant role in the decrease of heavy metal concentration in vermicomposts belongs to worms. The assumption is based on our preliminary studies of accumulation of heavy metals in worm's biomass during the SS vermicomposting process. The maximum level of accumulation was marked for Cr and Cd, their contents in worms' bodies after vermicomposting procedures having increased, respectively, by 27 and 12 times (unpublished data).

Table 2. Content of heavy metals (HM) in sewage sludge (SS), vermicompost of sewage sludge (VSS) and vermicompost of cattle manure (VCM), mg kg^{-1} 10^2

Organic matter	Cd	Cr	Cu	Ni	Pb	Zn
SS	3.10±0,40[a]	4.8±2.5	3.8±1.3	3.0±1.1	1.2±0.1	39±14
VSS	0.12±0,07	1.3±0.8	1.3±1.0	3.2±2.0	0.2±0.0	5.4±1.8
VCM	0.01	0.3	1.3±1.0	8.3	0.10	0.7
MAC of HM[b]	0.15	5.0	100	1.0	5.0	25

[a]Values are standard deviation of means of triplicates.
[b]Maximum allowable concentration of HM in the soil.

Fatty Acid Profile of Total Compost Community Biomass

GC-MS analysis showed the presence of 62 compounds belonging to normal, unsaturated, branched and hydroxy fatty acids, palmitic aldehyde and sitosterole. Some specific substances (markers) signify the presence of numerous microbial groups in the community. For example, microscopic fungi are determined by linoleic acid (octadecadienoic, 18:2). Tuberculostearic acid (10Me18) is an indicator of actinomycetes, presumably *Nocardia* and *Corynebacterium*. Another 10Me-branched FA refers to *Rhodococcus* (10Me16) (McNabb et al. 1997) or *Thermoactinomyces* (10Me14). The genus *Bacteroides* is suspected when hydroxypalmitic (h16) and hydroxy-*iso*-heptadecanoic (hi17) acids are detected. *Pseudomonas* and *Acetobacter* are determined by 2- and 3-hydroxydodecanoic (2h12,3h12) acids along with 3-hydroxydecanoic (3h10) acid. The specific proportion within these three markers helps to calculate microbial genera and species (*Weyant et al. 1996*). The presence *Sphingobacterium* is confirmed by 2--hydroxy-*iso*-pentadecanoic (2hi15) and 2-hydroxy-*iso*-heptadecanoic (2hi17), which have a marker rank. *Bacillus* is supposedly due to *iso*-pentadecanoic acid (i15) residual after balance fitting. *Caulobacter* genus could be estimated by *anteiso*-heptadecanoic acid (ai17) after equation fitting. The genus *Streptomyces* is supposed when *iso*-hexadecanoic (i16) acid excess is revealed. *Wolinella* genera could be estimated by hexadecanal (palmitic aldehyde, 16a), and *Vibrio* by hydroxylauric (h12) and hydroxypalmitic (h16) acids residue in FA balance equations after *Pseudomonas* and *Bacteroides* includs were substracted. Data for specific FA were taken from papers, cited above, or other numerous publications of the C.W.Moss laboratory, E.Yantzen papers, other scientists who measured cellular FA by gas chromatography and our own investigation.

Some substances are characteristic for the whole majority of microorganisms in a community. They are palmitic, stearic, mirystic, oleic, lauric acids and other substances. These acids are not specific and are used only for equation fitting. If community members were chosen correctly, the sum of the parts included from each member to a given substance should be equal to the measured concentration of this substance. Negative difference, when occurs, means that some microbe was not present in the community. A positive difference means that some microorganisms were not included in the community list, and the set of such positives formulates the fatty acid profile of forgotten microbe. Sometimes this procedure leads to the discovery of previously unknown species.

Microbial Diversity in Vermicompost

The following groups of microorganisms were found in the vermicompost microbial complex: Gram-negative aerobes of genera *Acetobacter, Pseudomonas, Cytophaga* and "stalked" bacteria of *Caulobacter*; facultative anaerobes of genera *Aeromonas*, *Wolinella* and *Vibrio*; anaerobic bacteria of genus *Bacteroides*. Gram-positive microorganisms with low G+C included spore-forming bacteria of the genera *Bacillus* and *Clostridium*; microorganisms of actinomycetes line (Gram+

bacteria with high G+C) *Nocardia, Micromonospora, Streptomyces* and membes of the *Maduromycetes* group. Microscopic fungi were also found (Fig.1). Microbial population densities were usually high (10^6–10^9 cells g^{-1} of compost).

Altogether, 21 genera of microorganisms were revealed in SS (with 29 species) and CM (26 species) vermicomposts communities in comparison with 19 genera from SS (25 species). Thus, species diversity appears to be higher in the presence of organic matter from SS with addition of CM manure in comparison with CM compost microbial composition. We speculate that it might be connected with substrate inhibition due to excess of CM organic matter.

The most common members of vermicompost are the genera *Pseudomonas, Bacillus, Aeromonas* and *Vibrio*, which have been shown in the intestines of *Eisenia foetida* (Toyota and Kimura 2000). Presence of *Streptomyces* is common to vulgar composts. The unexpected compost genus is *Wolinella*, although *Wolinella succinogenes* was reported to be frequent in the rumen. The abundance of microorganisms from genera *Acetobacter, Sphingobacterium, Aeromonas, Vibrio* and *Streptomyces* increased remarkably after SS vermicomposting procedure. We presume that the quality of organic matter should improve, for *Acetobacter diazotroficus* is known to fix nitrogen and some members of genera *Aeromonas, Vibrio* and *Streptomyces* can reduce nitrate, i.e. the activities of these bacterial genera improve nitrogen metabolism and thus plant productivity.

The interaction between various functional groups of microorganisms depends on nutrient resources and the biochemical mechanisms of organic and inorganic matter transformation. The ecolotrophical (Nikitin and Kunc 1988) approach to the analysis of microbial diversity can reveal a number of trophical groups in the compost community. Microorganisms with high hydrolyzing activities prevailed. They are spore forming (*Bacillus*), mycelial (*Streptomyces, Nocardia,* members of the *Maduromycetes* group) and cellulolytic (*Cytophaga*) microorganisms. Their products serve as substrates for fermenting organisms (*Bacteroides, Clostridium, Vibrio, Wolinella*). The bacteria *Acetobacter, Pseudomonas, Sphingobacterium* and *Sphingomonas* use simple organic material formed at hydrolysis and fermentation in aerobic conditions.

Dominant Microorganisms

Microorganisms are considered dominant when their number exceeds or approaches 30% of the common quantity of organisms in community (Dobrovol'skaya et al.1997). In analysis of the data on organic matter and VC microbial diversity we considered as dominating those species of microorganisms whose relative abundance exceeded or reached about 10%.

SS community had two genera of dominant microorganisms, *Pseudomonas* (27%) and *Bacteroides* (37%). The dominants came to 64% of total biomass and formed basic structural-functional association of bacteria in the primary organic matter.

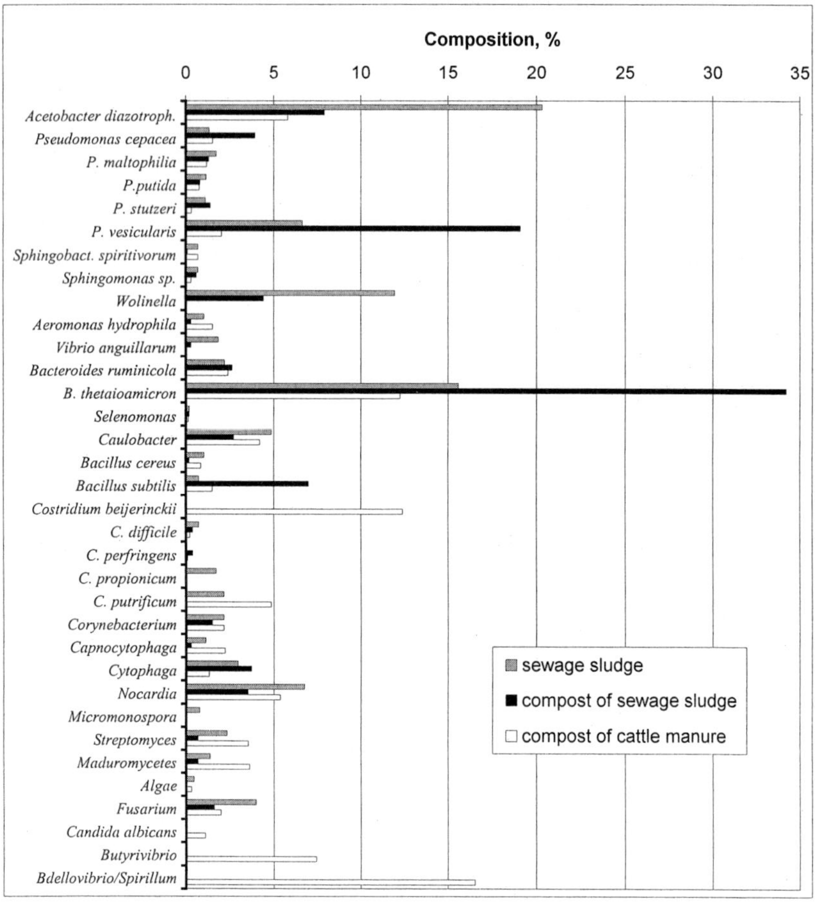

Fig.1. Composition of the microbiota in sewage sludge, vermicompost of sewage sludge and vermicompost of cattle manure

Thus, SS organic matter has been preferably used by aerobic bacteria with high hydrolytic activity (*Pseudomonas*) and anaerobic fermenting microorganisms (*Bacteroides*) contributing to the active development of these groups in SS; moreover, the anaerobic processes dominated.

The number of dominant microorganisms in VC of the SS community increased to four genera. They are members of genera *Acetobacter* (20%), *Pseudomonas* (11%), *Wolinella* (12%) and *Bacteroides* (10%), making totally 61%. Supposedly, the prevalence of genera *Acetobacter* and *Wolinella* may improve nitrogen metabolism in compost matter. Finally, the number of dominant microorganisms in VC of the CM community had three genera of dominant microorganisms. They are facultative anaerobic and obligate anaerobic

Bdellovibrio (and /or *Spirillum*), usual inhabitants in wastewater (17%), *Bacteroides* (15%) and *Clostridium* (17%), which are common for rumen as well. Consequently, the anaerobic processes dominated in this type of compost.

The redistribution of populations to a degree of domination and the development of the stable forms of microorganisms preserve the diversity of species in the structure of community and signify structural-functional changes, in reply to the worms' improvement of SS organic matter.

Conclusion

Our knowledge about the taxonomic composition of bacterial communities in vermicomposts is at an early stage. Nonetheless, some important conclusions can be derived so far:

- Microbial population densities are usually high (10^6–10^9 cells g^{-1} compost).
- 21 genera of microorganisms were found in SS (29 species) and CM (26 species) vermicompost communities compared to 19 genera in SS (25 species).
- Microorganisms with high hydrolyzing activity prevailed in the SS worm composts. They are spore forming (*Bacillus*), mycelial (*Streptomyces, Nocardia*, members of the *Maduromycetes* group) and cellulolytic (*Cytophaga)* microorganisms. Their products serve as substrates for fermenting organisms *(Bacteroides, Clostridium, Vibrio, Wolinella)*. Then members of *Acetobacter, Pseudomonas, Sphingobacterium, Sphingomonas* use simple organic material formed after hydrolysis and fermentation in aerobic conditions.
- The SS community had two dominating genera: *Pseudomonas* (27%) and *Bacteroides* (37%). The four following genera dominated in cenoses from SS composts: *Acetobacter* (20%), *Pseudomonas* (11%), *Wolinella* (12%) and *Bacteroides* (10%). Three genera dominated in CM vermicompost: *Bdellovibrio* (and/or *Spirillum*), usual inhabitants in wastewater (17%), *Bacteroides* (15%) and *Clostridium* (17%), which are common for rumen.
- The use of GC-MS for studying the complexity of a compost's microbial community allowed to detect different trophical groups and dominants in complex community with many species of useful properties.

References

Brondz J, Olsen J (1986) Microbial chemotaxonomy, chromatography, electrophoresis and relevant profiling techniques. J Chromatogr Biomed Appl 379:367-411

Brown GG (1995) How do earthworms affect microbfloral and faunal community diversity? Plant Soil 170:209-231

Carpenter-Boggs L, Kennedy AC, Reganold JP (1998) Use of phospholipid acids carbon source utilization patterns to track microbial community succession in developing compost. Appl Environ Microbiol 64:4062-4064

Dobrovol'skaya TG, Chernov IYu, Zvyagintsev DG (1997) Characterizing the structure of bacterial communities Microbiologia 66:408-414

Garland JL, Mills AL (1994) A community-level physiological approach for studying microbial communities. In: Ritz K, Dighton J and Giller KE (eds) Beyond the biomass: compositional and functional analysis of soil microbial communities. Wileys, New York, pp 77-83

Goodfellow M, Minnikin DE (eds) (1985) Chemical methods in bacterial systematics. Academic Press, London

Guo Yongcan (1995) Soil heavy metal pollution and earthworm isozyme. Chinese J Appl Ecol 6:316-322

Guo Yongcan, Wang Zhenzhong, Zhang Youmei, Mo Xiaoyang (1998) Bioconcentration effects of heavy metal pollution in soil on the mucosa, epithelia cell ultrastructure injuring of the eartworm's gastrointtestinal tract. Bull Env Contam Toxicol 60:280-284

Insam H, Amor K, Renner M, Crepaz C (1996) Changes in functional abilities of the microbial community during composting of manure. Microb Ecol 31:77-87

Madsen EL, Alexander M (1982) Transport of *Rhizobium* and *Pseudomonas* though soil. Soil Sci Soc Am J 46:557-560

McNabb A, Shuttleworth R, Behme R (1997) Fatty acid characterization of rapidly growing pathogenic aerobic actinomycetes as a means of identification. J Clin Microbiol 35:1361-1368

Marialigeti K (1979) On the community-structure of the gut-microbiota of *Eisenia lucens* (Annelida, Oligochaeta). Pedobiologia 19:231-220

Nikitin DI, Kunc F.(1988) Structure of microbial soil associations and some mechanisms of their autoregulation. In Vancura V, Kunc F (eds) Soil microbial associations: control of structures and functions. Academia, Praga

Osipov GA, Turova ES (1997) Studying species composition of microbial communities with the use of gas chromatography-mass spectrometry. Microbial community of kaolin. FEMS Microb Rev 20: 437-446.

Paul EA, Clark FE (1989) Soil microbiology and biochemistry.Academic Press, San Diego

Pedersen JC, Hendriksen NB (1993) Effect of passage though the intestinal tract of detritivore earthworms (*Lumbricus* spp.) on the number of selected Gram-negative and total bacteria. Biol Fertil Soils 16:227-232

Pennanen T, Frostegard A, Fritze H, Baath E (1996) Phospholipid fatty-acid composition and heavy metal tolerance of soil microbial communities along 2 heavy metal-polluted gradients in coniferous forests. Appl Environ Microbiol 62:420-428

Steward CC, Nold SC, Ringelberg DB, White DC, Lovell CR (1996) Microbial biomass and community structures in the burroms of bromophenol producing and nonproducing marine worms and surrounding sediments 133:149-165

Toyota K, Kimura M (2000) Microbial community indigenous to the earthworm *Eisenia foetida*. Biol Fertil Soil 31:187-190

Wang Zhehzhong (1994) Effect of heavy metals in soil on earthworms (*Opisthora*). Acta Sci Cirum 2:237-243

Weyant RS, Moss CW, Weaver RE, Hollis DG, Jordan JG, Cook EC, Daneshvar MJ (1996) Identification of unusual pathogenic Gram-negative aerobic and facultatively anaerobic bacteria. Second edition. Williams & Wilkins, Baltimore-Philadelphia

White DC, Davis WM, Nickels JS, King JD, Robbie RJ (1979) Determination of the sedimentary microbial biomass by extractable lipid phosphate. Ecology 40:51-62

Processes and Controls

Heat Production During Thermophilic Decomposition of Municipal Wastes in the Dano-System Composting Plant

J.E. Dziejowski and J. Kazanowska[1]

Abstract. Composting is a controlled thermophilic aerobic decomposition of organic solid wastes. The main part of the Dano-System composting plant is the biostabiliser. The biostabiliser was fed with 28 to 81 t of municipal wastes during 24 h. The temperature of composting wastes (28-56 °C) depended on the time of sampling and actual loading of the biostabiliser. Fresh compost from the biostabiliser was put in piles for a period of one to some months to obtain a marketable product. The samples of composting material were taken in 1998–1999 from the middle and the end part of the biostabiliser. Additionally, compost from the piles was sampled. A prototype isothermal calorimeter was used to determine the rate of heat production (RHP) of composting material. The method of closed jars for determining CO_2 production was used. The temperature of composting material in the biostabiliser was measured twice every day in 1998 and 1999. A few times the temperature was measured over a period of 24 h. The temperature inside about 2 m high piles was measured at the depth of 1 m.

Introduction

The Dano system is the most popular in Poland. Five composting plants work using this system with a possibility of utilisation from 80 to 580 tons per 24 h of organic, mainly municipal, wastes.

The investigations described in this chapter were performed on the material sampled from a production line working in the Dano system made by MACUM S.A. in Suwalki. The town Suwalki, situated in the northeastern part of Poland, has about 70,000 inhabitants. The region of Suwalki has the coolest climate in Poland with average temperatures of 4.9 °C in January and +17.7 °C in July. The town produces about 16-17 thousand tons of municipal wastes per year. With Dano technology 12.76 and 14.77 thousand tons of municipal wastes were utilised during these years. The average composition of municipal wastes undergoing utilisation is given in Table 1.

The thermophilic character of the composting process results from the heat production by microbes using organic and inorganic wastes as food during microbial growth. Several studies were performed on bench- or pilot-scale levels to evaluate the composting process parameters (Cambell et al. 1990; VanderGheynst

[1]Warmia and Mazury University in Olsztyn, Chemistry Department, Faculty of Agriculture, Plac Łódzki 4, 10-957 Olsztyn, Poland

H. Insam, N. Riddech, S. Klammer (Eds.)
Microbiology of Composting

et al. 1997). The knowledge of microbial heat production by composting material is very important for better understanding of mass and energy transport in this three-phase system.

The heat output of biological material reflects its metabolic activity. The calorimetric investigation of microbial processes can be applied for the determination of kinetic parameters of microbial biodegradation of organic substrate or determination of actual metabolic activity of biological material (Gustafsson 1991; Kemp 1999). The composting material undergoes continuous physical, chemical and microbial changes in bioreactors, e.g. in the Dano biostabiliser or in piles.

Table 1. Approximate composition of municipal wastes (Suwalki 1995)

Component	Content (%)
Fraction 0-10 mm	21
Glass	6
Metals	3
Paper	11
Textiles	4
Food wastes	30
Plastics	11
Organic residue	5
Inorganic residue	9
Total moisture	45
Fraction 0-10mm , moisture	35

For those reasons determinations of the changes of the rate of heat production during the composting process were calculated mainly for well-defined laboratory experiments (Wiley 1957; Griffis and Mote 1982; Seki and Komori 1984; Van Ginkel 1996). Reported maximum rates of heat production during the composting process are between 5 and 28 W kg^{-1} dry mass (Wiley 1957; Mote and Griffis 1982; Van Ginkel 1996). Heat production can be estimated on the basis of CO_2 production or O_2 consumption (Cooney et al. 1968; Van Ginkel 1996).

This chapter describes results of temperature and calorimetric measurements of composting material taken from the biostabiliser (Fig. 1) working in the Dano system and from piles.

Equipment destined for the utilisation of municipal wastes consists of a biostabiliser, conveying machines, screen, solid and ferromagnetic materials separators. The biostabiliser is 32 m long with a diameter of 3.6 m and it has a volume of 325 m^3. Up to 85 tons of municipal wastes during 24 h can be composted at the holding time about 36-48 h. The inside wall of the biostabiliser has equipment for shifting, breaking up and homogenisation of the composting wastes. A round screen for separation of large and hard materials from raw compost is fastened to the end of the biostabiliser. After passing through a vibration screen and separator of hard parts, raw compost was deposited in piles, where it was naturally composted for a few months to obtain a commercially availableproduct. To obtain better-quality compost some part of it is mixed with peat and kept in piles.

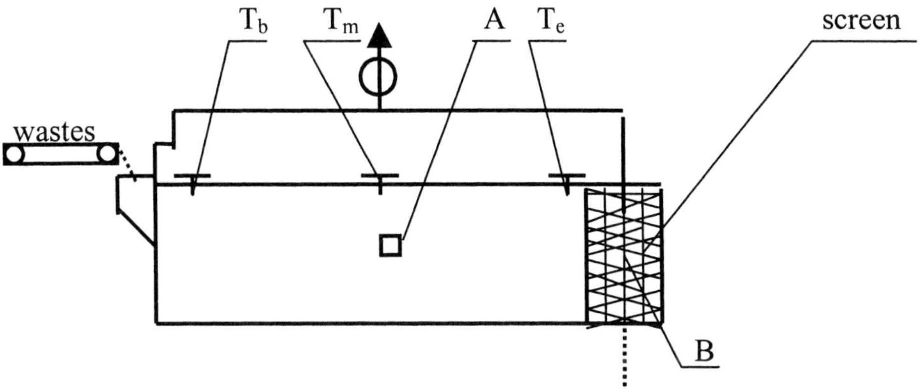

Fig. 1. Schematic diagram of the biostabiliser where: T_b, T_m, T_e – places of temperature measurement and *A*, *B* – sampling ports

Materials and Methods

The temperature of composting material in the biostabiliser was measured twice every day in 1998 and 1999. A few times the temperature was measured over a period of 24 h. Thermometers are fastened to the wall in the distance of 2.5, 16 and 29.5 m, respectively, along the biostabiliser for temperature measurements: T_b at the beginning, T_m in the middle and T_e at the end. The temperature inside about 2 m high piles was measured at the depth of 1 m.

A prototype isothermal calorimeter was used to determine the rate of heat production (RHP) of composting material. It contains four 25 cm^3 cells and sensitivity is about 6-7 $\mu W\ \mu V^{-1}$. Before measurement, each cell was electrically calibrated ($\pm 0.01\ \mu W\ \mu V^{-1}$). A convenient method for measuring metabolic heat rate at temperature $25\pm 0.01\ ^0C$ was used (Criddle et al. 1990). A technique of base line correction was applied for better reading of the calorimetric signal. The samples of composting material for calorimetric analysis were taken from the middle (A) and the end (B) part of the biostabiliser (Fig. 1). Additionally, compost from the piles was sampled. The method of closed jars for determining CO_2 production was used (Alef and Nannipieri 1995). Immediately after sampling the composting mass was put into self-opening 1 dm^3 Dewar flask and sent to the laboratory, where it was sieved on a 3 x 3 mm^2 screen. The composting waste was calorimetrically investigated 6–12 h after sampling for technical reasons. In this period the temperature of the compost decreased by about 5–10 $^{\circ}C$.

Results and Discussion

The temperature of composting wastes was read twice a day at 07.00 and 15.00 h every day during the work of the biostabiliser. Municipal wastes supplied to the compost plant have different temperatures, depending on the season of year. This can affect the composting process because some part of the microbially evolved heat is used to raise the temperature of composted wastes and the metal biostabiliser. For these reasons we paid attention to temperature changes in the biostabiliser during the chosen months: July 1999 and January 1999. Figure 2 shows the temperature changes inside the biostabiliser in July 1999. The observed large differences in the course of temperature changes depended on time, place of temperature measurement and actual loading of the biostabiliser with a new portion of wastes. Considerable changes in T_b are characteristic for the beginning stages of the biostabiliser because the temperature of the freshly loaded mass of wastes is lower in comparison to temperature T_m inside the bioreactor and temperature T_e. The changes in T_e probably result from shifting the composting mass from the middle to the end of the biostabiliser after loading. The changes in T_m are not so large as in the case of T_b and T_e. The average temperature T_b (n=60) measured twice a day in July was 44.1±6.9 ^{0}C, T_m 47.0±4.8 ^{0}C and T_e 42.2±10.8 ^{0}C at an average loading rate (n=30) of 42.4±12.0 tons day^{-1}.

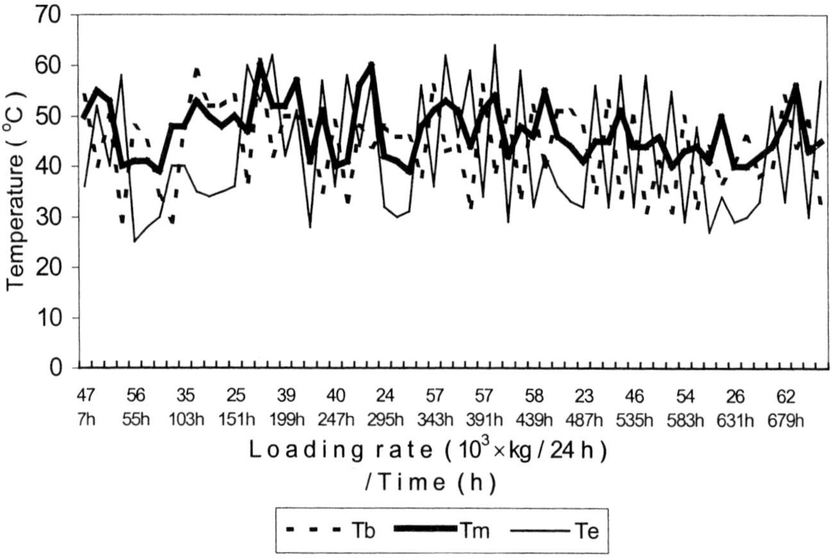

Fig. 2. The changes of temperature T_b, T_m and T_e in July 1999

Figure 3 shows the changes of temperatures T_b, T_m and T_e in January 1999. A very sophisticated course of temperature changes in the biostabiliser strongly depends greatly on the temperature of the freshly loaded wastes, particularly in winter. Violent fluctuations in temperature T_b were often greater than 30 ^{0}C. The average temperature T_b (n=62) was 31.7±12.9 ^{0}C, T_m 34.8±6.9 ^{0}C and T_e 33.1±10.5 ^{0}C during the period of measurements at average loading rate (n=31) 34.6±11.5 tons day^{-1}. The comparison of average temperature T_b, T_m and T_e in the biostabiliser shows differences in the composting process running in July 1999 and January 1999. The reported temperature changes indicate the mesophilic character of biodegradation of wastes inside the bioreactor in January 1999.

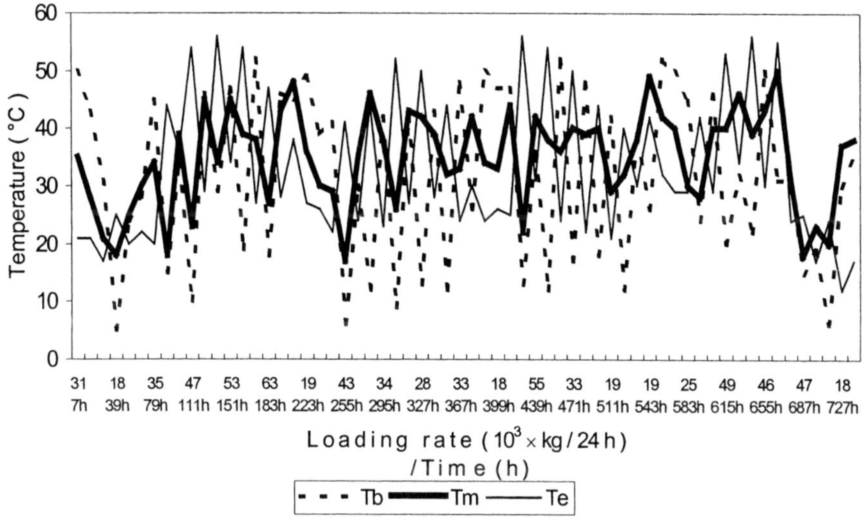

Fig. 3. The changes of temperature T_b, T_m and T_e in January 1999

Figures 4 and 5 show 24 h temperature measurements of the composting process in the biostabiliser during the chosen days of summer and winter.

The course of temperature changes was very characteristic. Almost immediately after feeding the biostabiliser with a new portion of wastes which took place between 08.00 and 14.00 h, the temperature T_b decreased. After reaching a minimum value of T_b, a period of temperature increase in relation to the values characteristic for T_m took place. The direction of temperature T_e changes was contrary to the changes of temperature T_b. The observed oscillating changes of T_b and T_e clearly show the temperature effect of fresh-loaded wastes on thermal parameters of the composting process in the Dano biostabiliser.

The results of calorimetric, respirometric, moisture and temperature measurements are presented in Table 2. There are divided into two groups depending on the period of sampling.

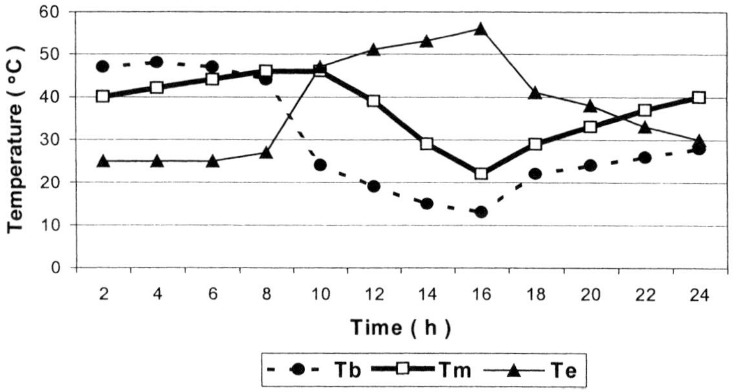

Fig. 4. Changes in temperature T_b, T_m and T_e inside the biostabiliser on 14 July 1999 at loading rate 57 tons 24 h^{-1}

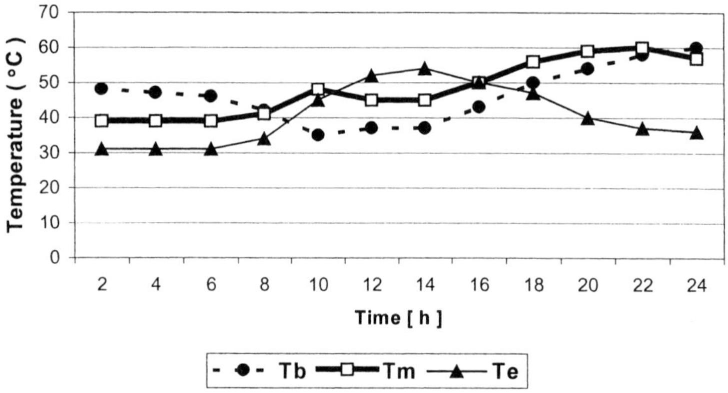

Fig. 5. Changes in temperature T_b, T_m and T_e inside the biostabiliser on 18 January 1999 at loading rate 55 tons 24 h^{-1}

The temperature was measured immediately in the place of sampling but other investigations were performed after supplying the samples (6-12 h) to the laboratory. For this reason there are only some approximations of real processes during the composting. The higher temperature of the composting material during summer induces the increase of metabolic activity expressed as RHP and rate of respiration of the compost mass. Enhanced water evaporation contributes to lower moisture in the composting material in the period June–October. The obtained RHP values are relatively low in comparison to maximum values obtained during stable laboratory experiments.

Table 2. The changes of temperature, rate of heat production (RHP), moisture and respiration of composting material in 1998–1999

Place of sampling	Temperature[a] (° C)	RHP[a] (mW 1 g $^{-1}$dry mass)	Moisture[a] (%)	CO_2 Production[a] (mg 1 g $^{-1}$dry mass 24 h^{-1})
June–October				
Middle of the biostabiliser	44.3 ± 5.5 (*n*=6)	0.64 ± 0.36 (*n*=6)	42.6 ± 7.8 (*n*=6)	3.02 ± 1.62 (*n*=6)
End part of the biostabiliser	41.5 ± 9.3 (*n*=6)	0.63 ± 0.46 (n=6)	44.6 ± 6.2 (*n*=6)	3.04 ± 1.32 (*n*=6)
From piles (1 and 2 months)	47.2 ± 7.2 (*n*=4)	0.56 ± 0.24 (*n*=6)	37.8 ± 3.1 (*n*=6)	0.67 ± 0.37 (*n*=6)
From piles compost with peat (1 : 1) (1 and 3 months)	56.2 ± 4.7 (*n*=4)	0.26 ± 0.15 (*n*=6)	27.4 ± 5.2 (*n*=6)	1.34 ± 1.03 (*n*=6)
November–March				
Middle of the biostabiliser	45.8 ± 5.2 (*n*=7)	0.53 ± 0.33 (*n*=7)	49.0 ± 3.2 (*n*=7)	2.00 ± 0.24 (*n*=7)
End of the biostabiliser	40.4 ± 7.8 (*n*=7)	0.43 ± 0.17 (*n*=7)	49.5 ± 2.7 (*n*=7)	2.55 ± 1.07 (*n*=7)
From piles (1 and 4 months)	47.6 ± 4.5 (*n*=5)	0.58 ± 0.41 (*n*=7)	39.5 ± 3.4 (*n*=7)	2.73 ± 0.37 (*n*=7)
From piles compost with peat (1 : 1) (1 and 6 months)	46.2 ± 7.1 (*n*=5)	0.33 ± 0.15 (*n*=7)	33.2 ± 6.7 (*n*=7)	1.76 ± 0.81 (*n*=7)

[a] ± SD

This indicates that decomposition of easily degradable compounds of municipal wastes takes place during the first h after mixing with innoculum at the beginning part of the biostabiliser. The composting process is still going on intensively in compost piles and also in piles containing peat.

The investigations were performed in the composting plant continuously working during the analysed period. They suggest the need for their extension in particular months of year, application of other techniques and methods of calorimetric measurements, i.e. DSC, combustion calorimetry and methods of sampling. Analysis of temperature changes and calorimetry of composting processes can be useful for the improvement of thermal parameters of bioreactors particularly working in low-temperature environments.

References

Alef K, Nannipieri P (eds) (1995) In: Methods in applied soil microbiology and biochemistry, Academic Press, New York pp 214-218

Cambell CD, Darbyshire JR, Anderson JG (1990) The composting of tree bark in small-scale reactors-adiabatic and fixed temperature experiments. Biol Wastes 31: 175-185

Cooney CL, Wang DIC, Mateles RI (1968) Measurement of heat evolution and correlation with oxygen consumption during microbial growth. Biotechnol Bioeng 6: 95-123

Criddle RS, Breidenbach RW, Rank DR, Hopkin MS, Hansen LD (1990) Simultaneous calorimetric and respirometric measurements on plant tissues. Thermochim Acta 172: 213-221

Griffis CL, Mote R (1982) A method of measuring the rate at which heat is generated by aerobic composting of wastes. Report Series 275, Nov 1982, Agricultural Experimental Station, Division of Agriculture, University of Arkansas, Fayetteville

Gustafsson L (1991) Microbial calorimetry. Thermochim Acta 193: 145-171

Kemp RB (ed) (1999) Handbook of thermal analysis and calorimetry, vol 4, From macromolecules to man. Elsevier Science, Amsterdam

Mote CR, Griffis CL (1982) Heat production by composting organic matter. Agric Wastes 4: 65-73

Seki H, Komori T (1984) Heat transfer in composting process. J Agric Meteorol 40: 37-45

Suwalki (1995) Atest. Okręgowa Stacja Chemiczno-Rolnicza w Bialymstoku, Poland

VanderGheynst JS, Gossett JM, Walker LP(1997) High-solid aerobic decomposition: pilot-scale reactor developement and experimentation. Process Biochem 32: 361-375

Van Ginkel JT (1996) Physical and biochemical processes in composting material. PhD Thesis. Agricultural University Wageningen, Wageningen, The Netherlands

Wiley JS (1957) II Progress report on high-rate composting studies. Eng Bull, Proc of the 12th Industrial Waste Conference, Purdue University, Series 94, pp 596-603, West Lafayette, IN

Composting of Different Horticultural Wastes: Effect of Fungal Inoculation

M.A. Elorrieta; M.J. López; F. Suárez-Estrella; M.C. Vargas-García and J. Moreno[1]

Abstract. A study was undertaken to evaluate the composting process with two different horticultural waste substrates: (1) a mixture of pepper, cucumber and bean plants and (2) pepper plants, and the effect of inoculation with *Trichoderma* sp. was analysed. Four piles at semi-pilot scale were built and periodically turned and aerated. Moisture content was maintained at 50–60%, and the temperature monitored daily. Microbial, physical and chemical parameters were followed during the process.

Thermophilic phase was reached within 2 days in all piles and maintained for 24–32 days. After this time, temperature decreased until its stabilization at environmental levels. Physical and chemical parameters evolution suggested a suitable process in all piles, although horticultural wastes mixture seemed to be slightly more appropriate as substrate. Inoculation of horticultural wastes mixture led to a faster decrease of C/N ratio and a higher formation of humic substances.

Introduction

Composting is a solid-state oxidative biotransformation process that requires a rich and varied microbiota. Microbial transformation of organic materials leads to a stable by-product that provides several benefits to plants when applied to the soil. Composting of green wastes has been shown to be a good system for the treatment and utilisation of agricultural residues, particularly vegetables. Composting contributes to the disposal of wastes, enhances the preservation of the environment and preserves soil fertility (García et al. 1993). Currently, there is a need for the adaptation of the process to particular substrates, together with the production of a good material for agricultural use.

Composting methods vary with regard to the process control mechanisms applied, such as aeration, turning and watering. The process usually needs quite long periods, from 6 to 12 months, to be accomplished (Kakezawa et al. 1990), but it can be accelerated by proper selection of control parameters. This would be feasible only with a detailed knowledge of the process according to the raw material used. Since composting is an aerobic process, conditions must promote efficient oxygen transfer and carbon dioxide removal from the medium. Turning and/or

[1]Unidad de Microbiología, Departamento de Biología Aplicada, Escuela Politécnica Superior, Universidad de Almería, 04120 Almería, Spain
e-mail: melorri@ual.es/ Tel.: 950015892/ Fax: 950015476

H. Insam, N. Riddech, S. Klammer (Eds.)
Microbiology of Composting

aeration are usually applied in an attempt to overcome this problem. It is also a common practice to mix substrates with materials comprising large pieces (bulking agents) to create interstitial spaces for the air to flow and diffuse uniformly (Bernal et al. 1996).

The successful outcome of the composting process greatly depends upon the presence of the necessary microorganisms and upon the conditions conducive to microbial activity and proliferation (Golueke 1992).

Many organic materials are susceptible to decomposition during composting. A wide range of organisms can easily use readily biodegradable materials (sugar, amino acids etc.). However, polymers such as lignin are natural barriers to microbial attack. Although non-external supplementation of microorganisms is necessary for the process to occur, some reports indicate the suitability of inoculation with proper microorganisms (Baca et al. 1992; Kakezawa et al. 1992). Some microorganisms active in the biodegradation of lignocellulose can accelerate the humification (Wani and Shinde 1978) or decrease the C/N ratio (Gaur et al. 1982; Yadav et al. 1982; Matthur et al. 1986). However, the efficacy of inoculation in natural composting processes is a subject of controversy among many authors. While Kakezawa et al. (1990, 1992) found an advantage in using *Coriolus versicolor* as inoculum, other authors report dubious effects (Finstein and Morris 1975; Solbraa 1984). Differences among results might be partially attributed to the various substrates employed in each study.

As composting becomes a significant part of solid waste management, it will be increasingly important to evaluate the fate of new solid waste material under composting conditions. This research focuses on the study of composting of horticultural plant wastes. Those wastes are generated in the Southeast of Spain (Almería) at rates greater than 1.000,000 t y^{-1} (Escobar 1998). This large amount of plant wastes represents a serious environmental, agronomic and sanitation problem. Solutions for crop waste recycling are being demanded urgently by the municipalities affected. However, there is a lack of information on the composting of horticultural wastes (Bernal et al. 1996). In the present study, the composting of horticultural wastes in turned and aerated piles is attempted. The effect of inoculation with a cellulolytic fungus (*Trichoderma* sp.) is also analyzed.

Material and Methods

Microorganisms and Inoculum Preparation

Trichoderma sp., isolated from horticultural wastes and identified according to Domsch et al. (1980), was used as inoculum for the composting processes. The fungus was cultured into test tube slants with YMPG (Yeast Malt Peptone Glucose) medium, in g l^{-1}: 10 glucose, 10 malt extract, 2 peptone, 1 yeast extract, 1 $MgSO_4.7H_2O$, 1 L-asparagine, 0.001 thiamine-HCl. Mycelia grown at 30 °C for 2 weeks were gathered in sterile saline solution. Fungal suspensions from three slants were inoculated into conical 2 l flasks with 400 ml of YMPG. Cultures were statically incubated in the dark at 30 °C for 2 weeks.

Preparation of Substrates and Inoculation

Sun-dried plants of green bean (*Phaseolus vulgaris* L.), pepper (*Capsicum annuum* L.) and cucumber (*Cucumis sativus* L.) and wood chips were chopped into 15—20-mm pieces. Wood chips were used as bulking agent.

Two different mixtures were prepared for composting purposes: mixture 1 or horticultural wastes mixture (HW) was composed of all plant wastes (1:1:1) and 25% wood chips (dry basis weight). Mixture 2 or pepper wastes (PW) was composed of pepper wastes only and the same ratio of wood chips. Moisture content was adjusted to 60% (w/w) by addition of water. Initial characteristics of the mixtures are given in Table 1.

When inoculation was applied (see below), 10 l of fresh inoculum (approximately 10^6 cfu ml^{-1}), prepared as mentioned above, were added to each pile. Water to adjust moisture and inocula (when supplied) were thoroughly mixed with raw material to ensure a homogeneous distribution. *Trichoderma* counts in raw material after mixing were approximately 10^{-4} cfu g^{-1}.

Table 1. Initial characteristics of mixed materials for composting.

	Waste mixtures[b]	
Characteristics [a]	HW	PW
C (%)	30.9	32.8
N (%)	2.1	1.8
C/N ratio	14.7	18.8
OM (%)	53.6	70.4
CEC (mEq 100 g^{-1})	26.7	24.8
Porosity (%)	92.7	92.0
WRC (%)	315.3	244.1
EC (mS cm^{-1})	12.1	15.4
Na (mEq 100 g^{-1})	13.2	14.4
Ca (mEq 100 g^{-1})	36.1	27.4
K (mEq 100 g^{-1})	22.3	24.0
P (mg 100 g^{-1})	378.0	242.5

[a]Values presented are averages among five replicates. OM, organic matter; CEC, cation exchange capacity; WRC, water retention capacity; EC, electrical conductivity.
[b]HW, horticultural wastes mixture; PW, pepper wastes mixture

Compost Piles and Treatments

Compost piles were maintained inside an open windrow. Piles were 1.2 m x 1.5 m at the bottom of the pile and were up to 1.2 m high. Air-forced composting of each pile was achieved through five perforated PVC tubing (5 cm diameter, 120 cm long) placed below a fine mesh screen near the bottom of the piles. Air was supplied by a blower S&P CBB-60.

The piles were periodically turned and aerated while fluctuations in temperature of the material were observed (active composting phase). Temperature of the piles was monitored daily during the process using composting thermometers (Dr. Friedrichs, Germany) inserted at different heights in the piles. Turning was applied weekly from the first 14 days of composting, until 42 days, at which time temperatures were stabilized at environmental levels. After first turning, air was injected twice a week for 10 min (2 $m^3 min^{-1}$), until the last turning. Water was added during turns to maintain the moisture content at 60%. During the curing phase (from 42 to 115 d) the piles were not turned or aerated.

Four different piles were prepared in order to evaluate the effect of substrates and inoculation in composting. HWI (HW mixture inoculated), HWNI (HW mixture not inoculated), PWI (PW inoculated) and PWNI (PW not inoculated).

Sampling and Analyses

Compound samples from 12 samples of ca. 500 g each were taken from each pile at different locations (top, middle and bottom levels). After homogenization, each compound sample was subdivided into five subsamples, which were frozen until analysis. Total carbon and nitrogen, cation exchange capacity (CEC), organic matter (OM) and humic extract (HE) were analysed on each dried (24 h at 105 °C) and homogenized sample by grinding through 1 mm sieve. Microbial analysis was conducted directly after sampling.

Counting of Microorganisms

Samples of 10 g (wet weight) were added of 90 ml of saline solution and shaken at 150 rpm for 30 min at 28 °C. Serial dilutions were obtained in saline solution until 10^{-10}. Bacteria were counted on Nutrient Agar (Difco, Michigan. USA) and fungi on Rose Bengal agar (Difco). Each analysis was performed in triplicate. Plates were counted after aerobic incubation at 30 °C for mesophilic microorganisms and 60 °C for thermophiles. Bacteria were incubated for 2–3 days while fungi were maintained until 5–7 days . Fungi plate counting was used to check the survival of Trichoderma sp.

Microbial population was also studied in HW and PW wastes left to decay on the ground. Microbial counting was done as previously indicated.

Chemical and Physical Analysis

Dry weight was determined by drying the material at 105 °C for 24 h. Samples were ashed at 600 °C for 8 h and organic matter was obtained by subtracting ash content to the whole sample. C and N contents were measured using a LECO CHNS-932 analyzer (St. Josephs, MI).

Cation exchange capacity (CEC) was determined by the method of Harada and Inoko (1980). Samples were mixed with 0.05 N HCl and soaked overnight. Mixture was centrifuged and washed with distilled water. The solid material was saturated with 1 N $Ba(CH_3OO)_2$ solution at pH 7 and soaked for 3 h. The suspension was filtered and washed with distilled water. The filtrate and washings were mixed and titrated with 0.1 N NaOH using phenolphthalein as indicator.

Extraction of humic-like fractions was done by the method proposed by Kononova (1966). Humic extract (HE) was obtained by shaking 1 g of sample overnight with 200 ml of 0.1 M $Na_4P_2O_7.10\ H_2O$ and 0.1 N NaOH mixture (pH 13). The dark-coloured extract was filtered until a transparent solution was obtained and elemental carbon was determined in the supernatant.

Statistical Analysis

Three to five replicates were used for each analysis, and data obtained throughout these studies were subjected to statistical analysis, using Statgraphics Plus 4.0 software (Manugistics, Inc. Rockville, MD).

One-way analysis of variance (ANOVA) was performed to compare the pile mean values for the different levels of sampling time, substrates and inoculation pattern, and to test whether there were any significant differences among the means at the 95% confidence level. In order to determine which means were significantly ($P<0.05$) different from which others, multiple comparison tests (Fisher's least significant difference) were used.

Results

Figure 1 shows mean temperature evolution through time in the four piles. The temperature patterns of each composting pile were very much alike, although significant differences between piles were observed when substrates and treatments were considered as main effect.

All piles showed typical composting phases. A mesophilic phase, shorter than 24–48 h, was observed in all piles. A thermophilic phase was reached in 24 h for HWI and HWNI piles in which maximum temperatures were around 60–65 °C. Four days were needed to reach these temperatures in PWI. The highest temperature, around 70 °C, was reached in the PWNI pile.

Thermophilic temperatures remained during approximately 28–32 days in all composting trials, due to turning and aeration facilities.

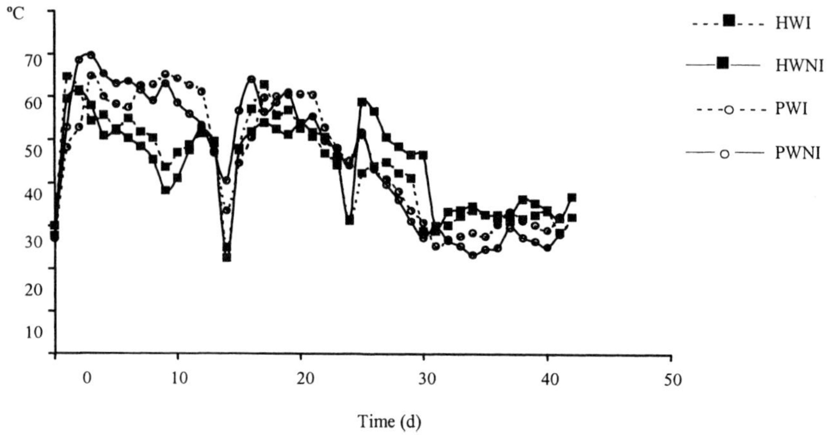

	Pile			
	HWI	HWNI	PWI	PWNI
Temperature	43.695 a	44.48 b	45.27 ab	44.03 c

Fig. 1. Time courses of temperatures during composting in HWNI, HWI, PWNI and PWNI piles. In the table, mean values for different piles during all process are shown. *Means with the same letter* are not significantly different.

The HWNI pile maintained higher temperatures than other piles after the second turning. After the thermophilic phase, temperature fell to the environmental level, and the curing phase began. Thus, the active phase of composting lasted around 32 days in all piles.

Thermophilic temperatures remained during approximately 28–32 days in all composting trials, due to turning and aeration facilities. The HWNI pile maintained higher temperatures than other piles after the second turning. After the thermophilic phase, temperature fell to the environmental level, and the curing phase began. Thus, the active phase of composting lasted around 32 days in all piles.

C/N ratio ranged between 14.6 ± 0.5 and 18.8 ± 0.9 for HW and PW mixtures, respectively, at the beginning of the process (Fig. 2a). However, all piles showed a similar trend for this factor. The closest pattern was observed within piles made with the same material, although there were significant differences between values obtained for HWI and HWNI piles. In HW mixture piles, C/N ratio decreased significantly during composting, mainly during the thermophilic phase. A faster decrease in C/N ratio was observed in HWI. Moreover, C/N ratio was stabilized after 32–42 days of composting in HWI and HWNI piles, respectively. A slight, but a significant decrease was observed at the end of the process, reaching final values around 10.6 ± 0.2 in both piles. In pepper piles, PWI and PWNI, a significant decrease was registered during the first 2 weeks.

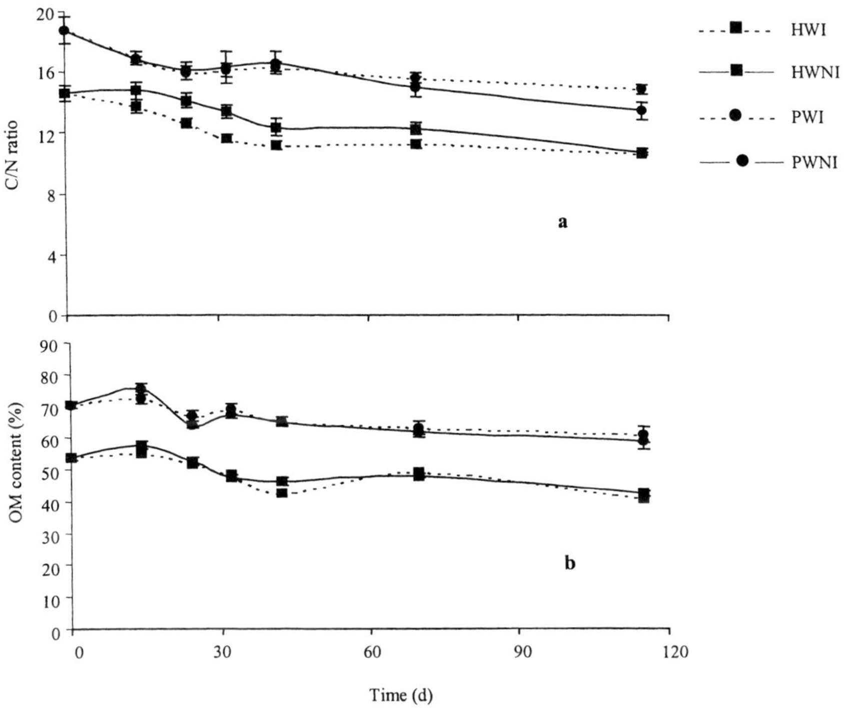

	HWI	HWNI	PWI	PWNI
C/N	12.22 a	13.16 b	16.13 c	16.35 c
OM	48.73 a	49.73 a	66.81 b	66.04 b

		Time (days)						
		0	14	24	32	42	70	115
HW	C/N	14.60 a	14.23 ab	13.35 bc	12.47 cd	11.73 d	11.76 d	10.67 e
PW	C/N	18.76 a	16.85 b	16.00 bc	16.21 bc	16.43 b	15.28 c	14.13 d
HW	OM	53.65 a	56.38 b	52.15 a	47.92 c	44.36 d	48.60 c	41.57 e
PW	OM	70.38 a	73.70 b	65.43 cd	68.15 ac	64.99 d	62.44 de	59.89 e

Fig. 2. Evolution of C/N ratios, ***a*** and organic matter (OM) contents ***b***, during composting in HWNI, HWI, PWNI and PWNI piles. In tables, mean values for different piles or for different sampling times are shown. Within a row, *means with the same letter* are not significantly different.

After this time, non-significant differences were observed until the 42nd day. However, during the curing phase (from 42 to 115 days), the C/N ratio dropped significantly, reaching final values of 14.8 ± 0.3 and 13.4 ± 0.5 for PWI and PWNI piles, respectively.

Initial OM values of the two raw materials tested were significantly different, 53.6 ± 0.5% and 70.3 ± 1.0% for HW and PW mixtures, respectively. In spite of this, OM evolution trend was similar in each pile during composting process (Fig.

2b). OM changes occurred mainly during the active phase (until 32–42 days). A small, although significant, decrease was observed after this time. Non-significant differences were observed between inoculated and non-inoculated piles made with same raw material. At the end of the process, lower OM content values were attained in HW piles (around 41%) than in PW piles (around 60%).

Figure 3 shows CEC and humic content evolution. Despite the similar initial CEC values in HW (26.7 ± 0.4 mEq 100 mg^{-1}) and PW piles (24.8 ± 1.3 mEq 100 mg^{-1}), a markedly significant difference in the CEC evolution through composting was noted for each material (Fig. 3a). HW piles registered the highest CEC values which increased mainly during the active phase of composting, reaching stabilization after 24 days. In contrast, the CEC hardly changed in pepper piles (there were no significant differences during the trial). On the other hand, there were no significant differences between inoculated and non-inoculated piles made with the same material.

HE content, as percentage of humic carbon of the total carbon, was 15.5 ± 0.8% and 15.5 ± 1.6% in HW and PW piles, respectively, at the beginning of the process (Fig. 3b).

An opposite trend in HE evolution was observed between HW and PW piles. In compost of HW mixture, HE continuously increased, approaching the highest value of 25% for the HWI pile, at 42 days. After this, no more noticeable changes were observed for these piles, although a significant slight drop in HE was quoted at the end of the process. The final HE content in these piles was significantly higher than the initial one. Moreover, significant differences between inoculated and non-inoculated piles were observed. A higher degree of humification was observed in HWI than in HWNI piles. The HE content of pepper compost decreased during the first 42 days of transformation. After this stage, PWNI exhibited a significantly increasing trend and approached to the initial value, while PWI maintained the low values reached at this time.

Changes of microbial population through the active phase of composting are showed in Fig. 4. Although there were non-significant differences in the pattern of microbial succession between HWI and HWNI piles, the number of thermophilic bacteria in HWI was larger than that in HWNI, at 32 days of composting. In contrast, mesophilic bacteria counts were lower in HWI than in HWNI between 14 and 32 days. However, both mesophilic and thermophilic bacterial counts were equalled at the end of the active phase in these piles. Mesophilic fungal populations were maintained, around 10^5 cfu g^{-1} during the whole process, but with significantly higher values in HWI than in HWNI when the active phase (32 days) was finished.

Significantly different microbial population values and patterns were obtained in piles with pepper when compared to HW piles. In pepper piles (PWI, PWNI), thermophilic bacteria fell during the first 14 days followed by a significant increase. Afterwards, thermophilic population decreased to the lowest values at the end of this active phase. Mesophilic bacteria significantly decreased while thermophilic temperatures were maintained in PWI and PWNI piles.

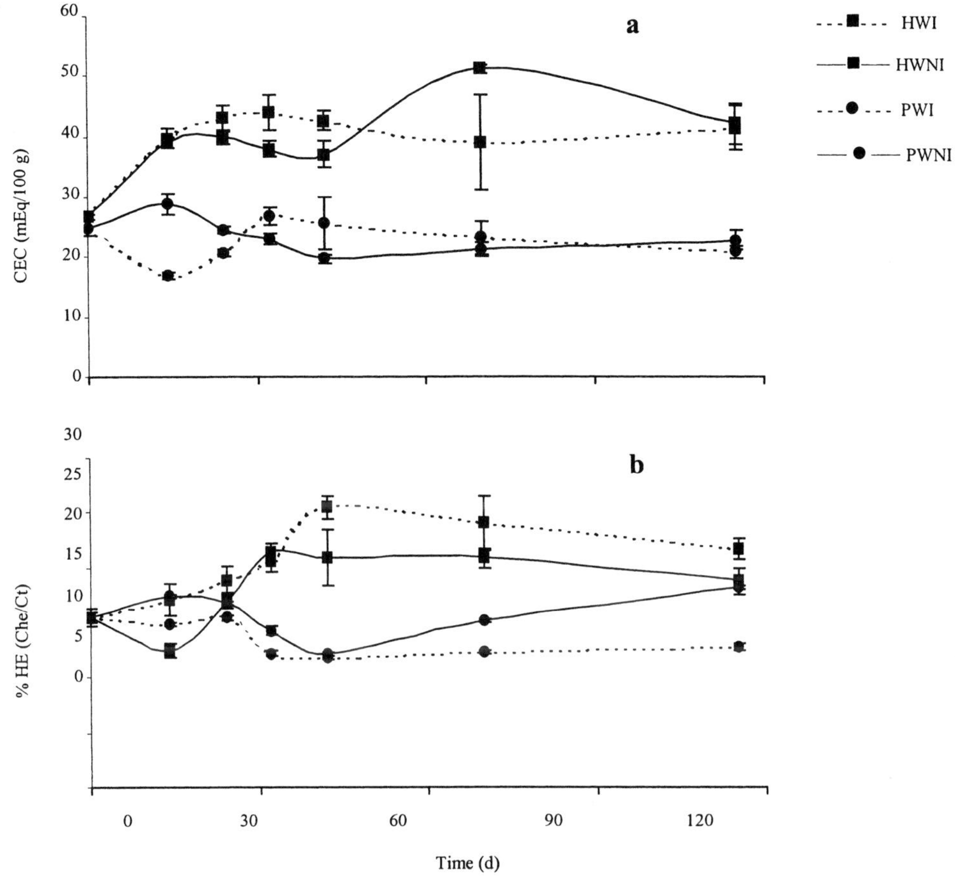

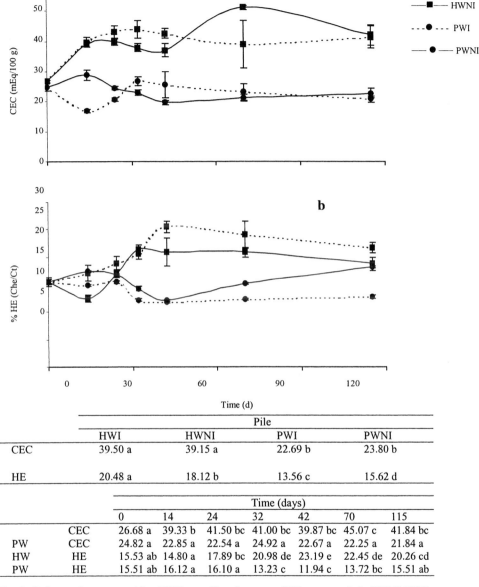

	Pile			
	HWI	HWNI	PWI	PWNI
CEC	39.50 a	39.15 a	22.69 b	23.80 b
HE	20.48 a	18.12 b	13.56 c	15.62 d

		Time (days)						
		0	14	24	32	42	70	115
	CEC	26.68 a	39.33 b	41.50 bc	41.00 bc	39.87 bc	45.07 c	41.84 bc
PW	CEC	24.82 a	22.85 a	22.54 a	24.92 a	22.67 a	22.25 a	21.84 a
HW	HE	15.53 ab	14.80 a	17.89 bc	20.98 de	23.19 e	22.45 de	20.26 cd
PW	HE	15.51 ab	16.12 a	16.10 a	13.23 c	11.94 c	13.72 bc	15.51 ab

Fig. 3. Cation exchange capacities (*CEC*) ***a*** and humic extract (*HE*) contents ***b***, in HWNI, HWI, PWNI and PWNI piles. In tables, mean values for different piles or for different sampling times are shown. Within a row, means with the same letter are not significantly different

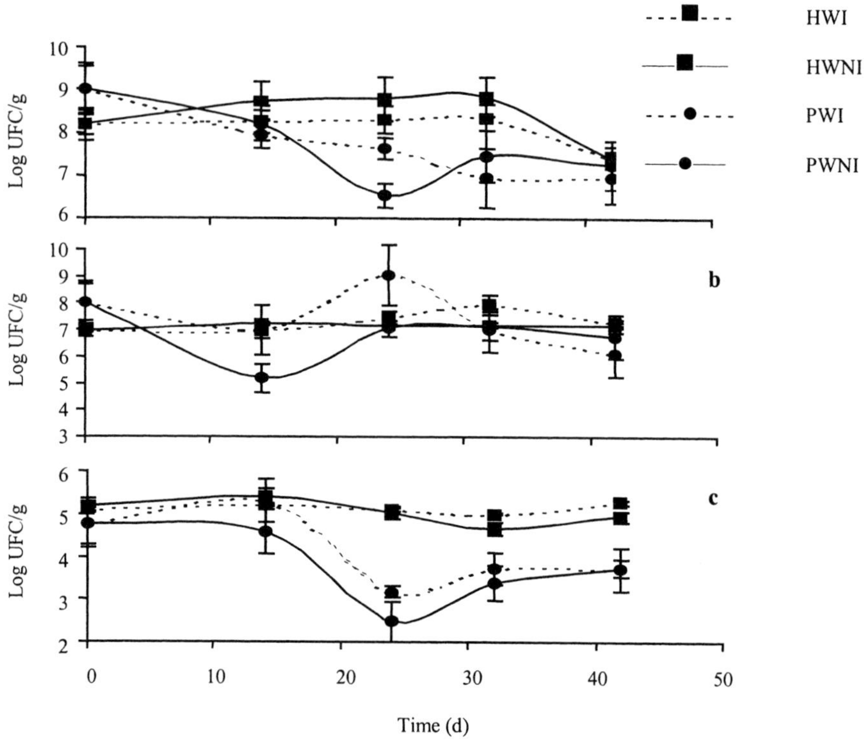

	Pile			
	HWI	HWNI	PWI	PWNI
MAM	8.09 ab	8.38 b	7.70 c	7.64 c
MAT	7.38 a	7.13 ab	7.23 ab	6.73 c
MFM	5.14 a	5.10 a	4.12 b	3.84 b

		Time (days)				
		0	14	24	32	42
HW	MAM	8.14 a	8.50 b	8.55 b	8.59 b	7.39 c
PW	MAM	9.01 a	8.08 b	7.08 c	7.19 c	6.99 c
HW	MAT	6.90 a	7.08 ab	7.46 cd	7.57 d	7.24 bc
PW	MAT	7.99 a	6.08 b	8.02 a	7.05 c	5.76 b
HW	MFM	5.14 a	5.31 b	5.16 ab	4.88 c	5.10 a
PW	MFM	4.77 a	4.94 a	2.76 b	3.54 c	3.87 c

Fig. 4. Microbial populations of mesophilic bacteria ***a***, thermophilic bacteria ***b*** and mesophilic fungi ***c***, during composting in HWNI, HWI, PWNI and PWNI piles. In tables, mean values for different piles or for different sampling times are shown. Within a row, means with the same letter are not significantly different

Mesophilic fungal counts in PW piles were significantly lower than in HW piles, at the beginning of the assay. This difference was maintained through the whole composting period. Thermophilic temperature led to a significant decrease in these mesophilic fungi counts in pepper piles. As soon as temperature decreased (after 28 days), cfu of mesophilic fungi began to increase slowly. Thermophilic fungi were not found in any case.

Trichoderma was detected during whole active phase of composting in the HWI mesophilic fungi plates. However, this fungus was isolated from PWI plates only until the first 14 days.The evolution of mesophilic bacteria and fungi in raw materials studied, left to decay on the ground, was also tested, in order to know the potential microbial colonization of these substrates under natural conditions. Mesophilic bacterial populations were maintained between 10^6–10^7cfu g^{-1} through time in HW mixture. PW wastes always gave lower microbial counts than HW mixture. The mesophilic bacterial population decreased strongly after 60 days in PW wastes (from 10^6cfu g^{-1} initial values to 10^2cfu g^{-1} final values). In respect to fungal population, main differences between both materials tested were obtained after 60 days' exposure. From this time, fungal counts were lower in PW than in HW mixture (around 10^6cfu g^{-1} in HW and around 10^4cfu g^{-1} in PW). Initial values were similar in both kind of wastes, around 10^3 cfu g^{-1}, and were maintained at this level until 60th day.

Discussion

Thermal evolution observed during the course of the active composting phase indicated a proper aeration in all heaps (Fig. 1). Golueke and Díaz (1990) pointed out that excessive aeration causes early cooling of the mass and accelerates moisture content reduction by water evaporation. They support the advantage of an aeration that should be properly applied depending on substrate. In this sense, turning practices in aerated piles improve the aeration, homogenization and sanitation of all materials.

Oxygen flow support allowed shortening residence time since the thermophile phase extended for 32 days. This effect has been previously reported (Lau et al. 1992; Sartaj et al. 1997). However, Golueke and Díaz (1990) indicated the risks of a faster process, since a poor mineralization of organic matter might occur. C/N ratio and OM content results obtained in the present work support the adequate mineralization of organic matter during the composting process carried out, similar to those reached by other authors (Iglesias and Pérez 1992). This transformation seemed to be more adequate in piles made with HW mixture (Fig. 2) since the lower OM and C/N values obtained could indicate a greater mineralization and stabilization of the product. In this sense, a C/N ratio of 10 to 12 is usually considered to be an indicator of stable and decomposed OM (Iglesias and Pérez 1992). A similar trend in OM and C/N evolution in PW and HW piles was observed despite the different initial values showed for these two kinds of raw materials. In this sense, several authors (Baca et al. 1992; Iglesias and Pérez 1992) indicated that the initial C/N ratio is not a good indicator of the subsequent transformation of the lignocellulose residues. Raw materials may have a wide range of lignin to cellulose ratio and, therefore, different biodegradability (Senesi 1989). It also depends on the relative nitrogen content of the original material (Morel et al. 1985), although in this study it was very similar for both materials (Table 1). Thus, the C/N ratio, in the range studied for the two raw materials tested, did not show noticeable influence in the way in which composting took place. On the other hand, the initial low C/N ratio (lower than 20 in both mixtures) was not found to reduce the effectiveness of the process. Lau et al. (1992) reported similar results for low C/N ratio materials.

Related to results mentioned above, humification of the two raw materials tested evolved in a different way (Fig. 3). Factors such as CEC and humic substances content inform about the course of humification of organic matter. As Gray et al. (1973) pointed out, it is possible that conditions which optimize the speed of mineralization might not coincide with those necessary for optimum humification. In agreement with data previously reported (Harada et al. 1981; Iglesias and Pérez 1992), results obtained in this work showed that an adequate maturation process was developed in the HW mixture, since a significant increment of CEC and humic substances content was observed, mainly in the inoculated pile. Harada et al. (1981) also found that CEC increases progressively to stability in the active phase of composting. CEC values and humus contents showed the worst evolution in pepper heaps. In this substrate, there were no changes of CEC through time and final humic substance contents were similar to the initial ones.

Significant different microbial population evolution was observed in HW and PW piles (Fig. 4) through the active phase of composting. This could also be related to the worse composting evolution of PW piles. Thus, trends of different microbial populations in PW counts were lower than in HW, perhaps as a consequence of the high temperatures reached. The PWNI pile reached 70 °C, which is considered lethal by many authors (Golueke 1992). However, there were no significant differences between temperature evolution observed in this pile and in the HW piles. On the other hand, maximum temperatures reached in PWI did not exceed 60–65 °C, similar to those reached in HWI and HWNI piles. Thus, it seems that the mixture of pepper wastes with other horticultural wastes offers some benefits to the raw material which improve the microbial development, showing the important role of the nature of raw material, also according to other reports (Guedes et al. 1991). Other evidence of worse microbial development in PW was found in microbial analysis carried out on plants decaying on the ground. Microbial counts in PW decreased through time to a greater extent than in the HW mixture. Since the differences between raw material tested did not seem very noticeable (Table 1), the low microbial counts observed in PW through time might be also a consequence of other factors, such as inhibiting substances. In this sense, the production of tannins and phenols at the beginning of composting has been described (Saveie and Gourbiere 1989; Kostov et al. 1996). The presence of this kind of substance could explain some differences in microbial growth observed in raw materials with similar C/N ratios (Kostov et al. 1994). Thus, the presence of toxic substances in PW could partially explain microbial differences observed between these two kinds of wastes, and consequently the differences in some of the physical and chemical parameters evolution observed.

All aspects discussed above could explain the results derived from testing the effect of fungal inoculation. The use of active cellulose-decomposing microorganisms for acceleration of the decomposition of composting raw materials assayed led to significant changes in HW piles. In this raw material the humification level was significantly higher than in the non-inoculated one. Besides, there were also differences in C/N ratio evolution, which decreased more quickly in HWI than in HWNI. Thus, evolution of these parameters was better in the inoculated than in the non-inoculated pile. Inoculation in the PWI pile did not improve the process. This could be related to the lower microbial development in pepper mentioned above, that could also affect growth of *Trichoderma* sp..

According to the results obtained, pepper characteristics can be improved for biotransformation by the addition of other horticultural wastes. Besides, inoculation with *Trichoderma* sp. offers benefits in HE content and C/N ratio evolution when a mixture of horticultural wastes is used as raw material.

Acknowledgements. This study was supported by the Comisión Interministerial de Ciencia y Tecnología (CICYT), Ministerio de Educación y Ciencia, Project AMB96-1171. The authors gratefully thank the Department of Chemical Engineering of the University of Almería, for the valuable use of the elemental analyzer.

References

Baca MT, Fornasier F, de Nobili M (1992) Mineralization and humification in two composting processes applied to cotton wastes. J Ferment and Bioeng 74 (3): 179-184

Bernal MP, Navarro AF, Roig A, Cegarra J, García D (1996) Carbon and nitrogen transformation during composting of sweet sorghum. Biol Fertil Soils 22 (1): 141-148

Domsch KH, Gams W, Anderson TH (1980) Compendium of soil fungi. Academic Press, London

Escobar A (1998) Residuos Agrícolas. In: Encuentro Medioambiental Almeriense: en busca de soluciones. Gestión de Residuos. Universidad de Almería, Almería, pp 23-46

Faure D, Deschamps AM (1991) The effect of bacterial inoculation on the initiation of composting of grape pulps. Bioresour Technol 37: 235-238

Finstein MS, Morris ML (1975) Microbiology of municipal solid waste composting. Adv Appl Microbiol 19: 113-151

García C, Hernández T, Costa F, Ceccanti B, Masciandaro G, Calcinai M (1993) Evaluation of the organic matter composition of raw and composted municipal wastes. Soil Sci Plant Nutr 39 (1): 99-108

Gaur AC, Sadavisam KV, Matthur RS, Magu SP (1982) Role of mesophilic fungi in composting. Agric Wastes 4: 453-460

Golueke CG, Díaz LF (1990) Understanding the basis of composting. Biocycle April: 56-59

Golueke GC (1992) Bacteriology of composting. Biocycle Jan: 55-57

Gray KR, Sherman K, Biddlestone AJ (1973) A review of composting, part1. Process Biochem 31: 32-36

Guedes RA, González CG, Neves OR, Sol MC (1991) Composting of pine and eucalyptus barks. Bioresour Technol 38: 51-63

Harada Y, Inoko A (1980) The measurement of the cation exchange capacity of compost for estimation of the degree of maturity. Soil Sci Plant Nutr 26: 127-134

Harada Y, Inoko A, Tadaki M, Izawa T (1981) Maturing process of city refuse compost for the estimation of the degree of maturity. Soil Sci Plant Nutr 26: 127-134

Iglesias E, Pérez V (1992) Determination of maturity indices for city refuse composts. Agric Ecosyst Environ 38: 331-343

Kakezawa M, Mimura A, Takahara Y (1990) A two-step composting process for woody resources. J Ferment Bioeng 70: 173-176

Kakezawa M, Mimura A, Takahara Y (1992) Application of two-step composting process to rice straw compost. Soil Sci Nutr 38 (1): 43-50

Kononova MM (1966) Soil organic matter. Pergamon Press, Oxford

Kostov O, Petkova G, Van Cleemput O (1994) Microbial indicators for sawdust and bark compost stability and humification processes. Bioresour Technol 50: 193-200

Kostov O, Tzvetkov Y, Petkova G, Lynch JM (1996) Aerobic composting of plant wastes and their effect on the yield of ryegrass and tomatoes. Biol Fertil Soils 23: 20-25

Lau AK, Lo KV, Liao PH, Yu JC (1992) Aeration experiments for swine waste composting. Bioresour Technol 41: 145-152

Matthur RS, Magu SP, Sadavisam KV, Gaur AC (1986) Accelerated compost and improved yields. Biocycle 27: 42-44

Morel JL, Colin F, Germon JC, Godin P, Juste C (1985) Methods for the evaluation of the maturity of municipal refuse compost. In: Gasser JKR (ed) Composting of agricultural and other wastes. Elsevier, London, pp 56-72

Sartaj M, Fernández L, Patni NK (1997) Performance of forced, passive and natural aeration methods for composting manure slurries. Trans of the ASAE 40 (2): 457-463

Saveie JM, Gourbiere F (1989) Decomposition of cellulose by the species of the fungal succession degrading *Abies alba* needles. FEMS Microbiol Ecol 62:307-314

Senesi N (1989) Composted materials as organic fertilizers. Sci Total Environ 81/82: 521-594

Solbraa K (1984) An analysis of compost starters used on spruce bark. Biocycle March: 46-48

Wani SP, Shinde PA (1978) Studies on biological decomposition of wheat-straw: II – Screening of wheat-straw decomposing microorganisms under field conditions. Mysore J Agric Sci 12: 388-391

Yadav KS, Mishra MM, Kapoor KK (1982) The effect of fungal inoculation on composting. Agric Wastes 4: 329-333

Backyard Composting: General Considerations and a Case Study

P. Illmer[1]

Abstract. A considerable part of the total decomposable organic waste is treated in small-scale backyard composters. Nevertheless, as the scientific community mainly cares about large-scale composting facilities (including their end products), very few publications dealing with the specific problems connected with household composting are available. The results discussed in the present chapter clearly demonstrate that (1) even problematic organic material can be composted successfully in a comparatively short time in household composters (2) the best results concerning the course of composting and the endproduct are achieved by frequent mixing of the compost, together with applying chopped wood at a rate of 10%, and (3) further investigations on backyard composting are urgently needed, especially as the results of the interactions between different treatments of compost are still hardly predictable.

Introduction

The beginnings of composting can be traced back to ancient times (Martin and Gershuny 1992) and it is today commonly defined as the biological decomposition of organic matter under controlled, aerobic conditions into a humus-like, stable product (Eppstein 1997).

For both ecological and economical reasons, an increasing part of the total solid waste is collected, divided up and delivered to recycling facilities of a varied complexity: from simple backyard compost heaps to highly sophisticated large-scale compost plants. About 60 to 80% of the total municipal solid waste is considered decomposable (USEPA 1992), depending mainly on country-specific differences like standard of living, recycling systems etc. Gathering and treatment of decomposable organic wastes has therefore turned out to be an important commercial factor that has led to the development of a prospering sector of the economy. Besides the production of soil and soil ameliorating substrata, the space necessary for landfills is markedly reduced.

In the beginning of scientific investigation of the innumerable interactions between microorganisms and several abiotic parameters (temperature, moisture, C and N contents, physical properties), mainly experiments were conducted at laboratory scale, not considering that these results are often far from the reality found in large-scale composting plants (Eppstein 1997).

---

[1]Institute of Microbiology (N.F.), University of Innsbruck, A-6020 Innsbruck, Austria
e-mail: : Paul.Illmer@uibk.ac.at; Tel: +512/507/6005; Fax: +0512/507/2928

H. Insam, N. Riddech, S. Klammer (Eds.)
Microbiology of Composting

While the differences between large-scale composting (LSC) and small-scale composting (SSC) have increasingly been taken into account during the past decades, the distribution of mass flows of decomposable wastes is not in the least met by the number of scientific investigations.

Fig 1 illustrates the distribution of organic matter which is treated in LSC facilities and backyard composting, respectively. While in the entire province of Tyrol about 25% of the total decomposable organic matter is treated in backyard composting, this percentage increases to as much as about 67% when a village with a comparatively high average income and mainly detached houses is taken as a basis (M. Mölgg, pers. commun. and M. Karbon, pers. commun.). No comparable data are available from other countries, but as Austria is very advanced in what the composting of organic wastes is concerned, the respective ratios of recycled organic matter should be lower in most other countries. The figures presented in Fig.1B C are in contrast to Fig 1A, showing a literature search (Science Citation Index, ISI® 1999) where the search term compost* resulted in 237 hits, only 3 of which were dealing with household composting. In this way nearly 100% of all investigations deal with LSC facilities or the application of their end products neglecting the real distribution of mass flows of organic waste. This and the remarkable number of backyard-specific conditions point to the necessity to intensify the efforts of investigating SSC.

The most important differences between large- and small-scale composting (the latter is sometimes divided into hot and cold SSC; USEPA 1992) are summarized in Table 1. Aerobe LSC is usually characterized by mixing resulting in aeration, a continuous supply of microorganisms with new substrate and oxygen, and a subsequent heat production. Hot SSC is very similar to LSC, but its success mainly depends on personal enthusiasm. These two types are characterized by typical textbook courses of temperature and microbial communities. However, textbooks often neglect that the major part of SSC is performed as cold composting, which is characterized by the continuous application of small amounts of fresh organic material to the top of the heap. This usually leads to distinct stratification, to anaerobe or microaerobe conditions, to only slightly elevated temperatures, and therefore to a very long duration of the process.

As stated above, these differences result in (1) different courses of composting inter alia characterized by flat temperature curves and rather unchanged microbial communities, (2) a very long duration and (3) sometimes problems concerning sanitation. Several ordinances stipulate that every part of compost material has to be exposed to a temperature of 65°C at a water content of at least 40% for 3 days running (e.g. ÖNORM S2200) - criteria that can hardly be achieved in backyard composting (Illmer et al. 1997). Therefore it is recommended not to supply SSC with potential problematic material like bones, meat, faeces, diseased plants etc. Besides aesthetical and hygienic aspects, this is also a necessary prerequisite to avoid the occurrence of pests like mice, rats, Calliphoridae etc. (Oberfeld 1996). However, this is in opposition to several laws (e.g. in Austria) which in some cases prohibit the disposal of these materials via the so-called residual waste. A modification of present laws is urgently needed.

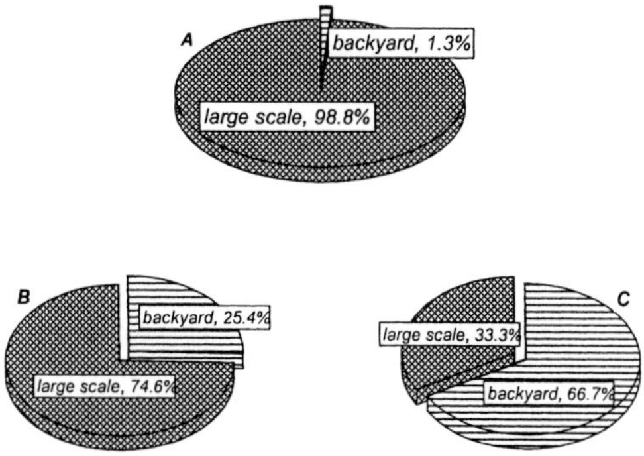

Fig 1A-C. Proportions of scientific publications dealing with large- and small-scale (backyard) composting as found in the Science Citation Index of 1999, *A*), distribution of mass flows of decomposable waste in the Austrian province of Tyrol (655,000 inhabitants, *B*), and in the village of Thaur (3,500 inhabitants, *C*).

Table 1. Important differences between large- and small-scale composting

Property/key factor	Large-scale composting, LSC	Small-scale composting, SSC	
		cold SSC	hot SSC
Quality of decomposable organic material (C/N, moisture, structure, etc.)	Fairly constant quality, balancing possible	Very different quality, balancing hardly possible	
Pollution (e.g. heavy metals)	Occurs	No[b]	
Mixing	Possible[a]	No	Possible[b]
Aeration	Possible[a]	No	Possible[b]
Temperature	High	Low	High
Sufficient sanitation	Yes	No	Possible[b]
Duration	Short	Long	Short
Expenditure of work	High[a]	Low	High[b]

[a]depending on the respective facility/system. [b]depending on personal enthusiasm.

The objective of the presented study was to investigate conditions which enable a successful small-scale composting despite the use of troublesome feedstock material.

Material and Methods

The investigation was carried out for 6 weeks using household composters of a volume of 700 l, described elsewhere (Illmer and Schinner 1997). All composters were equipped with covers to protect against rain, with sieve plates, and some of them with compost mixers (Fig. 2). Sieve plates and mixers were made of high-grade steel. A total of 400 l lawn clippings (water content 80%; C/N ratio = 10) were inserted into each composter within a few h after mowing on day 0 and day 7 (200 l each). The effects of mixing (two complete rotations per day vs. no mixing) and of the application of chopped wood as a structure material (SM) at a rate of 10% (v/v) vs. no application of SM were investigated. Treatments were studied in four replicates.

Sample preparation and determination of several physical and chemical parameters followed the methods given in ÖNORM S2023 and Illmer and Schinner (1997). The microbial respiration was determined according to Anderson and Domsch (1978). The parameters under investigation were: T temperature, WC water content, TOC total organic carbon, N_t total nitrogen, N_{sol} dissolved nitrogen in the compost extract, NH_4, NO_3 and NO_2 in the extract, CON conductivity, pH acidity, BIO dry biomass of test plants, CO_2 microbial respiration. Results were analyzed using ANOVA with a special focus on the interactions of the two independent factors, i.e. mixing and application of structure material (SM).

Results and Discussion

After 6 weeks of composting, the original volume of residual organic matter was distinctly reduced from the initial 400 l to about 145 and 110 l in the static and mixed variations respectively (Fig 3). This significant difference is of great importance, as one of the most apparent aims of consumers is the reduction of the volume of organic waste. Of course, the application of chopped wood (about 40 l) caused an increase in residual material. At the end of the composting still about 75% of the additional volume could be detected, and it seems very unlikely that this disadvantage can be overcome.

During the phase of composting with the highest temperatures (day 9-day 14; Fig 3) the temperature was significantly decreased from about 65 to about 40°C by mixing, whereas the addition of structure material caused only a slight, insignificant increase in temperature. On the one hand at least for hygienic reasons the temperature must not be too low (Eppstein 1997). On the other hand Bardos and Lopez-Real (1991) showed that too high temperatures may inhibit degradation and thus composting. These findings confirm our results, as all parameters indicating compost quality and maturity were significantly improved by mixing despite the decreased temperature. Therefore, the role of temperature as a central indicator for quality and progress in composting should probably be reconsidered.

The water content is a key factor in every composting process (Regan and Jeris 1970; DeSanto et al. 1993; Illmer et al. 1997). Water is released during microbial degradation of organic compounds, but enhanced microbial activity also induces a rise in temperature and thus evaporation. Consequently, the net increase or decrease in moisture content depends on the initial feedstock and the specific composting process (Eppstein 1997).

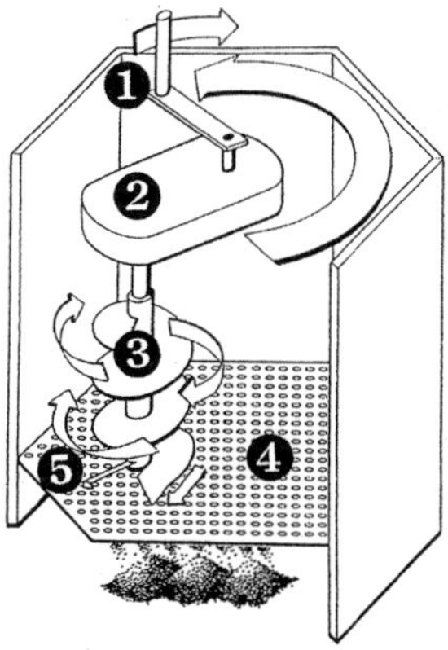

Fig 2. Household composter with an appliance for mechanical mixing. *1* Crank; *2* crank mechanism; *3* mixing worm; *4* sieve plate; *5* scraper for cleaning the sieve plate

Considering, the wet lawn clippings of the present investigation (WC ~ 80%) the significant decrease of WC in mixed composts (Fig 4) is a distinct improvement of the endproduct, although an even lower value would be preferable (Regan and Jeris 1970). The application of chopped wood also caused a significant reduction of WC, by far exceeding the nominal effect of the addition of dry wood.

A combination of both treatments led to synergistic effects illustrated by a highly significant interaction between the two factors (Fig.4.). For practical reasons (handling of material) the WC of the end product must not be too high. A WC of more than 60% was shown to distinctly reduce free air space and oxygen supply in composts (Eppstein 1997).

When taking the addition of wood (with a high content of C) into consideration, the significant reduction of TOC, which more than compensated the C addition to the system, can only be explained by an enhanced microbial activity (Canet and Pomares 1995; Illmer and Schinner 1997). Without the addition of structure material mixing had no effect on TOC, but - similar to the behaviour of WC - a synergetic accelerated degradation occurred when both treatments were combined.

As organic C is known to decrease during the composting of solid wastes (Canet and Pomares 1995) due to CO_2 evolution, both the decrease of C content and CO_2 production are useful for indicating the stage of composting (Eppstein 1997). Seen from this point of view our findings clearly indicate that both the application of structure material and mixing improve the process.

Total nitrogen (N_t) and dissolved nitrogen (N_{sol}) were significantly decreased by mixing, thus pointing to an accelerated N mineralization (Fig. 5; Illmer and Schinner 1997). The distinct reduction of all N fractions in mixed treatments in comparison to the controls (about minus 25%) is very important for reducing the problems caused by high N levels. Particularly NH_4 is associated with phytotoxicity, volatile compounds and reducing conditions, and often causes troubles when composting N-rich materials like lawn clippings (Illmer et al. 1997).

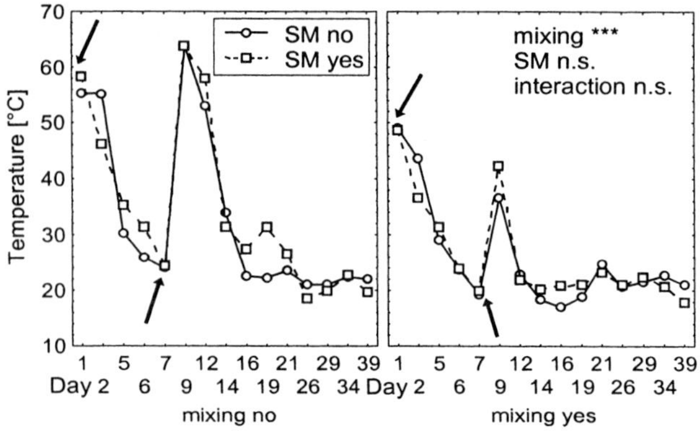

Fig 3. Temperature courses during household composting. *Arrows* indicate the addition of fresh lawn clippings. Plot of means – interaction effects from ANOVA analysis (abscissa scaling is not metric!). $p<0.001$ ***, $p<0.01$ **, $p<0.05$ *, $p\geq0.05$ n.s.

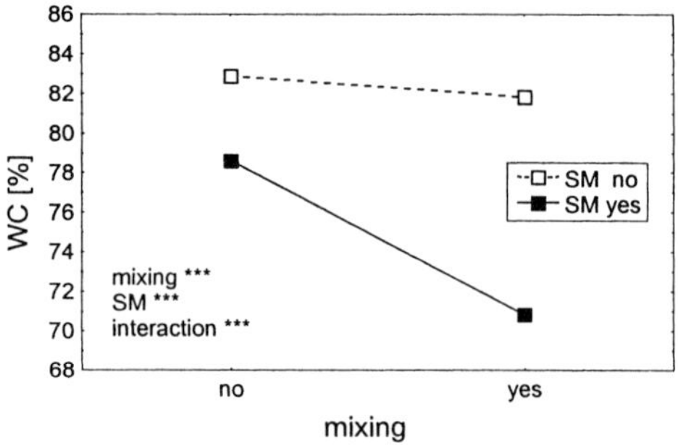

Fig. 4. Influences of both mixing and the addition of structure material (*SM*) on water content (*WC*). Plot of means – interaction effects from ANOVA analysis. $p<0.001$ ***, $p<0.01$ **, $p<0.05$ *, $p\geq0.05$ n.s.

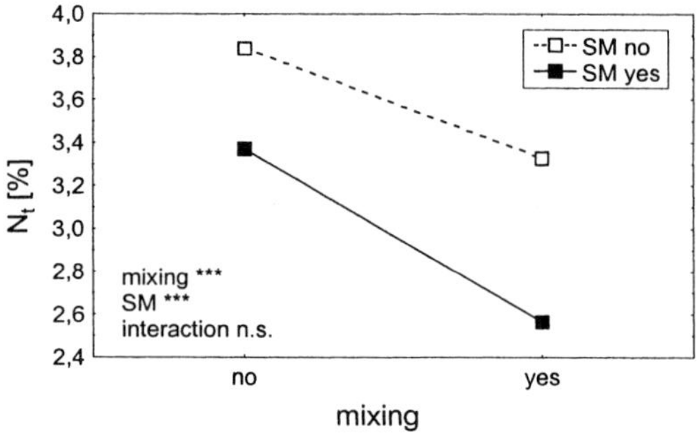

Fig. 5. Influences of both mixing and the addition of structure material (*SM*) on total N (N_t). Explanations see Figs. 3 and 4

Nitrite was detected in static composts but not in any sample of mixed compost, again indicating that oxygen supply in static composters was at least suboptimal (Forster et al. 1993). The application of structure material caused a significant decrease in N_t but - unlike to mixing - an increase of dissolved nitrogen (N_{sol}), mainly present in the form of NO_3-N was observed. Irrespective of mixing, NH_4 was almost completely eliminated by the adding of structure material (Fig. 6). On the other hand, mixing turned out to be a prerequisite of reducing the NH_4 concentration in treatments without SM addition, again indicating a distinct interaction between the two treatments. Bishop and Godfrey (1983) also found increases of available N in general and of NH_4-N in particular, when air supply was insufficient.

In the present investigation we found a distinct shift within the N_{sol} fraction. Due to the application of structure material, NH_4-N decreased whereas NO_3-N increased accompanied by a shift from anaerobe reducing to an aerobe, oxidizing environment. As nitrification mainly occurs in later phases of composting (Mathur et al. 1993) the observed shift is a sign of higher maturity (Kapetanios et al. 1993, Canet and Pomares 1995, Pare et al. 1998). However, as NH_4/NO_3 ratios can vary to a great extend (Mathur 1993), care has to be taken when interpreting them.

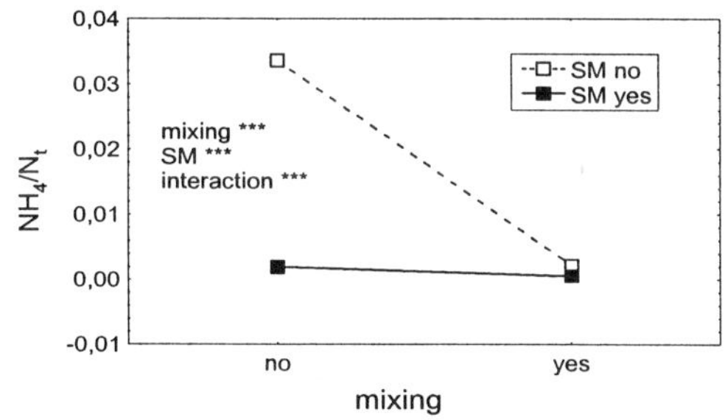

Fig. 6. Influences of both mixing and the addition of structure material (*SM*) on NH_4/N_t ratio. Explanations see Figs. 3 and 4

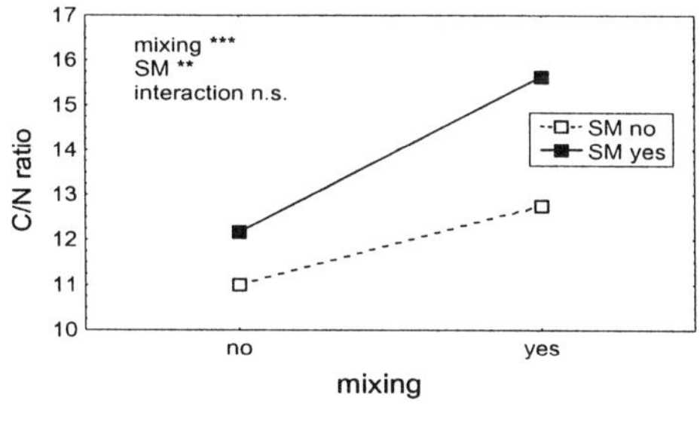

Fig. 7. Influences of both mixing and the addition of structure material (*SM*) on C/N ratio. Explanations see Figs. 3 and 4

Although both the total C and N contents decreased (Fig 5), these effects did not cancel one another in connection with the C/N ratios. The C/N ratio is significantly increased by mixing (Fig 7), which may be due to ammonia losses (Mathur 1993). The application of structure material also caused a highly significant increase in the C/N ratio - an expected effect when C-rich material was added to the system. C/N was found to finally reach a value of about 15, which corresponds to literature data (Kapetanios et al. 1993). This again indicates a higher maturity of the end product caused by the combined treatments. Many feedstocks for composting are dominated by wood or paper waste and are therefore characterized by high C/N ratios. These ratios tend to decrease during composting and are therefore a common indicator for maturation (Forster et al. 1993). However, C/N ratios of organic matter can vary to a great extent (from 10 for lawn clippings to about 500 for sawdust) and are therefore only suitable for the comparison of several composting processes and/or stages based on a similar feedstock (Inbar et al. 1990).

Conductivity was decreased from about 1.8 to 1.1 mS cm^{-1} due to mixing, pointing to a distinct reduction of the concentration of soluble ions in compost extracts, which confirms the above-mentioned results. Since overdoses of nutrients can cause phytotoxic effects (Iglesias-Jimenez and Alvarez 1993), the stated improvements were again beneficial. While mixing together with the application of chopped wood had the strongest positive effect, there was no difference in the absence of the structure material. This once again points to the necessity of a comprehensive investigation of the interactions between different treatments, investigations which have not been carried out so far.

When compost derived from mixed treatments was added to standard soil the biomass of test plants (*Lepidium sativum* L.) was significantly increased in comparison to controls, a result which in a way summarizes all the above-mentioned beneficial effects of mixing (Mathur et al.1993). While the application of structure material caused an additional positive effect when combined with mixing, the wood per se led to a reduced (n.s.) plant biomass when no mixing occurred (Fig 8). Phytotoxic effects of immature composts were shown to depend mainly on the presence of organic acids, ethylene oxide and ammonia (DeVleeschauwer et al. 1981; Wong 1985). High and harmful concentrations of organic acids were shown to occur in very young composts, and these acids are preferably degraded by yeasts under aerobe conditions (Choi and Park 1998). The degradation of these compounds, which depends on time and on the advance of composting, can be significantly accelerated by an increased O_2 supply (Eppstein 1997). The inhibition by SM addition alone may be caused by the additional NH_4 and CH_4 production accompanying wood degradation under O_2 limitation (Paul and Clarke 1996).

The outlook of the residual organic matter treated in two composters is presented in Fig 9. The end-product obtained without mixing and without SM addition is characterized by wet reducing conditions only 1 or 2 cm below the desiccated surface and did not show any further reduction in volume or change in consistency for several months. The end product of the other composting process (mixing, SM addition) showed an acceptable end product, especially when taking the short duration and the troublesome feedstock into consideration.

During the present investigation the best results concerning the course of composting and the quality of the end product were achieved by a combination of the two treatments tested, i.e. mixing and SM addition. However, many details and interactions between several influencing factors still remain unclear and demand further investigation.

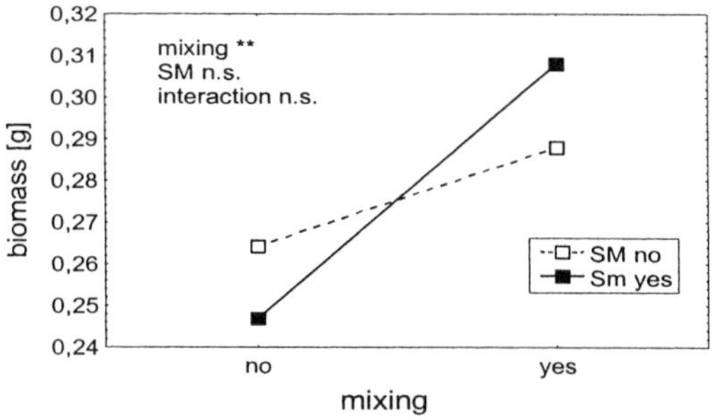

Fig. 8. Influences of both mixing and the addition of structure material (*SM*) on dry biomass of test plants (*Lepidium sativum* L.). Explanations see Figs. 3 and 4

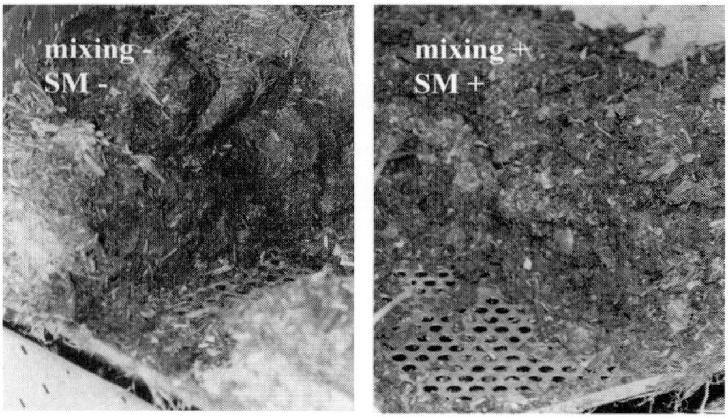

Fig. 9. Lawn clippings after 6 weeks of composting. *Left* without addition of SM and without mixing; *right* with addition of SM and mixing

Acknowledgements. This study was supported by Juwel H. Wüster, Austria.

References

Anderson JPE, Domsch KH (1978) A physiological method for the measurement of microbial biomass in soils. Soil Biol Biochem 10:211-221

Bardos RP, Lopez-Real JP (1991) The composting process: susceptible feedstocks, temperature, microbiology, sanitation and decomposition. In: Bidlingmaier W, L'Hermite P (eds) Compost process in waste management. European Community, Brussels, pp 179-190

Bishop PL, Godfrey C (1983) Nitrogen transformation during sludge composting. BioCycle 24:34-39

Canet R, Pomares F (1995) Changes in physical, chemical, and physicochemical parameters during the composting of municipal solid wastes in two plants in Valencia. Biores Technol 51:259-264

Choi MH, Park Y-H (1998) The influence of yeast on thermophilic composting of food waste. Lett Appl Microbiol 26:175-178

DeSanto AV, Berg B, Rutigliano FA, Alfani A, Fioretto A (1993) Factors regulating early-stage decomposition of needle litters in five different coniferous forests. Soil Biol Biochem 25:1423-1433

DeVleeschauwer D, Verdonck O, Van Assche P (1981) Phytotoxicity of refuse compost. BioCycle 27:44-46

Eppstein E (1997) The science of composting. Technomic Publishing, Lancaster

Forster JC, Zech W, Würdinger E (1993) Comparison of chemical and microbiological methods for the characterization of the maturity of composts from contrasting sources. Biol Fertil Soils 16:93-99

Iglesias-Jimenez E, Alvarez CE (1993) Apparent availability of nitrogen in composted municipal refuse. Biol Fertil Soils 16:313-318

Illmer P, Schinner F (1997) Compost turning – a central factor for a rapid and high-quality degradation in household composting. Biores Technol 59:157-162

Illmer P, Meyer E, Schinner F (1997) Thermic insulation and sieve plates – beneficial equipments for a rapid and high quality degradation in household composting? Bodenkultur 48:99-103

Inbar Y, Chen Y, Hadar Y, Hoitink HAJ (1990) New approaches to compost maturity. BioCycle 31:64-69

Kapetanios EG, Loizidou M, Valkanas G (1993) Compost production from Greek domestic refuse. Biores Technol 44:13-16

Martin DL, Gershuny G (1992) The rodale book of composting. Rodale Press, Emmaus, USA

Mathur SP, Owen G, Dinel H, Schnitzer M (1993) Determination of compost biomaturity. I literature review. Biol Agric Hortic 10:65-85

Oberfeld G (1996) Amtsärztliche Aspekte der Entsorgung biogener Abfälle. Mitt Österr Sanitätsverwalt 97/4:141-145

ÖNORM S2023 (1993) Untersuchungsmethoden und Güteüberwachung von Komposten. Österreichisches Normungsinstitut, Wien

ÖNORM S2200 (1993) Gütekriterien für Komposte aus biogenen Abfällen. Österreichisches Normungsinstitut, Wien

Pare T, Dinel H, Schnitzer M, Dumontet S (1998) Transformation of carbon and nitrogen during composting of animal manure and shredded paper. Biol Fertil Soils 26:173-178

Paul EA, Clarke FE (1996) Soil microbiology and biochemistry. Academic Press, New York

Regan RW, Jeris JS (1970) A review of the decomposition of cellulose and refuse. Compost Sci 11:17-20

USEPA (1992) Characterization of municipal solid waste in the United States. 1992 Update. EPA/530-S-92-019. US Environmental Protection Agency, Washington, DC

Wong MH (1985) Phytotoxicity of refuse compost during the process of maturation. Env Pollut Ser A 34:159-174

N-Dynamics During Composting – Overview and Experimental Results

I. Körner and R. Stegmann[1]

Abstract. Nitrogen plays an important role during composting since it may have valuable but also harmful effects when the compost is applied. An overview of the processes of N-dynamics is given. First, the theoretical courses of the processes most important for N-dynamics during composting are described, focusing on ammonification, nitrification, denitrification, immobilization and N released via leachate and exhaust gases. Experiments were carried out in a laboratory composting unit with 100-l bioreactors to evaluate their importance in composting. The composting parameters, such as water content, aeration, pH and temperature were evaluated with respect on their effect to the processes of N-dynamics.

Ammonification was lower during curing than during intensive rotting. Three-patterns could be detected; ammonia/ammonium formation during curing is significant, formation was negligible or immobilization is significantly higher than ammonia formation. Ammonification reaches higher rates for non-lignocellulosic substrates compared to lignocelluloses – independent of the N-content of the input material. Nitrification and denitrification occurred only during curing. Nitrification was limited by temperatures above 30 °C, pH over 8 and moistened substrate. Immobilization effects are assumed for the whole composting process and could be detected for the curing phase. They have not neccasarily been catalyzed by microorganisms. Chemical-physical processes may also be the reason. Releases take place via leachate (only if the process is insufficiently aerated) and via exhaust air. Ammonia releases via exhaust air were high for non-lignocellulosics and low/not detectable for lignocelluloses.

N-Turnover During Composting

The total amount of the N in composts can range widely. Only minor parts of the total N in compost, the water-soluble inorganic components, are directly available for plants. The major part of N is bound in organic forms and will slowly, over the course of many years, be mineralized and become available to the plant. In general, it is difficult to estimate mineralization processes and the fertilizing effect of compost. Compost N may have positive aspects in application, such as fertilization, but also result in damaging effects, such as over fertilization and pollution of ground water, when not used correctly.

[1]Department for Waste Management, Technical University of Hamburg Harburg, Harburger Schloßstr. 36, 21079 Hamburg, Germany

H. Insam, N. Riddech, S. Klammer (Eds.)
Microbiology of Composting

This paper gives an overview on the processes of N-dynamics during composting and shows the importance of the macronutrient N. The general processes of the natural N-cycle will be described, and their importance for composting will be shown with help of experimental results. The protein-N of waste can be transformed to inorganic and organic N-containing compounds during composting. Proteins or its degradation products may also be integrated into organic compounds. Furthermore, N-containing compounds can be discharged.

Ammonification and Ammonium-Ammonia Equation

The microorganisms responsible for ammonification are able to adapt to a wide range of milieu conditions. All conditions of the common composting process seem suitable. Ammonification or mineralization of N, is the complete process of protein degradation and takes place over several stages.

1. First, proteins resp. peptides are fragmented (polypeptides, oligopeptides, amino acids) in the course of proteolysis. This decomposition is catalyzed by microbial proteases (Schlegel and Schmidt 1985; Kleber and Schlee 1987; Morihara and Kohei 1992).

2. The following step – deamination – is the splitting-off of the amino group mainly from amino acids (Falbe and Regitz 1989), where ammonium or ammonia is produced. The most common type of amino acid degradation is oxidative deamination (Schlegel and Schmidt 1985), which can be divided into two different reaction types. Amino acids can be degraded to oxo acids, ammonia and hydrogen peroxide (H_2O_2) using molecular O_2 and unspecific aminooxidases. The splitting-off of ammonium from glutaric acid can also be catalyzed by NADP-dependent dehydrogenases (Sahm 1993).

3. Ammonium (NH_4^+) and ammonia (NH_3) are in chemical equilibrium in the aqueous phase via ammonium hydroxide (NH_4OH) (Fenn et al. 1981). Under conditions with high pH and high temperatures, the equilibrium is shifted towards ammonia, that can be released into the atmosphere (Gebel 1991).

Nitrification and Denitrification

The conversion from NH_4^+ to NO_3^- – nitrification – is a two-step oxidation which is carried out by two autotrophic bacterial groups:

1. Oxidation of ammonium ($NH_4^{+)}$ into nitrite (NO_2^-) by means of nitrosobacteria (*Nitrosomonas europeana ,-coccus, -spira, -cytis, -gloea*).
2. Metabolism of nitrite (NO_2^-) into nitrate (NO_3^-) through nitrite oxidation by nitrobacteria (e.g. *Nitrobacter winogradskyi, - agilis*).

Besides autotrophic nitrification, also fungal heterotrophic nitrification is known (Kiefer 1992).

Indications concerning the optimum temperature for nitrification vary (e.g. Kuntze et al. (1994): 15–35 °C; Scheffer and Schachtschabel (1992): 25–30 °C;

Beck (1979): 25–35 °C). The optimum pH for both nitrification steps lies at 8.5 but the steps differ with regard to their tolerance ranges (ammonium oxidation: 7.5 to 9.5; nitrite oxidation: 5.5 to 10.5). In an acidic milieu with pH values below 6, nitrification is weak. Nitrification needs a sufficient oxygen supply. Restricted aeration due to compaction or drenching impedes nitrification. O_2 availability determines the speed of the process (Beck 1979).

The microbial reduction of NO_3^- via NO_2^- to gaseous products NO–N_2O–N_2 is called biological denitrification. It is carried out by facultative anaerobic bacteria (e.g. *Pseudomonas, Alcaligenes, Achromobacter)* under conditions where the need for hydrogen acceptors is not sufficiently covered by O_2. NO_3^- is used as hydrogen acceptor for energy production. The gaseous intermediate products NO and N_2O can escape from the system before further reduction takes place.

The most important requirement for denitrification is an oxygen concentration lower than 1–5%. Higher concentrations inhibit denitrificating enzymes. Denitrification starts at approx. 5 °C (e.g. in soil) and increases with rising temperature. The optimum temperature lies at approx. 18–30°C (Kuntze et al. 1994). The optimum pH lies between 6 and 7 (Scheffer and Schachtschabel 1992).

Immobilization

The conversion of inorganic to organic N-compounds is called immobilization. Microbial proteins are formed during composting; in biological processes proteins are synthesized using ammonium as an N-source. If ammonium is not directly available, nitrate or nitrite may be transformed first. Furthermore, humic substances are produced during composting which can bind N-containing components. The bonding form and the chemical structure of the N-containing compounds have not been adequately investigated. However, the main N-components are compounds similar to proteins and amino acids (Knicker 1993). A significant part of some waste is lignin. The lignin of the living plant does not contain N. During composting lignin is partly transformed. During lignin transformation N-compounds can be built into the lignin molecule.

Releases

Release into the Gas Phase

During the conversions explained above, some gaseous N-containing compounds can be formed. NH_3 can be stripped into the gas phase. NH_3 release in the gas phase is mainly influenced by aeration rate, temperature and pH. It is also favoured by a high N-content of the substrate. Furthermore, N_2 , N_2O and NO_x, which are formed during denitrification, can escape into the gaseous phase.

Leaching

When the substrate contains large amounts of water, excess water settles as leachate, carrying water-soluble N-compounds and fine solid particles. Apart from the inorganic compounds, ammonium, nitrate and nitrite, organic N-compounds such as proteins, amino acids and humic-lignin compounds are leached.

Test Realization

The experiments were carried out in a laboratory composting unit using different substrates and different composting conditions. N compounds were measured in the substrate, in the leachate and in the exhaust air at different times during composting. Since it is not possible to give all the results here, some of the most important results were selected. A more detailed description of these experiments is given in Körner et al. (1999).

Experimental Setup and Analytical Methods

The experiments were carried out in a laboratory composting unit with air-tight 100-l bioreactors. They were equipped with an insulated, temperature-regulated, water-filled double wall. Air was pumped through the reactor bottom through sieve layers. The reactors were perforated with outlets for leachate and exhaust air. Exhaust air was cooled and condensed water was collected or returned to the substrate. The air stream was passed through bottles containing 0.5 M H_2SO_4 to dissolve ammonia. The following measurements of exhaust air for each reactor could be carried out in intervals: temperature; exhaust air rate; CO_2 content. Samples of the composting substrate were taken during agitation (varying between 1 and 4 weeks). Leachate was sampled once per week. Other liquid samples (condensed water, acidic solutions) were taken one to three times per week. The analyses used for this chapter are described shortly in the following:

- Aeration: continuous by a mass flow meter for fresh and exhaust air
- Temperature: continuous by PT 100 sensors at four places in the reactor
- Water content: sampling during turning and drying at 105 °C
- CO_2content: semicontinuous with an IR instrument
- Degree of degradation: Calculations using substrate mass, and water content
- PH: for the substrate – in the $CaCl_2$ extract
- NH_4^+/NH_3-N: for the substrate – in the $CaCl_2$ extract; for leachate, condensed water and acidic solution – in the sample; tirimetric determination
- NO_3^-, NO_2^- -N: for the substrate – measured in the $CaCl_2$ extract; for leachate – directly in the sample; determination by ion chromatography
- Total N: for the substrate – grinding of the dried sample, determination of Kjeldahl-N; for the leachate – in the sample; for both – disintegration of organic N and determination of NH_3 by distillation/titration.

Characteristics of the Experimental Runs

The list of experiments and some of the composting parameters are shown in Table 1. Different reproduceable biowastes were prepared from different ingredients:

- Green waste compounds: wood, bark, straw, leaves, grass
- Kitchen waste compounds: apples, pies, potatoes, wheat, meat powder
- Inorganic compounds: sand, lime.

Process conditions varied with respect to aeration, temperature, turning frequency and pH. Turning occurred mainly after 1 week during the thermophilic phase and every 2-4 weeks during curing. In some experiments, additives were used during composting, e.g. to regulate moisture, structure, and pH as well as to simulate recirculation of condensed water and to investigate N-transformation during the mesophilic phase.

Table 1. Characteristics of biowaste and composting conditions in terms of composition and initial pH of the substrate, aeration rate during thermophilic phase, temperature regulation during the thermophilic/curing phase and additive addition during composting

No. of exp.	Composition(% DM)			pH	Aeration rate ($l\ h^{-1}$)	Temp. regulation (°C)	Additives
	Green waste	Kitchen waste	Inorganic				
R5a	15.0	85.0	0	4.9	150	SH,65/ 45	W, CW
R8a	15.0	85.0	0	4.6	150	SH/ 30	W, CW
R3b	14.7	76.7	8.6	6.1	80	SH	-
R4a	14.8	76.7	8.6	6.5	150	SH	-
R9a	14.8	76.7	8.6	6.1	280	SH	Wa
R2d	24.2	56.2	19.6	11.5	150	SH	L
R3c	24.4	56.6	18.9	10.9	150	SH	L
R6a	15.0	85.0	0	5.0	150	SH/ 40	W, CW
R7a	15.0	85.0	0	4.3	150	SH/ 35	W, CW
R4b	24.5	56.6	18.9	9.4	150	SH	-
R5d	22.4	51.2	26.4	5.7	150	SH	-
R6g	22.5	51.0	26.5	5.5	150	SH	-
R5b	22.4	51.1	26.5	6.3	150	SH	NH_3
R6b	59.3	18.2	22.5	6.0	150	SH	NH_3, PP
R7b	64.5	16.2	19.3	6.4	150	SH	NH_3
R8c	60.0	17.0	23.0	10.8	150	SH	NH_3
R4d	22.5	51.0	26.5	5.8	80-290	SH	L, NH_3
R8e	22.5	51.0	26.5	5.7	110-170	SH	CW

DM dry matter, *SH* self-heating, *W* wood, *CW* condensed water, *Wa* water, *L* lime, NH_3 ammonia, *PP* protein-rich powder.

Results and Discussion

In the experiments, N-compounds in substrate, leachate and exhaust air were measured. The results, described in more detail below, are given in Table 2.

Table 2. N-related results for different composting experiments on the end of the thermophilic inclusive cooling phase and at the end of composting after curing

No. of comp. Exp.	End of thermophilic/ cooling phase				End of curing phase			
	Time (week)	NH_3 rel. (%DS)	NH_3/NH_4^+ sub. (%DS)	Degr. rate (%DS)	Time (week)	NH_3 rel. (%DS)	NH_3/NH_4^+ sub. (%DS)	Degr. rate (%DS)
R5	14	0.1	0.9	33	21	0.1	0.7	39
R8	11	1.1	1.0	26	21	1.6	0.8	33
R3	19	1.5	0.4	60	24	1.5	0.4	61
R4	11	1.7	0.2	61	24	1.8	0.1	62
R9	12	1.8	0.2	57	24	1.9	0.0	58
R2	7	1.0	0.1	39	10	1.0	0.1	39
R3	5	0.9	0.4	49	10	1.1	0.2	53
R6	11	0.8	1.1	26	21	2.1	0.4	33
R7	11	1.1	0.9	21	21	1.8	1.0	31
R4	5	1.2	0.1	49	10	1.4	0.0	55
R5	9	0.8	0.1	49	19	0.8	0.0	49
R6	4	0.7	0.3	34	19	1.0	0.0	47
R5	6	1.0	0.1	34	19	1.0	0.0	39
R6	5	0.0	0.1	33	19	0.0	0.2	49
R7	5	0.1	0.1	17	19	0.1	0.0	28
R8	5	0.0	0.0	25	17	0.0	0.2	21
R4	4	0.8	0.3	29	19	1.0	0.1	33
R8	4	0.5	0.5	21	19	0.8	0.0	38

NH_3rel. cumulative ammonia-N losses until the end of the period, *NH_3/NH_4^+-sub.* ammonia/ammonium-N in the substrate on the sampling day at the end of the period, *Deg. rate* degradation rate at the end of the period, *DS* dry substance.

Table 3. N-related results for different composting experiments–start conditions and significant values during composting

No. of comp. Exp.	N_0 (%DS)	Max. decrease N_{org}		Start NO_3^--production (week)	NO_3^- max		NO_2^- max	
		(%N_0)	(week)		(%DS)	(week)	(%DS)	(week)
R5a	2.3	48.3	21	No	0.01	0	0.00	-
R8a	2.2	50.0	21	No	0.00	-	0.00	-
R3b	1.8	58.4	17	No	0.00	-	0.00	-
R4a	1.8	67.3	23/24	19	0.10	24	0.00	-
R9a	1.9	51.7	24	18	0.15	20	0.00	-

Table 3 (cont.)

R2d	1.4	68.3	10	No	0.00	-	0.00	-
R3c	1.4	75.9	10	No	0.00	-	0.00	-
R6a	2.3	58.4	19	No	0.01	0	0.00	-
R7a	2.2	60.9	21	No	0.00	-	0.00	-
R4b	1.6	71.3	8	8	0.00	-	0.00	-
R5d	1.1	57.5	12	6	0.31	9	0.10	9
R6g	1.4	71.9	19	10	0.08	15	0.00	-
R5b	1.4	71.2	17	12	0.06	13	0.15	7
R6b	1.4	14.7	1	6	0.12	8	0.09	6
R7b	1.1	35.3	17	4	0.05	8	0.11	4
R8c	0.6	4.0	0	4	0.08	8	0.09	17
R4d	1.2	54.8	9	10	0.21	19	0.10	9/12
R8e	1.1	60.7	19	10	0.03	12	0.00	-

N_0.... total N content of substrate on the begin of composting, *Max. Decrease N_{org}* ... Maximuml decrease of organic N during composting in relation to the initial N (N0=100 %) and time when that was measured, *Start NO_3- production* .... week when first traces of nitrate were detected, *NO_3- max/ NO_2- max* ... maximal measured nitrate resp. nitrite value during the whole composting process and the week, when that did occur, *DS* dry substance.

Ammonification

In order to describe ammonification and its influencing factors, the ammonia/ammonium production was measured in the substrate as well as in the outlets. Its total production could not be measured since parts of ammonium were immobilized again. Significant parts of the ammonia were released. The cumulative ammonia releases and also the ammonia/ammonium found in the substrate on the respective measurement day can be considered as the minimal ammonia/ammonium amount produced due to the ammonification processes (Fig. 1.). The end of the thermophilic phase (including the cooling phase) and the end of the composting experiment (after material had enough time for curing) were selected as significant stages. The length of the periods (Table 2.) was different for each test due to different substrates and composting conditions (thermophilic phase including cooling phase: 4–19 weeks; end of the composting: 10–24 weeks).

With respect to ammonia/ammonium-formation three different patterns can be distinguished:

1 ammonium/ammonia formation increases significantly during curing;
2 ammonium/ammonia changes are negligible during curing;
3 ammonium/ammonia is significantly immobilized during curing

The reasons leading to the development of the three different types, apart from the different absolute ammonia releases need further investigation. It is obvious that a significant ammonia/ammonium production takes place in most cases. Only in 3 (R6b, R8c, R7b) of 18 experiments was the maximum determined ammonia/ammonium-N formation below 40% of the total initial N.

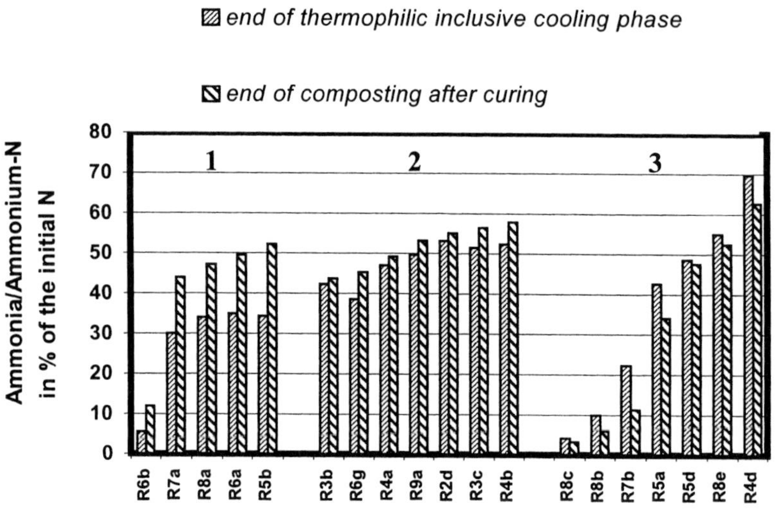

Fig. 1. Formation of ammonia (sum of cumulative ammonia-N releases and amount of the ammonia/ammonium-N contained in the substrate in relation to the initial total N of the substrate) at two significant stages of the composting process (end of the thermophilic including cooling phase; end of composting after sufficent curing). The experiments were grouped into the groups 1, 2 and 3 according to their patterns

Commonly, a minimum of between 40 and 70% of the initial N is ammonified. The absolute ammonification rate is probably higher since immobilization has to be considered. The decrease of ammonia/ammonium formation in group 3 clearly shows that during curing important immobilization effects occur. In these cases, immobilization was significantly stronger than the newformation of ammonium/ammonia from proteins during curing. In any case, ammonification, reflected by ammonium/ammonia formation, is significantly lower during curing than during intensive rotting.

The importance of the waste type on ammonification shall be clarified with Fig. 2. It shows the N-balances of wastes, which are very different in composition (Table 1.) but have a similar initial N-content (~ 1.4%DS, dry substance). Ammonification proceeded in an obviously different pattern in the different types of waste. The maximal decrease of organic N in the substrate during composting was marked in Fig. 2. The maximal decrease of organic N for each experiment is listed in Table 3. In the variants rich in green wastes (green waste content > ~ 60%DS) maximal decreases of only 4–35% of the initial total biowaste N (R6b: approx. 15 %) were observed. The variants rich in kitchen waste (kitchen waste content > ~ 50 %DS) showed much higher decreases (48–75 % of the initial total N; R2d: approx. 68 %). The low ammonification rates for substrates rich in green waste was linked with lower substrate degradation as well.

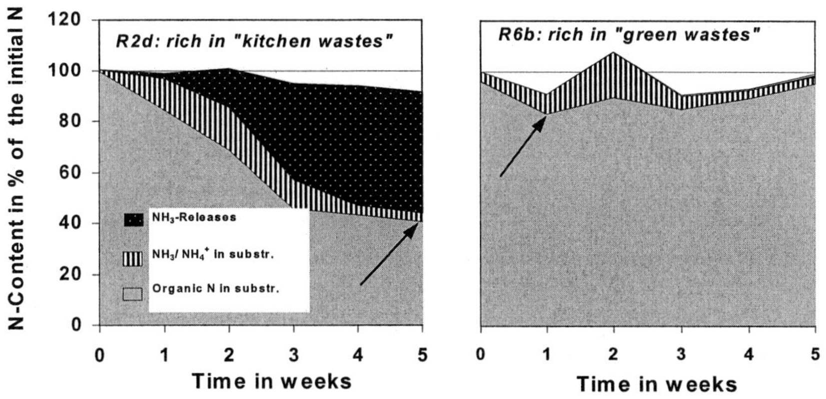

Fig. 2. Examples of N-balances (which show changes of N-containing compounds in the substrate as well as the cumulative ammonia releases) for composting experiments where different substrate types were used.

The maximum substrate degradation during the whole composting process varied from 21–49%DS when substrates with high green waste content were used, and from 33–53%DS for substrates with a high kitchen waste content. The low maximum decreases in organic N of green waste substrates led to the assumption that a significant part of the protein contained in fresh green waste is not ammonified at all during the composting process; but, theoretically, it would also be possible that ammonified compounds are immediately bound into humic fractions. Green waste contains chemical components different from kitchen waste (quantitatively seen, most important in green waste: cellulose, hemicelluloses, lignin; in kitchen waste: starch, other carbohydrates and to some extent protein). Considering the composition of the substrates it can be suggested that ammonification proceeds in a different way with regard to lignocellulosic and non-lignocellulosic substrate components.

Nitrification and Denitrification

The most important difference between the necessary conditions for nitrification and denitrification to occur is that nitrification needs aerobic conditions while denitrification is enhanced by anoxia. For detection of nitrification and denitrification processes during composting, their difference with respect to their metabolism products was used (nitrification products: NO_2^- and NO_3^-; denitrification product: N_2). Table 3. gives the maximal measured values for NO_2^- and NO_3^- during the composting process and the time, when detected. The detection of the denitrification product N_2 was not possible since N_2 was the main compound contained in the fresh air used for aeration. A potential N_2 increase was too small to be detected. The following discussion covers the results for nitrification and conclusions for denitrification during different composting phases.

Thermophilic Phase Including the Cooling Phase

NO_2^- and NO_3^- formation could not be observed during the thermophilic phase. Ammonia/ammonium was contained in the waste, so that the substrate was available for nitrification. The high temperatures probably inhibited nitrifying bacteria. Nitrification could only be ascertained with temperatures under 30 °C. Since no nitrate was formed during the thermophilic/cooling phase due to missing nitrification, denitrification was not possible either.

Curing

Nitrification was concluded through detection of low nitrate and partially nitrite formations in most of the tests during curing (Table 3.). However, in some experiments no traces of NO_2^- or NO_3^- were found:

- R5a, R6a, R7a: temperature over 30 °C due to temperature regulation
- R8a: temperature around and over 30 °C due to temperature regulation
- R3b: moistened substrate (reduces oxygen availability)
- R2d, R3c: PH over 8 due to pHregulation.

Immobilization

Immobilization of inorganic N compounds could only be detected by means of an increase in the organic N-fraction. This could only be visualized when the immobilization rate was higher than ammonification. During the thermophilic/cooling phase no signs of immobilization could be observed, probably since ammonification processes are quantitatively more important. However, this does not mean that immobilization processes did not occur. They definitely took place, since new microbial biomass was constant.

During curing immobilization could be made visible after addition of ammonia solutions. Figure 3. shows the respective details of two N-balances.

In R8c, almost 1/3 of the added ammonia was nitrified within the observation period (formation of nitrate and nitrite). However, more than 1/3 was immobilized. The most significant increase was found at the end of the observation period. In R6b the immobilized portion was even higher. The main difference between both tests was that the conditions in variant R8c allowed microbial activity and the conditions in R6b did not (limited water content). This means that an immobilization in microbial biomass did not take place and thus immobilization had to take place in the humic-lignin complex. It can be concluded that immobilization processes do not necessarily have to be catalyzed by microorganisms, but can take place by chemical-physical processes.

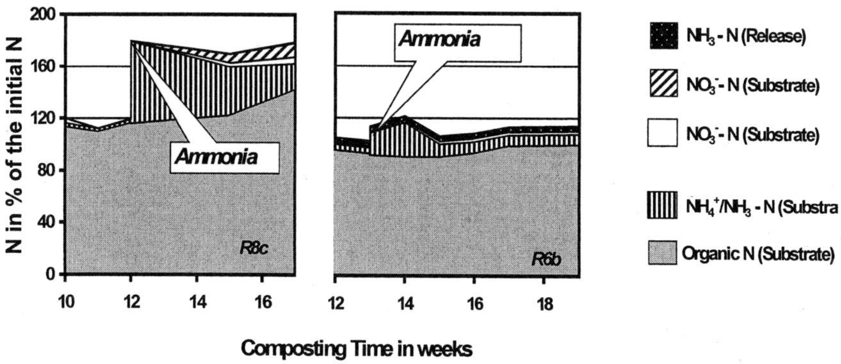

Fig. 3. Examples for experiments to visualize N-immobilization. Extracts from N--balances (which show changes of N-containing compounds in the substrate as well as the cumulative ammonia releases) for the period before and after ammonia addition

N-Releases

Quantitatively important N-releases could be measured via exhaust air and leachate.

NH_3-Releases via Exhaust Air

The releases were significantly different for different substrates. Following values could be determined over the whole composting period (Tables 2., 3.):

- Substrates with kitchen waste content > ~50% DS): 35–060 % of the initial N
- Substrates with green waste > ~60% DS: 0–10 % of the initial N

This can be attributed to the fact that, in general, green waste most likely involves lower ammonification. On one hand, the majority of the discharges stated took place during the thermophilic/cooling phase–due to the higher ammonification rate during these periods. On the other hand, generally higher temperatures and aeration rates allowed improved discharges of the ammonia formed compared with the curing phase. Apart from the kind of substrate, temperature, and aeration rate, a dependency on pH and on substrate structure could also be seen.

N_2-Releases via Exhaust Air

N_2 releases could only take place during curing as a basic condition with the availability of nitrate. Nitrate was only formed during curing. The maximum N_2--losses were approx. 20% of the initial N in some of the tests. They are concluded indirectly from the lack in N-balances. It has to be considered that the probability of errors is very high. Due to this, the results are not mentioned in detail.

N-Releases via Leachate

Leachate formation could only be observed for an insufficient process regulation. Due to this reason results are not mentioned in detail. In most cases where leachate formation could be found, N leaching amounted to under 10% of the initial N. The "worst-case" with a high leachate formation amounted to approx. 20% of the initial N.

References

Beck T (1979) Die Nitrifikation in Böden. Z Pflanzenernähr Bodenkd, 142: 344-364

Falbe J, Regitz M (1991) Chemie Lexikon M-Pk. In: Römpp Chemie Lexikon. Georg Thieme, Stuttgart

Gebel J (1991) Möglichkeiten einer umweltgerechten und wirtschaftlichen Aufbereitung von Gülle. Müll und Abfall 8: pp. 518-528

Kiefer C (1992) Charakterisierung der biologischen Stickstoffumsätze in der ungesättigten Bodenzone anhand der Bildung von Distickstoffoxiden. Dissertation, University Karlsruhe, Karlsruhe

Kleber H.-P, Schlee D (1987) Biochemie I. Gustav Fischer Verlag, Stuttgart

Knicker H (1993) Quantitative 15N- u. 13C- Festkörper- und 15N-Flüssigkeits-NMR-Spektroskopie an Pflanzenkomposten und natürlichen Böden. Dissertation, University Regensburg, Regensburg

Körner I, Brilsky H, Jensen U, Ritzkowski M, Stegmann R (1997) Possibilities for the regulation of the composting process to optimize the nutrient composition of compost. In: Stentiford E I (ed) Proceedings of International conference Orbit 97, Harrogate, United Kingdom, Zeebra Publishing, Manchester and Ritchie, Kilmarnock, UK, pp. 211-220

Körner, I, Ritzkowski M, Stegmann R (1999) Nährstofffreisetzung bei der Kompostierung und bei der Vergärung. In: Neue Techniken zur Kompostierung- Verwertung auf landwirtschaftlichen Flächen, Band II. Umweltbundesamt, Projektträger Abfallwirtschaft und Altlastensanierung des BMBF (ed), self published, Berlin

Morihara K, Kohei O (1992) Microbial Degradation of Proteins. In: Winkelmann G (ed) Microbial degradation of natural products, VCH-Verlag, Weinheim

Sahm H (1993) Biotechnology. In: Rehm H.-J; Reed G; Pühler A; Stadler P (eds.) Biological Fundamentals (Bd. 1, 2. Aufl.), VCH-Verlag, Weinheim

Scheffer F, Schachtschnabel P (1992) Lehrbuch der Bodenkunde. 13. Aufl. Ferdinand Enke, Stuttgart

Schlegel H, Schmidt K (1985) Allgemeine Mikrobiologie. Georg Thieme Verlag, Stuttgart

Unsuitability of Anaerobic Compost from Solid Substrate Anaerobic Digestion as Soil Amendment

H.M. Poggi-Varaldo[1], E. Gómez-Cisneros[1], R. Rodríguez-Vázquez[1], J. Trejo-Espino[1] and N. Rinderknecht-Seijas[2]

Abstract. This research aimed at evaluating the quality of anaerobic compost from solid substrate anaerobic digesters for potential use as a soil amender. A factorial experiment 4 x 2 x 2 was run. The factors were feedstock type, temperature in reactors (35 and 55 °C) and mass retention time (MRT, 16 and 23 days). The feedstocks used consisted of four mixtures of food waste (FW) and lignocellulosic fraction (LG): 100% FW, 67% FW-33% LG, 33% FW-67% LG and 100% LG which were designated as FS1, FS2, FS3 and FS4, respectively. The anaerobic compost coming from feedstocks with increasing proportion of lignocellulosic fraction (FS3 and FS4) were of higher quality and gave the lowest chemical oxygen demand, biochemical oxygen demand, volatile organic acids and total ammonia nitrogen concentrations in the extract and the highest germination indices. Heavy metal concentrations in all the anaerobic composts were lower than the maximum levels indicated in USA and European compost quality standards. However, high total oxygen uptake, moderate to high concentrations of volatile organic acids and ammonia, and germination indices under 60% indicated that the anaerobic compost was not suitable for direct use as a soil improver.

Introduction

In Mexico, nearly 52 000 tons of municipal solid wastes, 370 000 tons of industrial solid wastes and 1 900 000 m^3 of wastewater solids are generated on a daily basis (Poggi-Varaldo et al. 1997a). The Mexican pulp and paper industry produces approximately 110 000 dry tons $year^{-1}$ of non-hazardous solid wastes (primary sludge cakes, predominantly lignocellulosic wastes). The anaerobic co-digestion of municipal solid wastes, industrial solid waste and waste-activated sludges at high total solid contents (also known as solid substrate anaerobic digestion) has been shown to be an attractive and environmentally friendly alternative for dealing with the solid waste problem in Mexico (Poggi-Varaldo et al. 1995).

Our group has demonstrated the technical and economic feasibility of solid substrate anaerobic digestion of municipal and non-hazardous, industrial solid

[1]Centre for Advanced Studies and Research (CINVESTAV), Dept. of Biotechnology, Environmental Biotechnology and Anaerobic Processes R&D Group-GBPANAT, P.O. Box 14-740, Mexico D.F., 07000, Mexico E-mail: hpoggi@mail.cinvestav.mx
[2]ESIQIE-IPN, Division of Basic Sciences, México D.F., México

H. Insam, N. Riddech, S. Klammer (Eds.)
Microbiology of Composting

waste (Poggi-Varaldo et al. 1995; Poggi-Varaldo and Valdés-Ledezma 1995; Poggi-Varaldo et al. 1997a; Sparling et al. 1997; Valdés-Ledezma 1997). However, the quality of anaerobic compost related to its direct application as a soil amender or other alternative uses is an issue that has received relatively scarce attention in the literature. Therefore, the objective of this research is to evaluate the quality of anaerobic compost from solid substrate anaerobic digesters as a soil amender.

Methodology

Design of Experiments

A factorial experiment (4 x 2 x 2) was run. The factors were feedstock type, temperature in reactors (35 and 55 °C) and mass retention time (MRT, 16 and 23 days). The feedstocks used consisted of four mixtures of food waste (FW) and lignocellulosic fraction (LG): 100% FW, 67% FW-33% LG, 33% FW-67% LG and 100% LG which were designated FS1, FS2, FS3 and FS4, respectively. The lignocellulosic fraction consisted of paper mill sludge cake. The solid substrate 3 litre anaerobic digesters were fed twice a week in a draw-and-fill mode. Feedstock was prepared according to procedures reported in Valdés-Ledezma (1997) and Poggi-Varaldo et al. (1997a). Main process responses were monitored as described by Poggi-Varaldo et al. 1997a,b).

Monitoring and Analysis

Analytical methods used in this work can be found in Poggi-Varaldo and Valdéz-Ledezma (1995); Poggi-Varaldo and Rinderknecht-Seijas (1990); Poggi-Varaldo et al. (1997a,b). Germination index assays were carried out in compost extracts using raddish seeds, *Raphanus sativus* (Zucconi et al. 1981; Lozano-Vinalay and Poggi-Varaldo 1995). Respirometric tests of whole composts were performed in Bartha flasks (Alexander 1977).

Results and Discussion

Anaerobic digestion significantly decreased the C/N ratio of feedstocks 2 to 4 (Table 1). This was consistent with high volatile solid removal efficiencies and biogas productivities exhibited by the anaerobic digesters that received these feedstocks (Poggi-Varaldo and Valdés-Ledezma 1995). Table 2 shows the process removal efficiencies averaged per factor level. It can be seen that the C/N ratio parallels closely the results of the removal efficiencies. The C/N of FS1 did not improve very much because the digesters (four in total) fed with FS1 were acidogenic (Trejo-Espino 1999; Poggi-Varaldo 1999). The anaerobic composts coming from these digesters would not deserve such a name, rather, they were acidic pastes rich in soluble organic metabolites (mainly organic acids); i.e. the biodegradable organic matter in FS1 was not degraded, it was just fermented. Iglesias and Pérez (1989) recommend a C/N less than 20 for mature composts.

Table 1. Results of C/N ratios for feedstocks and anaerobic composts[a]

Factor	Main levels	C/N ratio[b]	
		Feedstock	Anaerobic compost
Mixture type	100% FW[c]-0%LG[d]	9.04	6.19
	67% FW-33%LG	22.53	10.94
	33%FW-67%LG	33.73	15.71
	0% FW-100%LG	38.63	17.29
Temperature	35° C	NA[e]	12.05
	55° C	NA	13.01
Mass retention time	16 days	NA	11.84
	23 days	NA	13.23

[a] Results are averaged per factor level, the standard error of the experiment was 1.14 (Montgomery 1991).
[b] g carbon (g^{-1} nitrogen Kjeldahl) or dimensionless.
[c] FW food waste fraction.
[d] LG lignocellulosic waste fraction.
[e] NA is not applicable.

The anaerobic composts generated from the digestion of FS 2 to 4 seemed to comply with this guideline; however, this sole feature is not enough to indicate compost maturity. Neither the temperature nor the mass retention time had a significant effect on the C/N of the anaerobic composts (Table 1).

The volatile organic acid content in the compost extracts from digesters fed with FS1 was very high, and was relatively low in the other composts (Fig. 1). Volatile organic acid concentrations as low as 1 mg HAc (g^{-1} dry compost) were obtained in digesters fed with FS4, and operated at 35 °C and 23-day mass retention time. However, this "low" value is higher than the recommended criterion of less than 0.3 mgHAc (g^{-1} dry compost) for mature composts (Iglesias and Pérez 1989). The anaerobic composts from thermophilic digesters showed higher volatile organic acid contents than those from mesophilic reactors (Fig. 2).

Table 2. Process removal efficiencies for solid substrate anaerobic digesters[a]

Factor	Main levels	Removal efficiency (% volatile solids)	
		η[b]	η_b[c]
Mixture type	100% FW[d] - 0% LG[e]	15.67	21.26
	67% FW - 33% LG	41.33	61.22
	33% FW - 67% LG	37.91	62.13
	0% FW - 100% LG	33.16	60.51

Table 2. (cont.)

Temperature	35 °C	33.19	NA [f]
	55 °C	32.89	NA
Mass retention time	16 days	29.26	NA
	23 days	34.77	NA

[a] Results are averaged per factor level, the standard error of the experiment was 4.61% and 6.42% for η and η_b, respectively (Montgomery 1991).
[b] On total volatile solids basis.
[c] On biodegradable volatile solids basis.
[d] FW food waste fraction.
[e] LG lignocellulosic waste fraction.
[f] NA is not applicable.

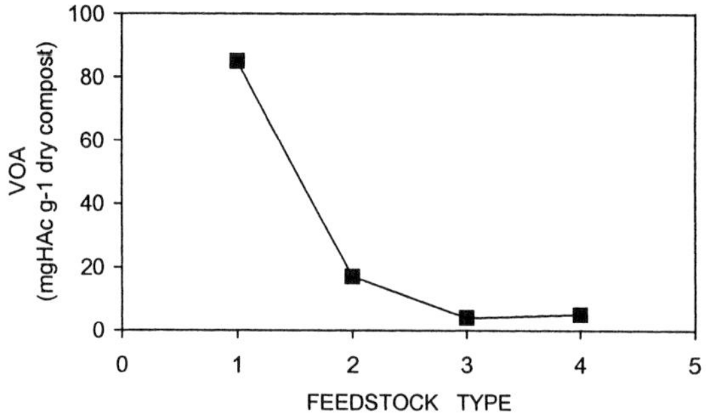

Fig. 1. Volatile organic acids contents (VOA) in the extracts of anaerobic composts: effect of the feedstock type

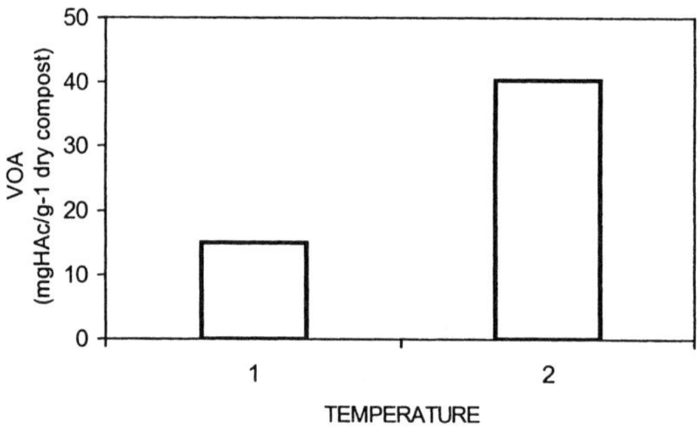

Fig. 2. Volatile organic acid contents (VOA)in the extracts of anaerobic composts: effect of the temperature. *1* 35 °C; *2* 55 °C

On average, the volatile organic acid content at 55 °C was nearly twofold the volatile organic acids at 35 °C; this effect was significant [P(F) < 0.01]. This trend agreed with previous findings of Poggi-Varaldo and Valdés-Ledezma (1995) and Poggi-Varaldo and Oleszkiewicz (1992), who reported instability of selected thermophilic, solid substrate anaerobic reactors at 14- to 16-d mass retention time, and higher volatile organic acid concentrations in thermophilic than in mesophilic digesters. Also, the increase of volatile organic acids with increase of temperature paralleled the increase in both chemical oxygen demand and biochemical oxygen demand of compost extracts (Gómez-Cisneros 1997). The ultimate oxygen demand (the long-term, cumulative O_2 uptake) of whole anaerobic compost significantly decreased with increase of the lignocellulosic fraction in the feedstock (Fig. 3). Again, and as was expected, the highest O_2 uptake corresponded to the acidogenic composts from digesters fed with FS1. The trend shown by the ultimate oxygen demand consistently mirrored the removal efficiency (Table 2). The anaerobic compost from digesters processing FS4 (that with the highest lignocellulosic fraction) showed the smallest ultimate O_2 demand. It is known that a large oxygen demand has deleterious effects on plants, microflora and soil quality when immature composts are applied onto soil. (Mathur et al. 1993). The mass retention time had a significant effect [P(F)< 0.10] on the ultimate oxygen demand (Fig. 4); the latter was lower at the longest mass retention time.

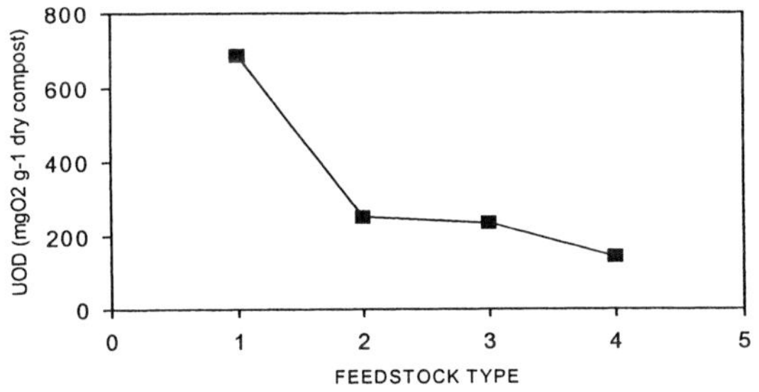

Fig. 3. Ultimate oxygen demand (UOD) of whole anaerobic composts: effect of the type of feedstock

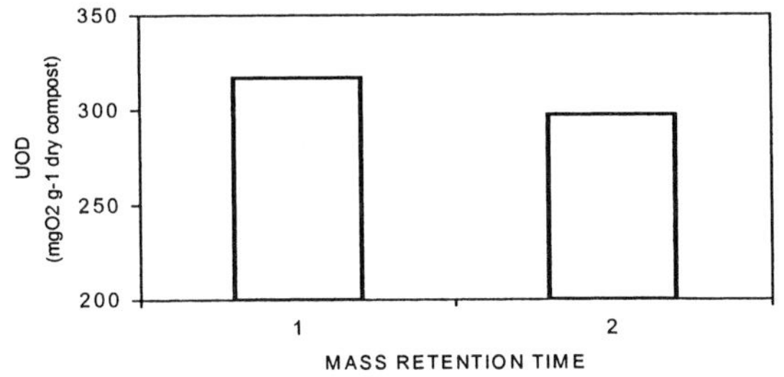

Fig. 4. Ultimate oxygen demand of whole anaerobic composts: effect of the mass retention time. *1* 16 day; *2* 23 day

This result was in line with the removal efficiency data (Table 2) obtained by Valdés-Ledezma (1997), who found higher removal efficiencies in digesters operated at 23-day mass retention time compared to those at 16-day retention time. Interestingly, the temperature did not have a clear effect on the ultimate oxygen demand of the anaerobic composts.

The immediate or initial oxygen demand rate (r_i) of anaerobic composts followed a similar trend to that reported for the ultimate oxygen demand. All the anaerobic composts had r_i values 4 to 20 times higher (Gómez-Cisneros 1997) than the recommended maximum value of 0.1 mg O_2 (h^{-1} g^{-1} dry matter) (Michel et al. 1995).

Heavy metal contents of the anaerobic composts were generally higher than the corresponding concentration in the feedstocks (Fig. 5). This result could be ascribed to the conservation of heavy metals and degradation of the organic matter of feedstocks, that would lead to an "increase" in the relative concentrations of the metals in the composts, since the same amount of metal is expressed on a smaller basis of dry matter. The profile of selected heavy metal concentrations of anaerobic composts obtained in this work seemed to comply with the international criteria (Table 3), including the most stringent guidelines of Canada and Germany.

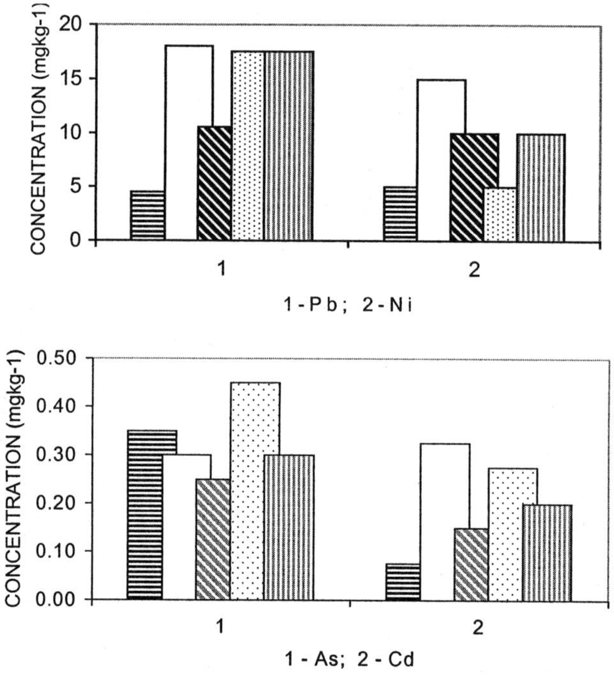

Fig. 5. Concentration of selected heavy metals in feedstock and anaerobic composts: *1st bar* feedstock 3; *2nd bar* anaerobic compost 3 (23 days/35 °C); *3rd bar* anaerobic compost 7 (16 days/35 °C); *4th* bar anaerobic compost 13 (23 days/55 °C); *5th bar* anaerobic compost 17 (16 days/55 °C).

Table 3. Maximum acceptable limits of heavy metals in composts: criteria from different countries selected results of this work. Concentrations in mg kg^{-1} dry compost [a]

Heavy metal	Germany	Belgium	Canada	France	Switzerland	Anaerobic compost[b]
Copper	150	100	100	---	150	50
Zinc	400	1000	500	---	500	106
Molibdenum	---	---	5	---	5	---
Cobalt	---	---	34	---	25	< 12.5[c]
Cadmium	2	5	3	8	3	0.22
Níckel	50	50	62	200	50	10
Lead	200	600	150	800	150	15.5
Mercury	1	---	0.8	8	3	< 20[c]
Chromium	150	150	210	---	150	---
Arsenic	---	---	13	---	---	1
Selenium	---	---	2	---	---	---

[a] Adapted from Chaney and Ryan (1993). [b] Anaerobic compost 3 from this study, from a digester operated at 35 °C and 23-day mass retention time. [c] Sensitivity level of the method.

The germination indices of the anaerobic compost significantly increased with the proportion of lignocellulosics in the feedstock (Fig. 6) [P(F) < 0.05]. A few germination indices as high as 55 and 77% were obtained in anaerobic composts from mesophilic reactors working at 23-day mass retention time and processing the feedstocks richest in lignocellulosic wastes. However, germination indices of composts from feedstocks 1 and 2 (rich in food waste) were dramatically poor because of the acidogenesis. The germination index significantly decreased with temperature, see Fig. 7 [P(F) < 0.05].

This pattern agreed with the results discussed above for the volatile organic acids (Fig. 3) and the ammonia nitrogen contents of the extracts (Gómez-Cisneros 1997), both recognized as phytotoxic substances (Mathur et al. 1993). As expected, the germination index increased with the mass retention time, Fig. 8 [P(F) < 0.05], which was consistent with the better performance of digesters operated at 23 day mass retention time (Table 2).

Zucconi et al. (1981) recommend a germination index not less than 60% for 30% v/v extracts of composts. In general, anaerobic composts in this work failed to comply with this criterion.

In spite of the fact that the solid substrate anaerobic digestion was able to improve several quality parameters of the solid wastes fed to the process, the overall quality of the anaerobic compost generated was not suitable for direct application as a soil amender and some type of postTreatment seems to be required.

Table 4 summarizes the quality profile of the best anaerobic compost in this work and its corresponding feedstock.

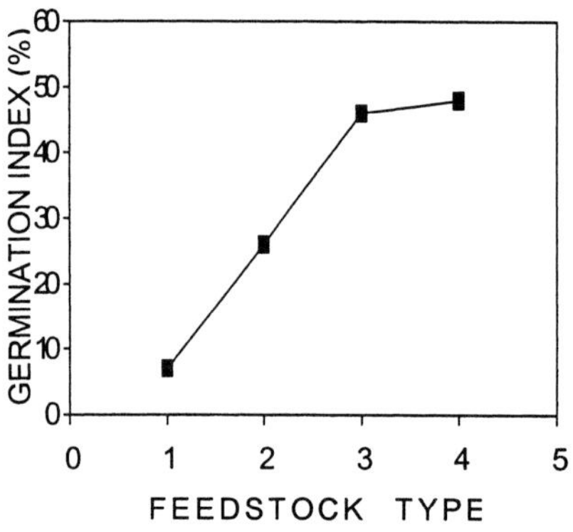

Fig. 6. Germination indices of anaerobic Composts: effect of the feedstock mixture

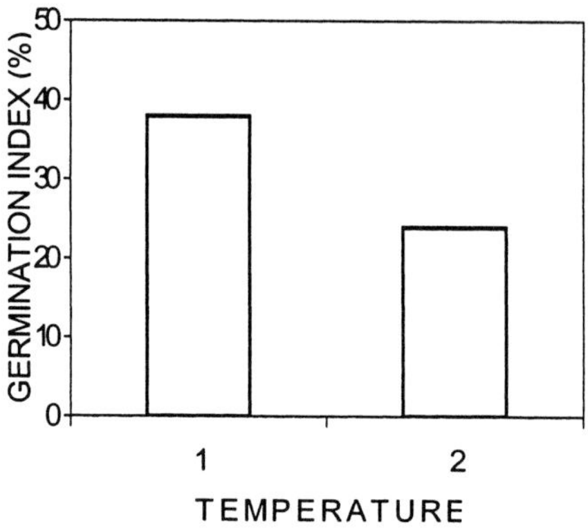

Fig. 7. Germination indices of anaerobic composts: effect of temperature. *1* 35 °C; *2* 55 °C

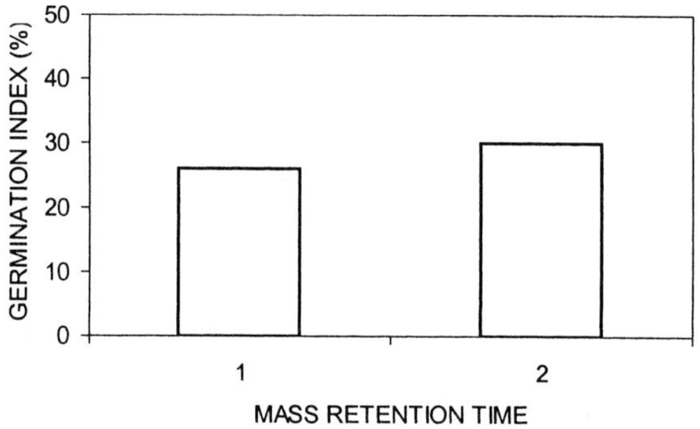

Fig. 8. Germination indices of anaerobic composts: effect of the mass retention time. *1* 16 days; *2* 23 days

The solid substrate anaerobic digestion process improved the compost quality (decreased C/N ratio, and decreased contents of volatile organic acids, total ammonia nitrogen, biochemical oxygen demand and chemical oxygen demand of extracts; and increased germination index). However, the values of volatile organic acid contents and total ammonia nitrogen in extracts were much higher than those indicated by compost quality guidelines. Furthermore, the initial oxygen uptake rate r_i of the anaerobic compost was threefold or higher than the criterion value of 0.10 mg O_2 $h^{-1}g^{-1}$ dry compost. Also, the germination index of the best anaerobic compost (55%) is still lower than the minimum value recommended (60%) for mature compost.

In conclusion, it is apparent that the anaerobic compost does not fulfill the criteria for direct use as a soil amender, and that some type of posttreatment and maturing is required.

Table 4. Summary of quality of the anaerobic compost

Parameter	Feedstock[a]	Anaerobic compost[b]	Recommended values
Chemical oxygen demand[c]	133.0	68.2	---
Biochemical oxygen demand[c]	40.0	28.5	---
Total ammonia Nitrogen[d]	1.07	0.82	<0.40[j]
Volatile organic acids[e]	>5	2.85	<0.30[k]
Germination index[f]	30.5	55.3	>60[l]
Ultimate oxygen demand[g]	> 500	162	---
r_i[h]	ND	0.26	<0.10[m]
C/N ratio[i]	28.35	14.83	<20[k]

[a] 67% paper mill sludge plus 33% food waste, amended with waste-activated sludge at a ratio 11 g TS (1 g TS waste activated sludge)$^{-1}$.
[b] Anaerobic compost from a digester at 35 °C and 23 - day mass retention time.
[c] Determined in the compost extract, in mg O_2 (g^{-1} dry compost).
[d] Total ammonia nitrogen of the compost extract, in mgN (g^{-1} dry compost).
[e] Volatile organic acids in the compost extract, in mgHAc (g^{-1} dry compost).
[f] Germination index of the compost extract, in % of the distilled water control.
[g] Ultimate oxygen demand of whole compost, in mg O_2 (g^{-1} dry compost).
[h] Initial oxygen uptake rate, mg O_2 h^{-1} g^{-1}.
[i] Dimensionless.
[j] Villar et al. (1993), Kapetanios et al. (1993).
[k] Iglesias and Pérez. (1989).
[l] Zucconi et al. (1981), Lozano-Vinalay and Poggi-Varaldo (1995).
[m] Michel et al. (1995).

Acknowledgements. This contribution is dedicated to the memory of Professor Herbert Wirth. Funding from CONACYT (Mexican Council of Science and Technology, Project No. 28071T and graduate scholarships to JL-T and EG-C), National Institute of Ecology (INE Ministry of Environment and Natural Resources of Mexico) and Dept. of Biotechnology and Bioengineering, CINVESTAV, is gratefully acknowledged. The authors wish to thank the excellent technical assistance and logistic support of Mr. Rafael Hernández-Vera, Mr. L. Valdés-Ledzma, Mr. Máximo Vázquez (all of them with the Environmental Biotechnology R&D Group, CINVESTAV del IPN), the help of Prof. Elvira Ríos-Leal with chromatographic analyses and Ms. Nelly Lozano-Vinalay with part of the germination tests.

References

Alexander M (1977) Introduction to soil microbiology. 2nd edn. John Wiley , New York

Chaney RL, Ryan JA (1993) Heavy metals and toxic organic pollutants in municipal solid wastes composts. In: Hoitink HAM, Keener HM (eds) Science and engineering of composting: design, environmental, microbiological and utilization aspects. Renaissance Publishers, Worthington, OH, pp 451-489

Gómez-Cisneros E. (1997) Aerobic post-composting of anaerobic compost from DASS process. MS Thesis, CINVESTAV Dept of Biotechnology, Mexico DF, México.

Iglesias E, Pérez V (1989) Evaluation of city refuse compost maturity: a review. Biol Wastes 27(2):115-142

Kapetanios E, Loizidou M, Valkanas G (1993) Compost production from Greek domestic refuse. Bioresour Technol 44(1): 13-16

Lozano-Vinalay N, Poggi-Varaldo HM (1995) Germination bioassay with seeds (*Raphanus sativus*) for the evaluation of waste biological treatment. 1st Int. Meet. on Microbial Ecology , México DF, México p. 43

Mathur S, Owen G, Dinel H, Schniter M (1993) Determination of compost biomaturity. I. Literature review. Biol Agric Hortic 10(1): 65-85

Michel FC Jr, Reddy AC, Forney LE (1995) Microbial degradation and humification of the lawn care pesticide 2,4-dichlorophenoxyacetic acid during the composting of yard trimmings. Appl Environ Microbiol 61(7): 2566-2571

Montgomery DC (1991) Design and analysis of experiments. John Wiley; New York

Poggi-Varaldo HM (1999) Interim report to CONACYT. Research project solid substrate anaerobic digestion of municipal and industrial solid wastes. DBB-CINVESTAV, Environ. Biotechnol. R&D Group, México DF, México

Poggi-Varaldo HM, Oleszkiewicz J (1992) Anaerobic co-composting of municipal solid waste and waste sludge at high total solids levels. Environ Technol 13(3): 409-421

Poggi-Varaldo HM, Rinderknecht-Seijas N (1990) Mini-handbook of solid waste analysis, CINVESTAV del IPN, México DF, México (in English) available upon request

Poggi-Varaldo HM, Valdéz-Ledezma L (1995) Optimization of dry anaerobic digestion start-up, Proc. 17th Canadian Waste Management Conference, Sept 11-14, Quebec City, QC, Canada. Proceedings on diskette

Poggi-Varaldo HM, Yu A, Rinderknecht N (1995) Dry anaerobic digestion: a review and feasibility study, R´95 Recovery, Recycling, Re-integration, Febr 1-4, Geneve, Switzerland

Poggi-Varaldo HM, Valdés L, Esparza-García F, Fernández-Villagómez G. (1997a) Solid substrate anaerobic co-digestion of paper mill sludge, biosolids and municipal solid wastes. Water Sci Technol 35(2-3): 197-204

Poggi-Varaldo HM, Rodríguez-Vázquez R, Fernández-Villagómez G, Esparza-García F (1997b) Inhibition of mesophilic solid-substrate anaerobic digestion by ammonia nitrogen. Appl Microbiol Biotechnol 47(3): 284-291

Sparling RS, Risbey D, Poggi-Varaldo HM (1997) Hydrogen production from inhibited anaerobic composters. J Hydrogen Energy 22 (6): 563-566

Trejo-Espino JL (1999) Determination of anaerobic compost quality and its post-treatment with ligninolytic fungi. MS Thesis. CINVESTAV del IPN, Dept of Biotechnology. México DF, México (in Spanish)

Valdés-Ledezma L (1997) Start-up of DASS reactors and the feasibility of the process for the co-digestion of municipal and industrial solid waste. MS Thesis, CINVESTAV del IPN, Dept of Biotechnology. México DF, México (in Spanish)

Villar MC, Beloso MC, Acea MJ, Cabaneiro A, González-Prieto SJ, Carballas M, Díaz-Raviña M, Carballas T (1993) Physical and chemical characterization of four composted urban residues. Bioresour Technol 45(1): 105-113

Zucconi F, Forte M, Monaco A, De Bertoldi M (1981) Biological evaluation of compost maturity. BioCycle 22(4): 27-35

Pile Composting of Two-phase Centrifuged Olive Husks: Bioindicators of the Process

G. Ranalli[1], P. Principi[2], M. Zucchi[2], F. da Borso[3], L. Catalano[3] and C. Sorlini[2]

Abstract. Composting is a process largely used to solve the organic material disposal problem, but composting of the olive husks produced by two-phase technology remains largely unknown. Because of their high content of salts, polyphenols, fatty acids and tannins, husk utilisation as fertiliser is not possible. Composting of the husks could permit the reduction of salt content and conversion of phenolic compounds into humic substances, but the knowledge about conditions of starting up, running the plant, degradation and microbiological characteristics of the cured compost has to be improved. Trials were carried out with uninoculated (A), and inoculated (B) piles and the process was monitored by ATP content and a pool of enzyme activities. The results showed that the performance of the process improved from pile A to pile B and ATP and mainly a pool of enzyme activities (alkaline phosphatase, esterase, esterase-lipase, phosphoamidase, β-galactosidase, β-gluco-sidase) showed effectiveness in the description of process. Finally, the molecular methods used to detect faecal contaminants in the cured compost resulted to be reliable.

Introduction

The present work was developed within the EC project FAIR5-CT97 3620 named Husks, whose aim is to solve in an energetically, economically and environmentally acceptable way the problem of the olive husk disposal. The Mediterranean EC countries contribute 96% of world production of olive oil (about 1.7 millions of tons year^{-1}). Olive oil can be produced by different extraction systems, the traditional discontinuous pressing process and the continuous centrifugation process being the most usual. In the continuous process a decanter is used, adding a high volume of water to the crushed olive and obtaining three phases: olive oil, olive oil wastewater and pulp with about 47% moisture. Recently, a new kind of centrifuge decanter was introduced; this new two-phase process produces olive oil and a husk mixed with olive oil mill wastewater. The new two-phase centrifugation technology solves the problem of olive milling wastewaters by dramatically reducing their production, but in the meantime shifts the problem to husks, which have more water content and higher phenols concentration.

---

[1]DISTAAM, University of Molise, 86100 Campobasso, Italy
[2]DISTAM, University of Milan, 20133 Milan, Italy
[3]DPVTA, University of Udine, 33100 Udine, Italy

H. Insam, N. Riddech, S. Klammer (Eds.)
Microbiology of Composting

Composting is a process largely used to solve the organic materials disposal problem, but very little is known about composting of the husks produced by two-phase technology. Because of their content of fatty acids, phenols and tannins (highly phytotoxic compounds), husk utilisation as fertiliser is not possible. Composting of the husks could permit the degradation of toxic compounds. Recent literature reports that the compost obtained by husks shows good characteristics in relation to both phytotoxicity and germination tests, and to agronomic evaluation (Sciancalepore et al. 1996; Epstein 1997). It is in any case necessary to extend the knowledge about the conditions of starting up, running of the plants, the degradation of toxic compounds and the fertiliser and hygienic characteristics of the mature compost.

The main purposes of this investigation were to compare simplified composting technologies and to utilise bioindicators other than microbial counts, for process monitoring. In particular, ATP content and several enzyme activities were tested as bioindicators of the process in a two-pile composting process on a pilot scale for olive husk treatment. The addition of separate solids or animal manure as inoculum to start up the process could introduce pathogen microorganisms to the raw material (Hussong et al. 1985). The safety of the cured compost (absence of *Salmonella* spp.) required by European regulations was evaluated by traditional culturable and biomolecular non-culturable techniques.

Materials and Methods

Plants

The experiments were carried out utilising two static piles (Fig. 1). The piles were set up in a greenhouse to be protected from atmospheric agents. Pile A was not inoculated, to the other (pile B) urea was added in order to balance the C/N ratio and corn stalks to soften the materials, finally it was mixed with separate solid from animal manure. To allow the aeration in the static pile, an electric blower with atmospheric air was used with a mean air flow of about 130 l h^{-1}. The trials in pile lasted for 130 days, until the pile temperature reached that of the environment.

The main characteristics of piles are reported in Table 1.

Fig. 1. Composting piles

Table 1. Main characteristics of composting piles technologies

Characteristics	Pile A	Pile B
Typology, shape	Static, trapezoidal	Static, trapezoidal
Dimensions (m) (width x height x length)	1.5 x 1.0 x 3.0	1.5 x 1.0 x 3.0
Weight (t)	1.7	1.7
Bulking agents (w/w)	Absent	Corn stalks, 10% on dry matter
Nitrogen source (w/w)	Absent	Urea, 0.03%
Inoculum (w/w)	Absent	Separate solid from animal manure, 1.0%

Sampling and Analyses

To obtain homogeneous samples, an amount of compost (2 kg) was collected from three different depths (bottom, middle, top) and mixed. The samples were stored at 4 °C during the transport and analyses were carried out within the following 24 h. Physical and chemical analyses (humidity, pH, ash, organic carbon, total N (TKN), organic carbon/total N, humification index, phenolic carbon) were performed according to DI.VA.P.R.A-IPLA (1992).

For the microbial viable counts tenfold serial dilutions were made from an initial suspension (30 g w/w of sample in 270 ml of phosphate buffer pH 7.0) and used as inoculum. Aliquots (0.1 ml) of each dilution were spread in duplicate onto different agarised media in Petri plates: total heterotrophic mesophilic and thermophilic bacteria and aerobic spore-forming bacteria on TSA (CM131, Oxoid) incubated at 28 and 55 °C, respectively, for 48 h; mycetes on RBC (CM549, Oxoid) at 28 °C for 48 h. All results are expressed on a dry weight basis (after drying aliquots of the samples at 105 °C until steady weight).

To determine the ATP content, the reaction between the enzymatic system luciferine-luciferase and the ATP present, was evaluated by the bioluminometer Biocounter Lumac 1500P (Lumac B.V., Landgraaf, The Netherlands). Values obtained as RLU (relative luminescence unit) were converted to ng of ATP/g d.w. by using a standard curve (Sciancalepore et al. 1996).

To determine the enzyme activities the API-ZYM kit (BioMérieux, Marcy l'Etoile, France) was used. With this assay it is possible to obtain semiquantitative evaluations of the activity of an enzyme pool (Humble et al. 1977; Hermann and Shann 1993). The 19 enzymes tested are: 1 control; 2 alkaline phosphatase; 3 esterase (C4); 4 esterase-lipase (C8); 5 lipase (C14); 6 leucine arylamidase; 7 valine arylamidase; 8 cistine arylamidase; 9 trypsin; 10 α- chymotrypsin; 11 acid phosphatase; 12 phosphoamidase; 13 α-galactosidase; 14 β-galactosidase; 15 β-glucuronidase; 16 α-glucosidase; 17 β-gluco-sidase; 18 N-acetyl-β-glucosaminidase; 19 α-mannosidase; 20 α-fucosidase. Briefly, from initial suspension of 20 g of compost sample in 180 ml of 0.9% NaCl, the enzyme activities were performed on 1/10 and 1/100 dilutions; after incubation at 37 °C for 4 h, the results were recorded by adding specific reagents and referring to a colorimetric standard table (Sciancalepore et al. 1996).

All data concerning chemical parameters, microbial counts, ATP and enzymes were submitted to statistical analyses using the SAS statistical software package (SAS Institute Inc. 1989). The values of each bioindicator were processed by ANOVA analysis; the relations between each couple of indicators were also tested by correlation and linear regression analyses.

To test the hygienic quality of cured composted materials, total genomic DNA was extracted from the matrix as recommended by Bethelet et al. 1996, purified with the Wizard DNA clean up system following the manufacturing instructions (Promega, Madison, USA) and used for PCR experiments or stored at minus 20 °C. Detection of *Salmonella* spp. was performed by amplification of the target sequence by means of the primers:

SAL-3 5-TATCGCCACGTTCGGGCAA and

SAL-4 5-TCGCACCGTCAAAGGAACC (Wang et al. 1997).

PCR Conditions

A final volume of 50 µl was used consisting of 5 µl of 10x Mg-free reaction buffer (Tris HCl 100 mM, KCl 500 mM, Promega, Madison, Wis, USA), $MgCl_2$ 1.5 mM (Promega), dNTPs 0.01 mM of each (Pharmacia Biotech AB) and 75 ng of each primer (Pharmacia), *Taq* polymerase, 1 U (Promega), and 1 µl of template (Ranalli et al. 2001). PCR mixture was processed with the following thermal protocol: one cycle consisting of 5 min at 94 °C; 34 cycles of 45 s at 94 °C, 45 s at 59 °C, and 1 min at 72 °C. Five µl of the PCR products were electrophorised (100 mV, Power Pac 300, Biorad) on a 2% agarose gel (Pharmacia Biotech AB) in 0.5x TBE buffer (89 mM Tris-boric acid and 2 mM EDTA), stained in a ethidium bromide solution, visualised and photographed under UV light.

Salmonella lili (DISTAM collection, I) was maintained on nutrient agar (Oxoid, Hampshire, UK) and utilised as positive control. The presence of *Salmonella* spp. was evaluated also by traditional culturable technique adopting Bacto selenite broth (Oxoid), incubation at 37 °C for 18h, and SS agar (Difco) at 37 °C for 24h.

Results

The results of this study are structured into sections: first, data related to the process by two piles and bioindicators of the composting of two-phase centrifuged olive husks are reported, and second, the hygienic quality of the cured compost obtained.

In Fig. 2 the changes of temperature (mean values) recorded for the two static piles experiment are reported.

In the first 70 days of the composting process, the temperature in the inoculated pile B increased more quickly than in the uninoculated one, in which it remained also lower.

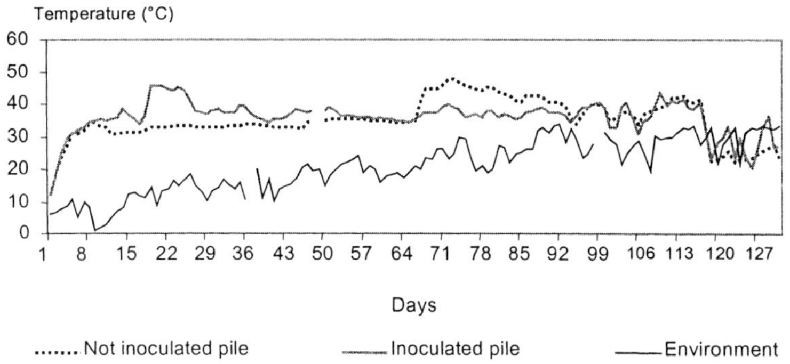

Fig. 2. Changes of process temperature in uninoculated (pile A) and inoculated (pile B) pile and environment temperature (*black line*)

In Table 2 main physical and chemical parameters determined for the two composting trials are reported.

Table 2. Physical and chemical parameters determined during the experimental trials

Parameters	Pile A uninoculated		Pile B inoculated	
	T=0	*T*=90	*T*=0	*T*=90
Moisture (%)	56	30	51	34
PH	5.9	6.6	6.3	7.1
Ash (%)	5.8	5.1	3.7	4.4
Total extr. carbon (mgC g^{-1})	45.13	37.53	55.05	39.8
Total N (%)	0.72	0.91	1.03	1.58
C/N	62.7	41.24	53.44	25.19
Phenolic carbon (mgC g^{-1})	36.42	18.60	35.32	14.82
Humification rate (%, HR)	22.1	25.3	24.5	29.7
Humification index (HI)	1.36	0.92	1.42	0.65

The results of 90 days of composting processes of crude olive husks showed a pH increase from subacidic (5.8 and 6.3) to pH 6.6 and fully neutral values in pile uninoculated and inoculated, respectively. Further, the moisture decreased pronouncedly, the total nitrogen values increase in two tests, to 1.5 times the initial content when the raw material was composted in pile B. Total extractable and phenolic carbon values show a decrease in all composting trials with the most favourable reduction in C/N ratio of 25 in pile B compost. Finally, humification parameters (HI) show positive results with decrease index (0.92 and 0.652) when uninoculated and inoculated piles of the composting process were adopted, respectively. These favourable results confirm that in our experimental conditions, fresh olive husks and raw organic residues (corn stalks) adopted as bulking agent can be submitted to composting and when idoneous technologies are employed, a composted material with a final humification index of good convenience can be obtained.

In order to verify the best bioindicator of the process, some microbiological parameters were monitored: viable microbial counts of total aerobic heterotrophic bacteria (mesophilic and thermophilic), mycetes and spore-forming bacteria are reported in Figs. 3 and 4.

On uninoculated pile A samples, total viable mesophilic bacteria counts showed some variations during the composting runs. In particular, they increased at day 10 after the start, retained a steady state until day 30, and decreased significatively at day 50. Total mesophilic aerobic spore-forming bacteria counts showed no sensitive variations during the first 30 days of composting process, but they decreased at the 50th day, followed by a significant increase at the 80th day. Mycetes and yeasts showed an unexpectedly high increase at the 30th and 30th day, followed by a subsequent decrease. Thermophilic microflora showed some variations but with no statistical significant value. These microbial changes are probably due to constant mesophilic conditions, instead of thermophilic ones, because the temperature remains low during the whole of the composting run.

In pile B samples total viable mesophilic bacteria counts showed no variations during the first 30 days of the composting run, followed by a significant decrease at the 50th day. Total mesophilic aerobic spore-forming bacteria and mycetes counts showed slight changes.

Thermophilic microflora showed some variations with a slight decrease at day 10 from the start, followed by an increase at day 30. In these cases, microflora changes are probably due to the early antagonism relationships among the microbial community of the raw materials, and the inoculum adopted, as well as the initial increase of the temperature, even if not higher than 50 °C, could be an index of a preliminary thermophilic condition supported by the microflora inoculated. However, considering all the data, the microbial cell counts in both piles (A and B) are heterogeneous and were not always related with other parameters.

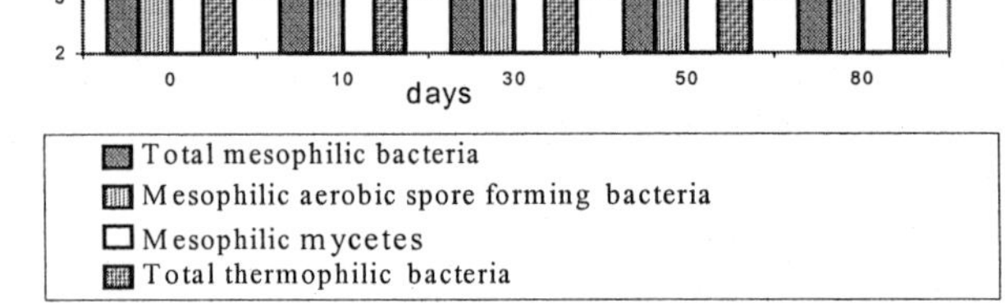

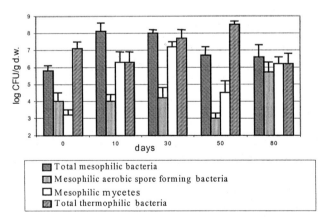

Fig. 3. Viable microbial cell counts on uninoculated pile A samples. Mean and error bars of three samples analysed separately

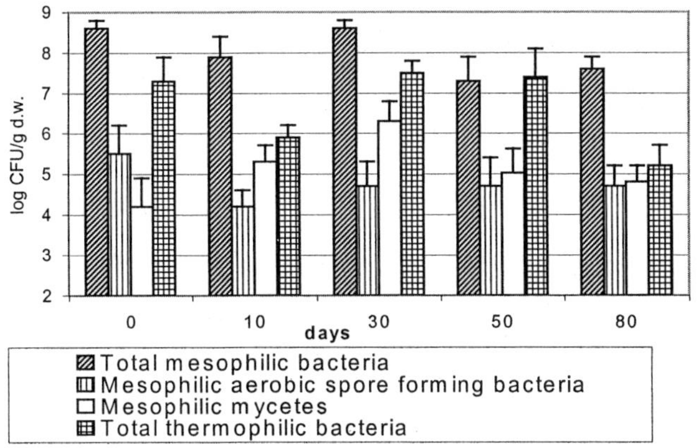

Fig. 4. Viable aerobic microbial counts on inoculated pile B samples. *Mean* and *error bars* of three samples analysed separately

The ATP content on composting samples from uninoculated pile A increased during the process up to 50th day, followed by a decrease with a final value at the 80th day of the same order as the early phase. This behaviour is directly related to that of the more important enzyme activities.

In the case of inoculated pile B, during an 80-day composting run, two peaks of ATP were observed: the first at the 10th day according to the peak of the enzyme activities (r=0.78); the second corresponds to the 50th day, as it occurred in the uninoculated pile, while the value at the 30th day is low.

In Fig. 5 the ATP content in the A and pile B composting samples are reported.

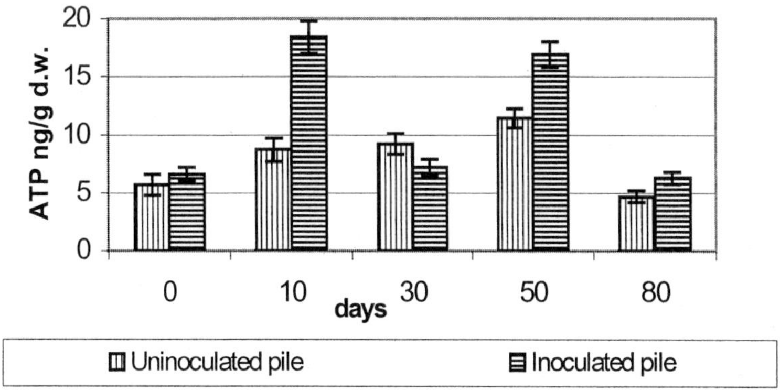

Fig. 5. ATP content in compost samples from uninoculated pile A and inoculated pile B during an 80-day composting run. Mean values calculated on five determinations and standard deviation

Enzyme activities are shown in Fig. 6 (pile A) and Fig. 7(pile B).

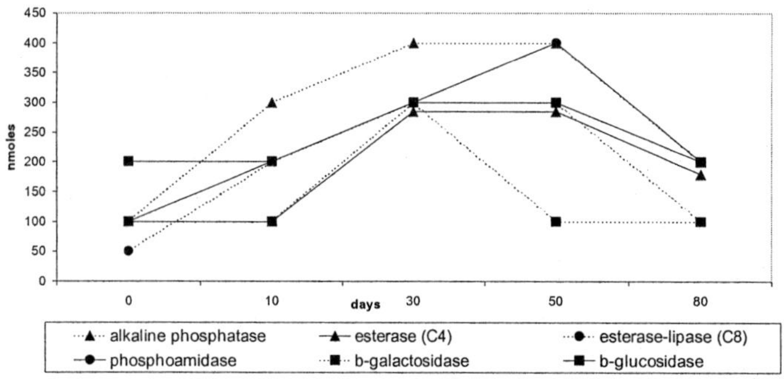

Fig. 6. Changes in enzyme activities during composting of uninoculated pile A

In pile A the highest activities were found for the phospho-hydrolases (alkaline and acid phosphatase) and for phosphoamidase enzymes; no activities of the β-glucoronidase, α-mannosidase and α-fucosidase enzymes were recorded. Of the 19 enzymes tested, remarkable changes were observed during the composting process for the following enzymes: alkaline phosphatase, esterase (C4), esterase-lipase (C8), phosphoamidase, β-galactosidase, β-glucosidase as reported in Fig. 7. Peaks of enzyme activities occurred at the intervals between 30 and 50 days after the start of the composting process, in relation with the highest temperatures recorded in the pile over the time (r=0.88) and with the mean value of ATP content (r=0.76).

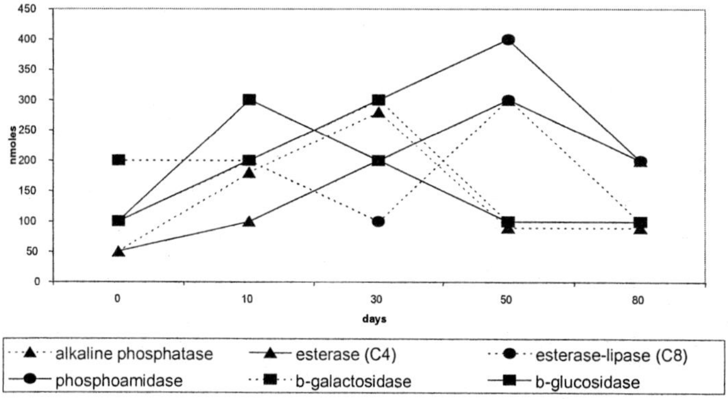

Fig. 7. Changes in enzyme activities during composting samples of inoculated pile B

In the case of the inoculated pile B, the highest activities for the phospho-hydrolase (alkaline and acid phosphatase) and phosphoamidase enzymes were registered; no activities of the β-glucoronidase, α-mannosidase and α-fucosidase enzymes were detected.

Total enzyme activities (mean of 9 over 19 tested) show a pattern similar to the temperature trend; this is the parameter commonly used to describe the trend of the composting process; moreover, the processed data confirmed this relation with a positive correlation value (r=0.89). The semiquantitative information obtained related to enzyme activities (high or low) could be useful for the characterisation of physiological and specific biodegradative conditions of microbial community in composting, as well as dynamic of microflora population over the time.

Hygienic Quality of the Cured Compost

Among potential pathogenic microorganisms in composts, according to the EC decision 1998/488/EC (7 April 1998) (Anonymous 1998) that establishes ecological criteria for the product group "soil improver", there are at present restrictions for *Salmonella* sp. (absent in 25 g) and *E coli* (< 1000 MPN g^{-1}). These limits must be verified before use and commercialisation in order to avoid any indiscriminate utilisation of non-stabilised materials. Moreover, it could represent a serious sanitary hazard for human and animal health. *Salmonella* spp. are not present in the fresh organic vegetable olive husk residues, but if other raw organic residues such as animal manure or municipal wastes are used as inoculum or to improve physical (porosity, density) and chemical composition (C/N ratio), the hygienic risk must be evaluated (Hussong et al. 1985).

In order to verify the presence of faecal bacteria, DNA extracted from cured compost samples was purified and amplified by specific primers (SAL-3 5-TATCGCCACGTTCGGGCAA and SAL-4 5-TCGCACCGTCAAAGGAACC) (Pfaller et al. 1994; Swamy et al. 1996; Wang et al. 1997) for detection of *Salmonella* spp. Positive control was prepared adding known suspensions of *Salmonella lili* to cured compost samples. Electrophoresis showed that the expected band of 275 bp was obtained only in the positive control, no bands was found in the cured compost sample from inoculated pile B (Fig. 8).

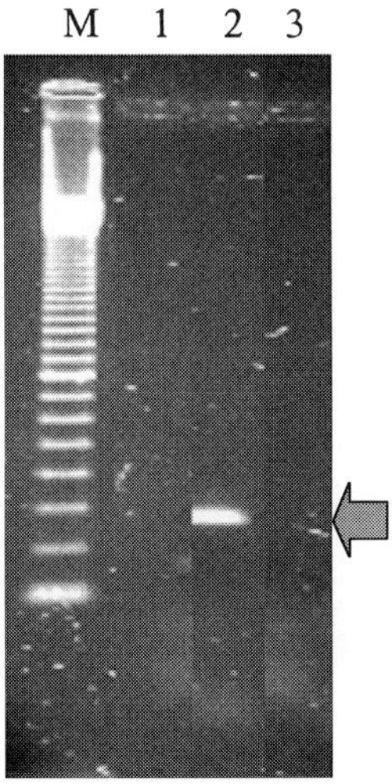

Fig. 8. Amplification of DNA target (in agarose gel 2%) extracted from cured compost samples of inoculated pile B, without and with addition of pure cultures of *Salmonella lili*. *M* Marker ladder 100 bp; *lane 1* C.n.; *lane 2* cured compost added with pure culture of *Salmonella lili*, diluted 1:100 and amplified with Sal3-Sal4 primers (specific for *Salmonella* spp.); *lane 3* cured compost diluted 1:100 amplified with Sal3-Sal4 primers

The results suggest that the procedure adopted for extraction, purification and amplification is effective for detection of *Salmonella* spp. Moreover, cured compost seems to be not polluted by *Salmonella* spp. or contaminated by a lower number below the detection limit. The data by traditional culturable technique confirmed the absence of viable cells of *Salmonella* spp. (value expressed as cfu 0 25 g^{-1} d.w.) on pile composting.

Conclusions

Microbial counts are not sensitive, it is also known that a very high number of microorganisms are not cultivable; as a consequence, by traditional culture methods it is possible to show only a lower number (10÷100 times) than is really present in the samples.

Rapid bioindicators for process monitoring were investigated, like ATP content and activity of a pool of enzymes. In fact, as bioindicators, the microbial viable cell counts are not reliable in describing the process due to their irregular behaviour, that only in the case of the peaks of thermophilic groups is related with the temperature. Both the bioindicators tested are easy to be assessed and can give information in a few h, while the cultural techniques need an incubation time of 2 or 3 days.

ATP content results appear to be a good bioindicator of the biomass, as proposed by other authors (Sciancalepore et al. 1996; Tseng et al. 1996). However, this indicator is not specific for a particular microbial group.

Enzyme activities are more sensitive indicators because they described microbial metabolism evidencing all the composting phases. In fact, there is activity enhancement for the starting up phase and the last phase when the microbial activities are more intense. Moreover, among the enzymatic hydrolytic activities, mainly those were detected that attack the molecules with a low or medium MW, that are more easily degradable by microorganisms. The biomolecular method to detect indicators of faecal pollution showed that the composted samples obtained from the pile enriched with animal manure are hygienically safe concerning the presence of faecal contaminants (*Salmonella* spp.) due to the inoculum.

Acknowledgements. Research supported by the EC fund FAIR5-CT97 3620.

References

Anonymous (1998) Directive 98/488 on eco-label soil improver. 7 April 1998

Bethelet M, Whyte LG, Greer CW (1996) Rapid, direct extraction of DNA from soils for PCR analysis using polyvinylpolypyrrolidone spin columns. FEMS Microbiol Lett 138: 17-22

DI.VAPRA-IPLA (1992) Metodi di analisi dei compost. Collana Ambiente, Torino

Epstein E (1997) The science of composting. Technomic Publishing, Lancaster, Pennsylvania

Hermann RF, Shann JF (1993) Enzyme activities as indicators of municipal solid waste compost maturity. Compost Sci Util 4: 54-63

Humble W, King A, Phillips I (1977) APIZYM: a simple rapid system for the detection of bacterial enzymes. J Clin Pathol 30: 275-277

Hussong D, Burge WD, Enkiri NK (1985) Occurrence, growth and suppression of Salmonellae in composted sewage sludge. Appl Environ Microbiol 50 (4): 887-893

Malik M, Kain J, Pettigrew C, Ogram A (1994) Purification and molecular analysis of microbial DNA from compost. J Microbiol Meth ods 20: 183-196

Pfaller SL, Vesper SJ, Moreno H (1994). The use of PCR to detect a pathogen in compost. Compost Sci Util 2: 48-54

Ranalli G, Bottura G, Taddei P, Garavani M, Marchetti R, Sorlini C (2001) Composting of solid and sludge residues from agricultural and food industries. Bioindicators of monitoring and compost maturity. J Environ Sci Health, 36 (4): 415-436

SAS Institute Inc (1989) SAS/STAT User's Guide, Version 6, 4th edn. SAS Institute, Cary NC

Sciancalepore V, Colangelo M, Sorlini C, Ranalli G (1996) Composting of effluent from a new two-phase centrifuge olive mill. Microbial characterization of the compost. Toxicol Environ Chem 55: 145-158

Swamy SC, Barnhart HM, Lee MD, Dreesen DW (1996). Virulence determinants invA and spvC in salmonellae isolated from poultry products, wastewater, and human sources. Appl Environ Microbiol 62: 3768-3771

Tseng DY, Chalmers JJ, Tuovinen OH (1996) ATP measurement in compost. Compost Sci Util 4 (3): 6-17

Wang RF, Cao WW, Cerniglia CE (1997) A universal protocol for PCR detection of 13 species of foodborne pathogens in foods. J Appl Microbiol 83: 727-736

Organic Acids as a Decisive Limitation to Process Dynamics During Composting of Organic Matter

T. Reinhardt [1]

Abstract. Composting in the presence of oxygen is generally defined to be an aerobic and thermophilic microbiological process. A delay in the development of temperature on a level of 40 - 50°C during the initial stages of the process is frequently reported and explained as a change in the bacterial population from mesophilic to thermophilic. In the course of the experimental trials presented here this effect is accompanied by a significant drop in pH due to the accumulation of organic acids. Even acetic acid can be found at concentrations hindering bacterial metabolism nonspecifically and completely, thermophilic populations seem to be affected in particular. Since acetic acid is a common product of various bacterial metabolic pathways, numerous explanations for its accumulation can be given. Recent experimental results support the anaerobic formation in terms of the mixed-acid fermentation as well as strictly anaerobic fermentations by members of the genus *Clostridia* i.e.; anaerobic conditions are likely to develop inside even very small particles due to limited transport capacities for oxygen from gaseous to liquid phases. Another reasonable explanation is given by the so called *bacterial crabtree-effect* as a response to excessive substrate supply.

Introduction

Spontaneous self-heating of organic matter in terms of composting and the relation of temperature and turnover is one of the classic and most frequently studied characteristics of this process (Finstein et al. 1986; Jäger 1997; Jeris and regan 1973; MacGregor et al. 1981; Miller and McCauley 1989, Niese 1969; Sikora and Sowers 1985; Suler and Finstein 1977; Waksman et al., 1939). Since the emission of excessive heat as the very cause for this phenomenon is predominantly a feature of the aerobic, oxygen-consuming microbial metabolism, composting is generaly defined to be an aerobic process. This rough definition is not taking into account the limitations of oxygen transport from the free air space into the heterogeneous solid particles of the organic matter, making this process more likely to be an anaerobic/aerobic co-process than a solely aerobic process (Hamelers 1993; Reinhardt 1999).

Development of temperature during the intitial stages of the composting process is usually not straight - up to the limiting temperature of microbial activity at about 75 - 80° C (Miller and McCauley 1989) - but is often characterized by a

[1]Dr.-v.-Brentanostr. 12, D-69518 Abtsteinach, Germany

H. Insam, N. Riddech, S. Klammer (Eds.)
Microbiology of Composting

retardation at a temperature level at about 40 - 50° C. An exemplary contribution to this phenomenon is given in a classic contribution by Carlyle and Newman (1941); the development of temperature, change in temperature per hour and the CO_2-formation-rate towards time in their experimental setup for self-heating is represented in Fig. 1.

It is well known, that during the initial stages of the composting process also a significant drop in pH, due to the microbial formation of organic acids may occur (Golueke et al. 1977; Haug 1980; Jäger 1979, 1997; Krogmann 1994; Jourdan 1988; Lechner 1991; Rüprich 1990), and the detrimental effects of organic acids on microbial activity have also been intensively studied (Russel et al. 1998; Cherrington et al. 1991). Although it seems to be obvious that formation and accumulation of organic acids on the one hand side and retardation in temperature development on the other hand side are interdependent, the latter phenomenon is even nowadays discussed solely in terms of the substitution of a mesophilic microbial population by a thermophilic population due to elevated temperatures (Smårs et al. 1999).

In the present work an experimental attempt is presented to investigate the dependence of microbial activity and formation / accumulation of acetic acid as well as the effect of this on the development of temperature.

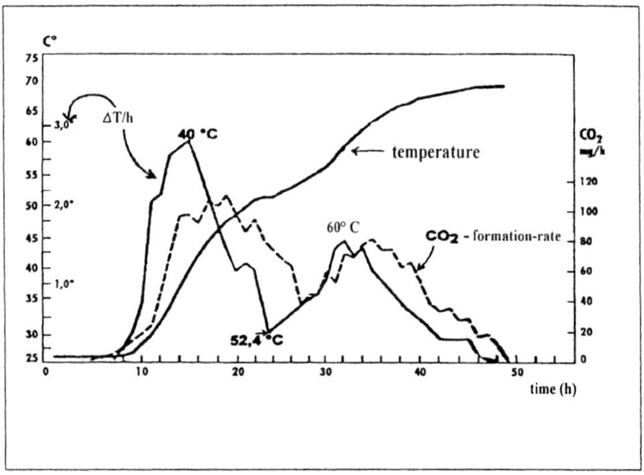

Fig. 1. Typical plot of temperature-development, change in temperature and CO_2-formation-rate for composting of organic matter (Carlyle and Newman 1941)

Materials and Methods

A laboratory reactor was designed to simulate large scale aerated piles at a waste treatment plant in the Bavarian town of Quarzbichl in Germany. The well insulated chambers could take up about 650 liters of organic matter at a heap-hight of about 180 cm. Compressed air was used to aerate the compost at a constant rate of about 4.5 m^3/ton of waste (fresh matter) per hour. Organic matter was derived also from the plant in Quarzbichl; a fraction < 40 mm after combined mechanical/biological pretreatment of residual household waste was used. Water content of the input material was adjusted to about 48%.

Samples were taken from the initial input material and from different levels of the central material column within the reactor every 24 h; solid samples were analyzed for dry solids (DS) and loss on ignition. A portion of 5 g DS of each solid sample was eluted in 50 ml of a 0.01 m $CaCl_2$-solution; pH-value was measured directly in the suspension and dilution-series were prepared. Microbial numbers were determined after spreading on Caso-Agar (Merck) and incubation at 35 and 50°C at aerobic and anaerobic conditions (Anaerocult P System by Merck). Clarified liquid samples after sedimentation of insoluble solids by centrifuge were further analyzed by GC-MS (Perkin ElmerR).

Gas samples from the air-supply and the exhaust-air and from 10 probes distributed over the height of the reactor were analysed for O_2, CO_2 and CH_4 at intervals of 1 h. Specific rates for O_2-consumption, CO_2 as well as CH_4-emission were calculated from the differences in the concentration at the different probe-positions.Temperature in the exhaust-air was recorded continously. Images of the temperature-distribution over the entire heap of organic matter were taken by an infra-red thermo-scan-system (AVIO TVS 2200) every 24 h.

From a series of 5 experimental trials the results of 2 representative examples are presented here.

Results

Initial characteristics of organic solid matter

The organic content (by loss on ignition) of the input material used for different experimental series ranged from 65 to 72 % (related to DS). The initial pH-value ranged from 6.2 to 6.4; the intial concentration of acetic acid ranged from 38 to 43 mg/g DS! Other organic acids like propionic and butyric acid were only determined in concentrations smaller than 1 mg/g DS. Initial microbial numbers ranged from 2.4×10^8 - 4.2×10^9 (35°C, aerobic), 3.1×10^8 - 4.6×10^9 (35°C, anaerobic), 1.4 - 8.5×10^6 (50°C, aerobic) and 2.2 - 6.3×10^6 (50°C, anaerobic).

Changes in the course of the composting process

pH-value

Starting at about pH 6.2 in both trials, in exp. 1 the pH-value increased to a level of about pH 8 throughout the entire heap after about 108 h (Fig. 2), whereas in exp. 2 only in the lower sections of the heap (that is near to the inlet of air) the same behavior could be observed; in the middle and upper sections in comparison pH-value remained slightly acidic, changing to neutral values towards the end of the experiment after 288 h (Fig. 3).

Acetic acid

Initial concentrations of acetic acid were about 40 mg/g DS in both experiments. After a slight increase at the beginning of the process in exp. 1 the concentration dropped to 0 mg/g DS almost throughout the entire heap after about 108 h (Fig. 4). In exp. 2 only in the lower sections of the heap a distinct decrease to a level of about 10 mg/g DS could be observed; in the middle and upper sections in comparison no definit change in the concentration of acetic acid could be detected until the end of the experiment after 288 h (Fig. 5).

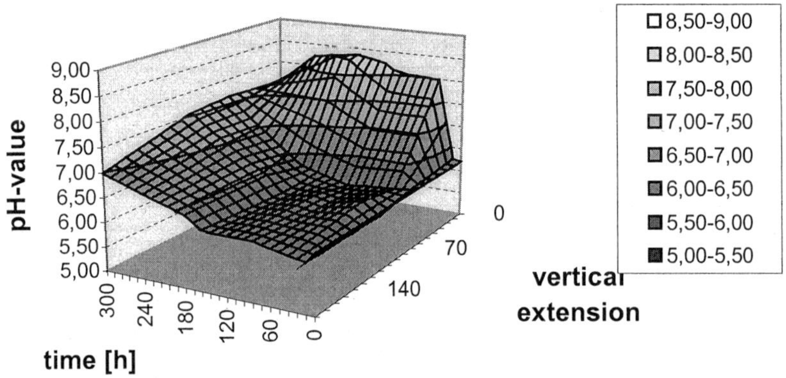

Fig. 2. Experiment 1; development of pH-value

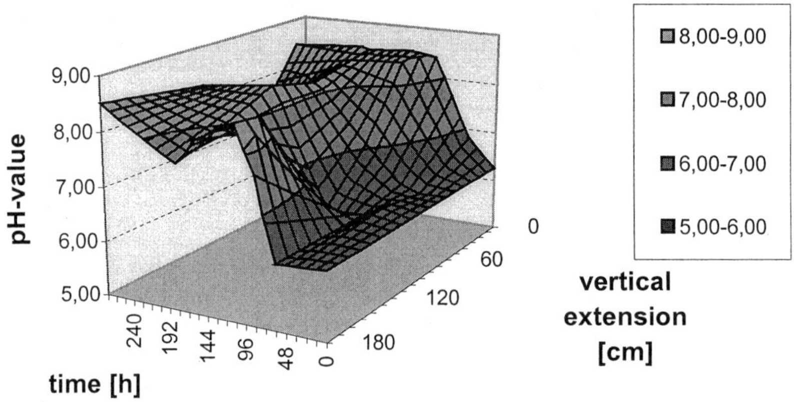

Fig. 3. Experiment 2; development of pH-value

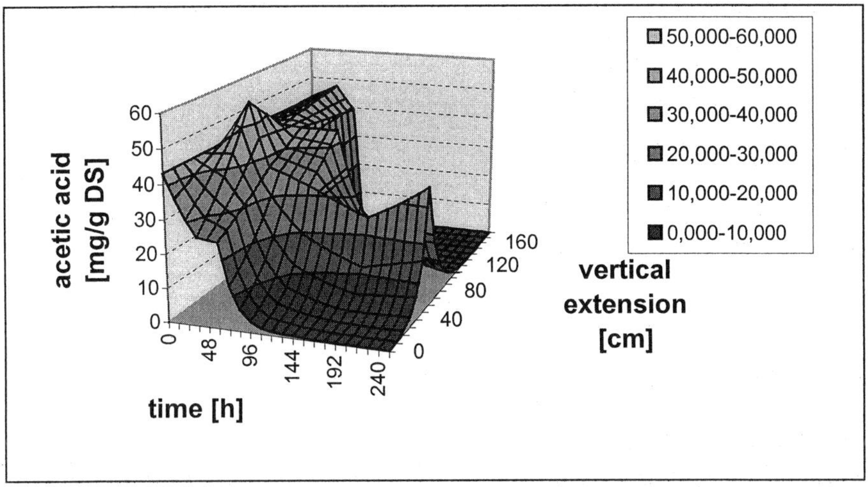

Fig. 4. Experiment 1; development of acetic acid concentration

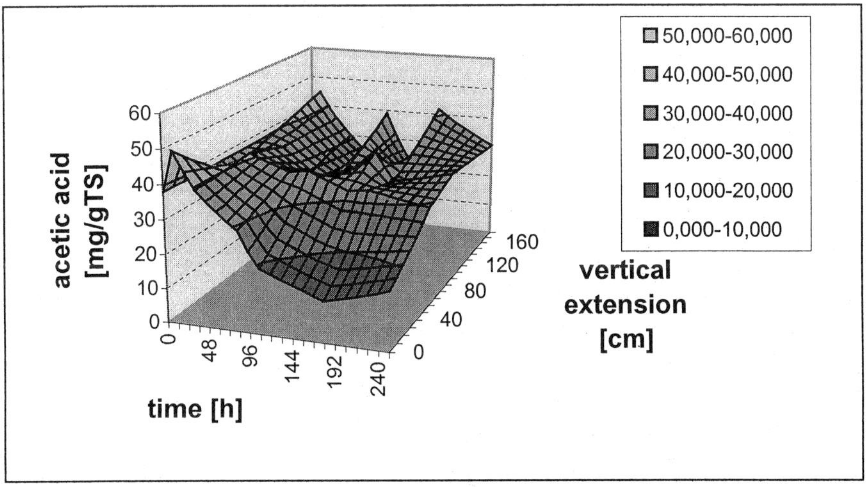

Fig. 5. Experiment 2; development of acetic acid concentration

Temperature

While the temperature of the exhaust air in exp. 1 showed an almost straight development up to an inhibitory level (that is >60° C) after about 72 h, the plot for exp. 2 shows the typical - and in that specific case drastic - retardation of temperature increase when reaching about 45° C after about 24 h (Fig. 6).

Distribution of temperature throughout the entire heap as recorded by infra-red thermo-scan-system correspond to the plots of the exhaust air temperature respectively (coloured images not presented here!). The graph of exp. 1 is characterized by a steady increase in temperature during the first 72 h, a steep gradient in temperature in the vertical dimension of the heap with a maximum ΔT of about 40° (temperature of the inlet air about 16°C) in the lower sections of the heap (from 0 - 30 cm) after about 108 h, an extended hot area (with respect to the temporal as well as spatial extension) with a temperature slightly above 60° C and a hot spot located in the upper sections of the heap showing a peak temperature of about 80° C after about 108 h. In contrast to this the graph of exp. 2 showed an uneven development of temperature with a first peak at about 45° C throughout a wide section of the heap after about 24 h, followed by a temporary decline in temperature. A second peak is reached after about 120 h on a level of about 50° C. Temperature gradients in this experiment are less distinct, but still give a maximum ΔT of about 30 ° (temperature of the inlet air about 20° C) in the lower sections of the heap (from 0 - 30 cm) after about 120 h. An extended hot area (with respect to the temporal as well as spatial extension) with a temperature at about 50° C was recorded, but no defined hot spot could be detected.

On-line gas analysis

Peak-uptake-rates of up to 8 mg O_2/g DS × h were achieved in several areas in the course of exp. 1 (Fig. 7), whereas in exp. 2 a peak-uptake-rate of about 9 mg O_2/g DS × h was achieved in a single, extended area in the lower sections of the heap (Fig. 8). The concentration of oxygen in the free air space was never below 14 % in both experiments.

Data for CO_2-emission (not given here) correspond to the results for oxygen-uptake. CH_4 was not detected in significant amounts.

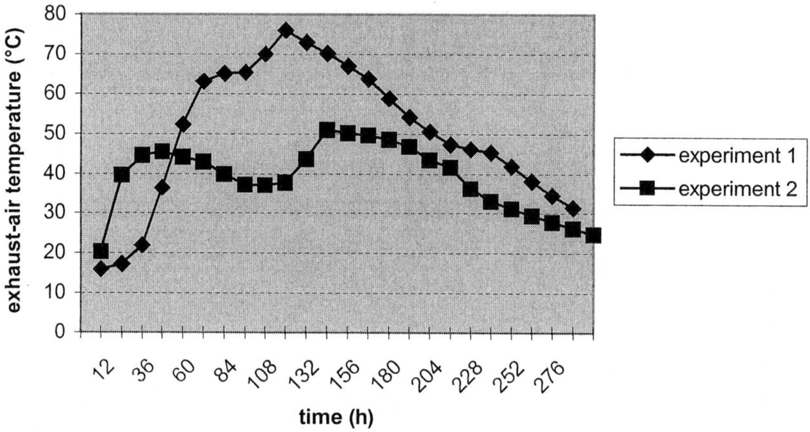

Fig. 6. Development of exhaust-air temperature

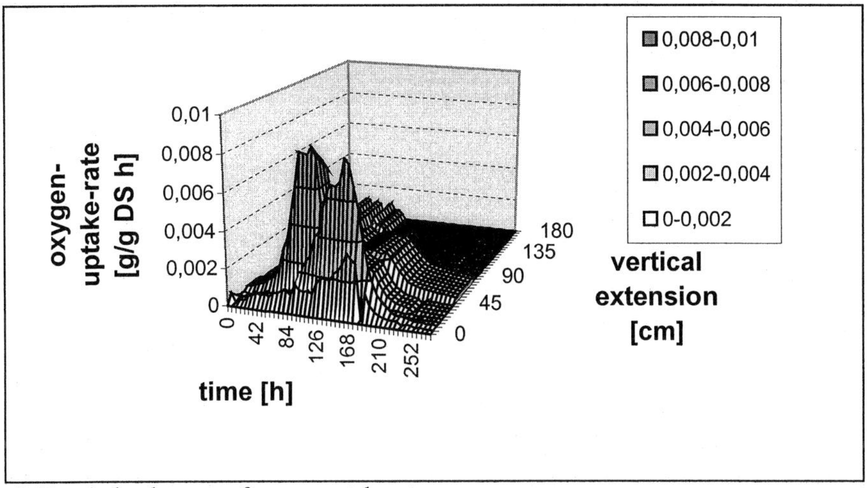

Fig. 7. Exp. 1; development of oxygen-uptake-rate

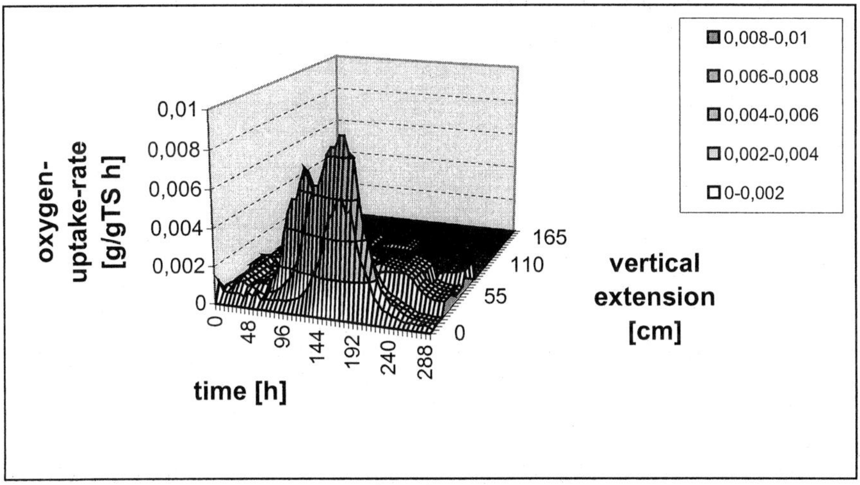

Fig. 8. Experiment 2; development of oxygen-uptake-rate

Microbial numbers

Since numbers for viable counts at aerobic and anaerobic incubation conditions were almost identical, only data for the former condition are given here.

The graphs for both experiments show quite similar characteristics, starting at similar numbers, with a drastic decrease for the mesophilic bacteria during the initial stages of the process - except the lower sections of the heap - and a slight increase of the thermophilic numbers in the lower sections of the heap after about 48 h in exp. 1 (Fig. 9) and 84 h in exp. 2 (Fig. 10) respectively.

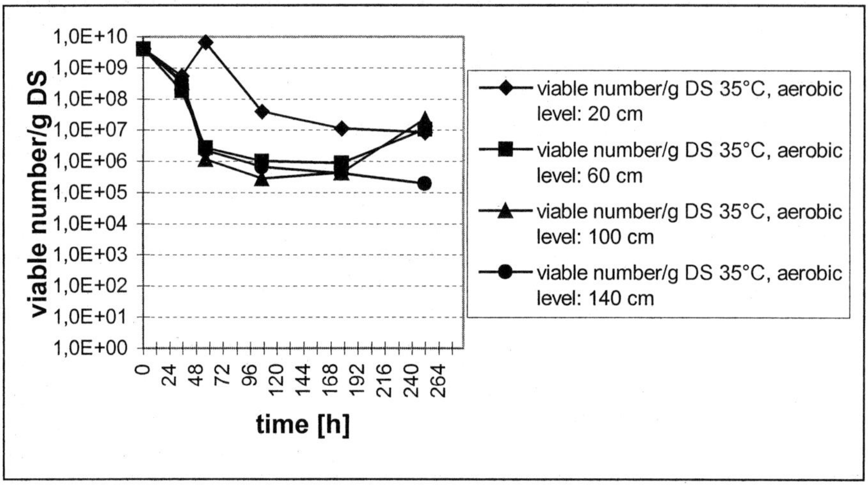

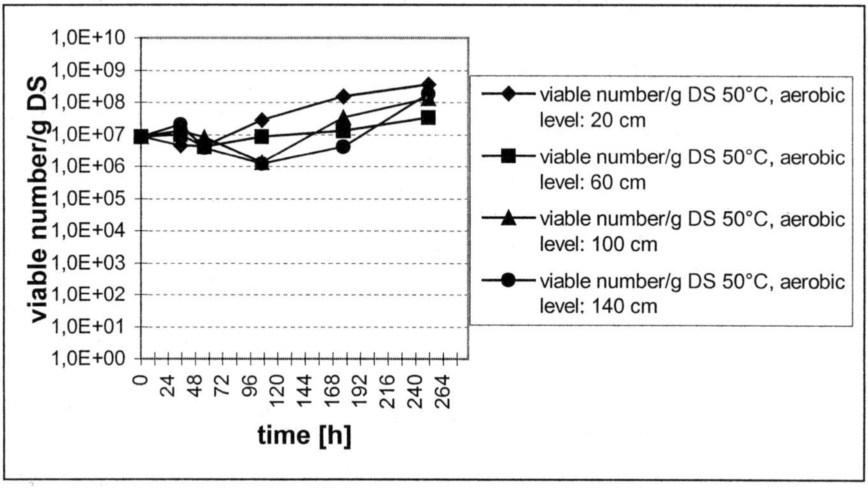

Fig. 9. Experiment 1; microbial numbers (mesophilic and thermophilic, aerob)

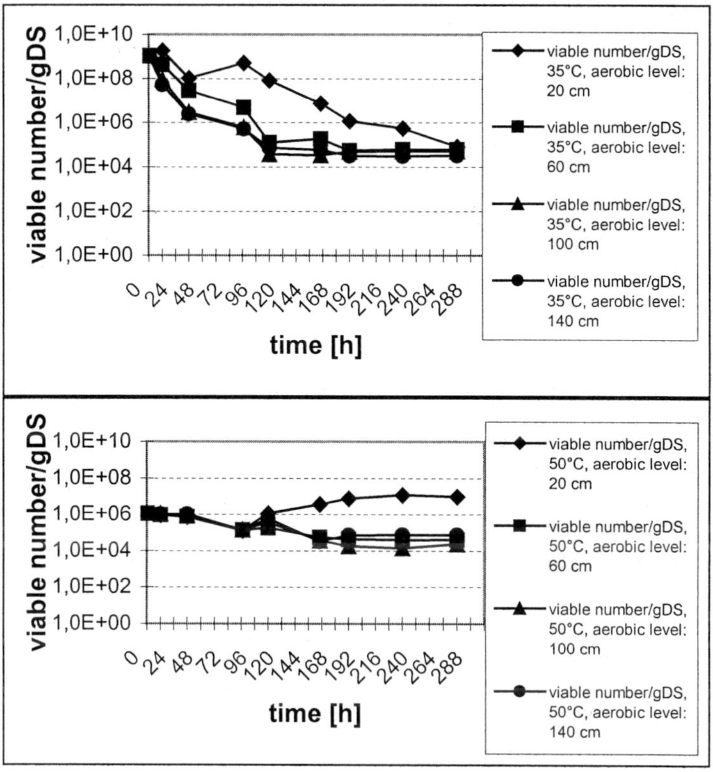

Fig. 10. Experiment 2; microbial numbers (mesophilic and thermophilic, aerob)

Discussion and Conclusions

With respect to the data presented an intimate correlation of the development of the pH-value, the concentration of acetic acid, the temperature and the microbial metabolic activity could be demonstrated.

In both experiments microbial activity - measured as oxygen-uptake-rate - establishes uniformaly throughout the entire heap even in the presence of very high concentrations of acetic acid or even accompanied by a further increase in acetic acid after start up of the process. Due to the emitted heat in the course of the developing microbial activity a rise in temperature can be detected throughout the heap after the start, exceeding optimal values to mesophilic microorganisms during 48 h (exp. 1) and 24 h (exp. 2). The development of intensive microbial activity - measured as oxygen-uptake-rate - was detected only in sections of the heap where acetic acid concentrations had already been diminished; that is suggested to be due to the metabolism of mesophilic bacteria. In exp. 1, maximum microbial activity was detected at temperature levels between 30 - 55° C, in exp. 2 a maximum could also be shown at mesophilic temperature levels. Spots of high microbial activity were more likely to establish in the cooler areas in the lower sections of the heap - due to the cooling properties of the constant air-stream of the forced aeration; also acetic acid concentrations are more likely to be depleted in this cooler sections. Temperatures exceeding 50° C could only be detected at extensively deminished concentrations of acetic acid in exp. 1. Temperatures up to 80° C are detected in exp. 1 but are not accompanied by detectable microbial activity.

Whereas in both experiments viable numbers for mesophilic bacteria are diminished due to elevated temperature in the middle and upper sections of the heap, only in lower sections a temporary increase in mesophilic numbers after about 36 h in both experiments can be shown. Numbers for thermophilic bacteria remain almost constant for the middle and upper sections of the heap, but also show a distinct increase in the lower section after 48 h (exp. 1) and 84 h (exp. 2) respectively. They are exceeding mesophilic numbers after about 108 h (exp. 1) and 180 h (exp. 2); in both cases after substantial decrease of acetic acid concentration.

Although process conditions for both experimental setups had been almost identical the two experiments presented here show different characteristics; no clear explanation for this differences can be given. It is assumed that a short delay in the establishment of the microbial activity at the beginning of exp. 1, which may be due to lower temperatures of the input material and the environment (exp. 1: 16° C; exp. 2: 20.5° C), is in some way responsible for that.

The present data suggest that a delay in temperature development frequently reported for composting processes is not solely a question of temperature and due to a change from mesophilic to thermophilic microbial populations, but is predominatly a question of acetic acid and the interdependencies of acetic acid concentration and microbial activity. Thus, it is worth to discuss the cause for the formation of acetic acid in the course of the composting process:

Usually there are two ways for the formation of acetic acid discussed in terms of composting: anaerobic formation by strictly anaerobic (i.e. by members of the genus *Clostridium*) or facultatively anaerobic (mixed acid fermentation) bacteria (Golueke 1977; Lechner 1991; Miller and McCauley 1988; Rüprich 1990) as well as aerobic formation by incomplete oxidation (Jäger 1979) of either ethanol or glucose by members of the genus *Acetobacter* or *Gluconobacter* respectively. No valid evidences for one or the other mechanism could be recieved from the experimental data presented here, but since no considerable amounts of ethanol, as a prerequesite for acetic acid formation in terms of incomplete oxidation as reported by Jäger (1979), and RQ-values > 1 for the represented experiments at the start of the process make an anaerobic mechanism - due to a limited oxygen transport from the gaseous phase into the liquid phase on the amorphous waste-particle (Hamelers 1993; Reinhardt 1999) - more likely.

At this point an alternative mechanism for the formation of acetic acid - under completely aerobic conditions - shall be mentioned. A vast number of bacteria produce acetic acid even when grown aerobicaly in the presence of easily degradable substrates - predominantly glucose. In this terms the formation of acetic acid is correlated to the growth rate of the population, starting when a specific critical growth rate is exceeded and increasing corresponding to the further increase of the growth rate up to the specific maximum. Since this phenomenon in some way resembles the so called *crabtree-effect* - an aerobic formation of ethanol by yeast - it is sometimes called the *bacterial crabtree-effect* (Doelle et al., 1982). Different explanations for this phenomenon have been given, including the saturation of the respiratory system (Chang et al., 1999) and a limited capacity of the TCA-cycle (Majewski et al., 1989). Recent studies suggest that acetate is excreted due to the pyruvate flux from the phosphoenolpyruvate-dependent phosphotransferase system (glucose uptake) since additional glucose is essential to balance the needs for metabolic precursors and NADPH (Chang et al., 1999).

Even most of these studies have been carried out on *E. coli*, a species not really relevant in terms of composting, this way of aerobic formation of acetate and related phenomena seem to be widespread among bacteria (Gottschalk, 1986). Since the phenomenon of the delay in temperature development is more likely to be observed on organic matter containing large quantities of easily degradable organic compounds - i.e. source separated organic waste – (Golueke, 1977), this gives some support to acetic acid formation in terms of composting due to the *bacterial crabtree effect.*

Since the formation and accumulation of acetic acid is a decisive limitation to process dynamics during composting of organic matter, in terms of process management sufficient strategies to avoid this limitations have to be developed - independent which mechanism the acetic acid derives from.

References

Carlyle RE, Newman AC (1941) Microbial thermogenesis in the decomposition of plant materials, J Bacteriol 41: 6-13

Chang DE, Shin S, Rhee JS, Pan JG (1999) Acetate metabolism in a *pta* mutant of *E. coli* W3110: Importance of maintaining acetyl coenzyme A flux for growth and survival. J Bacteriol 181: 6656-6663

Cherrington CA, Hinton M, Mead GC, Chopra I (1991) Organic acids: Chemistry; antibacterial activity and practical application. Adv Microb Physiol 32: 87-108

Doelle HW, Ewings NW, Hollywood NW (1982) Regulation of glucose metabolism in bacterial systems. Adv Biochem Eng 23: 1-35

Finstein MS, Miller FC, Strom PF (1986) Waste treatment composting as a controled system. In: Rehm HJ, Reed R (eds) Biotechnology, vol. 8, Microbial Degradations, VCH-Verlag, Weinheim, pp 363-398

Golueke CG (1977) Biological reclamation of solid waste. Rodale Press, Emmaus, Pennsylvania

Gottschalk G (1986) Bacterial metabolism. Springer, New York, Berlin, Heidelberg, Tokio

Hamelers HVM (1993) A theoretical model of composting kinetics. In: Hoitink HJ, Keener HM (eds) Science and Engineering of Composting, Columbus, Ohio, pp 36-58

Haug RT (1980) Compost Engineering (Principles and Practice). Ann Arbor, Michigan

Jager J (1979) About the chemical ekology of biological waste treatment processes. Dissertation, Naturwissenschaftliche Fakultät, Ruprecht-Karl-Universität Heidelberg

Jäger T (1997) Investigations on composting in lab-scale plants: material balance, microbiology and degradation of cellulose (in german). Dissertation, Department 10 (Biology), Technical University of Darmstadt, Germany

Jeris JS, Regan RW (1973) Controling environmental parameters for optimum composting. Compost Science, Jan./Feb:10-15

Jourdan B (1988) About the determination of maturity in compost derived from solid household-waste and mixtures of solid waste and sludge (in german). Stuttgarter Berichte zur Abfallwirtschaft 30, University of Stuttgart, Germany

Krogmann U (1994) Composting - basic evaluations for collection and treatment of organic wastes of various composition (in german). Hamburger Berichte 7, Economia, Bonn

Lechner P (1991) Recent experimental data relative to the Optimization of the composting process (in german). In: Umsetzung neuer Abfallwirtschaftskonzepte - Vermeidung und Recycling von Hausmüll, Bio- und Gewerbeabfällen - , Labor für Siedlungswasserwirtschaft der FH Münster, Band 3, Eigenverlag

MacGregor ST, Miller FC, Psarianos KM, Finstein MS (1981) Composting process control based on interaction between microbial heat output and temperature. Appl Env Microbiol 41: 1321-1330

Majewski RA, Domach MM (1989) Simple constrained-optimization view of acetate overflow in *E. coli*. Biotechn Bioengin 35: 732-738

Miller FC, McCauley BJ (1988) Odours arising from mushroom composting: a review. In: Austr J Exp Agric 28: 553-560

Miller FC, McCauley BJ (1989) Substrate usage and ddours in mushroom composting. Austr J Exp Agric 29: 119-124

Miller FG (1993) Composting as a process based on the control af ecologically selective factors. In: Metting FB (ed) Soil Microbial Ecology, Dekker, New York, pp 515-544

Niese G (1969) Determination of microbial activity in waste and waste-compost by measuring the consumption of oxygen and the emission of heat (in german). Habilitationthesis, Institut for Agricultural Microbiology, J. Liebig-Universität, Giessen, Germany

Reinhardt T (1999) About biological processes in three-phase-systems am: Exemplary studies on composting of residual solid waste with special attention on anaerobic coprocesses. Dissertation, Techn. Univ. Darmstadt, Germany, ISBN 3-923518-12-8,

Rüprich A (1990) Composting - process control and microorganisms. Studienreihe ABFALL-NOW, Band 5, Stuttgart, Germany

Russell JB, Diez-Gonzalez F (1998) The effect of fermentation acids on bacterial growth. Adv Microb Physiol 39: 205-234

Sikora LJ, Sowers MA (1985) Effect of temperature control on the composting process. J Env Qual 14: 434-439

Smårs S, Beck-Friis B, Jönsson H, Kirchmann H (1999) Influence of temperature during the initial phase of composting on biological activity and emission of ammonia and dinitrogen oxide. In: Bidlingmaier W, de Bertoldi M, Diaz LF, Papadimitriou EK (eds.) Organic Recovery and Biological Treatment, Rhombos, Berlin, pp 119-122

Suler DJ, Finstein MS (1977) Effect of temperature, aeration, and moisture on CO_2 formation in bench-scale, continously thermophilic composting of solid wastes. Appl Env Microbiol 33: 345-350

Waksman SA, Cordon TC, Hulpoi N (1939) Influence of temperature upon the microbiological population and decomposition processe in composting of stable manure. Soil Sci 47: 83-113

Effects of Interrupted Air Supply on the Composting Process – Composition of Volatile Organic Acids

M. Robertsson[1]

Abstract. The fate of potentially phytotoxic lactic acid, acetic acid, propionic acid and butyric acid during composting has been studied in self–heating bench scale bioreactors. To investigate the impact of air supply on these acids, the aeration was cut off the end of the mesophilic phase, at maximum temperature and in the cooling phase. Aeration was restored after 17 days and the control treatment was kept aerobic through out the whole experiment. Data on mass balances, temperature, pH, oxygen and carbon dioxide in the compost mixture are presented together with levels of volatile organic acids.

Introduction

Composting is an excellent way to utilise organic waste. The use of compost increases the content of soil organic matter and enables recycling of nutrients. However, the use of compost is not always beneficial to plants. Brinton (1998) analysed 712 compost samples from across the USA and found that 15% of the samples exceeded 10 000 ppm volatile organic acids (VOA). DeVleeschauwer et al. (1981) investigated levels of organic acids in fresh and mature compost and the effect on cress seed germination. Zucconi et al. (1981 a, b) evaluated the effects of immature compost on plants, and proposed the use of a cress seed germination index (Zucconi et al. 1985). The terms stability and maturity often appear in the literature. According to Epstein (1997), stability is a stage in the decomposition of organic matter and is a function of biological activity. Maturity is an organo–chemical condition of the compost which indicates the presence or lack of phytotoxic organic acids. Methods for determining compost maturity are reviewed by Morel et al. (1985), Jiménez and Garcia (1989), Inbar et al. (1990), Mathur et al. (1993) and Barberis and Nappi (1996).

The effects of organic acids on barley roots were studied by Lee (1977), who concluded that the leakage of ions from roots exposed to aromatic- and carboxylic acids was a result of a decrease in pH of the cytoplasm. The lipophilic nature of the respective compound determines the ability to enter through the lipid barrier of the plasma membrane. Low external pH increases the proportion of the undissociated form of the acid that can move into the cell. Once in the cytoplasm, with

[1]Department of Crop Science, Swedish University of Agricultural Sciences, P.O. Box 44, SE-230 53 Alnarp, Sweden

H. Insam, N. Riddech, S. Klammer (Eds.)
Microbiology of Composting

pH approximately 7, the weak acid dissociates and the pH can decrease so that uptake and retention of ions will be inhibited.

Potentially phytotoxic phenolic compounds are formed as a result of the degradation of lignin. However, the content of phenolic substances in the compost decreases as it takes part in the formation of humic substances (Sánchez–Monedero et al. 1999). The content of phenolic acids during composting of domestic waste and its effect on germination was studied by Lilja et al. (1996). Phytotoxic phenolic substances and long-chain fatty acids in compost with sawdust were examined by Tsuchida et al. (1984). Ortega et al. (1996) describe an example of phytotoxicity from naturally occurring phenolic compounds in bark from cork oak, when used as substrate without previous composting.

The study presented in this paper is focused on the potentially phytotoxic volatile organic acids (VOA) that can be accumulated in the compost during oxygen deficiency. The air supply to the bioreactors was turned off in three specific phases of the composting process and was resumed 7 days after the temperature had stabilised near ambient in the control treatment.

Material and methods

The composting process was studied under laboratory conditions in 12 bench scale bioreactors consisting of 3 l Dewar flasks (Isotherm); see Robertsson et al. (2001) for a more detailed description of the system. When aeration was running, 100 ml/min was supplied at the bottom of each bioreactor. No turning or mixing of the compost material occurred during the studied period. The process was totally self–heating, with no external heating or cooling included.

Material

Each bioreactor was loaded with 800 g raw material with the following composition (fresh weight, w/w percent): potato 36%, white cabbage 21%, carrot 21%, pine sawdust 11%, barley straw 11%. The same materials have been used in previous studies (Robertsson et al. 2001). All constituents were ground (screen size 3 mm). Unwashed entire vegetables were used. The water content of the raw material was 68% (w/w) and C/N was 57 at the beginning of the composting process.

Experimental layout

The first experiment was performed in three bioreactors with no repetitions. One of them was kept aerobic during the whole period. Air supply was turned off days 3–21 in the second bioreactor and days 6–21 in the last bioreactor. The total runtime was 30 days.

In a second experiment, the 12 bioreactors were divided into four treatments, each applied to 3 bioreactors. The air pump was turned off in three specific phases of the composting process. The first was day 3 when the temperature had reached mesophilic maximum temperature, 43, 45 and 46°C in the respective bioreactors. The second was day 4, just as the bioreactors had reached thermophilic maximum temperature, 62, 63 and 64°C. In the third treatment, the air supply was turned off at the cooling phase day 5, when the temperature had dropped to 47, 49 and 56°C. Aeration was resumed on day 17 in all treatments, except for the control that was kept aerobic the whole time. The experiment lasted for 31 days.

Measurements

Temperature was measured in the middle of each bioreactor with a thermocouple type T. Temperature data were sampled every 30 seconds and a mean value was stored at 6-minute intervals on a logger (Intab, AAC–2). Gas samples were taken manually at intervals for analysis of O_2, CO_2 and H_2S. Oxygen was measured with the galvanic cell principle and CO_2 by IR absorption with a gas analyser (Geo-technical Instruments, GA–94). The gas samples were pumped into the analyser from the air outlet. pH was determined in non–filtered extract of 5 g (fresh weight) compost sample from the top layer and 45 ml distilled water, after 30 min of agitation at room temperature. Analyses of aliphatic carboxylic acids were done with HPLC on the extract from the pH measurement after centrifugal filtration (Durapore 0.22 µm, Millipore). The HPLC analyses were performed under the following conditions: The pump (Waters 6000A) delivered 5 mN H_2SO_4, the flow was 0.7 mL/min. Sample (20 µL) was injected into a CAR–H column (Sarasep) at 80°C. Peaks were detected with a refractory index detector (Varian RI–4) operating at 50°C.

Results

Temperature

The course of temperature (Figs. 1–5) started with a lag phase for 2–3 days. After that, the temperature rose quickly to a plateau phase at the mesophilic maximum temperature 43–47°C. The thermophilic microbial population then took over and the temperature increased rapidly to maximum 62–63°C in the aerobic control treatment (Fig. 1.). When air supply was cut on day three at mesophilic maximum temperature, rapid cooling occurred approaching room temperature (the undermost serrated curve) (Fig. 2.). When the aeration was resumed on day 17, the temperature rose very quickly, reaching 55–57°C without a mesophilic plateau phase. After that, the temperature decreased but remained well above room temperature.

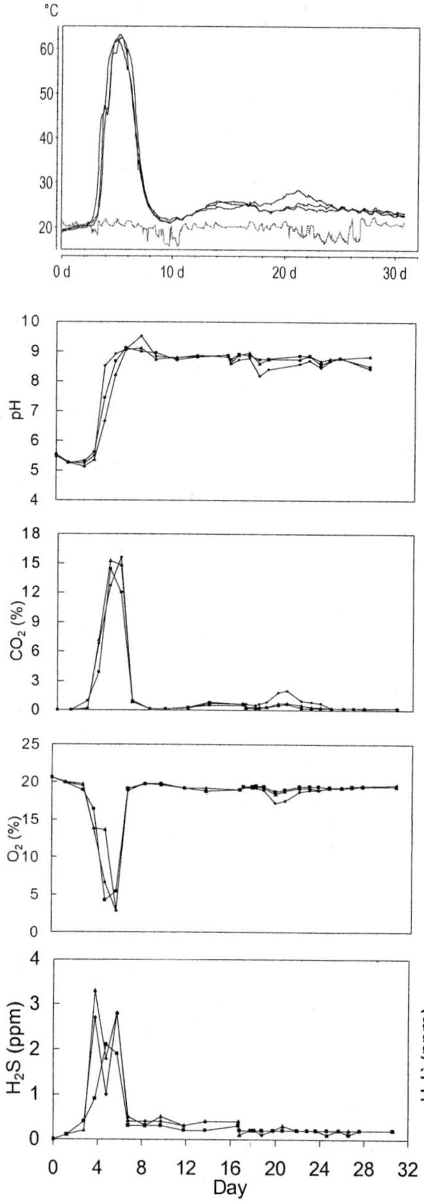

Fig. 1. Temperature, pH, CO_2, O_2 and H_2S in three bioreactors treated as aerobic control.

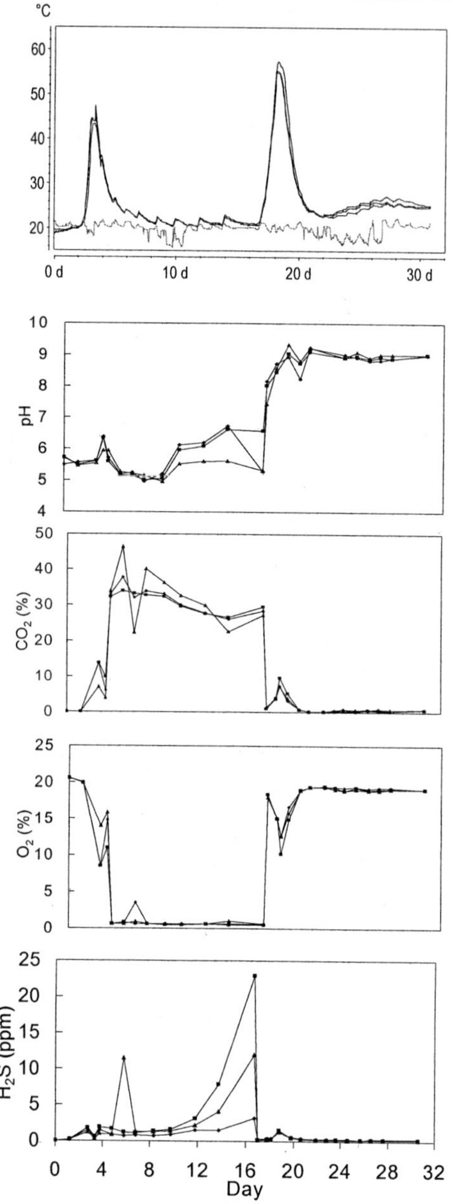

Fig. 2. Temperature, pH, CO_2, O_2 and H_2S in three bioreactors with aeration stopped on day 3 at mesophilic maximum temperature 43, 45 and 46°C. Aeration was resumed 17 days after the start of the experiment.

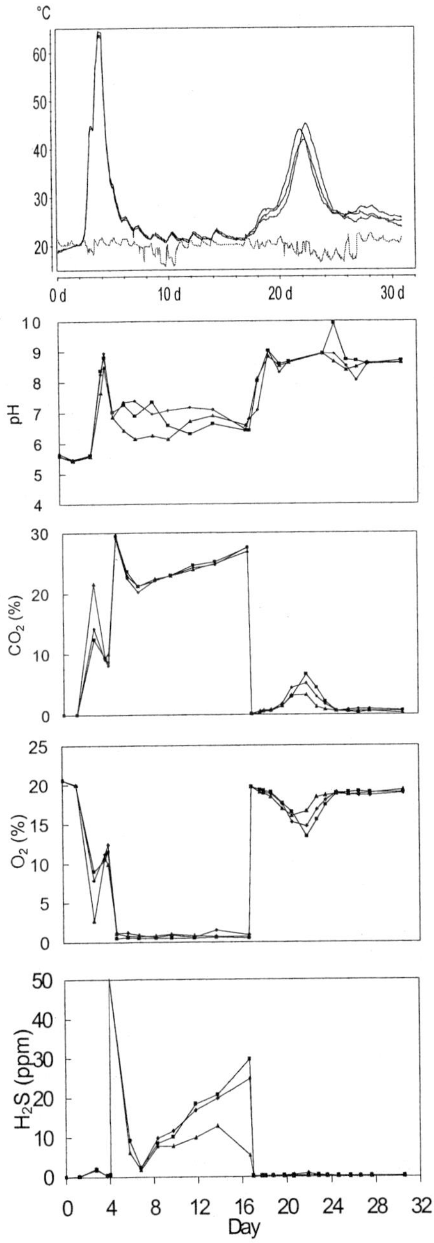

Fig. 3. Temperature, pH, CO_2, O_2 and H_2S in three bioreactors with aeration stopped on day 4 at thermophilic maximum temperature 62, 63 and 64°C. Aeration was resumed 17 days after the start of the experiment.

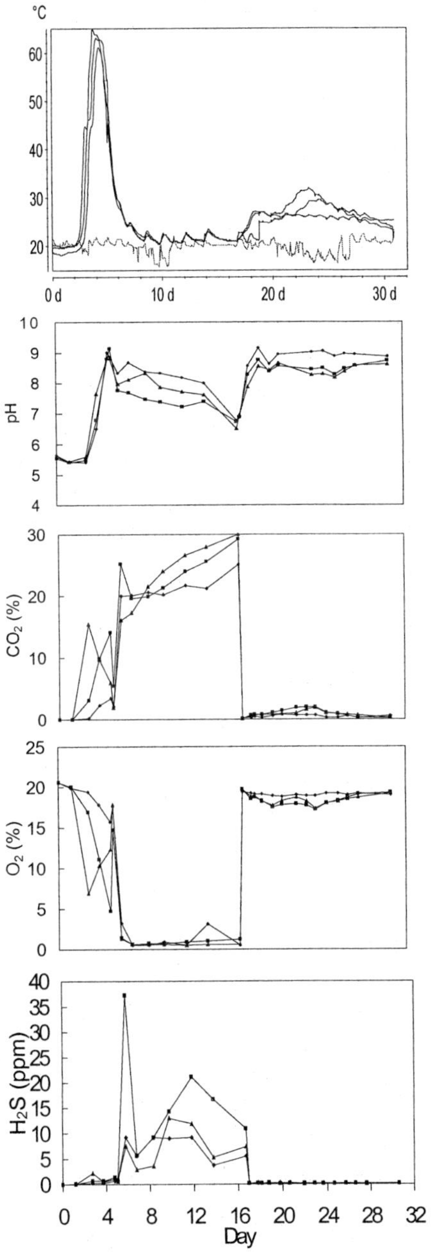

Fig. 4. Temperature, pH, CO_2, O_2 and H_2S in three bioreactors with aeration stopped in the cooling phase on day 5 when the temperature had dropped to 47, 49 and 56°C. Aeration was resumed 17 d after the start of the experiment.

When air supply was cut on day four at maximum temperature, the effect was an instantaneous cooling with temperatures close to ambient until aeration was resumed on day 17 (Fig. 3.). Then, the temperature increased relatively slowly to 42–45°C.

After day 25 it remained rather constant slightly above room temperature. Figure 4 shows the temperature in three bioreactors that were kept aerobic until the cooling phase day five. The temperature continued to decrease in the same way as in the aerobic control treatment (Fig. 1.).

However, the temperature did increase slightly after aeration was restored on day 17. Data from the first experiment with aerobic control, air stopped on day 3 at mesophilic maximum temperature (44°C) and on day 6 in the cooling phase (50°C), are shown in Fig. 5. The temperature was practically equal to ambient during the anaerobic period, but in the aerobic control it was a few degrees higher. No gas samples were taken during the anaerobic period in the first experiment. However, gas was sampled also when the aeration was cut off in the second experiment. The resulting effect was that air was pumped out of the bioreactor and replaced with air rich in oxygen. This resulted in a small rise in temperature after every sampling, giving the serrated appearance of the temperature curve.

pH

The pH was 5.5 at the start of the experiments. In the aerobic control (Fig. 1.) pH decreased to 5.2 the first days, but returned to pH 5.5 day 5. After that, pH increased rapidly and stabilised around 9 from day 7 towards the end of the experiment. When the air supply was cut off at mesophilic maximum temperature, the increase in pH had just begun (Fig. 2.). The effect of the oxygen deficiency was that pH gradually decreased towards pH 5 on day 7, after which it become slightly higher until aeration was reintroduced. Seven h after aeration was restored, pH increased from 5.3–6.6 to 7.7–8.2. The increase continued until it stabilised around pH 9. When aeration was stopped at the maximum temperature (Fig. 3.), the pH was 8.5–9.0 and after 18 h it was 6.8–7.0. pH was 6.5 just before aeration was reintroduced and after 2 days it stabilised around pH 9. Turning off the air supply in the cooling phase gave a less dramatic effect on pH (Fig. 4.). pH was about 9 when the aeration was stopped, after which it slowly decreased to pH 6.5–6.9 just before aeration was resumed. After 2 days, the pH had increased to the same level as before the air was stopped. Figure 5 shows the course of pH in all treatments from the first experiment. The response is equivalent to corresponding treatment in the second experiment. However, for some unaccountable reason, a low pH was measured on day 24 in the bioreactor where the aeration was turned off days 6–21.

Gases

In the aerobic control treatment, the highest CO_2 and lowest O_2 concentrations coincided with the highest temperatures (Fig. 1.). The sum of the CO_2 and O_2 concentration is approximately 20% as long as there is airflow through the bioreactor. However, a CO_2 concentration as high as 46% was obtained when the airflow was turned off at mesophilic temperature (Fig. 2.). The concentration of O_2 declined below 1% when aeration was turned off (Figs. 2–4). When aeration was restored on day 17, the CO_2 and O_2 concentrations first were close to the atmospheric content, but the respiration increased within one day.

The highest concentration of hydrogen sulphide in the aerobic control was obtained during the thermophilic phase and the maximum concentration was 3.3 ppm. When the air supply was stopped at mesophilic maximum temperature, the H_2S concentration slowly increased, and 23 ppm was recorded in one bioreactor just before the air was turned on. When the air was turned off at maximum temperature, the H_2S concentration exceeded the upper detection limit 50 ppm after 18 h, but the concentration decreased to 2 ppm day 7, thereafter it increased again, reaching 30 ppm in one bioreactor just before the air supply was restored. Turning off the air in the cooling phase gave a similar result to that obtained at maximum temperature, but the concentrations were lower. It is difficult to compare the figures obtained with and without aeration, since the airflow dilutes H_2S and the actual formation of H_2S can be much higher in spite of the lower concentration.

Mass balance

Mass balance from experiments 1 and 2 is shown in Tab. 1. and Tab. 2., respectively. The degradation expressed as loss in dry matter was clearly affected by the anaerobic period, especially when the air was cut off at the mesophilic maximum temperature on day 3. Although aeration was restored and a second active phase took place, it did not compensate for the inhibition of the composting process that occurred during the period without aeration. The reduction in the amount of fresh matter showed equivalent results. The reduction of the C content was close to the loss in dry matter. The loss of N was only 6% in the aerobic control, but 21% in the bioreactor with mesophilic air stop and 10% when the air supply was turned off in the cooling phase. The change in C/N ratio supports the observations. There was a slightly higher C/N than initially in the mesophilic air stop, as a result of the relatively high loss of N but a low C reduction. Note that the figures for C and N are based on only one sample from each treatment, and that previous investigations have shown a rather high variation in the N analyses.

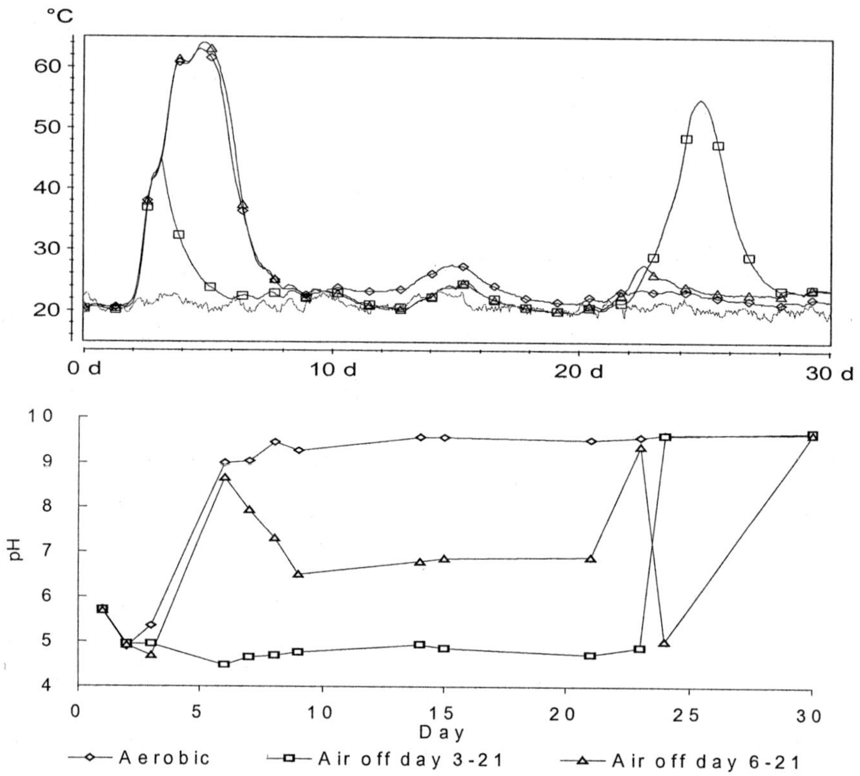

Fig. 5. Temperature and pH in three bioreactors. Aerobic control, aeration stopped at mesophilic maximum temperature 44°C on day 3 and in the cooling phase at 50°C on day 6. The aeration was resumed 21 days after the start of the experiment.

Table 1. Losses as % of initial fresh and dry matter, total C and N. Final C/N–ratio after 30 days of composting, initial C/N was 57 (first experiment, n=1).

Treatment	Fresh matter	Dry matter	Carbon	Nitrogen	C/N
Anaerobic days 3–21	10.4	16.7	17.0	20.7	59
Anaerobic days 6–21	11.9	22.6	24.1	10.3	48
Aerobic	20.5	24.3	25.3	6.5	45

Table 2. Changes as % of initial fresh and dry matter after 31 days of composting (second experiment). Mean and standard deviation (n=3).

Treatment	Air stopped day	Fresh matter	Dry matter
Mesophilic	3–17	12.6 ±0.5	17.4 ±0.3
Thermophilic	4–17	15.0 ±1.1	24.0 ±2.3
Cooling	5–17	16.0 ±0.1	22.3 ±3.5
Aerobic		20.2 ±0.9	27.0 ±2.1

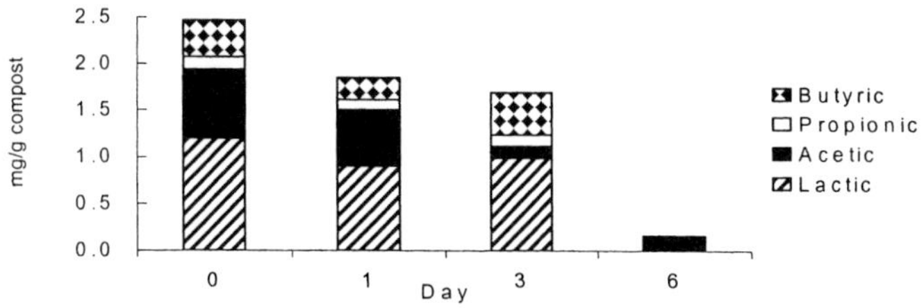

Fig. 6. Content of lactic acid, acetic acid, propionic acid and butyric acid in compost from the aerobic control treatment. All samples were under DL days 7–31. Mean with n=2 days 0–3 and n=3 days 6–31.

Volatile organic acids

Volatile organic acids (VOA) were analysed in the second experiment. The contents of VOA in the compost mixture at the beginning of the experiment and from the aerobic control are presented in Fig. 6. The concentration of acetic acid was 0.7 mg/g compost (fresh weight) at the start of the experiment.

The concentration decreased and was 0.1 mg/g after 3 days. In one sample from day 4 (data not shown) 0.2 mg/g was measured. After 6 days, 0.1–0.2 mg/g was detected in three samples (Fig. 6.). After that, no acetic acid was detectable in the aerobic control. The initial content of lactic acid was 1.2 mg/g and of propionic acid 0.1 mg/g compost. The content of lactic acid and propionic acid remained at the same respective levels during the first 3 days (Fig. 6.), but could not be detected after that. The concentration of butyric acid was initially 0.4 mg/g and decreased to 0.2 mg/g after one day, but increased again to 0.5 mg/g compost after 3 days. After that, no butyric acid was detected in the aerobic control treatment.

The highest total content of VOA was found in the composts where the aeration was turned off at mesophilic maximum temperature on day 3 (Fig. 7.). The concentration of lactic acid was 0.4 mg/g compost just before the air supply was turned off. The maximum content 3.0 mg/g was found 6–7 days after start. After that, the content decreased and no lactic acid was found after 14 days.

However, 0.5 mg/g was found in two of three bioreactors one day after aeration was resumed on day 18, but no lactic acid was found thereafter. The content of acetic acid was 0.1 mg/g compost just before the air supply was stopped. The effect of absent aeration was that the concentration of acetic acid gradually increased and maximum 2.9 mg/g compost was detected on day 17, just before the aeration was restored. Ten h later the composts contained 0.3 mg/g and one day after aeration was restored, the content of acetic acid was 0.1 mg/g compost. No acetic acid could be detected on the following days. The concentration of propionic acid was 0.1 mg/g compost just before the aeration was turned off. After 10 h the content was 0.03 mg/g (very close to detection limit) but it increased again and 0.3 mg/g was found on day 6; note that propionic acid was detected in only one of three bioreactors in this period.

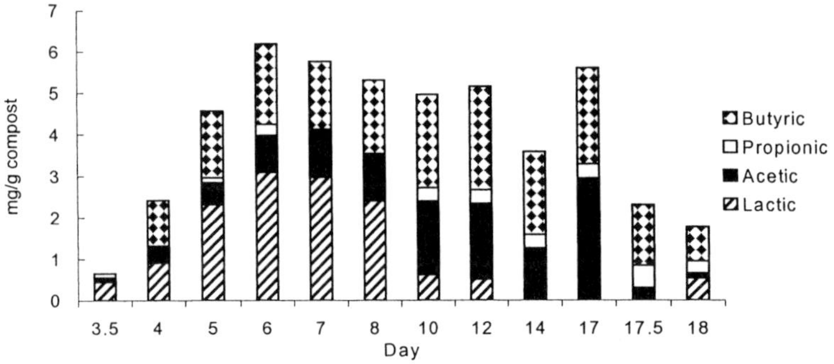

Fig. 7. Content of lactic acid, acetic acid, propionic acid and butyric acid in compost with aeration stopped at mesophilic maximum temperature, just after sampling day 3.5. Aeration was resumed just after sampling on day 17. All samples were under DL days 19–31. Mean with n=3 days 3.5–7, n=2 days 8–17 and n=3 days 17.5–31.

No propionic acid was detectable on days 7 and 8, but about 0.3 mg/g was found on days 10–17. Ten h after aeration was resumed the content increased to 0.5 mg/g, but after one day 0.3 mg/g compost was found. No propionic acid was found on the following days. Butyric acid was undetectable just before the airflow was turned off. Ten h later, 1.1 mg/g was found. The content of butyric acid was 1.0–3.1 mg/g on the following days without aeration. One day after the aeration was resumed 0.8 mg/g was found, but no butyric acid could be detected after that.

In the bioreactors where the air supply was turned off at thermophilic maximum temperature on day 4 (Fig. 8.), acetic acid dominated the content of VOA. Just before the aeration was turned off the concentration of acetic acid was very close to the detection limit (only 0.06 mg acetic acid/g compost was found in one of three bioreactors) and no other VOA was found. The concentration of acetic acid was 1.2 mg/g after 18 h without aeration and 1.6 mg/g the following day. However, the content decreased to 1.2 mg/g on day 7, but increased gradually from that level to 1.7 mg/g just before aeration was resumed.

Seven h later, the content of acetic acid was 1.5 mg/g and after one day with restored aeration 0.7–1.8 mg/g was found. No acetic acid could be detected after that. Lactic acid could not be found just before aeration was turned off, but after 18 h without aeration the content was 0.2 mg/g compost. Then, the concentration of lactic acid decreased, but at different rates in the three bioreactors, the last observation was on day 10. Propionic acid was found in only one of two samples just before aeration was restored, and 7 h later in two of three samples. The concentration was 0.15 mg/g compost. One of three samples contained butyric acid 0.2 mg/g, 18 h after the aeration was turned off. The same concentration was also found just before the aeration was resumed and 7 h later, but in only one sample per day.

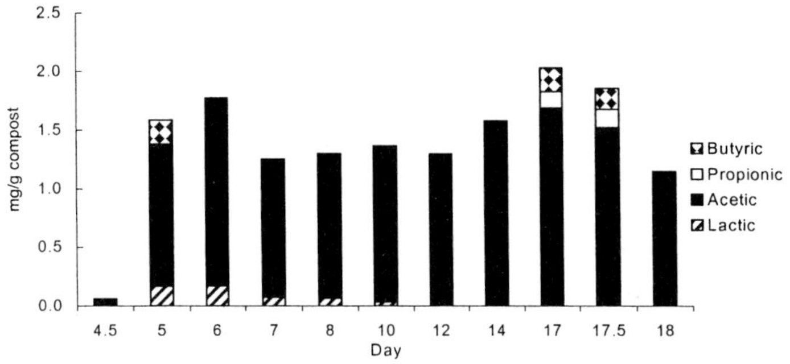

Fig. 8. Content of lactic acid, acetic acid, propionic acid and butyric acid in compost with aeration stopped at thermophilic maximum temperature, just after sampling day 4.5. Aeration was resumed just after sampling on day 17. All samples were under DL days 19–31. Mean with n=3 days 3.5–7, n=2 days 8–17 and n=3 days 17.5–31.

Figure 9 shows the content of VOA in the compost where the air supply was turned off during the cooling phase on day 5. No VOA could be detected just before the aeration was stopped, but after 1 day without aeration, 0.8 mg acetic acid per gram could be found. Then, in one of the bioreactors, the content of acetic acid gradually increased to 2.1 mg/g just before aeration was resumed. However, in another bioreactor the content decreased below the detection limit before it returned and was 1.8 mg/g just before aeration was introduced again. The concentration of acetic acid thereafter decreased and the last observation was one day after the aeration was resumed. Propionic acid first appeared after 10 d of composting. The concentration increased to 0.2 mg/g and remained at this level also 7 h after aeration was restarted. The final observation was in one sample one day after aeration was resumed.

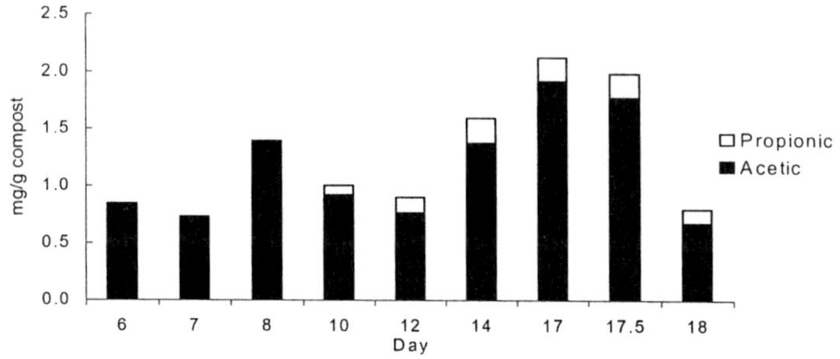

Fig. 9. Content of lactic acid, acetic acid, propionic acid and butyric acid in compost with aeration stopped at the cooling phase day 5.5. All samples were under DL just before aeration was stopped. Aeration was resumed just after sampling on day 17. All samples were under DL days 19–31. Mean with n=3 days 5.5–7, n=2 days 8–17 and n=3 days 17.5–31.

Discussion

It is clearly shown that the composting process requires oxygen and that the temperature declines immediately if oxygen is absent. However, if aeration is resumed, the composting process starts again (Figs. 2–5). The extension and maximum temperature of re–activated compost will be decided by how much microbial available energy is left in the compost material when the air supply is turned off. It is also interesting to notice that there is no observable change from mesophilic to thermophilic populations after restored aeration.

pH became lower and decreased faster the earlier the air supply was turned off. Michel and Reddy (1998) studied how different oxygenation rates effected pH and the total content of VOA. Their results cannot be directly compared with this study because they had constant aeration throughout the experiment. However, the change in pH was in line with the observations presented here.

The composition of VOA differed depending on the phase of the composting process in which the airflow was turned off. The highest total content of VOA was found in the composts where the air supply was turned off at the mesophilic maximum temperature (Fig. 7). Lactic acid was the first of the measured VOA to increase, showing its maximum after 3–4 days without aeration. As the content of lactic acid decreased, acetic acid and butyric acid increased until aeration was restored. This is in accordance with observations by Kirchmann and Widén (1994 a and b). In the insufficiently aerated windrow compost, they found high initial levels of lactic acid, which was replaced by acetic, propionic and butyric acid. When aeration was stopped later in the composting process, acetic acid dominated the content of VOA (Figs. 8 and 9). Relatively low concentration of lactic acid was found on the first days after the aeration was stopped at thermophilic maximum temperature (Fig. 8). Also butyric acid was found in low amounts the first day without aeration and just before and after aeration was resumed. In all treatments, propionic acid appeared in relatively low concentration after several days without aeration. No VOA could be detected in any treatment two days after aeration was restored. Note that the concentrations of VOA are presented per gram of fresh weight. Water content was 68% initially and was practically the same at the end of the experiment.

The consequences of interrupted air supply will be more seriously the earlier in the composting process it occurs. However, a resumed air supply can restore the compost surprisingly well in a short period of time.

References

Barberis R, Nappi P (1996) Evaluation of compost stability. In: de Bertoldi, Sequi, Lemmes, Papi (eds) The science of composting. Blackie Academic & Professional, London, pp 175-184, 1405 pp

Brinton W F (1998) Volatile organic acids in compost: Production and odorant aspects. Compost Science & Utilization Vol 6 No 1, 75-82

DeVleeschauwer D, Verdonck O, Van Assche P (1981) Phytotoxicity of refuse compost. Biocycle, January/February:44-46

Epstein E (1997) The science of composting. Technomic Publishing Company, Lancaster, USA. p 109, 483 pp. ISBN 1-56676-478-5

Inbar Y, Chen Y, Hadar Y, Hoitink H A J (1990) New approaches to compost maturity. Biocycle, December:64-69

Jiménez E I, Garcia V P (1989) Evaluation of city refuse compost maturity: A review. Biological Wastes 27:115-142

Kirchmann H, Widén P (1994) Separately collected organic household wastes. Swedish J. Agric. Res. 24:3-12

Kirchmann H, Widén P (1994) Fatty acid formation during composting of separately collected organic household wastes. Compost Science & Utilization Vol 2 No 1, 17-19

Lee R B (1977) Effect of organic acids on the loss of ions from barley roots. Journal of Experimental Botany 104:578-587

Lilja R, Hänninen K, Heikkinen Y (1996) Germination and phenolic acid content in determination of compost maturity. Publication in Science No. 8, University of Joensuu, Finland.18 pp. ISBN 951-696-603-9

Mathur S P, Owen G, Dinel H, Schnitzer M (1993) Determination of compost biomaturity. 1. Literature review. Biological Agriculture and Horticulture 10:65-85

Michel F C, Reddy C A Jr. (1998) Effect of oxygenation level on yard trimming composting rate, odor production, and compost quality in bench-scale reactors. Compost Science & Utilization Vol 6 No 4, 6-14

Morel J L, Colin F, Germon J C, Godin P, Juste C (1985) Methods for the evaluation of the maturity of municipal refuse compost. In: Gasser J K R (ed) Composting of agricultural and other wastes. Elsevier, London. pp 56-72

Ortega M C, Moreno M T, Ordovás J, Aguado M T (1996) Behaviour of different horticultural species in phytotoxicity bioassays of bark substrates. Scientia Horticulture 66:125-132

Robertsson M, Jensén P, Mårtensson A (2001) Small scale bioreactors – useful tools for studying compost process. Submitted to Compost Science & Utilization

Sánchez-Monedero M A, Roig A, Cegarra J, Bernal M P (1999) Relationships between water-soluble carbohydrate and phenol fractions and humification indices of different organic wastes during composting. Bioresource Technology 70:193-201

Tsuchida H, Azuma J, Ishida N, Nanjo I, Mizuno S (1984) Changes in phytotoxic components of sawdust barnyard manure during its rotting process. Rept Fac Agr Kobe Univ 16:277-290

Zucconi F, Monaco A, Forte M (1985) Phytotoxins during stabilization of organic matter. In: Gasser J K R (ed) Composting of agricultural and other wastes. Elsevier, London. pp 73-86

Zucconi F, Pera A, Forte M, de Bertoldi M (1981) Evaluating toxicity of immature compost. Biocycle, March/April:54-57

Zucconi F, Forte M, Monaco A, de Bertoldi M (1981) Biological evaluation of compost maturity. Biocycle, July/August:27-29

Reduction of Ammonia Emission and Waste Gas Volume by Composting at Low Oxygen Pressure

D.P. Rudrum[1], A.H.M. Veeken, V. de Wilde, W.H. Rulkens and B.V.M. Hamelers

Abstract. This chapter reports first results on the performance of a composting reactor with off-gas oxygen levels ≤10%. The low oxygen level is the result of uncoupling the two functions of the airflow: heat removal and oxygen supply. Uncoupling these functions is achieved by cooling and recirculating the gas, in this way reducing off-gas flow and emission of gaseous ammonia. The objective of the research was to identify a control strategy ensuring reactor performance stability, and a sufficient compost quality.

In an 80-li bench-scale composting unit three runs were made with oxygen concentrations of 10% (v/v) (run A), 5% (B) and 1% (C), respectively. Run A was considered the control object; this level is known to operate smoothly. The feed was a mixture of pig faeces and wheat straw 19:1(w/w). Temperature was regulated at 58 °C. It was not possible to reach the set temperature at oxygen levels below 10%, an initial level of 10% was necessary. Once the set temperature was reached, the oxygen level could be lowered.

The loss of ammonia and N_{tot} (g-N g^{-1}DM) from the solids was comparable for the three runs. At lower oxygen concentrations however, more ammonia was trapped in the condense: 53% for run A, 60% for B and 77% for C. Compared to run A, the reduction in off-gas was 33% for B and 50% for C. The loss of organic matter was 25% less for B and 33% less for C, but composting time may not have to be increased. The cumulative oxygen consumption after 450 h of composting was comparable for all runs. Apart from small differences in dry matter content, the composts were comparable for all three runs.

Introduction

In The Netherlands, there is a local excess of manure caused by an intensification of pig production. This means manure has to be transported over considerable distances, leading to higher production costs for the farmer. Export could be part of the solution, but due to EC regulations concerning animal manure, it cannot be exported unprocessed. Composting can be an option for manure management. If composted properly, the product complies with EC sanitation rules and can be exported. In addition, compost weighs only half as much as the fresh manure and

[1]Wageningen University, Subdepartment of Environmental Technology,
PO Box 8129, 6700 EV Wageningen, The Netherlands
e-mail: dale.rudrum@algemeen.mt.wag-ur.nl, tel.:+31(317)484993
fax:+31(317)482108

H. Insam, N. Riddech, S. Klammer (Eds.)
Microbiology of Composting

transport is consequently cheaper. Composting can expand the market for excess manure.

Traditional low-technology composting methods take a long time and require a lot of space. As space is scarce and expensive in The Netherlands, a high-rate composting process must be used. These high-rate processes emit a considerable amount of ammonia when composting low C/N materials. Therefore, Dutch law demands treatment of the wastegas emitted by high-rate composting. The costs of this treatment can outweigh the economic advantage of composting.

According to Finstein et al. (1983), the ventilation in temperature controlled high-rate composting processes has two functions: supply of oxygen and removal of excess heat. Veeken et al. (1999) showed that these functions can be uncoupled by introduction of gas cooling. In the uncoupled process fresh air is added to supply oxygen, and recirculated gas is cooled for heat removal. A part of the ammonia emitted from the compost is trapped in the condense that forms when cooling the recirculated gas. As a result, both thc off-gas amount and the ammonia emission to the environment are strongly reduced. Optimised to minimise emissions this system may be able to comply with Dutch legislation.

In the uncoupled system, the oxygen level and temperature can be independently controlled. There is a considerable literature on the choice of optimal temperature, see Richard and Walker (1998) for a review. It is generally accepted that the best composting results are obtained in the range from 55 to 60°C.

The major question regarding the uncoupled system is the optimal oxygen concentration of the off-gas. Model calculations showed that the lower this concentration is set the less air is required for composting and the less ammonia is emitted (Veeken et al. 1999). However, Richard et al. (1999) have shown that under low oxygen levels the composting rate can be limited. Any limitation of the process rate leads to a slower process and consequently higher land usage, thus higher costs.

The objective of this study is to investigate changes in the performance of the composting process at low oxygen levels and, if necessary, propose system changes to cope with the expected process limitations. This study is part of an integrated effort to modernise pig housing, the result will be used to design an on-farm composting unit.

Materials and Method

Three runs were made to compare the process at different oxygen concentrations ([O_2]). Run A had 10%, run B 5% and run C 1% (v/v) oxygen in the off-gas. Run A was run for 25 days, run B 21 days and run C for 19 days. Run C was done with a half-full reactor because of a shortage of suitable manure.

Manure was collected from a small pilot facility in Maartesdijk, in which it was separated on a convex conveyer belt as described by Kroodsma et al. (1998). The manure for the three runs was collected on the same day, and kept in cold storage (4 °C) until used. To improve aeration, wheat straw was added to the manure as a bulking material. The manure was mixed with the straw manually. The ratio manure to straw was 19 to 1 on a wet mass basis. Run A was mixed by hand once a week. The material was sampled and rewetted if necessary. Runs B and C were not opened during the run.

The experiments were performed using the 80-li composting vessel described by Veeken et al. (1999); the setup is represented in Fig. 1. The vessel was aerated with dry pressurised ambient air, entering at the bottom [Q_i (m^3/s)]. The gas at the top was recirculated to the bottom of the composting vessel [Q_{rec} (m^3/s)] via a water-cooled heat exchanger. Excess gas was led from the top of the composting vessel through an acid-filled water-lock to the gas analysing units. This off-gas flow was driven by the overpressure induced by the inflow. The water-lock kept the vessel slightly pressurised, preventing ambient air from leaking into the system.

The system was controlled by a computer with an input output system (RTI-820 board, 5B modules, Analog devices, USA) using Control EG. The following parameters were measured online every 15 minutes: oxygen, [O_2], (Oxi 219/90R WTW, Germany) and carbon dioxide, [CO_2], (Siemens Ultramat 1, Germany) concentration in the of gas (%(vv^{-1})); the compost temperature, T (°C), (thermocouples type T) at five points; and the airflow, Q_{in} ($m^3\ h^{-1}$), (5850S, AIR, thermal mass flow controller, Brooks, USA) into the reactor. The recirculation, Q_{rec} ($m^3\ h^{-1}$), was estimated from the pumping frequency, calibrated with a manually read gas flow meter (VEG .00094, G1.6, Schlumberger, The Netherlands).

The controls of the experiments were set to keep the average compost temperature at 58 °C, and [O_2] at a certain set oxygen level: run A at 10%, run B at 5% and run C at 1%. The temperature was controlled by adjusting Q_{rec}. The oxygen level was controlled adjusting Q_{in}. Both controls were step controls, increasing or decreasing the flows with a fixed step every 5 min.

The following characteristics of the material in and out of the reactor were measured using standard methods (APHA 1992): mass [M (kg)], dry matter [$DM_{110\ ^oC}$ (kg kg^{-1})] and pH. Ash (kg kg^{-1}), total nitrogen [N_{kj} (kg kg^{-1})], ammonium [N_{amm} (kg kg^{-1})] and volatile fatty acids [VFA (kg kg^{-1})] all on $DM_{110\ ^oC}$ basis.

The mass of condense formed [M_c (kg)], and mass of material trapped in the acid trap [M_a (kg)] were measured at intervals depending on the amount formed. After weighing condense and acid were analysed for ammonium [N_c, N_a, respectively (kg)] using the Kjeldahl method omitting the destruction step.

The VFA are nearly completely vaporised during the dry matter measurement (Derikx et al. 1994). The weight loss during drying consists not only of water, but also of organic matter, and the measured OM (DM - ash) does not include all substrate. For this reason, the DM and the OM were corrected by adding the VFA to the DM and the OM.

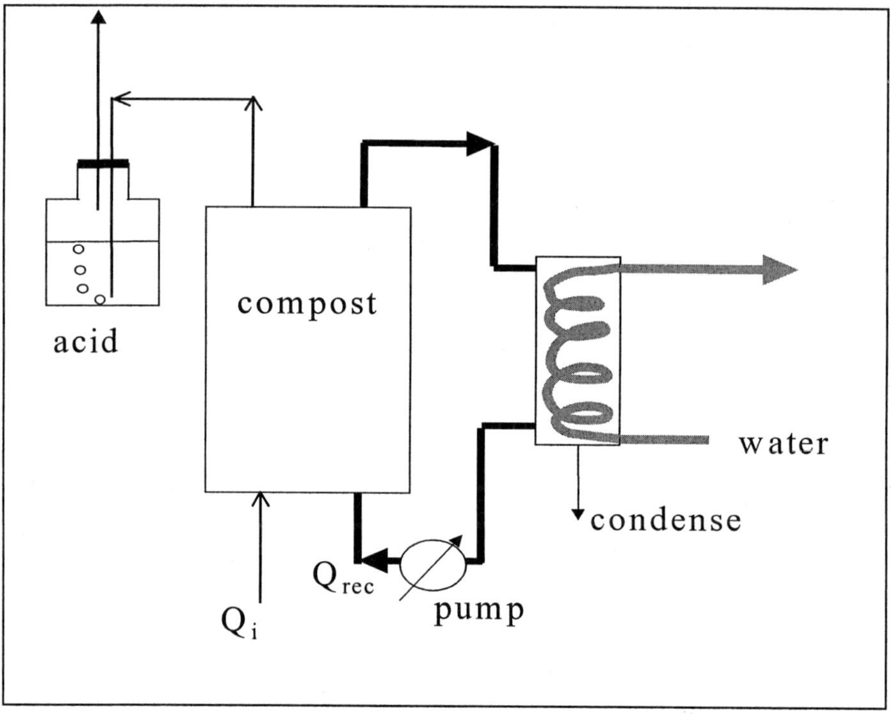

Fig. 1. Flow scheme of the batch-composting unit

From the acquired data the following parameters were obtained by calculation: DM and OM the corrected dry and organic matter content, OUR the oxygen uptake kg^{-1} initial OM, dOM the degraded organic matter, and dO_2, dC and dH_2O the "degraded" or formed amount of O_2, carbon and water, respectively.

The subscript in denotes the material used, the subscript out the compost.

- $DM = DM_{110\,°C} *(1+ VFA)$ $(kg\ kg^{-1})$
- $OM = 1\text{-}ash* DM_{110\,°C} /DM$ $(kg\ kg^{-1})$
- $OUR = 10*Q_{in} * ((20.9-[O_2]) /22.4)/(M_{in} *DM_{in} *OM_{in})$ $(mol\ (s*kg)^{-1})$
- $DOM = M_{in} *DM_{in} *OM_{in} - M_{out} *DM_{out} *OM_{out}$ (kg)
- $dO_2 = \int_{time} (OUR *OM_{in} * 32/1000)\ dt\ /dOM$ $(kg\ kg^{-1})$
- $dC = \int_{time} ((Q_{in} * [CO_2]) * 12/(22.4 * 100))\ dt\ /dOM$ $(kg\ kg^{-1})$
- $dH_2O = dOM + dO_2 - dC*44/12$ $(kg\ kg^{-1})$
- $N_{cap} = N_c/(N_c + N_a)$ (-)

The gas equations assume identical Q_{in} and Q_{out}. Although this proved not to be the case, corrections made for this did not influence the results. Therefore, the simpler formulas were used.

The runs were compared using the process parameters $[O_2]$, T, OUR, dO_2, dOM, dN and N_{cap}. The OUR reflects the biological activity during composting. The amount of organic matter lost, dOM, and the cumulative oxygen uptake, dO_2, are an indication for the progress in composting. The amount of nitrogen lost, dN, is important because in combination with the captured nitrogen, N_{cap}, this determines whether further gas treatment is necessary. The compost quality was compared using DM, OM, N_{kj}, N_{amm}, pH and C/N.

Results and Discussion

The graphs presented in this chapter are based on 5-h moving averages, that is over 20 data points. In this way, short-term fluctuations are smoothed out, and a clearer view of the process course is obtained.

During the startup of Run B it became clear that at an oxygen concentration of 5% the biological activity remained too low. Therefore, the oxygen concentration was temporarily raised to 10% until the desired 58 °C was reached. Because activity in run B did not start until the oxygen level was raised, this period of low activity was discarded. The point at which the oxygen level was raised to 10% is considered as time 0, allowing a better comparison between the runs. Run C was also started with an initial period during which the oxygen concentration was 10%.

Because it was expected that the start-up problem would repeat itself after each mixing, it was decided not to mix runs B and C. Run A had been mixed and rewetted once a week

On Line Measurements and Process Control

The oxygen concentrations are presented in Fig. 2. Apart from the start and the end of each run, the oxygen concentration is quite close to the intended concentration for all three runs. The step regulation used in these experiments is capable of maintaining the desired oxygen concentration.

The oxygen concentration rises towards the end of the runs because a minimal net airflow through the reactor was needed to have enough off-gas for measurements. If the oxygen supply through this flow is more than needed, the low concentration cannot be maintained. In run C the 1% level could be maintained only for a week.

As expected from the model calculations (Veeken et al. 1999), the amount of fresh air required for runs B and C was less than that required for run A (Table 1). A greater portion of the available oxygen was used, so less air is needed for the same amount of oxygen.

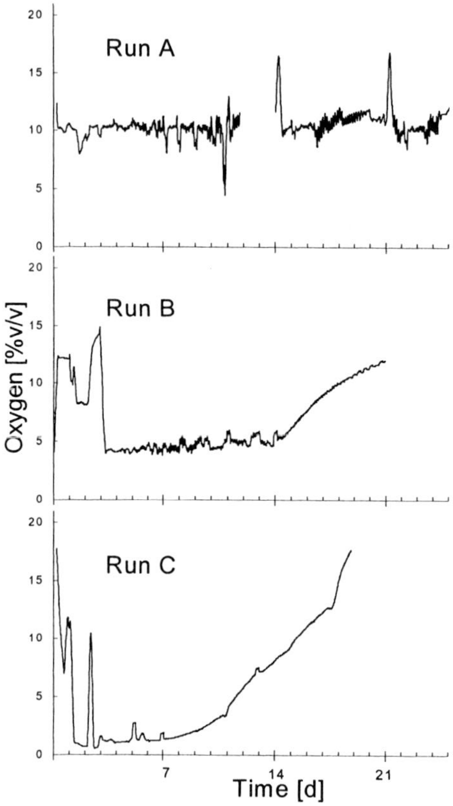

Fig. 2. Oxygen concentration of the off-gas

Runs B and C have a peak in recirculation flow that is half that of run A (not shown). The total recirculation flow, kg^{-1} organics, at the end of the runs are comparable, but the flow is more evenly spread in runs B and C.

The [CO_2] graph (not shown) is an exact supplement to the [O_2] graph (Fig. 2), the sum of the two constant around 20. The rate at which carbon dioxide was produced was less than equimolar to oxygen depletion in all experiments. This indicates that the degraded material was more reduced than carbohydrates. To correct for the CO_2 captured in the condense, it was assumed that an equimolar amount of CO_2 and NH_3 was dissolved, as $NH_4(HCO_3)$ is the main species in the condense (Veeken et al. 1999). This assumption changes the ratio. The corrected ratios are 0.96 for run A, and 0.95 for runs B and C. The carbon dioxide content of the condense should be analysed in future experiments. The ratio does show that even at 1% the OM was degraded by aerobic activity.

Table 1. The cumulative airflows during composting.

		Run A	Run B	Run C
Total inflow; Q_i	($m3\ kg^{-1}$ OM)	3.5	2.5	1.6
Total recirculation, Q_{rec}	($m3\ kg^{-1}$ OM)	10.1	8.7	7.1

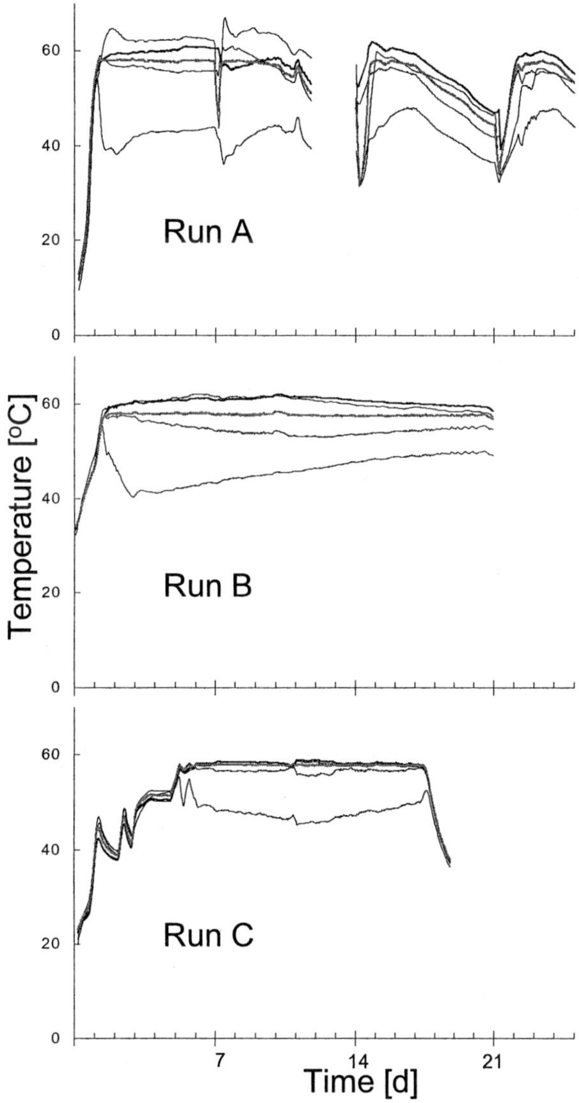

Fig. 3. Temperature at five points in the compost

In Fig. 3, maintaining the temperature at 58 °C was no problem at either 5% (run B) or 1% oxygen (run C). In fact, this temperature could be maintained longer than for the run at 10% oxygen (A), ensuring an excellent pathogen removal. At lower oxygen concentrations the temperature gradient in the reactor is smaller.

Fig. 4 shows the time course of the oxygen uptake rate (OUR). After an initial increase in the OUR, it drops in all runs at some point in time. Runs A and B show a sharp drop 2 days after start-up. In run B the drop begins before $[O_2]$ was lowered. After lowering the [O2] there is an almost instantaneous drop of the OUR. The OUR continues to decrease gently after the oxygen level is lowered. In run C the OUR is almost constant from day 4 through to day 10, and when the drop begins it does so despite the rising oxygen concentration.

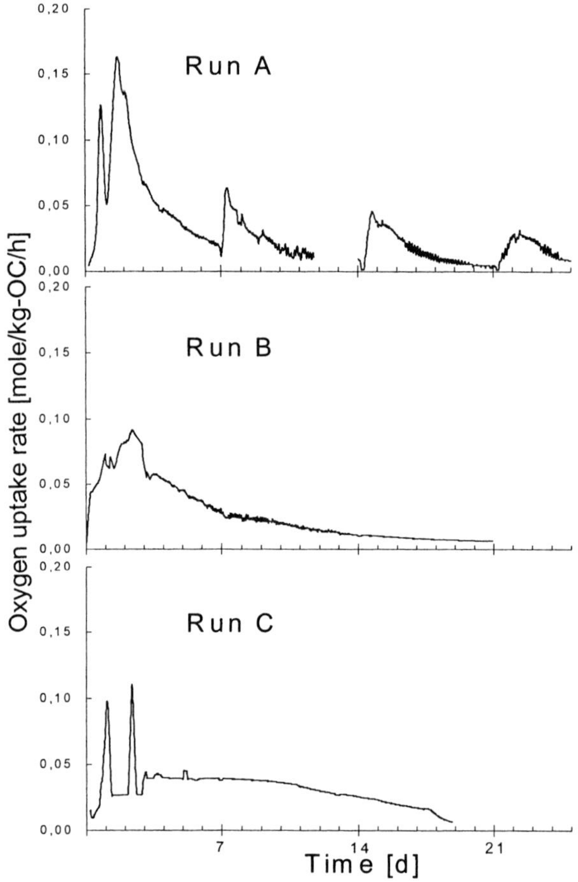

Fig. 4. The oxygen uptake rate of the compost

These results are in accordance with the predictions made by the kinetic model proposed by Hamelers (1993). In this model the microbial activity is concentrated in biofilms covering the compost material. The activity in these biofilms is limited by lack of either oxygen or substrate. Whether oxygen or substrate is limiting the composting process is determined by the size of the respective flows. At a fixed oxygen level and temperature the potential oxygen flow is almost constant during the whole composting process. Therefore, changes in OUR may be interpreted as changes in substrate flow.

Both the inability to raise the temperature initially high enough and the ability to maintain the high temperatures longer at low oxygen concentrations can be related to the significant differences in activity at the different oxygen levels. After the oxygen level was lowered in runs B and C, the activity is initially lower than in run A because of oxygen limitation. It consequently takes longer for the substrate to become limiting. The activity in runs B and C is therefore longer at a level high enough to compensate heat loss than in run A, but is too low to raise the temperature.

In the model (Hamelers 1993) the available substrate is a function of initial stock and hydrolysis. In time, the activity stabilises at a rate equal to the hydrolysis rate of the substrate. Once this rate is reached, all initial stock has been degraded. The graphs indicate a decreasing stock of readily degradable organics and a low hydrolysis rate.

The cumulative oxygen uptake of the three runs after nearly 3 weeks was similar, even though the OUR time course differed; run A 0.48 ($kg\text{-}O_2\ kg^{-1}\text{-}OM_{in}$), run B 0.44, run C 0.45. This comparison time was chosen because run C was stopped at this time.

Analysis

The analysis of materials in and out of the reactor is given in Table 2. The analysis of the condense and the acid trap in Table 6.

VFA evaporate during the dry matter analysis. As VFA are easily degradable organic compounds they were considered organic matter. Had this not been done the OM of all three materials would have been 0.77 kg kg^{-1}DM (Table 2). Apart from the VFA content the ingoing material is identical for the three runs. From the decrease in VFA it can be concluded that the material changed during storage.

The compost from the three experiments is very similar, except for dry matter content and organic nitrogen. In Table 3 there is a difference in organic matter decrease during the process. In run A organic matter degradation was 40%, 31% in run B and 25% in run C. Because of the difference in initial organic matter content and composting time it is difficult to compare these values.

A difference in oxygen demand kg^{-1} OM lost was found (Table 5). Runs A and B required 1.5 kg O_2 kg^{-1} OM, in run C this was 1.8 kg. This difference indicates why less organics were lost in run C despite a similar cumulative oxygen uptake to run A: biological activity can reduce the organic matter by consuming the more oxidised compounds. The VFA degradation during storage shows that there was biological activity. Therefore, the organic matter at the start of run C would have been more stable and reduced than the OM at the start of run A, which is consistent with the higher oxygen demand.

The difference in dOM between runs A and C may partly be explained by the difference in composting time. However, run A did not lose 15% of the initial organics in the fourth week of composting. The cumulative oxygen uptake during this time was 8% of the total. Because of the difference in the initial OM, the oxygen uptake is the more reliable measure for degradation.

Table 2. The analysis of composting materials before and after composting

		Run A in	Run B in	Run C in	Run A out	Run B out	Run C out
Dry Matter	(kg kg^{-1})	0.38	0.39	0.38	0.62	0.60	0.58
Organic Matter	(kg kg^{-1}DM)	0.81	0.78	0.77	0.72	0.71	0.73
Volatile Fatty Acids	(kg kg^{-1}DM)	46.20	9.30	0.94	0.00	0.00	0.00
Kjeldahl N	(kg kg^{-1}DM)	32.10	31.50	31.80	34.29	28.56	28.10
Ammonium N	(kg kg^{-1}DM)	14.28	14.10	14.20	8.93	8.67	9.70
Organic N	(kg kg^{-1}DM)	17.82	17.40	17.60	25.36	19.89	18.40
pH	-	7.20	7.20	7.20	8.50	8.40	8.50

Table 3. Mass of compost material components before and after composting

		Run A in	Run B in	Run C in	Run A[a] out	Run B out	Run C out
Mass	(kg)	42.00	42.00	22.10	17.68	20.70	11.60
Ash	(kg)	3.00	3.57	1.96	2.98	3.62	1.81
OM	(kg)	12.96	12.60	6.53	7.81	8.70	4.86
Water	(kg)	26.04	25.83	13.61	6.75	8.38	4.93
N	(kg)	0.51	0.51	0.27	0.37	0.35	0.19
Namm	(kg)	0.23	0.23	0.12	0.10	0.11	0.06
Norg	(kg)	0.29	0.28	0.15	0.27	0.24	0.12

a Values corrected for samples taken during composting.

The dry matter content of run A is highest at 62%, run B has 60% and run C 58% DM. This can be linked to the greater oxygen uptake in run A (Table 4). In our experimental setup relatively little drying occurs that is due to convectional physical drying (see also the discussion of the mass balance; Table 4). The recirculated air has approximately 100% moisture content, so that the air takes up water only when heated. This means that all the energy required for drying must be provided by degrading the organics.

The amount of organic N is in all cases almost constant as can be seen in Table 3. The relative increase (Table 2) is caused by the decrease in OM. The organic nitrogen content in the end product of run A is 25 g-N kg^{-1}-DM; this is much higher than the 20 and 18 g-N kg^{-1}-DM of runs B and C, respectively.

Mass Balance

The mass balances of the runs are presented in Table 4. The recovery is quite good. The mass that could not be accounted for is only 3% of either the ingoing material or Q_{in}. The leak in the mass balance is probably caused by water lost via the off-gas. The air used for Q_{in} is dry, and the acid trap does not completely dry the off-gas. The values found correspond with saturation at 25 °C. Because the off-gas does not pass through the cooler this is very well possible.

From the mass balance the stochiometric formula of the degraded organic matter was calculated. The result is presented in Table 5. The overall elemental composition of the degraded organics of runs A and B are similar, but in run C it contains only two thirds of the oxygen content.

Table 4. Mass balance of the composting runs

		Run A in	Run B in	Run C in	Run A out	RunB out	Run C out
Reactor	(kg)	42.0	42.0	22.1	14.0	20.7	11.6
Rewetting	(kg)	3.8					
Oxygen	(kg)	7.2	5.7	3.0			
CO2	(kg)				9.1	7.0	3.6
Condense	(kg)				17.9	14.4	8.5
Acid trap	(kg)				6.6	4.5	1.1
Samples	(kg)				3.8		
Total	(kg)	53.0	47.7	25.1	51.4	46.6	24.8
Imbalance as mass percentage of gas flow					2.7	2.7	1.6

Table 5. Calculation of the elemental composition of degraded organic matter

	formula		Run A	Run B	Run C
OMgone	dOM	(kg)	4.92	3.90	1.67
Oxygen uptake	dO_2 *dOM	(kg)	7.14	5.72	2.97
Carbon dioxide prod.	dC *dOM *3.7	(kg)	9.11	7.03	3.56
Water prod	dH_2O *dOM	(kg)	2.96	2.58	1.08
Carbon content dOM	dC * dOM	(kg)	2.48	1.92	0.97
Hydrogen content dOM	water*2/18	(kg)	0.33	0.29	0.12
Oxygen content dOM	OM-C-H	(kg)	2.11	1.69	0.58
Stochiometric formula			$CH_{1.6}O_{0.64}$	$CH_{1.8}O_{0.66}$	$CH_{1.5}O_{0.45}$
Mol CO_2 mol^{-1} O_2			0.93	0.89	0.87
%N of OM lost	100*dN/dOM	%	0.17	0.93	1.59

Ammonia Emission

In all 3 runs more nitrogen was retrieved in the condense and the acid trap (Table 6) than was lost from the composting material (Table 3). It seems unlikely that this is the result of nitrogen fixation. Probably it is a result of the N-Kjeldahl measurement.

Table 6. Mass of nitrogen components lost and ammonia retrieved

		Run A	Run B	Run C
Ncondense	(g)	106	131	95
Nacid	(g)	106	87	29
N loss of solids	(g)	143	159	83

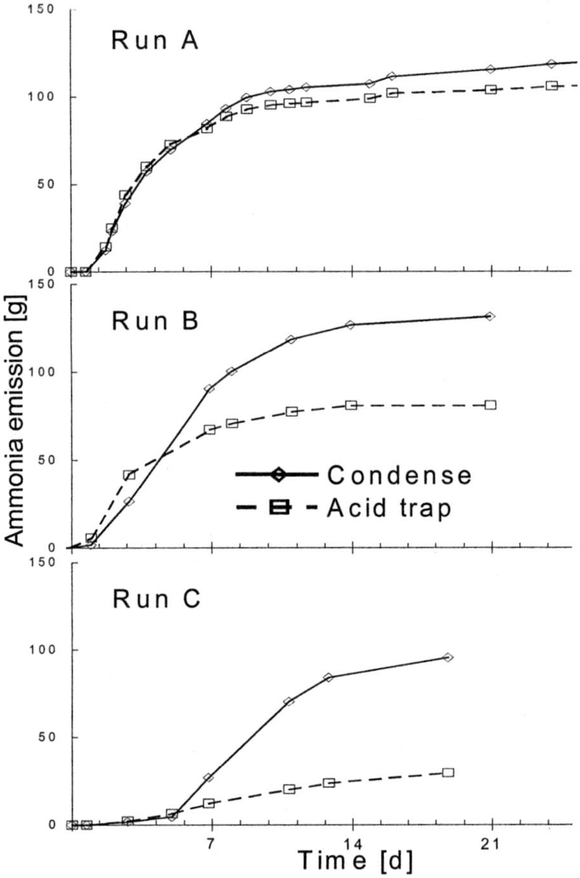

Fig. 5. Total amounts of ammonia-N in acid trap (N_a) and condense (N_c)

In accordance with the prediction by Veeken et al. (1999), more ammonia is dissolved in the condense at lower oxygen concentrations (Fig. 5). In run C only 23% of the emitted ammonia is released with the off-gas. In run B 40% is emitted, compared to 47% in run A and 100% in conventional systems.

There is no increase in the ratio of N_a to N_c toward the end of the runs. In this period the activity drops to the point that $[O_2]$ rises. By then the bulk of ammonia has been emitted and the rise in oxygen concentration does not effect ammonia emissions.

Conclusion

- As was predicted, the gaseous ammonia emission drops when composting at low oxygen concentrations. In run C only 23% of the evaporated ammonia was emitted via the off-gas.
- The drop in gaseous emissions is not due to retention in the compost, but due to a higher ammonia concentration in the condense. For the system to be applicable, a use for this condense must be found. In this study the condense, together with the separated urine, was intended as a nitrogen fertiliser.
- As has been shown, composting can take place at low oxygen concentrations. An initial concentration of 10% is needed for startup, but after this, the process can maintain itself. A step control is good enough to control the system.
- The excess airflow through the system, causing the rise in oxygen concentration in the latter part of the process when the activity has dropped, does not adversely affect ammonia emissions. The compost holds the ammonia not emitted by then.
- Confirmation was found to support the theory that at lower oxygen levels the activity drops. However, this does not change the compost if one considers a complete composting period of several weeks. This was shown by comparing the cumulative oxygen uptake. The organic matter degradation is, in this comparison, not a good indicator because of the different experimental times used and differences in starting material, especially VFA.

Acknowledgements. The Dutch Ministry of Economic Affairs via the Economy, Ecology, and Technology program financially supported the research. It is part of the Hercules project, a multidisciplinary effort to modernise pig farms.

References

APHA (1992) Standard methods for the examination of water and wastewater. 18th edn. American Public Health Association, Washington, DC

Derikx PJL, Willers HC, Ten Have PJW (1994) Effect of pH on the behaviour of volatile compounds in organic manures during dry-matter determination. Bioresour Technol 49(1): 41-45

Finstein MS, Miller FC, Strom PF, MacGregor ST, Psarianos KM (1983) Composting ecosystem management for waste treatment. Bio/Technology 1: 347-353

Hamelers HVM (1993) A theoretical model of composting kinetics. In: Hoitink AJ (ed) Science and engineering of composting: design, environmental, microbial and utilization aspects. Renaissance Publications, Ohio, pp 36-58

Kroodsma W, Ogink NWM, Satter IHG, Willers HC (1998) A technique for direct separation of pig excrements followed by on-farm treatment of the components. Int Conf on Agricultural Engineering, 24-27 Aug 1998, Oslo, Norway, pp 213-214

Richard TL, Walker LP (1998) Temperature kinetics of aerobic solid-state biodegradation. Proc IBE 1: A10-A30

Richard TL, Walker LP, Gosset JM (1999) The effects of oxygen on solid-state biodegradation kinetics. Proc IBE 2: A22-A39

Veeken AHM, de Wilde V, Hamelers HVM (1999) Reduction of ammonia emissions during composting by uncoupling of oxygen demand and heat removal. Orbit '99, Weimar, Germany, Rhombos, Berlin, pp 111-117

Review of Compost Process-control for Product Function

R.A.K. Szmidt[1]

Introduction

Composting processes for organic waste recycling have largely been designed to meet the needs of the waste management industry rather than the needs of users of recycled products. In particular, systems have been engineered and managed to maximise decomposition rate and reduction in volume or weight. The balance is usually between process-rate and cost. In a simple example the use of mechanisation such as straddle turners compared to bucket-loaders results in a higher throughput and lower labour costs but higher capital requirement. However, both approaches may result in a non-homogenous product and, because the process is predominantly carried-out outdoors, is subject to the influence of climatic variation such as temperature and rainfall.

For material destined for low- or nil-value markets the fact that composted end product is variable may be of little or no consequence. As a result, costs of production are the main driver. In some situations site restrictions, such as available space may also influence operational choices. Public perception of composting tends to be positive in so far as there is a general acceptance of the need to recycle or reduce waste but there are environmental issues which have become drivers or constraints, subsequent to the process-rate / cost equation. These include concerns over pollution in its various forms. In particular, odour, gaseous emissions, dust, microbial aerosols, run-off and aesthetics are important issues in the minds of planners, and neighbours.

Where there is a risk of offence these considerations have tended to result in a preference for contained systems. Containment can be in two forms. Firstly, the entire process may continue to operate indoors in the same way as it might outdoors, for instance with windrow turners or ventilated static piles, but within an air-conditioned building or, secondly, the process itself may be exclusively contained. The latter being so-called tunnel or in-vessel systems. There are also various hybrid systems such as agitated beds, drum-composters and bunker systems. Methods have been described by a number of authors, (de Bertoldi *et al.* 1996, Jackson *et al.* 1992, Stentiford 1992, 1993).

While containment of the process within a building may improve environmental management it does not intrinsically alter the process. For this the best solution, particularly for control of temperature and oxygen levels within compost

[1]Environment Division, SAC Auchincruive, Ayr, KA6 5HW, Scotland, UK.

H. Insam, N. Riddech, S. Klammer (Eds.)
Microbiology of Composting

is forced aeration (MacGregor *et. al.* 1981). This may be either as positive aeration, blowing into the stack, or sucking of air from composting material. The latter has the added advantage of focusing the exhaust air in one point where it may be treated to remove odour, for instance by using a biofilter (Psarianos *et al.* 1983, Stentiford 1993). However, in-vessel systems also offer the opportunity to apply a range of additional controls. While this can give environmental benefits there are also major opportunities for influencing product quality and performance (Gulliver *et. al.* 1991).

Control parameters

The range of parameters that may be controlled depends on the sophistication and location of the composting system. With regard to a simple outdoor windrow system the extent of control only extends to feedstock choice, initial moisture content, physical structure of the mix and the frequency of turning. This may be implemented on a routine basis (time), or on temperature within the stack. Nonetheless temperature within the stack may vary considerably and is broadly recognised as having a number of functional zones (Perrin & Macauley 1995).

In-vessel systems may have complex sensor and control systems (Fig. 1). The decision points for the process are, as for all compost systems in the choice of feedstocks but also in process management (Straatsma *et. al.* 1995). The various points at which sensors may be located are related to function in Table 1. Because of the number of control points that are possible in enclosed composting, process management is best done by computer control (Lindberg 1996). However the success of the system is dependent on the knowledge of the programmer and the set-points employed (Lamber 1991). As technology develops, other opportunities for control may arise, such as more complex integration of control (Bhurtun & Mohee 1996) or microbial probes (Ritz *et al.* 1994). Commercial computer control systems use protected software and little is published in respect of the algorithms employed by individual manufacturers. Adopting the waste management strategy of maximum decomposition rate may well miss the opportunity of fine control for particular product attributes. For instance control algorithms widely adopted in commercial glasshouse control systems employ a concept of 'ramp time' – that is, control of the rate of change between process phases. In the case of greenhouses this allows plants to adapt to change, for instance from night to day conditions without suffering shock. In some cases shock can have beneficial effects such as in plant growth regulation and a computer may be programmed to maximise this. Composting is also a dynamic process involving complex interactions. As feedstock choice becomes more complex, for instance in use of inoculants, biodynamic additives, and mineral supplements (Carpenter-Boggs *et. al.* 1999, O'Brien *et. al.* 1999, Balis & Tassiopoulou In Press) control of rate of changes during composting may become significant.

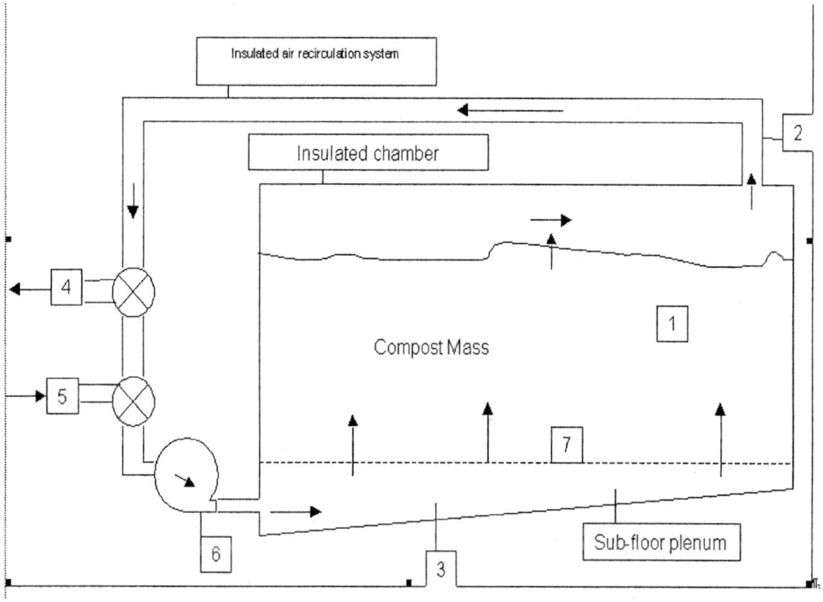

Fig.1. Control and sensor locations in a typical in-vessel system. Arrows denote the direction of air movement.
Key: Compost mass (1), Air recirculation system (2), Sub-floor plenum (3), Exhaust outlet (4), Fresh air inlet (5), Fan (6), Floor layout (7)

Table 1. Control choices for in-vessel systems

Feedstock	Process	Sensor point (see Fig.1)
Ingredients	Temperature	1,2,5
pH	Carbon dioxide	2
C:N ratio (total and available)	Oxygen	2,3
Conductivity	Odour	4
Nutrients	Ammonia	2
Inoculants	Moisture	2,3
Physical structure	Nutrient analysis	1
	Microbial activity	1
	Ventilation	
	Air flow rate	6
	Air speed	7
	Air mixing (recirculation / fresh air)	4,5

The relationship between feedstock and engineering

The engineering prerogative of ventilated compost systems is to carry out degradation by wholly microbial means rather than mechanical degradation of material.

How this can complement the product-related attributes needed by markets needs to be considered.

Simple composting processes such as turning inevitably result in a degradation of the substrate and therefore introduce an element of variability. However, by their nature agitated systems do give a high degree of mixing. By contrast, forced-air ventilation systems are usually unagitated and so the uniformity of the process reflects the uniformity of material loaded and the uniformity of the air distribution as engineered. It is therefore important to have pre-composting mixing or blending for such systems. Failure to achieve uniformity will result in uneven processing and in some cases system failure. While some parts of a compost mass may fail to compost properly others may be over active, resulting in hot-spots. In the absence of uniformity, probes such as for temperature, are unlikely to reliably reflect the mass. Failure to uniformly control a compost mass will result potentially in anaerobic zones and as a result odour discharge. Pre-composting uniformity may be achieved by physical destruction of feedstocks, for instance by grinding to a standard mesh size and then incorporation with a co-utilization material to provide the desired physical structure in the mass to ensure optimum air flow. Systems may use co-utilization material which will take either no part in the composting process, e.g. recycled tyre 'chips' , a limited part, e.g. hard wood 'chips' or will be degraded as part of the composting process, e.g. softwood 'chips' or shreddings. For material of small particle size and high moisture content, such as biosolids, co-composting with a bulking agent is essential (Stentiford 1993, Keener *et. al.* 1993). Co-composting with inert components to be screened out after composting more likely will give a uniform particle size distribution of compost than will co-utilization with degradable feedstock. Less sophisticated market products used for their aesthetic or fertiliser qualities may be satisfied with any particle-size distribution while those markets with specific requirements may require blending of composts to achieve an appropriate physical structure of end-product.

What are the target attributes?

The range of attributes required of a functional compost depend on the purpose for which it is to be used. In some markets, such as for mulch materials, aesthetics are important while for others only fertiliser value or physical structure are significant (Bragg 1990, Szmidt 1999). For proprietary peat-based materials, such as pot plant mixes, it is assumed that peat contributes no significant nutrients but is important in terms of providing the correct balance of air porosity and water availability, measured as air-filled porosity (AFP) (Bunt 1988, Bragg & Chambers 1988).

For less sophisticated agricultural markets there may be few clear targets. In many cases application of any fertiliser-grade material, including compost, is regulated and the target application rate is governed by crop requirements or environmental sensitivity of particular locations or soil types. For agriculture or for soil improvement, compost-use may be accompanied by application to land of other materials such as chemical fertilisers or manures to make up any short fall in nutrient content. In this context the nitrogen content of compost is important, as it is with other organic fertilisers. From a regulators point of view one should consider not just total fertiliser value but include a measure of availability to the crop and to the environment (Sikora & Szmidt 2001).

Compost processing influences the form of nitrogen present in compost. It can be present in ammonium, nitrate or organic forms. The latter consists of microbial protein and amino acids. Composting is a dynamic process with transition through a number of phases. For instance, mesophilic, thermophilic and secondary mesophilic phases take place and a sequence of microbial communities may live and die in each (Gerrits 1969, Gerrits 1988). Therefore the organic nitrogen fraction reflects both live and dead communities. The function of the live community will have implications for the subsequent rate of N-release and N-cycling from the dead. General agronomic advice has been to assume a release rate of 25% organic-N in year 1 and 10% in each of years 2 and 3 from compost applied to soil (Anon. 1994). However, this is not adequate for todays' precision farming and a better understanding of the interaction between climatic factors, soil and compost-type is essential. In reality N-mineralization rate may be considerably lower than these general guidelines. For instance, while chicken manure compost has been shown to have an N-mineralization rate of 28% (Castellanos & Pratt, 1981) other manure composts are typically in the range 5 – 15% (Wen *et al.*, 1995, Herbert *et al.*, 1991). There is also a need for improved routine analysis to include the organic-N fraction rather than just available forms. Tester *et. al.* (1977) showed that sequential screening of biosolids compost resulted in greater N mineralization in the smaller-sized fractions. Compost material that passed through 1 mm screen had a mineralization rate 3 times greater than compost that passed a 10 mm screen. The C:N ratio of the 10 mm pass compost was 14.7 and 10.4 for the 1 mm pass compost. A greater amount of woodchips (the bulking agent in the biosolids composting process) was removed by the 1 mm screen which reduced the C:N ratio. These data suggest screening of composts will produce fractions of different microbial potential that mineralize differently and can be marketed for various purposes.

In the literature some authors have linked composting-method to use of products (Lopez-Real & Vere 1992, Szmidt 1998). However, most tend to study development of particular systems and those papers dealing with use of composted material rarely are concerned with the choice of system to achieve a particular specification. In an analysis of more than 3000 papers and abstracts published on composting over the period 1990 – 2000, a clear majority do not examine the link between process and product. Those that do, represent only 27% of total. They can be split into a range of subheadings. Of these, the major area of linking is in the control of the end product (20%) and in the description of control methods (26.7%). More than half of these papers (66%) have been published since 1995. However, a substantial proportion of papers present information on the performance of a compost when used for a particular crop. Comparisons are predominantly against conventional cultivation such as soil or peat-based growing media rather than an investigation of management of compost as an element of management of the crop.

A number of papers have been presented that relate to specifications of compost quality (Verdonck 1998, Gabriels 1998, Folliet-Hoyte 1996, Ryan & Chaney 1993). However, there is an issue which has been largely overlooked in determining national or transnational standards. The processes of composting are biological and therefore potentially variable. In some cases such natural processes may be wholly acceptable as part of the local community. In the UK, providing no pollution occurs, there is a right to carry out natural processes, for which regulatory exemptions and guidelines are published. For instance, production of mushroom compost (Anon.1997). This typifies the role of quality standards. They tend to be tools for regulation, for instance of microbial hazard, rather than tools for optimising users confidence by specifying suitability and 'fitness-for-purpose'. One of the major advantages of composted materials over other products such as peat-based growing-media is that of plant disease suppression. The factors regulating this have been shown to vary between composts and are largely attributed to microbial dynamics rather than physical properties (Hoitink *et. al.* 1993). For instance, where chitin is present in the feedstock mix this may influence establishment of chitinolytic fungi and bacteria which may contribute particular disease suppressive attributes. Degradation of chitin is complex and specific to certain bacteria and fungi (Chen, 1997). Overall, there appears to be a link between feedstock and end-product potential in this respect (Table 2).

Table 2. Cited examples of compost type associated with plant disease suppression. (After Szmidt & Bragg, In Press)

Target Pathogen	Crop	Principal Compost Ingredients
Aphanomyces euteiches	Pea	Cattle manure
Botrytis cinerea	Strawberry Bean	MSW 'tea'
Erisyphe graminis	Barley Wheat	MSW 'tea'
E. polygoni	Phaseolus Bean	MSW 'tea'
Fusarium oxysporum	Nursery Stock Tomato Radish Sweet Basil Cotton	Cattle manure Chicken manure MSW Bark Poplar Bark Biosolids Sugarcane
F.conglutinans	Watermelon	SMS Papermill sludge
F.culmorum	Wheat	Greenwaste
Phoma medicaginis	Peas	Greenwaste Paperwaste

Table 2. (cont.)

Plasmodiophora brassicae	Cabbage	SMS Greenwaste Paperwaste
Phytophthora cinnamomi	Nursery Stock Sweet Basil Lupin	Pine Bark Vegetable waste Citrus
P.capsici	Sweet Pepper	Vegetable waste
P.fragariae	Strawberry	Greenwaste Paperwaste
P.infestans	Tomato Potato	MSW 'tea'
P.nicotianae	Citrus	Fresh MSW
Phytophthora sp.	Peas Cucumber Tomato Lupin	Greenwaste Biosolids MSW Citrus
Pythium ultimum	Cucumber Nursery Stock Chrysanthemum Peas Sugarbeet Iris	Pine BarkCattle manure / leaves Poultry manure / leaves Bark Grape marc MSW (organic fraction) Greenwaste Biosolids
P. graminicola	Creeping Bent Grass	MSW Brewery waste Biosolids Poultry manure
P.myriotylum	Cucumber Tomato Watermelon	MSW Greenwaste SMS Papermill sludge
P. irregulare	Cucumber	MSW Greenwaste
P.aphanidermatum	Cucumber	MSW Greenwaste Sugarcane Liquorice root
Rhizoctonia solani	Nursery Stock Cucumber Tomato Radish Bean Cotton	Cattle manure Hardwood Bark Poultry manure Vegetable waste Biosolids Greenwaste
Sclerotinia minor	Lettuce	Biosolids Grape marc
Sphaerotheca fuliginea	Cucumber	MSW 'tea'
Venturia inaequalis	Apple	SMS 'tea'

Microbial dynamics of compost

Metabolic rate and therefore energy as heat generated during composting is generally a function of microbial activity, physical and chemical properties of the substrate (Randle & Flegg 1978, Miller *et. al.* 1989, 1991, Hoitink & Keener 1993, de Bertoldi *et al.* 1996). Composting has been defined as "The controlled decomposition and subsequent stabilisation of blended organic substrates under aerobic conditions that allow the development of thermophilic temperatures as a result of biologically produced heat"(Szmidt 1999). Outside of these process-boundaries other events can occur. Because of the high rate of composting anaerobic activity may arise. Oxygen levels may fall rapidly to below 5% at which facultative anaerobes dominate (Burrage, personal communication). This can occur within 20 minutes following aeration by turning of a high rate system. Such variation in substrate utilization can confound process management, environmental management, particularly odour control, and product quality. Hillman (2000) observed that rate of enzymic degradation of starch under anaerobic conditions depends on the amylopectin / amylose ratio and that altering the botanical source of starch can have significant effects on bacterial populations. There were different responses in the microbial populations resulting from natural compared to processed (gelatinised) starch and so there would likely be differences between composting microbial dynamics, for instance in relation to starch from vegetable matter compared to food waste. However, the implications of this in terms of product quality are undetermined.

In the mushroom industry it was long considered that a proportion of anaerobic activity during the early stages of composting was useful in maximising compost quality (Van Griensven, 1988). Sharma *et. al.* (2000) observed that mushroom compost produced in a ventilated bunker system performed better than that produced from a windrow process. They observed that such composts differed in respect of mean straw length, dry matter, conductivity, nitrogen, ammonia, fibre and ash content. In addition there was a higher population of the thermophilic fungus *Scytalidium thermophilum,* believed to be a major factor in immature-compost quality. In-vessel systems, which guarantee maintenance of aerobic conditions throughout the mass, are now generally acknowledged as producing the highest yielding compost for mushrooms.

In some compost processes where temperature is allowed to climb excessively anaerobic activity can become predominant. This is likely to result in temperatures in excess of 70°C. The consequence may be that the process ceases to be biological and the compost mass becomes an exothermic biochemical reactor (Sinden & Hauser 1953). This type of high temperature phase may continue for some time although Miller *et. al.* (1989) considered that in compost, chemical oxygen uptake rates are minor compared to metabolic uptake. The microbial content of such compost is likely to be low after the continuous exposure of the microflora to pasteurisation conditions. When the temperature of the mass returns to near-ambient conditions there may be little re-establishment of a mesophilic population and therefore a risk of poor microbial attributes such as disease suppression. This latter point requires further investigation.

In-vessel systems with a high degree of containment, even where pasteurisation is controlled to a minimum period may still give similar results. In such a case, after pasteurisation the compost simply fails to maintain any microbial activity during the expected conditioning and maturation phases and temperatures simply return to ambient levels. To counter this there may be a need to 're-seed' pasteurised material to re-establish a mesophilic population. This can be done using a known supply of mature functional compost or by inoculation from culture.

Health and Safety

Safety of composted material is of critical importance for the long term future of the industry. Content of pathogens and zoonoses must be determined. The classic interpretation of safety lies in the attainment of pasteurisation on a time / temperature interaction (Folliet-Hoyte 1996, Ryan & Chaney 1993). The role of potential pathogens in the composting process itself was alluded to by Farrell (1993). The suggestion that organisms such as *Listeria sp.*, *Salmonella sp.* and *Cryptosporidium sp.* can actually be significant in the decompostion process implies that they not only remain a problem but can become niche colonists. This is supported by results of in-vessel composting in which the core areas achieved acceptable pasteurisation levels complying with EPA guidelines (Ryan & Chaney 1993) but where a failure in peripheral areas of the system allowed carry-over of pathogenic organisms. Subsequent determinations showed not only presence of pathogen levels in the core but higher levels than intitially present. Growth of pathogens in the core was considered to result from recolonisation from the unpasteurised zones. Consequently a combination of a biological void resulting from anaerobic / biochemical zones, described above, and unpasteurised material potentially is a hazardous combination. Hillman (2000) showed that in animal colons the presence of starch was correlated with a reduction in *Escherichia coli* populations and this may have implications in the generation of safe composts. Under such anaerobic conditions and at low-oxygen environments lipid and protein fractions were also considered to be important factors.

Product use

In respect of microbial function of products, there is little direct information on market demands. For agricultural and horticultural products specifications do not typically define microbial parameters. Fertilisers are typically assumed to have no functional microbial population although there is little doubt that nutrient levels will influence soil microbial dynamics (Ritz *et. al.* 1994). For intensive systems, such as greenhouse crop production, growing-media may have detailed physico-chemical requirements but are often assumed by users to have inconsequential microbial populations. However, growing-media can rapidly establish stable microflora once in use. Even in inorganic systems such as hydroponic production using mineral wool or perlite, substrate populations will readily reach $1E6g^{-1}$dw and $1E8g^{-1}$ dw for fungi and bacteria respectively (Szmidt *et. al.* 1988).

The level of microbial colonisation of compost depends on the level of maturity, feedstocks, and the type of composting process. Bess (1999) considered that a population of $1E8g^{-1}$dw and $1E4g^{-1}$dw of aerobic bacteria and fungi / yeasts respectively was appropriate for quality compost. In addition, a target in excess of $1E3g^{-1}$dw of Pseudomonad bacteria was recommended in view of the widely reported plant-related benefits that this group confer.

The phenomenon of suppressiveness of plant diseases is widely reported (Cohen *et. al.* 1998, Hoitink *et. al.* 1993) and there is a clear link between the approaches of classical biological control addressing one target with one organism and compost technologists attempting to maximise biological potential of the whole microbial compost community. There is a school of thought that the presence or absence of specific microorganisms is less relevant than is the community potential in respect of antifungal action or antibiosis. This approach is typified in the investigation of microbial communities using techniques such as Biolog™, described by Garland & Mills (1991) and Garland (1996). Similarly the organic-crop movement tends to adopt a holistic approach in the concept of promoting soil 'health' and the contribution that compost and other organic-based materials may have.

However, composts do have specific attributes related to process management and feedstock (Bahl 1991, Pardo *et. al.* 1995). In respect of plant disease suppression, activity does vary from source to source (Table 2). The challenge is to determine the optimum combinations of feedstock and engineering for compost activity and to examine the potential benefits of selecting compost blends for function.

Conclusions and recommendations

Although composting is a process that is well established there is little doubt from the literature here considered that important issues remain to be addressed. In some cases these are part of ongoing development while others offer new approaches which through research and development will increase the commercial value and scope of products.

In particular the following require further study:

- Uniformity of feedstock mixes and the relationship between feedstock and product quality;
- Design of compost control systems for influencing -
  - high-rate systems by stressing / de-stressing the microbial mass,
  - microbial functionality,

nitrogen form and availability,
product aesthetics;

- The colonisation of spatial and biological niches by pathogens ;
- The link between compost system and product function and
- Creation of standards for compost and compost production which not only address regulatory issues but also 'fitness for purpose' of the finished product.

Acnowledgements. SAC receives grant in aid from the Scottish Executive, Rural Affairs Department (SERAD) Advisory Activities programme. The author would like to acknowledge grant-assistance for compost technology development from Score Environmental and Fife Environment Trust.

References

Anon (1994) Compost Calculator USDA-CSRS 91-COOP-1-6159.The Compost Council Research Foundation

Anon (1997) Environmental Protection Act 1990, Part I: Secretary of State's guidance – Production of mushroom substrate. Pub.: Department of the Environment (PG6/30/97), London, UK

Bahl N (1991) Supplementation of nitrogen in *Agaricus* compost by agro wastes. Mushroom Science XII (1): 201-203

Balis C, V Tassiopoulou (in press) Triggering effect of hydrogen peroxide on composting and a new method for assessing stability of composts in a thermally insulated microcosm. Acta Horticulturae

Bess V (1999) Evaluating microbiology of compost. Biocycle. 40: 62-64

Bhurtun C, R Mohee (1996) Performance prediction of composting processes using fuzzy cognitive maps. In: The Science of Composting (2), Edit: de Bertoldi M, P Sequi, B Lemmes, T Papi Pub: Blackie, London, UK. 1083-1086.

Bragg NC (1990) Peat and its alternatives. Pub HDC, Petersfield, UK

Bragg NC, BJ Chambers (1988) Interpretation and advisory applications of compost air-filled porosity of potting substrates. Acta Horticulturae 294: 183 - 190

Bunt A C (1988) Media and Mixes for container grown plants. Pub Unwin Hyman, London. pp309

Burrage S (Personal communication) SW & WS Burrage Ashford, UK

Carpenter-Boggs, L, JP Reganold & AC Kennedy (1999) Effects of biodynamic preparations on compost development Biological Agriculture and Horticulture. 17: 313-328.

Castellanos JZ, PF Pratt (1981) Mineralization of manure nitrogen – correlation with laboratory indexes. Soil Sci Soc Am J 45: 354-357

Chen AA (1987) Chitin metabolism. Archives of Insect Biochem Physiol 6: 267-277

Cohen RB, Y Chefetz, Y Hadar (1998) Suppression of soil-borne pathogens by composted municipal solid waste. In: Brown S, JS Angle, L Jacobs (Ed) Beneficial Co-utilization of agricultural, municipal and industrial by-products, Pub Kluwer Academic, Dordrecht, Nl, 113-130

de Bertoldi, M, P Sequi, B Lemmes & T Papi (Eds) (1996) The Science of Composting (Parts 1&2) Pub: Blackie Academic, London, UK pp1383

Farrell JB (1993) Fecal pathogen control during composting. In: Hoitink HAJ, HM Keener (Ed) Science and Engineering of composting: Design, microbiological and utilization aspects. Pub: Renaissance, Worthington,Ohio, USA, 282-300

Folliet-Hoyte N (1996) Canadian national compost standards. In: de Bertoldi M, P Sequi, B Lemmes, T. Papi (Eds) The Science of Composting, Pub: Blackie Academic, London, UK 247-254

Gabriels R (1998) Quality of composts and EC legislation Acta Hort 469: 187-194

Garland JL (1996) Analytical approaches to the characterization of samples of microbial communities using patterns of potential C source utilization Soil Biol Biochem 28: 213-221

Garland JL, AL Mills (1991) Classification and characterization of heterotrophic microbial communities on the basis of patterns of community-level sole-carbon-source utilization. Appl Env Microbiol 57: 2351-2359

Gerrits JPG (1988) Nutrition and Compost. In: van Griensven LJLD (ed) The Cultivation of Mushrooms. Darlington Mushroom Laboratories. Rustington, UK 29 - 72

Gerrits JPG (1969) Organic compost constituents and water utilised by the cultivated mushroom during spawn run and cropping. Mushroom Sci 7: 111 - 126

Gulliver A, FC Miller, E Harper, BJ Macauley (1991) Environmentally controlled composting on a commercial scale in Australia. Mushroom Sci XIII: 155-164

Herbert M, A Karman, LE Parent (1991) Mineralization of nitrogen and carbon in soils amended with composted manure. Biol Agric Hortic 7: 349-361

Hillman K (2000) Modification of the intestinal microflora with starches. In: Animal, Food Science research report for 1999 Pub SAC 21-24

Hoitink HAJ, MJ Boehm, Y Hadar (1993) Mechanisms of suppression of soilborne plant pathogens in compost-amended substrates. In: Hoitink HA.J, HM Keener (Eds) (1993 Science and Engineering of Composting: Design, microbiological and utilization aspects. Pub: Renaissance, Worthington, Ohio, USA, 601-621

Hoitink HAJ, HM Keener (Eds.) (1993) Science and Engineering of composting: Design, microbiological and utilization aspects. Pub Renaissance, Worthington, Ohio, pp. 724

Jackson DV, J-M Merillot, PL'Hermite (Edit) (1992 Composting and compost quality assurance criteria. Pub: Commission of the European Communities, Brussels, B pp. 433

Keener HM C Marugg RC Hansen, HAJ Hoitink (1993) Optimizing the efficiency of the composting process. In: Hoitink HAJ, HM Keener (Eds) Science and Engineering of composting: Design, microbiological and utilization aspects. Pub: Renaissance, Worthington, Ohio, USA, 59-154

Lamber F (1991) Computer control in mushroom growing: An inventory of applications. Mushroom Sci XIII: 289-291

Lindberg C (1996) Accelerated composting in tunnels. In: de Bertoldi M P Sequi B Lemmes, T Papi (Eds) The Science of Composting, Blackie Academic, Professional, London, UK, 1205-1206

Lopez-Real J, A. Vere (1992) Composting control parameters and compost product characteristics In: Jackson DV J-M Merillot, PL'Hermite (Eds) Composting and compost quality assurance criteria. Pub: Commission of the European Communities 131-140

MacGregor ST FC Miller KM Psarianos, MS Finstein (1981) Composting process control based on interaction between microbial heat output and temperature Appl Env Microbiol 41: 1321-1330

Miller FC, ER Harper, BJ Macauley (1989) Field examination of temperature and oxygen relationships in mushroom composting stacks – consideration of stack oxygenation based on utilsation and supply. Austr J Exp Agric 29: 74-749

Miller FC, BJ Macaulay, ER Harper (1991 Investigation of various gases, pH and redox potential in mushroom composting phase I stacks. Austr J Exp Agric 31: 415-425

O'Brien TA, AV Barker, J Campe (1999) Container production of tomato with food by-product compost and mineral fines. J Plant Nutr 22, 445-457

Pardo J, FJ Gea, Pardo A, MJ Navarro (1995) Characterization of some materials derived from the grapevine industry and their use in traditional composting in Castilla-La Mancha. Mushroom Sci XIV: 213- 221

Perrin PS, BJ Macauley (1995) Positive aeration of conventional (Phase I) mushroom compost stacks for odour abatement and process control. Mushroom Sci XIV: 223-235

Psarianos KM, ST MacGregor, FC Miller, MS Finstein (1983) Design of composting ventilation system for uniform air distribution. Biocycle 24: 27-31

Randle P, PB Flegg (1978) Oxygen measurement in a mushroom compost stack. Scientia Horticulturae 19:315-323

Ritz K, J Dighton, K Giller (Eds) (1994) Beyond the Biomass: Compostitional and functional analysis of soil microbial communities. Wiley, Chichester pp.275

Ryan JA, RL Chaney (1993) Regulation of municipal sewage sludge under the clean water act section 503: A model for exposure and risk assessment for MSW-compost. In: Hoitink, H.A.J., H.M. Keener (Eds): Science and Engineering of composting: Design, microbiological and utilization aspects, Renaissance, Worthington, Ohio, USA, 422-450

Sharma HSS, G Lyons, J Chambers (2000) Comparison of the changes in mushroom (*Agaricus bisporus)* compost uring windrow and bunker stages of Phase I and II. Ann Appl Biol 136: 59-68

Sikora LJ, RAK Szmidt (2001) Nitrogen Sources, mineralization rates and plant nutrient benefits from compost. In: PJ Stofella, BA Kahn (Eds) Compost Utilization in Horticultural Systems. Pub: CRC Press, Boca Raton, USA 287 - 305

Sinden JW, E Hauser (1953) The nature of the composting process and its relation to short composting. Mushroom Sci 2, 123-131

Stentiford EI (1992) The composting process applied to sewage sludge and source separated refuse. In: Jackson DV, J-M Merillot, PL'Hermite (Eds) Composting and compost quality assurance criteria. Pub: Commission of the European Communities, Brussels, B 69-80

Stentiford EI (1993) Diversity of composting systems. In: Hoitink, HAJ, HM Keener (Edit): Science and Engineering of composting: Design, microbiological and utilization aspects. Pub: Renaissance, Worthington,Ohio, USA 95-110

Straatsma, G, TW Olijnsma, JPG Gerrits, LJLD van Griensven, HJM Op den Camp (1995) Inoculation of indoor compost with thermophiles. Mushroom Sci XIV: 283-291

Szmidt RAK, DA. Hall, GM Hitchon (1988) Development of perlite culture systems for the production of greenhouse tomatoes. Acta Hort 221: 371-379

Szmidt RAK (Edit) (1998) Composting and use of composted materials for horticulture. Acta Hort 469 pp.480

Szmidt RAK (Ed) (1999) Report of the National Waste Strategy Composting Task Group: Scottish Environment Protection Agency, Stirling, UK. pp. 60.

Szmidt RAK, NC Bragg (In press) Composted horticultural media. Pub Nexus Horticulture, Swanley UK

Tester CF, LJ Sikora, JM Taylor, JF Parr (1977) Decomposition of sewage sludge compost in soil. I. Carbon and nitrogen transformations. J Environ Qual 6: 459-463

Van Griensven LJLD (1988) The Cultivation of Mushrooms, Darlington, Rustington, UK

Verdonck O (1998) Compost specifications. Acta Hort 469: 169-178

Wen G, TE Bates, RP Voroney (1995) Evaluation of nitrogen availability in irradiated sewage sludge, sludge compost and manure compost. J Environ Qual 24: 527-534

Using Agricultural Wastes for *Tricholoma crassum* (Berk.) Sacc. Production

N. Teaumroong[1], W. Sattayapisut[1], T. Teekachunhatean[2] and N. Boonkerd[1]

Abstract. *Tricholoma crassum* (Berk.) Sacc. is usually found only once a year, particularly in the rainy season. The potential for commercialization of this mushroom was investigated. As an attempt to cultivate *T. crassum*, rather than collecting from nature, agricultural wastes (mungbean husks and potting soil) were used as substrates for cultivation. Prior to the fruiting step, sawdust and sucrose were used as main carbon sources without fermentation or decomposition during spawn run. The type of containers and substrates for cultivation were compared in relation to fruit body yield. The results suggest that potting soil in container with the least aeration exhibited the highest yield (44.5 g fresh weight of fruiting body per 800 g spawn). Total period of production took almost 3 months. Moreover, the cultivation approach developed here can control the appropriate size of fruit bodies for further commercialization.

Introduction

Tricholoma crassum (Berk.) Sacc. is, one of the wild edible mushrooms which belongs to the order agaricales, family Tricholomataceae (Chang and Hayes 1978). Only a few edible species of this family, such as *T. matsutake* and *T. crassum,* were found in the Asia region (in Japan, Thailand and Sri-Lanka) while several species such as *T. albobrunneum, T. flavovirens, T. paneolum, T. equestre, T. terreum* and *T. georgii,* were mostly found in Europe and America (Huffman et al. 1989). The general phenotypic characteristic of *T. crassum* is smooth with convex pileus with a diameter of 3–25 cm, thickness of 1–3 cm and white in colour. Gills are freely separated from the stalk with 20–25 grills cm^{-1}. Stalk lengths are in the range of 7–24 cm and are white in colour. The white spores are ovoid-shaped with size 5–6.5 x 6.5–7.6 μm. *T. crassum* grow naturally in various habitats such as meadow, rice field, forest and soil containing high organic matter.

It is edible and non-toxic (Chilton 1994). However, it is usually found only once a year, particularly in the rainy season as widely distributed in every region of the country, thus the various common names among local people, such as Hed-Tin-Raed (northeast), Hed-Jan (north), Hed-Hua-Sum (south) and Hed Yai or Hed-Tub-Tao-Khao (central) (Wason 1996). The nutritional value and taste are rather promising for commercialization, with 0.287 g fats, 10.02 g carbohydrates,

[1]School of Biotechnology and [2]School of Plant Production, Institute of Agricultural Technology, Suranaree University of Technology, Nakhon Ratchasima 30000, Thailand
E-mail : neung@ccs.sut.ac.th

H. Insam, N. Riddech, S. Klammer (Eds.)
Microbiology of Composting

2.91 g protein, 0.486 g fibre, 84.34 g water, 2.71 mg calcium, 115.75 mg phosphorus and, 3.35 mg iron, and contain only 54.3 cal per 100 g fresh weight (pers. data obtained from the Ministry of Health, Thailand, 1992). In addition, *T. crassum* is easy to handle particularly in the postharvest, step due to its hard tissue and slow autolysis properties. However, their growth under natural conditions was uncontrollable and usually of a gigantic size, which gave consumers the impression of being poisonous rather than edible. This study aims to develop an appropriate cultivation technology of *T. crassum* on the basis of natural imitation. Culturing was carried out with agricultural waste and compost soil used as cultivation materials under limited space to promote a smaller size of fruit-body for commercial purposes.

Materials and Methods

Preparation of the Inoculum

The mycelial form of *T. crassum* was isolated from the fruiting body available in local markets and further selected by a research unit, University Farm, Suranaree University of Technology. The mushroom tissue culture was cultivated on potato dextrose agar (PDA) and incubated at room temperature (28–30 °C) for 15–30 days. Mycelial growth rate was also determined by daily measurement of colony diameter. The mycelia from pure culture was used as the inoculum for preparing the spawn. The medium for incubating spawn was composed of 100 kg sawdust, 5 kg rice bran, 1 kg $CaCO_3$, 0.5 kg gypsum ($CaSO_4$), 0.2 kg and 200 g sucrose (commercial grade). The moisture was adjusted to 60%. Eight hundred g of the mixed medium was packed into clean plastic bags and partially by sterilized by steaming at 95 °C for 3 h. The completely cooled medium was then inoculated with a mycelial mat of *T. crassum*. The progress in spawning was measured every 2 weeks to determine mycelial growth rate in the spawning step.

Fruiting Condition

Two types of containers for fruiting were polypropylene baskets and earthenwear pots, both 42 cm in diameter and 35 cm in height. The substrates for cultivation were potting soil containing 1.4% organic matters, mungbean husks and potting soil mixed with mungbean husks (1:1, v/v). The complete spawning was removed from plastic bag after 4 weeks cultivation prior to bedded with the substrates. Each container was first bedded for 5 cm with substrate for cultivation then overlaid with four complete spawning with a distance of 10 cm between the spawn. The substrate for cultivation was then filled to the top of the container. A moisture content of approximately 60–70% was continually maintained throughout the cultivation period by watering.

Results and Discussion

Mycelial Growth Rate in the Spawning Step

PDA was chosen as the medium for pure culture inoculum. Several media were tested (Deprom 1976), but only malt extract and potato dextrose peptone yeast extract promoted higher growth rate when compared to PDA. However, the performance of mycelial growth of *T. crassum* on PDA was not much different from these letter media, and the ingredients of PDA were cheaper. Thus, PDA was chosen as medium. Mycelial growth rate was determined as depicted in Fig 1. Along with the colony diameter, the log phase was during 19–23 days of cultivation. The spawning rate was also measured by the growth of mycelium every 2 weeks (Fig.2).

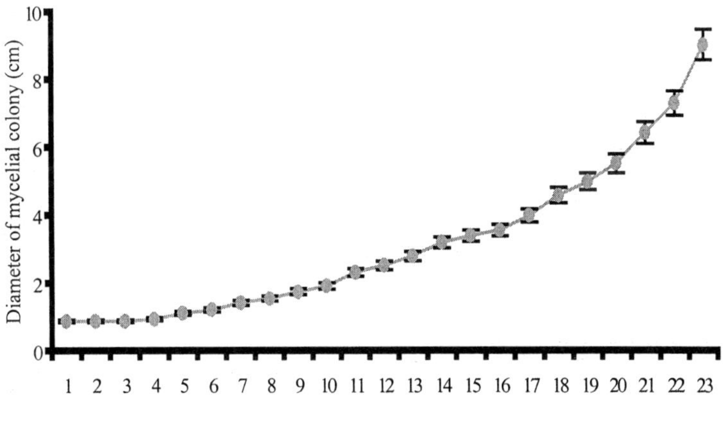

Fig. 1 Growth rate of *T. crassum* mycelial formation on PDA, incubated at 30 °C. Values represent means and standard diviation of three replications

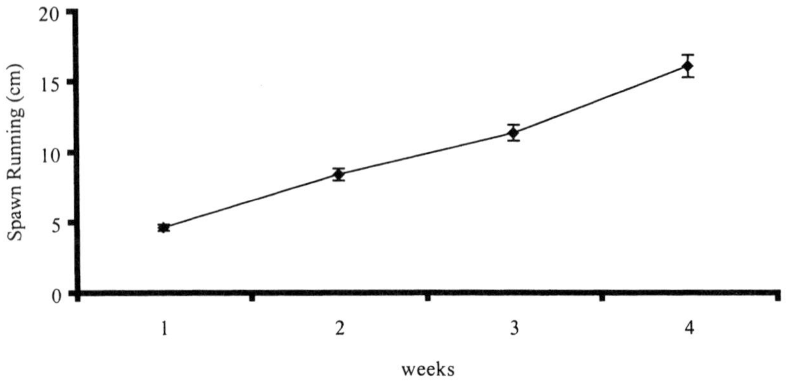

Fig. 2 Growth rate of *T. crassum* mycelial formation in spawning, incubated at room temperature 28-30 °C. Values represent means of strandard diviation of three replications

Full spawning could be achieved within 6 weeks under 60% moisture content. Previous works had used media either decomposed or fermented by microorganisms for incubating spawn. For example, encasing in sawdust fermented with horse or pig dung for 3 days previously, (Pantawee 1975), using peanut husks and sawdust fermented by horse dung for 1 week, or using corn waste:sawdust:blended corn (15:4:1) followed with fermentation for 3 weeks (Wason and Wijai 1987). However, in this study sawdust and sucrose were used as the main carbon sources without decomposting or fermentation. The results showed a promising spawning rate without the problems of a time-consuming process and chances of contamination.

Effect of Substrates and Containers to the *T. crassum* Yield

After spawns of *T. crassum* were added into containers, the full colonization of mycelia could be clearly seen within 17 days and first development of primordia was established in the next 12 days. The fruiting body of *T. crassum* could be harvested 9 days after primordia formation. The time from the start of preparation of pure culture till the fruiting step was almost 3 months. When compared with the yield obtained from various substrates, it was found that cultivation with potting soil alone promoted the highest yield at 44.5 g (fresh weight) $spawn^{-1}$ (Fig. 3).

With cultivation on mungbean husks and mungbean husks mixed with potting soil (1:1, v/v), a yield of only 12.6 and 16.1 g (fresh weight) $spawn^{-1}$, respectively could be gained. This might imply that to cultivate *T. crassum*, aeration provided by mungbean husks is somewhat unnecessary or in some way retards the fruitification. In addition, some other essential minerals and nutrients in high organic substances containing soil as potting soil were also critical in terms of both quantity and quality for *T. crassum* fruiting body development.

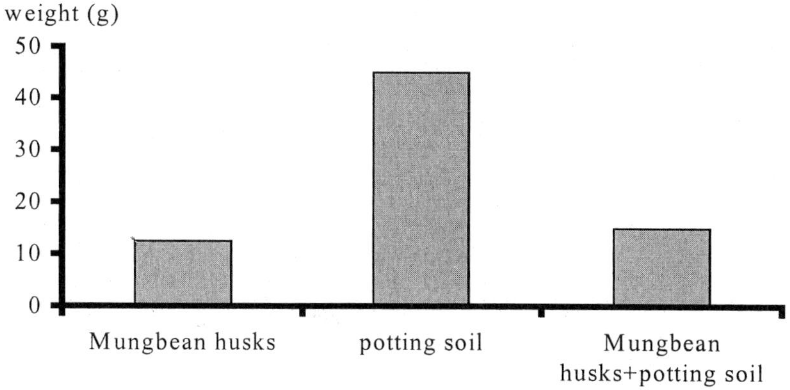

Fig. 3 Yield of *T. crassum* fruiting body grown on various substrates; mungbean husks, potting soil and mungbean husks mixed with potting soil (1:1, v/v)

Moreover, when compared with the yield obtained from different containers as baskets and earthenwear pot that used potting soil as substrate, the yield also corresponded to the result from the first experiment. The production yield from cultivation in baskets was 20% less than that from earthenwear pot cultivation. The average size of the cultivated fruiting body could be controlled as 10 cm of cap diameter, 8–10 cm for the length of stalk and 1.5—2.0 cm stalk diameter, while a naturally grown fruiting body varied in size, with cap diameter ranging between 5 and 25 cm and a height of 5–40 cm (Fig. 4) or even larger.

In the past, the approach to cultivation of a suitable or controlled size of *T. crassum* mushroom was conducted with plastic bag cultivation, which is rather labour-intensive for a large farming system. Therefore, this simple cultivation technique of *T. crassum* developed here would be an alternative appropriate technology to transferring to farmers.

A

Fig. 4 A. Legend see next page

B C

Fig. 4 A-C Comparison of size between naturally grown of *T. crassum* and from cultivation. A) Young fruiting body (6 days after primordia formation), B) Mature fruiting body (9 days after primordia formation) C) Naturally grown fruiting body

Acknowledgement. This work was fully supported by a Suranaree University of Technology grant.

References

Chang ST, Hayes WA (1978) The biology and cultivation of edible mushrooms. Academic Press, San Diego, California

Chilton WS (1994) The chemistry and mode of action of mushroom toxins. In: Spoerke DG, Rumack BH (eds) Handbook of mushroom poisoning: diagnosis and treatment. CRC Press, London

Deprom C (1976) Cultivation of *Tricholoma crassum* in plastic bags. J Thai Mycol 1:22-25

Huffman DM, Tifany LH, Knaphus G (1989) Mushrooms and other fungi of the midcontinental United States. Iowa State University Press, Ames, Jowa

Pantawee P (1975) The study of *Tricholoma crassum* cultivation. J Thai Mushroom 2:34-43

Wason P (1996) Cultivation of wild mushroom: VIII Hed Tin Raed [*Tricholoma crassum* (Berk.) Sacc.]. Songklanakarin J Sci Technol 4:397-406

Wason P, Wijai R (1987) Cultivation of *Tricholoma crassum*. J Thai Plant Pathogen 7:1-13

Microbial Transformation of Nitrogen During Composting

S.M. Tiquia[1]

Abstract. Microorganisms are fundamentally involved in important changes to the N compounds during composting. However, their role in composting systems is not well understood. Hence, this study was conducted to evaluate the microbial transformation of nitrogen during composting of spent pig litter-sludge and poultry litter in forced-aeration piles. Most N in spent pig litter sludge and poultry litter is in organic forms, which serve as a reservoir of N. During composting, some of the organic N was slowly converted to the much smaller inorganic N pools. The mineralization process was continued further by conversion of ammonium to nitrite/nitrate by nitrifiers. The ammonium- and nitrite-oxidizing bacteria are the microorganisms that gain their energy from these inorganic oxidations. Denitrification occurred during the early stage of composting, as indicated by a higher population of denitrifying bacteria. However, as composting proceeded, the population of denitrifying bacteria declined significantly, indicating that very little denitrification took place once the air was forced into the pile. The multiple regression analysis showed that the physico-chemical properties of the spent pig litter-sludge and poultry litter are the most critical factors affecting the changes in N and its different forms during composting. The equilibria and rates of N were affected by interactions between microbial biomass community structure and the physico-chemical properties of the manure such as temperature, water content, pH, and C:N ratio during composting.

Introduction

Nitrogen is an essential nutrient for all life on earth (Galloway 1998). Thus, its fixation into usable forms by microorganisms and subsequent transformations and recycling through organic and inorganic forms are of great interest. Indeed, N is the nutrient most limiting plant growth in terrestrial ecosystems (Wild 1988). Current concerns include high concentrations of nitrate in ground and surface waters and the contribution of gaseous nitrogen oxides, such as NO and N_2O, to large-scale environmental problems of acid rain, ozone depletion, and greenhouse warming (Galloway 1998). The large diversity of N-containing compounds, which exist in numerous oxidation states, and the wide array of microbial transformations, make the N cycle an extremely interesting intellectual knowledge.

[1]Environmental Sciences Division, Oak Ridge National Laboratroy, P.O. Box 2008, Oak Ridge, Tennessee 37831, USA
e mail: tiquias@ornl.gov/ tel: +1 865-5747302/ fax: +1865-5768646

H. Insam, N. Riddech, S. Klammer (Eds.)
Microbiology of Composting

Nitrogen is also the nutrient that has received the most attention in composting, as it could be lost significantly during the composting process. Recent studies have showed that about 20–70% of the initial N of the initial feedstock could be lost due to ammonia volatilization, leaching, and runoff during composting (Martins and Dewes 1992; Rao Bhamidimarri and Pandey 1996; Tiquia and Tam 2000a). These losses not only reduce the value of the composted product as an N fertilizer, but they could also lead to serious environmental pollution (Kirchmann and Lundvall 1998). Apart from N losses through ammonia volatilization, leaching, and run-off, there are a number of microbially mediated processes (ammonification, immobilization, nitrification, and denitrification) that are involved in other changes to the N compounds during composting. These processes are responsible for moving the fixed nitrogen from one form to another in the compost, and are vital in understanding the composting process.

Despite the unique role played by microorganisms in determining the characteristics of composts added to soil, very few studies have been published on the microbial communities of such residues, particular those populations that are involved in nitrogen transformation during composting (Finstein and Morris 1975; Diaz-Ravina et al. 1989; Nodar et al. 1990). Hence, this study was carried out to describe the evolution of microbial populations involve in N transformation during composting. Since the key processes in N dynamics are also affected by other controlling factors such as environmental conditions and physico-chemical properties of the compost, this study will also examine the most important physical and chemical factors affecting the transformation of N in the spent litter-sludge and poultry litter.

Materials and Methods

Composting Setup and Sampling

Spent litter-sludge (a mixture of spent pig litter and sludge; 2:1 litter: sludge, wet volume) and poultry litter (a mixture of poultry manure, wood shavings, wastes feed, and feathers) were composted using forced-aeration method (Tiquia and Tam 2000b). Three forced-aeration piles were setup for the spent pig litter-sludge and also for the poutry litter. Triplicate composite samples (approximately 1 kg each) were taken from each pile at day 0 and then weekly until the termination of the composting trial. Average temperature readings were three locations of the forced-aeration piles every 4 days.

Chemical and Microbial Analysis

The manure composts were characterized for the following parameters: water content, pH; electrical conductivity, organic matter (OM), ash; different forms of

N (Sparks 1996). The theoretical total N concentration of the compost samples was calculated by adding the Kjeldahl N with the NO_x^--N, whereas the organic N concentration was derived from subtracting the NH_4^+-N from the Kjeldahl N. Denitrifying bacterial population was quantified by inoculation of tubed liquid media (Tiedje 1994) using the most probable number (MPN) method (Woomer et al.1990). Total aerobic heterotrophs, and ammonium- and nitrite-oxidizing bacteria were quantified on appropriate media using the plate frequency technique (Tiquia et al. 1998).

Statistical Analysis

Pearson product-moment correlation coefficients were calculated to show the relationship between different forms of N and physico-chemical and microbial properties of the manure compost. To determine the most important factors (physical, chemical, or microbial) affecting the transformation of N in the spent pig litter-sludge and poultry litter, a stepwise multiple regression analysis was performed. Statistical analyses were computed using SigmaStat 1.0 for Windows statistical package.

Results and Discussion

The Composting Process

The temperature within the composting mass determines the rate at which many biological processes take place, and the material is considered mature if the declining temperature reaches ambient level (Golueke 1972). The temperature of the spent pig litter-sludge piles mass increased rapidly to between 55 and 65 °C, and persisted until the active decomposition was over, thereafter slowly decreasing (Fig. 1A).

The declining temperature of spent litter-sludge reached ambient level by day 77, which was 111 days earlier than the poultry litter piles (Fig. 1A and B). The temperature of the poultry litter took 168 days to reach ambient temperature (Fig. 1B). Higher peak temperatures (70–75 °C) also persisted during the first 21 days of composting (Fig. 1B). The poultry litter piles had higher peak temperatures and conserved heat longer than spent pig litter-sludge piles due to heat generated by on-going microbial activities (Fig. 1D), and possibly due to the higher insulating quality of the poultry litter (Mathur 1998).

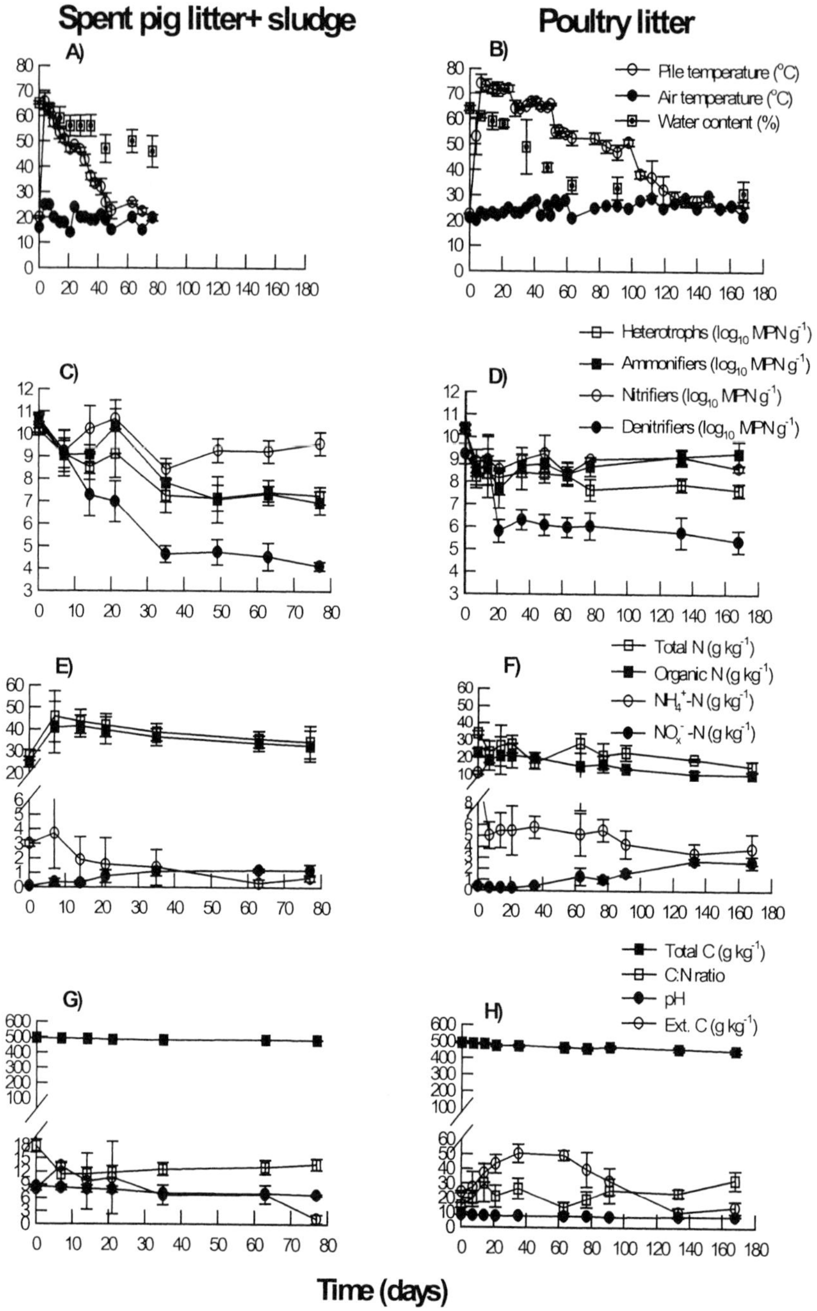

Figure 1. Changes in physico-chemical and microbial properties of manure composts.

The poultry litter also had higher concentrations of available nutrients and water-extractable C than the spent pig litter-sludge (Tiquia and Tam 2000a,b).

Nitrogen Dynamics During Composting

The changes in total N concentration of the manure composts were very similar to the organic N (Fig. 1E–F). This result relates to the facts that the inorganic fractions (NH_4^+-N and NO_x^--N) were low, and that organic N was the major nitrogenous constituent in both feedstocks. After the slight increase in organic N by day 14 (for spent pig litter-sludge) and day 21 (for poultry litter), the organic N gradually decreased (Fig. 1E,F). The initial increase can be attributed to a concentration effect as a consequence of degradation of organic C compounds (Fig. 1G,H), which reduced the dry mass. On the other hand, the subsequent decrease was a result of the ammonification process, which converted a fraction of the organic N to NH_3 and NH_4^+ ions.

The NH_4^+-N concentration decreased dramatically during composting. The dramatic decline in NH_4^+-N concentrations of the spent pig litter-sludge and poultry litter did not, however, correspond to a rapid increase in NO_x^- -N, which indicated that some of the NH_4^+-N in these piles had been lost through NH_3 volatilization and/or microbial denitrification. The increase in NO_x^--N concentration during composting was only 1.3 g kg^{-1} and 2.6 g kg^{-1} for spent pig litter-sludge and poultry litter, respectively. This means that about 1.24 g kg^{-1} and 4.60 g kg^{-1} of NH_4^+-N in the spent pig litter-sludge and poultry litter, respectively were lost during composting. Some of the NH_4^+ ions were lost through NH_3 volatilization as the pH of both piles were > 7.0 (Fig. 1G,H), giving conditions that favored NH_3 volatilization. Some of the NO_3^-/NO_2^- ions were lost through microbial denitrification as the population of denitrifying bacteria was high during the early stage of composting (Fig. 1C,D).

Evolution of Microbial Populations Related to N Transformation

It is generally believed that microorganims are fundamentally involved in biochemical transformations during composting (Mathur 1998). The ammonium-oxidizing bacteria are the microorganisms involved in the oxidation of NH_4^+ to NO_2^-, while nitrite-oxidizing bacteria are involved in the subsequent oxidation of NO_2^- to NO_3^-. There are also reports that heterotrophic organisms including bacteria, fungi, and actinomycetes are able to oxidize NH_4^+ to NO_2^-/NO_3^- (Wild 1988). Numbers of total aerobic heterotrophs, and ammonium- and nitrite-oxidizing bacteria were maintained at high population sizes during the active decomposition process (Fig. 1C,D), which suggests a rapid nitrification rate. Myrold (1999) estimated that about 3 x 10^5 nitrifiers g^{-1} soil is required to produce a nitrification rate of 1 mg N kg^{-1} day^{-1}. This population is about half the initial population of total aerobic heterotrophs (10.16 and 10.40 log_{10} MPN g^{-1}), ammonium-oxidizing bacteria (10.50 and 10.31 log_{10} MPN g^{-1}), and nitrite-oxidizing bacteria (10.70 and

10.29 $\log_{10}$ MPN g^{-1}) of the spent pig litter-sludge and poultry litter, respectively. Despite this, there was no dramatic increase in NO_x^--N concentration (Fig. 1E,F). This result relates to the facts that a fraction of the NH_4^+-N had been lost through ammonia volatilization and denitrification caused by other interacting factors such as aeration, pH, temperature, water content, and initial C:N ratio (Mathur 1998; Tiquia and Tam 2000a). If these factors are not optimized during composting, they could stimulate volatilization and denitrification of N in the compost piles.

Denitrification is the dissimilatory reduction of NO_3^- to N gases NO, N_2O, and N_2. This process is carried out by a wide variety of mainly heterotrophic bacteria that use NO_3^- as a terminal electron acceptor when O_2 is unavailable. Hence, denitrification occurs in composting where C and NO_3^- are available during periods of restricted O_2 availability (Bishop and Godfrey 1983). In this study, the numbers of denitrifying bacteria were highest at the beginning of composting (Fig. 1C,D) due to higher initial water content (65%) of the piles, which could hinder aeration and subsequently produce anaerobic or microaerophilic pockets in the compost piles. Under these conditions, the denitrifying bacteria use NO_3^- as an electron acceptor instead of O_2, leaving N and N_2O gases to be released from the compost piles to the atmosphere. As composting proceeded, the population of denitrifying bacteria declined significantly (Fig. 1C,D), indicating that very little denitrification took place once the air was forced into the pile. The NO_x^--N concentration also started to increase at a faster rate as composting progressed (Fig. 1C,D).

Relationship Between Different Forms of N and Other Compost Properties

Significant negative correlations were found between C:N ratio and total and organic N of the manure composts (Table 1). This result means that as the total and organic N decreased, the C:N ratio increased. The C:N ratio of the compost pile normally decreases during composting (Golueke 1972); however, due to vigorous losses of N, the C:N ratio increased (Fig. 1E,F). Morisaki et al. (1989), Tiquia and Tam (2000a), and Tiquia et al. (2000) also reported increases in C:N ratio during composting. These losses were attributed mostly through ammonia volatilization. Here, significant correlations were found between NH_4^+-N and temperature, water content, pH, and carbon, while significant negative correlations were found between NO_x^--N and these four parameters (Table 1). The NH_4^+-N was positively correlated with the microbial properties, with the exception of nitrite-oxidizing populations of spent pig litter-sludge where no correlation was found. On the other hand, significant negative correlations were found between NO_x^--N and microbial properties of these two manures (Table 1). The multiple regression analysis demonstrated that C:N ratio was the most important factor affecting the changes in total N during composting of spent pig litter-sludge and poultry litter.

Table 1. Pearson product-moment correlation coefficient (*r*) values between different forms of N and physico-chemical and microbial properties of spent pig litter-sludge and poultry litter

	Parameters									
Different forms of N	temperature	watercontent	pH	Carbon	ext. C	C:N ratio	hetero	ammoni	nitri	denitri
Spent pig litter-sludge										
Total N=	0.94***	0.08	0.12	-0.003	0.62*	-0.97***	-0.07	-0.02	-0.21	-0.01
Organic N=	0.88**	-0.05	-0.01	-0.14	0.52	-0.99***	-0.19	-0.06	-0.21	-0.15
NH_4^+-N	0.60*	0.94***	0.90***	0.87**	0.76*	0.02	0.83*	0.74*	0.29	0.91***
$(NO_3^-+NO_2^-)$-N	-0.59*	-0.93***	-0.97***	-0.93***	-0.68*	-0.13	-0.91***	-0.85**	-0.61*	-0.96***
Poultry litter										
Total N=	0.04	0.64*	0.56	0.70*	0.24	-0.66*	0.71*	0.19	0.63*	0.61*
Organic N=	0.35	0.91***	0.76*	0.89**	0.53	-0.60*	0.68*	-0.06	0.33	0.69*
NH_4^+-N	0.59*	0.66*	0.78**	0.68*	0.13	-0.51	0.98***	0.64*	0.74*	0.66*
$(NO_3^-+NO_2^-)$-N	-0.61*	-0.89***	-0.67*	-0.87**	-0.63*	0.31	-0.64*	0.60*	-0.61*	-0.64*

Ext. C= water extractable carbon; ammoni=ammonium oxidizing bacteria; nitri= nitrite oxidizing bacteria; denitri=denitrifying bacteria.
Correlations were based on 10 average data of the spent pig litter-sludge and poultry litter piles; *, ** and *** indicate significance at 0.05, 0.01 and 0.001 probability levels, respectively.

For the spent pig litter-sludge, the C:N ratio, water content, and pH are the most critical factors affecting the changes of organic N, NH_4^+-N, and NO_x^--N, respectively. For poultry litter, the changes in organic N and NO_3^--N were affected by water content, while change in NH_4^+-N was affected by population of total aerobic heterotrophs (Table 2).

Results of this study revealed that the microbial transformation of N during composting varies with the properties of the initial composting materials. The microbial biomass community structure as well as the physico-chemical properties of the manure during composting such as temperature, water content, pH, and C:N ratio would affect the transformation of N. The equilibria and rates of N dymanics were affected by interactions between microbial biomass community structure and populations, and the physico-chemical properties of the manure.

Table 2. Multiple regression analysis between different forms of N and microbial and physico-chemical properties of spent pig litter-sludge and poultry litter

Regression equation	Multiple R^2 value	F value	Significance of F
Spent pig litter-sludge			
Total N= 72.3 - (2.57 * C:N ratio)	0.92	70.30	0.0004
Organic N= 68.7 - (2.5 * C:N ratio)	0.98	285.00	<0.0001
NH_4^+-N= -7.82 + (0.12 * water content) + (0.35 * pH)	0.79	12.40	0.0192
(NO_3^-+NO_2^-)-N= 5.71 - (0.65 * pH)	0.93	81.50	0.0003
Poultry litter			
Total N= -37.6 - (0.62 * C:N ratio)	0.57	6.33	0.0360
Organic N= 3.33 + (0.30 * water content)	0.82	40.90	0.0002
NH_4^+-N= -16.2 + (2.6 * heterotrophs)	0.96	183.50	<0.0001
(NO_3^-+NO_2^-)-N= 3.86 - (0.06 * water content)	0.76	29.60	0.0006

Regression analysis was calculated based on four microbial (total aerobic heterotrophs, ammonium- and nitrite-oxidizing bacteria, and denitrifying bacteria) and six physico-chemical (temperature, water content, pH, total C, water-extractable C, and C:N ratio) parameters with stepwise method and PIN (probability of f-to-enter) = 0.050 limit. Heterotrophs=total aerobic heterotrophs.

Acknowledgements. This project was financially supported by Central Matching Fund (Postdoctoral Fellowship), City University of Hong Kong. The author is grateful to N. F.Y. Tam for providing meaningful suggestions for this project.

References

Bishop PL, Godfrey C (1983) Nitrogen transformations during sludge composting. BioCycle 24: 34–39

Diaz-Ravina M, Acea MJ, Carballas T (1989) Microbiological characterization of four composted urban refuses. Biol Wastes 30: 89–100

Finstein MM, Morris ML (1975) Microbiology of municipal solid waste composting. Adv Appl Microbiol 19: 113–151

Galloway JN (1998) The global nitrogen cycle:changes and consequences. Environ Pollut 102: 15–24

Golueke CG (1972) Composting: a study of the process and its principles. Rodale Press, Emmaus, Pennsylvania, 110 p

Kirchmann H, Lundvall A (1998) Treatment of solid manures: identification of low NH_3 emission practices. Nutr Cycl Agroecosyst 51: 65–71

Martins O, Dewes T (1992) Loss of nitrogenous compounds during composting animal wastes. Biores Technol 42: 103–111

Mathur SP (1998) Composting process. In: Martin AM (ed) Bioconversion of waste materials to industrial products, 2nd edn. Blackie Academic and Professional, Chapman and Hall, London, pp154–193

Morisaki N, Phae CG, Nakasaki K, Shoda M, Kubota H (1989) Nitrogen transformation during thermophilic composting. J Ferment Bioeng 67: 57-61.

Myrold D (1999) Transformations of nitrogen. In: Sylvia DM, Fuhrmann JJ, Hartel PG, Zuberer DA (eds) Principles and applications of soil microbiology. Prentice Hall, Upper Saddle River, New Jersey, pp 259–294

Nodar R, Acea, MJ, Carballas T (1990) Microbial populations of poultry sawdust litter. Biol Wastes 33: 295–306

Rao Bhamidimarri SM, Pandey SP (1996) Aerobic thermophilic composting of pigerry solid wastes. Water Sci Technol 33: 89–94

Sparks DL (1996) Methods of soil analysis. Part 3–Chemical methods. SSSA, Madison, Wisconsin, 130 pp

Stentiford EI (1996) Composting control: principles and practice. In: De Bertoldi M, Sequi. P, Lemmes B, Papi T (eds) The science of composting. Part I. Blackie Academic and Professional, Chapman and Hall, London, pp 49–59

Tiedjie JM (1994) Denitrifiers. In: Weaver RW, Angle JS, Bottomley PS. SSSA, Book Series No. 5. Madison, Wisconsin, pp 245–267

Tiquia SM, Tam NFY (2000b) Co-composting of spent pig litter and sludge with forced aeration. Biores Technol 72: 1–7

Tiquia SM, Tam NFY (2000a) Fate of nitrogen during composting of chicken litter. Environ Pollut 110: 535–541

Tiquia SM, Tam NFY, Hodgkiss I.J (1998) *Salmonella* elimination during composting of spent pig litter. Biores Technol 63: 193–196

Tiquia SM, Richard TL, Honeyman MS (2000) Effect of windrow turning and seasonal temperatures on composting of pig manure from hoop structures. Environ Technol 21: 1037–1046

Wild A (1988) Plant nutrients in soil: nitrogen. In: Wild A (ed) Russel's Soil conditions and plant growth, 11th edn.. Longman Group UK. Harlow, Essex, pp 652–694

Woomer P, Bennet J, Yost R (1990) Overcoming the flexibility of most-probable number procedures. Agron J 82: 349–353

Effect of Additives on the Nitrification-Denitrification Activities During Composting of Chicken Manure

H. Yulipriyanto[1], Philippe Morand[1], P. R.[2], G. Tricot[3] and C. Aubert[3]

Abstract. In order to know their effect on composting, we added ligno-cellulosic waste, microbial additive, or *Yucca* juice to chicken manure piles, which were composted in 2 m^3, temperature and air flux-controlled bins. The nitrification/denitrification activity during the composting process was tested in the different mixtures.

In the turned piles, microbial activity developed after 6 weeks. Denitrification occurred in the manure and manure-ligno-cellulosic waste mixture piles, while nitrification occurred in the manure-*Yucca* juice pile. In the static pile, which was composed of a manure-microbial additives mixture composted for 12 weeks, no nitrification/denitrification activity was detected.

Related to the data on temperature, moisture, gas emissions and the chemical analyses, the results obtained show that the additives used in the experiment have an obvious effect on composting. All the additives reduce the nitrogenous losses, owing to the modification of the composting conditions and microbial environment (microbial additive), the increase of the C/N ratio (ligno-cellulosic waste) or the regulation of the emission of ammonia (*Yucca* juice). The microbial additive leads to the more conservative process in terms of nitrogen, and the *Yucca* juice addition, to a larger proportion of dinitrogen in the released nitrogenous gases.

Introduction

In Europe, more than 80% of ammonia emissions are of agricultural origin (Asman 1992) and, of these, 95% come from animal waste (Buijsman et al. 1987), poultry breeding contributing 21% of these (Bline and Aubert 1998). Modern agriculture is also responsible for 50% of the emissions of CH_4 (Husted 1994) and for a significant share of N_2O (Mariotti 1997), both greenhouse gases.

Chicken manure is intensively produced in Europe. Composting is seen as a promising method of stabilizing chicken manure (Aubert and Guiziou 1997; Aubert 1998; Elwell et al. 1998). However, before composting of this manure is done on a large scale, it is necessary to ensure that the nitrogen is not transferred into an equally harmful gaseous form.

---

[1]UMR 6553, CNRS/Université de Rennes I, 35380 Paimpont, France
[2]INRA, Bioclimatologie, 35042 Rennes Cedex, France
[3]ITAVI, Zoopôle Beaucemaine, BP 37, 22440 Ploufragan, France

H. Insam, N. Riddech, S. Klammer (Eds.)
Microbiology of Composting

The transformation of nitrogen and the synthesis of nitrous oxide by nitrification/denitrification are linked to physical, chemical and biological factors. The purpose of this chapter is to determine if there are significant differences in nitrification/denitrification activities during composting of chicken manure given various conditions. These varying conditions include the use of additives such as ligno-cellulosic waste, *Yucca* juice and a microbial preparation. The piles also differ, as some are static piles, while others are aerated through turnings.

Material and Methods

Substrates

The chicken manure used in the experiments was obtained from commercial broiler buildings. The manure used in exp. 1 was either 35 or 50% dry matter (DM), depending on where it was collected in the band. The manure in experiment 2 had a DM content of 75%. Ligno-cellulosic waste was a mixture of wood shavings and sawdust. The firm COBIOTEX provided the microbial additive, which was fixed on a calcareous substratum. *Yucca schidigera* juice, containing 16.8% saponines, was supplied by the firm INOBIO.

Piles

In exp. 1, two static piles of manure were composted for 6 weeks, one at 50% water content (C1), and one at 65% water content (MA). The MA pile was spiked with the microbial additive. Two aerated piles (turned twice) were brought to 65% water content and composted for 6 weeks, one without additive (T), and one with a ligno-cellulosic waste additive (TLW1). The ligno-cellulosic waste included oak shavings (50% of the pile in volume) and sawdust (50% wood in dry mass of the pile).

In experiment 2, a control static pile was also composted, but this time with chicken manure at 25% water content (C2). Again as in exp. 1, an aerated pile with ligno-cellulosic waste was composted. However, this pile (TLW2) had a slightly different ratio of manure and waste (oak shavings and sawdust making up 50% of the pile in carbon mass), and was turned only once. Pile MA from exp. 1 continued without being disturbed, thus being composted for 12 weeks. The final pile (TY) was made with *Yucca* juice added to a concentration of 650 g t^{-1}, which gave 100 g saponins t^{-1}. Water was added to bring the moisture content to about 65% in piles TLW2 and TY.

Composting

The chicken manure was composted in four cells each 3 m^2 in area and 8 m^3 in volume, which were naturally ventilated and located in a room with controlled temperature (Fig. 1). The ground was covered with a PVC film, in order to prevent infiltration loss. The piles were carefully sampled at the beginning, at each turning, and at the end, and weighed with a precision weighing machine (Mettler). They were compacted with feet, at the beginning and at each turning, except the controls (C1 and C2).

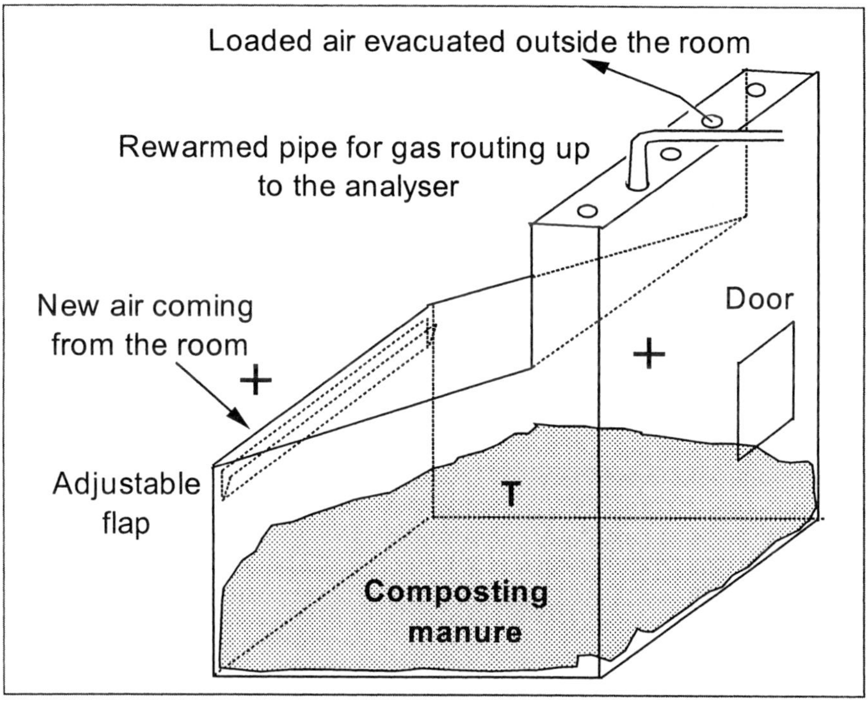

Fig. 1. Experimental cell. + Temperature and hygrometry measurements; T Temperature measurement (thermocouple)

Measurement of Composting Parameters and Analyses

Water, ammonia and nitrous oxide concentrations were measured continuously in the ambient air and in the cells air with a spectro-photo-acoustic gas analyser (Bruel and Kjaer 3426). Nitric oxide concentrations were checked regularly with Draeger tubes to ensure that they were negligible compared to ammonia. Gas lines were warmed to avoid water condensation. The air flow rates were measured continuously with a hot-wire anemometer (TSI 8450), dry and moist temperatures were measured by copper-constantan thermocouples. Inside the piles, the thermocouples were set at 0.5 and 0.2 m depth, in the centre and at the edge. All measurements were averaged, and stored on an AOIP datalogger (SA120 and SA70). Mixture samples were refrigerated and immediately delivered to the Côtes d'Armor Department Laboratory (France) for analysis.

Nitrification/Denitrification

Samples were taken in the centre of the piles (except otherwise specified), and put in sealed bags, then tested for activity measurements. The denitrifying activity was determined by measuring N_2O concentrations after blocking the reactional chain by adding acetylene (Yoshinari and Knowles 1976). For each sample, 30 g replicates were put in 250-ml bottles and a small part was used to calculate dry weight using a drying oven for 48 h at 105 °C (Kakezawa et al. 1990). In half of the bottles, the reaction was blocked by the injection of 20 ml of acetylene. All samples were incubated at 40 °C. Gas extractions were recorded at 0, 1, 2, 3 and 4 h by means of a 20-ml syringe of which 5 ml was injected in Venoject vacuum tubes. Between each extraction, the syringe was cleaned by purging it three times with ambient air and, before extraction, the interior of the bottle was homogenized in the same way. A chromatograph CP9001 was used in the analysis.

Statistical Tests

Differences between experimental conditions were tested by the non-parametric Mann-Whitney test. The level of significance was fixed at $p \leq 0.05$.

Results

Composting

The composting parameters and the composting results are given in Table 1. At all locations in piles C1 (control), T (turned), TLW1 and TLW 2 (turned with ligno-cellulosic waste added), and TY (turned with *Yucca* juice added), the temperature was very close to 60 °C, with a decrease at the time of turning and a rapid return to 60°C (Fig. 2). The temperature did not remain at 60 °C in pile MA (with microbial additive), where the "composting" process proposed by the furnisher of the microbial additive differed considerably from the standard composting techniques, and for pile C2 (control), which was very dry.

Table 1. Composition of the piles before and after composting

	Unit	C1	T	TLW1	MA	C2	TY	TLW2
Initial content								
Manure	kg	869	869	456	1501	592	609	287
Water	kg			214	778		307	599
Shavings	kg			200				229
Sawdust	kg			36				29
Initial weight	kg	869	869	906	2279	592	916	1144
Dry matter	%	47.7	28.5	35.9	31.5	73.4	35.0	30.5
Organic matter	%	38.9	24.0	33.2	26.0	63.5	30.1	27.7
Total nitrogen	%	2.19	1.46	0.87	1.54	3.48	1.63	0.84
C/N		n.d.	n.d.	n.d.	n.d.	9.0	8.4	16.5
Final weight	kg	511	375	511	1806	520	613	736
Dry matter	%	47.0	47.0	46.1	24.9	71.7	53.7	36.3
Organic matter	%	38.2	36.1	41.4	20.1	60.6	44.2	32.2
Total nitrogen	%	2.06	1.95	1.35	1.64	2.06	2.64	1.18
C/N		9.4	9.4	15.6	6.2	16.2	8.0	13.4

C : control pile, T : turned pile, TLW : turned pile with ligno-cellulosic waste added, MA : pile with microbial additive. 1 and 2 for experiments 1 and 2. n.d. : not determined.

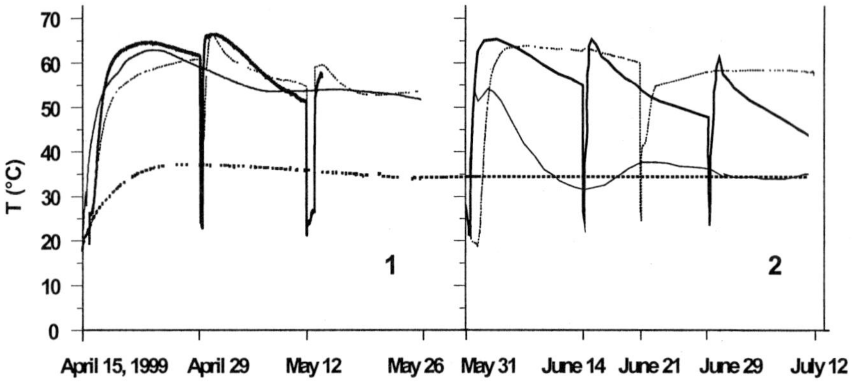

Fig. 2. Temperatures of the composting piles, at 50 cm depth in the centre for C1 (—), T (▬) and MA (...), and at 50 cm depth at the edge for TLW1 (...) in exp. 1 (1); at 50 cm depth in the centre for C2 (—) and MA (...), at 50 cm depth at the edge for TY (▬), and at 20 cm depth in the centre for TLW2 (...) in experiment 2 (2). The compost designation is as in Table 1

Nitrification/Denitrification

Nitrification/denitrification activities were tested by the production of N_2O in bottles containing compost samples, for the seven piles, and for the ligno-cellulosic waste substrate of exp. 1 (LW). Each time, six bottles were used, acetylene was added to three of the bottles to block N_2O production by denitrification. In Table 2, the results of N_2O production h^{-1} during the 2 (C1,MA and TY), 3 (TLW2 and C2) and 4 (LW,T and TLW1) first h of incubation are presented.

We assumed that there was denitrification activity if N_2O production was zero in bottles where acetylene was not added (-a) and positive in samples with acetylene added (+a), and nitrification activity in the inverse case. Positive figures in both situations indicate the presence of nitrification/denitrification activities, which are difficult to dissociate, and negative figures, with acetylene added at least, could indicate the consumption of N_2O, present at time=0 in the sample, by nitrifying bacteria.

Evolution in Time

Table 2 shows a significant microbial activity at time=0 in exp. 1 for mixtures C1 and MA, which were made from dry chicken manure, and a low microbial activity for mixtures T and TLW1, which were made from chicken manure coming from the same band, but in an area of higher moisture content. After composting for 42 days, the microbial activities (mainly denitrification in C1, T and TLW1, no activity in MA) were not related with the activities observed at the beginning.

Consequently, in experiment 2, the behaviour of the composts was studied only during composting. For C2 and MA, we distinguished top and bottom samples. The tests performed on C2 confirmed the results of the tests performed on C1, with the localisation of the denitrification activity at the bottom of the pile. No activity was observed in MA. Addition of *Yucca* juice (compost TY) had an obvious effect in directing the microbial activities towards nitrification : nitrification and denitrification can be present after 28 days; however, in our case only nitrification was present after 42 days.

The differences between samples +a and -a (with and without acetylene) were significant ($p \leq 0.05$) in TLW1, C2 at the bottom and TY at day 41/42. No significant differences were found between the different replicate sets at time=0 in exp. 1, or at the beginning of composting (day 14) for TY in experiment 2.

The differences between replicate sets related to different composting times were significant ($p \leq 0.05$) for MA and TLW1 between day 0 and day 42, as well for samples -a and +a as for these samples counted together (±a) (high microbial activity versus no activity). For pile T between day 0 and day 42, the difference was significant ($p \leq 0.05$) for denitrification and global activity. For pile TY, the differences were significant in all cases (-a,+a,±a) between days 14 and 29 as well as between days 29 and 41.

Effect of Microorganisms Addition

Comparison between C1 and MA showed no significant difference at time=0 (p contained between 0.51 and 0.87 on tests made with 2 or 4 h of incubation) and for replicates taken by a set of three (-a,+a), or by a set of six (±a), but a significant difference ($p \leq 0.05$) for denitrification or global activity at day 42. Otherwise, there was a significant difference between MA and T, at the beginning of composting and at day 42, in all cases tested (-a,+a, ±a).

The results show a difference between the composts with and without microbial additive, but do not allow us to ascribe this difference to the additive itself. When the chicken manure contained initially a sufficient quantity of active microorganisms, no difference was found at the beginning of composting, and the conditions of aeration (initial moisture or turning) and temperature during the course of composting could be sufficient to explain the final differences.

Effect of Ligno-Cellulosic Waste Addition

The two composts T and TLW1 did not show a clearly different evolution in the microbial activities tested.

Table 2. N_2O production by nitrification and denitrification at various stages in the composting process in various compost mixtures. Compost designation is as in Table 1

Substrate or compost	Sampling depth	Acetylene presence	Sampling date	ng N_2O g^{-1} compost h^{-1} Mean	Standard deviation
LW	Standard	-	April 16, 1999	-13	22
		+		-32	55
C1	Standard	-	April 15, 1999	1939	2629
		+		9434	10560
	Standard	-	May 26, 1999	21	34
		+		197	131
T	Standard	-	April 16, 1999	0	0
		+		0	0
	Standard	-	May 26, 1999	22	17
		+		77	80
TLW1	Standard	-	April 16, 1999	0	0
		+		13	16
	Standard	-	May 26, 1999	0	0
		+		148	32
MA	Standard	-	April 15, 1999	3868	6564
		+		9604	8974
	Standard	-	May 26, 1999	0	0
		+		0	0
	Top	-	July 12, 1999	0	0
		+		0	0
	Bottom	-	July 12, 1999	0	0
		+		0	0
C2	Top	-	July 12, 1999	0	0
		+		0	0
	Bottom	-	July 12, 1999	0	0
		+		72	23
TY	Standard	-	June 14, 1999	0	0
		+		-26	45
	Standard	-	June 29, 1999	54	25
		+		67	33
	Standard	-	July 12, 1999	12	8
		+		0	0
TLW2	Standard	-	July 12, 1999	0	0
		+		0	0

Effect of Moisture and Aeration

Although there was a significant difference at the beginning of composting between C1 and T (samples +a and ±a), at the end of composting there was no significant difference. Likewise, there was no significant difference between C1 and C2 at day 41/42 or between T and C2 at day 41/42, if we consider the samples of C2 taken out from the bottom of the pile, the microbiologically active part of the pile. The two parameters, moisture and aeration, may act in opposite directions and additional trials are needed to conclude the effect each has.

Effect of *Yucca* Juice Addition

TY differed significantly from TLW2 and C2 (top part), in the case of samples -a, as it showed a nitrification activity, absent in TLW2 and C2 (top part). It differed from C2 (bottom part), in both sets of samples -a and +a, as nitrification activity was detected in C2 and not in TY.

The microbial activity in the *Yucca* juice pile was obviously different from standard composts.

Nitrogenous Gas Emissions During Composting and Balance Sheet

The microbial activity in the samples was compared to total nitrogenous gases emissions (Figs. 3 and 4). Two correlations were noticed. First, gaseous emissions decreased rapidly in the MA pile, which means that the microbial populations were active in producing proteic biomass, and that NH_3 and N_2O production was avoided. Secondly, due to a continuous availability of ammonia, a higher nitrification activity was noticed in the TY pile, and NH_3 and N_2O emissions decreased over time.

On the contrary, the absence of nitrification and denitrification activities in samples of TLW2 at day 41 and increase in N_2O emissions could be explained by samples which are not representative of the heterogeneity of the pile. Compost mixtures using ligno-cellulosic waste as a bulking agent are often more heterogeneous than when using straw or manure.

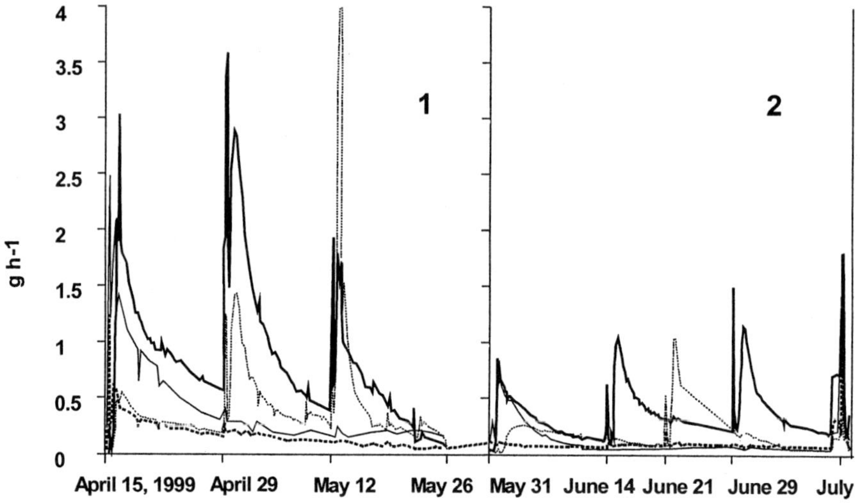

Fig. 3. Hourly mean flux of N-NH_3 during composting of piles C1 (—), T(▬▬), TLW1 (...), and MA (...) in exp. 1 (1), and C2 (—), TY (▬▬), TLW2 (...), and MA (...) in experiment 2 (2). Compost designation is as in Table 1

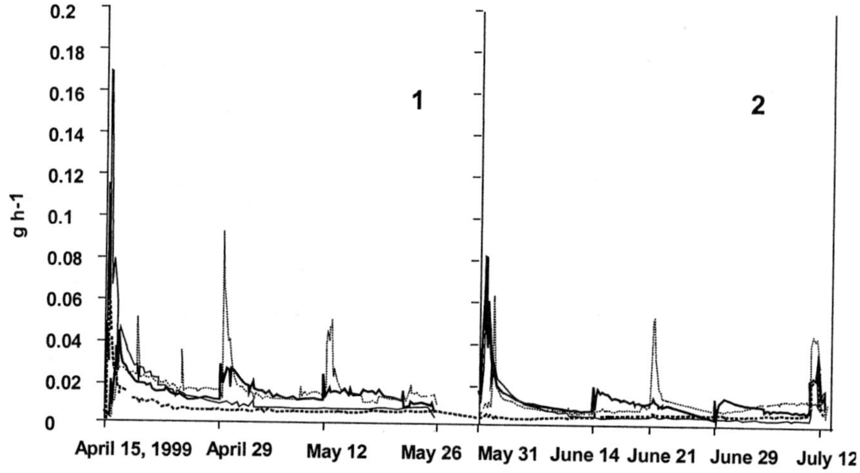

Fig. 4. Hourly mean flux of N-NO_2 during composting of piles C1 (—), T(▬▬), TLW1 (...), and MA (...) in exp. 1 (1), and C2 (—), TY (▬▬), TLW2 (...), and MA (...) in experiment 2 (2). Compost designation is as in Table 1

The balance sheet also shows up the relationships between gaseous emissions and nitrification/denitrification activities. Table 3 shows that the difference between nitrogen loss in the pile during composting and nitrogen found in the form of gaseous emissions (NH_3 or N_2O gas) was equal to 70, 0, 0, 60, 90, 20 and 5%

respectively for C1, T, TLW1, MA, C2, TY and TLW2, 100% meaning that the NH_3 and N_2O emissions were zero. It was observed that the higher percentages are in relation with denitrification at the end of composting in static piles (C1 and C2), and the lower percentages, with the release of ammonia during turning in piles (T, TLW1 and TLW2). TY and MA percentages are in the middle, this is due to the fact that MA was composted for 12 weeks and because denitrification activity was very low at times in TY.

Discussion

The way in which composting is managed has a great influence on the movement or sequestering of nitrogen. Gaseous emissions are linked directly to the physico-chemical and biological conditions of composting (Van den Weghe et al. 2000). If one seeks a fertilising compost, one will try to avoid both ammonia volatilisation and denitrification. On the other hand, if nitrogen abatement is looked for, one has to be careful to favour N_2 emission rather than N_2O or NH_3 release. The study of the behaviour of nitrogen is important today, due to the increase in greenhouse gas emissions and contamination of waters (Mariotti 1997).

Pile MA had the highest nitrogen conservation. This pile was kept at 35 °C throughout composting and was not aerated : it was the wettest, it was not turned, nor were bulking agents added. Its C/N was very low at the end of composting (6.2). During the first 15 days, the temperature curves of piles MA and T, which had the same moisture content and were not turned during this time, were obviously diverging. This can be explained only by microbial activities, which were directed differently. In MA, the microbial biomass was certainly high, as well as microbial activity, in terms of respiration, but not in terms of nitrification/denitrification. The results obtained during the initial trials of the COBIOTEX complex, on composts with high C/N, went along the same lines, as far as temperature decrease and nitrogen conservation are concerned (Anonymous 1996). The low oxygen demand was explained by the use of aerobic-anaerobic facultative bacterial strains (four *Lactobacillus* and five *Bacillus* strains). The final product from pile MA appears to be a fertilizer as much as an organic enriching agent, and should be tested for its agronomic value and compared with standard composts. In addition, it would be necessary to test the CH_4 emission of the process.

Table 3. Balance sheet for the chicken manure composting experiments. Compost designation is as in Table 1

	Unit	C1	T	TLW1	MA	C2	TY	TLW2
Initial content								
Water	% Manure DW	110	251	430	217	36	100	468
Nitrogen	% Manure DW	4.6	5.1	5.8	4.9	4.7	4.7	5.7
Pile losses								
DM	% Manure DW	42	29	66	37	14	28	48
Water	% Manure DW	44	171	226	28	2	38	192
Nitrogen	% Manure DW	2.1	2.2	0.7	0.8	2.3	1.1	1.0
Nitrogen	% Initial N	46	43	13	16	49	23	17
Gaseous emissions								
Water	% Manure DW	39	164	226	34	13	46	214
$N\text{-}NH_3$	% Manure DW	0.63	2.82	2.00	0.33	0.18	0.87	0.94
$N\text{-}N_2O$	% Manure DW	0.011	0.017	0.053	0.014	0.010	0.014	0.016
$N\text{-}NH_3$	% Initial N	14	55	34	6.7	3.8	19	16
$N\text{-}N_2O$	% Initial N	0.24	0.33	0.91	0.29	0.21	0.30	0.28

The other composts and the control piles are divided into two groups in terms of N conservation : TY, TLW1 and TLW2 with losses of about 20% of the initial N, and C1, C2 and T with losses from 40 to 50%.

In the first group, the most interesting in terms of N conservation, pile TY had the lowest emission of polluting gases, NH_3 and N_2O. The nitrogen conservation in pile TY is due to the absence of denitrification activity at the end of composting and the presence of nitrification activity. The *Yucca* juice saponines have the ability to fix ammonia reversibly (Clay 1993). It has been used in animal feeds, in order to decrease the production of ammonia emissions but the results obtained are contradictory (Tricot 1999). Bline (1998) is the first to have used it in composting. Nevertheless, using a dose of 100 g juice t^{-1}, the decrease in NH_3 and N_2O emissions was only temporary. In the TY pile, which had a sufficient quantity of saponines (100 g t^{-1}), the *Yucca* juice regularized the flux of ammonia, and limited its emission, thus favouring nitrification and reducing N_2O emissions. Piles TY and T were similar with respect to compost management, initial nitrogen, water content and temperature. However, pile T emitted three to four times more NH_3 and one and half to two times more N_2O. This additive seems therefore to decrease emissions of polluting gases during composting, without modifying the process itself.

N conservation was also observed in piles TLW1 and TLW2. This result is in agreement with the idea that an increase in the C/N ratio is linked to a decrease in relative N loss (Morand et al. 2000). Nevertheless, in these piles, significant NH_3 and N_2O emissions occurred, with respect to their N loss. Denitrification was incomplete, compared to C1 and C2, and did not lead to production of N_2. This can be explained by a better aeration of the piles, which favours NH_3 volatilisation and N_2O release. The emissions were higher in TLW1, which was turned twice, than in TLW2, which was only turned once. Pile T that was turned twice also presented a high level of emissions, higher than TLW1 did, because of the high N loss in the T pile, linked to a low C/N ratio (N contents were 2.4, 2.8 and 4.7% DW for TLW1, TLW2 and T, respectively). Misselbrook et al. (2000), in their study of gaseous emissions in farm yard manure, also concluded that mixing only once at 40 days led to the lower emission of polluting gases.

The piles with no additives were the most adept at maximizing nitrogen abatement. In this group, control piles, which were static, produced the highest ratio of N_2 emissions and the lowest ratios of ammonia and nitrous oxide emissions (comparable to those of TY) in relation to nitrogen losses. This had been seen in a previous experiment although total N loss was lower in that experiment (Bline 1998; Tricot et al. 1999). Piles C1 and C2 differed in water content (50% and 30%, respectively) and temperature. However, they maintained a similar rate of N loss. NH_3 emissions in pile C2 were very low, but, due to the dry matter content, sterilisation and composting conditions could not be reached.

In terms of quantity, the nitrification and denitrification activities in our experiments led to an N_2O flux comparable to what we had already found in a previous trial composting poplar bark-poultry dung. In this trial, N_2O flux ranged from 0 to 180 ng N_2O g^{-1} compost h^{-1}, at the end of a 9-month composting period in damp conditions, with two outlying points at 320 (sample -a) and 2700 ng N_2O g^{-1} compost h^{-1} (sample +a) (Morand et al. 2000).

Conclusion

After comparing seven different piles and recording nitrification/denitrification activities, temperature, N loss, NH_3 and N_2O emissions, correlations between composting management and nitrogenous gas emissions were observed. It was found that the use of additives (microbial additives, *Yucca* juice and ligno-cellulosic waste) leads to nitrogen conservation. However, if the objective is to abate nitrogen, it is best to compost chicken manure in a static pile without additives. In any case, turnings have to be minimized to limit ammonia volatilisation and release of nitrous oxide which makes denitrification incomplete.

The nitrification/denitrification activity seems to be a permanent feature in composting with different predominant parts following the initial inoculum, and the time and management of composting. Further studies should therefore include testing nitrogenous gas emissions over longer composting periods and a more complete study of microbial activities over time. These studies should also be done in larger piles to determine the effect of compaction on gas production. Finally, the compost products should be evaluated for their agronomic value.

Acknowledgements. We would like to thank P. Marmonnier for his help in the statistical analysis of the data, P. Perrin and N. Josselin for their technical assistance during the composting experiments and the chromatography analysis respectively, and Y. Picard for completing the figures. We are grateful to J.C. Poupa and J.Y. Buis for the supply of ligno-cellulosic waste. We are also indebted to N. Nicholson for checking and correcting the English manuscript.

References

Anonymous (1996) Tradicompost. Complexe fermentaire pour le compostage des végétaux et des matières organiques. Santel, Louverné, France, 28 pp

Asman WAH (1992) Ammonia emissions in Europe : updated emissions. Report n. 228471008. National Institute of Public Health and Environmental Protection. Bilthoven, The Netherlands, 88 pp

Aubert C (1998) Le compostage des fumiers de volailles. In: Le compostage à la ferme des effluents d'élevage. Actes du colloque. Paris, 15 décembre 1998. ACTA, Paris, pp 45-55

Aubert C, Guiziou F (1997) Le compostage des fumiers de volailles à la ferme. Sci Tech Avicoles 19 : 21-30

Bline D (1998) Ammoniac en aviculture : étude en bâtiments et lors du compostage du fumier. Mémoire d'ingénieur des techniques agricoles. Ecole Nationale d'Ingénieurs des Travaux Agricoles de Bordeaux, Bordeaux, 132 pp

Bline D, Aubert C (1998) Pertes d'azote et dégagements d'ammoniac par les élevages de poulets. Sci Tech Avicoles 25 : 11-16

Buijsman E, Maas HFM, Asman WAH (1987) Anthropogenic NH_3 emissions in Europe. Atmos Environ 21 : 1009-1022

Clay J (1993) Control of ammonia in poultry houses using deodorase. Zootech Int 10 : 34-37

Elwell DL, Keener HM, Carey DS, Schlak PP (1998) Composting unamended chicken manure. Compost Sci Util 6(2) : 22-35

Husted S (1994) Seasonal variation in methane emission from stored slurry manures. J Environ Qual 23 : 585-592

Kakezawa M, Mimura A, Takahara Y (1990) A two-step composting process for woody resources. J. Ferment Bioeng 70 : 173-176

Mariotti A (1997) Quelques réflexions sur le cycle biogéochimique de l'azote dans les agrosystemes. In: Lemaire G, Nicolardot B (eds) Maîtrise de l'azote dans les agrosystemes. INRA, Paris, pp 9-24

Misselbrook TH, Balsdon S, Pain BF, Gibbs PA, Parkinson RJ (2000) Gaseous emissions from on-farm composting of cattle farm yard manure. In: 9th international workshop of the European cooperative research network Recycling of agricultural, municipal and industrial residues in agriculture. Gargnano (BS), Italy, 6-9 Sept 2000. Abstr papers. University of Milan, Italy, p 101

Morand P, Baron S, Yulipriyanto H, Robin P (2000) Gaseous emissions during composting of poplar bark-poultry dung mixtures. First results. In: Warman PR, Taylor BR (eds) Proceedings of the International Composting Symposium (ICS'99), September 19-23, 1999, Dartmouth/Halifax, Nova Scotia, Canada. CBA Press, Truro, NS, Canada, pp 544-570

Tricot G (1999) Maîtrise des émissions azotées lors du compostage de fumier de volailles. Mémoire de fin d'études. Ecole Supérieure d'Agriculture d'Angers, Angers, France, 118-XLII pp

Tricot G, Aubert C, Robin P, Bline D (1999) Maîtrise des émissions azotées lors du compostage de fumier de volailles. Sci Tech Avicoles 31 : 25-31

Van den Weghe H, Zhou S, Zaied H (2000) Gaseous emission control by composting pig slurry in an aerated static reactor - influence of several physical parameters on gaseous emissions. In : 9th international workshop of the European cooperative research network Recycling of agricultural, municipal and industrial residues in agriculture. Gargnano (BS), Italy, 6-9 Sept 2000. Abstr papers. University of Milan, Italy, p 106

Yoshinari T, Knowles R (1976) Acetylene inhibition of nitrous oxide reduction by denitrifying bacteria. Biochem Biophys Res Commun 69 : 705-710

Biodegradability

Biodegradability Study on Films for Packaging Based on Isotactic Polypropylene Modified with Natural Terpene Resins

S. Cimmino[1], E. D'Alma[1], E. Ionata[2], F. La Cara[2] and C. Silvestre[1]

Abstract. This chapter discusses the influence of a natural terpene resin, poly (α-pinene) on tensile behaviour, gas permeability and biodegradability of isotactic polypropylene (iPP)-based films. The presence of poly(α-pinene) confers improvement of physical and mechanical properties of films, compared to those of plain iPP.

To test the biodegradable behaviour of these innovative films for general packaging uses, several microbial communities isolated from the polluted area were used. It was found that one microbial community, C.B.1, was able to erode the blend films and not the plain iPP film.

Introduction

The large-scale commercial use of synthetic polymers involves plastics being a significant part of waste in municipal landfills by weight and volume, and this fraction of waste is projected to increase. The disposal of synthetic polymers in the environment is a phenomenon less than half a century old, a minuscule duration in the evolutionary time scale required for microbial evolution on Earth. Consequently, the microorganisms are not biochemically equipped to catabolize synthetic polymers and this is the main reason for their "recalcitrance" in the environment. Nevertheless, data on the biodegradability of some polymers by bacteria, fungi and plants or animal cells have already been published (Albertson and Banhidi 1980; Wasserbauer et al. 1990; Krupp and Jewell 1992). Recently, a new polymer system, based on iPP modified with natural terpene resins, was formulated and proposed for general packaging uses, food packaging included (Albertson et al. 1987). The terpene resins are low molecular weight materials with a structure similar to that of hydrocarbon resins and are glassy at room temperature. The aims of this work are: (1) to test the biodegradable behaviour of iPP based films modified with poly(α-pinene); and (2) to select microorganisms and suitable conditions to obtain the degradation of the films.

[1]Istituto di Chimica e Tecnologia dei Polimeri (ICTP) CNR. Via Campi Flegrei 34. 80078 Pozzuoli (NA), Italy. (e-mail: cimmino@irtemp.na.cnr.it)
[2]Istituto di Biochimica delle Proteine (IBP) CNR. Via P. Castellino 111, 80128 Napoli,Italy. (e-mail: lacara@dafne.ibpe.na.cnr.it)

H. Insam, N. Riddech, S. Klammer (Eds.)
Microbiology of Composting

Materials and Methods

Materials

The polymer materials tested for their biodegradability were polypropylene (iPP), poly(α-pinene) (PαP) and their blends (iPP/PαP 95/5, 90/10, 80/20 in wt %). Isotactic polypropylene is a commercial product, Shell HY 6100, with MW=3.0 x10^5 g mol^{-1}. Poly(α-pinene) is a natural terpene resin and derives from the polymerization of α-pinene mono-terpene, which is the main constituent of the wood of coniferous plants. The sample used, kindly supplied by Hercules (The Netherlands), is the Piccolyte A115, with glass transition (Tg) at 61±3 °C, Mn=680 g mol^{-1}, MW=1075 g mol^{-1}. Binary blends of polyolefin/resin were obtained by mixing iPP and PαP in a Brabender-like apparatus (Rheocord EC of HAAKE Inc., Karlsruhe, Germany) at 210 °C and 32 rpm for 10 min. Specimens of each composition were sterilized for 15 min with 70% ethanol and dried under aseptic conditions before introduction into the incubation media.

Mechanical Tensile Tests

Dumbell-shaped specimens were cut from the compression-moulded sheets and used for mechanical tensile measurement. Stress-strain curves were obtained by using an Instron machine (Model 1122) at cross-head speed, V_t, of 10 mm min^{-1}, gauge length = 22 mm, and nominal strain rate = 0.45 min^{-1}. Modulus, stress and elongation at yield and rupture were calculated from such curves on an average of 15 specimens.

Permeability Measurements

The permeability measurement of the films (thickness 200 μm) was performed at 30 °C in presence of a gas mixture of CO_2, O_2 and N_2 (1:1:1) with a Lyssy isostatic analytic permeability tester, mod. GPM 200, and a gas chromatography apparatus with thermal conductivity.

Isolation of Bacteria

Soil samples collected from about ten different sites that were rich in plastic wastes were used.

One g of each soil sample was added to 100 ml of tap water and incubated at 25°C. Successive subcultures in the presence of plastic materials and decreasing glucose concentrations were then incubated under the same conditions. Final subcultures were grown with no addition of glucose. Evaluating the vial biomass of every culture by serial dilution on nutrient agar plating incubated at 28 °C, we have isolated one type of bacteria consortium, named C.B. 1, particularly able to grow with the polymer films as the only carbon source. This consortium was used for biodegradability tests.

Biodegradability Test

The microbial consortia were maintained on a minimal medium in shake flasks by using samples of plain iPP, plain PαP, iPP/PαP 90/10 and iPP/PαP 80/20 as the sole carbon source. The samples were incubated at 28 °C for 3 and 6 months in 500-ml flasks containing 200 ml of minimal salts medium and inoculated with 10 ml of microbial community.

The experimental assessment of biodegradability was carried out by different approaches, in samples collected at fixed times. The microbial growth over the polymer as sole organic substrate was evaluated by monitoring the increasing of biomass through measurement of pH, optical density at 600 nm (for total biomass estimation) and determination of the number of viable cells by serial dilution on nutrient agar plates. Oxidase activities, the most probable enzymes involved in the degradation, were then evaluated as described below. Finally, analysis by scanning electron microscopy (SEM) of the polymer surfaces was performed.

Oxidase Activity Assay

The assay was performed utilizing xantine 0.3 $mgml^{-1}$ as substrate in 1 ml final volume of phosphate buffer 0.05 M, pH 7.5, by adding 650 μl of culture supernatant and following the increasing of absorbance at 290 nm against a blank tube without substrate.

SEM Analysis

The surface analyses were performed by using a SEM Philips XL 20 series microscope on cryogenically fractured surfaces. Some samples were sonicated and used to investigate the formation of the erosion. Other samples were used without sonication in order to detect the microorganisms presence. Before the observation, all the surfaces were coated with Au/Pd alloy with SEM coating device (SEM Coating Unit E5150, Polaron equipment Ltd.).

Results and Discussion

Table 1 reports the tensile parameters obtained from stress-strain curves.

The iPP has a plastic behaviour characteristic of semicrystalline polyolefins, with yielding phenomenon, cold drawing, fibres drawing and break. The 95/5, 90/10 and 80/20 blends present a stress-strain behaviour like to plain iPP. From the data of Table 1 it is observed: (1) the Young modulus increases with the composition of resins; (2) the values of the stress and elongation at the yielding, rupture and cold drawing region of the 95/5 and 90/10 blends do not present significant differences; (3) the parameters at rupture (σ_r and ε_r) of the 80/20 blend are lower than those of the iPP, 95/5 and 90/10 blends.

Table 1. Stress-strain parameters of iPP/PαP films

IPP/PαP (% weight)	E (Mpa)	σ_r (MPa)	ε_y (%)	σ_r (MPa)	ε_r (%)
100/0	930±25	25±1	10±1	42±1	1020±45
95/5	1180±40	30±1	10±1	43±1	1030±35
90/10	1150±40	28±1	11±1	37±3	1030±100
80/20	1340±43	30±5	10±1	29±4	850±75

Gas permeability coefficients to CO_2 and N_2 are reported in Table 2. The data analysis shows that the permeability coefficient to CO_2 and O_2 decreases with the composition of PαP. Analyzing the N_2 permeability coefficient, the 90/10 and 80/20 blends present similar values, but lower than iPP.

Table 2. Gas permeability coefficient (P) of iPP/PαP films. [(cm^3(STP)·cm)/($cm^{2\cdot s}$·Pa)]

iPP/ PαP (% weight)	P (CO_2) x 10^{-13}	P (O_2) x 10^{-13}	P (N_2) x 10^{-13}
100/0	3.1±0.2	0.94±0.06	0.22±0.05
95/5	2.8±0.3	0.87±0.08	0.25±0.06
90/10	2.4±0.1	0.71±0.07	0.16±0.01
80/20	1.8±0.2	0.63±0.05	0.18±0.05

The biodegradability test was performed on all the samples utilizing a selected microbial community isolated from polluted area (C.B.1). The growth parameters (pH, OD at 600 nm and viable cells number) were monitored during the 6 months of incubation in the cultured microorganisms. The pH value was maintained between 7.0 and 8.0 and both the OD at 600 nm and the viable cells number values had a slight increase (data not shown).

The SEM analysis showed that the C.B.1 microorganism consortium is able to erode the films. As an example of the biodegradable activity of the C.B.1 consortium microorganism, Fig 1 and 2 show the surface of the 80/20 blend film after 6 months of incubation at 28°C without and in presence of the microorganisms, respectively. From the comparison of the Fig 1 and 2, it is evident that the formation of eroded zones is due to the activity of the microorganisms. It is very interesting to note that the microorganisms eroded the films and not the film formed by plain iPP.

In fact, Fig 3 is relative to the surface of plain iPP film after incubation at 28 °C for 6 months, practically identical to that before the incubation.

SEM analysis detected the presence of microorganisms on the polymer surface. In order to detect the microorganisms, the samples were not sonicated.

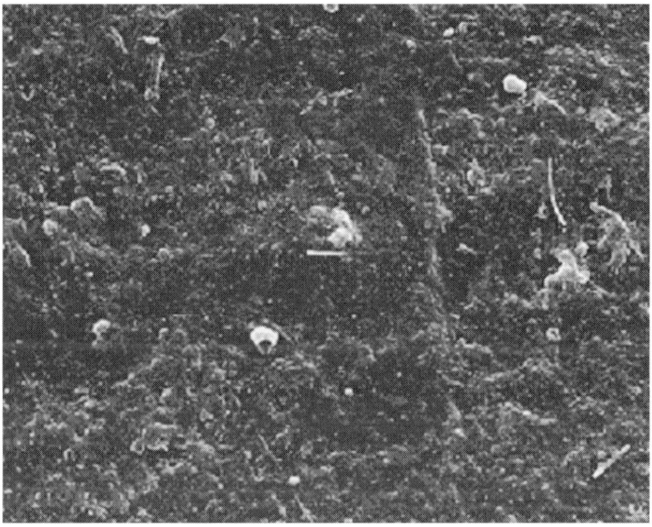

Fig.1. SEM micrograph of 80/20 blend film incubated without microorganisms (3000x)

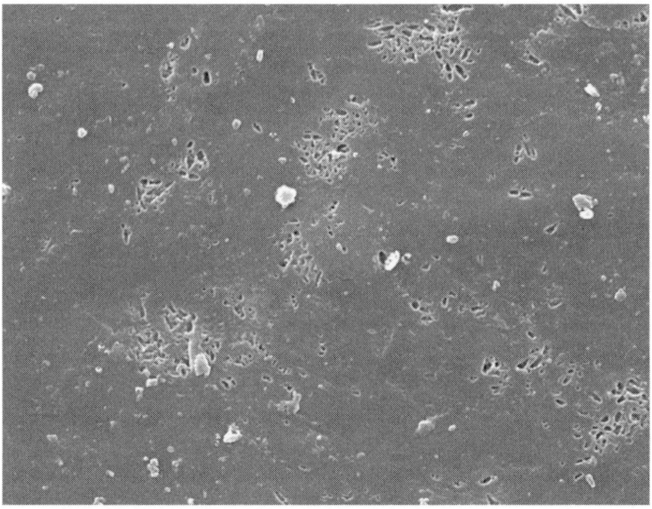

Fig. 2. SEM micrograph of 80/20 blend film incubated in presence of microorganisms. (3000x)

Figures 4 and 5, relative to the surface of 80/20 blend film, are shown as an example of the presence of microorganisms on the surface found for all the films.

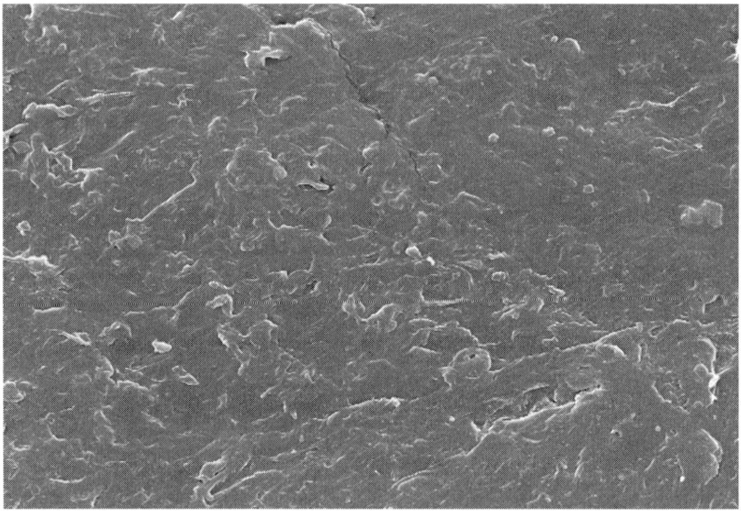

Fig. 3. SEM micrograph of plain iPP film incubated in presence of microorganisms (3000x)

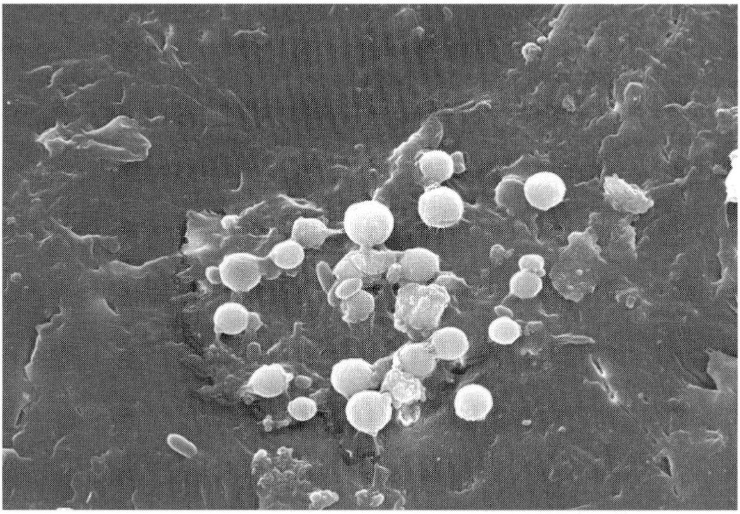

Fig. 4. SEM micrograph of 80/20 blend film in presence of microorganisms (3000x)

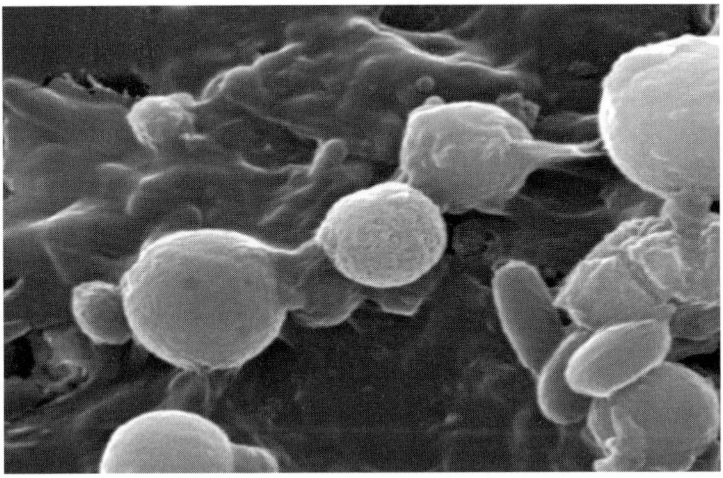

Fig. 5. SEM micrograph of 80/20 blend film in presence of microorganisms (12 000x)

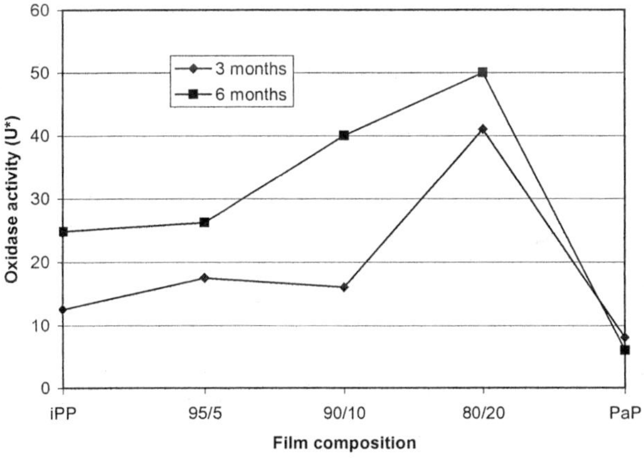

Fig. 6. Oxidase activity after 3 and 6 months. (One unit of oxidase activity is defined as $\Delta OD_{290\ nm}\ min^{-1}\ mg^{-1}$ protein)

Because polyolefin biodegradation was correlated by many authors to microbial extracellular enzymes like oxidases (Albertson and Banhidi 1980; Albertson et al.1987; Albertson and Karlsson 1988; Pometto et al.1992), such an activity was also tested in the culture supernatant at several times. The activity values, after 3 and 6 months of incubation, are shown in Fig. 6.

The 80/20 sample showed a high oxidase activity both at 3 and 6 months of culture supernatants, confirming the SEM observations.

From the data reported above it can be concluded that these aerobic species with different catabolic capabilities can act in close cooperation to degrade iPP/PαP films.

The evidence of biodegradation comes from the increasing oxidase activity in the culture supernatant of the incubated blend films, confirmed by SEM observation.

The finding that enzymatic attack of some polyolefin occurs as the enzymatic attack of trypsin on poly(ether urethane) (Bouvier et al.1991) suggests that synthetic polymers may be recognized by natural metabolic machineries and then transformed into low-molecular-weight compounds.

Moreover, the adhesion of bacteria on the film surfaces increases the susceptibility of the polymer to microbial attack (Iman and Gould 1990). Hence, it is suggested that the well-known metabolic flexibility and adaptability of microorganisms can result in the biodegradation of iPP/PαP films composed by two macromolecules that seem recalcitrant to biological metabolism.

References

Albertsson A C, Banhidi Z G (1980) Microbial and oxidative effect in degradation of polyethylene. J Appl Polym Sci 25: 1655-1671

Albertsson A C, Karlsson S (1988) The three stages in degradation of polymers: polyethylene as a model substance. J Appl Polym Sci 35: 1289-1302

Albertsson A C, Andersson S O, Karlsson S (1987) The mechanism of biodegradation of polyethylene. Polym Degrad Stabil 18:73-87

Bouvier M, Chawla A S, Hinberg I (1991) In vitro degradation of a poly(ether urethane) by trypsin. J Biomed Mater Res 25: 773-789

Cimmino S, D'Alma E, Di Lorenzo M L, Di Pace E, Silvestre C (1998) Blends of isotactic polypropylene and natural terpene resins. I. Phase structure, thermal, and dynamic-mechanical properties. J Polym Sci part B Polym Phys 37: 867-878

Iman S H, Gould J M (1990) Adhesion of an amylolytic Arthrobacter species to starch-containing plastic films. Appl Environ Microbiol 56: 872-876

Krupp L R, Jewell W J (1992) Biodegradability of modified plastic in controlled biological environments. Environ Sci Technol 26: 193-198

Pometto A L, Lee B, Johnson K E (1992) Production of an extracellular polyethylene degrading enzyme(s) by Streptomyces species Appl. Environ. Microbiol. 58: 731-733

Wasserbauer R, Beranovà M, Vancurovà D (1990) Biodegradation of polyethylene foils by bacterial and liver homogenates. Biomaterials 11: 36-40

Isolation and Characterization of Thermophilic Microorganisms Able to Grow on Cellulose Acetate

F. Degli-Innocenti[1], G. Goglino[2], G. Bellia[1], M. Tosin[1], P. Monciardini[3] and L. Cavaletti[3]

Abstract. Cellulose acetate (CA) can be successfully used for producing compostable plastics. It is biodegradable under composting conditions, but neither the mechanism of degradation nor the involved microorganisms are known. We isolated five thermophilic strains from compost which were identified by molecular characterisation as belonging to the family Thermomonosporaceae, probably to the genus *Actinomadura*. The strains are able to grow at 50 °C on minimal plates supplemented with CA (DS=2) or microcrystalline cellulose in powder. The degradation of CA is visible by the clarification of the area surrounding the mycelium.

The strains grow at 37 °C on cellulose but not on CA. At 20 and 58 °C no growth is detected on both substrates. The strains are viable after a 2-week incubation period at 65 °C. Neutral and basic pHs are the best for CA degradation, while poor results are obtained under acidic conditions. The best degradation of CA is obtained using nitrate as nitrogen source. Solid-state respiration tests, performed using sterile vermiculite as a matrix, confirmed that the strains are able to mineralise the CA. The strains failed to grow in CA under liquid conditions.

Introduction

Cellulose esters are considered very attractive in terms of cost, material applications and environmental persistence (Buchanan and Gardner 1993). They are already used, in combination with starch, to produce compostable cutlery for fast-food restaurants (Anonymous 1997). The biodegradation of cellulose esters and in particular of cellulose acetate (CA) has been extensively studied under different test conditions (Buchanan and Gardner 1993; Gardner et al. 1994; Ji-Dong Gu et al. 1993; Komarek et al. 1993). It has been shown that the degradation of CA is affected by the degree of substitution (DS). The DS is the degree of substitution of the three hydroxyls on each anhydroglucose unit. The complete substitution leads to a DS of 3.0. The cellulose triacetate (DS=3) is substantially recalcitrant to biodegradation, while the cellulose diacetates, having a DS of less

[1]Novamont S.p.A., via Fauser 8, 28100 Novara, Italy
[2]Università di Milano, Dipartimento di Genetica e di Biologia dei microrganismi, Via Celoria 26, 10133 Milano, Italy
[3]Biosearch Italia S.p.A., via R. Lepetit 34, 21040 Gerenzano, Italy

H. Insam, N. Riddech, S. Klammer (Eds.)
Microbiology of Composting

than 2.5, are susceptible to biodegradation (Buchanan et al. 1993). It is assumed that the synergistic action of esterases and cellulases is needed to degrade CA (Buchanan and Gardner 1993). In order to better understand the mechanism of biodegradation of CA, it would be important to isolate the active microorganisms and characterize their action. Cellulolytic enzymes from *Thermomonospora curvata* alone do not have the ability to attack CA with DS=2.5. Mixed cultures are expected to have both esterases and cellulases, and, therefore, to be capable of cleaving the ester bonds and degrading both cellulose and the acetic acid (Stutzenberger and Kahler 1986). On the other hand, the mesophilic fungus *Pestalotiopsis westerdijkii* during growth completely consumes the cellulose acetate (DS=0.76) but not cellulose triacetate (Reese, 1957). This organism appears to contain all the enzymes required for complete de-acetylation of low DS CA. More recently, a mesophilic bacterium has been isolated from enrichment cultures started with compost and incubated at 37 °C (Gross et al. 1993) by plating on pure CA filters having a DS of 1.9–2.3. The bacterium was identified as *Pseudomonas paucimobilis* and reported to be able to grow using CA of 1.7 and, to a lesser extent, CA of 2.5 and 3, as sole carbon source. This strain was not subjected to further characterization and, at present, it is not available because lost (R.A. Gross pers. comm..).

We were interested in performing a similar search focusing on thermophilic microorganisms, active during composting. Composting is considered the more appropriate system for recycling the solid organic fraction of the municipal waste (Gardner et al. 1994), and biodegradable plastics made with CA are supposed to be recycled into compost. Therefore, it is of interest to understand the mechanisms of biodegradation of CA under composting conditions from both an academic and a practical viewpoint, in order to determine which conditions can affect the degradation rates.

Materials and Methods

Growth Media

Minimal medium (MM): KH_2PO_4 (2.0 gl^{-1}); $MgSO_4$ (0.5 gl^{-1}); NH_4NO_3 (1.0 gl^{-1}); pH adjusted to 7.0 with 1 M NaOH. After sterilisation at 121 °C for 10 min, the following solutions are added: 1 ml of NaCl (10%), 1 ml of $CaCl_2$ (10%); 1 ml of trace elements solution (Bellia et al. 1999). For solidification, agar is added at a concentration of 15 gl^{-1} (MM agar medium) or 3 gl^{-1} for the top layer (MM soft agar medium). In the experiments of physiological characterization $(NH_4)_2SO_4$ (1.65 gl^{-1}) or $NaNO_3$ (2.125 gl^{-1}) were used instead of NH_4NO_3 as nitrogen source.

Saline solution: 8.8 gl^{-1} NaCl.

Nutrient agar (NA) and nutrient broth (NB) were purchased by Oxoid (UK), actinomycete isolation agar (ACTI) by Difco (USA), and Trypticase soy broth (TSB) by BBL (USA). Cellulose acetate was obtained by the Eastman Company, Kingsport, Tennessee. Cellulose microcrystalline Avicel was obtained by Merck (D). Both cellulose and cellulose acetate are supplied as a powder.

Enrichment Cultures

Some consecutive composting cycles were performed in order to enrich the microbial population capable of degrading the CA, and facilitate the isolation of the active species. Cellulose acetate was used as a cosubstrate and the final compost of one cycle was used to inoculate the subsequent composting cycle. The composting method we applied is derived from an already described laboratory test method (Tosin et al. 1996) and it is under standardisation at ISO level (Degli-Innocenti et al. 2000). It requires the use of a synthetic waste (Table 1) derived from a previous study (Palmisano et al. 1993).

Table 1. Composition of synthetic waste used in the composting process at laboratory scale

Material	Dry mass (g)
Sawdust	200
Rabbit-feed	150
Corn Ssarch	50
Saccharose	25
Corn seed oil	15
Urea	10
Total	450

Mature compost, used as source of thermophilic microorganisms, was sampled from the composting plant of Castelceriolo (Italy). In a first composting cycle, 50 g of mature compost were used to inoculate 450 g of the synthetic waste. CA (DS=2) was also added (100 g). The mixture was wetted and incubated at 50 °C, following the procedure already described (Tosin et al. 1996). After 69 days of composting, a sample of the resulting compost (90 g) was used as inoculum of a second trial (denominated trial A). In this case the synthetic waste (450 g, dry weight) was supplemented with 50 g of CA DS=2 and 50 g of CA DS=2.4. At the end of the composting cycle (after 127 days), one sample of compost was subjected to the isolation procedure described below. Another sample (100 g) was used as coinoculum, together with 50 g of mature compost from a composting plant, of a third composting cycle (trial B), which was performed using the same CA addition of trial A. After 72 days, a sample was withdrawn for microbial isolation. A liquid extract was obtained, as indicated below, with a further sample from trial B (100 g), and this extract was used to inoculate a fourth composting cycle (trial D). In this case, instead of the synthetic waste, a medium based on vermiculite was used. The composition of this artificial composting bed was the following: vermiculite (200 g), MM medium (300 ml), compost extract (300 ml), NB (3.9 g), urea (1.75 g), starch (6 g), cellulose Avicel (6 g) and CA DS=2 (6 g). This method has been already applied to simulate the composting process with an inorganic carbon-free matrix (Tosin et al. 1996). The mixture was incubated at 50 °C and, after a 25-day incubation, a sample (100 g) was subjected to the isolation procedure.

In parallel, a different approach has been used, that is the use of baits (trial C). Pieces (35x15 , 1.5 mm in thickness) of a material based on starch and CA DS = 2.4, were buried within a synthetic waste under composting. After 120 days of composting, a white-grey mould covered the baits which were clearly degraded. The biofilm was scraped from the surface of the specimens, suspended in 24 ml of saline solution, and used to inoculate plates for the isolation of CA-degrading microbes, as described below.

Compost Extract

A sample of compost (100 g) is mixed with 500 ml of saline solution (NaCl, 8.8 gl^{-1}) for 3 h. The slurry is filtered with a strainer, to remove the compost particles, and then with a paper filter. The suspension is finally used to inoculate bacteriological plates or other enrichment cultures.

Isolation of CA-Degrading Colonies

Twentyfour ml of the compost extract is mixed with 200 ml of the MM soft agar medium, maintained at about 50 °C and supplemented with 12 g of CA DS=2 (or cellulose). Samples of 8 ml are withdrawn with a pipette and poured on top of MM agar medium plates and carefully distributed on the top before solidification. During this operation the suspension is maintained in agitation with a magnetic stirrer. The plates are then wrapped in aluminium foil and incubated at 50 °C.

The area of soft agar showing a visible microbial growth and a clarification of the CA powder is recovered using sterile tools. The agar is then transferred into sterile tubes and mixed with a vortex along with glass spheres in order to disrupt the agar clumps. The suspension is then diluted and plated into NA and ACTI plates to obtain single colonies. Pure cultures were obtained through serial isolation and retested for growth in CA.

Characterization of Isolates

The microbial strains were streaked with a sterile loop into MM plates covered with a layer of MM soft agar medium supplemented with a powder of CA or cellulose and incubated under different conditions. The growth under liquid conditions was performed by inoculating 100 ml of MM medium supplemented with 1 g of CA DS = 2, in 1000-ml flasks, with the spores of the different strains. The flasks were incubated at 50 °C under vigorous shaking for several days.

Assessment of the Degree of Substitution (DS) After Degradation

Films of about 90 µm in thickness were prepared by casting, dissolving 1 g of CA DS = 2 in 100 ml of acetone. Specimens of 5x5 cm were washed, dried and UV-sterilised. These pieces were placed on the surface of MM plates, inoculated with spores of the strain C300 and incubated at 50 °C. At intervals, samples were withdrawn and the degree of substitution was measured using the method described in ASTM D 871-72 (1983).

Vermiculite Respirometric Test Method

The respirometric method used in this work is an adaptation of a test method recently developed to assess the biodegradability of plastics (Bellia et al. 1999). The method is a solid-phase fermentation in which the vermiculite is used as solid matrix. The carbon dioxide evolved through the mineralization of the CA supplemented to the medium is measured. Vermiculite is a mineral clay material particularly suitable as a microbial carrier, displaying very favourable properties for growth and survival of microorganisms (Pesenti-Barili et al. 1991). Vermiculite can be easily sterilised, supplemented with nutrients (Tosin et al. 1998), and inoculated with a culture of the strain of interest. The bioreactors applied in this work were 250-ml glass flasks, endowed with an air inlet and an air outlet. The bioreactors can be sterilised and maintained under sterile conditions placing 0.3-µm air filters (Hepa Vent, Whatman) to the two openings. The bioreactors are filled with: 8 g vermiculite , 12 ml minimal medium MM 2x (with double concentration of solutes); 100 mg nutrient broth. After sterilization, a 10% solution of urea, sterilized by filtration, is added (1.4 ml). The solid bed is inoculated with 2 ml of spores suspended in saline solution. The suspension is obtained by scraping cultures of the micro-organism under study grown in agar plates. The bioreactors are then incubated at 50 °C. At intervals, the bioreactors are withdrawn from the incubator, connected to a source of CO_2-free air for a period of time sufficient to replace the inner atmosphere several times. By doing this, the oxygen possibly consumed in the previous phase is restored and the accumulated CO_2 washed away. During this phase the outlet of each bioreactor is connected to a systems of CO_2 traps, containing $Ba(OH)_2$. The system of CO_2 measuring and the calculation of the biodegradation percentage are described in ISO 14852.

The CO_2 evolution usually almost stops after 5 days, an indication that the supplemented carbon source has been consumed; 100 mg of UV-sterilised cellulose or CA (or any other test material, i.e. nutrient broth) is added to the test reactors. After 9-day incubation at 50 °C continued. The CO_2 is measured every 2–3 days.

Procedure for Chromosomal DNA Extraction and 16S rDNA Amplification and Sequencing

The strains were grown in 50 ml of TSB supplemented with 1.28 ml of a 20% glycine solution in 500 ml flasks, continuously stirred at 50 °C for 3 days. Then the method of Pospiech and Neumann (1995) was used. 16S rDNA was amplified by PCR using primers F27 (5'-AGAGTTTGATCMTGGCTCAG-3') and R1492 (5'-TACGGYTACCTTGTTACGACTT-3') (Heuer et al. 1997). Amplification was performed in a final volume of 50 µl, containing 1 µl of genomic DNA (approximately 10 ng), 50 mM KCl, 10 mM Tris-HCl, pH 8.3 (20 °C), 500 nM of each primer, 1.5 mM $MgCl_2$, 0.2 mM of each dNTP and 1.5 U *Taq* DNA polymerase (Roche Molecular Biochemicals). The parameters were: denaturation for 5' at 95 °C; 30 cycles of denaturation for 45" at 94 °C; annealing for 45" at 61 °C; extension for 2' at 72 °C; final extension step of 10' at 72 °C. PCR was performed in a PE9600 thermal cycler (Perkin Elmer). The amplified products were purified with the JETQUICK PCR kit (Genomed), according to the manufacturer's protocol. PCR products were sequenced using primers F27 and Seq3F (5'-AACACCGGTGGCGAAG-3'), corresponding to *E.coli* pos. 715–730) for the forward strand and primers R1492, Seq2R (5'-GGGCATGATGACTTGACG-3'), corresponding to *E.coli* pos. 1210–1192) and R513 (5'-CGGCCGCGGCTGCTGGCACGTA-3'), (Heuer et al. 1997) for the reverse strand. For each primer, two reactions were performed, on independent amplification products, to account for possible misincorporation of nucleotides during PCR. Sequencing reactions were performed with the BigDye Cycle Sequencing Kit (Applied Biosystems), and sequences were run on an ABI Prism 310 automatic sequencer (Perkin Elmer). Individual sequences were assembled to give the full sequence using the fragment assembly tools of the Wisconsin Package Version 9.1, Genetics Computer Group (GCG), Madison, Wisc. The search for homologies was performed using BLAST (Altschul et al. 1997) against all the sequences available in GENBANK and EMBL and using the sequence match software of the Ribosomal Database Project II (Maidak et al. 2000) against the specific 16S database (over 32 000 sequences). Sequence alignments and genetic distances were calculated with the Wisconsin package software.

Results and Discussion

Three sequential enrichment processes were performed at laboratory scale. CA was used as a cosubstrate and the inal compost of one composting cycle was used to inoculate the subsequent one. The composting processes showed the typical sequence of phases (Tosin et al. 1996). The CA-degrading microorganisms were selected from the compost by using the clear-zone method, already applied in the assessment of biodegradability (Augusta et al. 1993). Five strains were identified as capable of growing using CA (DS=2) and clarifying the surrounding area (see Table 2 and Figure 1).

Table 2. List of microorganisms isolated from different sources

Strain denomination	Source
A001	Composting trial A
B001	Composting trial B
C300	From CA-based plastic items, used as baits and buried in compost
D004	Composting trial D
204	From CA-based plastic items, used as baits and buried in compost

All the strains were also capable of growing in minimal medium supplemented with cellulose, ACTI and NA. All the strains were able to grow, to some extent, also in the minimal agarized medium, without any carbon source. This was particularly evident with an expired batch of agar, which was then discarded, and also in a case of overheating during media sterilization, caused by a mismanaged vapour sterilization (at about 135 °C instead of 110–120 °C). Probably some agar degradation product, formed during prolonged storage or during sterilization at high temperature, can be metabolised by the microbial isolates and used as carbon and energy source. This problem was partially solved by using fresh batches of agar and sterilising at lower temperatures (115 °C for 10 min).

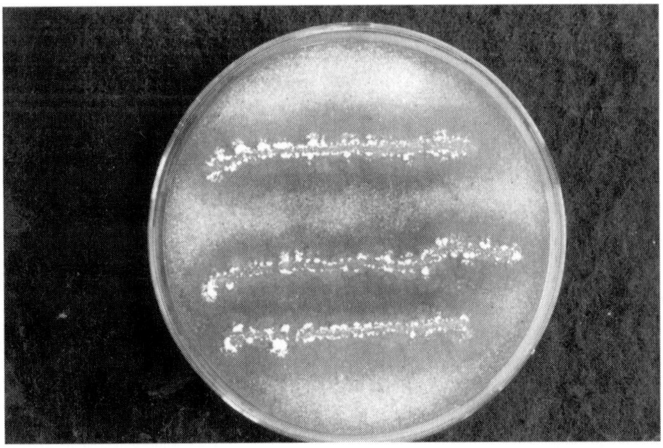

Fig.1. Growth of the strain C300 on a MM agar plate supplemented with CA DS=2 in powder

Identification of the Strains

The strains under investigation presented similar morphology, typical of the order *Actinomycetales*, i.e. vegetative hyphae with production of aerial mycelium. In order to identify the isolated strains, the chromosomal DNA of each was isolated. The 16S rDNA of the strains A001, D004 and C300 were amplified with primers F27 and R1492, partially sequenced with primers F27, Seq2R and R513, and the sequences compared. The three strains have an identical 16S rDNA sequence in the investigated region (about 1100 bp). We can, therefore, conclude that the three isolates are members of the same species or of very closely related species. The sequence of strain D004 was completed with primers Seq2F and R1492, and the almost complete 16S sequence obtained was compared with known sequences. Search in public databases revealed a high homology with the 16S rDNA sequences of several Thermomonosporaceae, the highest scores (98% overall identity) being with *Actinomadura rubrobrunea* (accession number AF134069) and *Excellospora viridilutea* (D86943).

The 16S sequence was also aligned with a small subset of Actinomycete 16S rDNA sequences, including representatives of all the Actinomycetales families. Genetic distances were calculated, based on the Jin-Nei algorhitm (Jin and Nei 1990), and a phylogenetic tree constructed. In this tree, strain D004 clusters together with *Actinomadura rubrobrunea*, *Excellospora viridilutea* and other *Actinomadura* species, confirming the results of the homology search (data not shown).

Based on 16S rDNA analysis, we can conclude that strains A001, C300 and D004 are members of the same species (or very closely related species) of the family Thermomonosporaceae, probably belonging to genus *Actinomadura* or to a closely related genus.

Characterizations of the Strains

The growth of the strains on plates of minimal medium using CA or cellulose as the only carbon source was tested at different temperatures, at different pH and using different nitrogen sources. The results of these tests are shown in Tables 3, 4 and 5.

Table 3. The different isolates have been grown in plates of minimal medium adjusted at different temperature. The level of growth obtained with the different carbon sources has been assessed by a qualitative judgement (symbols: - = no growth; from + to ++++ = increasing level of growth)

Strain	Incubation temperature (°C)	MM	MM +cellulose	MM +CA
A001	20	-	-	-
A001	37	+	++	-
A001	50	-+	++	+++
A001	58	-	-	-
A001	65	-	-	-
A001	65→50 [a]	++	++	++
B001	20	-	-	-
B001	37	+	++	-
B001	50	-+	++	+++
B001	58	-	-	-
B001	65	-	-	-
B001	65→50 [a]	-	-	+++

Table 3. (cont.)

D004	20	-	-	-
D004	37	+	+ +	-
D004	50	- +	+ +	+ + +
D004	58	-	-	-
D004	65	-	-	-
D004	65→50 [a]	+	+ +	+ +
204	20	-	-	-
204	37	+	+ +	-
204	50	- +	+ +	+ + +
204	58	-	-	-
204	65	-	-	-
204	65→50 [a]	+	+ +	+ + +
C300	20	-	-	-
C300	37	+	+ +	-
C300	50	- +	+ +	+ + +
C300	58	-	-	-
C300	65	-	-	-
C300	65→50 [a]	-	+ +	+

[a] After 10 days at 65 °C the plates were transferred at 50 °C and the growth evaluated after 14 further days of incubation.

Table 4. The different isolates have been grown in agar plates of minimal medium adjusted at different level of pH, at 50 °C. The level of growth obtained with the different carbon sources was assessed by a qualitative judgement (symbols: - = no growth; from + to ++++ = increasing level of growth)

Strain	PH	MM	MM + cellulose	MM + CA DS = 2.0
A001	5.0	+ -	+ +	+
A001	7.0	- +	+ +	+ + +
A001	8.5	+	+ +	+ +
B001	5.0	-	+ +	+
B001	7.0	- +	+ + +	+ + +
B001	8.5	+	+ +	+ +
D004	5.0	-	+	- +
D004	7.0	- +	+ +	+ + +
D004	8.5	+	+ +	+ +
204	5.0	-	+	-
204	7.0	- +	+ +	+ + +
204	8.5	+	+ +	+ +
C300	5.0	-	++	+
C300	7.0	- +	+ + +	+ + +
C300	8.5	+	+ +	+ +

Table 5. The different isolates have been grown in agar plates of minimal medium supplemented with different nitrogen sources, at 50 °C and pH 7. The level of growth obtained with the different carbon sources was assessed by a qualitative judgement (symbols: - = no growth; from + to ++++ = increasing level of growth)

Strain	Nitrogen source	MM	MM + cellulose	MM + CA DS = 2.0
A001	NH_4NO_3	- +	+ +	+ + +
A001	$(NH_4)_2SO_4$	+	+ +	+ +
A001	$NaNO_3$	+ -	+ +	+ + +
B001	NH_4NO_3	- +	+ +	+ + +
B001	$(NH_4)_2SO_4$	+	+ +	+ +
B001	$NaNO_3$	+ -	+ +	+ + + +
D004	NH_4NO_3	- +	+ +	+ + +
D004	$(NH_4)_2SO_4$	+	+ +	+ +
D004	$NaNO_3$	+ -	+ +	+ + +
204	NH_4NO_3	- +	+ +	+ + +
204	$(NH_4)_2SO_4$	+	+ +	+ +
204	$NaNO_3$	+ -	+ +	+ + +
C300	NH_4NO_3	- +	+ +	+ + +
C300	$(NH_4)_2SO_4$	+	+ +	+ +
C300	$NaNO_3$	+ -	+ +	+ + + +

The strains are capable of growing in CA DS=2 and in cellulose at 50 °C. At 37 °C the strains can grow on cellulose but do not grow on CA. This indicates that the cellulolytic enzymes and the esterases needed to cleave the ester bond between the glucose residues and the acetic acid are both produced and active at 50 °C while, at 37 °C the esterases are either not produced or inactive. The strains are not able to grow at 58 °C or higher temperature. This is a reversible effect: they can grow on CA and cellulose, even after a prolonged exposure at 65 °C, once the temperature is decreased to 50 °C. However, the CA is degraded, when blended with starch, under composting conditions at 58 °C (Degli-Innocenti et al. 2000). Probably other active species, present in the compost, may act together in a synergistic fashion (one providing the esterases, the other the cellulases). Our system of selection would probably miss these consortia. Another possibility is that the CA starch blend is more degradable than the CA alone. This seems to be the case (unpubl. data) but, at the moment, we cannot differentiate between the two hypotheses.

The best pH for growing the strains on CA is 7, while lower growth is obtained with a pH of 5.

Growth Under Liquid and Solid Conditions

No sign of growth of the stain 204 was observed under submerged, stirred, liquid conditions in MM supplemented with CA DS = 2 as only carbon source. No further trials were performed. The vermiculite test method was then applied in order to obtain a more quantitative measurement of the CA degradation caused by the different strains, and to overcome the apparent impossibility of performing growth tests under liquid conditions. The vermiculite test method, recently developed to study the biodegradability of plastics under solid conditions (Bellia et al. 1999), has been used in the present study, for the first time, to determine the degradation of a substrate (CA) when exposed to a single selected strain instead of a mixed population. For technical reasons, we could only apply a discontinuous air supply to the cultures. In a first trial the strain C300 was fed with NB or cellulose (Fig. 3). The final CO_2 conversion was low, for both NB and cellulose, compared to the CO_2 level usually found with continuously aerated systems and using mixed microbial populations (Degli-Innocenti et al. 2000). This cannot be considered as an indication that the strain C 300 is unable to degrade the substrates (NB and cellulose). As a matter of fact, the growth on cellulose plates and, in particular, on NA plates is very strong. In a second trial, the same strain was grown with CA DS=2 and cellulose (Fig. 4). In this case, the final levels of conversion into CO_2 were also low. It is interesting to remark that the biodegradation level of CA was very similar to the biodegradation of cellulose and NB, suggesting that the CA DS=2 is a carbon source as good as cellulose (and NB) for the strain C300.

The test has been repeated with another strain (A001), obtaining the same results The low biodegradation level could be a consequence of the discontinuous aeration system and better results could be obtained by using a continuous system.

The degradation of the CA DS =2 occurs under solid-state conditions, but fails under submerged stirred liquid conditions. This shows that the phase of the growth medium is an important factor in the degradation. Very poor degradation results were obtained in another study, where microorganisms from compost or soil were used as an inoculum for aqueous tests (Schäfe et al. 1998). Therefore, it is of great interest to set up a solid-phase method for testing the species isolated from compost or soil species.

The DS of CA films placed on MM plates and inoculated with the strain C300 and incubated at 50 °C has been followed for two months. The results, shown in Table 6, indicate that the strain is able to decrease the DS of the CA DS=2.

Table 6. Deacetylation of cellulose acetate during degradation caused by the strain C300

Incubation time (days)	DS
0	2.0
12	2.0
42	2.0
50	1.99
60	1.03

Further studies at biochemical levels will be necessary to better characterise the degradation mechanisms of CA by the isolated strains. Furthermore, it will be interesting to determine the minimum DS of CA which still allows the degradation by the strains.

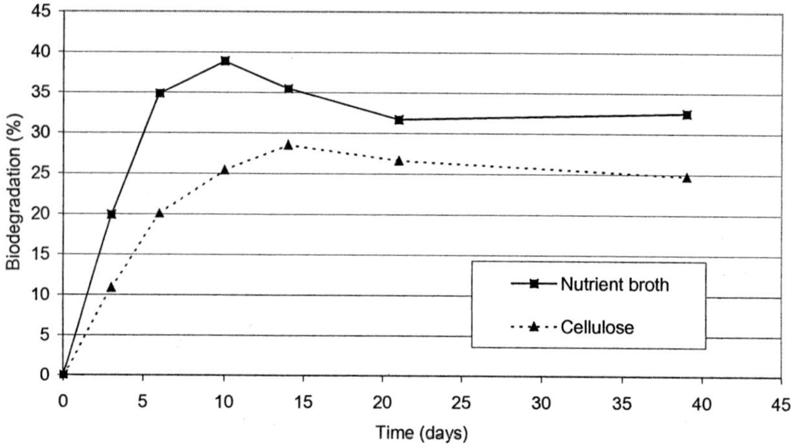

Fig.3 Biodegradation, measured through CO_2 evolution, of nutrient broth and microcrystalline cellulose caused by the strain C300. Each *point* is the average of two replicates

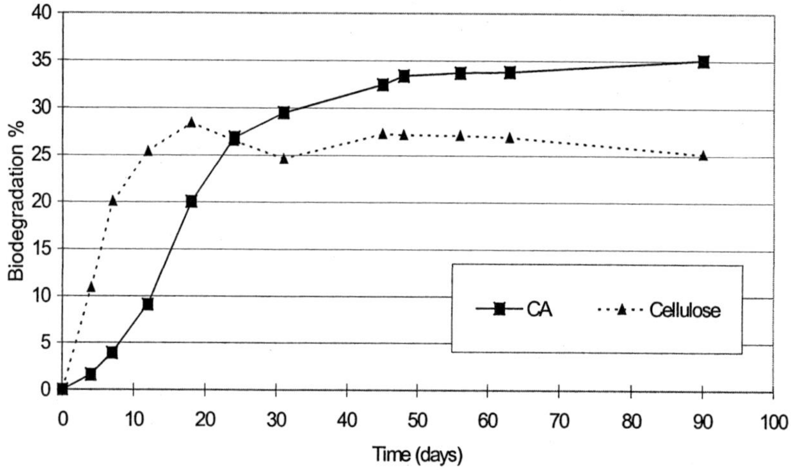

Fig.4. Biodegradation, measured through CO_2 evolution, of CA DS=2 and microcrystalline cellulose caused by the strain C300. Each *point* is the average of two replicates

Acknowledgements. Many thanks to Dr. Renato Fani (University of Florence), Dr. Anna Maria Puglia (University of Palermo), and Prof. Enrica Galli (University of Milan) for assistance and suggestions. Thanks also to Dr. Giulia Gregori and Dr. Sara Guerrini for reading the manuscript.

References

Altschul S.F, Madden TL, Schäffer AA, Zhang J, Zhang Z, Miller W, Lipman DJ (1997) Gapped BLAST and PSI-BLAST: a new generation of protein database search programs. Nucleic Acids Res 25:3389-3402

Anonymous (1997) Compostable cutlery. Perseco Focus 4:3

ASTM D 871-72 (1983) Annual book of ASTM standards. American Society of Testing and Materials, Easton, pp 172-187

Augusta J, Müller RJ, Widdecke H (1993) A rapid evaluation plate-test for the biodegradability of plastics. Appl Microbiol Biotechnol 39:673-678

Bellia G, Tosin M, Floridi G, Degli-Innocenti F (1999) Activated vermiculite, a solid bed for testing biodegradability under composting conditions. Polymer Degrad Stabil 66:65-79

Buchanan CM, Gardner RM, Komarek RJ (1993) Aerobic biodegradation of cellulose acetate. J Appl Polymer Sci 47:1709-1719

Degli-Innocenti F, Tosin M, Bellia G (2000) Degradability of plastics. Standard methods developed in Italy. Int Conference on Biodegradable polymers: production, marketing, utilisation and residues management Wolfsburg (Germany) 4-5 Sept 2000

Gardner RM, Buchanan CM, Komarek R, Dorschel D, Boggs C, White AW (1994) Compostability of cellulose acetate films. J. Appl. Polymer Sci 52:1477-1488

Gross RA, Gu JD, Eberiel DT, Nelson M, McCarthy SP (1993) Cellulose acetate biodegradability in simulated aerobic composting and anaerobic bioreactor environments as well as by a bacterial isolate derived from compost. In: Ching C, Kaplan D, Thomas E (eds) Biodegradable polymers and packaging. Technomic Publishing Company Lancaster PA, USA pp 257-279

Heuer H, Krsek M, Baker P, Smalla K, Wellington EMH (1997) Analysis of actinomycete communities by specific amplification of genes encoding 16S rRNA and gel–electrophoretic separation in denaturing gradients. Appl Environ Microbiol 63: 3233-3241

ISO 14852 (1999) Determination of the ultimate aerobic biodegradability of plastic materials in an aqueous medium – Method by analysis of evolved carbon dioxide. International Organisation for Standardisation, Geneva, Switzerland

Ji-Dong G, Eberiel DT, McCarthy SP, Gross RA (1993) Cellulose acetate biodegradability upon exposure to simulated aerobic composting and anaerobic bioreactor environments. J Envir Polymer Degrad 2 :143-153

Jin L, Nei M (1990) Limitations of the evolutionary parsimony method of phylogenetic analysis. Mol Biol Evol 7:82-102

Komarek RJ, Gardner RM, Buchanan CM, Gedon S. (1993) Biodegradation of radiolabeled cellulose acetate and cellulose propionate J Appl Polymer Sci 50:1739

Maidak BL, Cole JR, Lilburn TG, Parker CT Jr, Saxman PR, Stredwick JM, Garrity GM, Li B, Olsen GJ, Pramanik S, Schmidt TM, Tiedje JM (2000) The RDP (ribosomal database project) continues. Nucleic Acids Res 28:173-174

Palmisano AC, Maruscik DA, Ritchie CJ, Schwab BS, Rapaport RA (1993) A novel bioreactor simulating composting of municipal solid waste. J Microb Methods 18: 99-112

Pesenti-Barili B, Ferdani E, Mosti M, Degli-Innocenti F (1991) Survival of Agrobacterium radiobacter K84 on various carriers for crown gall control. Appl Environ Microbiol 57: 2047-2051

Pospiech A, Neumann B (1995) A versatile quick-prep of genomic DNA from Gram-positive bacteria. Trends Genet. 11: 217-218

Reese ET (1957) Biological degradation of cellulose derivatives. Ind Engin Chem 49:89-93

Schäfe A, Müller RJ, Pantke M, Pagga U (1998) Evaluation of the ultimate aerobic biodegradability of plastic materials in an aqueous medium. Report of a ring test. International Biodeterioration Research Group. Chester (UK) September 1998

Stutzenberger F, Kahler G (1986) Cellulase biosynthesis during degradation of cellulose derivatives by Thermomonospora curvata. J of Applied Bacteriol 61:225-233

Tosin M, Degli-Innocenti F, Bastioli C (1996) Effect of the composting substrate on biodegradation of solid materials under controlled composting conditions. J of Environ Polymer Degrad 4(1):55-63

Tosin M, Degli-Innocenti F, Bastioli C (1998) Detection of a toxic by-product released by a polyurethane-containing film using a composting test method based on a mineral bed. J Environ Polymer Degrad 6:79-90

PCB's Biotransformation by a White-Rot Fungus Under Composting and Liquid Culture Conditions

G. Ruiz-Aguilar[1], J. Fernández-Sánchez[1], R. Rodríguez-Vázquez*[1], H.M. Poggi-Varaldo[1], F. Esparza-García[1], R. Vázquez-Duhalt[2]

Abstract. The white-rot fungus *Phanerochaete chrysosporium* H-298 CDBB-500 was able to dechlorinate the polychlorinated biphenyl (PCB's) congeners. In this study, the extent of transformation and the ligninolytic enzymes involved in PCB transformation were compared between the composting and liquid cultures. Composting culture (CC) showed a higher CO_2 evolution when fungal inoculum was present than those containing only the native microflora. The fungus and the native community showed synergistic effects as demonstrated by the increase of CO_2 production and PCB transformation. The added fungi and the native microflora transformed around 70 % of the PCB congeners. In the other hand, the extent of PCB degradation by the fungus was 85 % in the liquid culture (LC). There was a direct correlation between PCB degradation and fungal growth. In both systems (LC and CC), the highly chlorinated congeners were transformated, while the less chlorinated congeners increased during the treatment, suggesting a possible dechlorination process. Mn dependent peroxidase (MnP) was detected in both systems, while lignin peroxidase (LiP) was detected only in LC. MnP is suggested to be involved in PCB transformation.

Introduction

Treatment systems using microorganisms for the degradation of toxic recalcitrant organopollutants hold the promise for detoxification of vast quantities of contaminated soil (Bumpus et al. 1985). The white-rot fungus *Phanerochaete chrysosporium* is able to transform a wide variety of environmental persistent organopollutans to CO_2 (Bumpus et al. 1985; Sasek et al. 1993). The bioaugmentation process permits to use mixed cultures in order to improve the overall transformation, in which each species occupies its own niche for growth and substrate degradation. Different lignin contents are found in many sediments and soils in which organopollutans are adsorbed. Thus, this is an ecological advantage for ligninolityc microorganisms that are able to grow and metabolize

[1]Depto. de Biotecnología y Bioingeniería, CINVESTAV-IPN, Av. Instituto Politécnico Nacional 2508, San Pedro Zacatenco, Deleg. Gustavo A. Madero, 07360, México, D.F. México. Phone +(52 5) 747 7000 Ext. 4351. Fax: +(52 5) 747 7000 Ext. 4305.
E-mail: rrodrig@mail.cinvestav.mx. gruiz@mailbanamex.com.
[2]Instituto de Biotecnología, UNAM, Apdo. Postal 510-3, Cuernavaca, Morelos 62250, México.

H. Insam, N. Riddech, S. Klammer (Eds.)
Microbiology of Composting

insoluble recalcitrant substrates. Most of the degradation studies have been performed on LC (Andersson and Henrysson 1996; Braun-Lülleman et al. 1997; Boochan et al. 2000). However, only few studies have addressed the potential of those fungi to degrade hydrophobic pollutants in contaminated soil (Tiedje et al. 1993). Also, liquid media may not represent the situation in the natural environment into which xenobiotics have been previously introduced (Neilson 1996). The fungal morphology and physiology in liquid and composting conditions are usually different. LC permits the establishment of the metabolic mechanisms involved in pollutant degradation more easily than in CC, because of the soil interferences. *P. chrysosporium* has the ability to remove individual PCB congeners from Aroclors 1242, 1254 and 1260 (Yadav et al. 1995). Different combinations of the ligninolytic enzymes are known to ensure efficient biodegradation of recalcitrant lignin (Hatakka 1994) and some xenobiotic compounds (Field et al. 1993), but most of these enzymes have so far been implicated in PCB degradation.

Ligninolytic conditions imply extreme environmental conditions, which affect either physiological or metabolic functions, including formation of other secondary metabolites that have an adverse effect towards production of the ligninolytic enzymes (Dosoretz et al. 1999). Several white-rot fungi may readily proliferate in lignocellulose substrate and therefore to sure their growth on soil, where is not their natural habitat (Lang et al. 1997). Some fungal spores may be mature and able to germinate immediately when they are introduced into a suitable environment. These spores may complete their maturation, or if the spores are already mature, they may require a special activation treatment involving conditions other than those that normally support somatic growth (Moore-Landercker 1996).

The secretion of ligninolytic enzymes by white-rot fungi has been demonstrated in LCs using defined media, complex broth or submerged lignocellulose substrates (Lang et al. 1997). Ligninolytic enzymes using white-rot fungi do the first transformation step of the pollutants. The prerequisite for transformation of xenobiotics by fungi in soil is the transport into, or the production of the ligninolytic enzyme production in soil (Lang et al. 1997).

The aim of this work is to compare the extent of PCB biodegradation by a white rot fungus and to monitor the ligninolytic activity in both LC and CC.

Material and Methods

Microorganisms and their cultivation

Phanerochaete chrysosporium strain H-298 CDBB-500 was obtained from the Microbial Cultures Collection of Department of Biotechnology and Bioengineering, CINVESTAV-IPN. The strain was maintained on malt extract agar (Merck-Mexico) slopes. This strain is able to degrade benzo(a)pyrene

(Rodriguez et al. 1999), pentachlorophenol (Mendoza-Cantú et al. 2000), PCB's (Fernández-Sánchez et al. 1999).

Media for Liquid Culture

For inoculum production a medium containing yeast-peptone-glucose (YPG) was used. A nitrogen-limited mineral medium (Morgan et al. 1991b) was used for treatments without glucose. An extract from PCB contaminated soil was used as carbon sources.

Support for Composting Culture

Sugarcane bagasse was used as support and it was treated according to Fernández-Sánchez et al. (1999). The bagasse was air-dried and sieved through an 860-590 μm mesh.

Soils

5 kg of soil from a plot of land contained PCB in Mexico were collected. The granulometric analysis of the soil reported 78% sand, 15 % silt and 4% clay and had a pH of 6.3. Total organic C content was 1.88%. Bi-, tri-, tetra-, penta-, hexa-, hepta-, octa- and nona-chlorobiphenyls were found in the soil.

Carbon Dioxide Analysis

The CO_2 production was determinated by gas chromatography (GC). This analysis was performed using a Gow-Mac chromatograph equipped with thermal conductivity detector and a concentric column CTR1 (Alltech, USA). The GC detector and injector temperatures were set at 100 °C and 40 °C respectively. The column was kept at room temperature and the detector potential was set at 125 volts. Helium was used as a carrier gas at a flow rate of 55-ml/min. Runtime for each sample was approximately 4 min. The CO_2 production was calculated according to Saucedo-Castañeda et al. (1994). 2-ml headspace sample was taken from microcosms and analyzed for CO_2 evolution.

PCB's Extraction and Analysis

For CC, PCB's were extracted and analyzed as explained by Fernández-Sánchez et al. (1999). The contents of microcosms were extracted in a Soxhlet apparatus with n-hexane-acetone mixture. Extracts were concentrated down to 2-ml in a rotary evaporator and treated as indicated in Fernández-Sánchez et al. 1999. In the case of LC, medium was washed with an n-hexane-acetone mixture. The new mixture was extracted with hexane in a separation funnel and treated as reported by Fernández-Sánchez et al. 1999.

The PCB's analysis was done by high-pressure liquid chromatography (HPLC) and gas chromatography-mass spectrometry (GC-MS). The HPLC system consisted of a Consta Metric 3200 pump (LCD Analytical), an UV Spectro Monitor 3200 detector (Thermo Separation Products) and a Chromoject integrator. The column was a C18 (Chromanetics, 150x4.6 mm, and particle diameter 5 μm). The solvents were acetonitrile-water at a flow rate of 1.0 ml/min. The quantitation of PCB's was performed by a standard procedure: calculation from the sum of characteristic peaks selected from chromatograms of samples and compared with a PCB standard.

The GC-MS system consisted of a Varian Saturn 3 GC/MS, with electron multiplier as detector and a spi injector. A DB-5 fused silica capillar column (30 m x 0.25 mm) was used. The operating conditions were the injector and detector temperatures 250 °C and 260 °C, respectively. Helium was used as a carrier gas at a flow rate of 1 ml/min; sample column 1 μl. Interpretation of the chromatographic data was carried out by integration software (Varian, Saturn 3 GC/MS version 4.1).

Enzyme Assays

The samples were immediately analyzed for ligninolytic enzyme activities. Mn dependent peroxidase (MnP) activity was measured according to Tien et al. (1988). Lignin peroxidase (LiP) was estimated with veratryl alcohol as the substrate (Kuwahara et al. 1984).

Treatments for Composting and Liquid Cultures

The CC was made in a microcosm with PCB-contaminated soil and bagasse that was inoculated with *P. chrysosporium*. Microcosms were amended with 1.5 g (dry weight) of baggasse and 8.3 mg of mycelium of *P. chrysosporium*. The material was moistened with 1.5-ml Kirk's medium (Kirk et al. 1976) without glucose, to reach 60 % of moisture content. The medium consisted of (in 1L): KH_2PO_4 (200 mg), $MgSO_4.7H_2O$ (50 mg), $CaCl_2$ (80 mg), NH_4NO_3 (48 mg), L-asparagine (79 mg), 0.5-ml of vitamins and 1-ml trace mineral solutions. The pH of the medium was adjusted to 4.5 with HCl, and the microcosms were acclimated at 39 °C for 11 days prior to PCB-contaminated soil addition (8.5 g). To improve the contact between soil and bagasse, 1 ml of sterile Ringer's solution (Morgan et al. 1991a) was applied to the soil surface. Then, the microcosms was sealed and incubated at 39°C in the dark for 20 days. The bagasse and soil were sterilized to eliminate microorganisms for the controls. To preserve aerobic conditions and avoid carbon dioxide cumulating, each flasks was flushed daily with sterilized and moistened air for 15 min. Cultures were aerated every day for 30 days.

The LC was carried out in flasks and inoculated with pellets of *P. chrysosporium* H-298 CDBB-500 previously grown in YPG medium. We added the PCB extract before the experiment started. Two controls cultures were carried out, in both systems, one without inoculum and other with inactive fungus (abiotic control). The percentage of PCB transformation was estimated by comparing the PCB concentrations at the beginning and at the end of the treatment.

Results and discussion

CO_2 Evolution

The effect of the addition of a foreign microorganism (fungus) to native microflora on the CO_2 evolution was evaluated (Fig. 1.). Aerobic heterotrophic activity showed high levels (about 20 mg CO_2) in the treatments with fungus plus sterile and non-sterile bagasse-soil. This value suggests that the fungus had a physiological activity in the microcosms. It is reflected in the treatment without fungus where CO_2 evolution was 13 mg. There was a possible synergism between *P. chrysosporium* and native microflora that might favor the PCB transformation. The competition of non-degrading microorganisms seems to be minimal, when lignocellulosic residue was used for fungal growth. In sterilized bagasse-soil without fungus microcosm a low level of CO_2 (4 mg of CO_2) was found. It is possible that some microorganisms resisted the autoclave process and produced CO_2 after 15 days. Sylvia et al. (1999) found that some endospores could resist environmental changes during soil sterilization and gave some CO_2 activity.

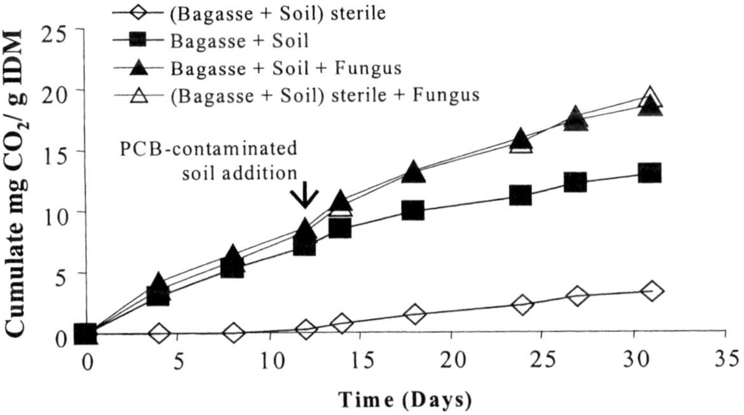

Fig. 1. Cumulative CO_2 production during composting culture of *Phanerochaete chrysosporium* H-298 CBDD-500 for PCB-contaminated soil treatment. Arrow indicates the time that PCB-contaminated soil was added (11 days). IDM: initial dry matter.

PCB's Transformation in Liquid and Composting Cultures

Two systems, LC and CC, were compared for their PCB metabolization and ligninolytic activity. Analysis by GC-MS and HPLC showed a reduction in highly chlorinated congeners and an increase in the proportion of less chlorinated biphenyls in both systems (Fig. 2. and 3.). LC's showed better levels of transformation (85 %) than CC (70 %).

Some congeners, as di-CB to nona-CB, were biotransformed during the process of bioaugmentation in CC (Fig. 2.b). It has seen that hexa-CB and hepta-CB congeners are difficult to degrade by microorganisms (Safe 1984). However, the interaction between fungus and native microflora indicated us that their degradation was possible. Certain bacterial species have been reported to degrade chlorinated biphenyls at concentrations comparable, with those examined in this research. Tiedje et al. (1993) observed that the aerobic microorganisms use either dioxygenases or oxygenases to metabolize PCB's, however, it is known that biotransformation mechanism for highly chlorinated congeners are performed by anaerobic community (Abramowicz 1990). It is possible that anaerobic microniches were present on soil and thus they were responsible for the degradation.

In addition, white rot fungi are known to produce nonspecific extracellular peroxidases. These ligninolytic enzymes attack chlorinated lignins in papermill effluents, polycyclic aromatic hydrocarbons, dyes, various insecticides and polychlorinated phenols (Thomas et al. 1992). *P. chrysosporium* is particularly attractive because it can metabolize pollutants completely to carbon dioxide (Thomas et al. 1992). This suggests that *P. chrysosporium* uses a dechlorination process to transform at PCB's. This is supported by the fact that LC's showed dismiss in the highly chlorinated congeners, where aerobic conditions prevail. PCB transformation was attributed to *P. chrysosporium* in LC since minimal PCB lost (5 %) was found in abiotic control (Fig. 3.b). Also, CC has more interferences than LC has. Then let PCB's to be more susceptible to be attacked by a fungus in LC. In this study, it was found that *P. chrysosporium* was effective in the PCB transformation. The ability to treat high concentrations of pollutants shows that white rot fungi should be considered seriously for the treatment of PCB-contaminated soil.

Enzymatic Activity

LiP and MnP activities of *P. chrysosporium* grown on CC and LC are shown in Fig. 4. LiP activity was detected only in LC; nevertheless MnP activity was detected in both culture systems. In LC's MnP activity preceded LiP activity. No LiP activity could be detected in CC in the presence of PCB's. However, LiP activity was found in CC with surgarcane bagasse and without PCB's (Cruz-Córdova et al. 1999), suggesting an inhibition effect of PCB on the LiP production.

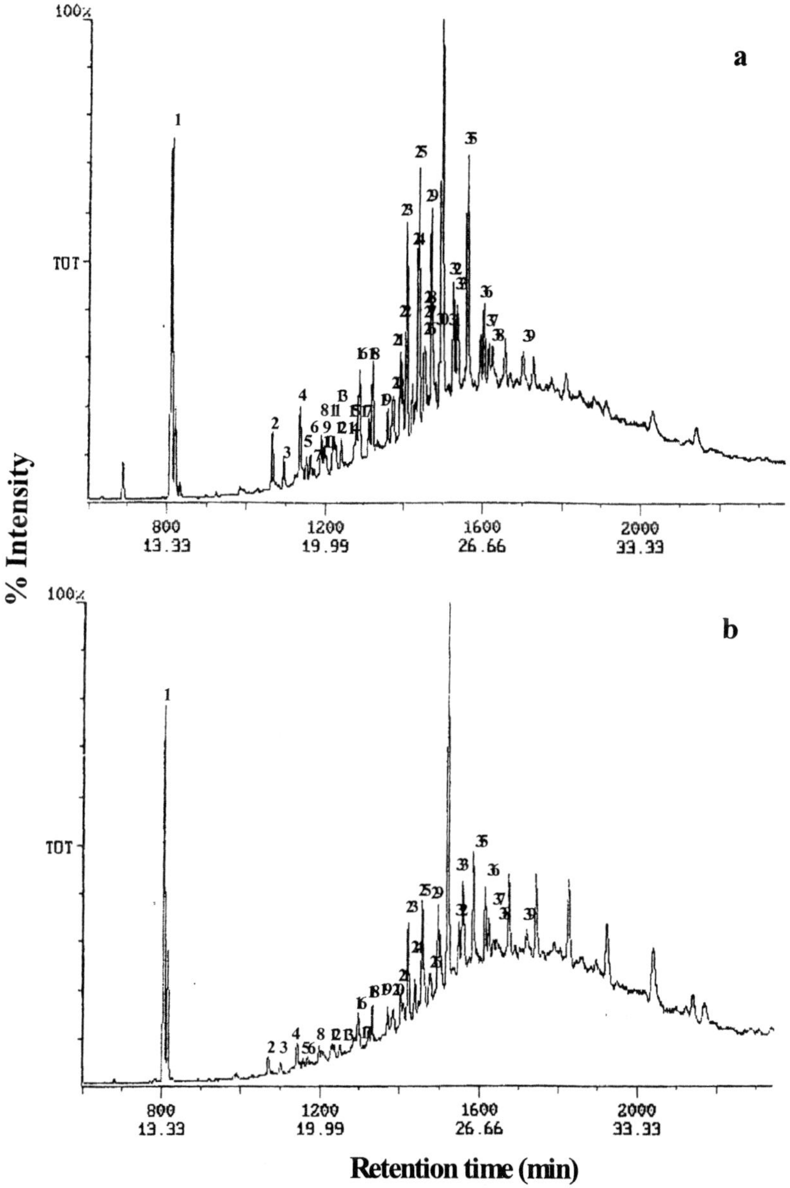

Fig. 2. GC-MS chromatograms of PCB transformation in composting culture by a consortium (fungus-native microflora). **a** Initial sample, **b** Bioaugmentated sample, after 20 days of treatment. Each number on the peaks represents congeners found in the samples.

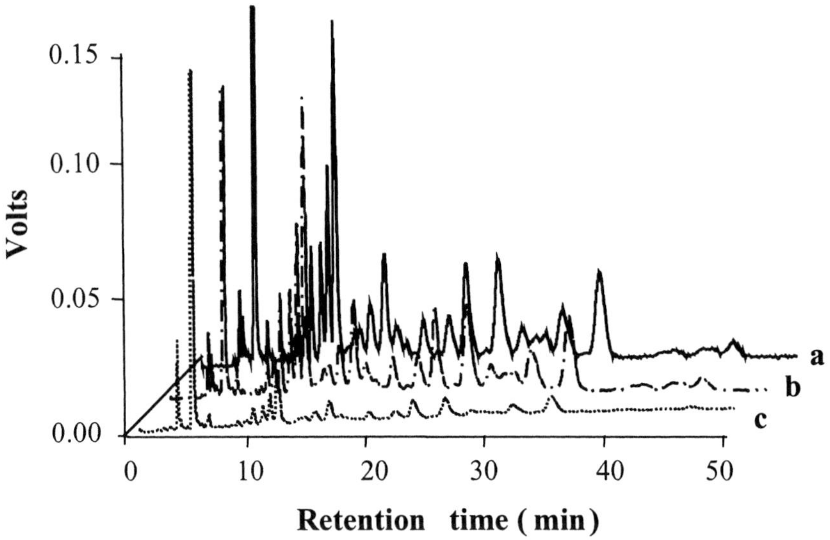

Fig. 3. HPLC chromatograms of PCB transformation in liquid culture by *P. chrysosporium* H-298 CDBB-500. **a** Initial sample, **b** Abiotic control, **c** Sample after 10 days of treatment.

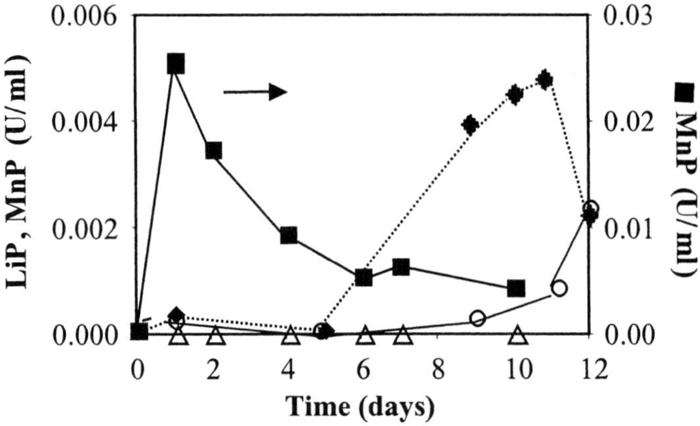

Fig. 4. LiP and MnP activities produced by *P. chrysosporium* H-298 CDBB-500 in composting and liquid culture, in the presence of PCB's. For liquid culture ◆ LiP and ■ MnP. For composting culture △ LiP and ○ MnP.

Ligninolytic activity from *P. chrysosporium* is affected by the culture conditions (pH, nutrient concentration, inductors, agitation, etc.) (Kirk et al. 1978). In addition, it is well known that levels of manganese in the medium have a drastic effect on the LiP production (Boominathan and Reddy 1991). These results show that the soil inoculation with *P. chrysosporium* improves the PCB metabolization, and that the fungal metabolism is able to dechlorinate the PCB molecules.

Conclusion

White-rot fungi are able to degrade several toxic compounds. Our results suggest that *Phanerochaete chrysosporium* have the ability to penetrate nonsterile soil and produce ligninolytic enzymes in a soil environment. Interaction between fungus and indigenous microflora was beneficial for PCB degradation. LC was useful as a reference and confirmed the data obtained by CC, because soil interference (soil microorganisms, organic and inorganic components, etc.) is avoided under these conditions. *P. chrysosporium* should be taken into account for remediation treatments.

Acknowledgments. We are especially grateful to Alfredo Medina-Dávila and Dolores Díaz-Cervantes for their excellent technical assistance. This work was sopported by IMP-FIES 95-106-VI. Graduate scholarships to José Manuel Fernández-Sánchez and Graciela Ruiz-Aguilar from CONACyT are gratefully acknowledged.

References

Abramowicz DA (1990) Aerobic and anaerobic biodegradaation of PCBs: a review. Critical Rev Biotechnol 10(3):241-251

Andersson BE, Henrysson T (1996) Accumulation and degradation of dead-end metabolites during treatment of soil contaminated with polycyclic aromatic hydrocarbons with five strains of white-rot fungi. Appl Microbiol Biotechnol 46(5/6):647-652

Boominathan K, Reddy CA (1991) Fungal degradation of lignins biotechnological applications. In: Arora DK, Bharat R, Mukerji KG, Knudsen GR (eds) Handbook of applied mycology, Vol. 4. Marcel Dekker Inc, New York, pp 786

Boochan S, Britz ML, Stanley GA (2000) Degradation and mineralization of high-molecular-weight polycylic aromatic hydrocarbons by defined fungal-bacterial cocultures. Appl Environ Microbiol 66(3):1007-1019

Braun-Lülleman A, Majcherczyk A, Hütterman A (1997) Degradation of styrene by white-rot fungi. Appl Microbiol Biotechnol 47(2):150-155

Bumpus AJ, Tien M, Wright D, Aust DS (1985) Oxidation of persistent environmental pollutants by a white rot fungus. Science 228:1434-1436

Cruz-Córdova T, Roldán-Carrillo TG, Díaz-Cervantes D, Ortega-López J, Saucedo-Castañeda G, Tomasini-Campocosio A, Rodríguez-Vázquez R (1999) CO_2 evolution and ligninolytic and proteolityc activities of *Phanerochaete chrysosporium* grown in solid state fermentation. Resources, Conservation and Recycling 27:3-7

Dosoretz CG, Chen AH-C, Grethlein HE (1999) Effect of oxygenation conditions on submerged cultures of *Phanerochaete chrysosporium.* Appl Microbiol Biotechnol 34:131-137

Fernández-Sánchez JM, Ruiz-Aguilar G, Rodríguez-Vázquez R (1999) Polychlorinated biphenyls transformation by bioaugmentation in solid culture. In: Alleman BC, Leeson A (eds) Bioremediation of nitroaromatic and haloaromatic compounds, Fifth International In Situ and On-site Bioremediation Symposium, Vol. 7. Battelle Press, Columbus, Ohio, pp 173-178

Field JA, deJong E, Foijoo-Costa G, deBont JAM (1993) Screening for ligninolytic fungi applicable to the biodegradation of xenobiotics. Trends Biotechnol 11:44-49

Hatakka A (1994) Lignin-modifying enzymes from selected white-rot fungi: production and role in lignin degradation. FEMS Microbiol Rev 13:125-135

Kirk TK, Connors WJ, Zeius TG (1976) Requeriment for a growth subtrate during lignin descomposition by two wood-rotting fungi. Appl Environ Microbiol 32(1):192-194

Kirk TK, Schultz E, Connors WJ, Lorenz LF, Zeikus JG (1978) Influence of culture parameters on lignin metabolism by *Phanerochaete chrysosporium.* Arch Microbiol 117:277-285

Kuwahara M, Glenn KJ, Morgan AM, Gold HM (1984) Separation and characterization or two extracellular H_2O_2-dependent oxidase from ligninolytic cultures of *Phanerochaete chrysosporium.* FEBS Lett 169(2):247-250

Lang E, Eller G, Zadrazil F (1997) Lignocellulose decomposition and production of ligninolytic enzymes during interaction of white rot fungi with soil microorganisms. Microbial Ecol 34:1-10

Mendoza-Cantú A, Albores A, Fernández-Linares L, Rodríguez-Vázquez R (2000) Pentachlorophenol biodegradation and detoxification by the white-rot fungus *Phanerochaete chrysosporium.* Environ Toxicol 15:107-113

Moore-Landecker E (1996) Spores. In Fundamentals of the fungi Part II (Physiology and reproduction). Fourth Edition. Prentice Hall, USA, pp 360

Morgan P, Cooper CJ, Battersby NS, Lee SA, Lewis ST, Machin TM, Graham SC, Watkinson RJ (1991a) An automated image analysis method for the determination of fungal biomass in soils and on solid matrices. Soil Biol Biochem 23(7):609-616

Morgan P, Lewis ST, Watkinson RJ (1991b) Comparison of abilities of white-rot fungi to mineralize selected xenobiotic compounds. Appl Microbiol Biotechnol 34:693-696.

Neilson AH (1996) An environmental perspective on the biodegradation of organochlorine xenobiotics. International Biodeterioration & Biodegradation. 37(1-2):3-21

Rodríguez RV, Montalvo CP, Dendooven L, Esparza FG, Fernández LL (1999) Degradation of benzo(a)pyrene in soil by white rot fungi. In: Alleman BC, Leeson A (eds) Bioremediation technologies for polycyclic aromatic hydrocarbon compounds, Fifth International In Situ and On-site Bioremediation Symposium, Vol. 8. Battelle Press, Columbus, Ohio, pp 93-98

Safe SH (1984) Microbial degradation of organic compounds. In: Degradation of polychlorinated biphenyls, Vol. 13. Marcel Dekker Inc, New York, pp 361-369

Sasek V, Volfová O, Erbanová P, Vyas BRM, Matucha M (1993) Degradation of PCBs by white-rot fungi, methylotropic and hydrocarbon utilising yeast and bacteria. Biotechnol Lett 15(5):521-526

Saucedo-Castañeda G, Trejo-Hernández MR, Lonsane BK, Navarro JM, Roussis S, Dufour D, Raimbault M (1994) On-line automated monitoring and control systems for CO_2 and O_2 in aerobic and anaerobic solid-state fermentations. Process Biochem 29:13-24

Sylvia DM, Fuhrmann JJ, Hartel PG, Zuberer DA (1999) Habitat and organisms. In Principles and applications of soil microbiology. Prentice Hall, Upper Saddle River, New Jersey, pp 72-92

Thomas DR, Carswell KS, Georgiou G (1992) Mineralization of biphenyl and PCBs by the white rot fungus *Phanerochaete chrysosporium*. Biotechnol Bioeng 40:1395-1402

Tiedje JM, Quensen JF, Chee-Sanford J, Schimel JP, Cole JA, Boyd SA (1993) Microbial reductive dechlorination of PCBs. Biodegradation 4:231-240

Tien M, Kirk KT (1988) Lignin peroxidase of *Phanerochaete chrysosporium*. Methods Enzymol 161:228-249

Yadav JS, Quensen JF, Tiedje JM, Reddy CA (1995) Degradation of polychlornated biphenyl mixtures (Aroclors 1242, 1254 and 1260) by the white rot fungus *Phanerochaete chrysosporium* as evidenced by congener-specific analysis. Appl Environ Microbiol 61:2560-256

Tests on Composting of Degradable Polyethylene in Respect to the Quality of the End Product Compost

B. Raninger[1], G. Steiner[2], D.M. Wiles[3] and C.W.J. Hare[4]

Abstract. The evaluation of degradable polyethylene as an input material for composting processes was carried out. The degradation of EPIs (Environmental Products Inc.) TDPA (total degradable plastic additives)-based polyethylene is accomplished using an additive approach such that both lifetime control and final total degradability are achieved. The influences on the rotting process and the compost produced were investigated in a one-to-one field scale research programme established at the composting plant of Wiener Neustadt in Austria. The EPI TDPA-based polyethylene (PE) was mixed to source-separated organic biowaste from households and yard waste into a tunnel-batch of approx. 60 t (approx. 1.1% FM of PE). The feedstock was processed over a total residence time of 26 weeks. As a result, EPI TDPA-based polyethylene is principally qualified as a composting input material and does not cause any negative effects on the parameters compost quality and composting process.

Introduction

Packaging materials made of olefins and recoverable through composting and degradation are going to become more popular in the future. No tendency can be observed for a decline in the use of products made from, e.g. polyethylene or polypropylene. Its popularity is based on the wide spectrum of useful physical and chemical characteristics of these ubiquitous thermoplastics and on the relatively low price, making them highly competitive with other materials.

Packaging waste represents about 17% by weight of the municipal solid waste (28 Mio t per year). Ten t of plastics are used for packaging purposes only, about 17% of specific packaging material are disposed of in the packaging waste bin (EC 1999). Disposal of plastics is still a very important problem to cope with in modern waste management. In particular, the addition of packaging material to source-separated organic waste is a major issue in waste management. Concerning composting of organic waste, impurities lower the quality of the compost

[1]Mining University of Leoben, MU-IED, Peter Tunnerstr. 15, 8700 Leoben, Austria & TB für Umweltschutz und Entsorgungstechnik, 5322 Hof/Salzburg, Austria
[2]GUA Corporation for Comprehensive Analyses Ltd., Sechshauserstrasse 83, A-1150 Vienna, Austria
[3]EPI (Canada) 802–1788 W. Broadway, Vancouver, BC, V6J1Y1, Canada
[4]EPI (Europe) Dunston Rd., Chesterfield, Derbyshire, S418Xa, UK

H. Insam, N. Riddech, S. Klammer (Eds.)
Microbiology of Composting

produced. Biodegradable packaging material may help to increase the quality obtained (Raninger and Steiner 2000).

For many years research activities have been going on to develop biodegradable plastic materials to reduce the problems caused after their use during waste treatment and disposal. The company EPI (Environmental Products Inc.) achieved an advance in biodegradable polymer development with their totally degradable plastic additive (TDPA[5]) approach to promoting the degradation of polyolefins. This technology enables the use of polyolefins in applications where environmental disposal, for example in composting and soil, can be practical. Polyolefins are generally not considered to be biodegradable on the time scale of composting. However, in small-scale simulation tests, TDPA-based polyethylene degraded in the compost cycle by oxidation to small fragments of lower molecular weight, which were then biodegradable either during the compost cycle or subsequently in soil. Degradation thus is recognised to occur in two stages, the first is promoted by oxidation and the second is biodegradation mediated by living organisms[6].

Contrarily, there was a lack of practical data of the application of degradable plastic to biowaste and a lack of investigations on the quality of the end-product compost. To provide reliable data, a one-to-one field scale research programme was established at the municipal composting plant of Wiener Neustadt in Austria.

Method

Composting Facility

The municipal composting plant of Wiener Neustadt is one of the most advanced composting plants of this size available in Austria, based on the tunnel technology. This plant was implemented in 1997. The design of the tunnels was established by AE&E / LINDE KCA, providing a fully developed method for composting of source-separated organic waste. Each of the three highly instrumented, automated tunnels in the composting plant holds 90 m^3 or 60 t of organic waste and green matter garden waste.

The municipal composting plant serves a population of about 100000 inhabitants. In 1998, about 6000 t of organic (household) waste and 4000 t of green matter/yard waste were treated at this facility. The composting procedure occurs in two stages: an in-vessel, forced aeration "tunnel" process for 1 week,

[5]TDPA (totally degradable plastic additive) is a registered trade mark of EPI Enviromental Products Inc.

[6]Further details on the principles of degradation of compostable plastics can be found in the literature listed in the references, e.g. Scott 1997; Cermak et. al. 1998, 1999. Comments should be sent to EPI – e-Mail: epieurope@epi-global.com.

followed by outdoor treatment on a paved area in "triangle" windrows, with weekly watering and turning for at least 10 weeks.

The compost product is intended for fertilisation (in accordance to Austrian National Standards) and used mainly in landscaping and gardening. The plant is approved with the Austrian Compost Quality Seal (ÖKGS), a legally defined quality certification system, executed by the Austrian Compost Quality Society (KGVÖ). The ÖKGS will be given to composting plants if the requirements on plant operation and end-product quality are permanently controlled and achieved.

Composting Test Procedure

First, the EPI TDPA compost sacks (granulate from EPI, which was extruded into bags by Unterland-Kunststoffverpackungen GmbH, Kufstein, Tyrol) were shredded together with screening rejects from yard waste composting; 10 000 single bags were used (660 kg EPI TDPA). By grinding the bags together with the screening rejects, the mean size of the bags was decreased to approx. 25 cm^2 (about hand-size). The reduction of the size goes along with the requirements of DIN V 54900-3 (1998).

In total, 10 300 kg FM (fresh matter) material containing rejects and TDPA-based sacks was stored outside and exposed to natural environmental conditions (UV, moisture of ambient air) for 2 weeks while the material was turned three times by a front loader. The concentration of EPI TDPA was 6.4% FM. This step simulated the phase of material handling and distribution (transportation, storage, delivering to consumer and use) prior to disposal. Simulating this process instead of distribution of the plastic bags to households in a field trial ensured a quantitative test procedure at the composting plant.

Afterwards, the material was mixed to source-separated organic biowaste from households into a tunnel-batch of approx. 60 t of biowaste and yard waste. The rotting material contained about 1% of EPI TDPA (according to DIN V 54900-3). The intensive tunnel maturation period lasted for 2 weeks instead of 1 week. The expansion was due to obtain more data about the behaviour of compostable polyethylene in indoor composting facilities. The postmaturation was carried out for approx. 6 months.

The compost was analysed before and after the intensive maturation stage (2 weeks), during the postmaturation stage (12, 18 weeks), and finally after 6 months according to ON S 2200. In addition, a total quantification of the process was guaranteed.

In order to compare the organic waste batch containing TDPA-based polyethylene with usual procedure and usual compost qualities, a standard compost batch was treated and observed parallel to the EPI batch. The standard batch was processed 2 weeks indoors at the same time, but in a different tunnel. In addition, the standard batch was treated accordingly 6 months outdoors beside the batch containing compostable PE. The standard batch did not contain any EPI TDPA-based PE (but contained other plastic impurities in an extent such as is always found in biowaste batches). There was no need to store the standard batch outside prior to bioprocessing because the biowaste containing non-degradable plastics and other impurities was delivered by the municipal garbage collection service. The batch containing TDPA-based polyethylene was stored outside in order to simulate these steps that the standard batch was exposed to (use, transport, disposal, etc.).

Results and Analyses

Composting Process

The TDPA-based plastic bags composted very well. Despite unfavourable conditions for a pioneering project during the 6-month practice-relevant trial, all requirements of ON S 2200 on the composting process conditions were met. Hence, the presence of 1.1% by weight of EPI TDPA-based PE (10 000 bags, 660 kg) in the test batch had no deleterious effect on the composting operation at the Wiener Neustadt plant. The amount of plastic pieces in the feedstock will be expected to be even smaller due to distributional reasons if established as standard biowaste sack, or consumer goods packaging.

The temperature profile reached more than 65 °C in the hygienic stage and did not significantly deviate from prior registered graphs in the indoor stage and in the postmaturation stage.

The temperature profiles of both batch containing TDPA-based PE and standard charge were similar. However, the batch containing TDPA-based PE was generally lower, because the charge contained dry, woody ballast added in the first exposure stage. The standard batch contained more easily biodegradable carbon. The indoor intensive maturation process of the charge containing compostable TDPA-based PE (indicated by a stop in the temperature profile) was stopped after 1 week to investigate the change in the organic waste and the PE and to undertake a homogenisation step. The indoor treatment of the batch containing TDPA-based polyethylene was continued for another week.

A specific loss of 36% DM (dry matter) of the input total solids and of 60% DM of the total input organic mass (volatile solids) could be observed (see Table 1).

The energy removed via waste air exhaust stream was 159 GJ (standard batch without EPI TDPA: 196 GJ). The energy released by biochemical oxidation was 72–89 GJ (38–47% of the theoretical potential) after the intensive stage (standard - no EPI content - batch: 80–100 GJ, 41–51% of the total possible energy).

Table 1. Mass balances of both compost batches, filled with EPI TDPA material and standard compost

Date	Action	EPI TDPA PE	Total		Specific loss of mass	Organic content[a]	Total organic content[a]	Specific loss of organic content[a]
		(kg)	(kg FM)	(kg DM)	(% DM)	(% DM)	(kg VS)	(% VS)
Compost batch containing EPI TDPA-based PE								
21.06.99	Shredder	660	10 300					
05.07.99	Input	640	57 075	26 939		60.3	16 244	
12.07.99	Mechanical intermed. stage		49 185	24 248	10.0			
19.07.99	End of intensive composting stage		51 460	23 414	13.1	51.7	12 105	25.5
27.09.99	End of post maturation stage I		46 000	18 032	33.1	49.8	8980	44.7
16.11.99	Mid-post matu. Stage II		35 540	19 440	27.8	34.6	6726	58.6
10.01.00	End of trial		34 120	17 299	35.8	37.1	6418	60.5
Standard compost batch								
05.07.99	Input	0	69 630	27 852				
19.07.99	End of intensive composting stage		49 460	22 752	18.3	53.1	12 081	
27.09.99	End of post maturation stage I		40 360	20 987	24.6			
16.11.99	Mid-post matur. Stage II		35 420	18 418	33.9			
10.01.00	End of trial		33 120	17 222	38.2			

VS Volatile solids; DM dry matter; FM fresh matter.
[a] Organic content measured as ignition loss IL (according to ON S 2200 TOC would be 0.58*IL).

Compost Quality

Ballast Matter

A steady degradation of the plastic ballast matter within the compost could be registered. Approx. 25% of the plastics degraded after 2 weeks of intensive composting. The degradation rate of total plastics at the end was assumed to be

43% (by weight of dry matter, fraction 2–20 mm and fraction > 20 mm). However, the amount of plastics calculated on the basis of the total impurity content was still about the amount of added TDPA-based PE. For a total quantitative report of degradation of TDPA-based PE detailed investigations by separating of the TDPA-based PE and "normal" plastics were carried out.

The identification of EPI TDPA-based PE within the compost was performed in two ways:

- Firstly, the impurities content of the investigated charge containing EPI TDPA-based PE was subtracted by the content of impurities of the standard compost batch.
- Secondly, the plastic pieces analysed were separated by hand according to their mechanical properties and visual characteristics.

However, due to several methodological failures both procedures were not found to be useful.

While sorting of ballast matter as required by ON S 2200 allows only general findings about plastics within the compost, a detail description of the types of plastics is necessary to obtain further information on the degradation rate of compostable plastic pieces. The parameter ballast matter plastics did not significantly reflect the degradation of EPI TDPA-based material as the visual controls of the compost do (especially in the first weeks of the test trial, the degradation effect was clearly visible and easy to observe). Furthermore, a logical change of the parameter (increasing, decreasing or constant) could not be observed to a sufficient extent. The high amount of usual impurities, statistical insufficiencies of compost analyses and the low content of EPI TDPA-based PE in the compost led to principle failures in assessing the final EPI TDPA-based PE fragment content in the compost.

As a matter of fact, the prEN 13432 (1998) admits that the limit values for disintegration and the test duration are based on present experience and do not reflect general scientific findings (see prEN 13432 drafting note to A.3, p. 11). The limit values may be confirmed or modified if results of findings are available. Correspondingly, the problems concerning both the ballast matter analyses and the high content of impurities in the biowaste have to be considered for requirements of practice-relevant tests. Moreover, the prEN 13432 (1998) states that biodegradation of the packaging material has not to be fully completed by the end of biological treatment in technical plants (see prEN 13432, Note, p. 8). Subsequently, it grants that the completion of biodegradation is also possible during the use of the compost produced. The DIN V 54900 approves the tests even with a grade of disintegration less than 90% if the tests are carried out according to the DIN V 54900, the maturing stage (Rottegrad) has a minimum level of IV and temperature is for 1 week above 65 °C and for 3 weeks higher than 55 °C. Besides, the Din V 54900 requires for the tests under practice-relevant conditions two treatment steps: the sorting of the impurities before mixing with the packaging material for both the biowaste and the structure material, and the filling of biowaste, structure material and plastic material together in specific bags to distinguish between the sample and the usual biowaste.

While these steps do not totally reflect the realistic conditions of composting processes in full-scale facilities (oxygen supply, mechanical stress, distribution, seasonally or occasionally input variations, weather conditions), it is certainly a better approach to receive reliable data for the assessment of compostable packing material than merely observing the total ballast matter.

Heavy Metals

The results from analytical investigation of the EPI TDPA test material showed that the content of heavy metals is very low. According to the actual limit values of EU-CEN prEN 13432 (11/98), the material is qualified as a biopackaging material recoverable by composting. Out of seven relevant heavy metals only zinc (6 ppm - limit value 150 ppm) and nickel (3 ppm - limit value 25 ppm) could be detected. Total carbon content (TC) is 80% DM, the ignition loss giving the volatile solids is 96.8% DM. The prEN 13432 (1998) requires a minimum of 50% of total solids (ignition loss) for packaging materials. Contents of nitrogen, phosphate and calcium are very low. Potassium and magnesium could not be detected. To conclude, the plastics material of EPI TDPA does not contain any hazardous substances that could directly influence the quality of the compost.

The heavy metals contents of the compost produced did not reach the limit values of the ÖKGS and ÖN S 2200 standards as well as of the draft Austrian compost ordinance. Moreover, all metal parameters except zinc and copper did not even reach the limit values of the standard compost type I (ON S 2200 - highest), but these parameters were significantly lower than the decisive limits for standard compost type II of ON S 2200.

Nutrients, Organic Content

The analyses of the test batch showed in a qualitative way the degradation process of the compost. While ignition loss and total organic carbon decreased, the content of nutrients, nitrogen, and calcium as well as magnesium increased significantly. Due to the content of nutrients the compost could be used for fertilisation applications even after the 12th week.

Table 2. Analytical results of EPI TDPA composting tests [Units in percentage of dry (% DM) or fresh matter (% FM)].

Parameter / sample		EPI TDPA PE	Test compost input tunnel	Standard compost output tunnel	Test output tunnel	Test post-mat.	ÖKGS / ÖN S 2200 Limit V. A^a/B^b
Bioprocessing week/date		0/7.99	0/7.99	2/7.99	2/7.99	26/1.00	
Water content	% FM	1.5	52.8	54.0	54.5	49.3	25 – 50
Dry matter content	% FM	98.5	47.2	46.0	45.5	50.7	
Ignition loss	% DM		60.3	53.1	51.7	37.1	>20

Table 2. (Cont.)

Total organic carbon TOC	% DM		35.0	31	30	18.6	>12
Total nitrogen - (N_{tot})	% DM	0.09	1.57	1.93	1.82	1.75	
Ammonia Nitrogen (NH_4-N)	% DM			0.001	n.p.	0.015	<0.1[a]
Nitrate Nitrogen (NO_3-N)	% DM			0.086	n.p.	0.004	<0.2[a]
PH-value	-		6.1	7.1	7.1	8.0	
Ballast matter > 25 mm	% DM		32.6	58.11	19.53	6.33	
Ballast matter Glass > 2 mm	% DM		0.002	-	-	0.24	
Ballast matter Metal > 2 mm	% DM			-	0.23	0.05	
Ballast matter Plast. 2--20 mm	% DM		1.13	0.07	2.05	2.12	<0.2
Ballast matter Plastic > 20 mm	% DM		1.97	1.36	0.79	0.83	0
Conductivity	mS/cm		3.80	2.8	2.7	2.81	<2/4
$P_2O_{5\ total}$	% DM	0.008	0.86	1.0	1.0	1.24	
K_2O_{total}	% DM	n.d.	1.69	2.08	2.11	2.22	
CaO	% DM	1.57	8.35	9.55	10.71	12.04	
Mg	% DM	n.d.	1.47	1.76	1.84	2.02	
Cu	ppm	n.d.				84[c]	70/100
Zn	ppm	6				279[c]	210/400
Ni	ppm	3				22[c]	42/60
Cr	ppm	n.d.				26[c]	70/70
Pb	ppm	n.d.				66[c]	70/150
Cd	ppm	n.d.				0.39[c]	0.7/1
Hg	ppm	n.d.				0.12[c]	0.7/1

n.p. Not possible; n.d. Not detected.
[a] Compost type A
[b] Compost type B
[c] Metal content related to an ignition loss of 30% DM.

Ecotoxicity

To check the Ecotoxicity of the compost produced, ecotoxicological tests like the plant tolerance test (Linzer Substrattest), determination of seeds and propagules were carried out according to DIN V 54900-3, ON S 2200 and ON S 2023 at the laboratories of BVA Linz and OWS Gent. In conclusion, it can be stated that, compared to the blank compost, there was no negative effect on the survival and live weight of earthworms in compost/artificial soil mixtures using compost obtained at the end of an aerobic composting test in which EPI TDPA-based PE was added in a 1 to 2% concentration at the start. This means that after composting EPI TDPA plastic, no residuals were left such as metabolites, non-degraded components and inorganic components, which exert a negative influence on the earthworms. A summary of the tests is given in Table 3.

Table 3. Results of Ecotoxicity test carried out at BVA in Linz (Austria) and OWS in Ghent (Belgium)

OWS Gent		
Test method	Used standard	Results
Cress test	OWS int. Standard	No negative effects
Summer barley plant growth test	OWS int. Standard	No negative effects
Daphnia test	OWS int. Standard	No negative effects
Earthworm test	OWS int. Standard	No negative effects
BVA Linz		
Test method	Used standard	Results
Plant tolerance test	ON S 2023 (required by	No negative effects (compost type A)
Seeds and propagules	ON S 2200)	No seeds detected (compost type A)

Conclusion

As a result, the following conclusions can be drawn:

1. The EPI TDPA-based PE is in principal a material containing a high amount of not readily biodegradable hydrocarbons like other material of natural origin (lignin, collagen, chitin, etc). Nevertheless, the presence of 1.1% by weight of EPI TDPA-based PE (10 000 bags, 660 kg) in the test batch had no deleterious effect on the composting operation at the Wiener Neustadt plant.
2. The process conditions have to be monitored carefully to adjust the system during the first implementation stage considering the physical conditions of the material.
3. The intensive indoor maturation process is not the decisive step for the degradability of EPI TDPA-based PE, but an effective initial process. Disintegration will be completed afterwards in the postmaturation stage. Consequently, the major breakdown of the material occurs in a stage that is usually not highly controlled and dependent on other circumstances like weather or education of the operator's staff.
4. An effective distinguishing feature (colour, hallmark, sign) of EPI TDPA-based material is definitely necessary.
5. Research is necessary on both achieving final disintegration and monitoring of disintegration.
6. No negative effects on the quality of the resulting compost were observed. After 26 weeks of bioprocessing, the end product < 12 mm fully meets the Austrian standard requirements ON S 2200 of the highest compost quality A (except for a low-level zinc and cooper contaminant, which has nothing to do with TDPA materials).
7. A final degradation rate of 63% is achieved in 26 weeks, failing to pass ballast matter limiting standards. EPI material easily disintegrates during composting, but does not meet all the requirements of CEN prEN 13432. EPI TDPA-based

olefins would not pass the pre-norm prEN 13432 (1998) requirements concerning biodegradability. Due to the environmental condition during the decomposition period of 6–12 months, the material will not release 90% of TOC as CO_2, to be classified as biodegradable. EPI TDPA-based plastics show behaviour like other products of natural origin such as chitin, lignin and collagen, materials that will not pass the prEN 13432.

8. The EPI product using polyethylene and TDPA technology meets the requirements to be classified as a degradable/compostable plastic and the compost end product is fully acceptable as land fertiliser. EPI products may be considered to degrade successively by abiotic and biological pathways in a manner that produces no harmful or toxic by-products.

References

BMUJF (1999) Verordnung betreffend Qualitätsanforderungen an Kompost aus Abfällen, Entwurf, Vienna

CEN prEN 13432 (1998) Requirements for packaging recoverable through composting and biodegradation – test scheme and evaluation criteria for the final acceptance of packaging, Brussels

Cermak BE, Gho JG, Hare CWJ, Wiles DM (1998) Controlled-lifetime, environmentally degradable plastics based on conventional polyethylenes. Presented at RAPRA Addcon 1998 conference, London

Cermak BE, Gho JG, Hare CWJ, Wiles DM, Tung J-F (1999) Biodegradable polyethylene. In: proceedings of the Int Conf Orbit 99. Rhombos, Berlin

DIN 54900 (9/1998) Prüfung der Kompostierbarkeit von Kunststoffen, Berlin

EC DG XI (1999) 99/31/EC Council directive on the landfill of waste, Brussels

EC (1994) The packaging and packaging waste directive 94/62/EC, Brussels

EC (1999) Discussion paper on the revision of the packaging and packaging waste directive 94/62/EC, Brussels

EPI (1998) Internal information for the General Assembly, Chesterfield

KGVOE (1996 -99) Regulation sheets No 1-12, Salzburg

ONORM S 2023 (1993) Analytical methods and quality control of compost, Vienna

ONORM S 2200 (1993) Quality requirements for biowaste-compost; Vienna

Raninger B (1999) Compost quality requirements related to the end product users, ISWA Symposium, Paris

Raninger B, Nelles M, Harant M, Steiner G, Staber M, Lorber K-E (1999) Long-term behaviour of mechanical biological pre-treated material under landfill conditions, 5th Int Landfill Symposium, Sardinia

Raninger B, Steiner G (2000) Biodegradability of EPI TDPA polyethylene recoverable through composting and biodegradation under practice-relevant conditions and testing of the quality of the compost. For EPI Europe and Sera Trade Switzerland. Salzburg

Scott G (1997) Abiotic control of polymer biodegradation. In: TRIP vol 5, no 11/97. Elsevier, Amsterdam

Microbial Degradation of Sulfonylurea Herbicides: Chlorsulfuron and Metsulfuron-Methyl

E. Zanardini[1], A. Arnoldi[2], G. Boschin[2], .A. D'Agostina[2], M. Negri[1] and C. Sorlini[1]

Abstract. Sulfonylurea herbicides are characterised by low field rates, high herbicidal activity, broad action spectrum, good crop selectivity and low human and animal toxicity. Their mode of action is highly specific since these herbicides inhibit acetolactate synthase (ALS), an enzyme of the biosynthetic pathway of the branched amino acids (valine, isoleucine and leucine) in plants and microorganisms, but not in animals.

In our research we selected from soil samples two microbial consortia able to degrade two sulfonylureas, chlorsulfuron and metsulfuron-methyl. From these mixed cultures, several bacterial strains, including *Pseudomonas fluorescens* B1 and B2, are able to degrade under growth conditions (with herbicides added as sole carbon and energy source at the initial concentration of 100 mgl^{-1}) 11 and 15% of the two compounds, respectively, while under co-metabolic conditions *Pseudomonas fluorescens* B2 degrades in 2 weeks about 21 and 32% of metsulfuron-methyl and chlorsulfuron, respectively.

This confirms that the degradation is more pronounced in the presence of cosubstrates.

For metsulfuron-methyl three intermediate metabolites were identified and characterised, suggesting two different degradative pathways by *P. fluorescens* B2: the demethylation of the triazinic ring and the cleavage of the sulfonilureic bridge.

PCR analysis reveals that the gene encoding the catechol 2,3 dioxygenase (C23O), a key enzyme involved in the degradative pathway of aromatic compounds, is present in all the isolates from the initial consortium, except in a rod Gram-positive bacterium, B3 strain.

Introduction

Sulfonylureas are a relatively new class of herbicides characterised by several properties that make them particularly interesting in comparison with other products. At first they were applied in medicine; the first study on their use as

[1]Department of Food Science and Microbiology (DISTAM), MAAE Section, University of Milan, Via Celoria 2, 20133 Milan, Italy.
[2]Department of Agricultural and Food Molecular Sciences (DISMA), Chemistry section, University of Milan, Via Celoria 2, 20133 Milan, Italy.

H. Insam, N. Riddech, S. Klammer (Eds.)
Microbiology of Composting

herbicides was performed in 1966 by DuPont, who registered the first sulfonylurea herbicide in 1977, followed by Ciba Geigy in 1982.

The main characteristics of these compounds are: low field rates (2–100 gha^{-1}), high herbicidal activity, broad action spectrum, good crop selectivity and low human and animal toxicity (DL50 on rat generally > 5000 $mgKg^{-1}$) (http://dns.agrsci.unibo.it/agro/agroeco/).

Their chemical structure is characterised by three portions: an arylic group (R_1), a sulfonylurea bridge and an heterocyclic portion (R_2) as shown in Fig 1.

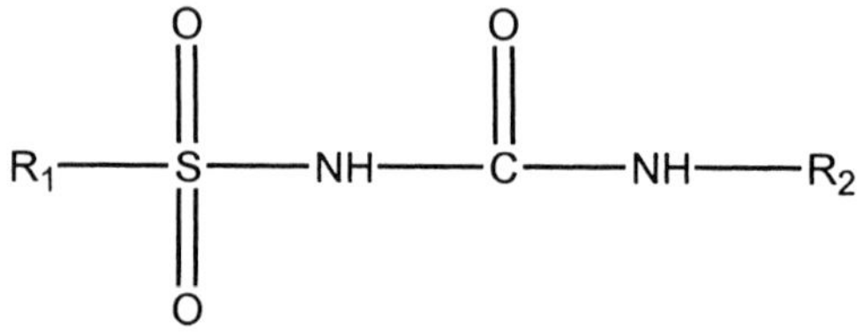

Fig. 1. Chemical structure of sulfonylurea herbicides

Their mode of action is highly specific on the protein synthesis, since these herbicides inhibit acetolactate synthase (ALS), an enzyme of the biosynthetic pathway of the branched amino acids (valine, isoleucine and leucine) in plants and microorganisms, but not in animals (Blair and Martin 1988; Brown 1990).

Soil adsorption depends on the pH and on the content of organic matter; sulfonylurea compounds do not have a relevant volatility, but some of them can have a high mobility in soil. They can mainly be degraded by chemical or microbial processes, while photodegradation is not relevant (Walker and Welch 1989; Blacklow and Pheloung 1992; Oppong and Sagar 1992).

Under acid conditions these compounds can be chemically hydrolysed via the cleavage of the sulfonylureic bridge (Blacknow and Pheloung 1992; Hemmanda et al. 1994; Strek 1998).

The study of the intermediate products, both chemical and microbial, is very important since in some cases they can be more toxic than the initial compounds, leading to environmental problems.

The general aim of this work was to study the degradation of two sulfonylurea compounds, chlorsulfuron and metsulfuron-methyl (Figs. 2 and 3), by soil microorganisms.

The choice of these compounds was suggested by their high persistence in soil under certain conditions that can cause problems to successive cultures (Blair and Martin 1988).

The research was performed in the following steps: (1) the selection, from different inocula, of mixed or pure cultures able to grow in the presence of the two herbicides added as sole source of carbon and energy; (2) the study of the microbial degradation by pure cultures under growth and cometabolic conditions; (3) the identification of formed intermediate metabolites.

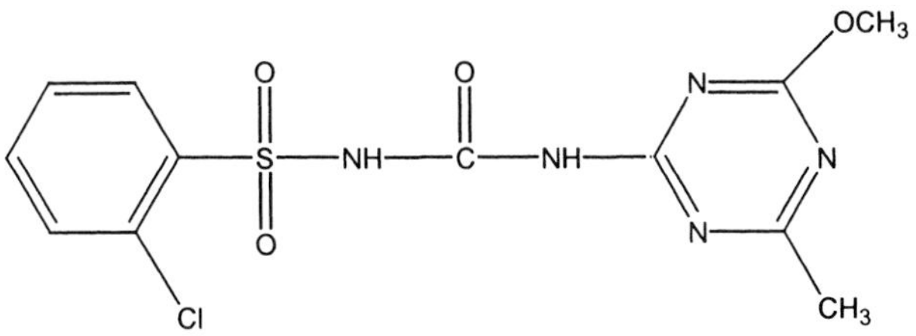

Fig. 2. Chlorsulfuron

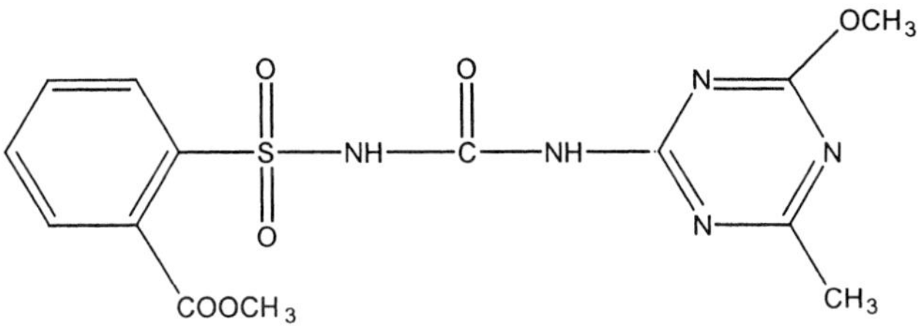

Fig.3. Metsulfuron-methyl

Further investigations were conducted on the total protein analysis and the detection in all the isolates of the gene encoding the catechol 2,3 dioxygenase (C23O), a key enzyme involved in the degradative pathway of the aromatic compounds.

Materials and Methods

Microbiological Methods

Enrichment cultures were performed in M9 mineral medium (Kunz and Chapman 1981); soil samples treated with sulfonylurea herbicides and a mixed culture obtained from soil after an enrichment for some years in presence of chlorsulfuron were used as inocula. The herbicides were added at various concentrations (ranging from 100 to 500 mgl^{-1}) in the different tests and will be reported during the discussion of the results. For cometabolic studies, M9 medium added with sodium acetate (2‰) was used or plate count broth, PCB, (Difco), were used as carbon sources. Plate count agar, PCA, (Difco) was used for conservation of the bacterial isolated strains. The strains were classified by API 20 NE System that permits the identification of Gram-negative rods, not enteric bacteria.

Protein Extraction

P. fluorescens B2 was cultivated for 24 h in PCB in presence of the separately added herbicides (500 mgl^{-1}) and in absence of these compounds. The cellular pellets obtained after centrifugation were analysed for total protein following the Laemmli protocol (Laemmli 1970); the extracted products were run (200 mV, Power Pac 300, Biorad) on an acrylamide gel, stained with a solution of Coomassie Brilliant Blue R250 (2.5 gl^{-1}) in acetic acid (10%) and ethanol (25%).

C23O Amplification

For the DNA extraction, an alkaline lysis following the Maniatis protocol was used (Sambrook et al. 1989). A set of specific primers XylF and XylR (XylF GTTGTATGCAGACAAGGAAT; XylR CCTCGAAACCTGGGAAGACT) was used for the amplification of the gene that encodes the C23O enzyme. The C23O available sequences were acquired in GeneBank (http:// www.ncbi.nlm.nih.gov/cgi-bin/genbank). For the amplification a thermal cycler Perkin Elmer 2400 was used and the thermal protocol was:

°C	95°	95°	60°	72°	95°	55°	72°	95°	50°	72°	72°	4°
Time	1'	45"	45"	2'	45"	45"	2'	20"	45"	2'	3'	∞
		x 5 cycles			x 5 cycles			x 25 cycles				

The amplified products were run (100 mV, Power Pac 300, Biorad) on an agarose gel (ranged between 0.8 and 2%) depending on DNA fragment dimensions, in 0.5x Tris borate EDTA (TBE) buffer, stained in ethidium bromide, visualised and photographed under UV light.

Chemical Analyses

Materials

Chlorsulfuron (analytical grade 99.5%) and metsulfuron methyl (analytical grade 97.4%) were kindly provided from DuPont, France. Samples for HPLC were filtered through disposable nylon 66 filters (0.45 μm, Alltech).

HPLC Analyses

Broth cultures were extracted with ethyl acetate (3 x 20 ml), the combined extracts were dried with anhydrous sodium sulphate and, the solvent was evaporated under vacuum at 30°C. The obtained residues were dissolved in methanol and analysed by HPLC-DAD. HPLC analyses were conducted on a HP-1050 quaternary pump fitted with a Rheodyne injector (20 μl, loop) and equipped with an HP-1050 diode array detector (HPLC-DAD). The system was controlled by a HP Chemstations

(DOS Series, Hewlett-Packard). Chromatograms were recorded at 235 and 254 nm. Spectral data were recorded from 220 to 550 nm for the peaks of interest. The column was a Lichrospher 100 RP-18 (5 µm, 250 x 4 mm, Merck, Darmstadt, Germany), the flow rate was 1 $mlmin^{-1}$ and the gradient from 20:80 methanol/water + 1% acetic acid to 100:0 methanol/water + 1% acetic acid over 50 min, then 5 min isocratic. The column used for the LC/MS analyses was a Luna C8 (3 µm, 50 x 20 mm, Phenomenex, Torrance, USA). The loop was 20 µl, the flow rate 0.2 $mlmin^{-1}$. The analyses were carried out with a linear gradient from 0:100 acetonitrile/water plus 0.1% acetic acid to 50:50 acetonitrile/water plus 0.1% acetic acid over 16 min. The detector was a mass spectrometer PE Sciex API 3000 and the source was a Turbo Ion Spray.

Results and Discussion

Biodegradation in Growth and Cometabolic Conditions

From the two microbial consortia selected by the enrichment cultures several bacterial strains were isolated. Corresponding to the API System 20 NE, most of them belonged to *Pseudomonas* genus (*P. fluorescens* and *P. putida*). To investigate the ability of the pure strains to degrade these two herbicides, the bacterial strains were inoculated in M9 mineral medium in the presence of chlorsulfuron or metsulfuron-methyl, added separately as sole carbon and energy source at the final concentration of 100 mgl^{-1}. Under these conditions, after 1 month of incubation, only a low biotic degradation was observed. The highest percentage of biodegradation, also very low, was obtained with *Pseudomonas fluorescens* B1 showing a removal of 15% of the chlorsulfuron and 11% of the metsulfuron-methyl (Fig.4).

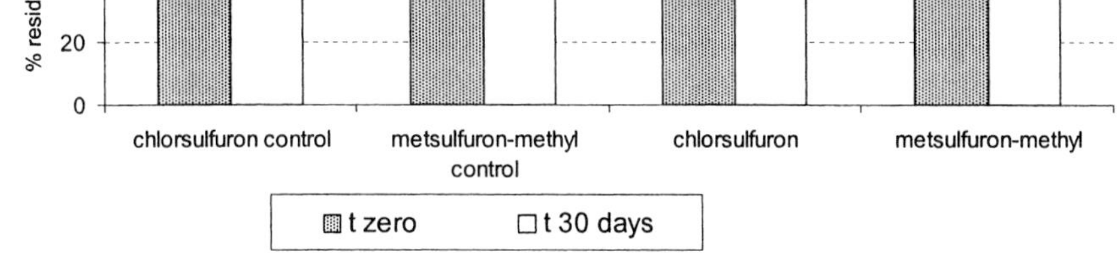

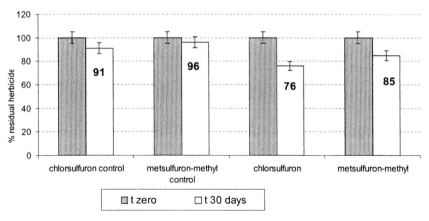

Fig. 4. Biodegradation in growth condition of chlorsulfuron and metsulfuron-methyl (100 mgl^{1}) by *P. fluorescens* B1 strain

Under cometabolic conditions (M9 medium added with Na-acetate 2‰), a higher degradation was observed after only a week of incubation; in fact for metsulfuron-methyl a removal of about 23% was found by *P. fluorescens* B2, while no significant removal was detected for chlorsulfuron (Fig.5).

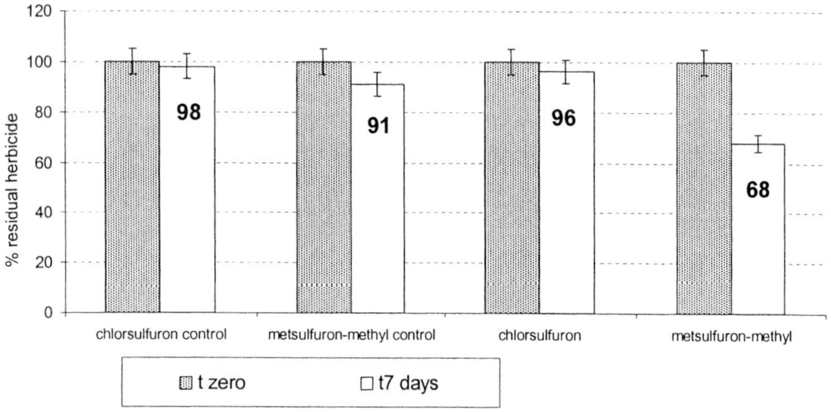

Fig. 5. Biodegradation in cometabolic condition (Na-acetate 2‰) of chlorsulfuron and metsulfuron-methyl (100 mgl-1) by *P. fluorescens* B2 strain

Under cometabolic conditions (in plate count broth), the highest percentages of biotic removal are obtained by *P. fluorescens* B2: 21% for metsulfuron-methyl and 32% for chlorsulfuron in 2 weeks (Fig. 6).

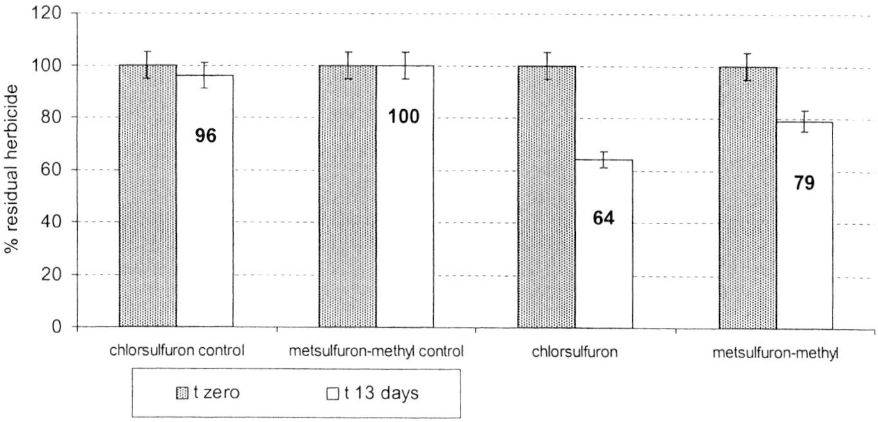

Fig. 6. Biodegradation in PCB of chlorsulfuron and metsulfuron-methyl (500 mgl-1) by *P.fluorescens* B2 strain

Therefore, the biodegradation of these herbicides by the isolated *Pseudomonas* strains is more evident and relevant under co-metabolic conditions.

Protein Analysis

The total protein analysis evidenced no differences in the patterns between the control without the herbicides and the results obtained in presence of the two herbicides added separately (Fig. 7). This suggests that the enzymatic attack, by *P. fluorescens* B2, occurs in no specific way by constitutive enzymes and not by enzymes induced by the compounds.

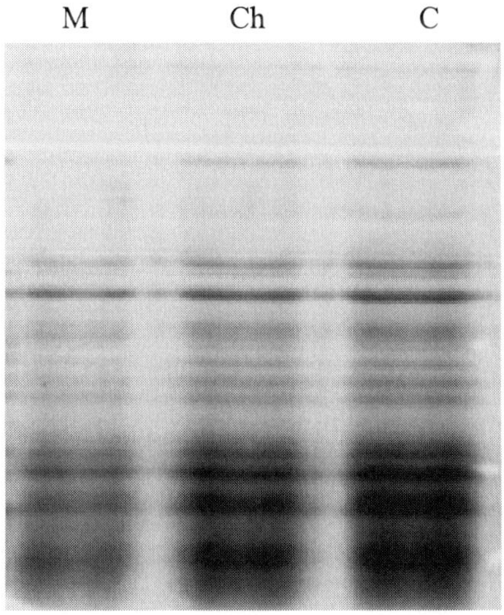

Fig. 7 Protein patterns obtained from broth cultures of *P.fluorescens* B2 grew in absence (control) and in presence (500 mgl-1) of the two herbicides added separately. M with metsulfuron-methyl; Ch with chlorsulfuron; C control

Identification of the Biodegradation Metabolites

The separation and identification of the intermediate products of the degradative pathway of *P. fluorescens* B2 was performed by HPLC and LC-MS analyses and permitted the characterisation of three metabolites for metsulfuron-methyl:

2-amino-4-methoxy-6-methyl-1,3,5-triazine, obtained via cleavage of the sulfonylureic bridge

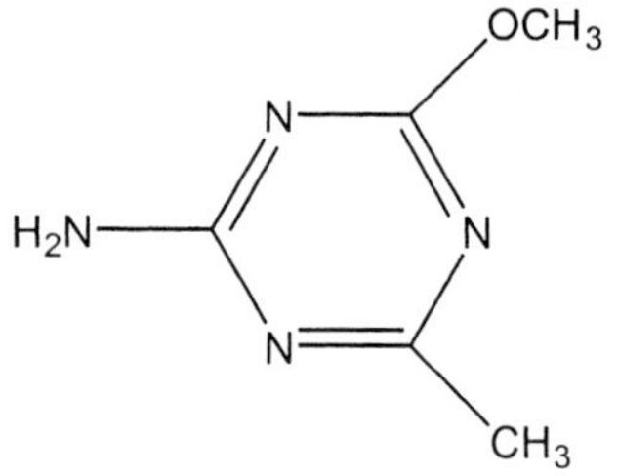

methyl-2-[[[[[4-hydroxy-6-methyl-1,3,5-triazin-2yl]amino]carbonyl]amino] sulfonyl] benzoate, obtained through demethylation of the triazinic group

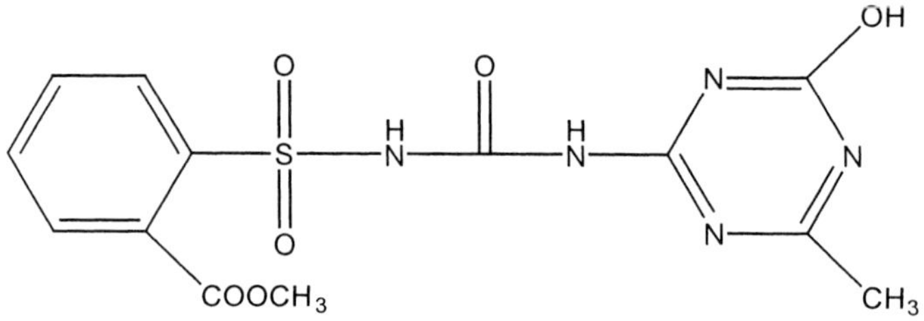

methyl 2-[(acetylamino)sulfonyl]-hydroxy-benzoate

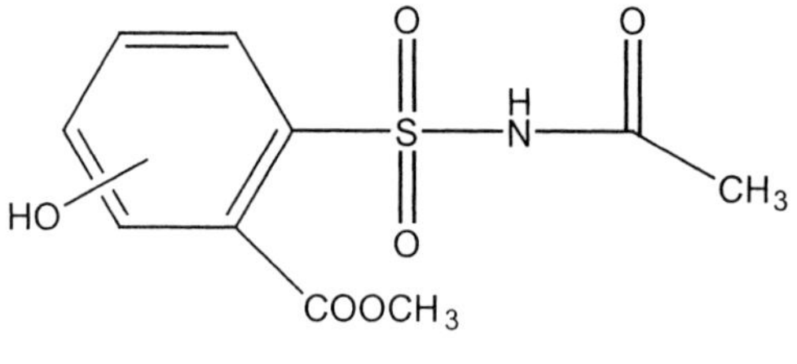

In this case, the structure is tentative and is probably obtained by hydroxylation of the aromatic ring and acetylation of the sulfonamide residue deriving from the cleavage of the sulfonylureic bridge.

These results lead us to hypothesise two separate and independent pathways by *P.fluorescens* B2 for degradation of metsulfuron-methyl, the demethylation of the triazinic ring and the cleavage of the sulfonylureic bridge.

Some of these metabolites were also obtained by other authors, but under different conditions: in fact, most of the studies were performed in soil or with mixed cultures and in presence of lower concentrations of the herbicides (10-15 mgl^{-1}) (Romesser and O'Keefe 1986; Cambon and Bastide 1992; Brusa and Ferrari

1997), while we used higher concentrations of the compounds and pure cultures belonging to *Pseudomonas* genus.

These metabolites found for metsulfuron-methyl have also been detected by other authors both by chemical (Braschi et al. 1997) and microbial degradation (Brusa and Del Puppo 1995; Li et al. 1999) in soil and not with pure culture of *Pseudomonas* genus.

Only Joshi et al. (1985) reported a study in which the biodegradation of these herbicides was obtained with pure cultures, but with other microorganisms, an *Actinomyces*, *Streptomyces griseolus*, and two fungal strains, *Aspergillus niger* and *Penicillium* sp.

Amplification of the Gene Encoding the C23O

In order to investigate the potential catabolic activities of the bacterial isolates, the presence of the gene encoding the catechol 2,3 dioxygenase, an enzyme that determines the cleavage the aromatic ring in *meta* position was researched. For the applied PCR, a set of specific primers (XYLF and XYLR) was used.

The result of the amplification of this gene was positive and the presence of the expected band (320 bp) was observed in all the isolates from the initial consortia, except in B3, a Gram-positive rod, not classified (Fig. 8).

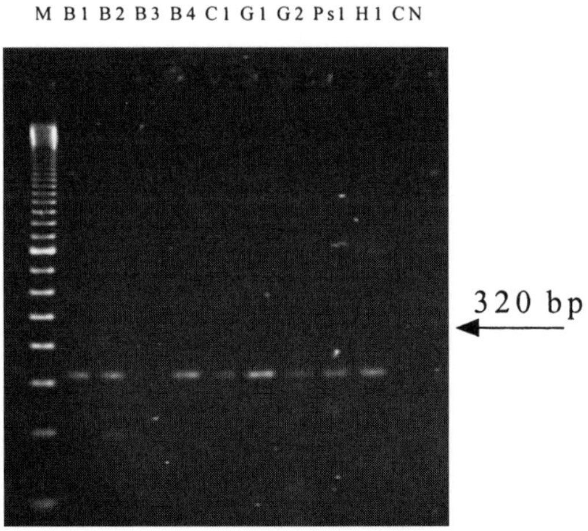

Fig. 8 Amplification of the gene encoding the C23O

Conclusions

The results obtained by this research permit the following conclusions:

1. The biodegradation of the two herbicides by the isolated bacterial strains belonging to *Pseudomonas* genus preferentially occurs under cometabolic conditions. In fact, a low removal of 15% chlorsulfuron and of 11% metsulfuron-methyl was observed when the herbicides are added as sole carbon and energy source, while a higher removal occurs under cometabolic conditions, 32% for chlorsulfuron and 21% for metsulfuron-methyl.
2. The total protein analysis suggests that the enzymatic attack probably occurs in no specific way by constitutive enzymes.
3. Three metabolites were identified for metsulfuron-methyl (analyses on chlorsulfuron are still in progress), and *P. fluorescens* B2 seems to degrade this herbicide following two separate and different pathways.

The presence in the isolated bacterial strains of the gene codifying the catechol 2,3 dioxygenase, gives further information about their potential metabolic activity.

This study contributes to the knowledge about the microbial degradation of sulfonylurea herbicides by the soil microflora. Beside this, it could permit the application of microorganisms in the treatment of wastewater containing residues of the tested compounds from working processes or in the bioremediation of soil accidentally contaminated by herbicides.

Acknowledgements. This research was supported by the Italian National Research Council (CNR), Target Project Biotechnology. We thank Dr. Fanelli (Istituto di Ricerche Farmacologiche Mario Negri, Milan) for LC-MS analyses and Dr. Bonacini (DuPont) for the herbicides.

References

Blacklow WM, Pheloung PC (1992) Sulfonylurea herbicides applied to acidic sandy soils: movement, persistence and activity within the growing season. Aust J of Agric Res 43: 1157-1167

Blair AM, Martin TD (1988) A review of the activity, fate and mode of action of sulfonylurea herbicides. Pestic Sci 22: 195-219

Braschi I, Calamai L, Cremonini MA, Fusi P, Gessa C, Pantani O, Pusino A (1997) Kinetics and hydrolysis mechanism of triasulfuron. J of Agric Food Chem 45: 4495-4499

Brown HM (1990) Mode of action, crop selectivity, and soil relations of the sulfonylurea herbicides. Pestic Sci 29: 263-281

Brusa T, Del Puppo E (1995) Microbial degradation of the sulfonylurea herbicides: current knowledge. Ann Microbiol Enzimol 45: 321-330

Brusa T, Ferrari F (1997) Bensulfuron-methyl: mixed culture microbial degradation maintenance. Microbiol Res 152: 137-141

Cambon JP, Bastide J (1992) Degradation chimique ou microbiologique des sulfonylurees dans le sol III. Cas du thifensulfuron methyle. Weed Res 32: 357-362

Hemmanda S, Calmon M, Calmon JP (1994) Kinetics and hydrolysis mechanism of chlorsulfuron and metsulfuron-methyl. Pestic Sci 40: 71-76

Joshi MM, Brown HM, Romesser JA (1985) Degradation of chlorsulfuron by soil microorganisms. Weed Sci 33: 888-893

Kunz DA, Chapman PJ (1981) Catabolism of pseudocumene and 3-ethyltoluene by Pseudomonas putida (arvilla) int-2: evidence for new function of TOL (pWWO) plasmid. J of Bacteriol 146: 179-191

Laemmli UK (1970) Cleavage of structural proteins during the assembly of the head of bacteriophage T_4. Nature 227: 680-685

Li Y, Zimmerman WT, Gorman MK, Reiser RW, Fogiel AJ, Haney PE (1999) Aerobic soil metabolism of metsulfuron-methyl. Pestic Sci 55: 434-445

Oppong FK, Sagar GR (1992) Degradation of triasulfuron in soil under laboratory conditions. Weed Res 32: 167-173

Romesser JA, O'Keefe DP (1986) Induction of cytochrome P-450-dependent sulfonylurea metabolism in Streptomyces griseolus. Biochem Biophys Res Commun 140: 650-659

Sambrook J, Maniatis T, Fritsch EF (1989) Molecular cloning. A laboratory manual 1, 2, 3 (II ed) Cold Spring, Harbour Laboratory, Cold Spring Harbor, New York

Strek H J (1998) Fate of chlorsulfuron in the environment.1. Laboratory evaluations. Pestic Sci 53: 29-51

Walker A, Welch SJ (1989) The relative movement and persistence in soil of chlorsulfuron, metsulfuron-methyl and triasulfuron. Weed Res 29: 375-383

The following web sites were consulted:

http://dns.agrsci.unibo.it/agro/agroeco/ for general information regarding the sulfonylurea herbicides

http://www.ncbi.nlm.nih.gov/cgi-bin/genbank for the data bank alignment of the sequences of the gene codifying the C23O enzyme.

Maturity Testing

Hydrogen Peroxide Effects on Composting and Its Use in Assessing the Degree of Maturity

C. Balis, V. Tassiopoulou and K. Lasaridi[1]

Abstract. In a series of laboratory-scale experiments using adiabatic reactors, addition of hydrogen peroxide to the compost was shown to exert a triggering effect on the composting process. On treatment with hydrogen peroxide, the temperature in the composting mass rose at a significantly faster rate and reached higher levels compared to the control trials. Triggering effects similar in response were observed in all trials performed so far, using a variety of substrates. The initial rise of temperature may be attributed to exothermic non-biological chemical reactions. The subsequent long-term rise, however, is considered to reflect intensification of microbiological activity in the catabolic processes, as indicated by control experiments using an antimicrobial agent (toluene).

The intensity of the triggering effect of the hydrogen peroxide treatment appears to depend on the hydrogen peroxide concentration, the substrate's moisture and its degree of decomposition, with fresh composts exhibiting a sharper temperature boost compared to stable ones. This property could be used for the assessment of the maturity/stability degree of composts, in a more sensitive modification of the self-heating test. A new thermal stability index (TSI) is proposed which could offer valuable information on energy balance and state of composting, especially if used in combination with respirometry to provide a calorimetric: respiration ratio in presence and absence of hydrogen peroxide.

Introduction

Compost quality is an intricate topic and its precise definition is often controversial. It can be described as the summation of the numerous characteristics of compost that are recognised as playing a role in sustaining soil quality and contributing to a plant's good health and yield. Although there may be disagreements on quantitative measures, this definition implies a response of a soil ecosystem, including its plant constituents, to a particular compost treatment. Specifications are usually determined as minimum admissible levels of required substances or maximum tolerable limits for unwanted ones, and the relevance of the various quality criteria will depend on the intended compost use. Among the multitude of characteristics that determine compost quality, the state of maturity or stability of the compost is amongst the most important, for both process control and product evaluation, as it affects a number of its agricultural properties:

[1]Harokopio University of Athens, 70 El. Venizelou, 176 71 Athens, Greece

H. Insam, N. Riddech, S. Klammer (Eds.)
Microbiology of Composting

phytotoxicity, nutrient availability, suppressiveness against soil-borne pathogens, etc.

Numerous parameters have been proposed for the evaluation of compost maturity/stability. These can be broadly classified into chemical tests, microbiological assays, higher plant bioassays and humified organic matter analyses (Forster et al. 1993; Avnimelech et al. 1996; Lasaridi and Stentiford 1998). Among them, oxygen respirometry and the self-heating test have gained the widest acceptance by both regulation authorities and the scientific community (LAGA 1984; ASTM 1996; US Composting Council 1997; Scaglia et al. 2000). The *self-heating test* is based on the fact that aerobic microorganisms decomposing organic substrates generate heat and the heat produced is proportional to microbial activity. The test is performed by placing the sample in a 1.5-l Dewar flask, under standard conditions of moisture, compaction and room temperature. The higher the temperature rise in the sample, the more unstable is the compost (FCQAO 1994). The intensity of microbial activity and the state of a compost process can be assessed through the *oxygen consumption rate* of the compost, using respirometric techniques (Manios and Balis 1983; Iannotti et al. 1993). The test is usually performed in a sealed environment under controlled conditions of temperature and moisture. A variation of this method is based on the estimation of a stability index over the entire thermal spectrum of composting using a thermogradient respirometric apparatus (Lasaridi et al. 1996). In both cases, two parameters are of importance in assessing stability: presence of the appropriate microbial consortium and a sufficient period of time for their adaptation into their new environment.

The aim of this study was to address some of the problems encountered during the application of the self-heating test. The rise of temperature was often erratic from sample to sample; the required time of incubation was often long and inconsistent; and the rise of temperature was invariably lower than the expected levels in apparently unstable composts. The method described was developed in the course of the IMPROLIVE project (FAIR CT96 1420) and is based on the finding that hydrogen peroxide has a dramatic triggering effect on the composting process.

Materials and Methods

A variety of agricultural wastes and composts were further composted in laboratory-scale experiments using a microcosm thermally insulated aerobic system with a capacity of 2 l. The materials examined include: composted olive oil mill sludge from two-phase decanters (*alpeorujo*); extracted olive press-cake compost (*orujo*) from three-phase decanters; orujo, cocomposted with the olive oil mill wastewater from three-phase decanters and presses (*alpechin*); cotton gin waste; and cotton gin waste cocomposted with olive pulp (i.e. the fraction of de-oiled alpeorujo de-stoned by centrifugation). Composts had been prepared in temperature feedback, aerated, laboratory-scale composting units of *ca.* 0.125 m^3 capacity. The active composting phase lasted for 70 days and was followed by a maturation period of at least 1 year. At the end of the maturation phase all the composts were alkaline (pH 7.3–8.5) and were not phytotoxic, according to the germination index test (Zucconi et al. 1981). In addition, commercial fresh and spent mushroom compost (FMC and SMC, respectively) from the Hellenic Mushroom Farm was examined.

The Hydrogen Peroxide Test

The microcosm composting experiments were carried out as follows: the mixture was air-dried and subsequently 1.5 l of it was placed in a 2-l capacity thermally insulated container (adiabatic reactor), where it was moistened to 60% of its water-holding capacity with either tap water for the controls, or with a hydrogen peroxide water solution. Concentrations of 1 to 5% H_2O_2 in the moistening solution were tested. The flasks were loosely closed, to keep them aerobic, and stored in a room with constant temperature, at 20 °C. The temperature in the composting mass was continuously monitored using a thermocouple.

To assess the nature of the triggering effect of hydrogen peroxide on the composting process, control experiments were performed by adding in the reactor 5 ml toluene, a weak volatile anti-microbial agent, while the system was already progressing in the thermophilic phase. The temperature evolution was recorded before and after the addition and the volatilisation of toluene.

Microbial Isolations

In order to investigate the nature of the triggering effects of H_2O_2 on the composting process, changes in the microbial populations of the orujo compost under different treatments were studied. *Mesophilic bacterial strains* were isolated with the dilution method on plate count agar medium after 4-day incubation at 25 °C. Isolates were transferred onto nutrient agar plates and examined regarding their morphological features, Gram stain reaction and endospore formation. Thermophilic endosporeformers were isolated following a selective method according to which vegetative cells are killed with the addition of hot water for a period of 5 min. After heat killing the samples are cooled quickly in cold water, plated on nutrient agar and incubated at 55 °C for up to 5 days. Thermophilic isolates were also tested for their ability to grow at 65 °C. A number of isolates has been identified with the use of classical microbiological methods, in addition to the BIOLOG identification system and the MATIDEN identification program for streptomycetes.

Fungal species were isolated with the aid of a modified soil-washing apparatus, which possesses sieves of different mesh diameters (Dhingra and Sinclair 1995). After washing the samples overnight, the smaller particles and most of the spores are removed, while the remaining particles are distributed on the three sieves. The different fractions are dried separately on sterile filter paper, placed on potato dextrose agar with 50 mg l^{-1} Rose bengal and incubated at 28 °C.

Results and Discussion

In all trials (Fig. 1–4) the addition of hydrogen peroxide in the moistening solution was found to exert a triggering effect on the composting process, as reflected by the rate and level of temperature rise in the composting mass compared with the control. This has also been observed in a range of other trials with similar materials, not reported here (Balis and Tassiopoulou 2001).

Figure 1a presents the temperature variation during the self-heating process of alpeorujo with and without the addition of hydrogen peroxide (at 5%). In the H_2O_2 treatment temperature increased faster, reaching 50 °C after 3 days, while the control did not exceed 36 °C. The nature, biological or chemical, of the processes causing the release of the excess heat was investigated by the addition of a weak, volatile antimicrobial agent in the H_2O_2 treatment. On day 13, while the system of the H_2O_2-treated alpeorujo was progressing in the thermophilic phase, 5 ml of toluene were added to the flask, resulting in a sharp temperature decline. The temperature drop after toluene addition followed the pattern of temperature losses in a hot inanimate body. As toluene is volatile, after day 18 its antimicrobial effect was attenuated, allowing again microbial heat production and the temperature started to rise again.

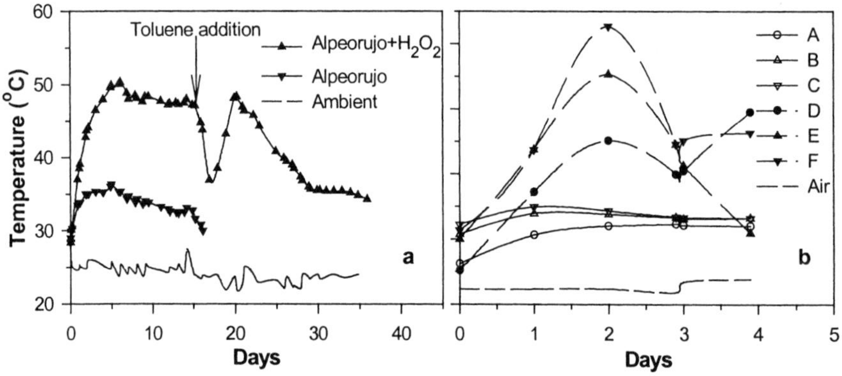

Fig. 1. a Effect of the addition of 5 ml toluene on the temperature course of alpeorujo compost triggered by hydrogen peroxide. **b** Temperature variation with time during the self-heating process of: *A* mature orujo compost (control); *B* mature compost +2.5% H_2O_2; *C* mature compost +5% H_2O_2; *D* mature compost +alpechin; *E* mature compost +2.5% H_2O_2; *F* mature compost +5% H_2O_2

This was a strong indication that the prolonged extra heat production in H_2O_2 treatments should be attributed to enhanced microbial activity in the compost and not to exothermic chemical oxidation of the organic matter by hydrogen peroxide.

In Fig. 1b the effect of different concentrations of H_2O_2 in mature orujo with and without the addition of alpechin was examined. Alpechin, which is high in easily biodegradable compounds, adds fresh organic matter to the mature compost resulting into increased microbial activity, expressed as higher recorded temperatures (45 °C in treatment D, compared with 35 °C for A). This has also been observed in pilot-scale systems and confirmed in respiration studies (Lasaridi et al. 1996). Hydrogen peroxide at both concentrations (2.5 and 5%) had only a minor effect on the self-heating process of the mature orujo compost, and this was restricted to the beginning of the process. On the contrast, in the alpechin reatments where fresh organic matter was available, H_2O_2 addition caused a distinct and prolonged temperature rise, reaching 55 and 63 °C for the 2.5 and 5% concentration, respectively. The addition of H_2O_2 amplifies the difference between the maximum temperatures recorded in samples poor and rich in fresh organic matter, i.e. between stable and unstable composts (compare the couples A and D, B and E, and C and F). This effect, also observed with other materials (Fig. 2, Fig. 4), could be exploited to develop a more sensitive version of the self-heating test for assessing compost maturity.

Similar behaviour was exhibited during the addition of H_2O_2 in cotton gin waste (Fig. 2a) and cotton gin waste amended with olive pulp (Fig. 2b). Cotton gin waste is relatively recalcitrant, as a large part of its organic matter is in the form of lignin and lignocellulose. Olive pulp, on the other hand, a smooth paste that contains the soft tissues of the olives including the water-soluble constituents, is rich in highly biodegradable compounds, although it cannot be composted on its own due to lack of the necessary porosity. The hydrogen peroxide treatment stimulated the self-heating process most dramatically in the case of cotton gin-pulp mixtures (Fig. 2b).

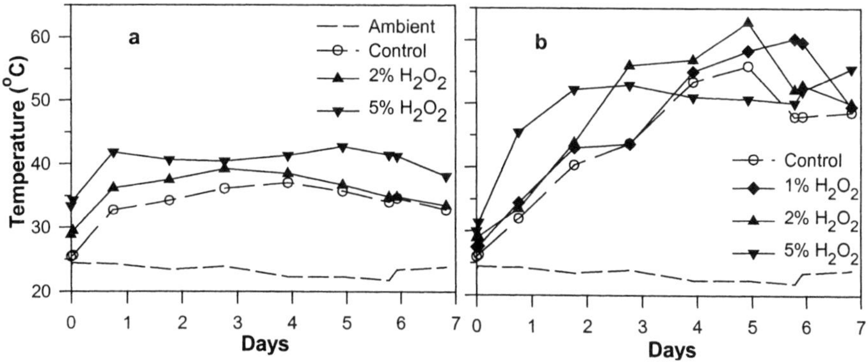

Fig. 2a,b Effect of different concentrations of hydrogen peroxide on the self-heating process of a cotton gin waste and b cotton gin waste cocomposted with olive pulp

In the cotton gin compost, temperature went just over 35 °C. Addition of 2 and 5% H_2O_2 caused a temperature increase of 4 and 9 °C, respectively. In all treatments of cotton gin-olive pulp mixtures, maximum temperature exceeded 55

°C, with H_2O_2 addition resulting in temperature boosts in excess of 12 °C at certain measuring events. Hydrogen peroxide addition, especially at high concentration, seemed to mainly accelerate the occurrence of the maximum temperature (T_{max}), rather than increase it in absolute terms. This is probably due to the high T_{max} recorded for the control, which was near the temperature ceiling for biological processes.

In the case of the heat-dried alpeorujo (Fig. 3), the triggering effect of the hydrogen peroxide was highly intensified when the substrate had been seeded with mature orujo compost. In absence of seeding matter and without hydrogen peroxide treatment (Fig. 3a, control), the temperature reached a peak value of 50 °C after 18 days of incubation. The hydrogen peroxide treatment, however, accelerated the self-heating process and the temperature rose to 45 °C in about 5 days, when the temperature of the control was about 30 °C. The acceleration effect of the hydrogen peroxide treatment was considerably more pronounced when the material was seeded with mature compost. In this case (Fig. 3b), the temperature overshot the 60 °C level and then declined sharply again, probably because the system reached pasteurising temperatures. The thermophilic process was reassumed and reached a second peak of 50 °C after 30 days.

The effect of hydrogen peroxide on fresh (before spawning: FMC) and spent (SMC) mushroom compost is illustrated in Fig. 4. FMC, when subjected to the hydrogen peroxide treatment indicated unstable compost, with H_2O_2 both accelerating and boosting the occurrence of T_{max}. On the contrary, the spent substrate did not respond to the hydrogen peroxide treatment, a kind of response typical of a stable compost. The high temperature reading, 2 h after incubation, is attributed to the exothermic chemical reactions occurring at the beginning of the hydrogen peroxide test, and does not affect stability assessment, which is based only on the biological triggering effects of H_2O_2.

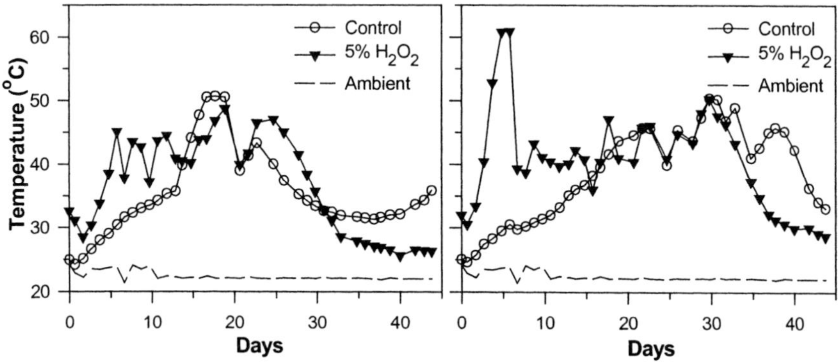

Fig. 3a,b Effect of 5% hydrogen peroxide on the self-heating process of **a** heat-dried Spanish alpeorujo and **b** heat-dried Spanish alpeorujo seeded with 10% mature orujo compost

In all trials, a usually slight increase in temperature is noted immediately after the hydrogen peroxide treatment. This was considered to be the result of exothermic, non-biological chemical reactions, as H_2O_2 oxidises certain compounds in the compost. The subsequent longrise of temperature is thought to reflect intensification of microbial activity in the catabolic processes, the intensity of which is probably triggered by the hydrogen peroxide and fuelled by the available organic substrate.

On the basis of the results reported here and elsewhere (Balis and Tassiopoulou 2001), a new stability parameter, the thermal stability index (TSI) was developed that exploits the triggering effect of H_2O_2 on composting. The test, which is in essence a variation of the self-heating test, is carried out as described in the Materials and Methods section, but with the addition of mature orujo compost as seed at a ratio of 10:1 on a volume basis, and at a standardised H_2O_2 concentration of 5% for the moistening solution. TSI is calculated as follows :

TSI = [60-peak temperature (°C) within 5 days after treatment]/[60-ambient temperature (°C)]

The TSI may yield values that range from one (fully stable compost) down to zero (unstable compost). On rare occasions it may yield negative values. This would occur when the peak temperature exceeds the level of 60 °C, an indication of a strongly unstable material.

The effect of hydrogen peroxide on the microbial population profile of the end product was also examined in order to gain an insight into the nature and mechanisms of the observed triggering effects. Population levels were counted (3 dilutions times 4 replicates per dilution) in the series of mature orujo composts described in Fig 1b and results were in accordance with the temperature variation of the different treatments. In all cases, H_2O_2 slightly increased the population counts, from 7×10^7 to 43×10^7 Cfu for the orujo compost and from 36×10^7 to 83×10^7 Cfu for the orujo + alpechin. In all treatments pH was neutral.

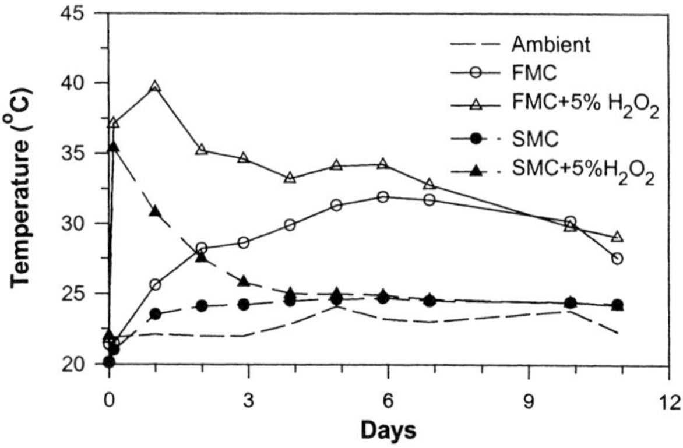

Fig. 4. Effect of 5% hydrogen peroxide on *FMC*: fresh mushroom straw compost, before spawning; and *SMC*: spent mushroom compost, after mushroom cropping

The various mesophilic and thermophilic isolates that were obtained from the different treatments are presented in Tables 1,2 along with some of their characteristics. A number of them have been identified, but the majority are still awaiting identification. The names of *Streptomyces* species reported refer to clusters arranged according to Williams et al. (1983).

Streptomycetes were the most frequently isolated group of microorganisms in the orujo compost treated with 5% H_2O_2 with or without alpechin (treatments C and F). Furthermore, when alpechin was added, the microbial population consisted mainly of endospore-forming species belonging to the genus *Bacillus*, while compost without alpechin contained different isolates of Gram-negative rods and coccobacilli belonging to the *Cytophaga/ Flavobacterium* group (Table 1). All the thermophilic bacterial isolates belonging to the genus *Bacillus* when cultivated formed filaments (Table 2).

Fungal isolates belong to the genera *Penicillium, Aspergillus* and *Rhizopus*. The soil-washing technique used in the study permitted the isolation of species existing in the compost as active mycelium, thus favouring the recovery of ligninocellulytic fungi. *Aspergillus* were the most frequently isolated species from the orujo containing alpechin (treatment D), especially when H_2O_2 was added (treatment F), while *Rhizopus* sp. was present only in the control (treatment A).

It has been reported that certain *Bacillus* as well as *Streptomyces* strains exhibit resistance towards oxidative stress caused by hydrogen peroxide (Murphy et al. 1987; Lee et al., 1993). Adaptation of *Bacillus subtilis* to treatment with H_2O_2 is followed by induction of several protective proteins (Murphy et al. 1987). These studies also indicate a similarity between the proteins synthesised by the bacterial cells when exposed to oxidative and thermal stress. It is not yet clear whether there is a close mechanistic relationship between various stress responses or whether the same proteins can simply be induced by a number of different mediators. A protective mechanism to treatment with H_2O_2 was also observed in the case of *Streptomyces coelicolor* (Lee et al. 1993). Adaptation was accompanied by induction of several enzyme activities involved in defence against oxidative stress such as catalase, peroxidase and glutathion reductase. In addition, mutants resistant to H_2O_2 exhibited greater resistance against heat shock compared to the wild type. *Aspergillus* and *Penicillium* species are also known to be resistant to oxidative stress (Emri et al. 1997).

Conclusions

Addition of hydrogen peroxide at concentrations 1 to 5% in the compost water content exerted a triggering effect on the rate of temperature rise and the maximum temperature reached, for a wide variety of composts. The response of various substrates and treatments to the hydrogen peroxide treatment appears to depend on the hydrogen peroxide concentration, the substrate's state of decomposition and its moisture content. The temperature boost cannot be attributed to exothermic chemical reactions, as it is inhibited by the addition of antimicrobial agents.

Table 1. Mesophilic microbial consortia isolated in the different treatments of the orujo compost

Isolate	Treatm.	Gram	Endospore	Morphology/genus	Colour	Genus/species
M4	A			Streptomyces	Beige	*Sphingobacterium thalpophilum*
M3	A	+	?	Rod	Beige	
M5	A	+	-	Rod	Beige	
M1	A	-	-	Cocci - coccobacilli	Beige	
252	B			Streptomyces	Beige	*Streptomyces rochei*
253	B			Streptomyces	Beige	
255	B			Streptomyces	Orange	
257	B			Actinomycete	White	
251	B	+	+, (X)	Rod	Beige	*Bacillus licheniformis*
254	B	-	-	Rod	Yellow	
256	B	-	-	Rod	Orange	
258	B	-	-	Rod	Orange	
53	C			Streptomyces	Beige	*Streptomyces rochei*
51	C	+	+, (X/T)	Rod	Beige	
52	C	-	-	Rod	Yellow	*Flavobacterium sp.*
54	C	-	-	Rod	Beige	
55	C	-	-	Coccobacilli	Orange	
56	C	-	-	Coccobacilli	Beige	*Sphingobacterium thalpophilum*
MK4	D			Streptomyces	Beige	
MK5	D			Streptomyces	Beige	*Streptomyces rochei*
MK7	D			Streptomyces	Orange	
MK8	D			Streptomyces	Beige	
MK3	D	+	+, (X)	Bacilli-Coccobacilli	Rose	*Bacillus firmus / B. subtilis*
MK1	D	+	+, (X)	Rod	Beige	
MK6	D	+	+, (T)	Rod	Beige	*Bacillus polymyxa*
MK2	D	+	?	Coccobacilli	Orange	
255K	E			Streptomyces	Beige	*Streptomyces rochei*
257K	E			Streptomyces	Beige	*Streptomyces rochei*
252K	E	-	-	Bacilli-Coccobacilli	Dark beige	*Pseudomonas aeruginosa*
251K	E	+	-	Rode	Orange	
253K	E	+	+, (X)	Rod	Beige	*Bacillus megaterium*
254K	E	+	+, (T)	Rod	Beige	
256K	E	+	+, (X)	Rod	Beige	
52K	F			Streptomyces	Beige	
55K	F			Streptomyces	Beige	*Streptomyces rochei*
53K	F	+	+, (T)	Rod	Beige	
51K	F	-	-	Coccobacilli	Orange	*Flavobacterium indologenes*

A: mature orujo compost (control). *B*: A+2.5% H_2O_2. *C*: A+5% H_2O_2. *D*: A+alpechin. *E*: D+2.5% H_2O_2. *F*: D+5% H_2O_2.
(X) = Central; (T) = Terminal.

Furthermore, it cannot be ascribed solely to the extra oxygen that is liberated shortly after the addition of hydrogen peroxide, as the conditions under which the experiments were performed precluded molecular oxygen limitations.

Table 2. Thermophilic microbial consortia isolated in the different treatments of the orujo compost

Isolate	Treatm.	Gram	Endospore	Morphology/Genus	Growth at 65 °C	Colour
ME1	A	?(-)	+, (T)	Rod	+	Beige
ME2	A	?(-)	+, (T)	Rod	+	Beige
ME3	A	+	+, (T)	Rod	-/+	Beige
ME4	A	+	+, (T)	Rod	+	Beige
251E	B	+	+, (T)	Rod	+	Cream
252E	B	+	+, (T)	Rod	+	Beigc
253E	B	+	+, (T)	Rod	-/+	Beige
51E	C	+	+, (T)	Rod	+	Cream
52E	C	+	+, (T)	Rod	ND	Beige
53E	C	+	+, (T)	Rod	ND	Cream
55E	C	?(-)	+, (T)	Rod	-	Beige
MKE1	D	+	+, (T)	Rod	+	Beige
MKE2	D	+	+, (T)	Rod	-	Beige
MKE4	D			Streptomyces	ND	Beige
MKE5	D	+	+, (T)	Rod	+	Beige
251KE	E	+	+, (T)	Rod	ND	Beige
252KE	E	+	+, (T)	Rod	+	Cream
253KE	E	?(-)	+, (T)	Rod	ND	Beige
254KE	E	?(-)	+, (T)	Rod	ND	Beige
51KE	F	?(-)	+, (T)	Rod	+	Beige
52KE	F	+	?	Rod	-	Beige
54KE	F			Streptomyces	-	White
55KE	F	?(-)	+, (T)	Rod	+	Beige

A: mature orujo compost (control). *B*: A+2.5% H_2O_2. *C*: A+5% H_2O_2. *D*: A+Alpechin. *E*: D+2.5% H_2O_2. and *F*: D+5% H_2O_2.
(T) = Terminal.

It is more likely that hydrogen peroxide, being a reactive oxygen intermediate by itself, elicits the formation of highly reactive hydroxyl radicals. The observed changes in the microbial consortia in the compost after the H_2O_2 treatments appear to support this hypothesis. Such mechanisms are well known to operate in the biological realm. The lignin moiety of lignocellulose is decomposed by *Phanerochaete chrysosporium* through a mechanism of co-metabolism. The formation of glucose from the cellulose yields hydrogen peroxide, hydroxyl and superoxide radicals that are needed to initiate a snowball reaction causing the breakdown of the lignin skeleton (Leisola and Waldner 1988; Schlegel 1992). Similar evidence has been reported in the case of the brown-rot fungus *Gloeophyllum trabeum*. The fungal chelator fosters the production of reduced metals which when in proximity to reactive oxygen species such as hydrogen peroxide or other oxidants, will react to form hydroxyl radicals which are capable of depolymerizing and oxidising lignocellulose compounds (Goodell et al. 1997).

The triggering effect of H_2O_2 on the initiation and advancement of composting can be exploited to assess compost stability in a modification of the standard self-heating test that was shown to offer increased sensitivity. For further study, the possibility of combining respirometry with the hydrogen peroxide effect on heat output to monitor the energy of the composting substrates is envisaged, from the point of view of determining optimum conditions for composting processes and assessing compost stability. Since the heat output and carbon dioxide can be measured continuously and may have different patterns, the calorimetric to respiration ratio in presence and absence of hydrogen peroxide could give valuable information on energy balances and stage of the composting process.

Acknowledgements. This work was supported by the EU IMPROLIVE project (FAIR CT96 1420). Dr. M. Kyriakou offered valuable comments on microbiological issues during the preparation of the chapter. Finally, it should be noted that this chapter reports some of the last work of the late Professor Costas Balis. Although it might have not been possible for the rest of the authors to fully complete and present his string of thoughts, we believe it presents some very interesting results and ideas that are worth further investigation.

References

ASTM (1996) Standard test method for determining the stability of compost by measuring oxygen consumption. American Society for Testing and Materials, D 5975-96, Philadelphia, PA

Avnimelech Y, Bruner M, Ezrony I, Sela R, Kochba (1996) Stability indexes for municipal solid-waste compost. Compost Sci. Util. 4(2):13-20

Balis C, Tassiopoulou V (2001) Triggering effect of hydrogen peroxide on composting and a new method for assessing stability of composts in a thermally insulated microcosm system. Acta Hortic. 549:61-70

Dhingra OD, Sinclair J (1995) Basic plant pathology methods. Lewis, CRC Press, Boca Raton, Florida

Emri T, Posci I, Szentirmai A (1997) Glutathione metabolism and protection against oxidative stress caused by peroxides in *Penicillium chrysogenum*. Free Rad. Biol. Med. 23(5):809-814

FCQAO (Federal Compost Quality Assurance Organization) (1994) Methods book for the analysis of compost, in addition with the results of the parallel interlaboratory Test 1993. Kompost-Information Nr 230. Bundesgütegemeinschaft Kompost e.V.

Forster JC, Zech W, Wurdinger E (1993) Comparison of chemical and microbiological methods for the characterization of the maturity of composts from contrasting sources. Biol. Fertil. Soils 16(2):93-99

Goodell B, Jellison J, Liu J, Daniel G, Paszczynski A, Fekete F, Krishnamurthy S, Jun L, Xu G (1997) Low molecular weight chelators and phenolic compounds isolated from wood decay fungi and their role in the fungal degradation of wood. J. Biotechnol. 53(2,3):133-162

Iannotti DA, Toth BL, Elwell DL (1993) A quantitative respirometric method for monitoring compost stability. Compost Sci. Util. 1(3):52-65

LAGA (1984) Quality criteria and application recommendations for compost obtained from refuse and refuse/sludge, Pamphlet M10. German Federal Department of the Environment, Bonn

Lasaridi KE, Stentiford EI (1998) Respirometric techniques for MSW compost stability evaluation: a comparative study. In: Stentiford EI (ed) Proceedings of the International Conference on Organic Recovery & Biological Treatment (ORBIT 97). Harrogate, UK, Zeebra Publishing, Manchester, pp 303-310

Lasaridi KE, Papadimitriou EK, Balis C (1996) Development and demonstration of a thermogradient respirometer. Compost Sci. Util. 4(3):53-61

Lee JS, Hah YC, Roe JH (1993) The induction of oxidative enzymes in *Streptomyces coelicolor* upon hydrogen peroxide treatment. J. Gen. Microbiol. 139:1013-1018

Leisola M, Waldner R (1988) Production, characterisation and mechanism of lignin peroxidases. In: Zadrazil F, Reiniger P (eds) Treatment of lignocellulosics with white rot fungi. Elsevier, London, pp 37-42

Manios V, Balis C (1983) Respirometry to determine optimum conditions for the biodegradation of extracted olive press-cake. Soil Biol. Biochem. 15:75-83

Murphy P, Dowds BCA, McConnell DJ, Devine KM (1987) Oxidative stress and growth temperature in *Bacillus subtilis*. J. Bacteriol. 169(12):5766-5770

Scaglia B, Tambone F, Genevini PL, Adani F (2000) Respiration index determination: dynamic and static approaches. Compost Sci. Util. 8(2):90-98

Schlegel HG (1992) General microbiology. Cambridge University Press, Cambridge

US Composting Council (1997) Respirometry. In: Leege PB, Thompson WH (eds) Test methods for the examination of composting and compost, The US Composting Council, Bethesda, Maryland, pp 9-165/9-194

Williams ST, Goodfellow M, Alderson G, Wellington EMH, Sneath PH, Sackin MJ (1983) Numerical classification of *Streptomyces* and related genera. J. Gen. Microbiol. 129:1743-1813

Zucconi F, Pera A, Forte M, de Bertoldi M (1981) Evaluating toxicity of immature compost. Biocycle 22(2): 54-57

Plant Performance in Relation to Oxygen Depletion, CO_2-Rate and Volatile Fatty Acids in Container Media Composts of Varying Maturity

W. F. Brinton and E. Evans[1]

Abstract. Compost performance in container media presents a challenge to the compost industry. Seedling and container composts represent high-value horticultural products, and many competing products are available. Therefore plants grown with compost container media must exhibit excellent properties. However, there are frequent reports of poor performance of container and seedling starter mixes. Many factors, including porosity, salt content and maturity of composts prior to starting container plants, may play a large role in the observed performance. Dilution and use of bark and peat can remove salt and porosity as factors. Immaturity has frequently been associated with poor plant performance, but it is not known what precise levels of maturity affect container performance. Ammonia and volatile fatty acids (VFA), known to be phytotoxic, are frequently found in immature composts. This study uses three composts of varying maturity in container mixes and examines plant performance over 21 days. The results indicate that immature and semicured composts reduce oxygen content very significantly in container media to levels that may directly and indirectly damage roots. The study evaluates which tests may be used to predict performance. The Dewar test which rates compost on a I - V scale, did not predict damage. Initial VFA, CO_2 respiration and Solvita tests all predicted the poor result of immature composts in container performance.

INTRODUCTION

Composts which are not fully stabilised are considered to be immature. Such composts may heat up in Dewar vessels, contain high levels of volatile fatty acids (VFA) and possess a high oxygen demand (Jourdan 1988; Manios et al.1989; Brinton et al. 1995, Ionaotti 1994a). A wide variety of tests have been proposed for compost maturity and toxicity (Itävaara et al. 1998; SEPA 1997).

Among many growth-suppressing traits in composts is the presence of VFA. The composting process involves unavoidable episodic oxygen depletion. This may result in temporary accumulation of short-chain volatile fatty acids (VFA). VFA have been previously shown to be responsible for poor plant performance in controlled studies (Devleeschauwer et al. 1981; Brinton 1998; Brinton and

[1]Woods End Research Laboratory, Box 297, Mt Vernon Maine 04352, USA
Email: wbrinton@ctel.net

H. Insam, N. Riddech, S. Klammer (Eds.)
Microbiology of Composting

Tränkner 1999; Lee 1977). However, under many circumstances, we have observed poor performance where little or no VFA is present or when the Dewar test results indicated grade IV and V composts, considered to be "finished" (LAGA 1984). A more sensitive test is needed for general usage. In order to be fully meaningful to compost users, laboratory tests used for maturity must be evaluated in relation to end uses of compost. Our study examines plant performance in relation to qualities of composts used in container media.

We previously reported a survey of 712 compost samples showing that 26% had VFA above 5000 mg kg dm^{-1} while 6% had VFA above 20000 mg kg dm^{-1}. The VFA correlated negatively with compost age and were highest in the first 20-35 days of composting (Brinton 1998). Prior work in plant growing media revealed that VFA levels as low as 500 mg kg dm^{-1} exert phytotoxic effects on plant seedlings (Lynch 1977). In liquid nutrient culture, VFAs of as little as 100 mg kg^{-1} cause 50% growth depression (Woods End 1997). Many factors, however, may be involved in growth depression. Some composts which have very little VFA and adequate nutrients still perform poorly in growing media. Oxygen depletion in the root zone may be one such factor. Oxygen is critical for root development, and adequacy of air governs important ion adsorption properties. Roots of plants growing under waterlogged or anaerobic conditions have extremely retarded respiration and low ion uptake rates (Salisbury and Ross 1978). It is likely, therefore, that immature composts used in containers may exert influence on oxygen supply traits. With composts being used in high-value markets such as for container media and starting of seedlings for vegetable culture, to better prediction of conditions that may cause poor performance is imperative.

Materials and Methods

In-vessel compost samples

Compost samples were obtained from an in-vessel biosolids compost facility in Rockland, Maine, in August of 2000. Representative samples were selected from three phases of the compost process; phase I, uncured compost discharged after 21 days; phase II, semi-cured compost cured for 60 days under cover and phase III discharged compost cured 250 days outside. The mixing formulae for these composts were held constant at the facility. The analytical traits of these composts are given in Table 1.

The test traits indicate a transition from high-ammonium, medium-high CN ratio composts to low CN, high nitrate composts. Although the Dewar test temperature indicates a wide range of heating, the rating scale employed in the Dewar places both the semicured and the cured in the same finished class.

Table 1. Physical / Chemical Traits of Biosolids Compost Examined in Container Study

Compost	pH 1:1	OM % (dm)	Total-N% (dm)	C:N	NH4-N ppm (dm)	NO3-N ppm (dm)	VFA ppm (dm)	Salt S m^{-1}
Uncured age 21 days	7.53	73.0	1.969	20.0	4872	1	2109	0.49
Semicured age 76 days	7.44	73.0	2.212	17.8	3295	1	993	0.42
Cured age 250 days	6.10	57.0	2.949	10.4	16	1734	319	0.45

Table 2. Biological Traits of Biosolids Composts

Compost	CO_2-C% of C	CO_2- C% (dm)	Solvita test unit	Wheat-Germination Rel%	Wheat Biomass Rel%	Cress Germination Rel %	Cress Biomass Rel%	Dewar C° (Grade)
Uncured age 21 days	0.53	0.20	4	93	62	45	41	31 (II)
Semicured age 76 days	0.59	0.23	4	93	56	35	37	10 (V)
Cured age 250 days	0.14	0.04	7	93	83	98	79	3 (V)

Moisture was added to the compost to reach optimal level prior to conducting Dewar and CO_2 respiration tests. Solvita tests correlated closely with CO_2 respiration whether reported as percent of carbon or percent of dry matter. VFA content was high for the uncured material and diminished as the material aged.

Simulated Container-Mix Formulation

In order to construct container mixes, we determined the needed dilution with peat moss to reach a suitably low conductivity of approximately 0.2 S m^{-1} . In addition, we determined that diluted compost mixtures would vary in air porosity, depending on the depth in pots. Air porosity of compost alone diluted with peat was higher (45 - 49%) than normally encountered with container mixes. Thus, we prepared a blend of compost / peat / washed sand (2:1:1) that resulted in a uniform air porosity, ranging from 16 - 18% throughout the container after packing. Two container volumes were selected: 3 and 12 l. No additional nutrients were provided.

To measure oxygen concentration in the container media, we inserted narrow 1.5-mm vinyl air tubes to specified depths at the top, middle and bottom or 4, 10.5 and 17.5 cm depths from the top, respectively. At the time of measurement, these tubes were attached to an O_2-sensitive electrode via a mini-air sampler that requires only 5 cm^{-3} of air to obtain a reading. A small suction syringe was used to extract sufficient air daily during the growth of the plants. The pots were seeded to sorghum-sudan grass at an equivalent rate of 400 kg ha^{-1} (approximately 1 seed /

3 cm^{-2}). Final harvests were made at 16 days after planting. Plants were held under Gro-Lux lights for 12-hr light/dark cycles at 22 °C.

Fig. 1. Oxygen measurement lines inserted container-media into pots at varying depths.

Analytical Methods

Volatile fatty acids (VFA) were determined after water extraction and distillation in H_2SO_4 at pH 1.8; the resulting distillate was titrated to a standard endpoint (SMM 1994). The CO_2 evolution rate was determined on 40-g samples after 1 day of equilibration after sampling and incubation temperature of 34 °C. CO_2 was trapped in a NaOH-barium, the Kjeldahl procedure for solid waste (EPA1996) and nitrate by water extraction followed by liquid ion chromatography (SMM 1994). Ammonia was determined by LiAC extraction followed by ion-electrode determination (Orion 1982). The presence of hydrogen sulfide was estimated by placing Merckoquant lead acetate indicator strips over samples of acidified compost (Merck 1996). Solvita maturity was determined with Solvita test kits (TMECC 2000). Phytotoxicity tests were conducted on each of the three composts and one control by 1:1 (v/v) dilution of limed, spaghnum peat (pH = 6.2) to obtain a conductivity of approximately 0.2 S m^{-1}. Subsequently, ten seeds each of garden cress (*Lepidium sativa*) and wheat (*Triticum aestivum* var. Rose) were sown into each of six 50-cc cells. Germination and growth was measured after 7 days by counting plants and cutting and weighing fresh material. Results are reported against a control of professional peat/nutrient media (Fafard 3-B Mix, Fafard, New Brunswick). To measure roots in large pots, plants were removed by cutting media cross-sectionally and then roots were washed with a gentle stream of water.

Results

Oxygen Content in Growing Media

Oxygen content of interstitial pores diminished with depth in media and was significantly affected by the apparent maturity of the composts (see Fig. 2). Surprisingly, the O_2 content did not vary appreciably over time but persisted near the levels shown in Fig. 2 throughout the growing period.

Performance of Plants as Seen in Tops and Root Development

At 21days, the entire plant/root mass was carefully removed from the media for visual examination (Fig. 3). These examinations correspond to the data collected for distribution of oxygen in the containers (Fig. 2). Root development and plant yield responded to increasing maturity of compost (Fig. 3). The differences between the uncured and semicured were more pronounced ($p < 0.04$) than between the cured and the control ($p < 0.10$).

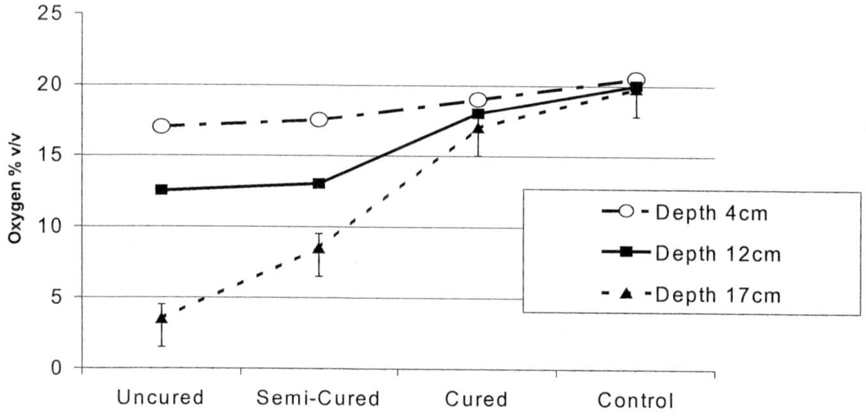

Fig. 2. Interstitial oxygen concentration in compost container media

It is evident that there is little root growth in the immature composts where rootlets were confined to the top and edges of the container. In the cured treatment, the roots extend to the bottom of the pot.

Fig. 3. Container media series, from left to right: uncured, semicured, cured, control

Effects of Compost Maturity on Root Development: Root Washings

Immediately after the root-ball of plants was removed from the pots, we also harvested the rootlets from the containers by washing them gently out of the compost-peat-sand mix. In Fig. 4 we provide evidence of the rootlet damage incurred as a result of the immaturity of uncured and semicured composts. Table 3 gives results from plant effects observed in Fig. 4.

We tabulated the root length and weight of tops. Table 3 gives the results for plant fresh weight and root length.

Table 3. Yield and root weight in relation to compost treatment. Means followed by the same letter in the row do not differ significantly at the $p < 0.05$ level

Variable	Uncured	Semicured	Cured	Control
Plant fresh weight (mg)	73 a	116 b	183 c	196 c
Root length (cm)	7.5 a	9.0 b	12 c	19 d

The plant effects observed from immature composts are more evident on closer inspection. A pronounced stiffening of the rootlets is evident, with tissue thickening above the hypocotyl, where the stems joins the root. There was also evident discoloration of the rootlets from immature composts. We detected hydrogen sulfide in uncured and semicured compost media at the 17 cm depth.

Fig. 4. Development of plant roots as affected my compost maturity: from left to right uncured, semicured, cured, control

Table 4. Correlations Observed Between Test Traits and Plant Growth

Relationship Examined	r factor
Solvita test : cress weight Solvita test : CO_2 rate	0.996* -0.992
CO_2 rate : cress weight (CO_2 rate as % of TS) CO_2 rate : cress weight (CO_2 rate as % of Carbon)	-0.998* -0.999**
CO_2-Rate : Wheat Weight (CO_2 as % of TS)	-0.998*
Root length : VFA content	-0.999**
Plant yield : ammonium	-0.997*
Plant yield : O_2 content (O_2 measured at 17cm depth)	-0.997*

Relationships Between Measured Parameters

We tabulated statistical correlations from averaged treatment effects. In Table 4 the significant correlations ($p < 0.05$) are listed.

The data suggest that a number of important, interrelated factors played a role in determining plant effects arising from immature composts. The causal mechanism was most likely elevated CO_2 evolution and VFA production, elevated ammonia levels, along with oxygen deprivation and hydrogen sulfide production in containers during growth. Early workers have showed that O_2 levels of 5% or less in the root zone may cause dramatic loss of potassium absorption potential (Vlamis 1944). We did not measure ethylene gas, but it is likely to have been produced under anaerobic conditions in the lower layers of the containers. Hydrogen sulfide was produced and is known to have damaging effects on root development.

Implications for Further Work: CO_2 Evolution Rate Versus Plant Growth

In view of the relationship of immaturity to plant growth we observed here, we decided to examine additional data on plant yield (cress test) in relation to compost CO_2-evolution rate. In Fig. 5 we present 155 compost analyses by regressing cress fresh weight at 7days against CO_2 rate. The linear correlation gave a very significant relation of $r = 0.37$ ($p < 0.001$) and the logarithmic scale CO_2 rate gave a very highly significant relationship at $r = 0.53$, $p < 0.0001$.

Germination results from the same study indicated that CO_2 evolution rate had little or no significant effect on seed germination for either cress or wheat. Germination is thus a poor predictor for compost maturity. The observed CO_2 rate relationship to growth may be direct, indirect, or both. CO_2 rate very likely affects other important parameters that in turn influence plant growth, such as levels of VFA, ammonia and C:N ratio. These observations suggest that maturity is indeed a complex phenomena.

Data in Fig. 5 show that about 1/3 of all composts examined achieved a growth performance comparable to a professional peat-based mix, while another 1/3 inhibited it slightly to significantly, and 1/3 very significantly. Additionally, several of the composts reduced yields even more than the unfertilized controls (50% growth), which suggests severe phytotoxicity traits. This may result indirectly from oxygen deprivation plus any one of the previously mentioned traits. More work should be focused on appreciating the significance of compost-induced growth suppression, but only in context of the intended use, since clearly maturity is relative to all factors encountered in the use of composts. Tests that act as predictors must show relevant relationships to the results of compost usage.

References

Brinton W (1998) Volatile organic acids in compost: production and odorant aspects. Compost Sci Util 75-82

Brinton W, Evans E (1997) Interpretation of compost analyses. Woods End Res J p1-4

Brinton W, Tränkner, A (1999) Compost maturity as expressed by phytotoxicity and volatile organic acids. Orbit-99 Conf Proceedings, University Bauhaus Weimar

Brinton, W, Evans E, Droffner M, Brinton R (1995) standardized test for evaluation of compost self-heating. Biocycle 3:64-69

Devleeschauwer D, Verdonck O, Van Assche P (1981) Phytotoxicity of refuse compost. Biocycle 22: 44

EPA (1996) EPA-600 Methods for chemical analysis of water and wastes. US EPA (RCRA)

Iannotti D A, Grebus ME; Toth BL, Madden LV, Hoitink HAJ (1994) Oxygen respirometry to assess stability and maturity of composted municipal solid waste. J Environ Qual 23: 1177.

Itävaara M., Venelampi O, Samsoe-Petersen L, Lystad H, Bjarnadottir H, Öberg L (1998) assessment of compost maturity and eco-toxicity. NT Tech Rep 404 Espoo, Finland

Jourdan B (1988) Zur Kennzeichnug des Rottegrades von Müll- und Müllklär-schlammkomposten. Erich-Schmid Stuttgart

LAGA (1984) Qualitätskriterien und Anwendungsempfelhungen für Kompost aus Müll und Müllklarschlamm. (Quality criteria and application recommendations for compost from waste and sludge) Merkblatt 10 der Länderarbeitsgemeinschaft Abfall, Erich Schmid Verlag Berlin

Lee R.B (1977) Effects of organic acids on the loss of ions from barley roots. J Exp Bot 28: 578-587

Lynch JM (1977) Phytotoxicity of acetic acid produced in the anaerobic decompostion of wheat straw. J Appl Bacteriol 42: 81-87

Manios VI, Tsikalas PE, Siminis HI (1989). Phytotoxicity of olive tree compost in relation to organic acid concentration. Biol Wastes 27: 307-317

Merck (1996) Merkoquant brand test strips. Postfach 4119, D-6100 Dsrmstadt 1, Germany

Orion (1982) Handbook of Electrode Technology. Orion Research, Cambridge Mass.

Salisbury FB, Ross CW (1978) Plant Physiology. Wadsworth, California

SEPA (1997) Report: compost quality and potential for use. Swedish EPA. AFR- 154, Stockholm Sweden

SMM (1994) standard methods for the examination of water & wastewater, 20th ED. Water Environment Federation

TMECC (2000) Test methods for examination of composts and composting. Draft Document. Composting Council of America, Alexandria Virginia

Vlamis J (1944) Effects of oxygen tension on certain physiological responses of rice, barley and tomato. Plant Physiol 19:33-50

Woods End (1997) Interpretation Guidelines for Compost Analyses. Woods End Res J V-I

Microbiological and Chemical Characterisation of Composts at Different Levels of Maturity, with Evaluation of Phytotoxicity and Enzymatic Activities

A.C. Cunha Queda[1], G. Vallini[2], M. Agnolucci[3], C.A. Coelho[4], L. Campos[1] and R.B. de Sousa[1]

Abstract. Composts from different vegetable residues as well as pig and horse manure at different levels of maturity were used in the present study. In order to characterise the different composts, microbial populations (namely, total aerobic bacteria, actinomycetes, filamentous eumycetes, aerobic cellulolytic fungi and bacteria), physicochemical and chemical parameters (moisture, organic matter, total nitrogen, ammonium nitrogen, C:N ratio, pH, electrical conductivity), self-heating capacity and phytotoxicity (measured by means of the germination test with *Lepidium sativum*), as well as enzymatic activities [cellulase, lipase (C10), protease], were evaluated. The research was designed to focus on a possible correlation between the enzymatic activities and the maturity level of composts. The results showed evidence that the protease:cellulase activity ratio was significantly correlated to the compost phytotoxicity.

Introduction

Compost quality is closely related to its stability and maturity. Quality is also mainly related to both the chemical characteristics of the initial substrates and the dynamics of the composting process itself. The latter is affected by several parameters such as aeration, temperature control, matrix structure (porosity and mechanical strength), moisture, pH and carbon:nitrogen ratio.

The composting process can be divided into two main phases: the active phase and the curing phase. During the active phase, the most easily biodegradable materials are transformed and partially mineralised; the organic matter becomes stabilised as a consequence of the intense microbial activity (Adani et al. 1997).

[1]Departamento de Química Agrícola e Ambiental, Instituto Superior de Agronomia, Tapada da Ajuda 1349-017 Lisbon, Portugal
[2]Dipartimento Scientifico e Tecnologico, Università degli Studi di Verona, Verona, Italy
[3]Consiglio Nazionale delle Richerche, Centro di Studio per la Microbiologia del Suolo, Pisa, Italy
[4]Departamento de Matemática, Instituto Superior de Agronomia, Tapada da Ajuda 1349-017 Lisbon, Portugal

H. Insam, N. Riddech, S. Klammer (Eds.)
Microbiology of Composting

The curing phase is characterised by the conversion of part of the stabilised organic matter into humic substances (Chen and Inbar 1993). Thus, while reaching compost stability can be regarded as the result of high-rate microbial reactions occurring throughout the active phase of the process, compost maturity appears, on the contrary, as the effect of the curing phase. Consequently, indicators that evidence compost stability and maturity are important tools to define the end point of both composting phases and to evaluate the compost quality. Furthermore, in order to decide about compost utilisation, parameters such as electrical conductivity, pH, available nutrients and heavy metals should be considered.

Different methods have been so far proposed to evaluate compost stability. In particular, monitoring of either the respiratory activity (Jodice 1989) or the ability of the material to heat up again once rehydrated (Harada et al. 1981; Jourdan 1988) have been often suggested. These methods aim at detecting residual microbial activity, which depends on the presence of easily decomposable compounds, possibly not fully transformed. Furthermore, studies have been carried out to establish a possible relation between enzymatic activities and compost stability (Herrmann and Shann 1993). Other parameters such as cation-exchange capacity (Harada and Inoko 1980; Saharinen et al. 1996), C:N ratio (Poincelot 1974; Juste 1980; Golueke 1981), and the degree of humification (Sugahara and Inoko 1981; Jodice 1989) have also been proposed to determine compost maturity. Germination index (GI), which is a measure of phytotoxicity, has been considered as a reliable indirect quantification of compost maturity (Zucconi et al. 1981a, b).

The present chapter deals with the characterisation of a range of different composts, attempting to relate their physicochemical, enzymatic and microbiological properties to their stability and maturity.

Materials and Methods

Composts

Composts used in this study were obtained from different organic wastes:

- vegetable residues (VR) collected at the centralised garden-produce market of the city of Lisbon, Portugal;
- pig slurry solid fraction (PSSF) coming from a pig-breeding farm;
- horse manure (HM) provided by the Instituto Superior de Agronomia (ISA), Lisbon, Portugal.

In order to improve the mechanical characteristics, initial organic substrates were mixed with shredded residues from corrugated cardboard production (RCCP) and/or straw (S), as reported below (all percentages are expressed by weight on a wet basis):

- C1: pig slurry solid fraction and straw (87.5% PSSF and 12.5% S);
- C2: horse manure and residues from corrugated cardboard production (95% HM and 5% RCCP);
- C3: horse manure and residues from corrugated cardboard production (90% HM and 10% RCCP);
- C4: horse manure and residues from corrugated cardboard production (85% HM and 15% RCCP);
- C5: vegetable residues, residues from corrugated cardboard production and straw (87.5% VR; 7.5% RCCP and 5% S).

All composts were produced in the same pilot composting unit, located at ISA, Lisbon, Portugal. Composting was carried out in windrows aerated by periodic turning for 2 months, on a roofed concrete platform. Once produced, composts were stored at room temperature for different periods: C1 for 36 months; C2, C3 and C4 for 18 months; C5 for 12 months. The composts were then screened through a sieve (5 mm) and carefully homogenised by hand. Sub-samples were finally prepared for analytical characterisation.

Analytical Methods

Moisture content was determined by drying 100-g samples of compost in an oven at 105 °C until constant weight was reached. Organic matter content was quantified by calculating weight loss of oven-dried samples on ignition in a furnace at 550 °C. Carbon was expressed by multiplying the organic matter content by 0.5 (Zucconi and De Bertoldi 1987).

Total and ammonium nitrogen contents were analysed through distillation (Commission of the EC 1978). Conductivity and pH measurements were performed on suspensions of the compost samples in water (1:5 w/v) by electrometric determination (Commission of the EC 1978).

Phytotoxicity was evaluated by means of a seed germination test using cress (*Lepidium sativum*). The germination tests were carried out for 24 h in the dark at 27 °C. Seeds were placed in Petri dishes on sterile filter paper wetted with 1 ml of either filter-sterilised aqueous extract from compost or distilled water. Fifteen plates, each with seven seeds, were prepared for both the control (germination only on sterile water) and the treatment (germination on the 30% dilution of compost aqueous extract). Compost extracts were obtained from each compost, adjusted to a moisture content of 60%, by maintaining the samples at 250 atm for 15 min with a hydraulic press (Zucconi et al. 1981 b, 1985).

The ability of the material to reheat was evaluated using a Dewar self-heating test. Vessels (1.5 l) were filled with compost samples with a standardised moisture content of 35%. Temperature evolution was measured until heat production had ceased. According to Laga-Merkblatt M10 (1995), the class of stability was attributed by taking into account the maximum temperature reached in each vessel: class 1 – maximum temperature >60 °C; class 2 – maximum temperature between

50 and 60 °C; class 3 – maximum temperature between 40 and 50 °C; class 4 – maximum temperature between 30 and 40 °C; class 5 – maximum temperature <30 °C.

Determination of enzymatic activities was performed on three separate samples of aqueous compost extracts, each one prepared by suspending 50 g of compost in 1000 ml of distilled water (Herrmann and Shann 1993). Before preparation of the aqueous extracts, each compost sample was normalised to a moisture content of 60% and then incubated at 27 °C for 48 h (Cunha Queda 1999). Total cellulase activity was assayed using a modified protocol of the method proposed by Hope and Burns (1987). The reaction mixture consisted of 5 ml of acetate-buffered solution [0.1 M, pH 5.5, containing 0.2% (w/v) azide], 0.5 g of Avicel (as substrate), and 1 ml of aqueous compost extract. The method is based on the determination of reducing sugars (expressed as μmol glucose g^{-1} compost dry matter) released into the reaction mixtures after incubation at 40 °C for 16 h.

Lipase (C10) activity was determined according to the method proposed by Cunha Queda (1999). The reaction mixture consisted of 3 ml of substrate solution [2.5 mM of *p*-nitrophenyl caprate in 50 mM potassium phosphate buffer at pH 8.0, containing 0.23% (w/v) sodium desoxycholate and 0.11% (w/v) gum arabic] and 1 ml of aqueous compost extract. This method is based on the determination of *p*-nitrophenol (expressed as μmol g^{-1} compost dry matter) released into the vials after the incubation of mixtures at 25 °C for 1 h.

Protease activity was evaluated by using a modified procedure of the method proposed by Ladd and Butler (1972). The reaction mixture contained 10 ml of aqueous compost extract, 5 ml of a Tris(hydroxymethyl)aminomethane buffered solution (12.5 mM; pH 8.1) and 5 ml of casein solution at 2% (w/v) as substrate. This method is based on the determination of amino acids (expressed as μmol tyrosine g^{-1} compost dry matter) released in reaction mixtures after incubation at 50 °C for 2 h.

Microbial populations (i.e. total aerobic bacteria, actinomycetes, filamentous eumycetes, aerobic cellulolytic fungi and bacteria) were quantified according to the methodologies described by Pochon and Tardieux (1962).

Data were analysed using one-way ANOVA designs. In all the analyses the factor being studied was the compost, with five levels (C1, C2, C3, C4 and C5). For all the parameters analysed, the factor effect was significant at the 0.05 level. The Tukey HSD test was then used to separate the means. All the samples were previously subjected to the normality Shapiro-Wilk and Lilliefors tests in order to check for the necessary assumptions for the ANOVA analyses. Whenever for each individual analysis no reference is made, all the samples passed the test at the 0.05 level.

Results and Discussion

All composts show marked differences with respect to their physicochemical characteristics (Table 1), in particular for the parameters of prominent importance such as ammonium nitrogen content, C:N ratio and electrical conductivity. For the analysis of moisture content, although the sample for C1 did not pass the normality test at the 0.05 level for both Lilliefors and Shapiro-Wilk tests, its presence in the ANOVA analysis does not affect the significance of the differences among the other composts, so that we only have to be cautious about the significance of the differences between C1 and the other composts. Also C4 did not pass the Lilliefors test in the analysis of C:N ratio, although its presence in the analysis does not affect the significance of the differences among the other composts. Results of organic matter, total nitrogen, ammonium nitrogen pH and electrical conductivity are significantly different at α=0.05 for all compost samples.

Table 1. Physicochemical and chemical characterisation of composts studied in the present research

	Compost				
Parameter	C1	C2	C3	C4	C5
Moisture (%)	13.81a	33.70b	27.40c	26.55c	26.65c
Organic matter (% dw)	56.32a	39.18b	63.17c	69.34d	67.68d
Total nitrogen (g kg^{-1} dw)	23.00a	15.10b	20.40c	18.20d	30.50e
Ammonium nitrogen (mg kg^{-1} dw)	531.93a	53.21b	1098.15c	264.84d	928.57e
C:N ratio	12.24a	12.97b	15.49c	19.05d	11.10e
PH	6.13a	7.14b	7.22c	7.91d	8.80e
Electrical conductivity (mS cm^{-1})	11.02a	5.70b	5.97c	3.13d	10.66e

dw, dry weight basis. Values presented are the mean of six determinations. Means of each parameter designated by the same letter are not significantly different at α=0.05.

The results from Dewar self-heating tests showed no differences among the composts studied (Table 2). The maximum temperatures reached by compost samples demonstrated that heat production was very low. Therefore all composts were highly stable according to Laga-Merkblatt M10 (1995). Conversely, the germination index (GI) showed a wide range of values (Table 2).

As described before, the germination bioassay with *Lepidium sativum* was performed by using the 30% dilution of compost extracts. Under these conditions, the germination index must be higher than 60% (threshold of phytotoxicity) in order to classify a compost as not phytotoxic (Pera et al. 1991). Only compost C5 had a GI<60% and could be considered phytotoxic. Composts C2, C3 and C4 presented germination indexes >60%, indicating absence of phytotoxicity. However, higher percentages of RCCP in the initial composting mixtures led to a reduction of the germination index in the final composts (Cunha Queda et al. 2000). Furthermore, it was worth noting that compost C1 reached a germination index >60% in spite of its high electrical conductivity (Table 1).

Table 2. Maximum temperature (Tmax) reached in Dewar self-heating tests, class of stability (Laga-Merkblatt M10 1995) and germination index of five different composts

Compost	Tmax (°C)	Class of stability	Germination index (%)
C1	29	5	84.60
C2	26	5	124.00
C3	28	5	92.00
C4	30	4	67.40
C5	30	4	46.00

The highest values of ammonium nitrogen content (Table 1) were found in composts C5 and C3, which resulted significantly different at α=0.05. Nevertheless, compost C3 did not exhibit phytotoxicity, despite its high ammonium content. References exist pointing out that factors such as low molecular weight fatty acids (Zucconi et al. 1985; Kirchmann and Widén 1994; Shiralipour et al. 1997), phenol compounds (Willson and Dalmat 1986), ammonium nitrogen (Wong et al. 1983; Jodice 1989; Grebus et al. 1994), aliphatic and aromatic amines (Jodice 1989), heavy metals (Wong et al. 1983; Anid 1986) and high electrical conductivity (Pera et al. 1991; Grebus et al. 1994) were responsible for inhibition of germination. Actually, Zucconi and coworkers (Zucconi et al. 1981a, b) reported that phytotoxicity depends on a combination of several factors rather than only one cause.

Another parameter traditionally considered to determine the degree of maturity of a compost and to define its agronomic quality is the C:N ratio (Jiménez and Pérez García 1989). Several authors report that a C:N ratio below 20 is indicative of an acceptable maturity (Poincelot 1974; Golueke 1981), a ratio of 15 or even lower being preferable (Juste 1980). However, Hirai et al. (1983) stated that the C:N ratio cannot be used as an absolute indicator of compost maturity, since the values of this ratio for well-composted materials present a great variability, mainly due to the characteristics of the initial matrix used. Actually, the compost C:N ratio depends on the ratio between relative contents in carbon and nitrogen in the initial matrix, on the presence of organic fractions refractory to biodegradation in the initial substrate and on the dynamic evolution of the composting process. Our results are in agreement with these authors (Hirai et al. 1983). In fact, composts C3 and C4 have a C:N ratio higher than 15 and GI values higher than 60% (threshold of phytotoxicity). On the other hand, compost C5 with a C:N ratio lower than 15, presents a GI lower than 60%, suggesting potential phytotoxicity.

Table 3 shows the results of enzymatic activities. All five composts have significantly different protease and lipase activities. For the analysis of protease activity, although the sample for C5 did not pass the normality test at the 0.05 level for both the Lilliefors and Shapiro-Wilk tests, its presence in the ANOVA analysis does not affect the significance of the differences among the other composts, so that we only have to be cautious about significance of the differences between C5 and the other composts. Similar comments apply to C4 in the analysis of lipase activity, which C4 did not pass the Lilliefors test. Concerning the results of cellulase activity, composts C1 and C4 did not show significantly different activities, as also composts C2 and C3. The compost C1 did not pass the Shapiro-Wilk test in the analysis of cellulase activity. The absence of compost C1 in the analysis of cellulase activity renders the difference between composts C2 and C3 significant at α=0.05 level.

Table 3. Enzymatic activities of five different composts

Enzymatic activity	Compost C1	C2	C3	C4	C5
Protease (μmol tyrosine g^{-1} dw 2 h^{-1})	10.80a	5.56b	20.26c	15.15d	113.12e
Lipase (C10) (μmol *p*-nitrophenol g^{-1} dw h^{-1})	21.47a	60.63b	28.09c	41.81d	163.68e
Total cellulase (μmol glucose g^{-1} dw 16 h^{-1})	19.02a	40.72b	45.66b	19.80a	94.44c

dw, dry weight basis. Values presented are the mean of three determinations, each one with three replications for lipase activity and four replications for protease and cellulase activities. Means of each parameter designated by the same letter are not significantly different at α=0.05.

Compost C5 showed the highest values of lipase (C10), total cellulase and protease activities (Table 3). However, when results of total microbial populations are considered (Table 4), it is difficult to establish any relations between those and specific enzymatic activities.

For instance, compost C5, which showed the highest lipase (C10) activity, also presented a number of total actinomycetes slightly higher than those measured in the other composts. Even the highest level of total cellulase activity, which is associated with compost C5, does not parallel the highest counts of cellulolytic microorganisms (bacteria and fungi) occurring in the other composts tested. These results may be due to the dilution-plate methodology used to estimate the total microbial populations. This methodology only records the number of viable cells, spores and mycelial fragments that are capable of growing on the agarised media. Furthermore, slow-growing microorganisms and microorganisms unable to proliferate on agar are quickly masked by opportunistic or spore-forming microorganisms (Hardy and Sivasithamparam 1989).

Table 4. Total counts of different groups of microorganims in five different composts [CFU (colony forming units) g^{-1} dry weight]

Group of microorganisms	Compost C1	C2	C3	C4	C5
Aerobic bacteria (x 10^9)	0.05	154.00	93.40	46.30	54.50
Actinomycetes (x 10^5)	1.13	2020.00	2080.00	1350.00	2260.00
Filamentous eumycetes (x 10^6)	0.16	9.62	9.64	62.60	5.45
Cellulolytic bacteria (x 10^5)	0.01	7890.00	5725.00	11600.00	1.06
Cellulolytic fungi (x 10^5)	0.75	170.00	413.00	2720.00	170.00

As mentioned above, one of the main objectives of the present study was to find possible correlations between enzymatic activities and compost maturity, evaluated through the GI in the composts tested. The correlation analysis was done with a total of n =15 data points, corresponding to each of three compost extracts for each of the five composts (Table 5). The highest correlations were found among the protease, total cellulase and lipase (C10) activities (Table 5).

Table 5. Correlation matrix between enzymatic activities and germination index (GI)

	GI	Lipases (C10)	Proteases	Total cellulases
GI	1			
Lipases (C10)	-0.5683	1		
Proteases	-0.7514	0.9328	1	
Total cellulases	-0.4594	0.9070	0.9173	1

The correlation coefficients with absolute value >0.5140 are significant at α=0.05; with absolute value >0.6412 are significant at α=0.01; with absolute value >0.7604 are significant at α=0.001 and with absolute value >0.8369 are significant at α=0.0001.

Cunha Queda (1999) characterised the enzymatic profiles during several composting trials. Evidence of high levels of protease, lipase (C10) and total cellulase activities throughout the active phase, of the process, with total cellulase activity still intense even after this phase was reported (Cunha Queda 1999). According to these enzymatic profiles, ratios of certain enzymatic activities were calculated (Table 6). Since when using the values of GI in both the Shapiro-Wilk and Lilliefors normality tests we are led to not reject the null hypothesis of normality for both tests at the 0.05 level, and since the F-test used for the correlations only needs the normality of one of the variables involved in the test (Kshirsagar 1972), all the correlation tests involving the variable GI are legitimate. The ratios lipase:total celullase activity and protease:total cellulase activity also passed the normality test but the ratio lipase:protease activity and all the three individual total cellulase, protease and lipase activities did not pass any of the normality tests. This may be taken as another point in favour of the use of the activity ratios rather than the enzymatic activities by themselves, although the enzymatic activities did pass the normality test under a different rearrangement of their values, when the ANOVA analyses were carried out, and this is enough to enable us to carry out the correlation test legitimately. Furthermore, the F-test is robust to non-normality when carrying out the correlation test (Seber 1997).

Table 6. Correlation coefficients between enzymatic activity ratios and germination index (GI)

Enzymatic activities ratio	r	Significance level
lipase/protease	0.7755	0.001
protease/cellulase	-0.9678	0.0001
lipase/cellulase	-0.3911	ns

ns, not significant.

Results show that protease:total cellulase activity ratio had the highest significant correlation with the germination index (Table 6). The estimated regression equation which corresponds to this correlation coefficient is:

$$GI = 126.7047 - 71.0816 \times (\text{protease/cellulase activity}),$$

$$r = -0.9678 \text{ and } r^2 = 0.9366 .$$

According to this regression model for GI=60, the estimated protease:cellulase activity ratio is 0.9384. This indicates that for values of the protease:cellulase activity ratio lower than 0.9384 GI values can be expected to be higher than 60% (i.e. absence of phytotoxicity). Thus, the highly significant correlation between protease:cellulase activity ratio and the germination index might be used as an indicator of compost maturity.

Conclusions

This research has shown that the five composts considered, although falling in the same class of stability, revealed different results for maturity evaluation. These results show that stability and maturity are distinct properties, both of prominent importance in the evaluation of compost quality along with several physicochemical parameters. Moreover, evidence has been gained that C:N ratio can not be used as an absolute indicator of compost maturity. Finally, data concerning enzymatic activities have showed that protease:cellulase activity ratio was significantly correlated to the germination index. However, further studies with a larger number of composts either from different starting matrices or obtained through different stabilisation strategies are needed in order to confirm the correlation found in this study.

Acknowledgements. We wish to thank referees for their helpful and valuable revision of this manuscript.

References

Adani F, Genevini PL, Gasperi F, Zorzi G (1997) Organic matter evolution index (OMEI) as a measure of composting efficiency. Compost Sci Util 5(2):53-62

Anid PJ (1986) Evaluating maturity and metal transfer of MSW compost. BioCycle 27:46

Chen Y, Inbar Y (1993) Chemical and spectroscopical analyses of organic matter transformations during composting in relation to compost maturity. In: Hoitink HAJ, Keener HM (eds) Science and engineering of composting: design, environmental, microbiological and utilization aspects. Renaissance Publ, Worthington, pp 550-600

Commission of the European Communities (1978) Standardisation of analytical methods for manure, soils, plants and water. Workshop organised by Commission of the European Communities, Gent, Belgium

Cunha Queda AC (1999) Dinâmica do azoto durante a compostagem de materiais biológicos putrescíveis (Nitrogen dynamics during putrescible biomass composting). PhD Dissertation, Instituto Superior de Agronomia, Universidade Técnica de Lisboa, Lisbon, Portugal

Cunha Queda AC, Almeida Duarte E, Campos L, Bruno de Sousa R (2000) Composting of horse manure enriched with paper board residue: study of physico-chemical and biochemical parameters during the composting process. In: Warman PR, Taylor BR (eds) Proceedings of the International Composting Symposium (ICS'99), Halifax (Canada), vol 1, pp 110-123

Golueke CG (1981) Principles of biological resource recovery. BioCycle 22(4):36-40

Grebus ME, Watson ME, Hoitink HAJ (1994) Biological, chemical and physical properties of composted yard trimmings as indicators of maturity and plant disease suppression. Compost Sci Util 2(1):57-71

Harada Y, Inoko A (1980) Relationship between cation-exchange capacity and degree of maturity of city refuse composts. Soil Sci Plant Nutr 26:353-362

Harada Y, Inoko A, Tadaki M, Izawa T (1981) Maturing process of city refuse composts during piling. Soil Sci Plant Nutr 27:357-364

Hardy GESJ, Sivasithamparam K (1989) Microbial, chemical and physical changes during composting of a eucalyptus (*Eucalyptus calophylla* and *Eucalyptus diversicolor*) bark mix. Biol Fertil Soils 8:260-270

Herrmann RF, Shann JR (1993) Enzyme activities as indicators of municipal solid waste compost maturity. Compost Sci Util 1(4):54-63

Hirai M, Chanyasak V, Kubota H (1983) A standard measurement for compost maturity. BioCycle 24:54-56

Hope CFA, Burns RG (1987) Activity, origins and location of cellulase in a silt loam soil. Biol Fertil Soils 5:164-170

Jiménez EI, Pérez García V (1989) Evaluation of city refuse compost maturity: a review. Biol Wastes 27:115-142

Jodice R (1989) Parametri chimici e biologici per la valutazione della qualità del compost. In: Proceedings of the COMPOST Production and Use International Symposium, S.Michelle all'Adige, 20-23 June, pp 363-384

Jourdan B (1988) Determination of the degree of decomposition for waste and waste/sludge derived compost. Stuttg Beri Abfallwirtsch, vol 30, 180pp

Juste C (1980) Avantages et inconvenients de l'utilisation des composts d'ordures ménagères comme amendement organique des sols ou supports de cultur. In: International Conference on Compost, 22-26 January, Madrid, Spain, Min. Obras Públicas

Kirchmann H, Widén P (1994) Fatty acid formation during composting of separately collected organic household wastes. Compost Sci Util 2(1):17-19

Kshirsagar AM (1972) Multivariate analysis. Marcel Dekker, New York

Ladd JN, Butler JHA (1972) Short-term assays of soil proteolytic enzyme activities using proteins and dipeptide derivatives as substrates. Soil Biol Biochem 4:19-30

Laga-Merkblatt M10 (1995) Qualitätskriterien und Anwendungsempfehlungen für Kompost. Müll-Handbuch Lfg. 5/95, Kennziffer 6856, Erich Schmidt, Berlin

Pera A, Vallini G, Frassinetti S, Cecchi F (1991) Co-composting for managing effluent from thermophilic anaerobic digestion of municipal solid waste. Environ Technol 12:1137-1145

Pochon J, Tardieux P (1962) Techniques d'analyse en microbiologie du sol. La Tourelle, Saint Mande

Poincelot RP (1974) A scientific examination of the principles and practice of composting. Compost Sci 15:24-31

Saharinen MH, Vuorinen AH, Hostikka M (1996) Effective cation exchange capacity of manure-straw compost of varying stages determined by the saturation-displacement method. Commun Soil Sci Plant Anal 27:2917-2923

Seber GAF (1997) Linear regression analysis. Wiley, New York

Shiralipour A, Mcconnell DB, Smith WH (1997) Phytotoxic effects of a short-chain fatty acid on seed germination and root length of *Cucumis sativus* cv. Poinset. Compost Sci Util 5(2):47-52

Sugahara K, Inoko A (1981) Composition analysis of humus and characterisation of humic acid obtained from city refuse compost. Soil Sci Plant Nutr 27(2):213-224

Willson GB, Dalmat D (1986) Measuring compost stability. BioCycle 27:34-37

Wong MH, Cheung YH, Cheung CL (1983) The effects of ammonia and ethylene oxide in animal manure and sewage sludge on the seed germination and root elongation of *Brassica parachinensis*. Environ Pollut 30:109-123

Zucconi F, De Bertoldi M (1987) Compost specifications for the production and characterization of compost from municipal solid waste. In: De Bertoldi M, Ferranti MP, L'Hermite P, Zucconi F (eds) Compost: production, quality and use. Elsevier, New York, pp 30-50

Zucconi F, Monaco A, Forte M (1985) Phytotoxins during the stabilization of organic matter. In: Gasser JKR (ed) Composting of agricultural and other wastes. Elsevier, New York, pp 73-88

Zucconi F, Forte M, Monaco A, De Bertoldi M (1981a) Biological evaluation of compost maturity. BioCycle 22(4):27-29

Zucconi F, Pera A, Forte M, De Bertoldi M (1981b) Evaluation toxicity of immature compost maturity. BioCycle 22(2):54-57

Monitoring of a Composting Process: Thermal Stability of Raw Materials and Products

M.T. Dell'Abate and F. Tittarelli[1]

Abstract. In the following chapter, thermal methods of thermogravimetry (TG) and differential scanning calorimetry (DSC) were utilised on agro-industrial raw materials and on samples collected during a 5-month composting period. The main objective was to investigate the energetic status of different raw materials with respect to that of compost samples at different stages of the composting process, in order to obtain a process monitoring. Two piles, based on wastes from citrus industrial processing, were produced according to the presence and absence of sludge in the initial mixture of matrices. Materials thermal characterisation showed that sludge organic matter was the most stabilised among the used starting materials, but its quantitative contribution to the final product was minor. DSC measurements on samples taken at different times of transformation demondtrated at a macroscopic level the evolution of organic materials towards more energetic organic fractions. TG data allowed the calculation of the thermostability index R1, able to quantify the relative amount of the thermally more stable organic matter fraction with respect to the less stable. Finally, data showed that organic matter stabilisation was accompanied by an increased water retention.

Introduction

Chemical composition and structure of raw materials used in the composting process are important factors in determining the potential degradation of organic wastes. The amount of lignocellulosic substrates and nitrogen are fundamental in determining the optimal starting conditions for microbial activity, when process parameters such as temperature, oxygen and water availability are not limiting factors to microbial growth. On a gross scale, the carbon to nitrogen ratio of raw materials is in practice considered one of the key parameters, even if no information on fibre components can be deduced. Furthermore, the wet chemical methods for determining the composition of lignocelluloses are complex and time-consuming (Sharma 1995), and few direct methods on whole samples are available so far. In this context, thermal methods of thermogravimetry (TG) and differential scanning calorimetry (DSC) provided estimation of lignin and fibre components (Sharma 1990, 1991, 1995; Kaloustian et al. 1997). On the basis of these TG and DSC applications, thermal techniques have also been successfully used for

[1]Istituto Sperimentale per la Nutrizione delle Piante, Via della Navicella 2, 00184 Roma, Italy; e-mail: mt.dellabate@isnp.it

H. Insam, N. Riddech, S. Klammer (Eds.)
Microbiology of Composting

maturity assessment in compost, as a comparative method in evaluating organic matter (OM) transformation of lignocellulose-based materials during the composting process (Blanco and Almendros 1994, 1997; Dell'Abate et al. 1998, 2000). While the most widespread chemical methods for characterisation of compost organic matter require extraction and fractionation procedures of C compound classes, the application of TG and DSC on whole compost samples allows the analysis of thermal stability in organic matter within the mineral matrix to which it is closely associated.

In the present study such a thermoanalytical approach to compost characterisation was used. It was aimed to investigate the energetic status of different raw materials with respect to that of compost samples at different stages of the composting process, in order to obtain a process monitoring and to verify the possible influence of the starting organic substrates on the compost thermal patterns. Samples from two different piles, obtained using citrus industrial processing wastes as input materials, were studied in the framework of a wider investigation carried out on the possible utilisation of citrus industrial process wastes for compost production. In particular, the aspects related to C turnover and humification were described in Tittarelli et al. (2002) and those regarding the microbial dynamics in Pinzari et al. (2002).

Materials and Methods

Raw Materials

The main typologies of wastes produced by the citrus industrial process are *pastazzo* and sludge. *Pastazzo,* a mixture of citrus pulp and skins, represents almost 60% of the initial fruit weight, while sludge, the final product of fruit purification treatment and factory washing waters, is produced in smaller quantity. Plant residues, mainly pruning materials from orange trees, were used as bulking agents in order to increase pile porosity and to adjust C/N ratio and moisture in starting mixtures.

Composting process and sampling

Two different composting processes (called Cp and Cps) were studied. Compost Cp was obtained by mixing pastazzo and plant residues, while compost Cps was prepared adding sludge, as input material, to those utilised for compost Cp. According to their physicochemical characteristics, raw materials were mixed in the following percentages (w/w):

Cp = pastazzo 60% + plant residues 40%;

Cps = pastazzo 40% + plant residues 40% + sludge 20%.

Initial pile dimensions were the following: 2 m wide, 6 m long and about 1.7–1.8 m high. Temperature and moisture were monitored in both piles in order to maintain optimal condition for microbial activity (Fig. 1). Piles were mechanically turned (two to three times a week during the first 4 weeks, once a week subsequently) to promote aerobic organic matter decomposition. Moisture level was maintained constant at 50–60% (w/w) by addition of the proper amount of water: moisture content determination was carried out weekly. The complete pattern of process management is schematically reported in Fig. 1. Compost samples from both piles were taken at 0 (T0), 29 (T1), 67 (T2), 89 (T3), 130 (T4) and 165 (T5) days from the beginning of the composting process. The temperature course during the two processes was similar up to 89 days, thereafter mean values were slightly lower in Cps than in Cp.

Each sample was made up of six subsamples, taken after turning the piles, then mixed until homogenisation. Samples were oven-dried at 50 °C, ground and sieved at 1 mm and finally stored for subsequent analysis.

Analytical Methods

Total organic carbon (C_{org}), total nitrogen (N) and ash contents were determined. For total organic carbon (C_{org}) determination, mineralization of the organic matter to CO_2 was carried out as follows: 100 mg of compost was treated with 20 ml of 0.33 M $K_2Cr_2O_7$ and 26 ml of 96% H_2SO_4 for 10 min at 160 °C (Springer and Klee 1954). Organic carbon content was determined by backtitration with a solution of 0.2 N $FeSO_4$.

Total nitrogen content (N) was determined by dry combustion (LECO FP 228), and the C/N ratio was calculated. Ash amount was deduced from thermogravimetric measurements, as the sample weight at the temperature corresponding to the end of organic matter combustion reactions. These temperature values ranged from about 680 °C for samples at time 0 to about 580°C for final composts.

DSC and TG experiments were carried out simultaneously with a Netzsch Simultaneous Analyser STA 409 (Netzsch-Geratebau, Selb/Bayern, Germany) equipped with a TG/DSC sample carrier supporting a type S thermocouple of PtRh10-Pt. This device, named heat-flux DSC, is capable of calorimetric measurements since it meets the requirement that the quantity M = area under peak/(latent heat of fusion per gram x sample mass) to be a constant for transformations in different standard materials. In particular, in a heat-flux DSC, the temperature difference between sample and reference material is recorded as a direct measure of the difference in the heat-flow rates to the sample; in TG the weight change of a sample is measured during the thermal program. The first derivative of TG trace (DTG) represents the weight loss rate (expressed as % min^{-1}): it was calculated in order to better distinguish among subsequent decomposition steps, which were evaluated on the basis of the DTG characteristic parameters, such as both the onset and the end peak temperatures. Samples were analysed without any pre-treatment, except for manual grinding in agate mortar. The following conditions were employed: heating rate of 10 °C min^{-1} from 20– 900 °C, static air atmosphere, alumina crucible, calcined kaolinite as reference, sample weight about 20–24 mg. The thermobalance was calibrated for buoyancy effects in order to obtain quantitative estimation of weight changes. Heat production in the heat-flux DSC was calibrated under the same conditions by using a sapphire standard and subtracting a baseline obtained by an additional run for the empty crucibles. The Netzsch applied software SW/cp/311.01 was used for data processing. Duplicate thermal runs were carried out in order to verify the measurement reproducibility.

Results and Discussion

The main process parameters, such as pile temperature and moisture, together with the indication of pile-turning operations are reported in Fig. 1.

Raw Materials

In Table 1 results of raw materials chemical characterisation and proximate analysis carried out by thermogravimetry are reported. The DSC and TG profiles of pastazzo, sludge and plant residues showed the presence of different main thermally active organic fractions, with increasing thermal stability (Fig. 2a, b, c). As general features, in all raw materials and composts, three main temperature ranges were recognised: the first one up to about 180 °C, mainly characterised by endothermic reactions of dehydration having peak temperature at about 80 °C; the second one up to 350–400 °C, in which progressively pectine, hemicellulose and cellulose pyrolysed together with microbial cell walls (Sharma 1995; Kaloustian et al. 1997); the third one at higher temperature (up to 680 °C for samples at time 0 and about 580 °C for final composts) is characteristic breakdown of more stable component, such as lignin. The char left at the end of organic material combustion was constituted mainly by ash with small amounts of pyrolysis by-products, which volatilised with increasing temperature. Finally, endothermic decomposition reaction of carbonates occurred in the range 750–830 °C. The thermal breakdown of pastazzo organic matter (OM) occurred in the temperature range from 134 °C to about 670 °C (Fig. 2a): it involved two labile pools and a third one more stable. From the DTG curve a poorly resolved two-step degradation process up to 400 °C was evident, maximum weight loss rates being 4.2 and 4.8% min^{-1} at 260 and 300°C, respectively (DTG peaks). About 65% of pastazzo OM was degraded in such reactions, being the overall weight loss associated to OM combustion 90% (w/w) of the initial moisture-free sample.

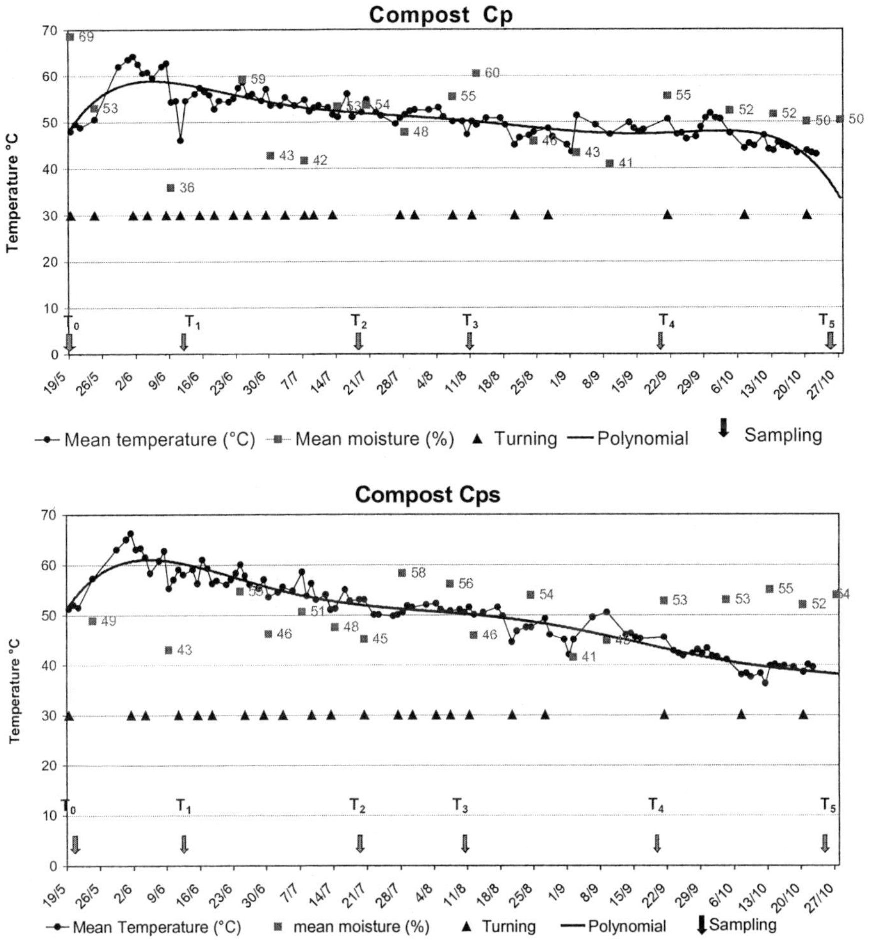

Fig. 1. Pile temperature and moisture during composting processes Cp and Cps

The peak temperatures on DSC trace were at 317.7, 332.4 and 463.5 °C, respectively.

Sludge showed a more complex thermal pattern (Fig. 2b), to which more fractions, having different thermal stability, contributed in the temperature range 187 to 644 °C: total weight loss was 51.3% on the basis of a moisture-free sample. In particular, the most thermal labile fraction of sludge OM was about 51% of OM. It was degraded in the temperature range of 187.3 to 385.7 °C, reaching the maximum weight loss rate of 2.2% min^{-1} at 270 °C. The more stable organic fraction pyrolised in two consecutive different steps, occurring, respectively, in temperature ranges from 385.7 to 469.9 °C and to 644.2 °C.

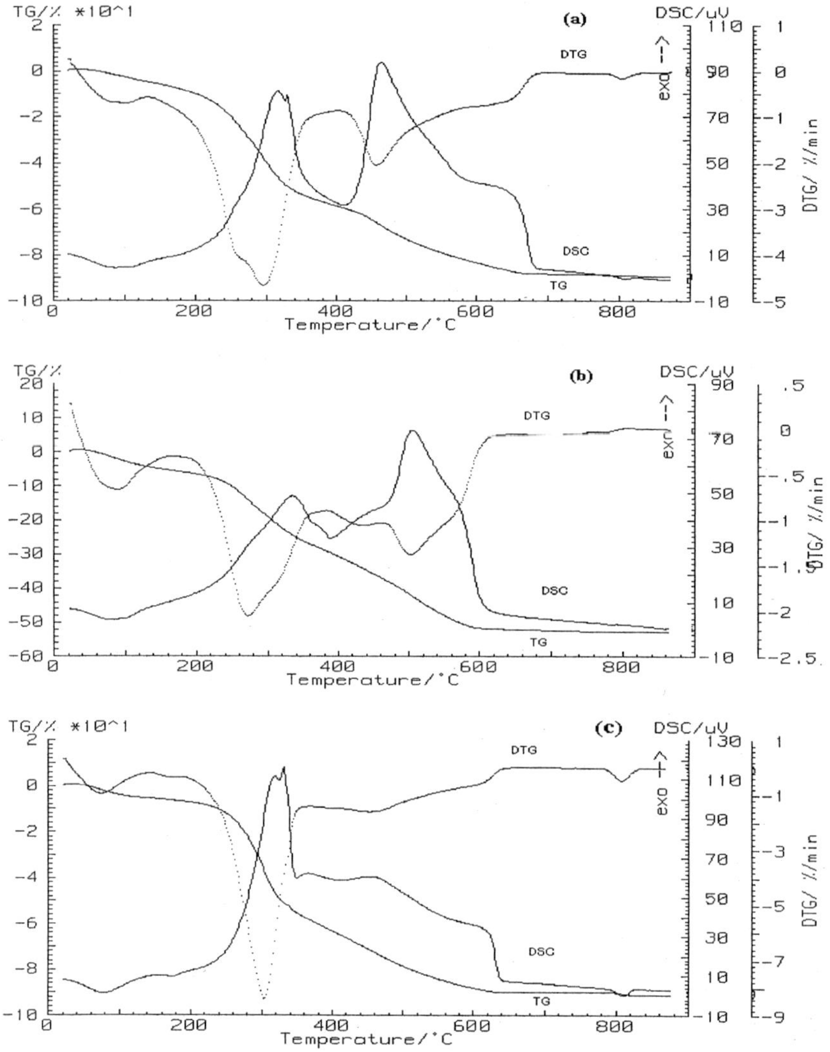

Fig. 2 a-c. DSC, TG and DTG traces of input materials: **a** pastazzo, **b** sludge and **c** plant residues

Maximum heat flows on DSC trace were recorded as peaks at 334.1 and 505.3 °C, respectively. The thermal behaviour of plant residues (Fig. 2c) showed the cellulose contribution, with an intense exothermic reaction at about 318 °C, associated to a weight loss on the TG curve in a narrow temperature range (Aggarwal et al. 1997). The rate of this weight loss reached its maximum value of 9% min^{-1} at about 300 °C (DTG peak). Thereafter, pyrolysis reactions without significant heat evolution occurred.

On the basis of both TG data discussed above for input materials (Table 1) and their percentages in the initial piles, it was possible to deduce the theoretical contribution of each material to the thermal behaviour of sample T0. In particular, since sludge contained a high level of ash at 850 °C, for the Cps process only 12% of the initial OM derived from sludge, whose contribution to the thermally more labile OM was 6%. The OM derived from pastazzo and plant residues was 43 and 45% of the initial OM, respectively, while in the Cp process their contribution to sample T0 was, respectively, 62 and 38%.

In evaluating thermal data related to samples from T0 to T5 for both Cp and Cps processes, two different levels were highlighted: the first was a qualitative approach, in which it was possible to demonstrate at a macroscopic level, by DSC measurements, transformations occurring in samples at different stages of composting; the second consisted in a quantitative estimation, deduced from TG data, of the size of the different pools of organic matter involved in the thermal degradation reactions.

Table 1. Total organic carbon content (C_{org}), C to N ratio and proximate analysis by TG: weight losses (% of total sample) and temperature ranges (°C) corresponding to the main exothermic reactions, and content (% of total sample) of sample residue at 850 °C

Raw materials	C_{org} (g kg^{-1})	C/N	Weight loss (%) ΔT (°C)	Weight loss (%) ΔT (°C)	Weight loss (%) ΔT (°C)	Residue (%) at 850 °C
Pastazzo	513	33.5	24.8 134.2-272	33.1 272-403	31.6 403-660	9.2
Sludge	305	8.6	23.6 187.3-385.7		22.89 458.8-644.2	45.0
Plant residues	510	51.5	49.66 189.3-353		33.9 353-663	6.1

Composting Process Cp

In Fig. 3 (a-f) thermograms of samples taken during the Cp process are reported. Mixed input materials before composting (T0) showed a thermal pattern that was characterised, after an endothermic reaction of dehydratation at low temperature, by an intense exothermic peak at 312.8 °C followed by a minor one at 329.7 °C, referable to cellulose and hemicellulose components of pastazzo and plant residues. Thereafter, undifferentiated pyrolysis reactions occurred in the temperature range 356–683 °C. Sample T1, taken after 29 days processing, showed some differences in its thermal pattern. The kinetic of thermal oxidation revealed on the DSC curve one exothermic peak at 301.1 °C with a shoulder at 333.4 °C, and the initial formation of a more thermally stable organic fraction, having a maximum of energy release at 465.9 °C. After 67 days from the beginning of composting, sample T2 was considerably transformed: its organic matter was characterised by two main fractions having different thermal stability, which released energy in two well-resolved peaks on the DSC curve, at 299.4 and

462.5 °C respectively. However, it was possible to differentiate a weak contribution of hemicellulose at about 330 °C. General features of sample T3, taken 22 days later than sample T2, did not differ from those of sample T2. On the contrary, a new change in thermal behaviour occurred during the curing phase in sample T4, 130 days from the beginning. First of all, it was possible to observe a significant increase of adsorbed water content, which demonstrates a deep change of material chemical-physical characteristics; moreover, the intensity of the second exothermic peak also increased showing a more symmetrical shape of the DSC peak. The residual pastazzo contribution was not identifiable. The extension of the curing phase up to 165 days from the beginning (sample T5) did not appreciably change the compost thermal behaviour.

Changes in thermal behaviour observed on DSC curves were associated with quantitative parameters deduced from thermogravimetric data and registered on the TG curve (Table 2). The maximum weight loss rate of the most thermal labile organic fraction, registered as a peak on the DTG curve, decreased from 8.2% min^{-1} for samples T0 and T1 to 6.8% min^{-1} for T2 and finally reached values about 4.4–4.9% min^{-1} for samples T4 and T5.

The weight loss observed in the overall temperature range for organic material thermal oxidation represents an estimation of compost organic matter content. At temperatures higher than 600 °C a little weight loss due to volatilisation of combustion byproducts occurred. Finally, a reaction of endothermic carbonate decomposition was observed at about 800 °C. As expected, the weight loss due to OM combustion gradually decreased in samples at different stages of composting, confirming the occurrence of the process of organic matter degradation up to 130 days from the beginning (sample T4). In the last part of the curing phase, a possible macromolecular rearrangement could explain the apparent slight increase of organic matter content in sample T5. This could be due to mobilisation of small amounts of carbon blocked as carbonates, formed during the composting process (Fig. 5). Instead, the amounts of thermal degradation byproducts reached a plateau value for samples T4 and T5.

The increased thermal stability of compost organic matter during the process was shown by the ratio between the weight losses associated with the second and the first exothermic reactions (R1). This thermostability index represents the relative amount of the thermally more stable organic matter fraction with respect to the less stable one, without regard to either sample moisture level or ash content (Dell'Abate et al. 1998, 2000). It increased from 0.68 to 0.93 for sample T3, while samples T4 and T5 showed a slight decrease of its values.

Two possible explanations can be hypothesised, the first being a possible macromolecular rearrangement, as suggested by the increased amount of absorbed water. The second is that the higher water content itself could have directly influenced the observed thermal pattern. On the other hand, literature reported evidence for the relationship between soil organic matter macromolecular state and water level content (Schaumann and Antelmann 2000); moreover, different forms of water have been identified and quantified in peat subjected to dehydration and re-hydration processes (McBrierty et al. 1996).

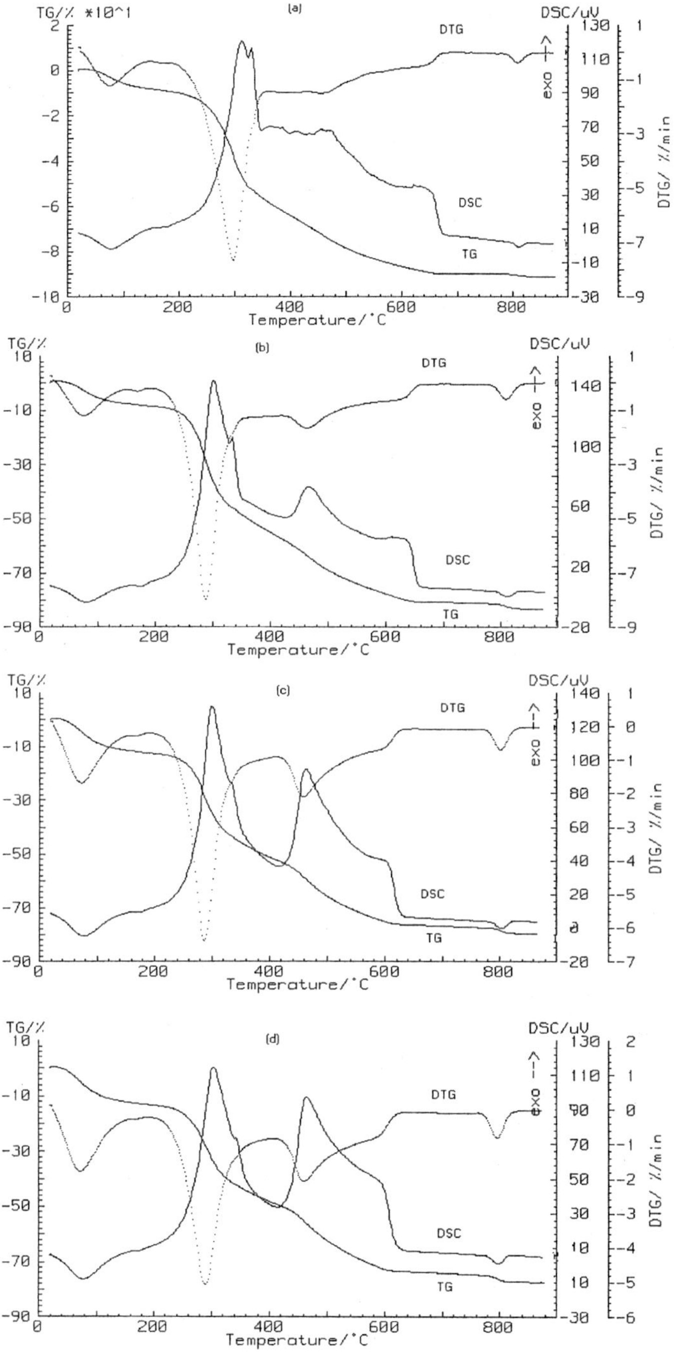
(a)
TG/% *10^1
DSC/uV
DTG
exo --->
DSC
TG
DTG/ %/min
Temperature/°C
(b)
TG/%
DSC/uV
DTG
exo --->
DSC
TG
DTG/ %/min
Temperature/°C
(c)
TG/%
DSC/uV
DTG
exo --->
DSC
TG
DTG/ %/min
Temperature/°C
(d)
TG/%
DSC/uV
DTG
exo --->
DSC
TG
DTG/ %/min
Temperature/°C

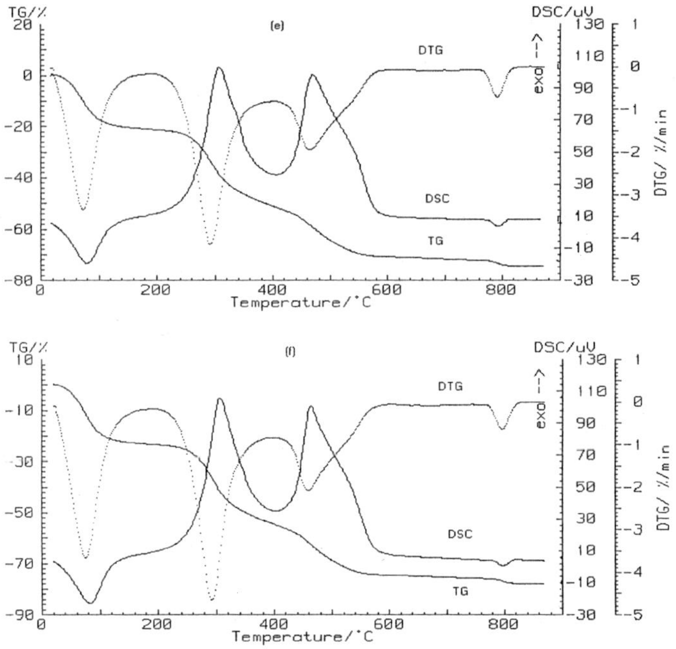

Fig. 3 a-f. DSC, TG and DTG traces of samples taken from the Cp composting process: **a** T0, **b** T1, **c** T2, **d** T3, **e** T4 and **f** T5

Composting Process Cps

The material obtained after mixing pastazzo, sludge and plant residues showed, before composting (T0), the thermal behaviour pictured in Fig. 4a. In addition to the strong exotherm at about 300 °C, due to cellulose combustion, on the DSC trace a more thermostable organic fraction, having a maximum of energy release at 474 °C, was evident, to which sludge OM possibly contributed. As in the Cp process, a shoulder at 340 °C was present. The sample resulting after 29 days of composting revealed that an intense degradation occurred: the shoulder (335 °C) of the main peak of cellulose combustion at 300 °C was still present. The thermal pattern at temperature higher than 400 °C was less complex than in Cp and the temperature level for complete degradation was reduced to 600 °C. Samples during the composting process showed a progressively more intense second exothermic reaction, due to both the depletion of labile organic fractions and the increased level of organic matter stability by means of humification-like processes. This finding was in line with results obtained from both soil organic matter and peats (Grisi et al. 1998).

From a quantitative point of view, a steep decline of organic matter content from T0 to T1 sample was observed (Table 2), thereafter the decline was slower up to T3. The organic matter composition, in terms of the relative presence of the more thermostable organic fraction, changed as well. The ratio R1 increased from 0.68 for sample at T0 to 0.84 after 89 days of composting (T3). The last part of the curing phase (samples T4 and T5) induced a further transformation: the DSC traces showed an intense endotherm at 82 °C, related to water release, and two sharp, well resolved exotherms having temperature maxima at 304.8 and 484.8 °C. The symmetrical shapes of these DSC peaks denoted the homogeneity of the reacting organic materials. The total organic matter amount, deduced from the thermogravimetric weight losses, did not change during this phase, indicating that organic matter biological degradation reached a plateau. On the contrary, the partitioning between the thermally labile and the more stable fractions changed; indeed the ratio R1 decreased to 0.73 (T4) and 0.68 (T5). Again, as in the Cp process, a possible macromolecular rearrangement or a direct influence of water content on thermal pattern can be hypothesised. The maximum weight loss rate of the more thermal labile organic fraction, registered as a peak on the DTG curve, instead, decreased progressively from 6.4- 6.2% min^{-1} for samples T0 and T1 to 4.2% min^{-1} for T2 and finally reached values about 3.5- 3.6% min^{-1} for samples T4 and T5.

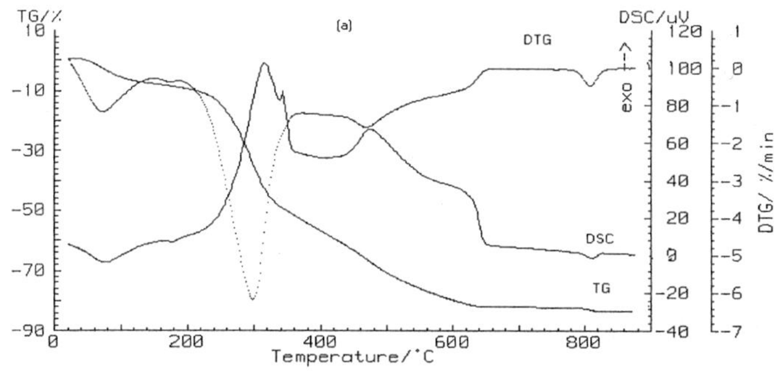

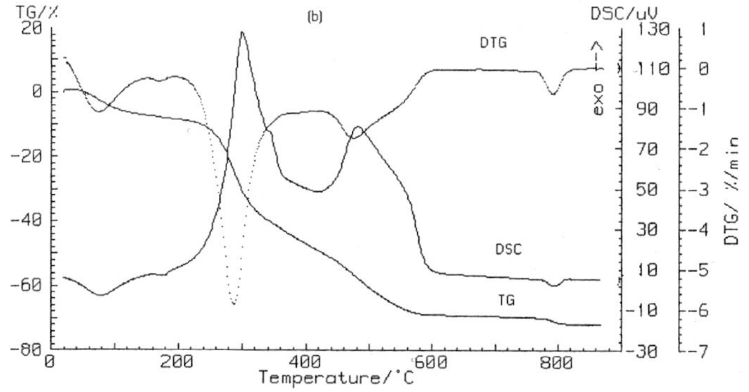

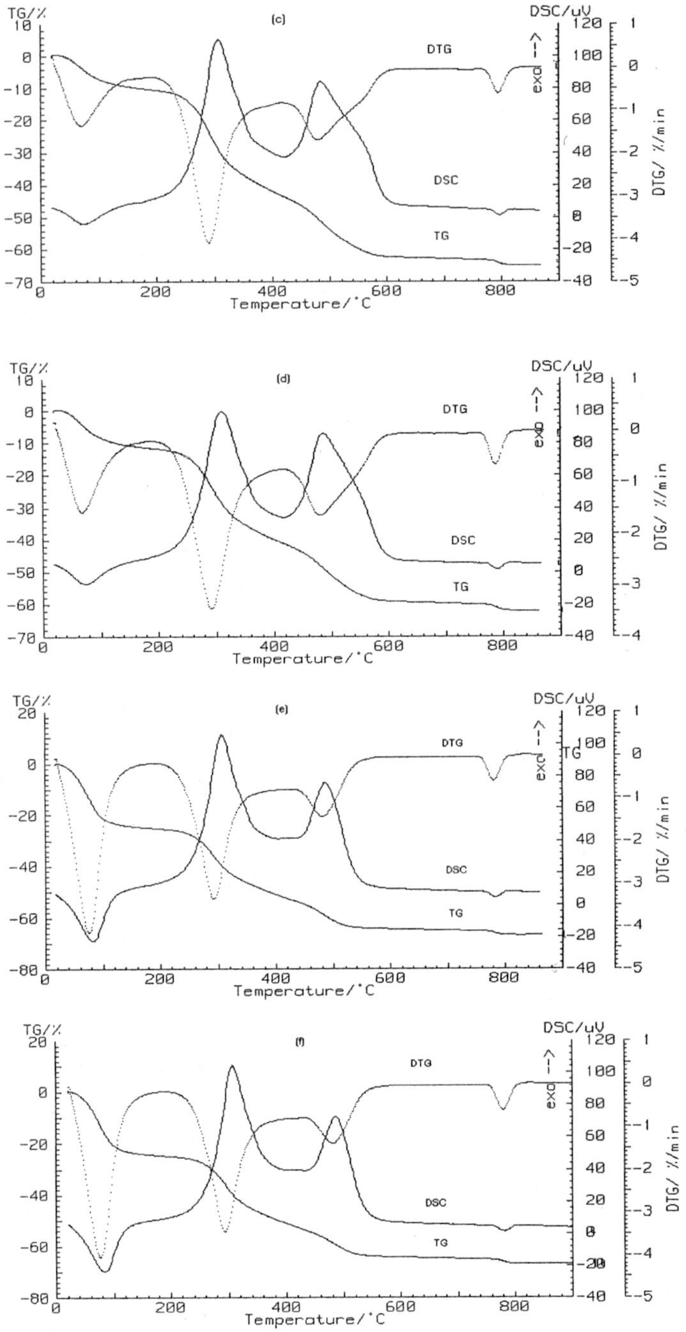

Fig. 4 a-f. DSC, TG and DTG traces of samples taken from the Cps composting process: **a** T0, **b** T1, **c** T2, **d** T3, **e** T4 and **f** T5

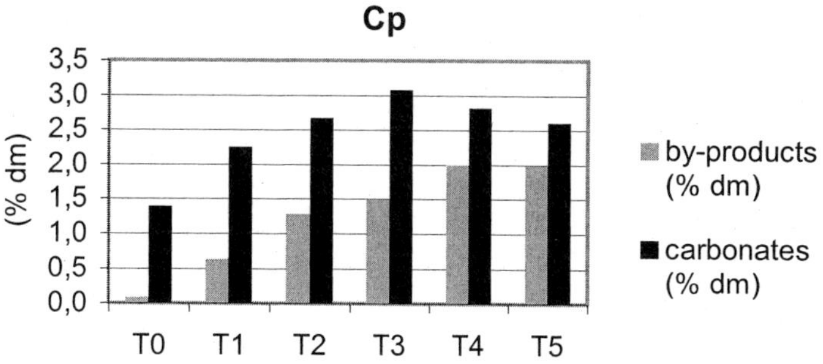

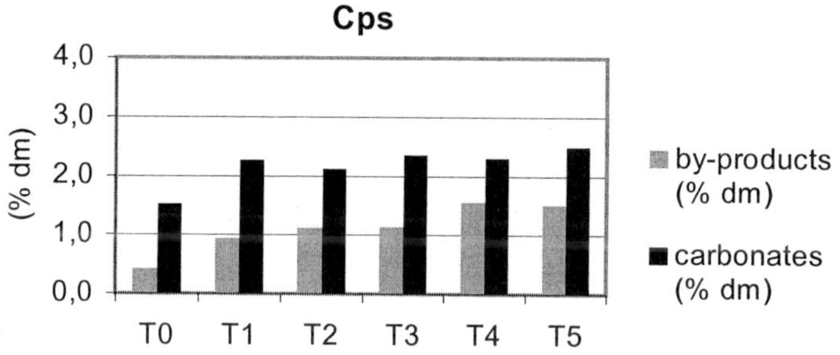

Fig. 5. Content of combustion by-products (% of dry matter sample) volatilised in the temperature range from the combustion end point and the beginning of carbonate decomposition reactions, and weight losses due to carbonate decomposition

Table 2. Main thermogravimetric weight losses (% of total sample) and temperature ranges in which they occurred. R1 refers to the ratio between mass losses associated with the second and the first organic matter (OM) exothermic reactions. 1st OM DTG peak is the temperature in which the rate of weight loss associated to the first exothermic reaction is at its maximum.

Process Cp	T0	T1	T2	T3	T4	T5
H_20 weight loss (%)	9.9	9.5	13.2	13.9	22.1	23.8
ΔT (°C)	26-146	26-152	26-152	20-156	24-163	25-166
Total OM weight loss (%)	81.0	72.8	64.5	60.8	50.2	51.9
ΔT (°C)	186-680	196-667	196-636	195-636	199-593	200-606
R1	0.68	0.78	0.89	0.93	0.90	0.88
1st OM DTG peak (°C)	299.2	289.6	289.1	290.5	292.0	292.2
Rate of weight loss (% min^{-1})	8.2	8.6	6.8	5.4	4.4	4.9
Process Cps	T0	T1	T2	T3	T4	T5
H_20 weight loss (%)	9.6	9.1	11.3	12.6	26.3	26.5
ΔT (°C)	26-151	28-152	26-159	25-162	25-170	25-169
Total OM weight loss (%)	73.6	61.2	52.7	48.0	39.4	39.4
ΔT (°C)	190-659	194-608	190-615	194-608	198-578	198-566
R1	0.68	0.72	0.78	0.84	0.73	0.68
1st OM DTG peak (°C)	301.0	288.5	290.9	291.8	292.9	294.5
Rate of weight loss (% min^{-1})	6.4	6.2	4.2	3.6	3.5	3.6

Conclusions

Results of thermal investigations carried out on raw materials and products of the composting process showed significant differences which gave a picture of both the initial and the final energetic status of organic materials. Material thermal breakdown showed the presence of organic fractions having different thermal stability. Among input materials, sludge OM showed the highest thermal stability, but from a quantitative point of view its contribution was minor. During the composting process both the depletion of labile organic fractions and the increased level of organic matter stability due to humification-like processes gave rise to characteristic thermal patterns. Several parameters, such as peak temperatures on DSC traces, weight losses due to different organic fractions on TG curves and maximum rate of weight loss calculated from the first derivative of TG (DTG peak), follow such changes in the chemical-structural composition of the organic matter. DSC measurements demonstrated at a macroscopic level the evolution of organic materials towards more energetic organic fractions: in particular, the thermally labile fractions of pastazzo and plant residues were detected on the DSC trace in the early phases of the composting process and progressively disappeared after 67 days in Cps and 89 days in Cp, while a progressively more intense exotherm at higher temperature range was evident. During the same period, the maximum rate of weight loss (DTG peak) related to cellulose combustion decreased progressively. The previously identified thermostability index R1 (Dell'Abate et al. 1998), which indicates the relative amount of the thermally more stable fraction of organic matter with respect to the less stable one, increased progressively during 89 days for both Cp and Cps processes. Thereafter, during the last period of the curing phase it decreased, especially in Cps compost. A possible macromolecular rearrangement was hypothesised, since the microbial degradation reached a plateau at this stage, as shown by Tittarelli et al. (this Vol.) on the same samples. A further indication supporting this hypothesis was represented by the occurrence of a large amount of absorbed water in samples T4 (130 days) and T5 (165 days) from both processes. This last aspect needs to be better investigated, since the water-holding capacity represents an important chemical-physical characteristic of composts. In particular, both a possible direct influence of the compost water content on the thermal pattern and the occurrence of a macromolecular state for compost organic matter have to be demonstrated.

In conclusion, DSC and TG measurements allowed deduction of a wide spectra of thermal parameters, which can be used for organic matter maturation monitoring during a composting process. In particular, the application of thermal methods on whole samples permits organic matter characterisation within the mineral matrix to which it is closely associated and to estimating its potential energetic status by means of kinetics of thermally induced organic matter oxidation.

Acknowledgement. Special thanks are due to Dr. Francesco Intrigliolo and his collaborators (Istituto Sperimentale per l'Agrumicoltura, Acireale, Catania) for matrice collection, pile preparation and technical assistance. Part of this research was funded by Italian Ministry of Agriculture.

References

Aggarwal P, Dollimore D, Heon K (1997) Comparative thermal analysis study of two biopolymers, starch and cellulose. Thermochim Acta 50:7-17

Blanco MJ, Almendros G (1994) Maturity assessment of wheat straw composts by thermogravimetric analysis. J Agric Food Chem 42:2454-2459

Blanco MJ, Almendros G (1997) Chemical transformation, phytotoxicity and nutrient availability in progressive composting stages of wheat straw. Plant Soil 196:15-25

Dell'Abate MT, Canali S, Trinchera A, Benedetti A, Sequi P (1998) Thermal analysis in the evaluation of compost stability: a comparison with humification parameters. Nutr Cycl Agroecosyst 51:217-224

Dell'Abate MT, Benedetti A, Sequi P (2000) Thermal methods of organic matter maturation monitoring during a composting process. J Therm Anal 61: 389-386

Grisi B, Grace C, Brookes PC, Benedetti A and Dell'Abate MT (1998) Temperature effects on organic matter and microbial biomass dynamics in temperate and tropical soils. Soil Biol Biochem 30: 1309-1315

Kaloustian J, Pauli AM, Pastor J (1997) Etude comparative par analyses thermique et chimique de quelques végétaux méditerranéens. J Therm Anal 50: 795-805

McBrierty VJ, Wardell GE, Keely CM, O'Neil EP, Prasad M (1996) The characterization of water in peat. Soil Sci Soc Am J 60: 991-1000

Pinzari F, Tittarelli F, Benedetti A, and Insam H (2002) Use of CLPP for evaluating the role of different organic materials in composting. In: Insam H, Riddech N, Klammer S (Eds) Microbiology of Composting, Springer, Heidelberg, pp.383–396

Schaumann GA, Antelmann O (2000) Thermal characteristics of soil organic matter measured by DSC: A hint on a glass transition. J Plant Nutr Soil Sci 163: 179-181

Sharma HSS (1990) Analysis of the components of lignocellulose degraded by *Agaricus bisporus* and *Pleotorus ostreatus*. Thermochim Acta 173:241-252

Sharma HSS (1991) Biochemical and thermal analyses of mushroom compost during preparation. Mushroom Sci 13:169-179

Sharma HSS (1995) Thermogravimetric analysis of mushroom (*Agaricus bisporus*) compost for fibre components. Mushroom Sci 14:267-273

Springer U, Klee J (1954) Prüfung der Leistungsfähigkeit von einigen wichtigeren Verfahren zur Bestimmung des Kohlenstoffs mittels Chromschwefelsäure sowie Vorschlag einer neuen Schnellmethode. Z Pflanzenernähr Düng Bodenkd 64:1-8

Tittarelli F, Trinchera A, Benedetti A, Intrigliolo F (this volume) Evaluation of organic matter stability during the composting process of agro-industrial wastes. In: Insam H, Riddech N, Klammer S (Eds) Microbiology of Composting, Springer, Heidelberg, pp.397–406

Compost Maturity – Problems Associated with Testing

M. Itävaara, O.Venelampi, M. Vikman and A. Kapanen[1]

Abstract. Safe use of compost in plant cultivation requires the utilization of mature compost. The complex composition of the organic matter and the changes occurring during biodegradation make maturity assessment a difficult task. In the present study we give an overview of the test methods applied, and some scientific background for the immaturity and toxicity of the compost samples.

The results of the acute toxicity test, Flash bioluminescent bacteria test *Vibrio fisheri*, correlated well with those of the plant growth assays. The immature composts studied were toxic in the Flash test and plant growth assays when processed for less than 3 months, but nontoxic after maturing during six months of composting.

In the present study we showed that oxygen deficiency during composting processing resulted in the development of toxicity in the compost. It was also confirmed that oxygen is a requirement for good quality compost, insufficient aeration during processing resulting in poor quality and retarded growth of the plants.

In order to evaluate the maturity of compost, stability and toxicity at least should be studied. Special attention should also be paid to the moisture content when testing stability with respirometric tests. This avoids the problem of false evaluation due to the lack of water needed for microbial respiration. Large-scale composting facilities very often run their composting processes with insufficient moisture content and aeration, resulting in reduced biodegradation and an increase in the length of time required to reach maturity. The moisture content of samples taken from such a composting facility should be balanced before testing for maturity.

Introduction

The tightening up of European legislation has increased the composting of biologically degradable waste. However, the use of compost as a valuable product for plant cultivation has been very restricted. Most forms of compost are still mainly used as landfill cover owing to their poor quality.

At the present time there is considerable international and national activity in a number of countries to develop standards for promoting the composting industry. There is a great need to find solutions for the quality control of composts, and to

[1]VTT Biotechnology, P.O. Box 1500, 02044 VTT, Finland

H. Insam, N. Riddech, S. Klammer (Eds.)
Microbiology of Composting

develop cheap and easy maturity tests that can be used by the composting operators. The Nordic countries, Finland, Denmark, Norway, Sweden and Denmark, are collaborating to find solutions to compost quality problems (Itävaara et al. 1998). Several methods have been evaluated by researchers, and the final phase is to ring test a number of selected tests. The aim is also to connect ongoing national research on maturity and ecotoxicity test development. The collaboration has been supported by Nordtest (1997 - 1998) and the Nordic Council of Ministers (1998 - 2000).

As compost continuously undergoes biological degradation, scientists have been faced with the problem of finding suitable tests that reliably determine the maturity of compost. Many chemical changes occur during maturation, such as changes in cationexchange capacity (Harada and Inoko 1980; Inbar et al. 1991), a decrease in dissolved organic carbon (Inbar et al. 1991), and a decrease in the ammonium and increase in the nitrate content of the compost (Spohn 1978; Standards Australia 1997). In addition, the dramatic increase in pH that occurs in the initial stages of degradation owing to the release of ammonia gradually drops to a level suitable for plant growth (Itävaara et al. 1997). The carbon/nitrogen ratio (Jimenez and Garcia 1989; Hue and Liu 1995), and the activities of several enzymes such as glucosidase, lipase, phosphatase (Herrmann and Shann 1993), dehydrogenase (Saharinen and Vuorinen 1997) and arginine ammonification (Foster et al. 1993), as well as microbial activities (Tseng et al. 1996), show clear changes during the maturation of compost. Maturity tests can be roughly classified into physical, chemical, plant and microbial activity assays as described in Table 1 (Itävaara et al. 1998).

Table 1 . Classification of maturity tests

• Physical - Temperature, odour, colour, structure - Rottegrade (self-heating test) • Chemical - pH - Reduction of organic matter - C/N ratio - Cation exchange capacity - NH_4^+, NO_3^- - Humification parameters - Organic compounds (e.g. acetic acid)	• Plant - Germination - Root elongation - Growth • Microbial activity - Carbon dioxide - Solvita - ATP - Enzyme

The abundance of chemical and biological changes that occur during composting, and the range of methods suggested in the literature, have made it difficult to agree on methods which should be accepted for the practical assessment of maturity. The following criterion should, however, be fulfilled: the compost should be stable enough to prevent the formation of anoxic conditions, caused by microbial degradation activity, when used as a growth substrate. Non-toxicity is another requirement, and means that the compost should not contain organic degradation products that will affect the growth of plants. The main compounds that have been shown to be phytotoxic are ammonia and a number of volatile organic acids (VOA), such as acetic acid and butyric acid formed during anaerobic degradation (Lynch 1977; Manios et al. 1989; Brinton 1998). Ammonia is formed in the initial stage of composting when proteins are degraded and nitrogen compounds released (Itävaara et al. 1997). Later ammonia is converted into nitrite and nitrate by nitrifying microorganisms. Volatile fatty acids are mainly formed under anaerobic conditions if the structure of the compost (e.g. porosity) is insufficient to maintain gas exchange and there is insufficient oxygen for the aerobic biodegradation of organic constituents.

Because of the complicated nature of maturity, our approach is not to identify one test but instead a pattern of test methods that can be used to verify the safe use of compost, i.e. maturity. The main idea is to determine maturity by utilising a combination of tests that confirm stability, anaerobicity and toxicity.

Materials and Methods

The FLASH bioluminescent bacteria test and correlation with the plant growth assay

The FLASH bioluminescent bacteria test, is based on kinetic measurements utilising the bioluminescent microorganism *Vibrio fischeri* (BioTox Kit) as a test organism, and has been developed to determine the toxicity of coloured samples (Lappalainen et al. 1999). Kinetic measurement was performed with 1251 Luminometer (Bio-Orbit, Turku, Finland) at 20 °C. Luminescence was measured throughout the 30-s exposure time. The maximum value of luminescence was obtained within the first 5 s and it was followed by a reduction in the case of toxicity of the sample. On the other hand, no bioluminescence decrease was recorded in absence of toxicity. Inhibition percentage was calculated as the ratio of the maximum light production (0–5 s) against the light production after 30-s exposure time. The negative inhibition values represent activation in light production. 2% NaCl solution was used as the test control in all experiments performed.

The plant growth assay was performed according to the OECD plant growth test (1984) using barley (*Hordeum vulgare)* and radish (*Raphanus sativus).*

The correlation between the FLASH bioluminescent bacteria test and the plant growth assay was studied in an experiment with samples of composted municipal food waste. The samples originated from a composting plant at the Ämmässuo biowaste treatment area, Espoo, Finland. The compost samples were classified into three groups according to the processing time: up to 3 months, 3–6 months and over 6 months composted municipal waste.

The Impact of Oxygen Deficiency on Temperature, Microbial activity and Toxicity

The effect of a lack of oxygen on heat generation development, microbial activity and toxicity of the compost samples was studied by varying the aeration flow into the composter bins. Vegetable-based biowaste collected from grocery stores was mixed with bark and wood chips in the ratio of 1:1, and composted in four bins with insulated walls (Biolan 200 liter vol.). The initial moisture content of the compost was 79.4% and the C/N ratio 26.8. Water content, volatile solids, pH, electrical conductivity, redox -potential, ammonium-N and dissolved organic carbon are routine analyses that are always performed, but are not presented in this chapter. All the bins were filled with the same biowaste mixture and the aeration level was set at 0.5 and 8 l min^{-1}. There were two parallel bins for each treatment. Aeration took place through the bottom of the bins, and the temperature was monitored by placing a probe, connected to a computer, in the middle of the bins. The contents of the composter bins were turned over once a week. The composting time was about 80 days. Samples were taken after 12, 20 and 36 days for toxicity analysis by the FLASH bioluminescent test, and after 12, 20, 36, 47 and 78 days for ATP and microbial activity analysis. ATP was extracted and determined according to Tseng et al. (1996).

Stability Test and Moisture Balancing

The biodegradation phase of the composting process is an important issue when evaluating maturity. We developed a small-scale carbon dioxide evolution test for testing the stability of compost samples.

Five grams of fresh compost was weighed out into a flask (vol. 120 ml) and sealed with a rubber cap. The compost was incubated at +37 ^{o}C and the carbon dioxide evolution determined after 4 hours by an infrared analyzer (Servomex PA404). Carbon dioxide evolution in the head-space test was evaluated by analysing samples from a composting facility. The samples were taken from the static piles after 3, 6, 8, 10, 13, 15 weeks of composting, and the water content and carbon dioxide evolution were measured immediately. In addition, all the samples were moistened to 80% of the maximum water-holding capacity (WHC), after which carbon dioxide evolution was again determined.

Results and Discussion

The main reason for determining toxicity is to prevent phytotoxicity when the compost is used for cultivation purposes. The results of the FLASH bioluminescent bacteria test were in good agreement with those of the plant growth assay (Fig.1). The samples analysed from the municipal composting plant remained extremely immature for up to 3 months according to both tests, after which the toxicity of the samples composted for 3–6 months gradually decreased. The samples of the municipal waste composted for more than 6 months were clearly non-toxic in the FLASH test and barley test. However, even though some of the samples were not toxic according to the FLASH test, they did retard the growth of radish seedlings. This is an indication of possible injury caused by these composts to sensitive plant species when mixed into the growth medium in large quantities.

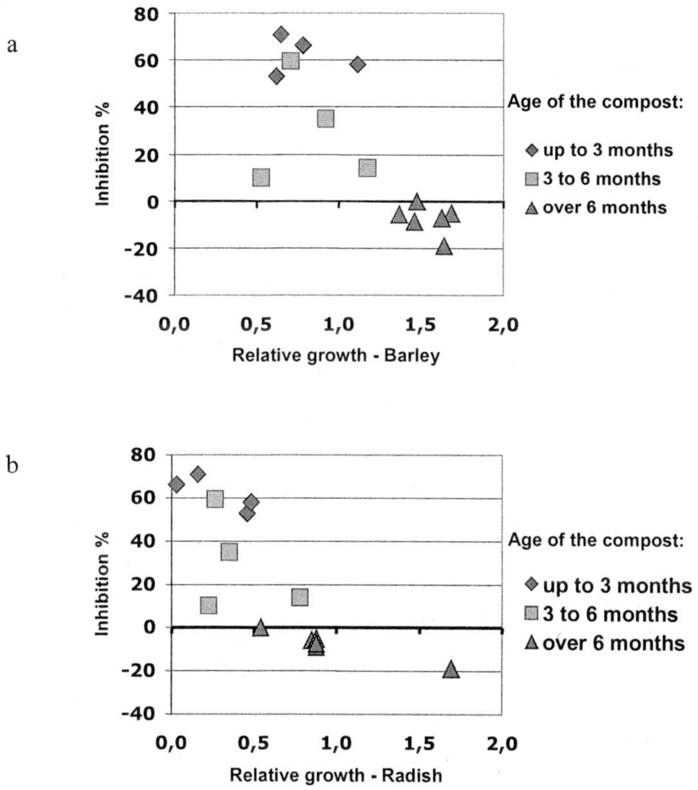

Fig. 1a b. Toxicity of municipal waste compost (MWC) samples analysed by the FLASH bioluminescent test (inhibition %) and the plant-growth assay using **a** barley and **b** radish as test plants

In this work we also demonstrated the effect of aeration during composting on microbial activity, heat generation and toxicity. Heat generation was considerably slower in the two composter bins aerated by forcing air at a rate of 0.5 l min^{-1} into the bins. In contrast, the temperature in the composts receiving abundant air immediately increased and reached a temperature of 70 °C during the first 4 days, while the bins with low aeration reached the maximum temperature of 70 °C after 17 days (Fig.2). The microbiological basis for the enhancing effect of aeration on heat generation was confirmed by the increased ATP concentration (20 μg ATP g^{-1} dry weight) after 12 days in the well - aerated compost, and the lower ATP content (1 μg ATP g^{-1} dry weight) in the poorly aerated bins (Fig.2). The temperature and microbiological activity, however, also subsequently increased in the poorly aerated bins due to the fact that weekly turning of the compost resulted in sufficient oxygen for microbiological degradation. The lack of oxygen in the 0.5 l min^{-1} aerated bins induced a strong toxic response, which was reflected in the FLASH bioluminescent test as a clear decrease in bioluminescence. The inhibition of bioluminescence was 23–55% in the immature composts after 12 and 20 days of processing. However, no inhibition was detected after 36 days. Inhibition was not detected in the immature composts aerated with 8 l of air min^{-1} (Fig.3).

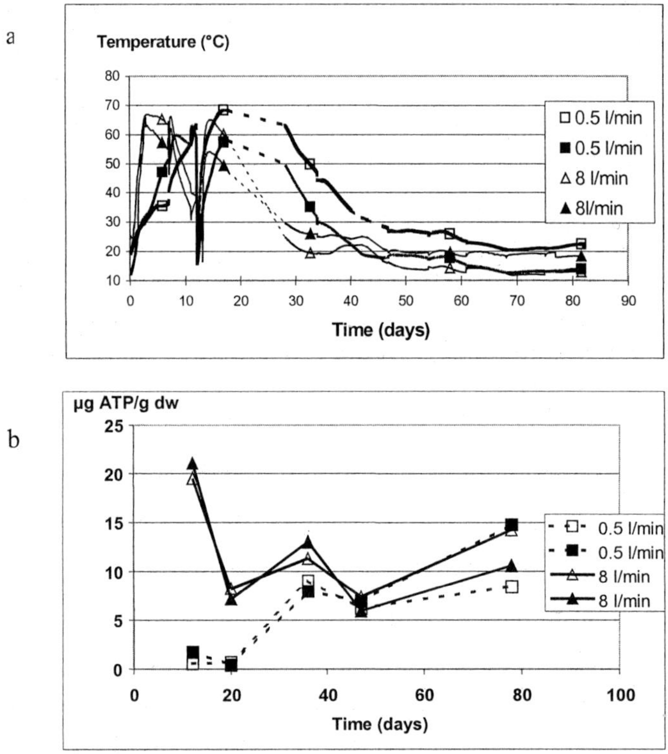

Fig. 2a, b. The effect of aeration on a temperature and b microbial activity determined as ATP content

The benefits of using the *Vibrio fischeri* based FLASH test as a fast screening method to evaluate the toxicity of immature composts and correlation with plant assays were clearly demonstrated. The use of marine bacteria for detecting toxicity in soil or compost has been critised but, according to our results, marine bacteria can be used if the salt and pH requirements of this species are taken into account and stabilised before measurement. The FLASH test has also been successfully used to determine the toxicity of oil-contaminated soils (Juvonen et al. 2000).

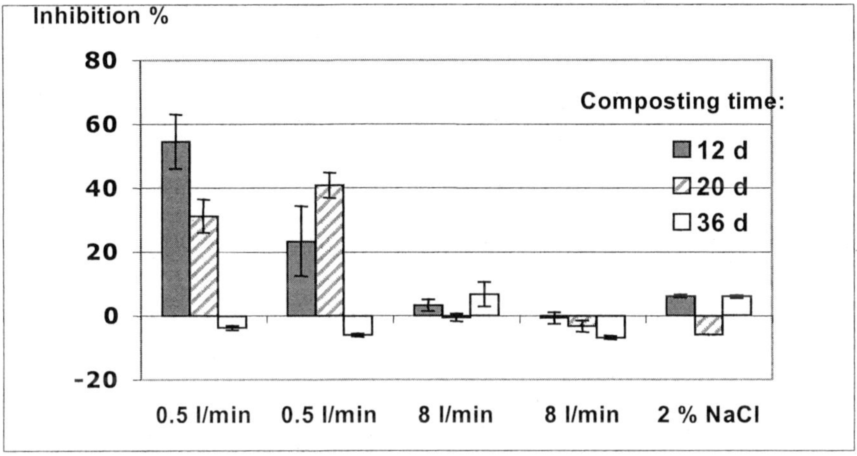

Fig. 3. Toxicity determined using the FLASH bioluminescent test on samples taken from 0.5 l min^{-1} and 8 l min^{-1} aerated composts. 2% NaCl was used as a control in order to verify the validity of the test

Our results also showed that toxicity was in this case induced by anaerobicity, and no toxicity was formed when aerated with 8 l of air min^{-1} even though the compost was immature. However, if immature compost is used as a plant growth medium, oxygen consumption in the medium may be excessive and cause poor plant growth. The formation of toxic anaerobic biodegradation metabolites may also be induced by high microbial activity under anoxic conditions. These are the main reasons for determining the stability of the compost. The stability test is very sensitive to the moisture content and this should be adjusted before measurement. One of the main problems in the composting plants seems to be achieving the optimum water content for microbial degradation. The samples taken at different times from the composting plant showed a decreasing moisture content as a function of time (Fig.4). After adjusting the moisture content to 80% water-holding capacity (WHC), the stability test gave a considerably higher carbon dioxide generation (Fig.5). If a stability test based on microbial activity such as Rottegrade or carbon dioxide evolution is applied directly to such a compost sample, a low temperature or carbon dioxide evolution due to drought may mask compost immaturity. When water is subsequently added in a practical application, the microbes will be activated, causing the problems mentioned earlier. If the compost is very dry it may also take several days to adjust the water content before maximum carbon dioxide generation is reached (Fig.6).

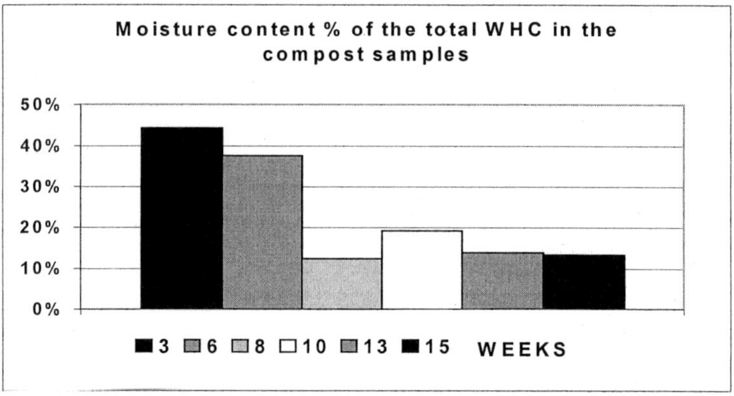

Fig. 4. The moisture content of the compost samples determined as % of the total water-holding capacity (WHC)

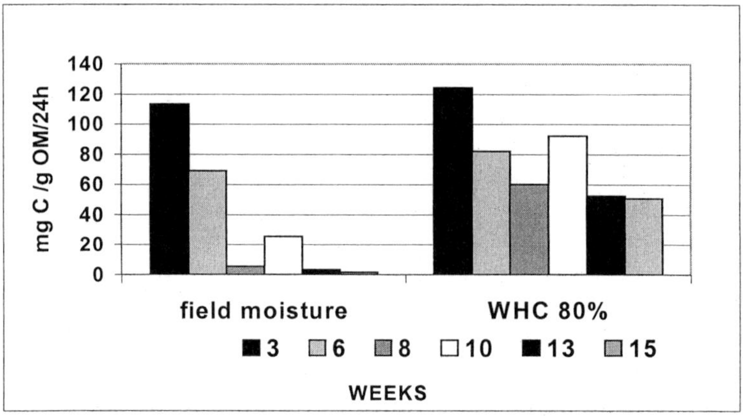

Fig. 5. The effect of adjusting the moisture content on carbon dioxide evolution in the stability test

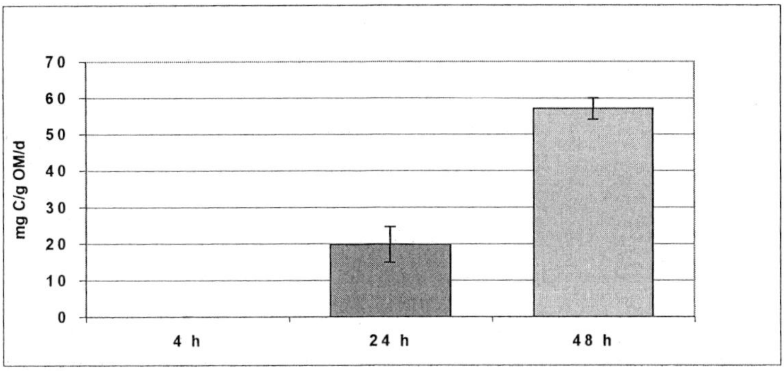

Fig. 6. CO_2 production in the VTT CO_2test after 4, 24 and 48 h

Concluding Remarks

Poor processing conditions or poor structure of the compost may cause anaerobic conditions, leading to the formation of toxic compounds. However, it may also reduce heat generation and microbial activity. When testing the stability of compost that has been processed in an oxygen-deficient environment, use of the carbon dioxide evolution test or the Rottegrade test based on a temperature increase may erroneously result in the compost being classified as mature. In addition, an insufficient moisture content will decrease the microbial activity and biodegradation rate of the composting process in several composting facilities. The moisture content will also affect the determination of stability, and should therefore be adjusted. Attaining the optimal water content and activating microbial degradation may take several days if the compost has been completely inactivated and dried out. This should be taken into consideration when determining stability, i.e. the respiration activity of the microbes, and is connected to the availability of easily utilisable nutrients.

Acknowledgements. This research was supported by the National Technology Agency of Finland, Biolan Oy, Vapo Oy, Kemira-Agro Oy and YTV.

References

Brinton W (1998) Volatile organic acids in compost: production and odorant aspects. Compost Sci Util 6 (1): 75-81

Foster JC, Zech W, Würdinger E (1993) Comparison of chemical and microbiological methods for the characterization of the maturity of composts from contrasting sources. Biol Fertil Soils 16: 93-99

Harada Y, Inoko A (1980) The measurement of the cation-exchange capacity of composts for the estimation of the degree of maturity. Soil Sci Plant Nutr 26: 127-134

Herrmann RF, Shann JR (1993) Enzyme activities as indicators of municipal solid waste compost maturity. Compost Sci Util 1 (4): 54-63

Hue NV, Liu J (1995) Predicting compost stability. Compost Sci Util 3 (2): 8-15

Inbar Y, Chen Y, Hadar Y, Hoitink, HAJ. (1991) Approaches to determining compost maturity In: Biocycle, the art and science of composting. JG Press, Emmaus, Pennsylvania, pp 183-186

Itävaara M, Vikman M, Venelampi O. (1997) Windrow composting of biodegradable packaging materials. Compost Sci Util 5 (2): 84-92

Itävaara M, Venelampi O, Samsøe-Petersen L, Lystad H, Bjarnadottir H, Öberg L. (1998) Assessment of compost maturity and ecotoxicity. Nordtest, Espoo, NT Techn Report 404. 84 pages. NT Project No 1363-97

Jimenez EI, Garcia VP (1989) Evaluation of city refuse compost maturity: review. Biol Wastes 27: 115-142

Juvonen R, Martikainen E, Schultz E, Joutti A, Ahtiainen J, Lehtokari M (2000) A battery of toxicity tests as indicators of decontamination in composting of oily wastes. Ecotoxicol Environ Safety 47: 156-166

Lappalainen J, Juvonen R, Vaajasaari K, Karp M. (1999) A new flash method for measuring the toxicity of solid and colored samples. Chemosphere 38 (5): 1069-1083

Lynch JM (1977) Phytotoxicity of acetic acid produced in the anaerobic decomposition of wheat straw. J Appl Bacteriol 42: 81-87

Manios VI, Tsikalas PE, Siminis HI, Verdonck O (1989) Phytotoxicity of olive tree leaf compost in relation to the organic acid concentration. Biol Wastes 27 (4): 307-317

OECD Guideline for Testing of Chemicals 208 (1984) Terrestrial plants, Growth test.

Saharinen MH, Vuorinen AH (1997) Evolution of microbiological and chemical parameters during manure and straw co-composting in a drum composting system. Agric Ecosyst Environ 66 (1): 19-29

Spohn E (1978) Determination of compost maturity. Compost Sci 3 (19): 26-27

Standards Australia (1997) Composts, soil conditioners and mulches. Homebush, NSW, Australia (AS 4454-1997)

Tseng DY, Chalmers JJ, Tuovinen OH (1996) ATP measurement in compost. Compost Sci Util 4 (3): 6-17

Use of CLPP to Evaluate the Role of Different Organic Materials in Composting

F. Pinzari[1], F. Tittarelli [1], A. Benedetti[1] and H. Insam[2]

Abstract. In this study, changes in the microbial functional diversity during composting were analysed. Seven samples corresponding to (1) three different residues used in composting (sludges, citrus industrial processing waste, green wastes), and (2) four different stages of compost maturity were compared by their "community level physiological profiles" (CLPP). Ecoplates (Biolog) were inoculated with compost and matrices and incubated at 30 °C; optical density was measured every 8 h for 8 days. The samples were compared on the basis of the kinetics of the curves produced in each well. The data were used to calculate the area under the curve, and to study the curve kinetics. As a whole, the CLPP succeeded in addressing the role of different residues in the functional diversity of the mature compost. The microbial physiological profile of the sludge was found in the final product, while no "metabolic" trace of the other two matrices was found. The samples corresponding to different stages of the composting process were well characterised by their CLPPs, showing an increase in the metabolic diversity during the stabilisation process.

Introduction

"A rational understanding of composting ecology provides the best basis for developing process control strategies" (Miller 1993), which means that successful composting requires the establishment of rational processing control criteria. Among the most important requirements for a good composting process are both the good quality of the residues used (C/N; pH; contents of heavy metals), and the physiological characteristics of the microbial species already present in the residues. The organic matrices used to produce composts are at times unfavourable for a good start of the composting process (Canali et al. 1998). The microbial community in certain residues may be lacking metabolic diversity and per se is inadequate for priming the composting process. This is the case of sludges and solid wastes obtained during the production and the transformation of citrus fruits and groves, which typically have high concentrations of aromatic compounds and a very low pH (3–4) (Correia et al. 1995). These residues are used to produce compost within a project on agro-industrial wastes recycling.

[1]Istituto Sperimentale per la Nutrizione delle Piante di Roma, Via della Navicella 2, 00184 Rome, Italy, e-mail: flavia.pinzari@tiscalinet.it
[2]Institut für Mikrobiologie, Innsbruck Universität, Technikerstr. 25, 6020 Innsbruck, Austria

H. Insam, N. Riddech, S. Klammer (Eds.)
Microbiology of Composting

In this study, changes in the microbial functional diversity during composting were analysed, and compared with the microbial pattern of the matrices used to produce the compost. Seven samples corresponding to: (1) three different residue types (sludges, citrus industrial processing waste, green wastes), and (2) four different stages of compost maturity were compared by their community level physiological profiles (CLPP) (Garland and Mills 1991; Garland 1997).

CLPPs are increasingly used for studying microbial communities of various ecosystems and environments, composts and sludges included (Garland 1997; Insam and Rangger 1997). The method involves direct inoculation of environmental samples into Biolog microtitre plates, and uses colour formation from reduction of a tetrazolium dye to assess utilisation of sole C sources during a 2–7 day incubation period.

Many options are available for evaluating CLPPs, such as overall rate of colour development (which is a function of inoculum density and incubation time), diversity (function of richness and evenness), pattern (analysed with multivariate statistics) and kinetics of substrate utilisation.

With the current techniques, the CLPP method is suitable for the study of bacteria and actinomycetes. Many solid wastes and industrial residues need bacteria to be transformed in the early stages of the composting process, while fungi become more important in the later stages. In sewage sludge composting, for example, about 40% of solids are decomposed by bacteria (Strom 1985). The use of CLPP, in evaluating the value of the matrices from the bacterial point of view, is of scientific and practical interest in order to develop process control strategies and optimise organic residue recycling.

Materials and Methods

Samples: Matrices and Compost

The composting process was carried out at the Azienda Sperimentale Palazzelli (Lentini) of the Istituto Sperimentale per l'Agrumicoltura- Acireale (Catania, Sicily, Italy). The windrow was prepared on a suitable cement platform by mixing three types of residues:

1. pastazzo, a mixture of citrus skins and pulps, which represents the main residue produced during the citrus-processing industry;
2. the citrus-processing industry sludge; obtained from the purification treatment of fruits and plants washing waters;
3. the green residues, straw and branches, produced in the management of grasses and trees in the orchards.

The weight percentages in the mixture composition were defined according to the C/N ratio and to the chemical, and physical characteristics of each residue (sludge 20%; pastazzo 40%; green residue 40%) (Tittarelli et al., this Vol.).

Temperature and moisture were controlled during the whole composting process in order to maintain optimal conditions for microbial activity (Dell'Abate and Tittarelli, this Vol.). Daily measurements of temperature and humidity were taken, and the pile was mechanically turned to aerate the inner layers. Water was added to the pile to maintain a humidity of 60–70% (w/w). Samples were taken four times, at the beginning of the process, and every month during the following 90 days (Table 2). Each sample, taken after turning, was made up of six subsamples subsequently mixed until homogenisation. The seven samples (Table 1) were slightly air-dried, ground and sieved to 2 mm; the sieving was repeated for each sample singularly, using conditions of sterility for all the instruments and containers used in the procedure.

Table 1.

	Sample	Sampling data (1999)	°C at sampling	% Total N	% Total organic C	C/N ratio
1	Sludge	13/02	n.m.	3.6	30.5	8.6
2	Pastazzo	13/02	n.m.	1.4	46.7	33.6
3	Green residues	13/02	n.m.	0.6	50.8	80.6
4	Compost time zero (T0)	13/02	42	1.4	44.2	30.7
5	Compost time one (T1)	17/03	61	1.8	40.0	21.7
6	Compost time two (T2)	19/04	56	1.9	35.8	18.6
7	Compost time three (T3)	14/05	53	2.1	35.2	16.8

The sample notation corresponds to: *1* sludge; *2* pastazzo; *3* green residues; *4* compost at time zero, when the matrices are only mixed and the composting process has not started yet; *5* 1-month-old compost, collected when the windrow was in the thermophilic phase; *6* 2-month-old compost, collected just after the thermophilic phase, when the windrow was cooling down; *7* 3-month-old compost, collected at the end of the composting process; this sample is considered as the "mature" product.

The C and N contents and the temperatures measured at the sampling times are shown in Table 1. Detailed descriptions of the chemical and biochemical characteristics of the matrices and the methods used in the monitoring of the pile are reported in Tittarelli et al. (this Vol.), and Dell'Abate and Tittarelli (this Vol.).

Samples Extraction, Inoculation and Incubation of Microtitre Plates

To obtain a bacterial suspension we used the first two steps of the Hopkins et al. (1991) extraction method, with some slight modifications as follows: 2.5 g dry weight of sample were transferred to a centrifuge tube (screw cap 50-ml tubes) with the addition of 20 ml of sodium cholate 0.1% (w/v) solution (sodium cholate $C_{24}H_{39}NaO_5$ Merck), 10 ml Na^+ form chelating resin (corresponding to 8.59 g of DOWEX Fluka Chemika Na^+, form 20–50 mesh) and 30 glass beads (Macdonald 1986).

Solutions, tubes, resin and the equipment for the inoculation procedure were sterilised before use by autoclaving at 120 °C for 20 min. The centrifuge tube with the sample was then shaken for 2 h at 4 °C with a shaker, and centrifuged at 2200 rpm for 2 min (Sorvall SS-34). The supernatant was collected and preserved at 4 °C. The pellet was resuspended in 10 ml Tris buffer pH 7.4 (Tris-hydroxymethyl--aminomethan; Merck), shaken for 1 h at 4 °C, centrifuged at 2200 rpm for 2 min (Niepold et al. 1979). The second supernatant was recollected with the first one, the pellet was discarded. Before further steps, the cell suspension was centrifuged again at 3000 rpm for less than 1 min at 4 °C, to allow precipitation of any remaining large light-weight particle (frequently present in compost extracts). The supernatant was used to prepare the desired dilutions.

The choice of the correct dilution of each suspension obtained from the compost samples required special care because of the high organic matter content resulting in the coloration of the extract. The amount of substrate used to obtain the suspension (2.5 g dry weight) was standardised in order to use the same extraction ratio for all the samples. The density of the extracts before the further dilution was therefore proportional to the relevant samples: a higher density of the cell suspension corresponds to a higher content of (extractable) microbial biomass in the sample.

In order to compare the samples in both quantitative and qualitative responses, the suspensions were diluted according to the same procedure 100-fold. For the dilutions a sterile Ringer solution (Ringer's tablets; Merck) was used (Collins et al. 1989).

In our experimental trials we used the Biolog Ecoplates (31 C substrates plus a control well, repeated three times on a 96-well plate) which contain substrates particularly suitable to discriminate between environmental samples (Insam 1997).

Replicates of the samples were separately extracted (six separate extractions for each sample), and suspensions diluted and used to inoculate the Ecoplates. Each well of the Ecoplate was inoculated with 120 μl of suspension. Plates were put singularly into plastic bags, to prevent drying in the external wells (the bags were not tightly closed in order to allow O_2 and CO_2 to circulate), and incubated at 30 °C. Optical densities of the wells were measured, reading absorbances at 592 nm with a microtitre plate reader (TECAN, Grödig, Austria). Measurements were effected immediately after the inoculation (time zero), and then every 8 h for 8 days (up to 184 h).

Data Analysis

Raw data were transferred to an Excel (Microsoft) sheet according to sample (7 samples), replicate (6 replicates each), and reading time (24 reading points). The optical density for each substrate at a given time was adjusted according to Insam et al. (1996). The absorbance value at time zero (background correction), and the smallest absorption value (SAV) that was found on the plate were both subtracted from the optical density of each well. The SAV was not always coincident with the control well.

From each plate, at each reading time the average well colour development (AWCD) was calculated as follows:

$$\text{AWCD} = \Sigma_{(i=1,31)} (\text{R}i - \text{C})/31, \qquad (1)$$

where R is the optical density at 592 nm (OD_{592}) of the *i*-well, and C is the SAV at that reading time.

The samples were compared on the basis of the kinetics of the curves produced in each well. The curves were used, at first, to calculate an integral with Riemann's sum (Sharma et al. 1997), and then to study the kinetics according to Lindstrom et al. (1998).

The technique suggested by Lindstrom et al. (1998) enables the comparison of "samples with different initial populations by relying on analysis of kinetic parameters, which are invariant with respect to inoculum density". The model contains three curve parameters (r, s and k):

$$y = \text{OD590nm} = k/[1+e^{-r(t-s)}], \qquad (2)$$

where k is the asymptote of the colour development sigmoidal curve, r is the exponential rate, and s is the time to the midpoint of the curve (t in the formula is the time after inoculation). The adjusted data for each substrate showing colour development were fit to Lindstrom's kinetic model and the three parameters estimated with Origin 6.0 (OriginLab Corporation, Northampton, MA, USA). The analysis was performed with a template written in Origin script language, which used the Levenberg-Marquardt algorithm for fitting the curve and produced estimates of the model parameters for each substrate. The standard errors and an estimate of the goodness of fit (χ^2) were also produced by the template. Kinetic parameters thus obtained, as suggested by Lindstrom et al. (1998), were used in place of the single-point datum in evaluating substrate use patterns.

The Gini coefficient (Harch et al. 1997), known to be well correlated with the functional diversity, was also calculated.

Analysis of variance (ANOVA), factorial analysis (canonical - discriminant) were employed to investigate the differences in compost microbial communities. The ANOVA was employed with the pairwise method of Bonferroni (SPSS 1998). The significance of separation of groups' centroids identified by the canonical functions was defined by Wilks' lambda test.

Results

Average Well Colour Development and Total Microbial Biomass

Average well colour development (AWCD, Garland 1997) data were calculated for each sample at each reading time. The maximum values and the times after inoculation at which they occurred were 1.8 at 136 h for sample 1, 0.12 at 184 h for sample 2; 1.2 at 184 h for sample 3; 1.8 at 128 h for sample 4; 1.7 at 120 h for sample 5; 1.9 at 136 for sample 6; 2.02 at 96 for sample 7. A one-way ANOVA was performed on the AWCDs of the seven samples (six replicates each) for each reading time (α=0.05) in order to evaluate the statistical significance of the difference between the averages (Table 2).

Samples 1, 4, 6 and 7 do not differ significantly from each other, whereas samples 2, 3 and 5 differ both within them, and from samples 1, 4, 6 and 7. All the compost samples but one (sample 5, corresponding to the thermophilic phase) show the same development of average well colour in the plates as the sludge (sample 1).

Table 2. Multiple comparison of the AWCD of the seven samples (six replicates each) for each reading time (α=0.05; test used: Bonferroni) are reported. Both significant and insignificant differences between samples are listed. Sample notation: see Table 1

Sample	AWCD	
	Significant differences with samples:	Insignificant differences with samples:
1	2, 3, 5	4, 6, 7
2	1, 3, 4, 5, 6, 7	-----
3	1, 2, 4, 5, 6, 7	-----
4	2, 3, 5	1, 6, 7
5	1, 3, 4, 5, 6, 7	-----
6	2, 3, 5	1, 4, 7
7	2, 3, 5	1, 4, 6

Particularly Distinctive Substrates

When considering the overall colour development of each of the substrates among the samples, some compounds appeared more informative than others. Phenolic compounds (2- and 4-hydroxybenzoate) and polymers (glycogen, Tween 40 and 80, α-cyclodextrin) were therefore separately used to characterise samples by comparison on the basis of a one-way ANOVA (α=0.05; post-hoc test: Bonferroni) (Tables 3 and 4).

The 2-hydroxybenzoate clearly distinguished sample 6 from all the others: compost at time 3 (sample 6) was in fact the only one able to use this compound as a substrate. The 4-hydroxybenzoate differentiated many samples, but on the basis of a different degree of substrate utilisation (the highest in sample 7, the lowest in sample 2). The comparison between samples based on the use of polymers showed, as a whole, a good ability of this guild of substrates to distinguish among the microbial communities examined.

The behaviour of the microbial communities of the samples against Tween 40, α-cyclodextrin and glycogen was identical (Table 4, one-way ANOVA on the single substrates). Sample 1 (sludge) and samples 4, 5, 6, and 7 (compost at different sampling time) showed the same metabolic versatility against these substrates. Samples 2 and 3 (green residues and pastazzo, respectively) differed both among each other, and from samples 1, 4, 5, 6 and 7.

Table 3. In the table the results of the one-way ANOVA performed in the multiple comparison of the seven samples based on the use of the phenolic compounds (2-hydroxybenzoate, and 4-hydroxybenzoate) (α=0.05; test used: Bonferroni) are reported. Both significant and insignificant differences between samples are listed. Sample notation: see Table 1

Sample	2-Hydroxybenzoate		2-Hydroxybenzoate	
	Significant differences with samples:	Insignificant differences with samples:	Significant differences with samples:	Insignificant differences with samples:
1	6	2, 3, 4, 5, 7	2, 5, 6, 7	3, 4
2	6	1, 3, 4, 5, 7	1, 3, 4, 5, 6, 7	-----
3	6	1,2, 4, 5, 7	2, 5, 7	1, 4, 6
4	6	1,2, 3, 5, 7	2, 5, 7	1, 3, 6
5	6	1, 2, 3, 4, 7	1, 3, 4, 5, 6, 7	-----
6	1, 2, 3, 4, 5, 7	-----	1, 2, 5, 7	3, 4
7	6	1, 2, 3, 4, 5	1, 3, 4, 5, 6, 7	-----

Table 4. Multiple comparison of the seven samples based on the use of the polymers (Tween 40, Tween 80, α-cyclodextrin, glycogen) (α=0.05; test used: Bonferroni) are reported. Both significant (S) and insignificant (NS) differences between samples are listed. Sample notation: see Table 1

Sample	Tween 40		Tween 80		α-cyclodextrin		Glycogen	
	S	NS	S	NS	S	NS	S	NS
1	2, 3	4, 5, 6, 7	2, 3	4, 5, 6, 7	2, 3, 4	5, 6, 7	2, 3	4, 5, 6, 7
2	1, 3, 4, 5, 6, 7	----	1, 3, 4, 5, 6, 7	----	1, 3, 4, 5, 6, 7	----	1, 3, 4, 5, 6, 7	----
3	1, 2, 4, 5, 6, 7	----	1, 2, 4, 7	5, 6	1, 2, 4, 5, 6, 7	----	1, 2, 4, 5, 6, 7	----
4	2, 3	1, 5, 6, 7	2, 3, 5, 6	1, 7	1, 2, 3	5, 6, 7	2, 3	1, 5, 6, 7
5	2, 3	1, 4, 6, 7	2, 7	1, 3, 6	2, 3	1, 4, 6, 7	2, 3	1, 4, 6, 7
6	2, 3	1, 4, 5, 7	2, 7	1, 3, 5	2, 3	1, 4, 5, 7	2, 3	1, 4, 5, 7
7	2, 3	1, 4, 5, 6	2, 3, 5, 6	1, 4,	2, 3	1, 4, 5, 6	2, 3	1, 4, 5, 6

Gini Coefficient: the Unequal Use of Carbon Sources

The Gini coefficient (G) (Harch et al. 1997) is considered a measure of the unequal use of C sources. G has a minimum value of 0.0 when all wells have equal absorbance and a theoretical maximum of 1.0, when all wells but one have a value of 0.0. Thus the G value does not change if each well's absorbance is raised in the same proportion. The value of G was calculated for each replicate, time and sample. The differences between samples were analysed with ANOVA (Table 5). The Gini coefficient for 966 plate readings ranged from 0.07 to 0.380.

According to the Gini coefficient the most separate sample was number 3 (green residue), while sample 4 could not be distinguished from most other ones (compost at time 0).

Table 5. Multiple comparison of the seven samples based on the Gini coefficient (α=0.05; test used: Bonferroni) are reported. Both significant and insignificant differences between samples are listed. Sample notation: see Table 1

	Gini coefficient	
Sample	Significant differences with samples:	Insignificant differences with samples:
1	2, 3, 7	4, 5, 6
2	1, 3, 4, 5, 6	7
3	1, 2, 4, 5, 6, 7	-------
4	2, 3	1, 5, 6, 7
5	2, 3, 7	1, 4, 6
6	2, 3, 7	1, 4, 5
7	1, 3, 5, 6	2, 4

Riemann's Sum

The calculation of the Riemann's sum consists of a procedure able to summarise the "time" variable into a coefficient that preserves the information given by the multiple readings approach. Riemann's sum for each substrate was calculated from all the readings taken during 8 days of incubation.

Stepwise discriminant analysis was applied to the "new" data set (42 rows = 7 samples x 6 replicates; 32 columns = substrates).

The first two canonical variables accounted for 97% of the variance within the original data. The scatterplot of the second canonical variable (function 2) against the first (function 1) is shown in Fig. 1. The Wilks' lambda test rejected the equality hypothesis for the centroids. The samples are distinguished along both the first and the second axis.

The second canonical variable (function 2) corresponds to the total extent of colour development in plates: sample 2 showed the lowest score, and sample 7 the highest. The first canonical variable distinguished mainly between sample 1 and sample 5, on the basis of the different use of some substrates (for example, L--asparagine showed the highest correlation value with the first axis).

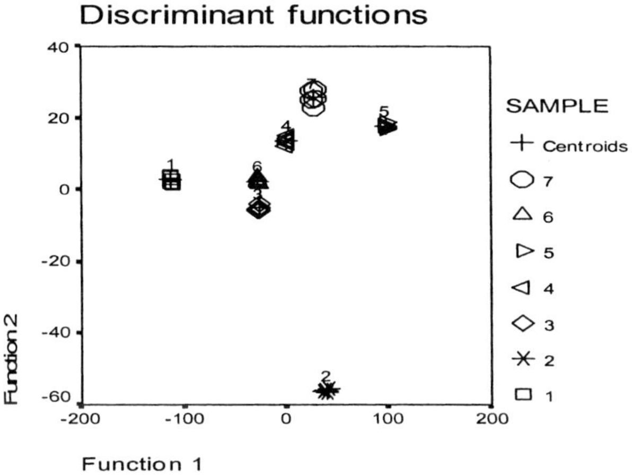

Fig. 1. Canonical discriminant functions calculated for the seven samples on Riemann's sum data set scatterplot of the second canonical variable (function 2) against the first (function 1). Sample notation: see Table 1

Lindstrom's Kinetic Parameters

Kinetic parameters obtained by the Lindstrom et al. (1998) model were used in place of single-point readings. This logistic model relates to a population of individuals. In the model, k, r and s are variables, where k represents the "carrying capacity" of the system and depends on the metabolic characteristics of the microbial community; r describes the shape of the curve; s is the time to the midpoint of the exponential portion of the curve. Generally, s values are negatively correlated with initial cell number in the inoculum. Stepwise discriminant analysis was applied separately to the three sets of data (r, k, s each forming a matrix of 42 rows = 7 samples, 6 replicates; and 32 columns = substrates).

The discriminant analysis was able to separate groups for all the parameters. The scatterplot of the second canonical variable (function 2) against the first (function 1) is showed for k, r and s in Figs. 2, 3, and 4, respectively.

The plot obtained with k-values (Fig. 2) showed a separation of groups 2 and 3 (green residues and pastazzo, respectively) both among them, and from samples 1, 4, 5, 6 and 7 along the function 1. Samples 1, 4 and 6 were not significantly separated along function 2. The plot obtained with r-values (Fig. 3) showed a significant separation of samples 7 and 5 both among them, and from samples 1, 2, 3, 4, and 6 along function 2. A better separation among all the groups was obtained along function 1, but the centroids of groups 1, 4 and 6 were not different according to the Wilks' test.

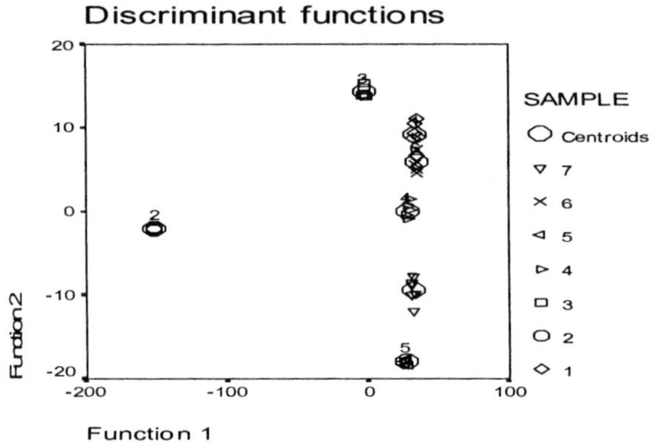

Fig. 2. Canonical discriminant functions calculated for the seven samples on the data set obtained calculating the k-parameter the Lindstrom's logistic model. Scatterplot of the second canonical variable (function 2) against the first (function 1). Sample notation: see Table 1

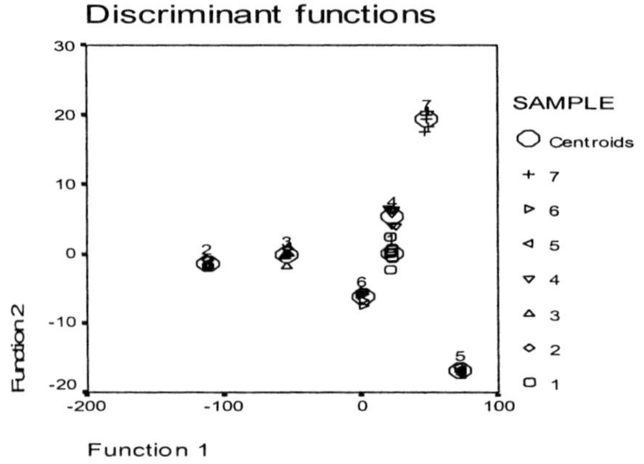

Fig. 3. Canonical discriminant functions calculated for the seven samples on the data set obtained calculating the r-parameter the Lindstrom's logistic model. Scatterplot of the second canonical variable (function 2) against the first (function 1). Sample notation: see Table 1

The plot obtained with s-values (Fig. 4) showed a separation graphically resembling that obtained for k-values: function 2 (used, in this plot, as horizontal axis) distinguished between sample 2 (with the lowest inoculum density) and all the other samples. function 1, separated samples 3, and 6 both between them, and from samples 1, 2, 4, 5, and 7. Both 2-hydroxybenzoate and Tween-40 showed the highest correlation values with the function 1 (correlation data are not reported).

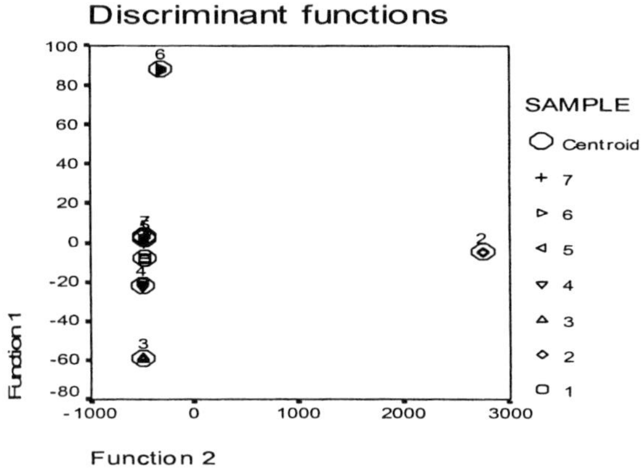

Fig. 4. Canonical discriminant functions calculated for the seven samples on the data set obtained calculating the **s**-parameter the Lindstrom's logistic model. Scatterplot of the second canonical variable (function 2) against the first (function 1). Sample notation: see Table 1

Discussion

According to McKinley and Vestal (1984) and Strom (1985), temperature is the dominant parameter controlling microbial activity during composting. Since the plates were incubated at 30 °C, thermophiles were not represented in our analysis. A simple "picture" of the matrices, and of four steps of the composting process, was taken. Despite this limitation, significant differences were observed between compost samples taken before, during and after the thermophilic phase. Lower values of AWCD and Riemann's sum was observed in sample 5 (compost sampled at the thermophilic phase) with respect to samples 4, 6 and 7. Moreover, the metabolic behaviour of sample 5 was peculiar, as showed in the plots obtained by the Lindstrom's coefficients (k and r) (Figs. 2 and 3). Sample 7 (mature compost) showed the maximum extent of metabolic activity both in the comparison based on AWCD values and on the Riemann's sum values. In most of the qualitative comparisons (i.e. Lindstrom's parameters and multiple comparisons based on single substrates) sample 7 was hardly distinguishable from samples 1, 4 and 6, thus suggesting a dominant role of sample 1 (sludge) in the composting process. The "metabolic fingerprint" of the sludge was extremely different from that observed for the two other matrices (samples 2 and 3: pastazzo and green residue, respectively). Although only a low percentage of sludge was used (20%), its effect as microbial inoculum was dominant. A confirmation came from the metabolic patterns obtained both for sample 2 and 3. Most of the data (AWCDs, Riemann's sum and Lindstrom's parameters) documented a behaviour of pastazzo (sample 2) and green residue (sample 3) significantly different from those of other samples (sludge and compost at different sampling times).

Phenolic compounds (2- and 4-hydroxybenzoate) and polymers (glycogen, Tween 40 and 80, α-cyclodextrin) were able to distinguish the microbial communities. The 2-hydroxybenzoate, for example, discriminated clearly between sample 6 and all the others: compost at time 3 (sample 6) was in fact the only one able to use this compound as a substrate.

According to Sparling et al. (1981), addition of *p*-hydroxybenzoic acid to soil stimulates microbial biomass. Williams (1962) observed a 100% inhibition of fungal spore germination by *o*-hydroxybenzoic acid, while *p*-hydroxy benzoic acid did not inhibit germination. The extent of the ability in using phenolic compounds can be related, for compost samples, to the presence of microorganisms involved with the synthesis or decomposition of humic substances (Sparling et al. 1981).

The inhibiting effect of *o*-hydroxy benzoic acid (=2-hydroxy benzoic acid) on fungal spores suggests a peculiar attitude of the microbial community of sample 6. It can be hypothesised that some bacteria from sample 6 are able to cope with this substance because they are competitors of fungi. Some authors (McKinley et Vestal 1984; Hellman et al. 1997) reported an increase in fungal activity in compost after the temperature decline (sample 6 was collected just after the thermophilic phase).

In summary, the CLPPs were shown to be sensitive for monitoring community changes during the composting process. The samples corresponding to the different stages of the composting process were well characterised by their CLPPs, showing an increase in the metabolic diversity during the stabilisation process. Moreover, the method succeeded in defining the role of different residues in the functional diversity of the mature compost. The microbial physiological profile of the sludge was found in the final product, while no "metabolic" traces of the other two matrices were found. The major question regarding this approach is whether the biological significance of the observed differences defines compost maturity or not. The increase in the use of different substrates by the microbial community cannot itself be considered as an indicator of maturity and quality of the product itself, but it can be used in evaluating the quality of the process. Among the available tools to analyse CLPPs during composting, the most powerful appeared the one that couples the kinetic analysis of the colour produced in each well with the multivariate analysis of variance. The indicators of biodiversity or of inequality of use of C sources, like the Gini coefficient, allow to define important differences between the examined communities. Using more than one approach, both differences in the pattern and in the catabolic efficiency can be evaluated.

Acknowledgements. This work was realised in the framework of the Action COST 831, and thanks to Dr. Francesco Intrigliolo (Istituto Sperimentale per l'Agrumicoltura- Acireale CT, Italy) who provided the compost samples. A particular thank is due to Scott Bertrand (Applications Project Manager of the OriginLab Corporation, Northampton, MA, USA) for his invaluable help in the development of the template used with Origin 6.0 Microcal software package. The authors thank the reviewers for their comments and recommendations.

References

Canali S, Trinchera A, Pinzari F, Benedetti A (1998) Study of compost maturity by means of humification parameters and isoelectric focusing technique. CD-Rom, Proc 16th World Congr of Soil Science. Montpellier, France 20-26 August 1998

Collins CH, Lyne PM, Grange JM (1989) Collins and Lyne's microbiological methods, 6th edn Butterworths,London

Correia Guerrero C, Carrasco de Brito J, Lapa N, Santos Oliveira F (1995) Re-use of industrial orange wastes as organic fertilizers. Biores Technol 53: 43-51

Garland JL (1997) Analysis and interpretation of community-level physiological profiles in microbial ecology. FEMS Microb Ecol 24: 289-300

Garland JL, Mills AL (1991) Classification and characterization of heterotrophic microbial communities on the basis of patterns of community-level sole-carbon-source utilization. Appl Environ Microbiol 57: 2351-2359

Harch BD, Correl RL, Meech W, Kirkby CA, Pankhurst CE (1997) Using the Gini coefficient with Biolog substrate utilisation data to provide an alternative quantitative measure for comparing bacterial soil communities. J Microbiol Methods 30: 91-101

Hellmann B, Zelles L, Palojarvi A, Bay Q (1997) Emission of climate-relevant trace gases and succession of microbial communities during open-windrow composting. Appl Environ Microbiol 63: 1011-1018

Hopkins DW, MacNaughton SJ, O'Donnell AG (1991) A dispersion and differential centrifugation technique for representatively sampling microorganisms from soil. Soil Biol Biochem 23: 217-225

Insam H (1997) A new set of substrates proposed for community characterisation in environmental samples. In: Insam H, Rangger A (eds) Microbial communities: functional versus structural approaches. Springer, Berlin Heidelberg New York, pp 259-260

Insam H, Rangger A (1997) Microbial communities: functional versus structural approaches. Springer, Berlin Heidelberg New York

Insam H, Amor K, Renner M, Crepaz C (1996) Changes in functional abilities of the microbial community during composting of manure. Microb Ecol 31: 77-87

Lindstrom JE, Barry RP, Braddock JF (1998) Microbial community analysis: a kinetic approach to constructing potential C-source utilisation pattern. Soil Biol Biochem 30: 231-239

Macdonald RM (1986) Sampling soil microfloras: dispersion of soil by ion exchange and extraction of specific microorganism by elutriation, Soil Biol Biochem 18: 399-406

McKinley VL, Vestal JR (1984) Biokinetic analysis of adaptation and succession: microbial activity in composting municipal sewage sludge. Appl Environ Microbiol 47: 933-939

Miller FC (1993) Composting as a process based on the control of ecologically selective factors. In: Metting FB (ed) Soil microbial ecology. Marcel Dekker, New York

Niepold F, Conrad R, Schlegel HG (1979) Evaluation of the efficiency of extraction for the quantitative estimation of hydrogen bacteria in soil. Antonie van Leewenhoek 45: 485-497

OriginLab Corporation (1999) Origin version 6.0 - Northampton, MA

Sharma S, Piccolo A, Insam H (1997) Different carbon source utilisation profiles of four tropical soils from Ethiopia. In: Insam H, Rangger A (eds) Microbial communities: functional versus structural approaches. Springer, Berlin Heidelberg New York, pp 132-139

Sparling GP, Ord BG, Vaughan D (1981) Changes in microbial biomass and activity in soils amended with phenolic acids. Soil Biol Biochem 13: 455-460

SPSS (1998) Statistical package of the Social Sciences. SPSS, Chicago

Strom P (1985) Effect of temperature on bacterial species diversity in thermophilic solid-waste composting. Appl Environ Microbiol 50: 899-905

Williams AH (1962) Enzyme inhibition by phenolic compounds. In: Pridham JB (ed) Enzyme chemistry of phenolic compounds. Pergamon Press, New York. pp 87-95

Evaluation of Organic Matter Stability During the Composting Process of Agroindustrial Wastes

F. Tittarelli[1], A. Trinchera[1], F. Intrigliolo[2] and A. Benedetti[1]

Abstract. Composting of wastes from citrus industrial processing (pastazzo and sludge) was studied in order to evaluate the evolution of organic matter during the process and to individuate chemical and/or biochemical techniques able to set the stability of the final product. Composts from two open-air piles of different composition were sampled every month during the whole period of composting (5 months) and the organic matter of each sample was characterised by chemical and biochemical techniques. Humification rate (HR%) and humification index (HI) were determined. Extracted organic matter of the six samples collected for each compost was investigated by isoelectric-focusing technique (IEF). The biochemical analysis were based on the study of C-mineralisation after the addition of each collected sample to soil. Results obtained clearly demonstrated organic matter evolution during composting processes. Humification rates increased and humification indexes decreased over time, while extracted organic matter showed electrophoretic behaviour typical of stabilised organic compounds. Moreover, mineralisation patterns confirmed the increased level of organic matter stability during the composting process.

Introduction

During the last years the citrus-processing industry, in Italy, has been increasing its economical importance. Nowadays, about 800 000 tons year^{-1} of citrus fruits are processed, with a production of 500 000 tons year^{-1} of pastazzo, a mixture of citrus pulp and skins, which represents the main residue of citrus-processing squeezing treatment. In addition to pastazzo, citrus-processing industry produces also a high amount of effluents, mainly constituted by fruits and factory plant-washing waters, which undergo a purification treatment with the production of sludge. The use of these residues as matrices for the production of a quality compost could constitute a typical example of transformation of a waste into a resource with a market value (Sequi and Tittarelli 1998).

One of the key issues in defining compost quality is the evaluation of its organic matter stability. For this reason, it is necessary to individuate and set up analytical methods able to follow the organic matter transformation during the composting process.

[1]Istituto Sperimentale per la Nutrizione delle Piante, Roma, Italy
[2]Istituto Sperimentale per l'Agrumicoltura, Acireale (CT), Italy

H. Insam, N. Riddech, S. Klammer (Eds.)
Microbiology of Composting

During composting, for a combination of biological and chemical transformations, total organic carbon content decreases, but the relative content in humic (or humic-like) compounds increases. Determination of the humification rate (HR%) and the humification index (HI%) demonstrated to be effective indicators of the formation of humic-like substances during compost maturation, being able to establish with accuracy the moment of complete stabilisation (Ciavatta et al. 1990).

Another chemical technique able to characterise organic matter of composted amendments is isoelectric focusing (IEF) (Ciavatta et al. 1993). This technique allows one to fractionate the organic compounds on the basis of their isoelectric point and their electrophoretic mobility. IEF was usefully utilised in order to obtain information on the qualitative characteristics of organic matter in soils (Ciavatta and Govi 1993), amendments and organic fertilisers (Ciavatta et al. 1997; Canali et al. 1998).

An effective biochemical method able to define, for a compost, the level of organic matter stability is the study of organic carbon mineralisation after addition of compost to soil (Kirchmann 1991; Tittarelli et al. 1998). Since organic matter resistance to soil microbial degradation is directly correlated to its chemical stabilisation, it is possible to follow organic matter evolution, during a composting process, by detecting evolved CO_2 after soil amendments with compost sampled from the heap at prefixed time.

The aim of this work is to study organic matter evolution of two composts from agroindustrial wastes (the first obtained mixing pastazzo, citrus-processing sludge and green residues, the second mixing only pastazzo and green residues), by means of chemical and biochemical techniques able to set the different degrees of organic matter stability during the composting process.

Composts considered in this work were further investigated by thermal analysis in Dell'Abate and Tittarelli (in press).

The reported study was carried out in the framework of the Research Program Perspectives in Organic Farming Fertilisation, funded by the Italian Ministry of Agriculture.

Materials and Methods

Composting trials were performed in Azienda Sperimentale Palazzelli (Lentini) of the Istituto Sperimentale per l'Agrumicoltura, Acireale (CT, Italy).

Two piles were prepared, the first by mixing pastazzo, coming from citrus-processing industry, with green residues (Cp) and the second by adding sludge from the citrus-processing industry to the matrices mentioned above (Cps).

Input matrices were mixed taking into account the chemical analysis of pastazzo and sludge (Table 1), and the C/N ratio of green residues (which was 77). The weight percentages of the used matrices are reported in Table 2.

Temperatures of the two piles were detected during the whole composting period and moisture maintained at the optimum values (50- 60%). Piles were turned more frequently during the thermophilic phase and less frequently during the mesophilic composting phase. At mixing time (T0) and after 29 (T1), 67 (T2), 89 (T3), 130 (T4) and 165 (T5) days, compost samples were collected from each pile. Each sample was constituted by six subsamples, taken after turning the heap, then mixed until homogenisation. The samples were oven-dried at 50 °C, ground and sieved at 1 mm and finally stored for subsequent analysis. All analysis were performed in three replicates.

Table 1. Main chemical-physical parameters of pastazzo and sludge

Parameter	Unit	Pastazzo	Sludge
Moisture	%	88.6	65.1
pH	-	3.2	8.2
N	%	1.39	3.55
P_2O_5	%	0.27	3.66
K_2O	%	0.90	0.54
C_{org}	%	51.4	28.0
C/N	-	37	8
Ca	%	1.06	6.25
Mg	%	0.12	0.72
Cd (total)	mg kg^{-1}	<0.5	5.0
Hg (total)	mg kg^{-1}	<0.1	0.12
Cu (total)	mg kg^{-1}	7.0	105
Zn (total)	mg kg^{-1}	12	680
Ni (total)	mg kg^{-1}	<0.5	37.5
Pb (total)	mg kg^{-1}	<0.5	27.0
Cr(VI)	mg kg^{-1}	n.d.[a]	n.d.[a]

[a] n.d. = not detectable

Table 2. Weight ratios of the matrices used for the production of composts Cp and Cps

Compost	Pastazzo (%)	Sludge (%)	Green residues (%)
Cps	40	20	40
Cp	60	-	40

Extraction and Fractionation of Organic Matter, Determination of C/N Ratio

The extraction was carried out on 2 g of each compost sample with 100 ml of a solution $NaOH/Na_4P_2O_7$ 0.1 N for 48 h at 65 °C. Samples were centrifuged at 2500 rpm and supernatant solution filtered through a 0.45 μm Millipore filter. Extracts were stored at 4 °C under nitrogen atmosphere.

Humic and fulvic acids were fractionated by acidification of 25 ml of the extract with H_2SO_4 50 %, separating humic-like acids (HA) (precipitated) from fulvic-like acids (FA) (in solution). The last ones were purified on a polyvynilpyrrolidone (PVP) column, resolubilised with NaOH 0.5 N and then joined to the humic portion. Combined fractions (HA+FA) were quantitatively transferred into a calibrated 50-ml flask, brought to volume with NaOH 0.5 N and stored at 4 °C under nitrogen atmosphere. Total organic carbon (C_{org}) in compost samples was determined according to Springer and Klee (1954), while total extractable carbon (C_{extr}) and humic and fulvic acids carbon (C_{HA+FA}) were determined following the procedure proposed by Ciavatta et al. (1990). Humification parameters for assessment of organic matter stabilisation in compost were so calculated:

$$\text{Humification rate (HR) \%} = (C_{HA+FA} \times 100)/C_{org}$$
$$\text{Humification index (HI)} = C_{not\ humified}/C_{HA+FA} = (C_{extr} - C_{HA+FA})/C_{HA+FA}$$

Total nitrogen content (%) of each compost sample was determined by the dry combustion method, using a LECO FP 228 nitrogen determinator, and the C/N ratio was then calculated.

Isoelectric Focusing (IEF)

Isoelectric focusing separations for T0- T5 samples of Cps and Cp were carried out in a Multiphore II, LKB electrophoretic cell, according to Govi et al. (1994). Ten mL of $NaOH/Na_4P_2O_7$ extract were dialysed in 6,000- 8,000-Da membranes, lyophilised and then separated on a 5.06% T and 3.33% C polyacrylamide slab gel, in which a pH range 3.5- 8.0 was created using a mixture of carrier ampholytes (Pharmacia Biotech): 25 units of ampholine pH 3.5- 5.0; 10 units of ampholine pH 5.0- 7.0; 5 units of ampholine pH 6.0- 8.0.

A prerun (2 h; 1200 V; 1 °C) was performed and the pH gradient formed in the slab was checked by a specific surface electrode. The electrophoretic run (2 h 30'; 1200 V; 1 °C) was carried out loading the water-resolubilised extracts (1 mg C 50 μL^{-1} $sample^{-1}$). The bands obtained were stained with an aqueous solution of Basic Blue 3 (30%) and scanned by an Ultrascan-XL Densitometer. In order to compare isoelectric focusing data of Cps and Cp samples, the same operative conditions were applied for both IEF separations.

Organic carbon mineralisation

Organic carbon mineralisation of samples taken during composting processes (T0-T5) from Cps and Cp piles was followed after their addition to soil, according to Kirchmann (1991), partially modified by Tittarelli et al. (1998). After incubation at constant temperature (30 °C) and moisture (60% WHC), determination of evolved C-CO_2 was performed at prefixed time (1, 2, 4, 7, 9, 11, 14, 16, 18, 22, 28, 35, 42, 49, 56 and 63 days). Soil respiration was used as control. The amount of organic carbon mineralised from the samples taken during the composting process was calculated as difference between the C-CO_2 evolved by the system soil + compost and the C-CO_2 evolved by the not amended soil, assuming that added material does not have priming effect on soil organic matter decomposition (Kirchmann 1991).

Results and Discussion

Composting Process

For both piles, after a first thermophilic phase, during which 60 °C were exceeded, the temperature was maintained over 50 °C for about 60 days. During the following period, a slow decrement of this parameter was detected in both piles, until reaching 40 °C in the curing phase. Cps and Cp composts, at the end of the processes, did not contain visible residues of the raw starting materials and appeared brown-coloured and homogeneous.

Organic Carbon, C/N Ratio and Humification Parameters

In Fig. 1, total organic carbon content (%), C/N ratio, humification rate (HR%) and humification index (HI) for each sample of Cps and Cp are reported.

For both the piles, C/N ratio decreased constantly, confirming the correct trend of organic matter transformation. Humification rates sensibly increased during the composting process, for both Cps and Cp, reaching quite high values. The final humification rate was the same for the two composts (about 46%), while a difference was noted in the final value of the humification index, that was 0.35 for Cps and 0.1 for Cp. The difference in the asymptotic value of the HI for the two composts could be determined by the different qualitative characteristics of the organic matter in the raw starting materials, since a difference in this humification parameter was detected also between the mixtures at T0 (0.75 for Cps against 0.58 for Cp).

Considered chemical parameters affirm that organic matter of the two composts, from a quantitative point of view, underwent a relevant stabilisation during the composting process, because of the mineralisation of labile fractions (decrease of total organic C and consequent decrease of C/N ratio) and the formation of humified or humic-like compounds, as attested by high values of final HR and low values of final HI (Govi et al. 1994; Sequi 1995; Dell'Abate et al. 1998).

Isoelectric Focusing (IEF)

In Fig. 1, the IEF profiles of extracted organic matter for Cps and Cp composts from T0 to T5 are reported.

The IEF profile of the extractable organic matter at the mixing phase (sample T0) for Cps compost showed one high peak at pH 3.5 and a group of peaks of lower intensity, not resolved, in the pH range between 4.0 and 4.7. At T1 time, the IEF profile resulted more defined and characterised by a group of bands between pH 4.2 and 4.7.

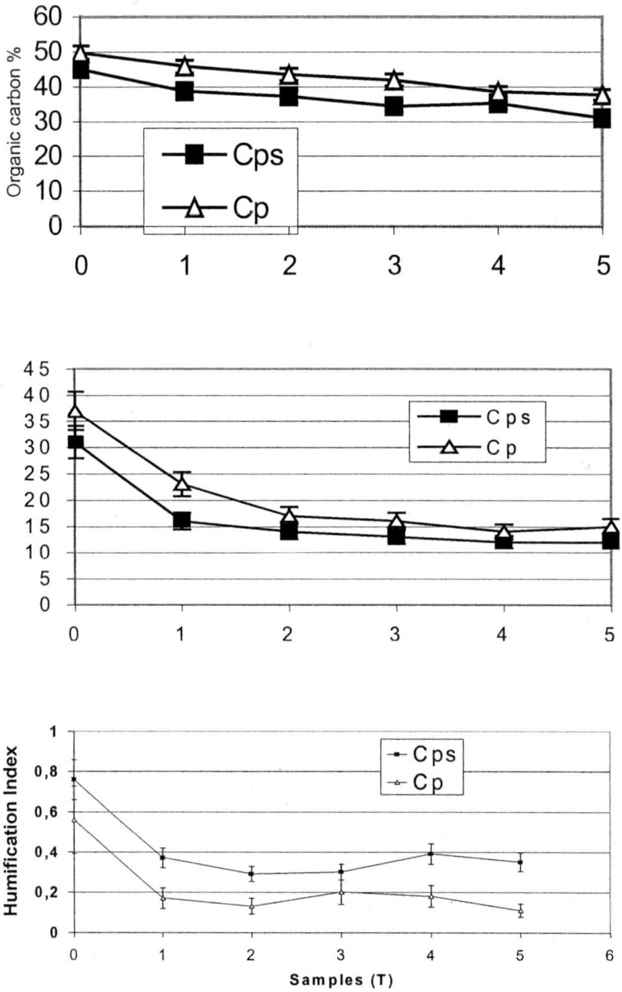

Fig. 1. Changes of Cps and Cp chemical parameters (organic carbon, C/N ratio and humification index) during the composting process. Data reported are means of three replicates.

Profiles relative to the subsequent samples (T2- T5) globally maintained the same configuration revealed at T1 time, even if at pH 4.7 a little increase of the last peak intensity was detected for the final T5 sample.

In the case of Cp, the IEF profile of the sample T0 was sufficiently resolved, with one high intensity peak at pH 3.5, one peak of medium intensity at pH 3.9 and a series of peaks well defined in the pH range 4.2- 4.7 with decreasing intensity. In T1- T5 profiles, some not-resolved peaks with low intensity appeared at pH >4.8, together with a reduction of peak focused at pH 3.5 and a gradual increase of the band focused at pH 4.7.

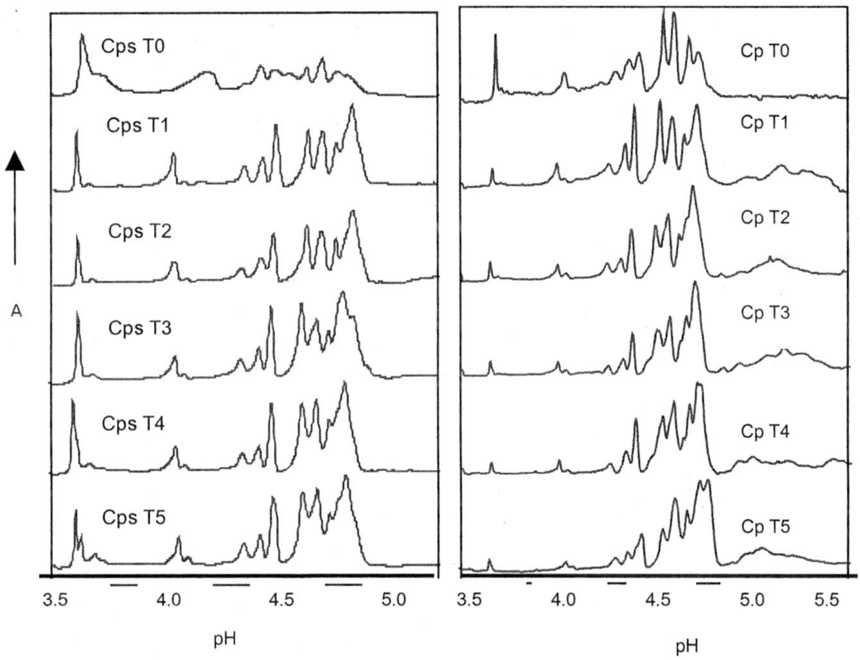

Fig. 1. IEF profiles of extracted organic matter for Cps and Cp composts (T0÷T5); pH ranges below reported are referred to pH values corresponding to the obtained IEF peaks after scansions between pH 3.5 and 8.0

IEF traces of the two composts showed that, during the composting process, extractable organic matter underwent a qualitative transformation, determined by the increase or appearance of bands in correspondence of pH>4.5 values. As reported in literature (De Nobili et al. 1989; Govi et al. 1994; Canali et al. 1998), more humified organic compounds focus at high pH values (4.5- 7.0). The increase of peaks intensity at pH higher than 4.5 during both the composting processes could correspond to the presence of more stabilised organic compounds. This result confirmed what arose from the determination of humification parameters.

Organic Carbon Mineralisation

Organic carbon mineralisation of Cps and Cp samples (T0- T5) added to the soil showed similar trends (Fig. 2).

In both cases, during 64 days of trial, about 30% of added organic carbon of composts was mineralised at T0, while about 16% was mineralised at T1. At time T2, mineralised organic matter from Cps compost was around 6% of added organic carbon, reaching the level of stability which was detected in the later samples T3, T4 and T5. In the Cp compost, at T2 the amount of mineralised organic carbon from compost was 10% and reduced to 6% for sample T3, whose level of stability resulted comparable to those of T4 and T5.

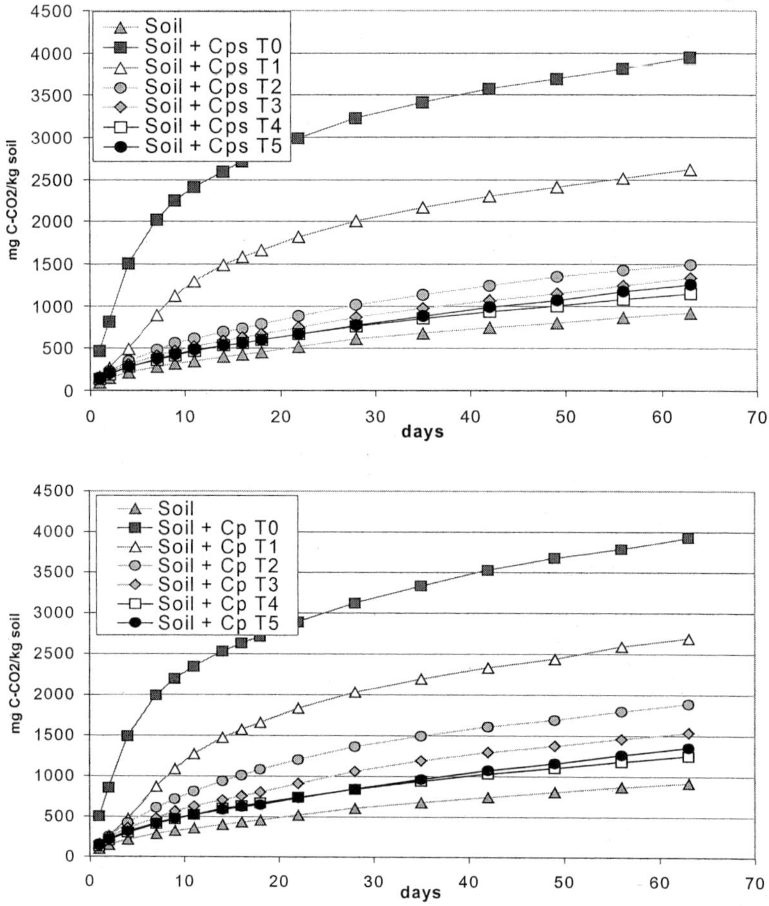

Fig. 2. Carbon mineralisation curves for Cps and Cp composts added to the soil. Values reported are means of three replicates. Calculated standard deviations are not graphically detectable

For both the typologies of composted amendments, the level of organic matter stability to the mineralising activity of soil microrganisms at T4 and T5 resulted the same (4% of total added carbon was mineralised during the whole period of incubation). The cumulative curves related to the considered composts evidenced that organic matter reached, at the end of the composting process, similar stability in presence or absence of sludge in the starting mixture. Moreover, while in Cps this stabilisation was reached after 67 days (T2), in Cp 90 days (T3) were necessary to obtain comparable level of organic matter evolution.

Conclusions

This study demonstrated that considered techniques were able to detect the organic matter transformation during the composting process. Humification parameters revealed a similar increase in organic matter stability for Cps and Cp composts, also confirmed by the study of organic matter mineralisation and the isoelectric focusing technique. Moreover, IEF traces of the two composts showed some qualitative differences in more stable organic matter fraction, probably determined by the difference in composts starting mixtures. Apparently Cp compost, produced without sludge, seemed to be better humified, as attested by the presence of some bands focused at pH>4.8, not revealed for Cps compost.

Humification rates were in accordance with data obtained by carbon mineralisation tests, while humification indexes corresponded to information arising from the isoelectric focusing technique. These findings were probably due to the different organic matter fractions considered in the analytical procedures. Indeed, mineralisation trials and HR determinations were based on total organic carbon contained in the studied composts, while HI and IEF were referred to total extractable carbon fraction. So, if mineralisation curves together with the HR could furnish quantitative information on the organic matter stability, the HI and the IEF are able to demonstrate changes in compost organic matter quality.

In conclusion, the composting processes developed in a regular way both in presence and absence of sludge in the starting mixture. Moreover, mineralisation patterns indicated that organic matter stability to soil microbial degradation activity was reached in advance in presence of sludge in respect to the compost produced without this starting material. The potential technological and economical consequence of this observation should deserve further investigation.

Acknowledgements. Thanks are due to the Cost Action 831 for financial support offered for the participation to the Conference Microbiology of Composting held in Innsbruck on 18-20 October, 2000.

References

Canali S, Trinchera A, Benedetti A, Pinzari F (1998) Study of compost maturity by means of humification parameters and isoelectric focusing technique. Proc 16th World Congr of Soil Science. Symp 40, Montpellier 20-26 August (CD-ROM)

Ciavatta C, Govi M (1993) Use of insoluble polyvinylpyrrolidone and isoelectric focusing in the study of humic substances in soils and organic wastes. J Chromatogr 643:261-270

Ciavatta C, Govi M, Vittori Antisari L, Sequi P (1990) Characterization of humified compounds by extraction and fractionation on solid polyvynilpyrrolidone. J Chromatogr 509:141-146

Ciavatta C, Govi M, Sequi P (1993) Characterization of organic matter in compost produced with municipal solid wastes: an Italian approach. Compost Sci Util 1:75-81

Ciavatta C, Govi M, Sitti L, Gessa C (1997) Influence of blood meal organic fertiliser on soil organic matter: a laboratory study. J Plant Nutr 20(11):1573-1591

Dell'Abate MT, Canali S, Trinchera A, Benedetti A, Sequi P (1998) Thermal analysis in evaluation of compost stability: a comparison with humification parameters. Nutr Cycl Agroecosyst 51:217-224

De Nobili M, Ciavatta C, Sequi P (1989) La valutazione del grado di maturazione della sostanza organica del compost mediante la determinazione di parametri dell'umificazione e per elettrofocalizzazione. In: Proc Int Symp on Compost Production and Utilisation, S. Michele all'Adige, pp 328-342

Govi M, Ciavatta C, Gessa C (1994) Evaluation of the stability of the organic matter in slurries, sludges and composts using humification parameters and isoelectric focusing. In: Senesi N, Miano TM (eds) Humic substances in the global environment and implications on human health. Elsevier, Amsterdam, pp 1311-1316

Isermayer H (1952) Eine einfache Methode zur Bestimmung der Bodenatmung und der Karbonate in Böden. Z Pflanzenernaehr Bodenkd 56:26-38

Kirchmann H (1991) Carbon and nitrogen mineralization of fresh, aerobic and anaerobic animal manures during incubation with soil. Swed J Agric Res 21:165-173

Sequi P (1995) Evolution of organic matter humification during composting processes. In: Lemmes B. (ed) The challenge. Fitting composting and anaerobic digestion into integrated waste management. ORCA Techn Doc n°5, Bruxelles, pp 153-159

Sequi P, Tittarelli F (1998) Outlook on perspectives for compost in Italy. In: Federal Ministry for the Environment, Youth and Family Affairs, (ed.) EU Symposium Compost–Quality Approach in European Union. Wien, 29-30 October 1998, pp 161-167

Springer U, Klee J (1954) Prüfung der Leistungsfähigkeit von einigen wichtigeren Verfahren zur Bestimmung des Kohlemstoffs mittels Chromschwefelsäure sowie Vorschlag einer neuen Schnellmethode. Z Pflanzenernähr Düng Bodenkd 64:1

Tittarelli F, Dell'Abate MT, Piazza P, Varallo G (1998) Effect of fly ash addition on organic matter stabilisation of composts. Proc 16th World Congr of Soil Science, Symp 40, Montpellier 20-26 August 1998 (CD-ROM)

Characterization of Organic Substances in Stabilized Composts of Rest Wastes

A. Zach[1]

Abstract. Composting of organic material is primarily a mineralization process but also a stabilization process accompanied by the formation of humic substances. Wastes from households (rest wastes) which undergo a sufficient mechanical-biological pretreatment (MBP) before deposition are, according to biological tests (gas generation, respiration activity), stable but still have a high content of about 30 % of organic matter. Different parameters were used to characterize stable organic matter showing that the content of humic substances seems to be an important one. Humic substances contribute up to 1/3 to the total amount of the organic material. It was found that after deposition of MBP-wastes humic substances remain stable or their amount even increases. Structural analysis by means of FTIR indicate increasing – age depending – stability of humic acids. Thus they represent a sink for carbon and nitrogen.

Introduction

A major goal of waste treatment before deposition is reduction and stabilization of the remaining organic substances in order to minimize emission potential to a harmless level. Mechanical-biological pretreatment (MBP) is one measure to reach this goal, yet, MBP-wastes still contain about 30 % of organic matter. However, biological tests show that these materials are largely stable and non-reactive leading to at least 90 % reduction of emissions compared to untreated wastes (Binner et al. 1999; Müller 1995).

The question arises if the remaining organic content is long term stable. And, basically, how do we define stability when we talk about pretreated wastes from households (rest waste)?

To define stability in general, above all, the factors time and environmental conditions must be considered. It is important to define the period, within deposited wastes from households remain stable. Thus, stability might be understood as no sudden release of emissions (gaseous or liquid) when waste is landfilled. MBP must anticipate the 10 to 20 years continuing phase of intensive emissions (which occur when fresh waste is landfilled). Considering this, stability in this regard may not to be seen as "always-lasting zero emission", but rather as

[1]Universität für Bodenkultur (University of Agricultural Sciences),
Department of Waste Management, Nußdorfer Lände 29-31, 1190 Wien

H. Insam, N. Riddech, S. Klammer (Eds.)
Microbiology of Composting

"delayed, apparently not noticeable emission", which does not affect the environment.

Material and methods

Landfill simulation tests were carried out with two MBP-wastes (both mixed with sewage sludge). Materials were treated in different treatment plants (Siggerwiesen and Oberpullendorf) and had a different degree of stabilization. At the MBP-treatment plant Siggerwiesen rest wastes were ground to a particle size < 100 mm and pre-decomposed for app. 3 days in a so-called DANO-drum. Before the active decomposition phase sewage sludge was added. Material 1 was directly taken after adding the sludge. At the MBP-treatment plant Oberpullendorf rest waste and sewage sludge were pre-decomposed in a so-called DANO-drum for 3 days, after separation of metals, bulk materials and plastics. Then material was sieved < 25 mm, mounted to piles of 2 m height and aerated for 5 weeks. After 5 weeks sample was taken (Material 2).

App. 100 kg of material 1 were further ground to a particle size < 20 mm. In adapted laboratory boxes material was aerated and decomposed under controlled conditions. After 8 weeks of decomposition – when stability criteria according to Binner et al. (1999) were reached – aeration was stopped and the material was put into a landfill simulation reactor (LSR, Figure 1), where leaching processes, degradation and humification can be measured under defined boundary conditions (e.g. landfill-like conditions). Material 2 was directly taken from the Oberpullendorf pretreatment plant after 5 weeks of decomposition and put into a LSR without any further treatment.

During the test, leachate was recirculated periodically. Landfill gas was trapped in bags and quantity and composition were measured in order to monitor the anaerobe environment. Investigation temperature was 35°C. Sampling was carried out irregularly depending on the status of the degradation process. To prevent oxygen diffusion solid samples were taken out of the LSR by flushing the reactor surface with N_2. Samples were analyzed for ignition loss, TOC and total nitrogen. Furthermore, analyses of substance groups was carried out according to a method from Van Soest (1963) to determine easy degradable substances, cellulose and non degradable substances. Yet, in most cases only cellulose was determined.

Humic substances were determined according to a method of Danneberg and Schaffer (1974). The authors prescribe a 4-step extraction of an air-dried sample with 0,1 M $Na_4P_2O_7$-solution. According to the color of the extracted sample, results lead to the relative parameter "optical density" for fulvic and humic acids. A quantitative determination of humic acids was carried out by precipitation with concentrated HCl and – after washing -redissolving the precipitate with 0,1 M $Na_4P_2O_7$. The solution was measured by a spectrophotometer at 400 nm. Calibration was done by a humic acid standard (Fluka No. 53680).

This method was also used to calculate the E4/E6-ratio, which is the quotient of the absorption at 400 nm and 600 nm. While high E4/E6-ratios indicate substances with lower molecular weight (fulvic acids, low-molecular humic acids), low E4/E6-ratios characterize more complex substances with higher molecular weight (humic acids, humine) and indicate a more aromatic character rather than an aliphatic one (Tan 1993; Schnitzer 1991).

For FTIR analysis precipitated humic acids were freeze dried and 2-3 mg were mixed with KBr, homogenized and pressed. Analysis were carried out on a Bruker Equinox 55 in a wave number range of 400 – 4000 cm^{-1}. Spectra were baseline-corrected and standardized.

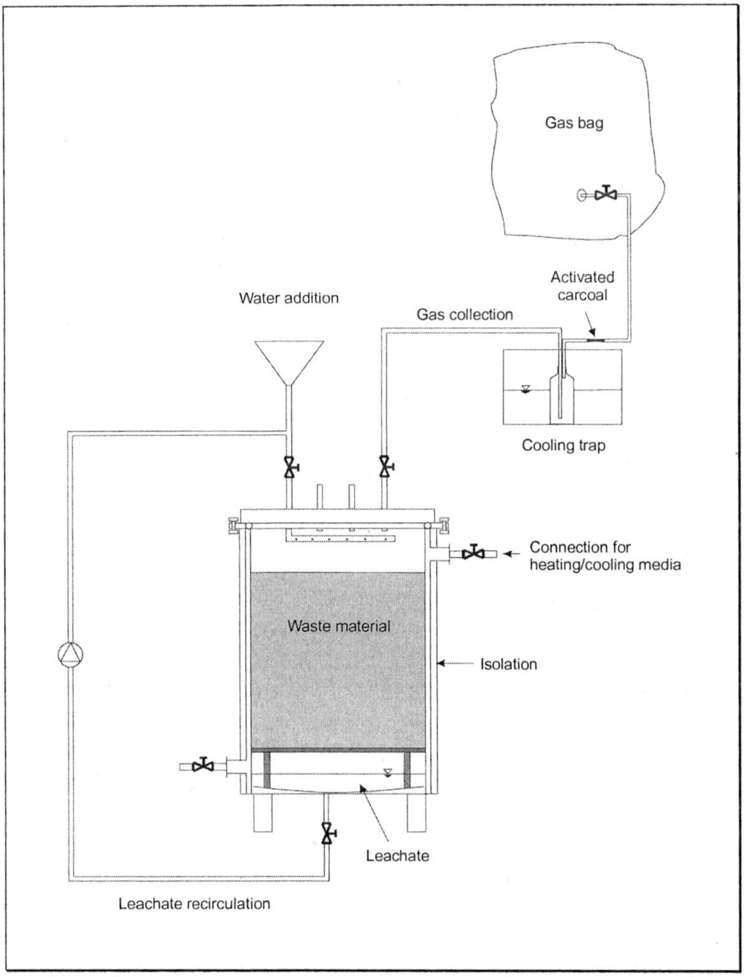

Fig. 1 Scheme of a landfill simulation reactor

Results and discussion

Development of organic matter

Results indicate, that MBP-waste from Siggerwiesen (material 1) was already well stabilized, TOC decreased only slightly within 80 weeks (table 1). Results vary due to material inhomogenities (high amount of plastic). Cellulose decreases to some extent. Nitrogen remains constant more or less at a value of app. 1 % DS.

For MBP-waste from Oberpullendorf (material 2) TOC decreases clearly. Although both materials were differently pretreated, TOC values after 80 weeks were nearly equal. Cellulose decreases clearly and thus contributes significantly to the gas generation. Nitrogen shows strong increase tendencies at the beginning, at the end thc content varies.

Table 1: Development of organic substances of MBP-wastes during a landfill simulation test

No.	Week	TOC (%DS)		Cellulose (%DS)		Nitrogen (%DS)	
		Sample 1	Sample 2	Sample 1	Sample 2	Sample 1	Sample 2
1	Start	18.9	27.5	7.6	14.4	0.93	0.78
2	3	n.d.	23.6	n.d.	15.1	n.d.	0.86
3	7	17.3	22.4	7.9	7.7	1.96	0.97
4	14	19.9	19.1	10.1*	6.8	1.04	1.00
5	20	19.4	18.4	7.8	6.0	0.95	1.12
6	30	17.5	18.3	7.6	6.0	0.89	1.00
7	55	16.6	19.3	7.0	4.1	1.02	0.90
8	80	16.8	18.4	5.3	3.8	0.90	1.04

*: outlier; n.d.: not detected

Van Soest analysis were carried out from the input samples of the landfill simulation tests and the output samples after 80 weeks, in order to investigate the change of easy degradable, cellulose and non degradable organic substances. This analysis provides sum parameters that can be correlated with a specific degradational behavior. In both samples cellulose seems to be the most important parameter to describe long term emission behavior (figure 2). The amount of non degradable organic matter, containing humic substances, lignin and plastics, remains to a large extent constant.

Degradation of cellulose

Stabilized material 1 showed only very small cellulose degradation, after 80 weeks still 5.3 % cellulose were determined. Material 2, however, showed a clear reduction of more than 70 %. For both materials the contribution of cellulose degradation to the gas generation was calculated. For material 1 more than 80 % of the gas generation result from cellulose degradation, for material 2 still 60 % were found, which indicates that in reactive material also other components contribute to the gas generation. However, to a large extent remaining gas generation of MBP-waste can be explained from cellulose degradation.

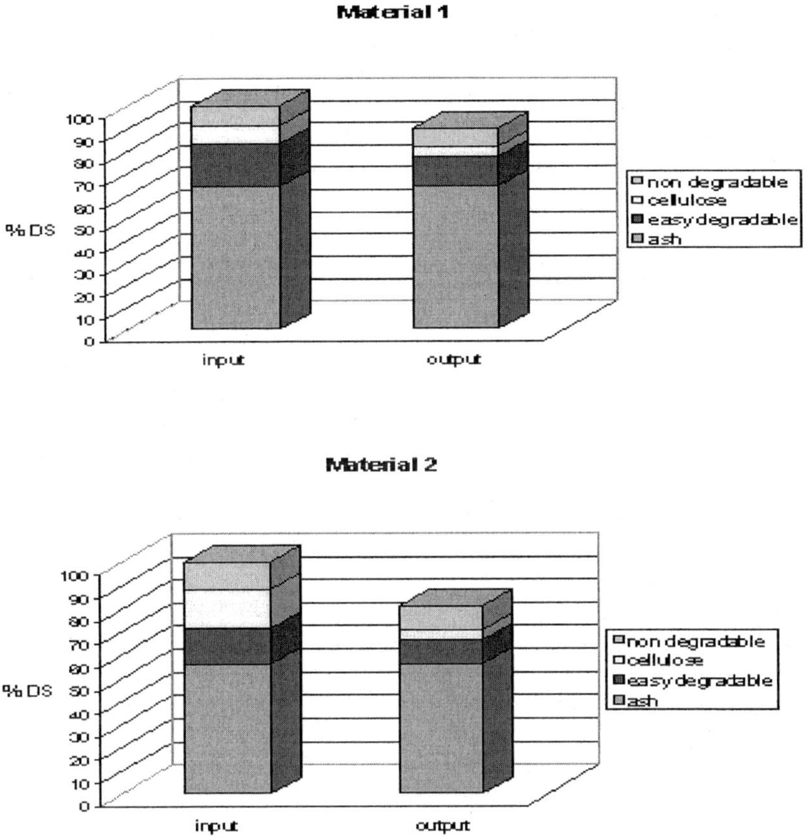

Fig. 2. Development of organic substances of MBP-waste during 80 weeks in a landfill simulation test

Figure 3 illustrates the good linear correlation ($r = 0.977$) between the cellulose content and the gas generation (in 240 days) of different MBP-materials.

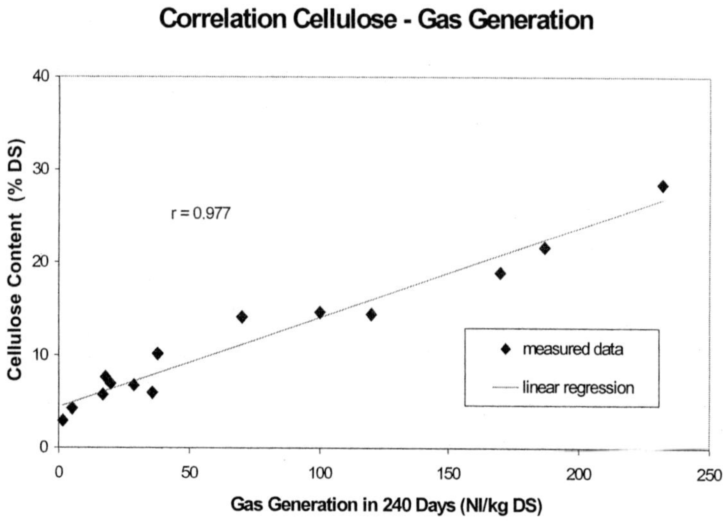

Fig. 3. Correlation between gas generation in 240 days (measured in incubation tests and LSR-tests) and content of cellulose (BINNER et al. 1998, ZACH 1996)

Development of stable organic matter

Humification

Figure 4 and figure 5 show the development of humic substances during the landfill simulation tests of MBP-materials. Results are given in optical densities related to the content of dry organic substance at the time of reactor input. Furthermore, the amount of humic acids is given percentually.

Already the amount of humic substances of the input material 1 – after 8 weeks of aerobic treatment – must be classified quite small. Compared to investigations of other materials from MBP-wastes amounts up to three times higher were found (Zach and Schwanninger 1999; Heisz-Ziegler 2000). The quantitative determination of humic acids underlines these results, yet showing an increase from week 20 – 60 even if finally the amount decreases again.

Figure 5 shows the development of humic substances of material 2. A significant increase of humic substances, especially humic acids, can be observed. However, this increase does not prove to be linear and indicates a specific dynamics of increase and decrease. At the beginning of the anaerobic phase app. 3 % of dry substance were humic acids, after 80 weeks 4.2 % - related to the input sample - were found.

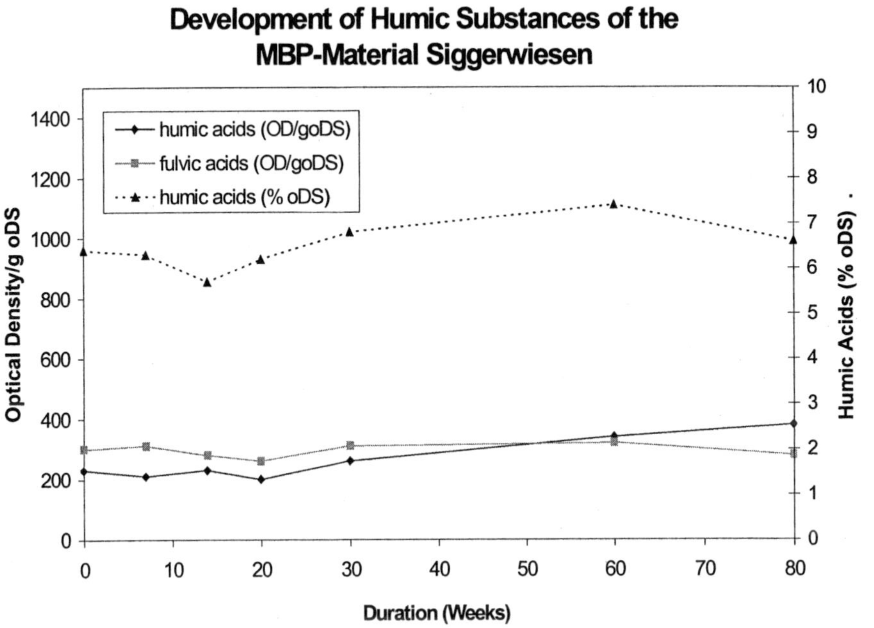

Fig. 4. Development of humic substances of material 1 (OD: optical density; oDS: organic dry substance)

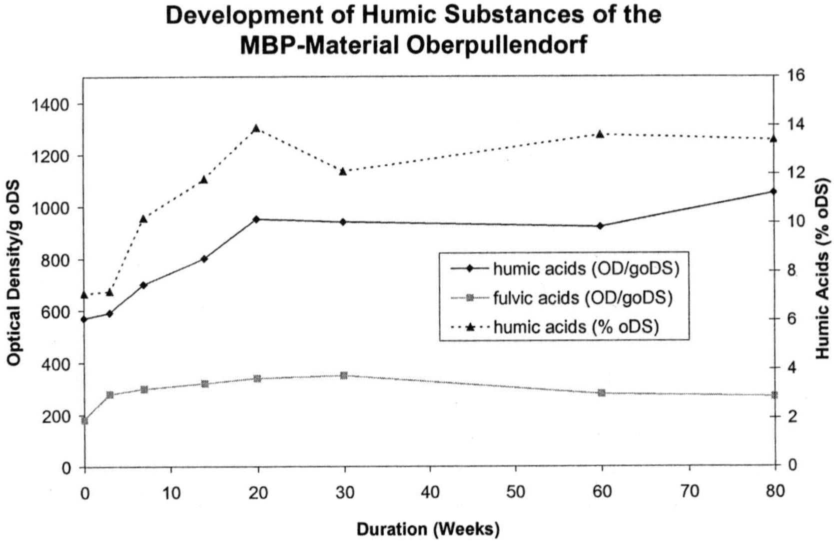

Fig. 5. Development of humic substances of material 2 (OD: optical density; oDS: organic dry substance)

The last value corresponds to 13.4 % of humic acids in the dry organic substance. The strongest increase of humic acids was found in the most active gas generation phase. Obviously, the presence of intermediate degradation products contributes to the humification. Towards the end of the active gas generation phase, a decrease of the humic acids was observed. Similar development was also discovered by Heisz-Ziegler (2000).

These investigations show that stability and formation of humic substances is dependent on the residual reactional behavior. They can be built up under anaerobic conditions. In aerobic stabilized MBP-wastes organic substance stability can be uphold under landfill-like conditions. Filip (1993) examined municipal solid waste (MSW), which was landfilled without pretreatment and determined a clear increase of humic acids during deposition. Within one year the humic acid content doubled itself. However, Filip determined also a reduction of humic substances. In table 2 analysis results of humic acid contents of MBP-wastes are compared with samples from old landfills and literature data.

Table 2. Humic acid contents of various waste materials and composting products

Sample	Age, Treatment	Humic acids (% OS)
Sample 1 (LSR 1 input – MSW/sewage sludge) Sample 1 (LSR 1 output – MSW/sewage sludge)	8 weeks aerobe + 80 weeks anaerobe	6.4 6.6
Sample 2 (LSR 2 input – MSW/sewage sludge) Sample 2 (LSR 2 output – MSW/sewage sludge)	5 weeks aerobe + 80 weeks anaerobe	7.1 13.4
Old landfill (Rautenweg)	7 years anaerobe	2.7 – 3.2
Old landfill (Haag/Hausruck)	5 years anaerobe	5.7
Old landfill (Haag/Hausruck)	15 years anaerobe	22.0
Old landfill (Haag/Hausruck)	20 years anaerobe	11.8
MSW (GRÜNEKLEE and KLEIN, 1993)	6 – 9 months	13.9
Paper/sewage sludge (GRÜNEKLEE and KLEIN, 1993)	90 days	3.2
MSW (GRÜNEKLEE and KLEIN, 1993)	2 years	23.8
MSW/sewage sludge (GRÜNEKLEE and KLEIN, 1993)	106 days	4.85
Biowaste (GRÜNEKLEE and KLEIN, 1993)	3 months	7.23 – 10.64
Biowaste (GRÜNEKLEE and KLEIN, 1993)	24 months	7.58 – 12.71
MSW (GRÜNEKLEE and KLEIN, 1993)	147 days	9.1
MSW/sewage sludge (GRÜNEKLEE and KLEIN, 1993)	239 days	13.7
MSW/sewage sludge(GRÜNEKLEE and KLEIN, 1993)	289 days	11.3
MSW <25 mm/sewage sludge (HEISZ-ZIEGLER, 2000)	24 weeks	6.8
Biowaste compost (GRUNDMANN, 1991)	3 months	7.23/10.64 *
Biowaste compost (GRUNDMANN, 1991)	24 months	7.58/12.71 *
MSW-compost (JIMENEZ & GARCIA, 1992)		12 – 14
Peat substrate (Fruhstorfer Erde)		7.1

*: different results from the use of various extracting agents

E4/E6-ratio

Table 3 shows the E4/E6-ratios for fulvic and humic acids in samples of both LSR's. Both materials behave more or less equal, E4/E6-ratios of fulvic acids were found between 8 and 14 which is – as expected – higher than the E4/E6-ratios of humic acids between 5 and 6. A significant change over the duration of anaerobic treatment could not be detected, while differences between fulvic acids and humic acids indicate a more intensive degree of polymerization and stabilization of the second ones. Similar observations were made by Grassinger (1998) for humic acids from composts. Grundmann (1991) found for humic acids values between 5 and 6 and Fricke et al. (1990) indicate that for humic acids from composts values under 5 are hardly achieved, due to their younger age compared to soil humic acids. For soil usually lower E4/E6-ratios are found. Results point out that humic and fulvic acids from mechanically-biologically pretreated rest wastes are – even if they are 1.5 years exposed to anaerobic conditions –structurally different from soil humic substances.

Table 3. E4/E6-ratio of humic substances

Sample No.	Week	E4/E6 Fulvic Acids		E4/E6 Humic Acids	
		Sample 1	Sample 2	Sample 1	Sample 2
1	Start	8.3	9.0	6.3	5.7
2	3	8.5	8.9	6.1	6.2
3	7	8.2	9.1	6.2	5.9
4	14	11.0	10.7	5.3	5.3
5	20	10.9	8.5	5.2	5.6
6	30	8.0	11.7	6.4	6.0
7	55	10.3	14.0	6.5	6.1
8	80	9.3	13.5	5.1	6.6

Elementary analysis of humic acids

To evaluate the ability of binding C and N, some of the humic acids were analyzed for their elementary composition. The results are indicated in Table 4.

The carbon-content of humic acids remains to a large extent constant, tendencies of decrease might, however, be indicated. Results for carbon can be compared with literature data for compost humic acids as well as for soil humic acids (Filip 1993; Haider 1996; Scheffer and Schachtschabel 1998). Nitrogen varies between 5.3 and 6.4 % which is higher than usually found in humic acids from soil (Fig. 6). Chen et al. (1996) point out as well, that the N-content of humic acids from composts tends to be higher than the one in soil. This might result from a higher N-offer in waste materials at the beginning of decomposition.

Table 4. Elementary analysis of humic acids of MBP-wastes

Sample	Age	C (%)	N (%)	S (%)
LSR 1:				
MHS A (sample No.1)	input	52.0	6.2	1.2
MHS B (sample No.7)	55 weeks	51.4	5.3	1.7
MHS C (sample No.8)	80 weeks	not analyzed		
LSR 2:				
MHS A (sample No.1)	Input	53.7	5.5	1.4
MHS B (sample No.7)	55 weeks	52.4	6.4	2.0
MHS C (sample No.8)	80 weeks	49.6	5.9	n.a.

n.a.: not analyzed

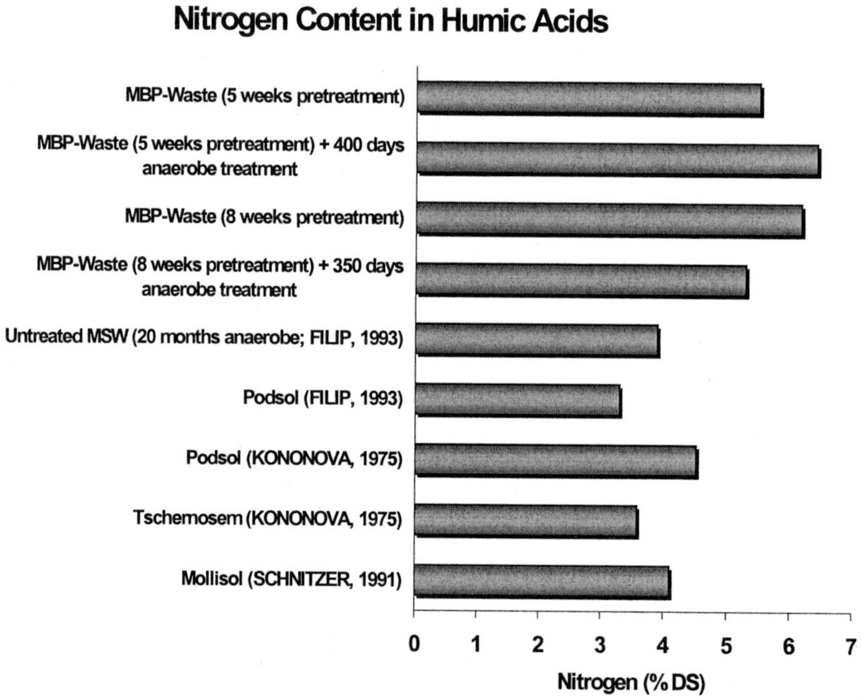

Fig. 6. N-content of humic acids originating from different materials

FTIR-characterization of humic acids

The Fourier Transform Infrared Spectrophotometry (FTIR) is a wide-spread procedure for the characterization of humic substances in soils and composts. So far, only few data exists about samples of MBP-waste (Filip 1993). In this investigation FTIR-spectra of humic acids extracted from differently pretreated MBP-wastes were recorded. A humic acid of a peat-containing substrate was used as a reference.

From both MBP-materials samples were taken at 3 different anaerobic stages. Spectra of humic acids are shown in figure 7 and figure 8. In general, spectra were similar and can be compared to the results of Filip (1993). Chen et al. (1996) also found a similarity of spectra from humic acids of composts amongst each other.

Characteristic for all samples are peaks at 2960 cm^{-1}, 2920 cm^{-1} and 2850 cm^{-1}, representing C-H vibrations of aliphatic methyl and methylene groups ($-CH_3$, $-CH_2$). The intensity of these peaks (which is stronger in MHS 2) decreases with the period of landfill like conditions, indicating a reduction of the aliphatic character. These peaks are weaker in the humic acid of the peat-containing substrate (figure 9). In MBP-humic acids the carboxylic peak (C=O) at 1720 cm^{-1} forms only a shoulder whereas in the peat-containing substrate a clear peak can be detected. All MBP-samples have a broad peak at 1655 cm^{-1}, which is typical for aromatic C=C bonds and C=O groups. It may also indicate amide I groups. An increase in this region – as it can be found for the samples after 55 weeks – might indicate a stronger aromatic structure; however, LSR-output samples (after 80 weeks) showed different development. A variation in this region was also observed during microbial modification and degradation of soil humic acids by Filip et al. (1999). In the reference material absorption at 1655 cm^{-1} is missing but an intensive peak appears at 1615 cm^{-1}. According to TAN (1993) this peak can be attributed to aromatic C=C bonds, thus indicating a highly aromatic structure.

A slight shoulder at 1533 cm^{-1} can be attributed to amide II groups (Chen et al. 1996; Filip 1993) as well as a vibration at 1518 cm^{-1}, which is nearly missing in the output sample. The latter might also represent C=C bonds. The peaks from 1455 cm^{-1} to 1370 cm^{-1} can be assigned to CH- and NH-vibrations. According to Chen et al. (1996) the peak at 1420 cm^{-1} represents molecule skeleton vibrations of the aromatic ring. A broad area around 1230 cm^{-1} indicates amide III bonds (Filip et al. 1999). At 1035 cm^{-1} $O-CH_3$ vibrations absorb, which is very intensive in the reference substrate.

Generally, both MBP-samples show strong changes in the area between 2960 cm^{-1}, 2920 cm^{-1} and 2850 cm^{-1} as well as at 1720 cm^{-1}, 1650 cm^{-1}, 1516 cm^{-1} and 1130 cm^{-1} during the anaerobic period. This indicates a structural change of humic acids over the time of deposition from an aliphatic to a more aromatic character. Results may also indicate a certain nitrogen dynamic. Very interesting results can be found by comparison of MBP-humic acids with a humic acid from a peat substrate. The spectrum of the second ones shows characteristic sharp peaks pointing out a highly aromatic structure. It seems, that with increasing age humic acids from MBP-wastes structurally approach this peat humic acid (of course, there are still differences, especially in the area between 1420 cm^{-1} and 1535 cm^{-1}).

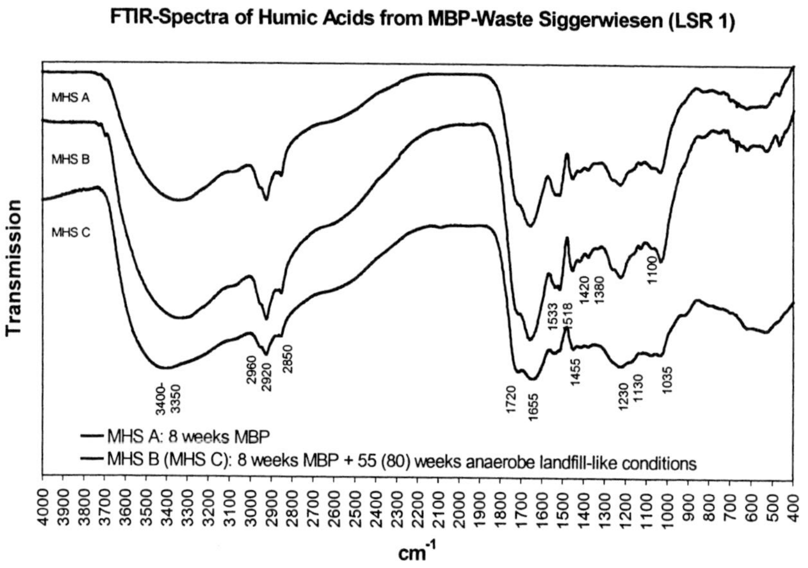

Fig. 7. TIR-spectra of humic acids from MBP-waste after 8 weeks aerobic treatment (MHS A), additional 55 weeks (MHS B) and 80 weeks (MHS C) respectively under landfill-like conditions

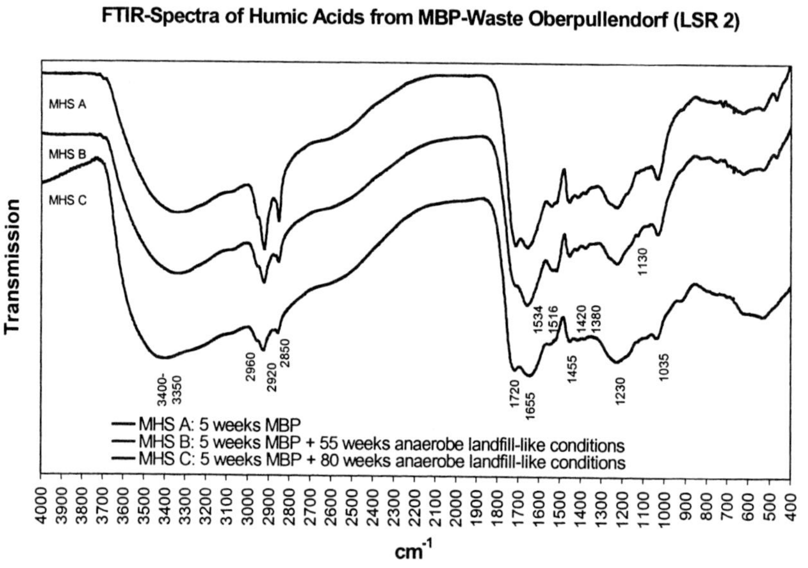

Fig. 8. FTIR-spectra of humic acids from MBP-waste after 5 weeks aerobic treatment (MHS A), additional 55 weeks (MHS B) and 80 weeks (MHS C) respectively under landfill-like conditions

However, similarity of MBP humic acids and peat humic acids could help to predict stability of organic matter.

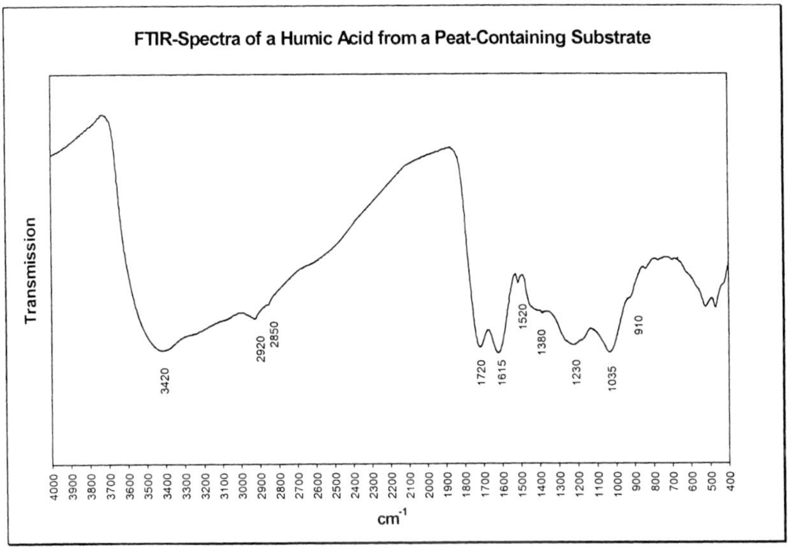

Fig. 9. FTIR-spectrum of a humic acid of a peat-containing substrate

Conclusions

The behavior of organic substances of two different mechanically-biologically pretreated wastes was examined under landfill-like conditions. Special focus was put on the development of stable organic substances, expressed by the parameter humic substances. By means of different analysis quality and quantity of the stable organic substance was studied and compared with analysis of several years deposited organic waste and data from literature.

Investigations show that organic substance of MBP-wastes changes under landfill-like conditions, depending on the degree of stabilization. MBP-waste, which is already stable, remains to a large extent stable in the anaerobic landfill environment. Only negligible emissions were detected which were less than 10 % compared to untreated waste. The most converted and degraded organic component is cellulose which contributes to 60 - 80 % to the remaining gas generation potential.

Formation of humic substances by means of mechanical-biological pretreatment leads to a fixation and immobilization of carbon and nitrogen. The investigations of the MBP-materials show that humic substances are either stable or can be built up, depending on the reactivity of the deposited material. This underlines their sink-function for carbon and nitrogen. Still reactive MBP-material shows an increase of the amount of humic substances. A total amount of humic acids of about 13 % of organic substance, which are a part of the humic substance system, were analyzed, while samples from an old landfill showed even 22 %. Compared with soil humic acids a tendency of increased nitrogen fixation could be observed.

FTIR-studies turned out to be a highly suitable investigation to examine structural changes of the humic substance system during the time of deposition (in this study only humic acids were analyzed). Changes in the areas between 2960 cm^{-1}, 2920 cm^{-1} and 2850 cm^{-1} as well as at 1720 cm^{-1}, 1650 cm^{-1}, 1516 cm^{-1} and 1130 cm^{-1} indicate a structural change of humic acids from aliphatic to more aromatic character. Many peaks show carboxylic groups which point out their acidic character and thus their ability to interact with metals and ions. Results may also indicate nitrogen fixation in $-NH_2$ and -NH- groups of MBP-humic acids. However, a more clear conclusion has to be underlined by further analysis (e.g. elementary analysis). With increasing age humic acids tend to higher structural stability. Interesting results can be found by comparison of MBP-humic acids with a humic acid from a peat substrate. This stable, old organic substance could act as a reference material in characterizing structural stability of humic acids, structural similarity may indicate stability of organic matter.

So far, FTIR-characterization of the total organic substance seems to be even more promising than characterizing humic compounds alone (Smidt 2000). A direct application of FTIR or NIR analysis to characterize organic substance in wastes might lead to an analysis procedure without any fractionating and extraction or other chemical analysis of the sample. This would also be a great step in simplifying analysis of a complex organic matrix. Yet, further investigations have to be carried out on this subject.

References

Binner E, Zach, A, Widerin M, Lechner P (1998) Auswahl und Anwendbarkeit von Parametern zur Charakterisierung der Endprodukte aus mechanisch-biologischen Restmüllbehandlungsverfahren. In: Lechner P (ed) Waste Reports No 07/1998.

Binner E, Zach A, Lechner P (1999) Stabilitätskriterien zur Charakterisierung der Endprodukte aus MBA-Anlagen. Projektbericht im Auftrag des Bundesministeriums für Umwelt Jugend und Familie.

Chen Y, Chefetz B, Hadar Y (1996) Formation and properties of humic substance originating from composts. In: de Bertoldi M, Sequi P, Lemmes B, Papi T (eds) The Science of Composting. Blackie Academic & Professional.

Danneberg O H, Schaffer K (1974) Eine einfache kolorimetrische Analyse des Huminstoffsystems. Sonderdruck aus „Die Bodenkultur", Band 25, Heft 4.

Filip Z (1993) Mikrobiologische, biochemische und stoffliche Beurteilung des Stabilisierungprozesses in einer Hausmülldeponie. In: Wiemer K, Kern M (eds) Biologische Abfallbehandlung. M.I.C Baeza-Verlag.

Filip Z, Pecher W, Berthelin J (1999) Microbial Utilization and Transformation of Humic Acids Extracted from Different Soils. Z. Pfanzenernähr. Bodenk. 2/99, Vol. 162, 215ff.

Fricke K, Pertl W, Mielke C, Schridde U, Vogtmann H (1990). Rottesteuerung und Qualitätssicherung. In: Fricke K, Turk T, Vogtmann H (eds) Grundlagen in der Kompostierung. EF-Verlag für Energie und Umwelttechnik, Berlin.

Grassinger D (1998) Einfluss von Temperatur und Sauerstoffgehalt auf die Humifizierung und die Mineralisierung bei der Verrottung von Bioabfall. Dissertation an der Abteilung Abfallwirtschaft, Universität für Bodenkultur

Grundmann J (1991) Reifegradbestimmung von Komposten durch Huminstoffanalytik – Eignung und Methode. Müll und Abfall 5/91, 268ff.

Grüneklee C E, Klein T (1993) Bedeutung der Huminstoffbildung für die biologische Restmüllstabilisierung, In: Wiemer K, Kern M (eds) Biologische Abfallbehandlung; S 905ff.

Haider K (1996) Biochemie des Bodens, Ferdinand Enke Verlag Stuttgart.

Heisz-Ziegler C (2000) Verhalten von Stickstoff bei der Ablagerung von Abfällen. Dissertation an der Abteilung Abfallwirtschaft, Universität für Bodenkultur.

Jimenez E I, Garcia V P (1992) Relationships between Organic Carbon and Total Organic Matter in Municipal Solid Wastes and City Refuse Compost. In: Biores. Techn. 41, 265ff.

Kononova, M. M. (1975). Humus of Virgin and Cultivated Soil. In: Giesking E (ed) Soil Components Vol. I, Organic Components. pp.475-526.

Müller W (1995) Leistungsfähigkeit der biologischen Restmüllbehandlung und Auswirkungen der biologischen Vorbehandlung auf die Stabilität des zu deponierenden Materials, Studienreihe Abfall Now, 14, Abfall Now e. V., Stuttgart.

Scheffer F, Schachtschabel P (1998) Lehrbuch der Bodenkunde. Ferdinand Enke Verlag, Stuttgart, 14. Aufl.

Schnitzer M (1991) Soil Organic Matter – The Next 75 Years. Soil Science Vol. 151, No. 1, pp. 41-55.

Smidt E (2000) Entwicklung der abfallbürtigen organischen Substanz in anthropogenen Ablagerungen. Dissertation an der Universität für Bodenkultur. in Bearbeitung.

Tan K H (1993) Principles of Soil Chemistry. Marcel Dekker Inc., New York, Basel, Hong Kong.

Van Soest P J (1963) Use of Detergents in the Analysis of Fibrous Feeds. II. A rapid Method for the Determination of Fiber and Lignin, J. Assoc. Off. Anal. Chem. (A.O.A.C.), 46 (5), 829-835.

Zach A (1996) Stoffgruppenanalyse - modifiziert nach Van Soest (Modified sequential fibre analysis - based on method developed by Van Soest). In: Lechner P (ed) Waste Reports No. 04

Zach A, Schwanninger M (1999) Charakterisierung von Huminsäuren in Abfallstoffen mittels FTIR. Österreichische Wasser- und Abfallwirtschaft, 11/12. Springer Verlag.

Application and Environmental Impact

Composting of *Posidonia oceanica* and Its Use in Agriculture

P. Castaldi[1] and P. Melis*

Abstract. The extraordinary accumulation of deposits of the seaweed *Posidonia oceanica* on the beaches of Sardinia has given rise to a series of problems linked to its disposal. Leafy deposits of *Posidonia oceanica* are treated as refuse and disposed of in waste dumps. This is an enormous waste of organic material.

Bearing this in mind, some composting experiments have been carried out using washed *Posidonia oceanica,* sludge and woody residues. The selected substrates were categorised and then mixed homogeneously, according to the previously established parameters, to obtain different mixtures which would satisfy the preliminary physical and chemical conditions for correct composting. The experiment was conducted using an open system with natural ventilation.

The final assessment of the chemical parameter values of the compost showed a good carbon, nitrogen and phosphorus content and a balanced C/N ratio. The degree of humification was raised appreciably, which indicated that the material had reached a high level of stability. The parameters for the physical characteristics (porosity, apparent density, easily available water and reserve water, etc.) were within the acceptable limits. No pathogenic microorganisms were found, which indicates that the compost was healthy, and the heavy metal content, in particular that of thallium, cadmium, chromium and lead, was below the levels established by law. These results offer an innovative solution to the disposal of *Posidonia oceanica* which would otherwise be taken to waste dumps, and offers undoubted advantages for the coastal communities interested in this problem.

Introduction

The disposal of the annual accumulation of *Posidonia oceanica* on the beaches of the Mediterranean causes a series of economic and environmental problems (Thelin and Giorgi. 1984). In this particular situation, the leafy deposits of *Posidonia oceanica* on the beach can be considered refuse, and at present they are transported to waste dumps, with the resulting loss of this enormous mass of organic material. It therefore seems of value to suggest an alternative system for recycling this waste in a way which satisfies the most recent U. E. directives.

[1]Dipartimento di Scienze Ambientali Agrarie Biotecnologie Agro - Alimentari, Sez. Chimica Agraria ed Ambientale,. Università di Sassari, Viale Italia 39, 07100 Sassari, Italia
*Corresponding author. Tel: +39 – 079 – 229214, Fax: +39 – 079 – 229276
e – mail: pmelis@ssmain.uniss.it

H. Insam, N. Riddech, S. Klammer (Eds.)
Microbiology of Composting

Refuse management places particular importance on the methods of recovery and recycling. Composting resolves this problem to a great extent and allows the part of the waste which is recyclable and useful for agriculture to be recovered (Melis et al. 1999 a). *Posidonia oceanica* is a material which in chemical terms, can be compared to other vegetal waste biomass (Baldissera Nordio et al. 1967; Melis et al. 1999 b). It is particularly rich in structural carbohydrates (C/N ratio > 65%) and thus suitable to combine in the right proportions with mainly nitric natural residues such as sludge, a potentially compostable waste material (Melis and Cattivello 1999).

Materials and Methods

The matrices used for the preparation of the experimental trial, sludge from urban sewage, *Posidonia oceanica* from the beach and pruning wastes were analysed using the appropriate methods (Min. Agricoltura e Foreste 1989, 1991) to find the ideal mixture which would satisfy the physical and chemical parameters needed as a preliminary condition for correct composting (Szmidt 1997). The initial characteristics also took into consideration the quantity of heavy metals.

Posidonia oceanica leaves have been collected from Alghero beach, Sassari (Sardinia), using a particular machine equipped with a rake. This has guaranteed a reduced removal of sand. Before of the preparation piles we provided for washing of the leaves with running water. This operation was done to prevent a possible negative influence of NaCl. After washing, the leaves were air dried and then ground in a crushing machine. The pruning wastes were treated in the same way. This allowed to be obtain ground material with dimensions between 5 to 10 cm. Because the leaves of *Posidonia oceanica* on the beach have a high structural carbohydrate content, and thus, as is well known, are not easily degraded, the mixtures were prepared with variable carbon residue content. One mixture (A) was prepared with a C/N ratio of 18, a second (B) with a C/N ratio of 22 and a third (C) with a C/N ratio of 36. One composting pile of each mixture was built on a concrete platform sheltered with a suitable roof. The piles had a rectangular base, 3 x 1.5 m, and were 0.9 m high. The total volume was about 4 m^3. The development of the process was monitored using a probe digital thermometer. Readings were taken daily at three depths in the piles (20-50-80 cm) during the active phase and weekly during the management phase. The piles were turned and the humidity level restored, depending on their temperature and humidity. Three samples were taken at 15, 30 and 90 d from the beginning of the process. The representative samples were characterised in conformity with the methods laid down by the Italian laws governing the analysis of compost (Min. Agricoltura e Foreste 1989, 1991). The humic and fulvic acid content (HA + FA), the non-humified portion (NH) and the humification parameters (percentage, rate and humification indices) were determined following the methods proposed by Sequi et al. (1986), Ghabbour et al.(1994) and Michaelson and Ping (1997). The heavy metal content was determined by mineralisation using the Milestone MLS 1200 microwave oven. The solutions were analysed by spectrophotometry with a Beckman plasma (Regione Piemonte 1992). The transformation of the lignocellulose component was measured by Van Soest's method for the determination of cellulose, hemicellulose and lignin (Martilotti et al. 1987). The biological assays (germination) were carried out following the methods of Nappi et al. (1990) to determine the phytotoxicity of the composts. In addition, tests for *Salmonella* and infesting seeds were carried out (Regione Piemonte 1992).

The physical characteristics of the substrate (apparent density, real density, porosity, easily available water, reserve water, bound water, etc.) were evaluated using the method proposed by De Boodt and Verdonck (1974).

Results and Discussion

Tables 3.1, 3.2 and 3.3 show the chemical characteristics of the matrices and the composition of the experimental mixtures.

Table 3.1. Chemical parameters of the matrices

Chemical Parameters	Sludge	*P. oceanica*
pH	7.16	-
Humidity (%)	76.41	10.64
Electric conductivity (mS cm^{-1})	0.70	-
Ashes (% d.m.)	49.45	9.52
Total organic carbon (% d.m.)	31.50	47.93
Total nitrogen (% d.m.)	2.43	0.73
C/N ratio	12.96	65.66
Total phosphorus (% d.m.)	1.28	0.07

Table 3.2. Heavy metals in the initial material (mg kg^{-1} d.m.)

Metal (mg kg^{-1} d.m.)	Sludge	*P. oceanica*
Zn	812.36	7.74
Cd	5.05	0.22
Cu	334.00	3.50
Pb	86.46	1.43
Cr	37.81	0.95

Table 3.3. Percentage makeup of the mixtures

Initial material	Mixture A	Mixture B	Mixture C
P. oceanica (%)	30	40	50
Sludge (%)	50	40	30
Lignocellulose residues (%)	20	20	20
C/N ratio	18	22	36

The composting process proceeded with an initial increase in temperature up to 60 °C, which was sufficient to guarantee the sanitation of the mass. After about a month the temperature began to fall due to a slowing down of the oxidising process carried out by microflora present in the organic matter. This followed a pattern influenced by the frequency of turning, with the temperature rising steadily and dropping abruptly when the piles were turned. The composting process and the resulting product was assessed from the results of the chemical analysis (Tables 3.4, 3.5 and 3.6). The total organic carbon content (TOC) fell in all the samples, above all in the first 30 d. This was due to the mineralisation to CO_2 and H_2O carried out by microorganisms. It then stabilised, indicating that the degradation process of the organic molecules was slowing down and that synthesis of the humic molecules was beginning. The total organic carbon content of the three piles after 90 d was similar to that reported in the literature for composts obtained from biological sludges and vegetal wastes (Tables 3.4, 3.5 and 3.6) (Adani et al. 1999). The nitrogen content followed a similar pattern, albeit to a different extent. The microorganisms use carbon and nitrogen, in a average ratio of 30:1, as an energy and structural resource for the synthesis of the proteins that they need to reproduce. During the phase of intense degradation of the organic substances there was a reduction in ammonia nitrogen, presumably due in part to volatilisation as NH_3 and/or oxidation to NO_3 (Tables 3.4, 3.5 and 3.6). Nonetheless, comparison of the final data with the initial situation shows that the ratio of organic to total N increased in all the piles. This was to associated with the reorganisation carried out by the microorganisms to facilitate their reproduction and the insertion of nitrogen in the humic molecules. The C/P ratio tended to fall in a way similar. The content in certain samples of the three different forms of total, extractable and humified carbon, enabled the humification parameters to be established. The carbon initially present in the experimental mixtures was in part mineralised and in part transformed into humic substances (Watanabe and Kuwatsuka 1992).

Table 3.4. Chemical characteristics of mixture A

Chemical characteristics	Initial material	15 days	30 days	90 days
pH	7.86	7.90	7.90	7.76
Humidity (%)	44.63	43.87	45.70	44.72
Salinity (me 100 g^{-1})	28.21	14.68	20.00	27.38
Total organic carbon (% d.m.)	38.91	24.15	22.74	22.49
Total nitrogen (% d.m.)	2.09	1.28	1.20	1.31
C/N ratio	18.62	18.87	18.95	17.17
Ammoniac nitrogen (% d.m.)	1.01	0.24	0.13	0.09
Nitric nitrogen (% d.m.)	0.18	0.31	0.41	0.47
Total extracted carbon (% d.m.)	10.08	8.30	8.76	8.80
HA + FA (% d.m.)	3.90	4.43	6.10	7.04
Humic acids (% d.m.)	1.98	2.56	4.00	5.18
Fulvic acids (% d.m.)	1.92	1.87	2.10	1.86
Degree of humification (DH %)	38.69	53.37	69.63	80.00

Table 3.4. (cont.)

Humification rate (HR %)	10.00	18.34	26.82	31.48
Humification indices (HI)	1.58	0.87	0.44	0.25
Total phosphorus (% d.m.)	1.22	1.16	1.19	1.30

Table 3.5. Chemical characteristics of mixture B

Chemical characteristics	Initial material	15 days	30 days	90 days
pH	7.86	7.95	8.00	7.84
Humidity (%)	52.26	50.65	54.22	50.79
Salinity (me 100 g^{-1})	31.91	21.21	22.62	32.23
Total organic carbon (% d.m.)	40.30	30.11	26.30	23.16
Total nitrogen (% d.m.)	1.85	1.16	1.18	1.22
C/N ratio	21.78	25.96	22.30	18.98
Ammoniac nitrogen (% d.m.)	1.08	0.18	0.11	0.11
Nitric nitrogen (% d.m.)	0.18	0.36	0.45	0.49
Total extracted carbon (% d.m.)	13.11	11.76	10.36	10.12
HA + FA (% d.m.)	3.95	4.67	5.86	7.00
Humic acids (% d.m.)	1.99	2.69	3.58	5.08
Fulvic acids (% d.m.)	1.96	1.98	2.28	1.92
Degree of humification (DH %)	30.13	48.21	56.56	69.17
Humification rate (HR %)	9.80	18.83	22.28	30.22
Humification indices (HI)	2.32	1.52	0.77	0.45
Total phosphorus (% d.m.)	1.13	1.11	1.16	1.02

Table 3.6. Chemical characteristics of mixture C

Chemical characteristics	Initial material	15 days	30 days	90 days
pH	7.95	8.00	8.47	8.10
Humidity (%)	54.49	50.05	50.26	50.00
Salinity (me 100 g^{-1})	35.52	22.50	23.62	36.50
Total organic carbon (% d.m.)	42.00	27.64	23.86	23.35
Total nitrogen (% d.m.)	1.17	0.96	0.96	0.93
C/N ratio	35.90	28.79	24.85	25.10
Ammoniac nitrogen (% d.m.)	0.69	0.23	0.12	0.15
Nitric nitrogen (% d.m.)	0.11	0.26	0.32	0.39
Total extracted carbon(% d.m.)	13.48	12.78	10.31	10.51
HA + FA (% d.m.)	3.99	4.98	6.03	6.71
Humic acids (% d.m.)	2.00	2.70	3.32	4.68
Fulvic acids (% d.m.)	1.99	2.28	2.71	2.03
Degree of humification (DH%)	29.60	46.80	58.49	63.84
Humification rate (HR %)	9.50	21.63	25.27	28.74
Humification indices (HI)	2.38	1.57	0.71	0.57
Total phosphorus (% d.m.)	0.98	0.92	0.96	0.98

The percentage of humic carbon increased considerably during the process until the stabilisation phase, and reached values which were near the limits established by Italian norms. In all the piles the HA/FA ratio tended to increase. This gave the compost greater resistance to degradation in the soil and thus extends the time when the additive is effective (Ciavatta et al. 1988; Chefetz et al. 1998).

The values for the humification rate (HR%) in the final product are typical of highly humified matrices, comparable with those found in organomineral fertiliser from humified peat (Fig. 3.1A) (Piccolo et al. 1997). The degree of humification (DH%) is an estimate of the proportion of organic matter humified compared to the total which could be humified. The values were very high in piles with 30% and 40% P. oceanica (Fig. 3.1B). A higher proportion of sludge in the compost mixture seems to have a positive influence on this parameter. The humification index values (HI) vary between 0 and 0.6, and are typical of humified material such as organic matter in the soil (Fig. 3.1C). The overall results for humification of the matrices used can be considered extremely satisfying.

The heavy metals content in the three composts was particularly low and below the acceptable limits established by law. There was no substantial difference between the three composts in this respect, although that with the highest proportion of sludge (compost A) had higher average levels (Table 3.7).

Table 3.7. Total heavy metals content (mg kg d.m.$^{-1}$)

Metal (mg kg d.m.$^{-1}$)	Compost A	Compost B	Compost C
Cr	19.05	16.85	14.28
Mn	285.7	359.6	255.1
Cu	80.95	76.40	40.82
Zn	369.0	359.6	214.3
Cd	0.95	0.79	0.51
Tl	0.05	0.03	0.20
Pb	9.52	9.21	7.14
Ni	59.52	58.43	53.06

The results of the analysis for variations in hemicellulose, cellulose and lignin, showed particularly significant variations in the case of polysaccharide. Hemicellulose fell by 95, 98 and 67% and cellulose by 80, 68 and 74%, in composts A, C and B respectively. The transformation level for lignin was also high, reaching 20% in compost C.

Biological analysis found no evidence of *Salmonella* in the three composts, while the enterobacterial levels in all cases were below those established by law.

The toxicity test, using germination and growing samples from annual crucifer watercress (*Lepidium sativum*), which was germinated in watery substrata of the different experimental composts, showed no evidence of phytotoxicity (Table 3.8) (Helfrich 1998). Indeed, the germination index was always higher than 70%. In addition, germination and root growth were stimulated, especially in compost C. This was may be attributed to the presence of hormone-like substances in the exctracts.

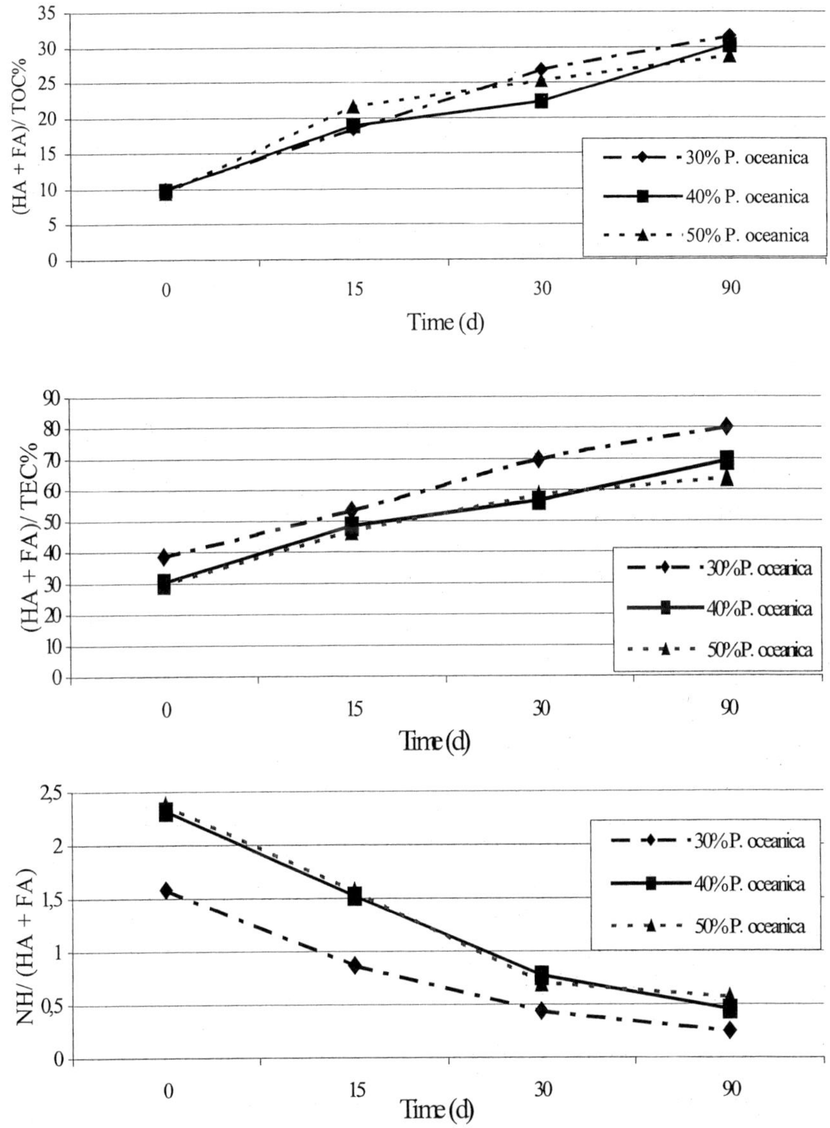

Fig. 3.1. A – C. Top: Humification rate (HR %) during the composting process. **Center**: Degree of humification (DH %) during the composting process. **Bottom**: Humification index (HI) during the composting process

To evaluate the compost's suitability for agricultural use as substrate, the physical properties of the compost were determined, as this governs the capacity of the substrate to supply sufficient water and air to the roots of the plants (Illera et al. 1999).

Table 3.8. Phytotoxicity test

	Average n° of germinated seeds	Average root Length (cm)	Germination index (Ig)
Test	8.80	7.35	-
Compost A 50%	7.20	6.33	70.46
Compost A 75%	8.00	8.37	103.53
Compost B 50%	8.20	7.74	98.19
Compost B 75%	6.50	7.85	78.89
Compost C 50%	8.80	9.38	127.62
Compost C 75%	8.80	8.25	112.30

Reducing the percentage of sludge in the mixture has a proportional effect on its density (Table 3.9). All experimental tests had good results for real and apparent density yardsticks such as porosity. It is not certain if the different composition of mixtures A, B and C in this experiment affected the porosity values. The parameters which measure the capacity of the absorbent substrate to retain water with different energy (easily available water, reserve water, and bound water) had excellent results in all the composts.

Table 3.9. Physical characteristics of the compost

	Compost A	Compost B	Compost C
Apparent density ($g\ l^{-1}$)	428.14	402.13	341.14
Real density ($g\ l^{-1}$)	2201	1642	1469
Porosity (%)	80.55	75.51	76.78
Easily available water (%)	23.87	26.58	29.30
Reserve water (%)	8.54	8.83	6.99
Water not easily available (%)	67.59	64.59	63.71

Conclusions

The results confirm the agricultural value of compost produced using the organic matrices of *Posidonia oceanica* from the beach.

In particular, the results meet the parameters for quality compost and the chemical characteristics of the initial material used are such that the risk of the presence of heavy metals is markedly reduced. All the composts have good physical characteristics, which means that root structures have water constantly available and are well aerated. The germination samples confirm the technical suitability of the compost as a component for substrate.

The three composts were produced using different percentages of the leaves of *Posidonia oceanica* from the beach in the composting mixture. They had optimal amendment characteristics and are thus suitable for use both in sustainable agriculture in open fields, to improve the physical and chemical qualities of the soil, and in horticulture, floriculture and nurseries as a principal component in the preparation of substrate.

References

Adani F, Genevini P L, Gasperi F, Tambone F (1999) Composting and humification. Compost Sci Util 7: 24 – 33

Baldissera Nordio C, Gallarati Scotti G C (1967) Sulla composizione chimica delle foglie di *Posidonia oceanica* raccolte in varie epoche località e profondità. Atti Soc Ital Sci Vet XXI: 1 – 5

Chefetz B, Adani F, Genevini P L, Tambone F, Hadar Yitzhak, Yona Chen (1998) Humic–acid transformation during composting of municipal waste. J Environ Qual 27: 794 – 800

Ciavatta C, Antisari L V, Sequi P (1988) A first approach to the characterisation of the presence of humified materials in organic fertilisers. Agrochimica 32: 510 – 517

De Boodt M, Verdonck O (1974) Method for measuring the water release curve of organic substrates. Acta Hort 37: 2054 – 2061

Ghabbour E A, Khairy A H, Cheney D P, Gross V, Davies G, Gilbert T R, Zhang X (1994) Isolation of humic acid from the brown alga *Pilayella littoralis*. J Appl Phycol 6: 459 – 468

Helfrich P, Chefetz B, Hadar Y, Chen Y, Schnabl H (1998) A novel method for determining phytotoxicity in composts. Compost Sci Util 6 (3): 6 – 13

Illera V, Walter I, Cuevas G, Cala V (1999) Biosolid and municipal solid waste effects on physical and chemical properties of a degraded soil. Agrochimica XLIII: 178- 186

Martilotti F, Antongiovanni M, Rizzi L, Santi E (1987) Metodi di analisi per la valutazione degli alimenti di impiego zootecnico. Quad Metodol 8 IPRA: 122 - 134

Melis P, Cattivello C (1999) Valutazioni analitiche e colturali su alcuni substrati di coltivazione. Colt Protette 4: 65 – 70

Melis P, Castaldi P, Melis A (1999a) Caratteristiche fisiche e chimiche di substrati di coltivazione a base di compost derivato da rifiuti dell'attività agroindustriale e civile ed effetti sulle specie coltivate. Ricicla '99: 476 – 481

Melis P, Mulè P, Senette C, Castaldi P (1999b) The use of quality compost from olive milling waste in the preparation of growing media in floriculture. Proc Int Composting Symp 19 –23 September 1999 Halifax/Dartmouth Nova Scotia, Canada: 905 – 916

Michaelson G J, Ping C L (1997) Comparison of 0.1 N sodium hydroxide with 0.1 M sodium pyrophosphate in the extraction of soil organic matter from various soil horizons. Commun Soil Sci Plant Anal 28: 1141 – 1150

Min Agricoltura e Foreste D M (1989) Metodi Ufficiali di analisi per i fertilizzanti. Suppl n 1 G U della Repubblica Italiana del 23/ 8/ 1989: 40 – 47

Min Agricoltura e Foreste (1991) Metodi ufficiali di analisi per i fertilizzanti. Suppl n 2 G U della Repubblica Italiana del 4/ 2/ 1991: 31 - 39

Nappi P, Vincenzino E, Barberis R (1990) Criteri per la valutazione della qualità dei compost. Acqua Aria 3: 261 – 268

Piccolo A, Pietramellara G, Mbagwu J S C (1997) Use of humic substances as soil conditioners to increase aggregate stability. Geoderma 75: 267 – 277

Regione Piemonte (1992) Metodi di analisi dei compost. Collana Ambiente 6: 29 - 49

Sequi P, De Nobili M, Leita L, Cercignani G (1986) A new index of humification. Agrochimica, 30: 175

Szmidt RAK (1997) Composting processing residuals of seaweed (*Ascophyllum nodosum*). Compost Sci Util 5: 78 – 86

Thelin I, Giorgi J (1984) Production de feuilles dans un herbier superficiel a *Posidonia oceanica*, evaluée par une methode derivée de la methode de Zieman. Int Workshop *Posidonia oceanica* 1: 271 – 276

Watanabe A, Kuwatsuka S (1992) Chemical characteristics of soil fulvic acids fractionated using polyvinilpyrrolidone (PVP). Soil Sci Plant Nutr 38: 31 – 41

Practical Use of Quality Compost for Plant Health and Vitality Improvement

J.G. Fuchs[1]

Abstract. High quality composts were produced and applied at practical scale. The effects of these compost treatments on the health of potted plants were tested in a series of five laboratory trials. (1) Cress inoculated with *Pythium ultimum* was grown in pots containing 30% of two compost from two different origins. The disease developed with one compost, and was inhibited by the other. (2) Seedlings of cucumber were grown in substrate containing 0, 33, 67 or 100% of compost, and were inoculated with *P. ultimum*. With increasing proportions of compost, the disease was reduced. (3) Freshly steamed soil was amended with 10% compost (control: no compost). In the treated soil, far less nitrite was measured than in the control soil, and cucumbers were less attacked by *P. ultimum.* (4) Soil from a field treated with compost for five years was compared with soil not treated with compost. In the treated soil, cucumber seedlings were less attacked by *P. ultimum*, and lettuce seedlings were less attacked by *Rhizoctonia solani*, than in the control soil. (5) Barley was grown in pots containing 0, 10, 30 or 50% of composts from four different origins. With increasing proportions of compost, the incidence of powdery mildew was reduced, and there were significant differences between the compost origins.

The composition and the maturity of the compost influence the potential for suppression of plant disease. The management of the composting processes, and the oxygen supply in particular, seem to affect compost quality strongly. This has consequences on the storage management of the end products, as inappropriate storage measures can lead to reduction of product quality.

Introduction

The positive effects of high-quality composts on plant growth and health are manifold. They influence plant development by an improved soil structure and an elevated soil humus content (Zebarth et al. 1999) as well as by supplying macro- and micronutrients. A special and interesting effect of quality composts is their direct influence on plant pathogen interaction due to their potential to activate and stabilise the soil microflora (Hoitink et al. 1997). The potential for plant disease suppression is a criterion of quality composts and is not a general attribute of everything that is called "compost". The objective of the presented work was to study the potential of selected high-quality composts in improving plant health

[1]Biophyt Ltd., Institute for Agricultural and Ecological Research and Consultation, Schulstrasse 13, 5465 Mellikon, Switzerland; e-mail: biophyt@pop.agri.ch.

H. Insam, N. Riddech, S. Klammer (Eds.)
Microbiology of Composting

and vitality and to solve some important problems in horticulture on a practical level

Materials and Methods

Microorganisms

Pythium ultimum Trow strain 67-1 and *Rhizoctonia solani* Kühn strain 160 were cultivated on malt agar plates at 20 °C for 7 days (*R. solani* 14 days). For soil inoculation, three 0.8-cm plugs of the malt agar culture were placed in Petri dishes containing 25 g of autoclaved millet seeds and 10 ml sterilized distilled water. After 10 days (*R. solani* 20 days) the colonized millet was broken into small pieces and mixed into the soil. *Blumeria (Erysiphe) graminis f.sp. hordei* was cultivated on leaf segments of the barley cultivar Igri. They were placed in Petri dishes on bendimidazole agar for 10 days at 17 °C with a 16-h light period per day. The inoculation of the barley leaves was carried out in a 2.5-m-high tower with 1 m diameter (Wolfe and Schwarzbach 1978).

Composts

Unless specified, the composts used for the described experiments came from an industrial compost plant at Fehraltorf (CH). These composts are composed of a mixture of green manure (about 60%), horse manure (about 30%) and organic residues from a commercial vegetable grower. Two composting systems were used: windrows with intensive turning and aerated boxes (12 m x 4 m x 3 m).

Disease Suppression Tests

All tests for plant disease suppression by composts were made in 200-ml plastic pots. First, defined quantities of compost and pathogens were added to the soil. Two days later, cucumber (germinated, germ length 3 - 5 mm) or cress seeds were sown and moistened. After 10 days the quantity of living plants per pot was counted and root and shoot weight measured. Growth and development of plants cultivated in soils with compost and with pathogens was compared with plants grown in soil without compost and without pathogens, as well as with the plants kept in soil without compost but with pathogens.

Disease Receptivity Test

For comparison, soils or substrates were inoculated with increasing quantities of pathogens and used to fill in 200-ml plastic pots. Cucumber or lettuce seeds were sown immediately afterwards. Ten days (lettuce 25 days) later, the disease incidence in the different soils and pathogen densities were compared.

Results and Discussion

Disease Suppression with Composts

In a first test the capability of different composts to protect plants from disease was investigated. Extreme variability was found between compost batches of the same industrial plant. Whereas some composts showed no reaction to *Pythium ultimum* on cress plants, other composts show full protection against this disease (Fig. 1). After a heat treatment (1 day at 90 °C), the suppressive composts lose their potential for disease suppression (Fig. 1). This indicates that disease suppression is linked with the microbiological activity of the composts, although physiochemical and biological properties of composts could also influence suppression capacity (Boulter et al. 2000).

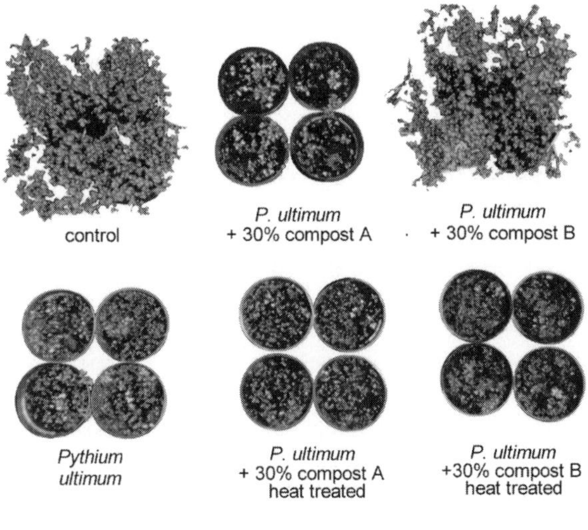

Fig. 1. Potential of two composts to protect cress against *Pythium ultimum* in comparison with untreated and heat-treated (1 day at 90 °C). The two composts come from the same industrial compost plant

The considerable variation in the potential of disease suppression of composts of the same industrial compost plant was found to have several causes. The control and management of composting processes, the stage of compost maturity and the storage conditions of the final product, all have major inputs on the biological quality of composts. Therefore, measuring the biological quality of a compost batch just before application is of high importance. On the one hand, phytotoxicity of composts was characterised with five plant growth tests according to Fuchs and Bieri (2000), on the other, the disease suppression potential of composts was determined with the *Pythium ultimum*-cucumber test. Composts showing phytotoxicity were excluded in the presented work.

Use of Quality Compost in Culture Substrate

Industrially used peat substrates are microbiologically inactive. There, a very small quantity of pathogen inoculum is sufficient to obtain a high disease incidence (Fig. 2). In these cases adding quality compost (Fig. 2) can efficiently protect plants. The addition of compost stabilises the substrate microbiologically, thus reducing the establishment of the pathogen. A drastic reduction in disease incidence occurs and the possibility for safe plant production is given.

This effect is of great importance in commercial production, especially in cultivation systems where the use of pesticides is impossible or not allowed, e.g. in organic plant seedling production or in sprout production on thin substrate layer (Fig. 3). In the latter, plants are produced on a thin layer of peat substrate and after sprouting packed in cardboard boxes with cellophane film windows (Fig. 3). They have to maintain their freshness for some days in the shop. The high humidity in the package is very favourable for the growth of mould fungi, which develop easily and quickly in such an environment. Infected sprouts have to be discarded. Adding quality compost (30% volume of substrate volume) solved this problem very efficiently. The compost microbiologically stabilised the environment surrounding the sprouts and the moulds could not develop further.

Use of Compost after Soil Steaming

Soil steaming is a very efficient but radical measure to eliminate soil-borne plant pathogens, microorganisms and weed seeds in horticulture and vegetable production. This technique has two crucial points: the building up of phytotoxic compounds in the soil, due to the degradation of dead biomass after soil heating, and the non-selective destruction of the whole soil flora and fauna complex, irrespective of whether these organisms are beneficial or harmful. "Biologically empty" soils therefore are highly susceptible to microbial colonisation after steaming.

Incorporation of quality compost (10% of the soil volume) immediately after soil steaming (Sterilo steamer, 6 h at 105 °C) prevents the accumulation of phytotoxic compounds such as nitrite (Fig. 4). This measure allows an earlier transplanting of seedlings without risk of losses by phytotoxic effects (Fig. 5), and second, the compost microbiologically stabilises the soil, and prevents soil recolonisation with pathogens like *Pythium ultimum* (Fig. 6).

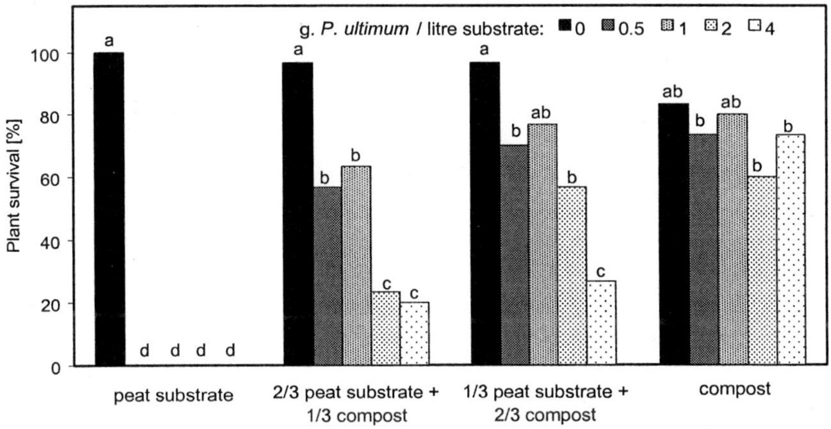

Fig. 2. Influence of compost on the incidence of cucumber damping off, causal agent *Pythium ultimum*, in peat substrate. Plant survival assessed 2 weeks after sowing. *Pythium ultimum* inoculum: 7-day-old culture on autoclaved sorghum grains. Mean of three independent experiments with six pots per repetition and four cucumber seeds per pot. Columns with the same letter do not differ significantly at P=0.05 (Multiple t-test), comparing each mean with each other mean considers one experiment as a repetition

Fig. 3. The living space of sprout culture on thin substrate layer is highly favourable for mould invasion if peat substrate is used only. By adding quality compost to the peat, the system can be microbiologically stabilised and mould growth is suppressed

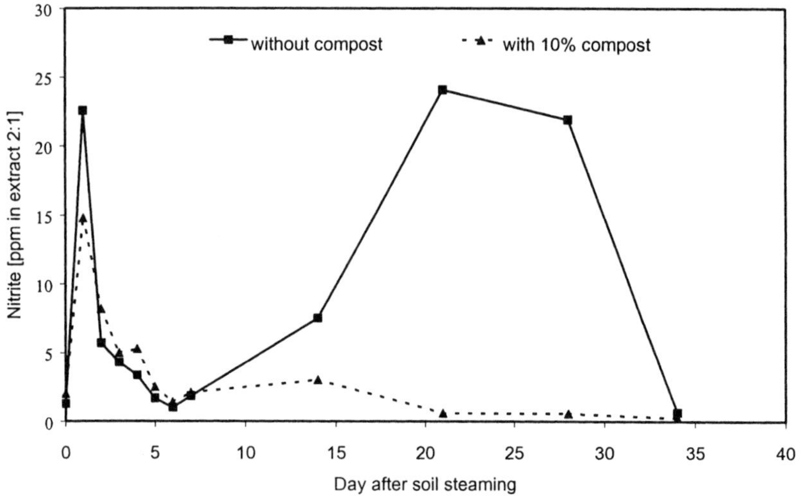

Fig. 4. Influence of compost on the development of nitrite in the soil after steaming. The soil was steamed for 6 h at 100 °C. Compost was added to the soil (10% of soil volume) when the soil temperature reached 40 °C in cooling down. Nitrite was analysed in a 2:1 water extract of the soils with the Spectroquant nitrite test of Merck (D-Darmstadt)

The application of quality compost enhances the efficacy of steaming and provides a profitable long-term effect to growers, and this technique can be a real and effective alternative to methyl bromide (De Ceuster and Hoitink 1999).

Fig. 5. Influence of compost on soil phytotoxicity towards tomato seedlings. The soil was steamed for 6 h at 100 °C. Compost was added to the soil (10% of soil volume) when the soil temperature reached 40 °C in the cooling phase

Use of Compost in the Field

For the assessment of long-term effects of compost on plant disease, fields were divided into two plots. On one half of the field each year compost was applied (10 tons dry weight ha^{-1}), while the other half was used as control. After 5 years, soil samples were taken on the different field plots. The disease receptivity of the soils was tested in the laboratory. The receptivity of the soils for *Pythium ultimum* or *Rhizoctonia solani* was lower in the plots with compost applied every year compared to those without compost (Figs. 7, 8). It has to be emphasised that suppressive effects of compost can still be clearly observed 1 year after compost application. This proves that compost enhances and stabilises soil fertility in a sustainable way. We found a clear negative correlation between more intensively worked and cultivated fields and disease receptivity (data not presented here). This is no surprise, as the biological equilibrium in intensively worked fields is more disturbed, and therefore the positive influence of compost became more distinct.

Influence of Quality Composts on the Whole Plant

The effects of quality composts added to soil are more complex and are not uniquely restricted to soil-borne disease suppression. They also negatively affect the development of leaf pathogens such as *Blumeria* (*Erysiphe*) *graminis* f.sp. *hordei*, the causal agent of barley powdery mildew (Fig. 9).

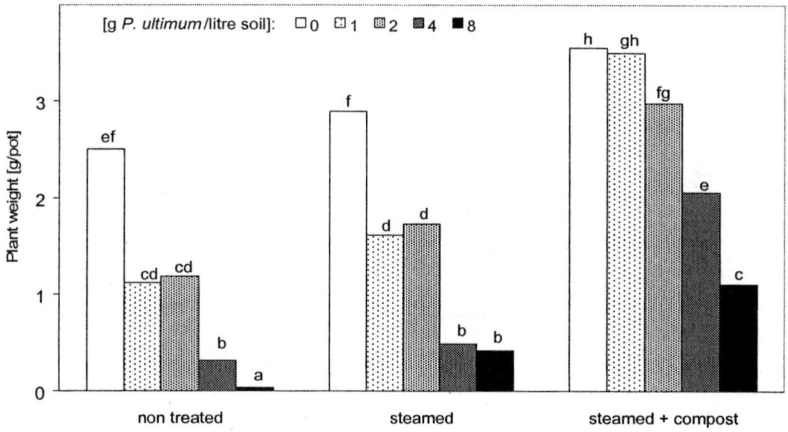

Fig. 6. Influence of compost on the receptivity to *Pythium ultimum* causing damping off on cucumber in steamed soil. The soil was steamed for 6 h at 100 °C. Compost was added to the soil (10% of soil volume) when the soil temperature reached 40 °C at the cooling phase. *Pythium ultimum* (7-day-old culture on autoclaved sorghum grains) was added to the soil 5 weeks later. Plant weight per pot was assessed 2 weeks after sowing. Means of three independent experiments with six pots and four cucumber seed per pot. Columns with the same letter do not differ significantly at P=0.05 (Multiple t-test), comparing each mean with each other mean considers one experiment as a repetition

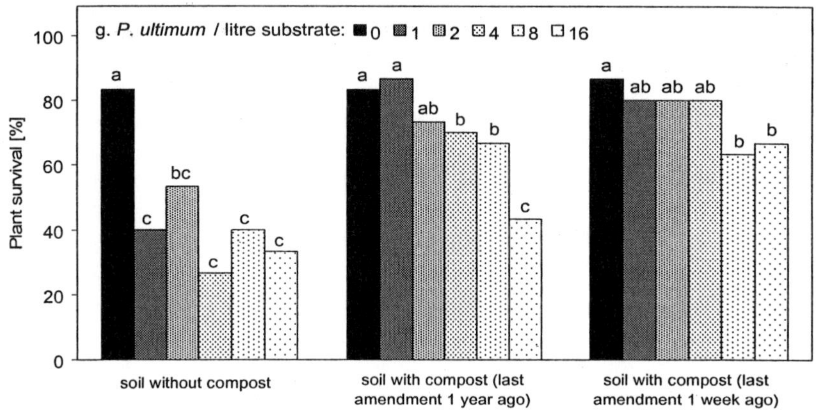

Fig. 7. Influence of quality compost amendments on the receptivity of soil to *Pythium ultimum*, pathogenic agent of cucumber damping off. Twenty tons of compost were given each year to one half of a vegetable field in Fehraltorf (CH), on the other half of the field no compost was applied. Soil samples were taken after 5 years on both field halves before and after amendment of new compost. Plant survival is assessed 2 weeks after sowing. *Pythium ultimum* inoculum in laboratory tests: 7-day-old culture on autoclaved sorghum grains. Each value is the mean of three independent experiments with six pots with four cucumber seeds per pot. Columns with the same letter do not differ significantly at P=0.05 (Multiple t-test), comparing each mean with each other mean considers one experiment as a repetition

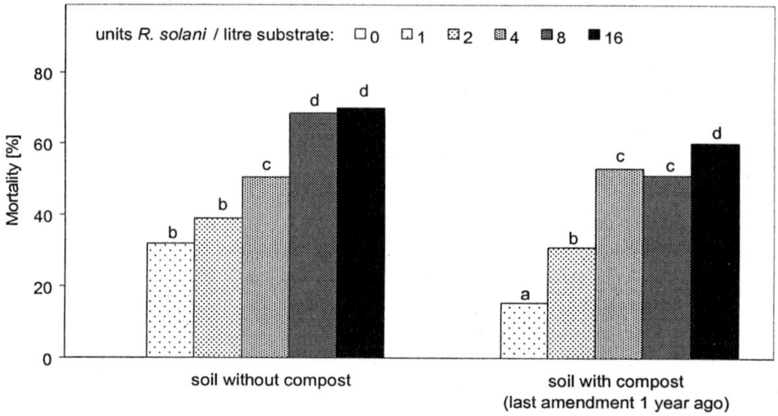

Fig. 8. Influence of quality compost application on the receptivity of soil to *Rhizoctonia solani*, pathogenic agent of lettuce damping off. Twenty tons commercial compost were given each year to one half of a vegetable field in St Sulpice (CH), the other half of the field received no compost. After 5 years, soil samples were collected on both field halves before the new compost amendment. Plant survival assessed 2 weeks after sowing. *Rhizoctonia solani* inoculum in laboratory tests: 21-day-old culture on autoclaved sorghum grains; 1 unit *R. solani* = 1 sorghum grain pot^{-1}. Each value is the mean of three independent experiments with six 200-ml pots with four cucumber seeds per pot. Columns with the same letter do not differ significantly at P=0.05 (Multiple t-test), comparing each mean with each other mean considers one experiment as a repetition

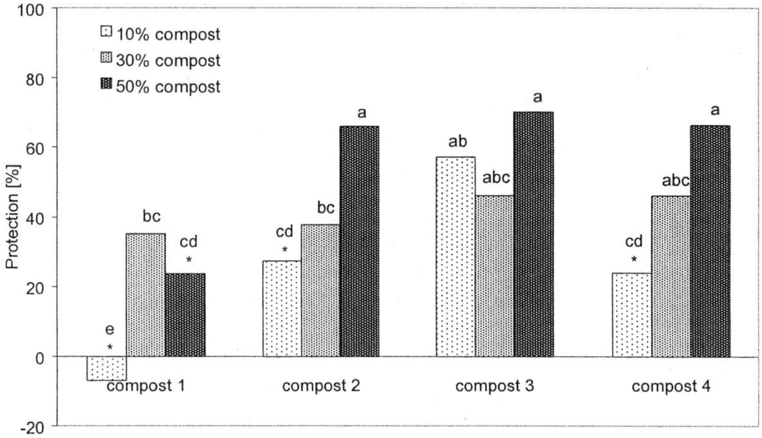

Fig. 9. Influence of composts on the incidence of the barley powdery mildew, caused by *Blumeria graminis* f.sp. *hordei*. Compost from different batches were added to the substrates (Brill No. 5, Gebr. BRILL, D-Georgsdorf) before sowing (10, 30 or 50% of substrate volume). After 10 days the first leaves of barley were inoculated with *B. graminis* conidia. Protection: reduction on the colonies units formed on the leaves 1 week after inoculation. Each value is the mean of three independent experiments with four pots with ten barley plants per pot. Columns with the same letter do not differ significantly at $P=0.05$ (Multiple t-test), comparing each mean with each other mean considers one experiment as a repetition. Columns with a * do not differ significantly at $P=0.05$ compared to the control

The potential to induce disease resistance in the barley plants, varied considerably from compost to compost, depending on their biological quality (Fig. 9). No correlation between nitrogen content or nitrogen availability of the composts with pathogen suppression was found in this and in all other experiments. The fact that defined quality compost can induce resistance to plant diseases is of high interest. It enables the development of new and sustainable plant protection strategies. The induced resistance effects of suppressive composts are not restricted to one plant - pathogen interaction only (Zhang et al. 1996, 1998).

Conclusions

On a practical application level we were able to demonstrate the positive effects of quality composts on crop growth and health status, disease suppression in field crops and in growing media, reactivation of the biological activity of soils after steam treatment and its detoxification, inducing diseases resistance on plant. Not all composts, however, show such positive characteristics. By our observations, we found some factors correlated with these effects. Amongst others, the composition and maturity of composts influence the potential of plant disease suppression, and our experiments (data not presented here) confirm the results of

Tuitert et al. (1998). In our studies we could also show that the guiding and management of the composting processes, in particular the oxygen supply, seem to be the most important factors affecting compost quality. This is also of importance in compost storage (data not published), but all mechanisms and factors that influence the biological compost quality are not yet known. At the moment we are obliged to work on two ways. First, we have to use biotests to measure the quality of different compost batches to give the growers the necessary informations for choosing the composts with the needed quality. Second, research for a better understanding of compost biology and its effect on plant health and vitality has to be continued. In this case, basic and applied research have to co-operate very closely. We now have the evidence that it is possible to control the composting processes for improving the biological quality and hence the biological efficacy of the composts.

Acknowledgements. The author expresses his appreciation to Cyrill Rogger and Jan Werthmüller for technical assistance and Markus Bieri for reviewing the manuscript.

References

Boulter JI, Boland GJ, Trevors JT (2000) Compost: a study of the development process and end-product potential for suppression of turfgrass disease. J Microbiol Biotechnol 16: 115-134

De Ceuster TJJ, Hoitink HAJ (1999) Prospects for composts and biocontrol agents as substitutes for methyl bromide in biological control of plant diseases Compost Sci & Util 7: 6-15

Fuchs JG, Bieri B (2000) New biotests to measure the biological qualities of composts. Agrarforschung 7: 314-319

Hoitink HAJ, Stone AG, Han DY (1997) Suppression of plant diseases by composts. Hortscience 32: 184-187

Tuitert G, Szczech M, Bollen GJ (1998) Suppression of *Rhizoctonia solani* in potting mixtures amended with compost made from organic household waste. Phytopathology 88: 764-773

Wolfe MS, Schwarzbach E (1978) The recent history of the evolution of barley powdery mildew in Europe. In: Spencer DM ed. The powdery Mildews. Academic Press, London, pp 129-157

Zebarth, BJ, Neilsen GH, Hogue E, Neilsen D (1999) Influence of organic waste amendments on selected soil physical and chemical properties. Can J Soil Sci 79: 501-504

Zhang W, Dick WA, Hoitink HAJ (1996) Compost-induced systemic acquired resistance in cucumber to *Pythium* root rot and anthracnose. Phytopathology 86 (10): 1066-1070

Zhang W, Han DY, Dick WA, Davis KR, Hoitink HAJ (1998) Compost and compost water extract-induced systemic acquired resistance in cucumber and *Arabidopsis*. Phytopathology 88 (5): 450-45

Environmental Impacts of Cattle Manure Composting

P.A. Gibbs[1*], R. J. Parkinson[2], T.H. Misselbrook[3] and S. Burchett[2]

Abstract. Measurement of the losses of nutrients during composting is needed to understand the composting process, to implement methods of nutrient conservation and highlight potential environmental impacts. Cattle manure was aerobically composted under ambient conditions to evaluate composting as a method of manure handling. Composting was conducted using 3-4-tonne heaps of fresh cattle manure in 5 x 5-m concrete bunkers. Temperatures inside the manure heaps were -monitored during the composting process. Total mass and nitrogen losses such as ammonia volatilisation, nitrous oxide emission and leaching were measured. Differences in temperature between composting treatments were small. Temperature in the composting manure piles reached in excess of 70 °C and remained above 55 °C for up to 70 days. Total mass loss ranged from 23% for the unturned static treatment to 67% of the initial mass for the indoor treatment turned three times. Nitrogen losses from the manure heaps ranged from 8 to 68% of the initial total manure N content. Leaching losses accounted for between 0 and 1.5% of the initial total manure N content. Gaseous N losses, primarily as NH_3, accounted for between 7 and 67% of the initial manure N content. The decision on whether to compost cattle manure needs to consider all of the issues, and not just nutrient losses. Some of the issues to consider are the effects of temperature on weed seeds and pathogens, the supply of plant nutrient following land spreading and the quantity and friability of the end product.

Introduction

Aerobic composting of organic wastes is a well-established method for stabilising materials before returning to cropping systems. There is considerable interest and activity in the UK in composting products from domestic waste streams (Groenhof 1998), and using the product on agricultural land (Parkinson et al. 1999). Less attention has been paid to the potential agronomic and environmental benefits that might accrue from the controlled composting of organic material generated from livestock production.

In the UK, approximately 90 million tonnes of animal manure are collected annually from farm buildings and yards. This material requires handling, storage

[1]ADAS Gleadthorpe Research Centre, Meden Vale, Mansfield, Nottinghamshire NG20 9PF,UK; email: paul.gibbs@adas.co.uk
[2]Department of Agriculture and Food Studies, University of Plymouth, Newton Abbot, Devon TQ12 6NQ, UK
[3]Institute of Grassland and Environmental Research, North Wyke, Devon EX20 2SB, UK

H. Insam, N. Riddech, S. Klammer (Eds.)
Microbiology of Composting

and subsequent land spreading (Williams et al. 2000). In the region of 50% of the manure are solid-based (cattle, pig, sheep and poultry) with the remainder handled as liquid slurries (cattle and pig). Organic manures are important sources of organic matter and major plant nutrients: nitrogen, phosphorus and potassium. Efficient utilisation of the nutrients, in particular nitrogen, contained in organic manure is essential if diffuse pollution from agriculture is to be reduced (Jarvis and Pain 1990; Smith and Chambers 1993; Smith et al. 1998). Nitrogen is the nutrient most susceptible to transformations affecting plant availability following field application. These transformations include mineralisation, nitrification, immobilisation, denitrification, ammonia volatilisation and nitrate leaching. Active composting, where heaps of manure are actively turned to improve oxygen supply, can stabilise N into more recalcitrant forms, thus slowing down N transformations after field application, but composting may be associated with increased emissions of ammonia (NH_3). There is increasing concern about the detrimental environmental effects due to NH_3 emissions and subsequent deposition to sensitive ecosystems. For this reason there is pressure on the agricultural industry, which is estimated to account for *ca.* 80% of the total NH_3 emission from the UK, to reduce emissions.

The major benefits to be gained from composting are the destruction of weed seeds and pathogens through elevated temperature, stabilisation of the ammonium-N content of manure through incorporation into the microbial biomass, a reduction in mass and volume through the loss of water and carbon and reductions in odour emissions during and after land application.

The aim of the work described here was to evaluate composting of cattle farm yard manure (FYM), which forms approximately 30% of the UK manure production total, as a method of efficient manure handling. Work was focused on composting as a low-technology solution. The project investigated manure quality, gaseous and leachate losses and the whole farm economics of the process. This chapter reports the results from five experimental treatments concentrating on temperature and nitrogen balances.

Materials and Methods

Fresh piles of FYM were created from straw-based material generated from the housing of young beef cattle in the University of Plymouth dairy unit at Seale-Hayne, southwest England. Composting was conducted in (purpose-built) bunkers, each 5 x 5 m, divided by low walls on three sides, open at the front and no roof. One set of trials was conducted in a lean-to shed to prevent rainfall addition (indoors). Each bunker had an outlet to a reservoir for collection of leachate, which was sampled as and when required. Manure heaps of approximately 3 to 4 tonnes of FYM were established at the start of the experiments. Measurements were conducted for approximately 90 days. Turning of the treatments took place after 14, 42 and 70 days in order to encourage microbial activity through aeration. Heaps were turned using a front end loader

(FL); one treatment was passed through a rear discharge muck spreader (MS) during its first turn. The experimental treatments are shown in Table 1. Four treatments involved turning to encourage O_2 supply and one was left as a static pile. All treatments used conventional FYM in the condition it was removed from the cattle housing, unless otherwise stated.

Table 1. Composting treatments

Treatment	Turning events	Time of turning (days)	Turning method	Straw addition
Static pile	0		FL	No
Turned pile	3	15, 42 and 75	FL	No
FYM + straw	3	15, 42 and 75	FL	Yes
Spreader turned	3	17, 50 and 80	MS	No
Indoors	3	15, 42 and 75	FL	No

FL front end loader, *MS* rear discharge muck spreader.

Manure Composition

During the establishment of each manure heap, ten grab samples were taken from each heap. These were bulked to form a heap sample and three subsamples were taken from each heap sample for laboratory analysis. This method follows current UK sampling recommendations for obtaining a representative solid manure sample (Chambers et al. 2001). Manure sampling was repeated at the end of the experiment to assess losses of manure nutrients during composting. All manure samples were analysed for dry matter, total C and N, NH_4-N, NO_3-N, total P and K using standard methods (MAFF 1986). Leachate samples were analysed for total N, NH_4-N and NO_3-N. To determine initial heap mass and mass loss during composting, manure piles were weighed at the start and end of the experiment.

Temperature

Temperature sensors (thermocouples) were installed near the centre of each heap at a height of approximately 50 cm above the bunker floor. The sensors were connected to a data logger which recorded hourly means of four sensors.

Gaseous N emissions

Gaseous N emission measurements were made at intervals throughout the storage period by enclosing each heap in a large polytunnel, with a fan at the front to draw air across the heap via an inlet at the back. Ammonia emission was calculated as the product difference in NH_3 concentrations in outlet and inlet air (measured using absorption flasks containing 0.01 M orthophosphoric acid) and the volume flow rate across the heap. N_2O emission was calculated in the same way, with concentrations being determined by gas chromatography.

Results

Manure composition

Measured characteristics of the conventional fresh manure and manure with an addition of extra straw are given in Table 2.

Table 2. Concentration of nutrients (on a dry weight basis) in the manure before composting. Data represent mean values (values in parenthesis represent standard deviations)

Treatment	Total C	Total N	NH_4-N ($g\ kg^{-1}$)	NO_3-N	Total P	Total K	C:N ratio
FYM	434 (9.3)	17.9 (2.30)	2.7 (1.24)	0.14 (0.23)	7.6 (0.26)	17.5 (0.88)	24
FYM + straw	425 (10.4)	17.0 (2.03)	2.2 (1.19)	0.20 (0.08)	5.9 (0.21)	10.2 (0.15)	25

The primary aim for the addition of extra straw was to increase the C:N ratio, as a result of the increased C addition. From the analysis conducted, no difference in manure C:N ratio between the conventional manure and the plus straw material was determined. The most notable changes in manure characteristics are the reduction in the total P and K concentrations following straw addition. These reductions probably result from the diluting effect of the straw, increasing the proportion of straw to animal excreta.

Temperature

Patterns of temperature inside the manure heaps are presented in Fig. 1.

The temperature within the manure heaps for all five treatments reached in excess of 60 °C within 2 days. Following the initial high temperatures the temperature within the static pile declined over time. In comparison, the turned treatments maintained temperatures between 55 and 65 °C for around 45–60 days. Turning the piles after 14 and 42 days stimulated temperature rises in the four turned piles. The highest temperatures were recorded within the heap turned once through a rear discharge muck spreader.

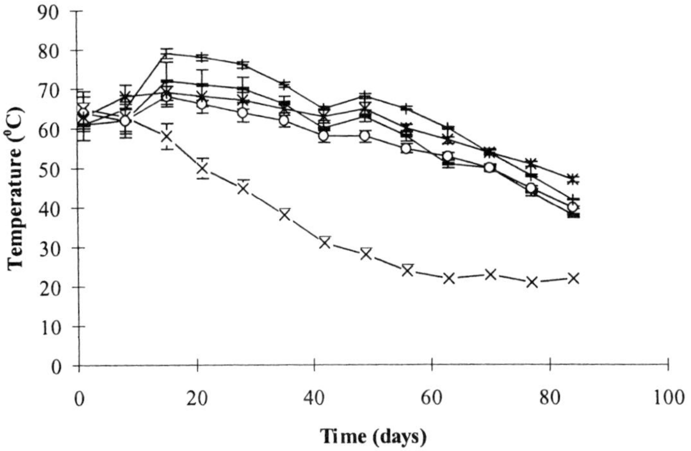

Fig. 1. Mean temperature at 50 cm depth in the manure heaps during composting. Static pile (x), three turns FYM (-), three turns plus straw (o), three turns, first through muck spreader (+), three turns indoors (*)

Gaseous Emissions

A steep increase in NH_3 emission rate in the first few days after initial heap stacking was noted for all treatments *ca.* 200–300 g NH_3-N heap^{-1} day^{-1}, followed by a rapid decline to minimal emission rates after 20 days if heaps were left unturned. Turning the heaps 2 weeks after heap creation promoted an increase in NH_3 emission to *ca.* 500 g NH_3-N heap^{-1} day^{-1} and 100 g NH_3-N heap^{-1} day^{-1} for the conventional FYM heaps turned by MS and FL, respectively. The FYM treatment receiving extra straw had NH_3 emission rates of *ca.* 60 g NH_3-N heap^{-1} day^{-1}. Accumulated N losses due to NH_3 volatilisation during cattle manure composting, for four of the treatments, are shown in Fig. 2. Due to location, it was not possible to erect the polytunnel over the indoor treatment.

The decline to minimal NH_3 emission after 20 days is very pronounced in Fig. 2 with very flat NH_3 accumulation curves after this point in time for all treatments.

The emission of N_2O declined over the storage period, with initial emission rates of *ca.* 18 mg N_2O-N heap^{-1} day^{-1}. A sharp increase in N_2O emission was recorded following the first turn by the rear discharge muck spreader *ca.* 40 mg N_2O-N heap^{-1} day^{-1} (data not shown).

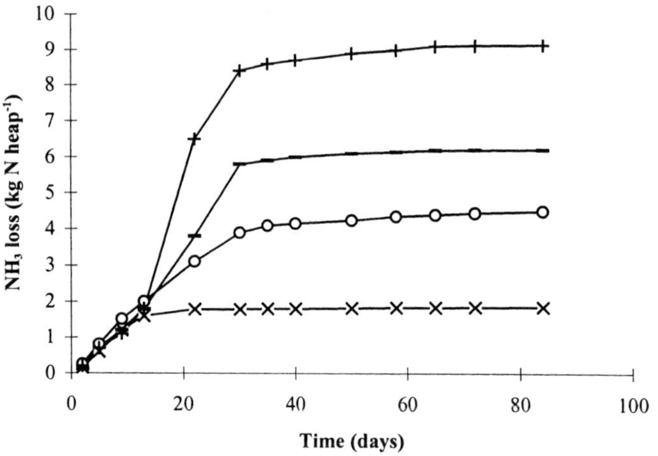

Fig. 2. Cumulative NH_3 volatilisation from manure heaps during composting. Static pile (x), three turns FYM (-), three turns plus straw (o), three turns, first through muck spreader (+)

Leachate Nitrogen

Accumulated N losses due to N leaching are shown in Fig. 3. The highest rates of N loss occurred during the first 2 weeks after heap creation. The highest leachate N loss occurred from the plus straw treatment *ca.* 0.5 kg total N was measured. No leachate was measured from the indoor treatment, thus no leachate N loss was measured.

Mass Losses

Total heap mass losses, on a dry matter basis, ranged from 23 to 67% (Table 3). The lowest losses were measured from the static treatment, highest from the indoor treatment. Mass losses from the FYM and the FYM plus straw treatments were similar, *ca.* 34% of the initial manure mass.

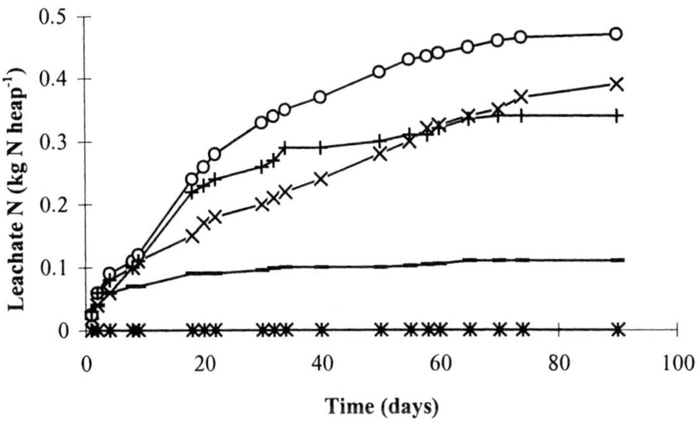

Fig. 3. Cumulative total N leached from manure heaps during composting. Static pile (x), three turns FYM (-), three turns plus straw (o), 3 turns, first through muck spreader (+), three turns indoors (*)

Table 3. Balances for nitrogen in the five composting treatments (numbers in parenthesis represent pool sizes as % of total N at start), and overall mass loss

Treatment	Initial total N	Final total N (kg N heap^{-1})	Leachate loss	Gaseous emission	Mass loss (% initial fresh weight)
Static pile	35.6	32.5 (91.3)	0.39 (1.1)	2.7 (7.6)	23.1
Turned pile	21.5	15.4 (71.6)	0.11 (0.5)	6.2 (28.0)	33.7
FYM + Straw	33.6	28.6 (85.1)	0.47 (1.4)	4.5 (13.5)	34.3
Spreader turned	28.6	19.1 (66.8)	0.34 (1.2)	9.2 (32.2)	52.0
Indoors	39.0	12.7 (32.5)	0.00 (0.0)	26.3 (67.4)[a]	67.3

[a] Gaseous loss calculated by difference, owing to no leachate

Discussion

Composting of solid cattle manure was characterised by high temperatures, the core temperatures of all five treatments were in excess of 60 °C within the first 2 days. These increased temperatures are related to heat liberated through aerobic respiration of microorganisms decomposing organic substrates. In order to destroy pathogens, temperatures in excess of 55 °C are required for 3 days, while a critical temperature of 63 °C is required to kill weed seeds (Eghball et al. 1997). All five treatments achieved these temperature thresholds, including the static pile (Fig. 1). However, it is unlikely that the outside layer of the static pile reached such high temperatures. Insam et al. (1996) demonstrated that the outer zone of unturned manure heaps can remain *ca.* 20 °C cooler than the heap core. This highlights the need for thorough pile mixing if complete heap sanitation is to be achieved. The high temperatures recorded for the four composting treatments over the first 40–60 days are encouraging from a manure sanitation standpoint. This is particularly true for organic systems where the return of weed seeds within manure can be a major problem. High temperatures will also encourage the loss of water vapour, resulting in greater mass reduction, the end result being less material needing to be handled during store emptying.

Ammonia volatilisation was highest during the first 2–3 weeks (Fig. 2); this has been found by others *e.g.* Petersen et al. (1998) and Amon et al. (1999). Petersen et al. (1998) attributed this to the production of extra gaseous products as a result of increased microbial activity and the extra heat generated during this activity. The resulting pressure gradient gives rise to a strong convective movement of gases towards the surface of the manure heap. As the centre of the manure pile is hotter than the surface this means NH_3 could be lost from all layers of the pile, and not just the surface. It is hypothesised that the treatment of turning a manure heap with a rear discharge muck spreader resulted in the highest pile temperature, and the potentially increased porosity of this treatment gave rise to the highest NH_3 emission (Fig. 2).

Loss of material during composting is principally as CO_2 and water vapour (Groenhof 1998). The greater mass loss measured for the spreader-turned and indoor treatments can be associated with faster rates of microbial decomposition (increased CO_2 production) leading to higher temperatures (evaporation of excess water vapour). Losses of manure mass in the region of 50–60% could have distinct benefits in terms of labour and time when the manure is returned to the land through field application. Theoretically, land spreading will take less time, thus labour and machinery can be used for other farm operations, although the increased concentration of nutrients in the end product, as a result of water and carbon losses, could counteract this time-saving. The increased nutrient concentrations could mean the material needs to be spread over a greater area to avoid applying excess nutrients.

Data on N pools and losses have been combined in Table 3. Leaching losses accounted for between 0 and 1.5% of the initial total manure N content from the five treatments. The greatest N leaching loss resulted from the plus straw treatment. It is hypothesised that the extra straw provided channels within the heap through which water could flow, picking up particulate and dissolved N and removing it through the base of the heap. However, leaching losses were a small fraction of the total N loss when compared to gaseous emissions. Gaseous losses accounted for between 7.6 and 67% of the initial total manure N content. Total N losses from the composting treatments in this study appear to be closely correlated with aeration and temperature. Treatments inducing greater aeration caused the highest rate of microbial activity and hence the highest temperatures (indoor and muck spreader turned), and as a result the highest NH_3 emissions. The higher aeration levels in these treatments resulted from the indoor treatment being protected from precipitation, hence water saturation was not a problem and turning the manure with a muck spreader produced a uniform material with a lower density allowing O_2 to infiltrate the whole heap.

The rates of NH_3 loss recorded during this study reflect the distribution of gaseous N losses recorded in the literature for studies monitoring cattle manure composting. Petersen et al. (1998) recorded as little as 5% of the total N lost as NH_3 from composting cattle manure, Martin and Dewes (1992) measured an N loss, as NH_3, from cattle manure of 47% of the total manure N while Eghball et al. (1997) measured, by difference, 41% of the total N lost from composting beef manure in gaseous forms.

The total NH_3 emission from the FYM plus straw treatment was *ca.* 30% lower than from the conventional FYM treatment, despite the mass loss from the two treatments being similar *ca.* 34%. (Table 3). The similarity in mass loss can be associated with similar rates of microbial decomposition while the reduced NH_3 emission from the FYM plus straw treatment can be related to the addition of extra C. The C addition will have increased NH_4 immobilisation by the microbial biomass and hence a lower rate of NH_3 emission, a factor observed by Kirchmann and Witter (1989), although the extra straw addition in this study was not confirmed by an increase in the C:N ratio (Table 2); this is probably due to manure sampling errors.

Gaseous nitrogen losses from agriculture have ecological importance. Emissions of NH_3 from the production of livestock accounts for *ca.* 80% of the total NH_3 emission in the UK (Misselbrook et al. 2000). Such emissions represent not only a potential ecological problem, but also inefficient use of manure N, which could potentially be available for crop growth. Composting manure in this study was associated with increased NH_3 emission, although the extent of the increase was dependent on the turning regime. A number of workers have concluded that cattle manure should not be composted due to the very high gaseous N losses unless high temperatures are desirable for sanitation (Eghball et al. 1997; Kirchmann and Lundvall 1998). Kirchmann and Lundvall (1998) found the best storage technique for livestock manure to restrict NH_3 emission was to keep the heap completely anaerobic, although this form of manure storage would not lead to such high temperatures and could lead to increased NH_3 loss during land spreading. In considering the merits of composting FYM, it is worth considering total NH_3 emission over both storage and land spreading. The emission of NH_3 following composting will be negligible, whereas up to 100% of the ammoniacal N content of fresh FYM can be lost after spreading, if the manure is not rapidly incorporated into the soil. Misselbrook et al. (2000) estimated that the storage of solid and liquid manures accounted for $< 10\%$ of the total NH_3 emission. Therefore, a small increase in NH_3 from manure storage might be far outweighed by the composting benefits, *e.g.* pathogen and weed seed destruction and mass loss.

One of the principle aims of manure storage is to retain essential plant nutrients so they may be returned to land and utilised for crop growth. It was highlighted above that the loss of N during manure storage, particularly as NH_3, is a waste of a valuable resource. The application of composted manure is more than simply providing plant nutrients, the major benefit is the supply of organic matter which can increase agricultural production through improvements in soil health. The issue for farmers is whether composted manure offers the same if not better, fertiliser replacement value than fresh FYM. Gibbs et al. (1999) highlighted that ryegrass yields following the application of fresh or composted cattle manure were not significantly different. Grain yields over a 4-year experimental period showed that the application of fresh and composted cattle manure resulted in similar yields (Eghball and Power 1999). Whilst composted cattle manure might not result in crop-yield benefits compared with uncomposted manure, it does have significant benefits over the use of fresh cattle manure (Gibbs et al. 2000). The use of composted manure poses a lower risk to the surrounding environment through atmospheric or hydrological pollution; composted manure is more friable; resulting in a more even spread with a lower risk of scorching; composting lowers the mass of material needing to be spread on land leading to lower transport costs and less trafficking on the fields.

Despite apparent large N losses during the composting process, it can be shown that composted cattle manure does have a fertiliser replacement value similar to that of uncomposted manure. Whether the fertiliser value is a direct relationship to nutrient content, or through improvements in the overall soil condition, is not confirmed. The major task faced by scientists is to highlight the fertiliser replacement value of animal manure, fresh or composted, to farmers and their advisors (Chambers et al. 2000).

Conclusions

Composting cattle FYM resulted in temperatures sufficiently high and over a time period long enough to kill manure-associated pathogens and weed seeds. Losses of N, particularly as NH_3, can be a major problem from composting cattle manure. As much as 67% of the total manure N was lost through gaseous emission during composting, compared to less than 1.5% in leachate. Total manure mass losses of up to 67% of the initial weight were measured.

The decision whether to compost is complex and depends on the desired result. If the primary aim is to restrict nutrient loss during manure storage then composting is perhaps not the best practice to follow. However, if you consider manure storage in the wider picture, where nutrient losses during storage are small compared to housing and spreading, the benefits of aerobic composting can be justified. The decision whether to compost cattle manure is site specific. Consideration needs to made of the benefits; sanitation, mass reduction, greater friability and less odour during spreading against the drawbacks of enhanced NH_3 emission.

Acknowledgements. Funding of this work by the UK Ministry of Agriculture, Fisheries and Food, the Environment Agency and the BOC Foundation (Project WA 0519) is gratefully acknowledged. Technical support was provided by Peter Russell. Frances Vickery and colleagues at Seale-Hayne Faculty, University of Plymouth.

References

Amon B, Amon T, Boxberger J, Pollinger A (1999) Emission of NH_3, N_2O and CH_4 from composted and anaerobically stored farmyard manure. 8th RAMIRAN Int Conf on Management Strategies for Organic Waste Use in Agriculture, Rennes, France, 26-29 May 1998, pp 209-216

Chambers BJ, Smith KA, Pain BF (2000) Strategies to encourage better use of nitrogen in animal manures. Soil Use Manage 16: 157-161

Chambers B, Nicholson N, Smith K, Pain B, Cumby T, Scotford I (2001) Managing Livestock Manures Booklet 3 - Spreading systems for slurries and solid manures (2nd Edition). Available from ADAS Gleadthorpe Research Centre, Medan Vale, Mansfield, Notts. NG20 9PF, UK

Eghball B, Power JF (1999) Composted and noncomposted manure application to conventional and no-tillage systems: corn yield and nitrogen uptake. Agron J 91: 819-825

Eghball B, Power JF, Gilley JE, Doran JW (1997) Nutrient, carbon and mass loss during composting of beef cattle feedlot manure. J Environ Qual 26: 189-193

Gibbs PA, Parkinson RJ, Fuller MP (1999) Response of grass following the application of composted and untreated cattle manure. In: Corrall AJ (ed) Accounting for nutrients – a challenge for grassland farmers in the 21st Century. BGS Occasional Symposium no33, Malvern, UK, pp 153-154

Gibbs PA, Parkinson RJ, Fuller MP, Misselbrook T (2000) Enhancing the effective utilisation of animal manures on farm through compost technology. In: Petchey AM, D'Arcy BJ, Frost CA (eds) Agriculture and waste management for a sustainable future. SAC/SEPA Conference, Edinburgh, UK, pp 63-72

Groenhof A (1998) Composting: renaissance of an age-old technology. Biologist 45: 164-167

Insam H, Amor K, Renner M, Crepaz C (1996) Changes in functional abilities of the microbial community during composting of manure. Microb Ecol 31: 77-87

Jarvis SC, Pain BF (1990) Ammonia volatilisation from agricultural land. Fertil Soc Proc 298: Fertiliser Society, Peterborough

Kirchmann H, Lundvall A (1998) Treatment of solid animal manures: identification of low NH_3 emission practices. Nutr Cycl in Agroecosyst 51: 65-71

Kirchmann H, Witter E (1989) Ammonia volatilisation during aerobic and anaerobic manure decomposition. Plant Soil 115: 35-41

MAFF 1986. The analysis of agricultural materials. Reference Book 427. HM Stationary Office, London.

MAFF 2000. Fertiliser recommendations for agricultural and horticultural Crops (RB209). HM Stationary Office, London.

Martins O, Dewes T (1992) Loss of nitrogenous compounds during composting of animal wastes. Bioresour Technol 42: 103-111

Misselbrook TH, van der Weerden TJ, Pain BF, Jarvis SC, Chambers BJ, Smith KA, Phillips VR, Demmers TGM (2000) Ammonia emission factors for UK agriculture. Atmos Environ 34: 871-880

Parkinson RJ, Fuller MP, Groenhof AC (1999) An evaluation of greenwaste compost for the production of forage maize (*Zea mays L.*). Compost Sci Util 7: 72-80

Petersen SO, Lind A-M, Sommer SG (1998) Nitrogen and organic matter losses during storage of cattle and pig manure. *J Agric Sci* 130: 69-79

Smith KA, Chambers BJ (1993) Utilising the nitrogen content of organic manures on farms- problems and practical solutions. Soil Use Manage 9:105-112

Smith KA, Chalmers AG, Chambers BJ, Christie P (1998) Organic manure phosphorus accumulation, mobility and management. Soil Use Manage 14: 154-159

Williams JR, Chambers BJ, Smith KA, Ellis S (2000) Farm manure land application strategies to conserve nitrogen within farming systems. In: Petchey AM, D'Arcy BJ, Frost CA (eds) Agriculture and waste management for a sustainable future. SAC/SEPA Conference, Edinburgh, UK, pp 167-179

Agronomic Value and Environmental Impacts of Urban Composts Used in Agriculture

S. Houot[1], D. Clergeot[2], J. Michelin[1], C. Francou[1], S. Bourgeois[1], G. Caria[3] and H. Ciesielski[3]

Abstract. Agronomic value and environmental impacts of three composts (a biowaste compost, BIO; a municipal solid waste compost, MSW and a compost made from green wastes co-composted with sewage sludge, GWS) are compared to those of farmyard manure (FYM) in a long-term field experiment located in Feucherolles (Yvelines, France) and initiated in 1998. The first compost spreading occurred in October 1998 and maize was sown in spring 1999.

Short composting time (in MSW) induced a large residual biodegradability of the organic matter and a transitory nitrogen (N) immobilisation after compost addition to soil followed by a faster organic N mineralisation than in more stabilised compost. The lowest concentrations of heavy metals were observed in the BIO compost in relation with the sorting of the composted wastes. Very low concentrations of polycyclic aromatic hydrocarbons (PAHs) and polychlorinated biphenyls (PCBs) were detected in the composts. Soil (loamy clay) was analysed before starting the experiment. All heavy metals concentrations were lower than the average measured in French soils. Lead and mercury presented the largest variability (25 and 16%, respectively). Organic pollutant (PAHs and PCBs) concentrations were also low. A gradient of concentration was observed with up to 126% variability of the initial concentrations.

The heavy metal input associated with compost spreading represented less than 5% of the initial stock present in the soil. The inputs of organic pollutants were proportionally larger but the evaluation of the other sources of contamination (aerial origin, for example) would be necessary to quantify the impacts of composts. Less than 5% of compost nitrogen and phosphorous was used by the plants. No cadmium, lead or chromium were found in the grains. No significant effect of the organic amendments was observed on the heavy metal content in grains and stems.

Introduction

Composting represents a valuable alternative to incineration for organic waste elimination. Actually, less than 10% of the municipal wastes, including the green wastes are composted in France (IFEN 1999). However, this proportion should

[1]INRA, Environment and Arable Crops, 78850 Thiverval-Grignon, France
[2]CREED, 291 Avenue Dreyfous Ducas, 78520 Limay, France
[3]INRA, Laboratory of Soil Analysis, 62000 Arras, France

H. Insam, N. Riddech, S. Klammer (Eds.)
Microbiology of Composting

increase since landfills are to be closed to recyclable wastes after 2002. Composts are mainly used as organic amendments rather than as fertilisers. Nevertheless, the availability of compost nitrogen and phosphorous must be known to adjust the mineral fertilisation to the crop needs and avoid nitrate leaching or excess phosphorous in soils. Many studies have been done on the agronomic values and the environmental impacts of composts on soil and plant quality (Stratton et al. 1995). Nevertheless, very few studies relate the results to the process of composting and compost physicochemical characteristics. This explains apparent contradictory results such as variable N availability in different composts, from 0 to 40% (Bernal et al. 1998; Mamo et al. 1999). On the other hand, maximum concentrations of heavy metals have been defined in France for sewage sludge but not for composts as in other European countries. The quality of the composts depends on the initial composted wastes and on the composting process. For example, the heavy metal concentration in compost is related to the waste sorting (Morvan and Carré 1994). Finally, very few data are available on the organic pollutants in composts (Berset and Holzer 1995; Niederer et al. 1995; Grossi et al. 1998)

To provide guarantees on the safe use of urban composts to French farmers, a program has been initiated in which both the agronomic value and the environmental impacts of various composts are studied. This program includes the initiation of a long-term field experiment to point out effects of agricultural practices on soil and plant quality under conditions representative of the farmer practices (Chaney and Ryan 1993; Werner and Warnusz 1997). Long-term observations are also necessary to determine the capacity of compost to increase soil organic matter and modify all soil properties related to soil organic matter such as water retention, structure stability, N and P availabilities. The first results of the field experiment are presented including: (1) compost characterisation, (2) soil initial analysis and evaluation of its homogeneity, necessary for the subsequent evaluation of the effects of composts and (3) results related to the first compost spreading, including an evaluation of the potential risk of soil quality degradation.

Materials and Methods

The Composts

Three urban composts were compared to a farmyard manure (FYM) as reference organic amendment: (1) a biowaste compost (BIO) issued from the co-composting of green wastes and the source separated organic fraction of municipal solid wastes, (2) a compost issued from the co-composting of green wastes and sewage sludge (GWS), (3) a municipal solid waste compost (MSW). The composting processes are summarised in Table 1.

Table 1. Summary of the composting processes

	Final Screening	Fermentation		Maturation	
		Process	Duration	Process	Duration
MSW	≤12 mm	Enforced aeration	24 days	Covered windrows	5 days[a]
BIO	≤20 mm	Enforced aeration	1 month	Covered windrows	2 months
GWS	≤20 mm	Covered windrows	3 months	Covered piles	2 months

[a] The compost had to be used immediately after fermentation, which happened frequently in the farmers'practice.

The composts were air-dried and ground before analysis (<250 µm). All analyses were done in triplicate. Total organic carbon (C) was determined by sulfochromic oxidation and total nitrogen (N) using the Kjeldhal method. Compost pH was measured in water (1/5, v/v). The residual biodegradability of the compost organic matter and the N availability were measured during laboratory incubations under controlled conditions with fresh soil (equivalent to 50 g of dry soil, sampled in the field experiment) added to 1 g of fresh compost. Mineral N was added at the beginning of the incubations (50 mg N kg^{-1} soil). Carbon mineralised in CO_2 was trapped in 10 ml of 1 M NaOH and analysed by colorimetry on a continuous flow analyser (Skalar, The Netherlands). Mineral N evolution during the incubations was analysed by colorimetry on a continuous flow analyser after KCl extraction (200 ml of 1 M KCl, 45 min shaking, decanting and filtration of the supernatant on glass fibre filters). The method of Berthelot was used for ammonium and that of Griess - Ilosvay for nitrates after reduction in nitrites on a cadmium column.

Total heavy metals (Cd, Cu, Cr, Zn, Hg, Pb, Ni) were solubilised in *aqua regia* and analysed by emission spectrometry (ICP) in line with mass spectrometry for Cd, Cr, Cu Zn, Pb and Ni and fluorescence spectrometry for Hg.

Total polycyclic aromatic hydrocarbons (PAHs) and polychlorinated biphenyls (PCBs) were extracted by accelerated solvent extraction (Dionex ASE 200, CA, USA). Mixtures of 2 g ground compost and 5 g celite were extracted with acetone/dichloromethane (1/1) for 20 min at 100 °C under 103 10^5 Pa for PAHs and with acetone/hexane (1/1) for 20 min at 150 °C under 103 10^5 Pa for PCBs. The recovered volumes after extraction were 30 ml for PAHs and PCBs. After extraction, 0.5 ml of n-dodecane was added to the PAH extracts to prevent volati-lisation of the molecules, then the extracts were evaporated to 1 ml under N_2. The residues were recovered in 2 ml of hexane, purified on a column filled with Al_2O_3 and eluted by hexane. The hexane solutions were concentrated as above and recovered in 2 ml of acetonitrile. After filtration on cellulosic filters (0.45 µm, Sartorius, Minisart RCK), PAHs were analysed by HPLC with fluorescence detection (Ternary pump Varian 9012, Autosampler Varian 9100, column Hypersil Green PAH-5 µm, 100 x 4.6 mm, detector Varian 9075). Volumes of 10 µl were injected. The mobile phase (2 ml min^{-1}) was a mixture of MilliQ water and acetonitrile. The gradient chromatography started with 50/50 water/acetonitrile (v/v) for 5 min, then gradually reached 100% acetonitrile during 25 min.

PCB extracts were evaporated as above, recovered in 2 ml of hexane and purified by shaking in 10 ml of 18M H_2SO_4. The hexane solution was recovered, then rinsed with 20 ml of MilliQ water, then filtrated on anhydrous Na_2SO_4. Sulphur was eliminated from the extracts on 2 g of reduced Cu. Then the extracts were evaporated and recovered in 2 ml hexane. The extracts were analysed by gas chromatography (Varian 3400, autosampler Varian 8200, injector split/splitless at 300 °C Varian 1077 in splitless mode) on a capillary column DB-5 (0.25 µm, 30 m * 0.25 mm), with electron capture detection (300 °C). The volume injected was 1.5 µl and the carrier gas was He. The temperature was programmed as followed: 70 °C during 1 min, gradient of 18 °C min^{-1} up to 250 °C, then 250 °C during 5 min, gradient of 5 °C min^{-1} up to 280 °C, 280 °C during 8 min.

The Field Experiment

The experiment was located at Feucherolles (Ile de France, 50 km west of Paris) and included four blocks of ten treatments. The three composts and the farmyard manure (BIO, MSW, GWS, FYM) were compared to a control treatment without organic amendment. These five organic treatments were crossed with two mineral N treatments: added mineral N fertilisation or no N fertilisation. The plots were 10 x 45 m and were separated from the adjacent plots by 6-m bands. The four blocks were 25 m apart to prevent contamination during compost spreading. The surface soil was sampled (average of three individual samples in each plot) in all the plots before compost spreading to determine initial variability. Deeper horizons were sampled in three holes opened between the plots. Samples were air-dried and 2-mm sieved before analysis. Total C, N and pH were analysed as described above. Soil samples were ground to 250 µm before heavy metal and organic pollutant analysis. Total Cu, Ni, Zn, Cr, Pb, Hg, Cd were solubilised in fluorhydric acid and analysed as above. PAHs and PCBs were analysed as described for composts.

The composts were spread during the first week of October 1998 on wheat stubble (10.7, 10.0, 16.2 and 13.1 t dry matter ha^{-1} for GWS, MSW, BIO and FYM, respectively) and ploughed into the soil the week after. The field experiment was cultivated with a wheat-maize rotation. Maize was sown in March 1999. Mineral N fertilisation was done in May 1999 (80 kg N ha^{-1}) in the fertilised plots. No mineral P or K was added. The maize was harvested in October 1999, dried and analysed for major elements and heavy metals after grinding.

Results and Discussion

Compost Characteristics

The GWS and BIO composts showed lower residual biodegradability than the MSW compost and the farmyard manure and 5 to 12% of the organic C was mineralised during the incubations, respectively (Fig. 1). This was probably due to their longer duration of composting. The composting of sludges increased the phosphorous and N content of the compost in GWS (Table 2). It also increased initial mineral N content in the compost (Fig. 1). The MSW compost was characterised by a large residual biodegradability related to its short composting time. Such compost decreased the availability of mineral N in soil immediately after compost addition because of N immobilisation during compost degradation. However, after this short period of N immobilisation, organic N mineralisation was faster than in the other composts. As compared to the composts, FYM had an intermediate residual biodegradability but a larger N availability (Fig. 1).

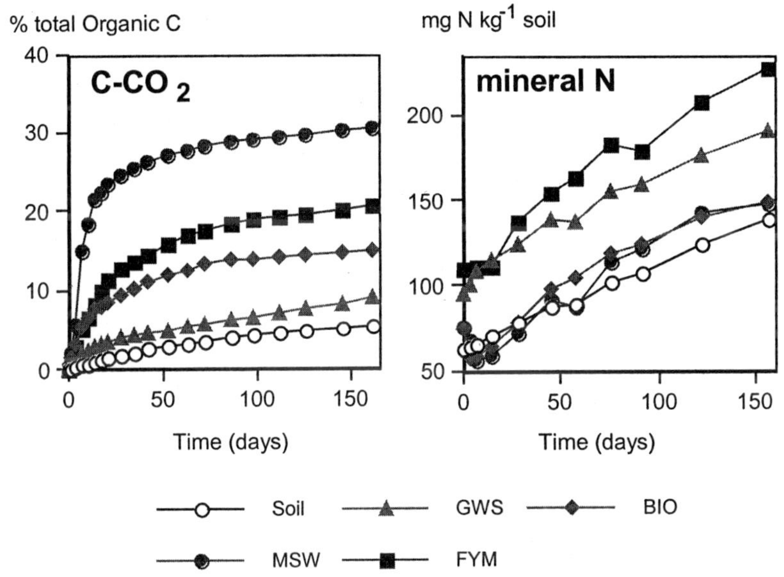

Fig.1. Kinetics of C and N mineralisation during incubations of soil–compost mixtures in laboratory controlled conditions. Results are expressed in percentage of total organic C mineralised during the incubations and in mg mineral N kg^{-1} soil for C and N, respectively

All composts had heavy metal concentrations lower than the maximum concentrations allowed in France for sewage sludge use in cultivated soils, but the values sometimes exceeded the concentrations proposed in the European Ecolabel (Table 3). The concentrations decreased in the order MSW> GWS> BIO, confirming that waste sorting results in decreasing the heavy metal contents of composts. However, even the BIO compost did not fit the Ecolabel criteria (too high Pb concentration). The concentrations of heavy metals in the BIO and GWS composts corresponded to the average concentrations found in similar composts in France (ADEME 2001), except for Pb in the BIO compost (average of 106 mg kg^{-1} dry matter) and Pb and Cd in the GWS compost (average of 62 and 1.6 mg kg^{-1} dry matter, respectively). Most of the heavy metals were more concentrated in the

MSW compost compared to the average for French MSW compost, but higher concentrations have also been reported for such composts (Pinamonti et al. 1997; Breslin 1999). Surprisingly, the FYM had higher heavy metal contents than usually observed (Juste 1993; Pinamonti et al. 1997).

Table 2. Physicochemical characteristics of the composts and farmyard manure. Dry matter is expressed as percentage of fresh matter. The other results are on a dry weight basis. Standard errors are indicated

	MSW	BIO	GWS	FYM
Dry matter (% fresh matter)	71.5	77.3	53.5	23.8
Organic C (g kg^{-1} DM)	304 ± 18	158 ± 13	273 ± 11	287 ± 5
Organic matter (% DM)	52.5 ± 3.1	27.3 ± 1.1	47.2 ± 1.9	49.6 ± 0.9
Total N (g kg^{-1} DM)	20.2 ± 0.4	16.3 ± 0.9	28.3 ± 0.7	23.9 ± 0.3
Total P_2O_5 (g kg^{-1} DM)	9.0 ± 0.5	7.0 ± 0.2	18.9 ± 1.1	14.3 ± 0.6
P_2O_5 Olsen (g kg^{-1} DM)	0.28 ± 0.01	0.79 ± 0.02	1.16 ± 0.02	4.02 ± 0.2
K_2O (g kg^{-1} DM)	13.1 ± 0.4	19.1 ± 0.3	21.1 ± 0.2	41.5 ± 1.1
C/N	15.3 ± 1.0	9.7 ± 1.0	9.6 ± 0.5	11.9 ± 0.1
Electrical conductivity (mS cm^{-1})	2.1 ± 0.1	0.9 ± 0	0.7 ± 0	3.0 ± 0.1
pH (in water)	8.1 ± 0.2	8.6 ± 0.1	8.5 ± 0.1	8.8 ± 0.1

Table 3. Heavy metal concentrations (mg kg^{-1} dry matter) in the three composts and the farmyard manure. Standard errors are indicated. Maximum concentrations allowed in sewage sludge applied on cultivated soil by the French legislation and proposed in the European Ecolabel for composts are given for comparison

Heavy metal (mg kg^{-1})	MSW	BIO	GWS	FYM	Allowed in sewage sludge	European Ecolabel
Cd	2.3 ± 0.7	0.7 ± 0.03	2.9 ± 0.2	3.7 ± 0.1	20	1
Cr	162 ± 22	45 ± 6	30 ± 1	78 ± 7	1000	100
Cu	312 ± 76	58 ± 1	154 ± 13	207 ± 4	1000	100
Hg	2.4 ± 0.2	0.5 ± 0.01	1.1 ± 0.1	0.2 ± 0.01	10	1
Ni	69 ± 10	19 ± 3	18 ± 2	35 ± 2	200	50
Pb	245 ± 47	125 ± 8	85 ± 11	404 ± 58	800	100
Zn	574 ± 50	211 ± 7	383 ± 15	716 ± 14	3000	300

All composts presented low concentrations of PAHs and PCBs, much lower than the concentrations measured in sewage sludges (Berset and Holzer 1995; Jauzein et al. 1995) or allowed by the French legislation on sludge use in agriculture (Tables 4 and 5). The location of the composting plants has been shown to influence PAH and PCB concentrations in the composts with large concentrations found in urban areas compared to rural areas (Berset and Holzer 1995; Niederer et al. 1995; Grossi et al. 1998). This could explain the larger concentrations of PAHs detected in the BIO compost, coming from an urban area. In the three composts, the concentrations were rather low and ranged from 900 to 2780 μg kg^{-1} dry matter for the total of 15 PAH and 30 to 100 μg kg^{-1} for the total of 7 PCB, comparable to the previously mentioned references.

Table 4. PAH concentrations (μg kg^{-1} dry matter) in the three composts and the farmyard manure. Standard errors are indicated

PAH	Code	MSW	BIO	GWS	FYM
	(-----	-----	μgkg^{-1} dry matter	-----	-----)
Fluoranthene	FLT	170 ± 46	481 ± 34	312 ± 11	54 ± 6
Benzo(b)fluoranthene	BaF	46 ± 15	215 ± 17	214 ± 22	28 ± 1
Benzo(a)Pyrene	BaP	17 ± 5	148 ± 25	81 ± 8	21 ± 2
Naphtalene	NAP	27 ± 13	10 ± 6	5 ± 1	4 ± 0.2
Acenaphtene	ANA	7 ± 2	41 ± 5	5 ± 0.5	2 ± 0.4
Fluorene	FLU	23 ± 4	61 ± 3	14 ± 1	6 ± 3
Phenanthrene	PHE	235 ± 65	456 ± 16	270 ± 6	53 ± 6
Anthracene	ANT	19 ± 5	56 ± 3	14 ± 1	6 ± 1
Pyrene	PYR	153 ± 47	315 ± 36	296 ± 8	43 ± 4
Benzo(a)anthracene	BaA	28 ± 7	182 ± 17	153 ± 69	23 ± 4
Chrysene	CHR	72 ± 20	239 ± 46	351 ± 13	49 ± 4
Benzo(k)fluoranthene	BkF	23 ± 8	118 ± 11	73 ± 4	13 ± 1
Dibenzo(ah)anthracene	DBA	15 ± 7	49 ± 12	33 ± 8	14 ± 2
Indeno(1,2,3-cd)pyrene	IPY	15 ± 7	137 ± 17	77 ± 6	21 ± 3
Benzo(ghi)perylene	BPE	47 ± 14	271 ± 14	245 ± 22	44 ± 10
Total of 15 PAH		897 ± 261	2780 ± 203	2144 ± 80	381 ± 32

In the French legislation on sewage sludge spreading on soil, the concentrations of three PAHs (FLT, BbF and BaP) and the total of seven PCB congeners (28, 52, 101, 118, 138, 153, 180) are limited to 5, 2.5, 2 and 0.8 mg kg^{-1} dry matter, respectively. Other PAHs are about to be added in the European regulation (ANA, FLU, PHE, ANT, PYR, BkF, IPY and BPE). These other PAHs presented similar concentrations in the composts (Table 4). The concentrations of PAH were lower in the cattle manure than in the composts, but higher than in pig and cattle slurries (Berset and Holzer 1995). The highest concentrations of PCB were detected in the MSW compost. PCB concentrations were similar in the BIO and GWB composts and in the FYM. Berset and Holzer (1995) also found similar concentrations in composts and animal slurries. In the three composts, the congener 138 was among the most concentrated but the concentrations of other congeners not included in the legislation were similar and should be included in future legislation. In FYM, PCB 118 was the most concentrated.

Initial Soil Characteristics

The soil was a loamy clay with a low organic matter content (Table 6). The pH was slightly acidic (6.9 in water). Heavy metals were analysed in the whole profile. In the surface horizon, the concentrations were below the average values of French soils (Baize 2000). They were three to ten times smaller than the maximum concentrations in soils defined for sludge application (Table 6). The concentrations measured in the carbonated loess (150-180 cm) were representative of the metal concentrations of soil substratum. Soil decarbonation increased the metal concentrations (110-150 cm). Concentrations of Cu, Zn, Cr and Ni in surface soil were smaller compared to deeper horizons. These elements were partly exported in plants. Only Cd, Cr and Hg presented larger concentrations in the surface horizon and could be of anthropogenic origin related to P fertilization or atmospheric deposition (Robert and Juste 1997). The initial variability of heavy metal contents varied from 3% for Cu to 25% for Pb without significant differences among the blocks. The soil was also homogeneous for all the agronomic parameters (less than 10% variability).

PAHs and PCBs are ubiquitous pollutants and have been detected in many soils (Alcock et al. 1995; Berset and Holzer 1995; Lichtfouse et al. 1997). PAHs are mainly of pyrolitic origin, coming from fossil fuel combustion (Lichtfouse et al. 1997). Atmospheric depositions largely contribute to their presence in soils (Alcock et al. 1995; Lichtfouse et al. 1997). Soil initial concentrations in PAHs (Table 7) and PCBs (Table 8) were equivalent to other concentrations determined in soils from rural areas (Niederer et al. 1995; Berset and Holzer 1995; Webber and Wang 1995). Among the PCBs, the congener 138 was much more concentrated than the others. Alcock et al. (1995) found that low chlorinated congeners were predominant in soils sampled during the 1960s and 1970s, being later progressively replaced by more chlorinated congeners.

Table 5. PCB concentrations (μg kg^{-1} dry matter) in the three composts and the farmyard manure. Standard errors are indicated

Congener n°	MSW	BIO	GWS	FYM
	(------------------------------μg kg^{-1} dry matter------------------------------)			
PCB 28	20.1 ± 2.7	3.4 ± 0.6	4.6 ± 1.6	4.5 ± 0.5
PCB 52	10.3 ± 1.7	3.1 ± 0.2	7.3 ± 3.7	1.2 ± 0.2
PCB 101	10.2 ± 1.6	4.0 ± 0.3	4.7 ± 2.4	0.7 ± 0.8
PCB 118	2.4 ± 0.8	2.3 ± 0.1	3.8 ± 1.7	24.3 ± 6.4
PCB 138	38.9 ± 14.9	8.8 ± 0.6	9.4 ± 3.7	4.0 ± 1.9
PCB 153	4.2 ± 0.9	1.9 ± 0.2	2.3 ± 0.9	0.9 ± 0.2
PCB 180	14.6 ± 3.0	6.1 ± 0.7	5.5 ± 1.3	2.9 ± 0.9
Σ 7 PCB[a]	100.7 ± 8.9	29.7 ± 1.1	37.1 ± 15.2	38.4 ± 9.7
PCB 8	27.3 ± 2.8	3.4 ± 1.1	2.6 ± 1.3	5.0 ± 1.2
PCB 18	14.0 ± 3.0	5.3 ± 3.2	4.2 ± 0.4	6.2 ± 2.0
PCB 44	13.7 ± 1.8	4.4 ± 0.4	8.1 ± 3.6	2.7 ± 0.4
PCB 66	19.0 ± 2.9	12.8 ± 4.4	13.6 ± 5.2	5.0 ± 0.3
PCB 77	14.1 ± 1.8	0.6 ± 0.0	5.5 ± 3.2	1.9 ± 0.5
PCB 105	19.0 ± 2.9	5.7 ± 0.2	10.2 ± 4.3	1.9 ± 0.1
PCB 128	23.8 ± 6.1	10.1 ± 0.6	10.3 ± 4.3	3.0 ± 0.2
PCB 126	0.8 ± 0.3	0.3 ± 0.1	0.7 ± 0.2	0.3 ± 0.0
PCB 170	7.7 ± 1.8	2.7 ± 0.2	2.4 ± 0.8	1.3 ± 0.4
PCB 187	7.7 ± 1.5	3.0 ± 0.1	2.7 ± 0.6	2.8 ± 0.4
PCB 195	1.8 ± 0.3	0.9 ± 0.2	0.6 ± 0.1	1.2 ± 0.5
PCB 206	4.2 ± 0.9	1.3 ± 0.2	0.9 ± 0.0	1.3 ± 0.5
PCB 209	1.2 ± 0.2	1.3 ± 0.2	1.2 ± 0.1	1.0 ± 0.3

[a] total of PCB 28, 52, 101, 118, 138, 153, 180.

Table 6. Soil characteristics. Each of the 40 plots was sampled in the surface soils and results presented correspond to the mean values of the 40 analysis. For the deeper horizons, three holes were opened and means of three analyses are given.

	Depth (cm)					
	0-25	30-45	45-90	90-110	110-150	150-180
Clay (g kg^{-1})	150 ± 10	223 ± 20	319 ± 19	286 ± 11	234 ± 12	219 ± 28
Silt (g kg^{-1})	783 ± 20	727 ± 29	646 ± 17	693 ± 19	741 ± 18	701 ± 36
Sand (g kg^{-1})	67 ± 9	50 ± 7	36 ± 4	22 ± 4	25 ± 10	80 ± 20
Org. C (g kg^{-1})	11.0 ± 0.8	3.6 ± 0.3	2.0 ± 0.2	1.6 ± 0.1	1.5 ± 0.1	0.8 ± 0.2
N_{Kj} (g kg^{-1})	1.08 ±0.06	0.50 ± 0.02	0.36 ± 0.02	0.30 ± 0.01	0.27 ±0.01	0.21 ±0.01
pH (in water)	6.9 ± 0.2	7.4 ± 0.4	7.7 ± 0.2	7.9 ± 0.2	7.9 ± 0.2	8.5 ± 0.1
$CaCO_3$ (g kg^{-1})	<1	<1	<1	<1	<1	151 ± 13
Cu[a] (mg kg^{-1})	11.6 ± 0.9	9.5 ± 0.6	14.2 ± 1.6	13.5 ± 0.2	12.8 ± 1.6	9.9 ± 0.4
Zn[a] (mg kg^{-1})	49.2 ± 3.4	52.3 ± 3.3	70.6 ± 4.5	67.2 ± 3.1	60.7 ± 2.0	47.9 ± 2.3
Cr[a] (mg kg^{-1})	45.3 ± 2.7	52.3 ± 3.5	67.3 ± 3.3	65.4 ± 1.2	61.4 ± 1.9	53.7 ± 3.5
Ni[a] (mg kg^{-1})	15.0 ± 0.9	21.4 ± 1.8	32.8 ± 3.7	33.1 ± 0.8	33.3 ± 1.5	23.7 ± 1.1
Pb[a] (mg kg^{-1})	25.3 ± 6.1	18.3 ± 1.2	19.8 ± 1.1	18.4 ± 0.0	18.5 ± 1.4	17.4 ± 0.7
Cd[a] (µg kg^{-1})	232 ± 13	98 ± 33	87 + 27	72 ± 26	112 ± 34	95 ± 28
Hg[a] (µg kg^{-1})	91 ± 16	30 ± 0	35 ± 5	30 ± 10	23 ± 6	20 ± 0

[a] Maximum concentrations of metals in soils for the possibility of sludge spreading (in mg kg^{-1}): Cu,100; Zn, 300; Cr, 150; Ni, 50; Pb, 100; Cd, 2; Hg, 1.
Average concentrations in French soils (815 samples were analysed, results in mg kg^{-1}): Cu,14.9; Zn, 149; Cr, 75; Ni, 41.3; Pb, 64.8; Cd, 0.42.

Table 7. Mean PAH concentrations (µg kg^{-1} dry soil) in the surface horizons of the ten plots of each block of the field experiment. Standard errors are indicated. Within a row, data followed by the same letter are not statistically different ($p<0.01$).

PAH	Block 1	Block 2	Block 3	Block 4
	(µg kg^{-1} dry matter)			
FLT	22.0 ± 8.1 (a)	27.5 ± 15.7 (a)	51.2 ± 27.7 (a, b)	83.6 ± 65.3 (b)
BbF	22.5 ± 4.6 (a)	26.1 ± 19.0 (a)	37.5 ± 21.5 (a, b)	61.1 ± 46.1 (b)
BaP	16.1 ± 3.7 (a)	17.2 ± 14.9 (a)	18.6 ± 15.6 (a, b)	43.1 ± 35.1 (b)
NAP	1.0 ± 0.6 (a, c)	0.7 ± 0.3 (c)	2.8 ± 2.1 (b)	2.6 ± 1.6 (a, b)
ANA	0.2 ± 0.2 (a, c)	0.1 ± 0.1 (a)	0.4 ± 0.2 (b, c)	0.5 ± 0.1 (b)
FLU	1.3 ± 0.3 (a)	0.5 ± 0.1 (b)	1.2 ± 0.6 (a)	1.7 ± 1.0 (a)
PHE	8.9 ± 3.8 (a)	7.4 ± 4.7 (a)	16.8 ± 10.0 (a, b)	29.0 ± 19.7 (b)
ANT	1.0 ± 0.6 (a)	1.0 ± 0.9 (a)	1.8 ± 2.0 (a, b)	4.5 ± 4.0 (b)
PYR	18.3 ± 5.3 (a)	20.1 ± 13.5 (a)	35.4 ± 21.3 (a, b)	61.4 ± 47.0 (b)
BaA	11.2 ± 2.9 (a)	14.1 ± 13.3 (a)	23.9 ± 16.4 (a, b)	47.0 ± 42.0 (b)
CHR	17.1 ± 2.8 (a)	21.6 ± 10.9 (a)	26.1 ± 13.0 (a, b)	46.2 ± 31.1 (b)
BkF	7.8 ± 1.7 (a)	10.0 ± 7.1 (a)	16.2 ± 9.9 (a, b)	28.3 ± 23.4 (b)
DBA	3.3 ± 1.5 (a)	6.4 ± 2.4 (a, c)	8.2 ± 3.9 (a, b)	9.4 ± 7.3 (b)
IPY	18.7 ± 2.9 (a)	24.4 ± 16.7 (a, c)	31.4 ± 18.6 (a, b)	48.2 ± 33.7 (b)
BPE	15.7 ± 3.7 (a)	28.7 ± 16.3 (a, c)	38.1 ± 21.4 (a, b)	62.3 ± 49.8 (b)
Total 15 PAH	165 ± 33 (a)	206 ± 131 (a)	310 ± 174 (a, b)	529 ± 400 (b)

Most PAHs were detected in similar quantities except NAP, ANA, FLU and ANT that had very low concentrations. The concentrations defining natural or uncontaminated soils varied between 0.02 to 0.1 mg kg^{-1} from one country to another

(Jauzein et al. 1995), pointing out that natural concentrations are poorly defined and mostly depend on the analytical capacity of laboratories. On the other hand, concentrations defined for contaminated soils are 30 times higher than measured. A gradient of PAH concentration was observed among the different plots (Fig. 2) which explained the large variability of their concentration (90% for FLT). Initial variability (35%) was also important for PCBs. Such variability would probably be due to the aerial origin of these molecules, coming from nearby sources.

Table 8. Mean PCB concentrations (μg kg^{-1} dry soil) in the surface horizons of the ten plots of each block, and mean PCB concentrations in the 40 plots of the field experiment ± Standard errors. Within a row, data followed by the same letter are not statistically different ($p<0.01$)

Congener	Block 1	Block 2	Block 3	Block 4
	(μg kg^{-1} dry matter)			
PCB 28	0.8 ± 0.8 (a)	0.6 ± 0.5 (a)	0.8 ± 1.1 (a)	0.4 ± 0.3 (a)
PCB 52	1.6 ± 3.7 (a)	0.4 ± 0.4 (a)	0.3 ± 0.3 (a)	0.2 ± 0.2 (a)
PCB 101	2.2 ± 3.9 (a)	1.2 ± 1.2 (a)	1.3 ± 0.5 (a)	0.2 ± 0.3 (a)
PCB 118	0.6 ± 0.2 (a)	0.7 ± 0.2 (a)	0.9 ± 0.2 (a)	1.0 ± 0.2 (a)
PCB 138	16.7 ± 7.6 (a)	24.3 ± 12.9 (a)	26.1 ± 7.8 (a)	30.1 ± 9.5 (a)
PCB 153	2.8 ± 1.0 (a)	3.7 ± 1.6 (a, b)	3.9 ± 1.2 (a, b)	4.9 ± 1.3 (b)
PCB 180	0.7 ± 0.4 (a)	1.4 ± 2.5 (a)	0.6 ± 0.2 (a)	0.8 ± 0.1 (a)
Σ 7 PCB[a]	25.3 ± 10.2 (a)	32.4 ± 13.2 (a, b)	33.9 ± 8.7 (a, b)	38.6 ± 10.1 (a)
PCB 8	0.9 ± 0.4 (a)	0.9 ± 0.4 (a)	0.9 ± 0.6 (a)	1.8 ± 3.2 (a)
PCB 18	1.4 ± 1.0 (a)	0.8 ± 0.5 (a, b)	1.0 ± 0.3 (a, b)	0.6 ± 0.1 (b)
PCB 44	0.6 ± 0.3 (a)	0.9 ± 1.4 (a)	0.5 ± 0.3 (a)	0.5 ± 0.1 (a)
PCB 66	0.7 ± 0.3 (a)	0.9 ± 1.1 (a)	1.2 ± 1.0 (a)	0.8 ± 0.1 (a)
PCB 77	1.0 ± 0.7 (a)	1.0 ± 0.4 (a)	1.2 ± 0.9 (a)	1.1 ± 0.3 (a)
PCB 105	1.3 ± 0.6 (a)	1.6 ± 0.6 (a, b)	1.9 ± 0.6 (a, b)	2.1 ± 0.4 (b)
PCB 128	1.9 ± 1.1 (a)	2.3 ± 1.2 (a)	2.4 ± 1.0 (a)	2.7 ± 1.1 (a)
PCB 126	2.7 ± 1.0 (a)	4.1 ± 2.4 (a)	3.7 ± 1.4 (a)	3.6 ± 1.2 (a)
PCB 170	1.1 ± 0.5 (a)	1.8 ± 1.2 (a)	1.4 ± 0.3 (a)	1.7 ± 0.4 (a)
PCB 187	0.2 ± 0.1 (a)	0.6 ± 1.3 (a)	0.2 ± 0.2 (a)	0.2 ± 0.1 (a)
PCB 195	0.1 ± 0.1 (a)	0.2 ± 0.2 (a)	0.1 ± 0.1 (a)	0.04 ± 0.02 (a)
PCB 206	0.1 ± 0.1 (a)	0.1 ± 0.1 (a, b)	0.2 ± 0.2 (a, b)	0.4 ± 0.5 (b)
PCB 209	0.04 ± 0.07 (a)	0.02 ± 0.05 (a)	0.05 ± 0.03 (a)	0.06 ± 0.03 (a)

[a]total of PCB 28, 52, 101, 118, 138, 153, 180.

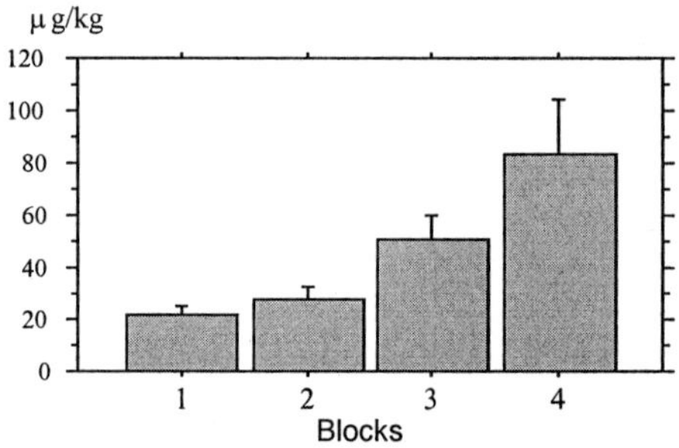

Fig. 2. Average concentration of fluoranthene in the four blocks of the field experiment before the first compost application, pointing out the initial gradient of concentration

First Compost Application

The doses of organic amendments to be applied were calculated on the basis of the rate of organic C application. The objective was to bring 4 t ha^{-1} of organic C in each treatment. The actual amounts applied varied between 3 to 4 t ha^{-1} (2.9, 3.0, 2.6 and 3.8 for GWS, MSW, BIO and FYM, respectively) because of difficulties in adjusting the spreading. The corresponding amount of dry matter applied varied from 10 - 16 t ha^{-1}, representative of farmer practices. Fluxes of heavy metals and organic micropollutants entering the soil were calculated on the basis of 5 compost spreadings (equivalent to 10 y of experiment) to be able to compare them to the maximum fluxes allowed in France for sludge spreading within 10 y. All fluxes of heavy metals were lower than the maximum allowed except for Cu applied with the MSW compost and Pb and Zn with the FYM (Table 9). The theoretical final soil concentrations after 10 y of experiment (five compost applications) were calculated by adding the initial stocks in soils (g m^{-2}) and the fluxes of elements brought with composts (g m^{-2}), then dividing by the weight of soil in 1 m^2 with the hypothesis that soil density did not vary during the experiment (Table 10).

All calculations were done based on 30 cm tillage and a soil density of 1.27. Final concentrations varied little although cumulated fluxes of metals represented up to 30% of the initial stocks at the beginning of the experiment (Table 9). They all remained below the average values of French soils. Considering the maximum concentrations of heavy metals in soils allowed for sludge application (Table 6), at least 100 applications of BIO, GWS and FYM would be possible. For MSW, Pb would limit the number of applications to 50. These maximum concentrations have been defined considering the adverse effects of metals on plants (MacGrath et al. 1995). On the other hand, considering the initial variability of the concentrations for some elements (e.g., Pb), the calculated increase remained within the variability of their initial concentrations and would be hard to detect.

Table 9. Fluxes of heavy metals calculated for 10 y (5 compost spreadings) compared to initial soil stocks. Maximum fluxes for sludge application allowed in France are given for comparison.

	MSW	BIO	GWS	FYM	Max	Soil initial stock
	(g m^{-2})					
Cd	0.01 (11%)[a]	0.01 (11%)	0.02 (22%)	0.02 (22%)	0.03 (33%)	0.09
Cr	0.8 (5%)	0.4 (2%)	0.4 (2%)	0.5 (3%)	1.5 (9%)	17.2
Cu	1.6 (36%)	0.5 (11%)	0.8 (18%)	1.4 (32%)	1.5 (34%)	4.4
Hg	0.012 (34%)	0.004 (11%)	0.006 (17%)	0.001 (3%)	0.015 (3%)	0.035
Ni	0.3 (5%)	0.2 (4%)	0.2 (4%)	0.2 (4%)	0.3 (5%)	5.7
Pb	1.2 (12%)	1.0 (10%)	0.5 (5%)	2.7 (28%)	1.5 (15%)	9.7
Zn	2.9 (16%)	1.7 (9%)	2.1 (11%)	4.7 (25%)	4.5 (24%)	18.7

[a] In parentheses, the percentage of soil initial stock brought with the five compost applications.

Table 10. Initial and calculated final heavy metal concentration after 10 y of treatment (5 compost applications). The coefficients of variation (in %) are indicated for the initial concentrations

	Final concentration after five compost spreadings				Initial
	MSW	BIO	GWS	FYM	Concentration
	(mg kg^{-1})				
Cd	0.26	0.25	0.27	0.30	0.23 (5.4%)
Cr	47.3	46.1	46.2	46.5	45.2 (6.1%)
Cu	15.7	12.8	13.8	15.2	11.6 (2.3%)
Hg	0.12	0.11	0.11	0.10	0.10 (15.6%)
Ni	15.8	15.3	15.4	15.5	14.9 (5.5%)
Pb	28.7	28.1	26.6	32.4	25.5 (24.8%)
Zn	56.6	53.5	54.4	61.3	49.0 (6.9%)

Table 11. Fluxes of some organic micropollutants calculated for 10 years (five compost spreadings) compared to the soil initial stocks of the same molecules (in parenthesis: % variation coefficients). Maximum fluxes for sludges allowed in France are given for comparison.

PAH	MSW	BIO	GWS	FYM	Max. sludge	Initial soil stock
	(mg m^{-2})					
FLT	0.85	3.90	1.67	0.36	7500	13.82 (93%)
BbF	0.23	1.75	1.15	0.19	4000	11.04 (82%)
BaP	0.10	1.20	0.43	0.14	3000	7.13 (96%)
NAP	0.13	0.08	0.03	0.02		0.54 (89%)
ANA	0.04	0.33	0.03	0.01		0.09 (67%)
FLU	0.11	0.49	0.08	0.04		0.35 (58%)
PHE	1.18	3.70	1.44	0.35		4.66 (90%)
ANT	0.10	0.45	0.07	0.04		0.62 (124%)
PYR	0.77	2.55	1.59	0.28		10.14 (92%)
BaA	0.14	1.48	0.82	0.15		7.21 (111%)
CHR	0.36	1.94	1.88	0.32		8.33 (74%)
BkF	0.11	0.95	0.39	0.09		4.66 (97%)
DBA	0.08	0.40	0.18	0.09		2.05 (71%)
IPY	0.08	1.11	0.41	0.14		9.20 (75%)
BPE	0.24	2.20	1.31	0.29		10.86 (89%)
Total 15 PAH	4.5	22.5	11.5	2.5		90.7 (86%)
Σ7 PCB	0.50	0.24	0.20	0.25	1200	9.75 (84%)

After 10 years of sewage sludge spreading on similar soil, metal concentrations did not vary significantly in another long-term field experiment (Michelin et al. 2001). Thus, it can be concluded that no adverse effects would be detected on soil quality due to heavy metal inputs after spreading of composts of similar quality, when they are applied in representative amounts. Nevertheless, it is necessary to keep looking for indicators of this potential degradation. Heavy metals extractable in DTPA or EDTA are considered as more representative than total contents to evaluate heavy metal accumulation in soils (Brendecke et al. 1993).

Only maximum fluxes entering the soils have been defined for PAHs and PCBs in sludge legislation. As for heavy metals, the calculated fluxes for 10 years of treatment were lower than the maximum allowed (Table 11). Fluxes of FLT, BbF and BaP represented 2 to 28% of the initial stocks measured in soils. For other

PAHs such as FLU and ANA applied with the BIO compost, the fluxes were larger than the initial quantities measured in the soil (140 and 366%, respectively). These two molecules will be added in the future European legislation on sludge use in agriculture. However, all organic pollutants presented large initial variability in the soil (up to 95%) and their detection in soils are partly due to aerial deposits even after many sludge applications (Alcock et al. 1995). After repeated applications of sludges according to recommended practices, minor increases in PAHs and PCBs were detected in other studies (Webber and Wang 1995).

Crop Results

Only FYM significantly increased maize yield probably because of a larger input of mineral N than with the composts (74, 34, 22 and 20 kg N-NH_4 ha^{-1} with FYM, GWS, MSW and BIO, respectively) but the influence of mineral N fertilisation was much larger (Table 12). Two to three hundred kg of N were applied with composts and FYM, mainly in organic forms. The apparent use of compost N and P by plants was evaluated by subtracting total N and P assimilated by the maize in the control and amended treatments. Only 5% of total N were used in the FYM

Table 12. Field experiment, results of the first maize crop. Yields and element concentrations in grains in the different treatments. Results are expressed on a dry matter basis. Within the same line, results followed by the same letter are not significantly different (p<0.05)

	Control	FYM	GWS	BIO	MSW
With N Yield (q ha^{-1})	103.7 ± 2.1 (a)	110.4 ± 2.9 (b)	103.0 ± 3.9 (a)	101.5 ± 2.8 (a)	103.4 ± 2.8 (a)
N (g kg^{-1})	12.8 ± 0.4 (a)	13.9 ± 0.4 (b)	13.0 ± 0.6 (a)	12.6 ± 0.3 (a)	12.9 ± 0.3 (a)
P (g kg^{-1})	2.39 ± 0.04 (a)	2.42 ± 0.14 (a)	2.44 ± 0.06 (a)	2.44 ± 0.09 (a)	2.37 ± 0.18 (a)
K (g kg^{-1})	2.84 ± 0.19 (a)	2.79 ± 0.15 (a)	2.76 ± 0.03 (a)	2.90 ± 0.17 (a)	2.74 ± 0.12 (a)
Cd (μg kg^{-1})	0	0	0	0	0
Cr (mg kg^{-1})	0	0	0	0	0
Cu (mg kg^{-1})	2.9 ± 0.2 (a)	2.8 ± 0.5 (a)	2.8 ± 0.4 (a)	2.8 ± 0.4 (a)	2.8 ± 0.3 (a)
Ni (mg kg^{-1})	0.33 ± 0.07 (a)	0.33 ± 0.07 (a)	0.39 ± 0.18 (ab)	0.31 ± 0.09 (a)	0.42 ± 0.10 (b)
Pb (mg kg^{-1})	0	0	0	0	0
Zn (mg kg^{-1})	13.9 ± 0.7 (a)	14.4 ± 1.0 (a)	14.3 ± 1.5 (a)	14.6 ± 1.3 (a)	14.9 ± 2.4 (a)
Hg (μg kg^{-1})	0.55 ± 0.16 (a)	0.78 ± 0.46 (a)	1.05 ± 0.5 (a)	0.55 ± 0.24 (a)	0.54 ± 0.11 (a)
Without N Yield (q ha^{-1})	83.3 ± 8.0 (c)	90.1 ± 4.0 (d)	84.1 ± 1.4 (c)	81.1 ± 3.0 (c)	85.3 ± 3.4 (c)
N (g kg^{-1})	11.4 ± 0.4 (c)	11.9 ± 0.5 (d)	11.3 ± 0.2 (c)	11.2 ± 0.2 (c)	11.4 ± 0.2 (c)
P (g kg^{-1})	2.76 ± 0.07 (b)	2.77 ± 0.15 (b)	2.88 ± 0.11 (b)	2.80 ± 0.05 (b)	2.78 ± 0.11 (b)
K (g kg^{-1})	3.06 ± 0.06 (b)	2.91 ± 0.16 (c)	3.16 ± 0.13 (b)	3.03 ± 0.10 (bc)	2.99 ± 0.10 (bc)
Cd (μg kg^{-1})	0	0	0	0	0
Cr (mg kg^{-1})	0	0	0	0	0
Cu (mg kg^{-1})	2.8 ± 0.2 (a)	2.7 ± 0.4 (a)	2.8 ± 0.2 (a)	2.7 ± 0.2 (a)	2.6 ± 0.4 (a)
Ni (mg kg^{-1})	0.43 ± 0.04 (c)	0.42 ± 0.04 (c)	0.38 ± 0.03 (ca)	0.44 ± 0.08 (c)	0.55 ± 0.15 (d)
Pb (mg kg^{-1})	0	0	0	0	0
Zn (mg kg^{-1})	16.4 ± 0.1(b)	16.7 ± 0.8 (b)	16.9 ± 1.1 (b)	16.9 + 1.1 (b)	17.4 ± 0.6 (b)
Hg (μgkg^{-1})	0.45 ± 0.25 (a)	1.06 ± 0.71 (a)	0.48 ± 0.12 (a)	0.53 ± 0.12 (a)	0.89 ± 0.42 (a)

treatment, 3% for MSW and 0% for BIO and GWS compost. Similarly very low values were found for P (0 to –4%). Mineral N fertilisation did not increase the use of P from amendments. This was probably due to the large initial availability of P in soil (81 mg kg^{-1} of P extracted using the Olsen method). Results on plant use of compost N vary largely in the literature: no valorisation of compost N (Abad Berjon et al. 1997), use of compost mineral N only (Gagnon and Simard 1999); availability of 5 to 40% of organic N (Hadas and Portnoy 1994).

Usually, more metals are accumulated in stems than in grains (Gardiner et al. 1995; Hue 1995). This was confirmed by our results. No Cd, Pb and Cr were found in the grains. All the metals were detected in the stems (results not shown). No significant effects of the organic amendments were observed on the heavy metal concentrations and total content both in grains and stems. On the other hand, mineral N fertilisation increased metal exportation in both grains and stems, because of the increased of dry matter production.

Conclusion

Our results confirmed how necessary it is to relate the effects of compost addition on soil and crop to the composting process. Short composting time was associated to a large residual biodegradability, which can provoke mineral N immobilisation immediately after spreading (example of MSW). However, such reactive organic matter could then have a larger potential N availability. The maize results pointed out an apparent very low availability of N and P from composts during the year after their application. Nevertheless, very large amounts of N have been brought which should mineralise during the following years. This confirmed the importance of long-term field experiment to study the agronomic value of composts. Other aspects are considered in the program: compost effect on soil structure stability, capacity of soil to accumulate organic matter and thus to contribute to decreasing of CO_2 in the atmosphere. Such result acquisitions need several years of observations and simulations that are under progress.

The fluxes of metallic and organic pollutants brought with the organic amendments were lower than the maximum allowed in French legislation on sludge use in agriculture. The calculated evolutions of metal concentrations resulted in possible applications during more than 50 years of composts of similar quality. Probably, no negative effects related to heavy metal accumulations would be observed. Nevertheless, total element analysis should not be enough to detect soil-quality evolution and it remains necessary to elaborate sensitive indicators of soil-quality evolution.

The input fluxes of PAHs and PCBs were also lower than the maximum allowed in sludge legislation. Organic pollutants in soils can be degraded or interact with natural organic matter which could decrease their detected concentration (Barriuso et al. 1996). The evolution of soil quality cannot be predicted with simple arithmetic addition of these molecules. On the other hand, the initial gradient in the field experiment indicated their possible atmospheric origin. This is also true for heavy metals entering the soils via vapour or dust depositions. Thus, compost effect on their concentrations in soil cannot be evaluated without measuring aerial depositions on soils, which will be done in the future in the field experiment. Finally, if waste sorting would allow to decrease metal contents in composts, the organic pollutant concentration could be more difficult to control since they can be of atmospheric origin in the composting plant as in the field experiment. More results will be acquired during the next years.

Acknowledgements. The authors would like to thank the compost producers for their help for compost sampling and their interest in the experiment and Jean-Noel Rampon for his help in the field experiment.

References

Abad Berjon M, Climent Morato MD, Aragon Revuelta P, Camarero Simon A (1997) The influence of solid urban waste compost and nitrogen-mineral fertilizer on growth and productivity in potatoes. Comm Soil Sci Plant Anal 28: 1653-1661

ADEME (2001) Les éléments traces métalliques dans les composts en France, Angers, France (in press)

Alcock RE., McGrath SP, Jones KC (1995) The influence of multiple sewage sludge amendments on the PCB content of an agricultural soil over time. Environ Toxicol Chem 14: 553-560

Baize D (2000) Teneurs totales en métaux lourds dans les sols: résultats généraux du programme ASPITET. Le courrier de l'Environnement de l'INRA 39: 39-54. (in French)

Barriuso E, Calvet R, Schiavon M, Soulas, G (1996) Les pesticides et les polluants organiques des sols, transformations et dissipation. Etude Gestion Sols 3-4: 279-296

Bernal MP, Navarro AF, Sanchez-Monedero MA, Roig A, Cegarra J (1998) Influence of sewage sludge compost stability and maturity on carbon and nitrogen mineralization in soil. Soil Biol Biochem 30: 305-313

Berset JD, Holzer R (1995) Organic micropollutants in Swiss agriculture: distribution of polyn uclear aromatic hydrocarbons (PAH) and polychlorinated biphenyls (PCB) in soil, liquid manure, sewage sludge and compost samples; a comparative study. Int J Environ Chem 59: 145-165

Brendecke JW, Axelson RD, Pepper IL (1993) Soil microbial activity as an indicator of soil fertility: long-term effects of municipal sewage sludge on an arid soil. Soil Biol Biochem 25: 751-758

Breslin VT (1999) Retention of metals in agricultural soils after amending with MSW and MSW-biosolids compost. Water Air Soil Pollut 109: 163-178

Chaney RL, Ryan JA (1993) Heavy metals and toxic organic pollutants in MSW-composts: research results on phytoavailability, bioavailability, fate, etc. In: Hoitink HAJ, Keener HM (eds) Science and engineering of composting. Renaissance Publications, Worthington, Ohio, pp 451-506

Gagnon B, Simard RR (1999) Nitrogen and phosphorus release from on-farm and industrial composts. Can J Soil Sci 79: 481-489

Gardiner DT, Miller RW, Badamchian B, Azzari AS, Sisson DR (1995) Effects of repeated sewage sludge applications on plant accumulation of heavy metals. Agricult Ecosyst Environ 55: 1-6

Grossi G, Lichtig J, Kraub P (1998) PCDD/F, PCB and PAH content of Brazilian compost. Chemosphere 37: 2153-2160

Hadas A, Portnoy R (1994) Nitrogen and carbon mineralization rates of composted manures incubated in soils. J Environ Qual 23: 1184-1189

Hue NV (1995) Sewage sludge. In: Rechcigl JE (ed) Soil amendments and environmental quality. Lewis, Boca Raton, pp 199-247

IFEN (1999) L'environnement en France. La Découverte, Paris (in French)

Jauzein M, Feix I, Wiart J (1995) Les micro-polluants organiques dans les boues résiduaires des stations d'épuration urbaines. Ademe, Angers (in French)

Juste C (1993) Matières organiques et comportement des éléments traces dans le sol. In: Decroux J, Ignazi JC (eds) Matières organiques et agricultures. GEMAS et COMIFER, Blois, pp 115-123 (in French)

Lichtfouse E, Budzinski H, Garrigues P, Eglinton TI (1997) Ancient polycyclic aromatic hydrocarbon in modern soils: ^{13}C, ^{14}C and biomarker evidence. Org Geochem 26: 353-359

MacGrath S, Chaudri AM, Giller KE (1995) Long-term effects of metals in sewagesludge on soils, microorganisms and plants. Journal of Industrial Microbiology 14: 94-104

Mamo M, Rosen CJ, Halbach TR (1999) Nitrogen availability and leaching from soil amended with municipal solid waste compost. J. Environ Qual 28: 1074-1082

Michelin J, Bourgeois S, Wiart J, Bermond A (2001) Evaluation de l'état de contamination d'un sol agricole par les éléments traces métalliques lié à des apports réguliers de boues résiduaires urbaines depuis 1985. In: Baize D, Tercé M (eds) Les éléments traces métalliques dans les sols, approche fonctionnelles et spatiales. INRA, Paris (in press) (in French)

Morvan B, Carré J (1994) Oligo-éléments et micro-polluants dans les composts. Symposium international sur le traitement des déchets, Pollutec, Lyon, France, pp 169-179 (in French)

Niederer M, Maschka-Selig A, Hohl C (1995) Monitoring polycyclic aromatic hydrocarbons (PAHs) and heavy metals in urban soil, compost and vegetation. Environ Sci Pollut Res 2: 83-89

Pinamonti F, Stringari G, Gasperi F, Zorzi G (1997) The use of compost: its effects on heavy metal levels in soil and plants. Resourc Conserv Recycl 21: 129-143

Robert M, Juste C (1997) Stock et flux d'élément traces dans les sols. In: Ademe (ed), Epandage des boues résiduaires, aspects sanitaires et environnementaux, Ademe, Paris, pp 192-204, (in French)

Stratton ML, Baker AV, Rechcigl JE (1995) Compost. In: Rechcigl JE (ed) Soil amendments and environmental quality. Lewis, Boca Raton, pp 249-309

Webber MD, Wang C (1995) Industrial organic compounds in selected Canadian soils. Can J Soil Sci 75: 513-524

Werner W, Warnusz J (1997) Ecological evaluation of long-term application of sewage sludges according to the legislative permission. Soil Sci Plant Nutr 43: 1047-1049

Composting in the Framework of the EU Landfill Directive

M. Kranert[1], A. Behnsen[1], A. Schultheis[2] and D. Steinbach[2]

Abstract. The EU directive on the landfill of waste will initiate far-reaching changes in the field of waste management. In most of the member states drastic measures will become necessary to comply the required limit values of biodegradable wastes in landfills within the given time frame (2006 -2016). On the basis of a research project on behalf of the European Union, the paper gives an overview of the state of composting and biowaste treatment in the European Union. Aerobic and anaerobic technologies are described and the quantities of treated biowaste shown. The main target of this project was to calculate scenarios for wastes from waste treatment operations with regard to future waste management measures. The effects of the biological treatment on the waste management and the quantity and quality of waste are described.

Beside the status quo of composting, this chapter also shows measures which must be taken by the EU Member States to meet the EU limit values. Both the mechanical-biological pretreatment of residual waste and the composting of separately collected organic waste or backyard composting will play a decisive role.

Modes of Biological Treatment Processes

The aim of biological treatment can be recycling as well as pre-treatment before final disposal of treated waste (Fig. 1). In both cases, aerobic methods (composting, decomposition) as well as anaerobic methods (digestion, methanizing) are possible.

If the aim is recycling of biological waste, a composite manure is generated which, as compost or digestate, may be returned into the material flow (Bundesgütegemeinschaft Kompost 1998). In addition, anaerobic digestion can have a positive influence on the greenhouse effect, when the generated biogas is used for an energetically self-sufficient operation. The input material is normally separately collected biowaste to generate a product with only a small content of harmful substances.

[1]University of Applied Sciences Braunschweig/Wolfenbüttel, Institute for Waste Management and Environmental Monitoring, Wolfenbüttel, Germany
[2]University of Stuttgart, Institute for Sanitary Engineering, Water Quality and Waste Management, Stuttgart, Germany

H. Insam, N. Riddech, S. Klammer (Eds.)
Microbiology of Composting

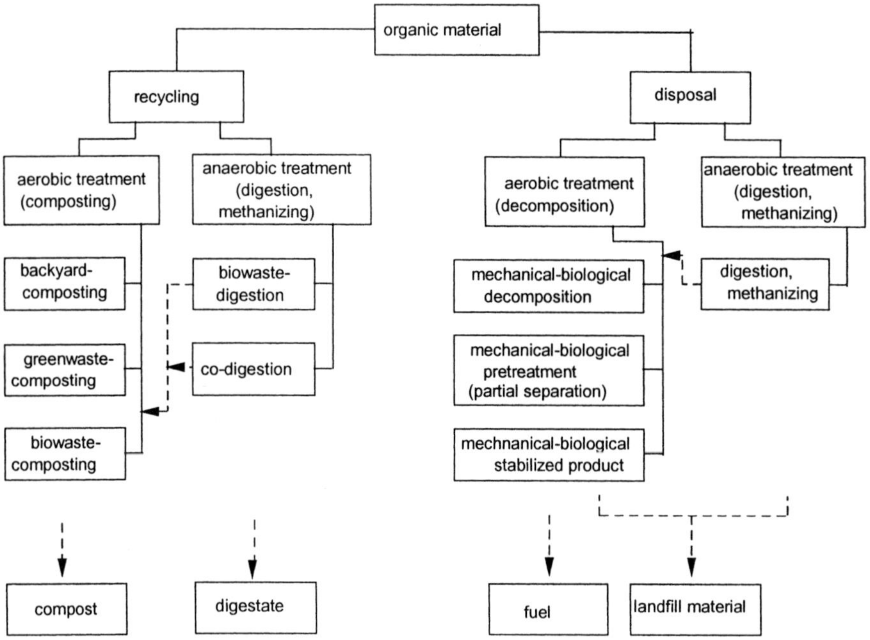

Fig. 1. Processes for the biological treatment of MSW

The aim of mechanical-biological treatment, in addition to the reduction of volume, is the generation of a low-emission landfill material. Mechanical-biological decomposition processes with an output which is landfilled directly or with a partial separation of material for recycling may be used. Effect of the treatment, beside the reduction of the organic substances, is the improvement of the separability and the reduction of water content. This leads to material which can be better landfilled and better used for the production of secondary fuel (refuse energy fuel, REF).

In Germany in the Technical Instruction on Municipal Solid Waste (TASi) the parameter volatile solids and total organic carbon (TOC) are defined as verification parameters for a low-emission landfilling material which needs little aftercare. This is acceptable for the output of incineration plants, but is not appropriate for the evaluation of mechanical biological pretreated waste. In the latter case, parameters which describe the biological activity (e.g. oxygen consumption, anaerobic gas production) are much more suitable. To address this, in Germany a landfill directive for thermal and mechanical-biological pretreated waste came into force in spring 2001.

Waste Amounts

The member states annually generate 180 million tonnes of municipal solid waste (MSW). The percentages concerning the waste amounts of the member states (Fig. 3) and the generated amounts per capita (Fig. 2) can be calculated on the basis of the total amount.

In individual states the definitions for MSW are quite different. Particularly commercial waste is not gathered commensurately in the statistics. Furthermore, the statistical data regarding waste amounts can vary considerably (e.g. Ireland 430 to 700 kg $capita^{-1}$ $year^{-1}$). It is apparent that the specific annual amounts per capita are typically in the range of 400 kg to 600 kg.

Analysis of the data reveals the great influence of the central European states as Germany, France, Italy and UK, which generate more than 72% of the whole MSW. Germany alone generates nearly a quarter of the whole MSW produced in the European Union.

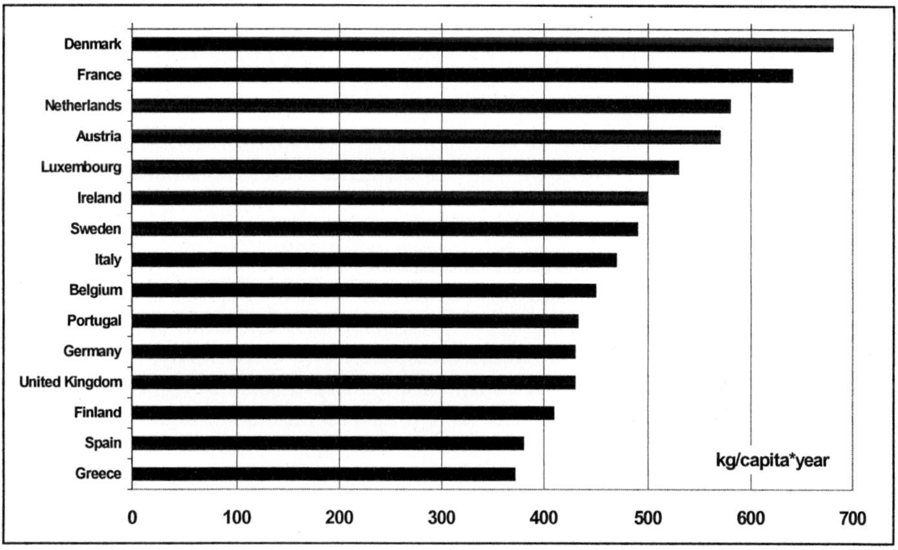

Fig. 2. Annual per capita generation of MSW in the member states of the EU (DHV Environment and Infrastructure EC 1997; Barth 1999; Stuttgart University, University of Applied Sciences Braunschweig/Wolfenbüttel, IBGE 2001)

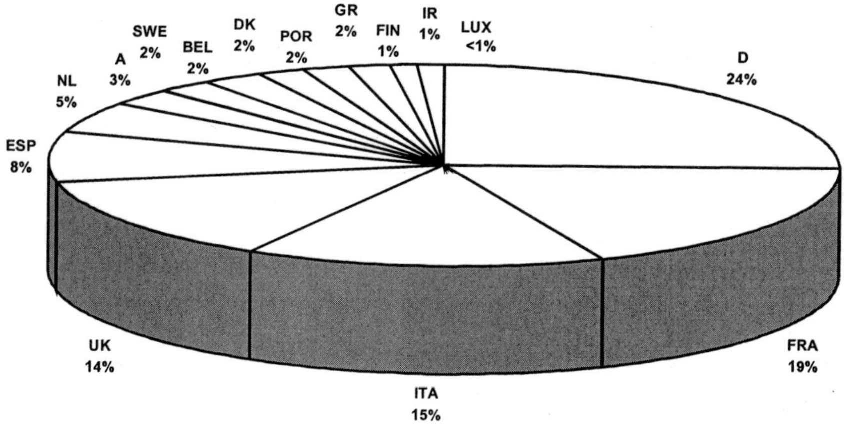

Fig. 3. MSW in the EU (proportion of whole MSW (%) of 180 million tonnes/a) (DHV Environment and Infrastructure EC 1997; Barth 1999; Stuttgart University, University of Applied Sciences Braunschweig/Wolfenbüttel, IBGE 2001)

The proportion of biowaste in household waste is ranges from 25% (France) to 50% (Greece) (Fig. 4). This demonstrates the influence of the separate collection of biowaste and home-composting as well as the separate gathering of valuable material (Lasaridi 1997).

In some cases (e.g. Greece) the high amount of biowaste is ascribed to the fact that backyard composting of biowaste or garden waste does not have any tradition and is not carried out. Additionally, in some countries, some special organic wastes (e.g. wood, leather) are gathered separately (e.g. Germany, Austria, UK, Italy). Biowaste which is generated by the commercial sectors (e.g. cultivation of vegetables or flowers) has a considerable influence on the waste amounts in The Netherlands and Belgium (Dewey 1998).

The four central European countries (Germany, France, Italy and the UK) greatly influence the biowaste potential in the EU. This may be calculated on the basis of the absolute MSW amounts and the relative national proportion of biowaste. More than 63% of the European biowaste potential is to be found in these countries (Fig. 5).

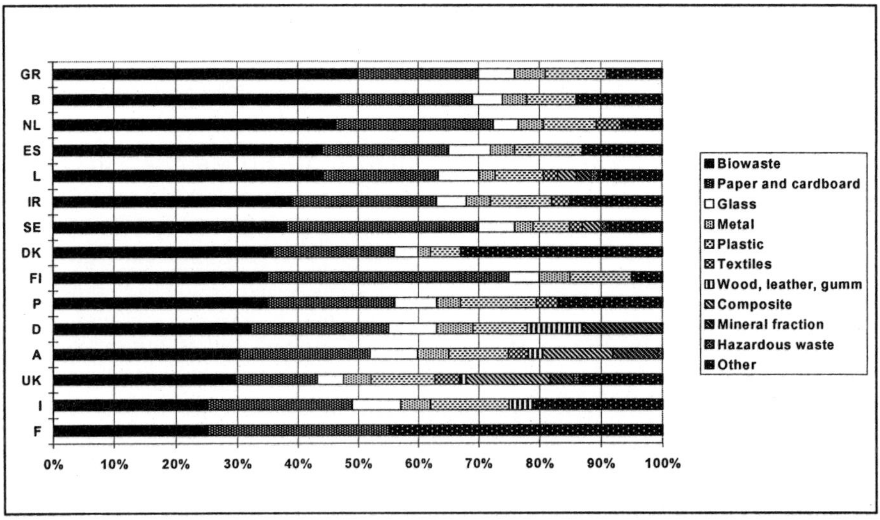

Fig. 4. Composition of the household waste in the states of the EU (Barth 1999; Stuttgart University, University of Applied Sciences Braunschweig/Wolfenbüttel, IBGE 2001)

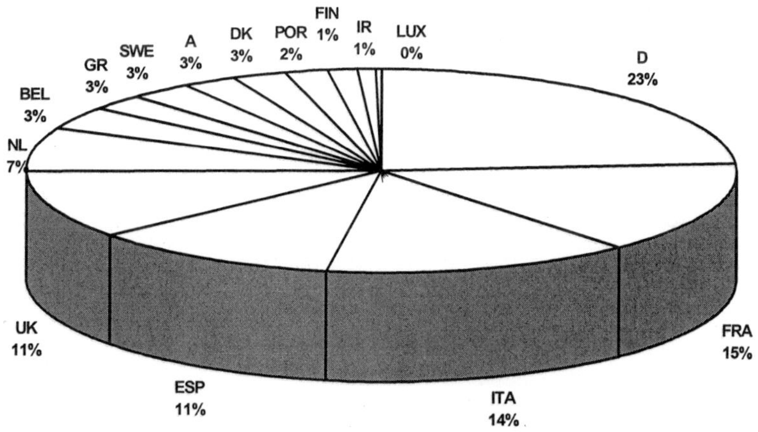

Fig. 5. Biowaste potential in MSW in the EU (quota of the total biowaste amounts (%) of 57 million tonnes/a) (DHV Environment and Infrastructure EC 1997; Barth 1999)

Legislative Framework

The Biological Waste Treatment Situation in the EU Member States

The regulation of waste management in most countries is determined by national concepts. Thus, the aim of the European Union to come to a coherent legislature concerning the separate collection and the biological waste treatment is relatively far away.

Some member states [Spain, Portugal, Greece, Ireland and Belgium (without Flanders)] have not implemented any national legislature to control separate collection and biological waste treatment (Lasaridi 1997, Dewey 1998). France and the United Kingdom are now in a test phase regarding separate collection (< 5%). Italy has legally regulated the separate collection of waste in its legislature, but has implemented it only in a qualified sense (approx. 1%) (Ragazzi, Gheser, Raninger 1998). By contrast, in Sweden separate collection is not regulated legally but is generally accepted by society and policy. In Denmark, Finland and Flanders (Belgium) the implementation of separate collection and treatment of biowaste is partly realised and will be pushed ahead. In Austria, Germany, Luxembourg and The Netherlands, the separate collection and recycling of waste is regulated in real terms and by parts substantial implemented (> 50%) (Barth, Kroeger, Bidlingmaier 1998, Kern, Funda Mayer, 1998,1999). The Netherlands hold an exceptional position because the regulations were implemented, nearly completely, within 10 years (1990: about 1% of the households were connected with the separate collection; 1996: approx. 92 %).

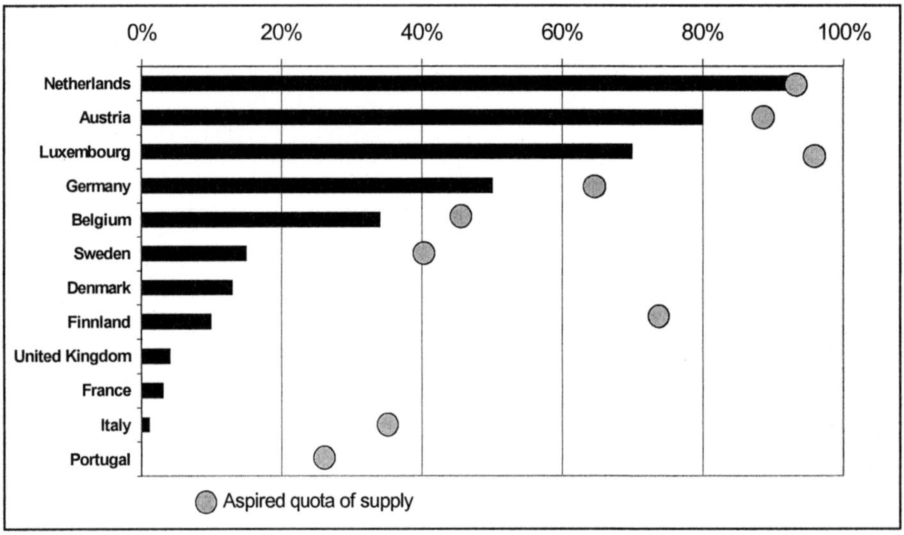

Fig. 6. Quota of supply on a separate collection system for biowaste

In Austria, Belgium, Denmark, Germany and The Netherlands, special quality assurance systems govern compost quality. Requirements for landfilling waste which are above the EU Landfill Directive are regulated or demanded in Germany, Austria, Belgium, The Netherlands and Sweden.

Landfill Directive

The European Directive on the Landfill of Waste is in force since July 1999 and has the aim of improving considerably the environmental standards of landfills in the member states of the European Union. Through this, the negative environmental impact coming from landfills shall be avoided or reduced. To reach this aim, strict standards on technical equipment and operation of landfills will be implemented. In addition, waste will be pretreated before final disposal, to reduce hazards and emissions from landfill sites. The advancement of measures to utilise organic waste by composting, anaerobic digestion or energy recovery is here well to the fore.

On the basis of EUROSTAT Data from 1995 (EUROSTAT, 1999) biodegradable waste has to be reduced. Member states, which have landfilled more than 80% of their MSW in 1995 can postpone the completion of the reduction figures by up to 4 years (Table 1).

Table 1. Reduction of biodegradable MSW required to comply with EU Landfill Directive (EU Commission 1999)

Years after implementation	Reduction
5 years (2006 / 2010)	25 %
8 years (2009 / 2013)	50 %
15 years (2016 / 2020)	65 %

The German technical instruction on MSW and the German landfill directive have very strict demands for the minimisation of organic matter to be landfilled. These demands are going far beyond the targets of the European Landfill Directive. Nevertheless, for some member states great efforts will be required to reduce the biodegradables in MSW going to landfill. According to our scenario results, Austria, Germany and The Netherlands are already achieving the last stage of the EU-Landfill Directive (Stief 1998, Wagner 1999).

Scenarios

The scenarios were developed to predict the effects of waste treatment operations on quantity and quality of wastes for recycling, residual wastes and residues from treatment operations. Changes resulting from legislation (e.g. national and EU Directives) and from different waste management operations, were also considered. Scenarios were calculated only for six EU member states (Austria,

The Netherlands, United Kingdom, Ireland, Germany, France EU-6), as data of other member states were not available or unsuitable for this kind of calculations. The EU-6 generate about 64% of the whole waste generated in the 15 member states (EU-15).

In view of the Landfill Directive, the data situation in some member states is a very important problem. Exact data on the amounts of waste, waste composition, residues, recycling and recovery rates, etc., is hard to obtain or is ambiguous. Moreover, data on the amount of waste, waste composition, residues, recycling, recovery rates, etc. are not comparable.

In addition, there is a serious definition problem of MSW, household waste, commercial waste, etc. Despite the European Waste Catalogue (EWC) the member states use their own waste definitions, which are often not congruent with either the EWC or other member states. France and the UK, for example, have almost the same number of inhabitants, but the MSW differs by several million tons. This is probably due to definition problems, because the countries subsume different waste fractions under the term MSW. Thus, in some member states sewage sludge is covered under MSW, but not in others.

In order to check whether the targets of the Landfill Directive are being achieved, detailed knowledge about waste quantities and composition is essential. This means that the biodegradable parts of MSW (e.g. paper, cardboard, organic kitchen waste, wood, textiles, garden and park waste, etc.) have to be known quantitatively. In addition, it is essential to know the quantities and biodegradability of wastes going directly to landfill, of wastes going to recycling and of the residues from recycling processes.

Because of data uncertainties and data gaps, scenario calculations could be performed only by making some assumptions. The waste management scenarios were developed in agreement with representatives of EU-DG Environment. The basis for the scenarios are population dynamics and waste streams of each member state, combined with waste input in selected treatment operations.

The scenarios consider collection of wastes for recycling and energy recovery, residual waste and the corresponding treatment operations. Materials which are not recyclable are pretreated to meet the requirements of the Directive on the Landfill of Waste and environmentally sound disposal, respectively.
Table 2 shows the five basic waste management scenarios.

Table 2. Waste management scenarios (Stuttgart University, University of Applied Sciences Braunschweig/Wolfenbüttel, IBGE 2001)

S1	Business as usual (autonomous development) Constantly developing waste quantities and adjusted recycling capacities
S2	EU-harmonising Transferring EU Directives (packaging and packaging waste, landfill of waste, ELV, WEEE) into national law. The scenario describes a minimum standard for Europe

Table 2 (cont.)

S3	National targets Comprises national waste plans and political targets exceeding the European Standards (S2) at least in some items. If the national standards do not Exceed the European ones, S3 is equivalent to S2
S4	Optimisation of waste treatment I – (intensive/high recycling and MBPT as treatment option) Extensive, low cost separate collection (bring systems) Material recovery (paper/cardboard, packaging waste, plastic, glass, metal, org. kitchen waste, garden/park waste) Treatment operations: Material recovery facility (MRF) and composting/fermentation MBPT as important treatment option for residual waste and residues from waste treatment operations before going to landfill
S5	Optimisation of waste treatment II – (maximum recycling and incineration as treatment option) Intensive separate collection (kerbside collection systems) Material recovery (paper/cardboard packaging waste, plastic, glass, metal, org. kitchen waste, garden/park waste) Treatment operations: MRF and composting/fermentation Incineration (incl. flue gas cleaning) as the only treatment option for residua l waste and (partly) for residues from waste treatment operations before going to landfill

Scenario 1 represents the current situation projected until 2021. Calculations until 2021 are made by assuming that recovery rates, as well as incineration and MBPT capacities (input of the reference year), are constant. This scenario is calculated for EU-6.

Scenario 2 considers EU targets [Packaging and Landfill Directive, directives on waste from electrical and electronic equipment (WEEE) and end-of-life vehicles (ELV)]. This scenario places special emphasis on recovery and recycling. Thus, the Landfill Directive must be achieved by increasing recovery and recycling rates, especially for paper/cardboard, organic waste, park/garden waste and wood. This scenario is relevant for France, the UK and Ireland.

Scenario 3 considers national targets (when exceeding EU targets). This scenario is important for the UK, setting high recovery and recycling targets. The most important national target for Austria, Germany and The Netherlands aims at banning untreated waste from landfill (preferring incineration as a treatment option).

Scenario 4 combines high recycling and recovery rates with pretreatment (mechanical-biological pretreatment alone or in combination with incineration) of waste to reduce the biodegradables in the waste stream. This scenario is valid for EU-6 except The Netherlands. Because in Austria and Germany mechanical-biological pretreatment (MBPT) is still under discussion as a treatment option, maximum 30% (Germany 35%) of residual waste is going into MBPT, while the rest is incinerated (to meet national targets). For calculating scenario 4, in France and the UK incineration input is going to zero by 2021, while MBPT input increases to 100%. For The Netherlands, MBPT is out of discussion, therefore it is not considered in this scenario.

Scenario 5 requires maximum recycling and recovery rates. In order to reduce the biodegradables in the residual waste stream, incineration is chosen as a pre-treatment option (MBPT is not considered).

The scenario results are presented in 5-year steps. This interval was chosen to meet most of the different stages of EU targets. In addition, changes in quantities typically become more obvious within a period of 5 years. Time horizons are: reference year (the year providing best waste data), 2001, 2006, 2011, 2016 and 2021.

In view of the Landfill Directive, the relevant scenario results show waste quantities and the portion of biodegradables going to landfill.

Waste Quantities Going to Landfill

Landfill input originates from different sources:

- residual waste that is not recovered;
- residues from treatment facilities (recycling plants, MBPT, incineration plants);
- impurities (collected with separately collected waste) that are not suitable for recycling.

The amount of waste projected as landfill input under the five scenarios is shown in Fig. 7.

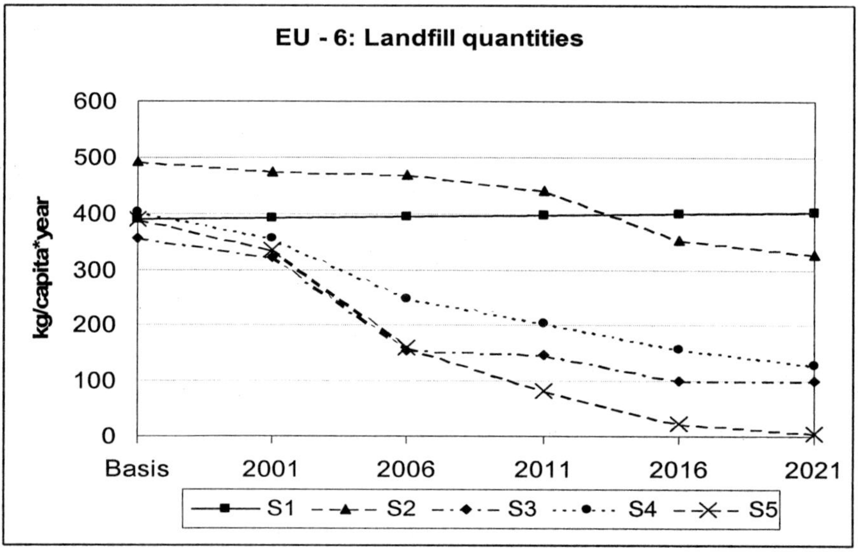

Fig. 7. Projected input to landfill (EU-6) of five scenarios (S1-S5) (Stuttgart University, University of Applied Sciences Braunschweig/Wolfenbüttel, IBGE 2001)

Because of positive population dynamics in the EU-6, except Germany, scenario 1 predicts steadily increasing waste quantities going to landfill.

In order to meet the EU targets in scenario 2, recycling and recovery rates must be increased considerably. These efforts result in decreasing landfill input.

The graphs of scenario 3 and 5 are almost congruent until 2006 (focus is on incineration of residual waste), then the scenario 3 profile flattens while scenario 5 decreases more steeply because of maximum recycling and recovery rates.

Under scenario 4, the landfill quantities decrease due to higher recycling and recovery rates and mechanical-biological pretreatment (and incineration) of residual waste.

Biodegradables Going to Landfill

The priority of the Landfill Directive is not the amount, but rather the portion of biodegradables being landfilled. Figure 8 shows the percentage of biodegradables for landfill compared to the biodegradables produced in the reference year. Although the reference year fixed in the Landfill Directive was 1995 or earlier, it was not possible to obtain any reliable information on this subject. Therefore, the reference year under this study was the year providing the necessary waste data and may therefore vary from country to country.

Since UK and Ireland are countries which sent more than 80% of collected MSW to landfills in 1995, they are assumed to postpone the attainment of the targets by 4 years.

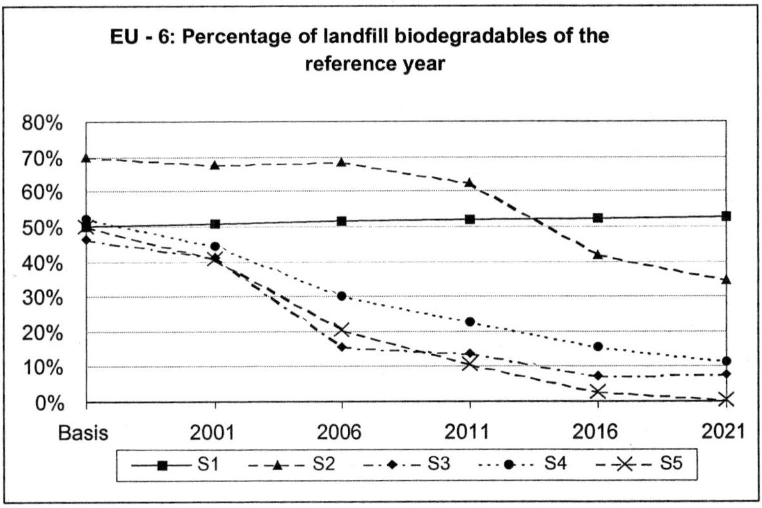

Fig. 8. Biodegradables going to landfill (EU-6) under five scenarios (S1-S5) (Stuttgart University, University of Applied Sciences Braunschweig/Wolfenbüttel, IBGE 2001)

In the reference year of scenario 1, the portion of biodegradables was 50%. The percentage of biodegradables increase slightly, due to constant incineration and MBPT capacities but increasing residual waste quantities (positive population dynamics).

Scenario 2 was calculated for France, Ireland and the UK. In the UK and Ireland, the targets of the Landfill Directive will probably be achieved with a postponement of 4 years. This does not apply for France, because France seeks to meet the targets as scheduled in the directive. The average of the three countries was projected. Thus, the targets of the Landfill Directive are achieved between the years fixed in the directive and the years of postponement.

The projection of scenario 3 predicts achievements of the last stage of the Landfill Directive in 2006. This is due to Austria, Germany and The Netherlands, which already attain (according to our calculations) the targets of the Landfill Directive.

Because of high recovery and recycling rates and MBPT (and partly incineration) as pretreatment before landfilling, in scenario 4, the single stages of the Landfill Directive can be achieved earlier than required.

In scenario 5, the percentages of biodegradables fall considerably beyond the target required in the last stage of the Landfill Directive. This is due to predicted maximum recycling and recovery rates, on the one hand, and incineration of residual waste and of some part of recycling residues, on the other hand.

Scenario Results

The Landfill Directive requires a decrease of biodegradable waste in MSW. This target was simulated for EU-6 by scenario calculations in several ways:

S1 The targets cannot be achieved by existing waste management systems in France, Ireland and the UK. Austria, Germany and The Netherlands already attain the targets (according to our programme calculation).

S2 High recovery and recycling rates will be necessary to meet the targets of the Landfill Directive when no additional treatment of residual waste is planned. This will mean great efforts in collection and recycling, especially for organic waste, waste paper and plastic.

S3 National targets vary widely. Austria, Germany and The Netherlands prefer incineration while the UK sets high recovery and recycling targets. France and Ireland are open for different kinds of waste treatment.

S4 Intensive recycling and recovery on the one hand and mechanical-biological pretreatment of residual waste (with and without additional incineration) on the other hand may be suitable treatment options to meet the Landfill Directive. Because of pretreatment of residual waste, collection, recycling and recovery rates can be lower than in S2.

S5 Maximum recovery and recycling efforts and incineration of residual waste may achieve the highest reduction of biodegradables.

The Landfill Directive requires above all directives, the strongest efforts from the member states. However, because of a definition gap of the term "MSW" and "biodegradable", each member state has various possibilities for interpretation. This influences, to a high degree, the amounts of MSW and, hence, the biodegradables stated in the official statistics.

Conclusion

Organic waste management and regulation is a very important element in the European waste policy. According to the Landfill Directive, organic waste going to landfill must be minimised. This will lead to intensified activities in separate collection and recovery of biodegradable waste (e.g. composting). Biodegradable residual waste which cannot be collected separately must be oxidised in residual waste treatment plants by incineration or mechanical-biological pretreatment.

The scenario calculations carried out under this study showed that the Landfill Directive may lead to significantly smaller amounts of waste and, hence, biodegradables going to landfill. The scenarios showed theoretically that a maximum of recycling and recovery activities (e.g. composting of organic wastes and recycling of paper), together with consequent residual waste treatment, can lead to very low landfill amounts and thus reduced landfill emissions.

References

Barth J (1999): Biological treatment in the EU; 4/99 (unpublished)

Barth J, Kroeger B, Bidlingmaier W (1998) Bioabfallkompostierung und Vermarktung in anderen Ländern Europas, Korrespondenz Abwasser 10/98; GFA Hennef

Bundesgütegemeinschaft Kompost (1994) Kompost für den Garten- und Landschaftsbau, Köln

Dewey W.-J. (1998) Stand der Kompostierung in Belgien und Frankreich; 58. Informationsgespräch des ANS in Viersen, ANS e.V., Mettmann

DHV Environment and infrastructure, European Commission (1997) Composting in the European Union; Final report; MdG/VS/AT973090, Ammersfoort

EUROSTAT (1999) Eurostat data shop Berlin, Statistisches Bundesamt, Zweigstelle Berlin

Kern M, Funda K, Mayer M (1998, 1999) Stand der biologischen Abfallbehandlung in Deutschland; Teil I, II; Müll und Abfall 11/98, 2/99; Erich Schmidt, Berlin

Kommission der EU (1999) Richtlinie 1999/31/EG des Rates vom 26. April 1999 über Abfalldeponien

Lasaridi K E (1997) Biological treatment of organic waste in Greece: history, present status and future perspectives, ORBIT 97, Weimar

Ragazzi M, Gheser D, Raninger B (1998) Abfallwirtschaft und Kompostierung in Italien; 58. Informationsgespräch des ANS in Viersen; ANS e.V., Mettmann

Stief K (1998) Die Europäische Deponierichtlinie ausgewählte Anforderungen an Deponien;
Korrespondenz Abwasser 10/1998, S 1827-1840; GFA Hennef

Stuttgart University; University of Applied Sciences Braunschweig/Wolfenbüttel; Institut Bruxellois pour la Gestion de Lénvironnement (IBGE) (2001) Final report Waste from Waste treatment operation DG Environment; E3/ETU/970089, Stuttgart

Wagner, K. (1999) Die europäische Deponierichtlinie zu erwartende Auswirkungen auf die TA-Abfall. In: Wiemer, Kern (Hrsg.) Bio- und Restabfallbehandlung III, S 497-513, MIC Baeza, Witzenhausen

Occurrence of *Aspergillus fumigatus* in a Compost Polluted with Heavy Metals

E. López Errasquín, B. Patiño, R.M. Fernández and C. Vázquez[1]

Abstract. An isolate of *Aspergillus fumigatus*, present at high frequency in compost samples from sewage sludge from Madrid, was analysed for tolerance and adaptation to copper. Although high copper concentrations (350 mg l^{-1}) inhibited fungal growth, training the fungus by successive sub-culturing in media containing progressively higher copper concentrations resulted in the fungus being able to grow in a medium containing 50 mg l^{-1} more copper than when it was not trained. Changes in some morphological features were also observed following training. Copper negatively affected two parameters of fungal growth, specific growth rate (μ) and lag-phase. Protein patterns obtained by SDS-PAGE of control and copper-amended cultures of *A. fumigatus* were compared and one protein of 20.2 kDa appeared to be specifically associated with the presence of copper in cultures.

Introduction

Sewage sludge disposal is an immediate and increasing problem for many municipalities because municipal wastewaters can contain potentially pathogenic microorganisms. Conventional wastewater treatment does not completely destroy these microorganisms, many of which are concentrated in the sludge (Clark et al. 1984). Composting, an alternative disposal treatment that is very effective for sludge treatment, has been investigated from engineering, economic, bacteriological and viral aspects (Burge et al. 1976; Colacicco et al. 1977; Willson et al. 1977). It is a thermophilic aerobic decomposition process (Clark et al. 1984) that can stabilize sludges prior to application on land. Temperature is a very important factor in composting, as both a consequence and a determinant of activity. Elevated temperatures are a fundamental characteristic of composting ecosystems, and temperatures of 50–70 °C are usually attained in the composting process provided the mixing or aeration is efficient (Cristóbal 1998). Under these conditions, the composting process will inactivate most microorganisms (Miller 1992). Although many microorganisms are eliminated during the composting process, many others, such as actinomycetes, mucorales and, in particular, *Aspergillus fumigatus*, are able to proliferate at the high temperatures generated during the process (Clark et al. 1984). Consequently, there are health risks

[1]Dpto. de Microbiología III, Facultad de Biología, Universidad Complutense de Madrid, Madrid 28040, Spain

H. Insam, N. Riddech, S. Klammer (Eds.)
Microbiology of Composting

associated with the composting of sewage sludge because compost workers may be exposed to fungi and their products, which may be allergenic or toxic.

The ability of microbial communities to adapt to anthropogenic chemicals is well described. The most important example is related to the widespread use of antibiotics. Similar adaptation processes may occur for a variety of other contaminants (Aelion et al. 1987). Wastewater often contains excessive amounts of metals such as cadmium, cobalt, copper and chromium. The impact of these metals on the environment and on the food chain has promoted research in developing alternative, efficient and low-cost wastewater purification systems (Wilhelmi and Duncan 1995). On the other hand, some microorganisms able to live in such conditions can be used to reduce and remove heavy metals from contaminated samples.

The objectives of this study were to investigate tolerance and adaptation to copper in a strain of *A. fumigatus* isolated from compost following the composting process.

Material and Methods

Organism and Culture Conditions

One strain of *A. fumigatus* was isolated from a sample of compost from piles of windrow in a composting plant located in the Comunidad of Madrid (Spain). This plant makes compost from sewage sludge from wastewater treatment plants. We used the dilution plate method for detection and isolation of fungi: compost (25 g) was added to 225 ml of sterile distilled water, dilutions of up to 10^{-1} or 10^{-3} were prepared and 1 ml of these dilutions were plated with Sabouraud dextrose agar--gentamicin-chloramphenicol (Biomedics). Duplicate samples were analysed. The fungus was maintained on solid media potato-dextrose agar (PDA) at 25 °C and, as stock cultures, on PDA slants at 4 °C or as microconidial suspensions in 15% glycerol at –80 °C.

The isolated fungus was cultured in 250-ml Erlenmeyer flasks containing 50 ml of liquid Sabouraud medium supplemented with concentrations of copper between 0 to 350 mg l^{-1}. The cultures were inoculated with mycelial discs cut from margins of 7-day-old colonies and incubated at 25 °C in an orbital shaker at 150 rpm. The mycelium was harvested by filtration through a nitrocellulose filter with a pore size of 0.45 μm (Millipore) and dried overnight at 65 °C. All assays were analysed in triplicate.

We used a rapid method to establish growth curves in the presence of copper by measuring optical density with an iEMS (Labsystem). Microtitre plates with 96 wells were used. Each well was inoculated with 50 μl of a 10^{6}-spore ml^{-1} suspension, 150 μl liquid Sabouraud medium, and different copper concentrations, and incubated at 28 °C and 500 rpm. Growth was followed over a period of 24 h at

20-min intervals. Agitation was triggered 1 min before each measurement at 620 nm. We also used controls of the copper solution and the liquid Sabouraud medium. The results given were the mean of six measurements.

Adaptation (Training)

Aspergillus fumigatus was trained by progressively increasing levels of copper using serial transfers on Petri dishes of PDA medium. The medium was initially amended with 50 ppm (mg l^{-1}) of copper and maintained at 25 °C. When the fungus was growing, it was transferred to fresh media amended with concentrations of copper increasing in increments of 50 ppm each. After three transfers and a final copper concentration of 150 mg l^{-1}, the fungus was still able to grow and was termed Cu-trained. During the process, growth and morphology of the fungus were also examined.

Study of the Tolerance of the Cu-Trained Strain

The Cu-trained and control fungi were analysed for copper tolerance in liquid Sabouraud medium amended with concentrations of copper between 0 to 400 mg l^{-1}, as described previously. The results obtained were compared with the curve of dry weight in presence of copper of the non-trained strain.

Gel Electrophoresis

SDS-PAGE was performed by the method of Laemmli (1970) with Tris-Tricine ready gels for peptides and small proteins with 10–20% linear gradient (Bio-Rad). A Bio-Rad kit containing the following polypeptides was used as standard: carbonic anhydrase (39.1 kDa), soybean trypsin inhibitor (29 kDa), lysozyme (18.2 kDa), aprotinin (8.5 kDa) and insulin (3.5 kDa). Gels were stained for total protein using the Coomassie method (Coomassie 0.002%, methanol 46%, acetic acid 7.6%).

Results and Discussion

Aspergillus fumigatus was the most abundant species present in the compost samples analysed. Its occurrence at high frequency could indicate a higher tolerance to the conditions of composting. As previously noted, high temperatures are generated during the composting process, resulting in an important reduction of the number of pathogens and also possibly species diversity. However, *A. fumigatus* is a thermophilic fungus that grows well at, or even above, 45 °C. For example, Millner et al. (1977) detected *A. fumigatus* in compost samples obtained from temperature zones of 63 °C or less, making it one of the most common microorganisms at compost sites. Given the presence of high concentrations of heavy metals in composts, among which copper is one of the most representative, we studied the tolerance of *A. fumigatus* to copper.

In vitro assays with *A. fumigatus* confirmed a high tolerance to copper, consistent with its dominance in heavily polluted compost, from where it was isolated. However, a reduction in growth was observed when high metal concentrations were used. The isolate examined survived at copper concentrations between 0 and 150 mg l^{-1} with similar levels of biomass (Fig. 1). Growth dramatically decreased at 150 mg l^{-1}, with a reduction of 50% in growth observed at 200 mg l^{-1} and no detectable growth at 350 mg l^{-1}.

Inhibition of sporulation by microorganisms in the presence of heavy metals is a well-known phenomenon. *A. fumigatus* also displayed low sporulation and other phenotypic changes such as abnormal pigmentation (Fig. 2). The colour of mycelia was altered in the presence of copper, with colonies growing in the presence of high concentrations of copper (100 mg l^{-1}) being white-ochreous while controls remained yellow-green. This observation is in agreement with the work of Lilly et al. (1992) and other authors (Rózycki 1992; Mandal et al. 1998), who reported heavy metal-induced changes in the mycelial morphology of several fungi.

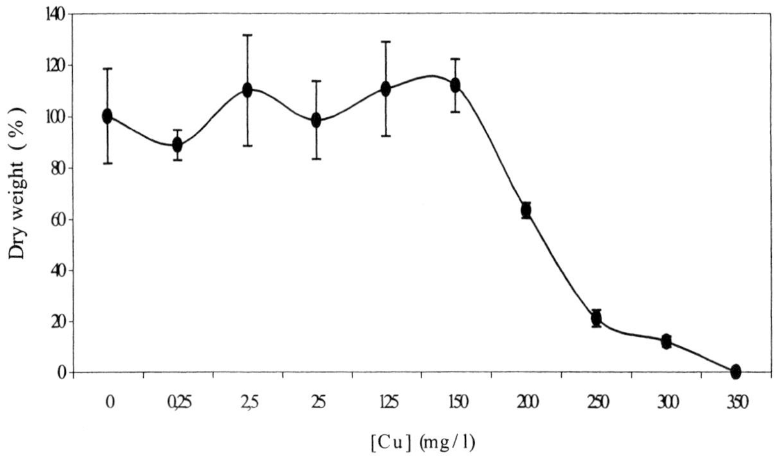

Fig. 1. Copper toxicity in *Aspergillus fumigatus*. The dry weight of 72-h-old mycelia is expressed as percentage growth

The effect of copper on specific growth rate (μ) and the lag-phase was tested in cultures with copper concentrations ranging from 0 to 350 mg l^{-1}. Our results indicated that μ, for this isolate, decreased from 0.24 to 0.005 h^{-1} when the copper concentration increased from 0 to 250 mg l^{-1} (Fig. 3). Lag-phase increased from

0.51 to 5.20 h when the concentration of copper in the medium was raised from 0 to 250 mg l^{-1}.

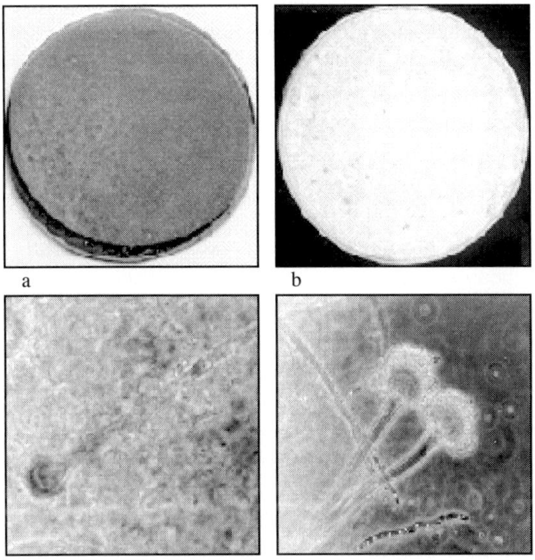

Fig. 2a-d. Effect of copper on morphology and physiology of *Aspergillus fumigatus*. **a** and **c** (40x): control cultures; **b** and **d** (40x): copper cultures

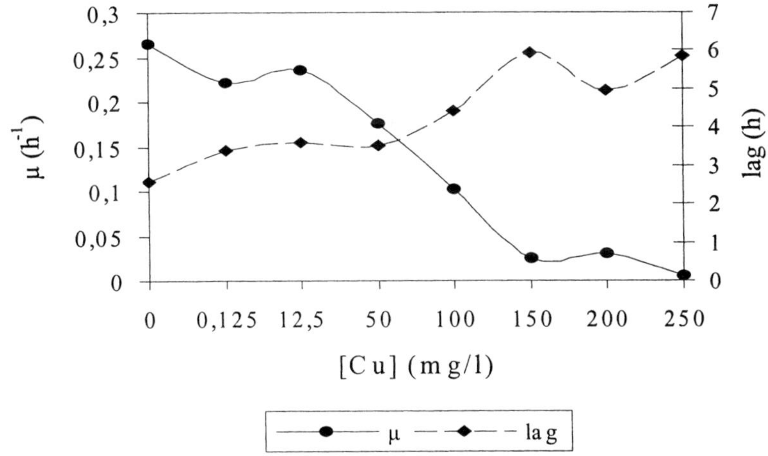

Fig. 3. Effect of copper on the specific growth rate (μ) and on the lag-phase in *Aspergillus fumigatus*

Microorganisms are capable of adapting to increasing metal concentrations. We observed that tolerance to copper could be developed by repeatedly sub-culturing parental strains on media containing progressively increasing toxic levels of metals. The ability of microbes to adapt to a heavy metal stress can also be acquired in natural environments and this adaptation has limits.

The influence of copper on the growth of non-trained and copper-trained *A. fumigatus* was determined on liquid Sabouraud medium supplemented with 0 to 400 mg l^{-1} Cu. Whilst the non-trained fungus did not grow on a medium with 350 mg of copper l^{-1}, the Cu-trained one exhibited some growth. None of the cultures was able to grow in presence of 400 mg l^{-1} Cu (Fig. 4). Therefore, the level of tolerance of the Cu-trained *A. fumigatus* increased slightly. This could be explained because microorganisms isolated from natural environments contaminated with heavy metals often exhibit tolerance to multiple pollutants. Heavy metals are usually present at elevated concentrations in these environments and microorganisms isolated from them are already adapted. Similarly, García-Toledo et al. (1985) demonstrated in *Rhizopus stolonifer* that the growth rate of a non-trained culture was lower than that of a Cu-trained culture. The non-trained culture did not grow on a medium with 450 ppm Cu, whereas the Cu-trained fungus exhibited some growth on media with 500 ppm but none with 600 mg l^{-1}. A more important adaptation to copper was noted among other classes of fungi, for example, a trained *Botrytis cinerea* isolate was able to grow in the presence of 750 mg l^{-1} Cu, whereas a non-trained parental isolate tolerated only 300 mg l^{-1} (Parry and Wood 1958).

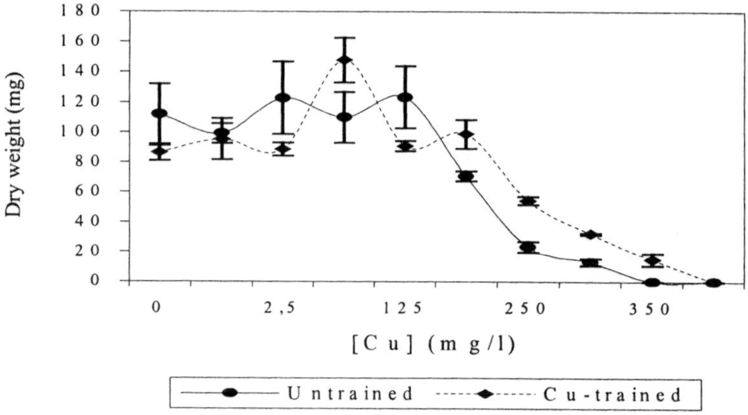

Fig. 4. Influence of copper on growth of non-trained and Cu-trained cultures of *Aspergillus fumigatus*

Experiments where fungi, particularly yeasts, are trained by successive transfers on media containing increasing concentrations of heavy metals may involve physiological adaptation (Macara 1978) and/or the selection of genetically stable variants from the original population. Adaptation can result in the expression of a resistance mechanism that can ensure avoidance or detoxification of the metal. However, adaptation may not be necessary if survival depends on the intrinsic properties of the organism (Gadd and Griffiths 1978) or if metal detoxification occurs through environmental factors. These specific mechanisms of resistance that involve adaptation should be clearly distinguished from those that do not.

In order to identify specific proteins that could be involved in copper tolerance, we analysed the protein pattern, initially from mycelia grown on copper (200 mg l^{-1}) and control conditions. Using SDS-PAGE and Coomassie staining, a specific protein was identified in the presence of copper, with an estimated molecular mass of 20.2 kDa. This protein was not detected in control mycelia (Fig. 5). Further analyses are currently being performed in order to purify and identify this protein to establish its possible role in copper stress.

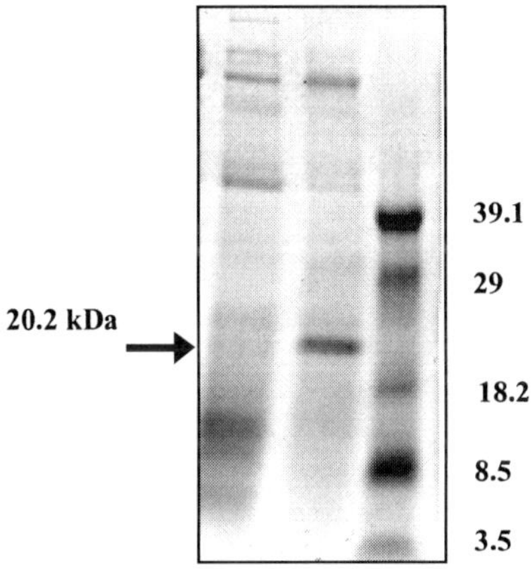

Fig. 5. SDS-PAGE. *Lane P* Polypeptide standards; *lane C* control culture; *lane Cu* Cu--amended cultures

Acknowledgements. This research was supported by Comunidad Autónoma de Madrid (CAM): Project 07M/0280/1997 and 07M/0029/1999, and E. López Errasquín by a CAM fellowship.

References

Aelion CM, Swindoll CM, Pfaender FK (1987) Adaptation to and biodegradation of xenobiotic compounds by microbial communities from a pristine aquifer. Appl Environ Microbiol 53: 2212-2217

Burge WD, Cramer WN, Epstein E (1976) Pathogens in sewage sludge and sludge compost. In: American Society for Agricultural Engineering Symposium on biological impact of waste application to land (paper no 76-2559). American Society for Agricultural Engineering, Chicago

Clark CS, Bjornson HS, Schwartz-Fulton J, Holland JW, Gartside PS (1984) Biological health risks associated with the composting of wastewater treatment plant sludge. J WPCF 56 (12): 1269-1276

Colacicco D, Epstein E, Willson GB, Parr JF, Christensen LA (1977) Costs of sludge composting, vol 79. US Department of Agriculture, Agricultural Research Service-Northeast, Washington, DC

Cristóbal F (1998) Destino final de fangos de depuración. In: XVI Curso sobre tratamiento de aguas residuales y explotación de estaciones depuradoras, chapter 19. Centro de Publicaciones de la Secretaría General Técnica del Ministerio de Medio Ambiente, Madrid

Gadd GM, Griffiths AJ (1978) Microorganisms and heavy metal toxicity. Microb Ecol 4: 303-317

García-Toledo A, Babich H, Stotzky G (1985) Training of *Rhizopus stolonifer* and *Cunninghamella blakesleeana*. Can J Microbiol 31: 485-492

Laemmli UK (1970) Cleavage of structural proteins during the assembly of the head of bacteriophage T4. Nature 227: 680-685

Lilly WW, Wallweber GJ, Lukefahr TA (1992) Cadmium absorption and its effects on growth and mycelial morphology of the basidiomycete fungus, *Schizophyllum commune*. Microbios 72: 227-237

Macara IG (1978) Accommodation of yeast to toxic levels of cadmium ions. J Gen Appl Microbiol 104: 321-324

Mandal TK, Baldrian P, Gabriel J, Nerud F, Zadrazil F (1998) Effect of mercury on the growth of wood-rotting basidiomycetes *Pleurotus ostreatus, Picnoporus cinnabarinus* and *Serpula lacrymans*. Chemosphere 36: 435-440

Miller FC (1992) Composting as a process based on the control of ecologically selective factors. In: Applications in agricultural and environmental management. M Dekker, New York, pp 515-544

Millner PD, Marsh PB, Snowden RB, Parr JF (1977) Occurrence of *Aspergillus fumigatus* during composting of sewage sludge. Appl Environ Microbiol 34: 765-772

Parry KE, Wood RKS (1958) The adaptation of fungi to fungicides: adaptation to copper and mercury salts. Ann Appl Biol 46: 446-456

Rózycki H (1992) Effect of heavy metals (Pb, Zn, Cu and Cd) on germination of conidia of *Cylindrocarpon bari* f. *destructans (Zinssm.) Scholten*. Zbl Microbiol 147: 261-269

Wilhelmi BS, Duncan JR (1995) Metal recovery from *Saccharomyces cerevisiae* biosorption columns. Biotechnol Lett 17: 1007-1012

Willson GB, Epstein E, Parr JF (1977) Recent advances in compost technology. Proc of the Nat Conf on Sludge Management, Disposal and Utilization. Information Transfer, Rockville, MD, pp 162-177

Important Aspects of Biowastes Collection and Composting in Nigeria

K. T. Raheem[1], K. I. Hänninen[1] and M. Odele[2]

Abstract. Many of the areas that generate organic waste that could be biotreated are inaccessible for collection because of the lack of accessible roads, poverty of the residents and general neglect of such places by the government authorities. In Lagos in 1999 it was observed that trucks collecting wastes cannot get into 40% of the inner part of the municipalities because the roads are too narrow or blocked by illegal structures. So one-third of the population received no refuse collection service. There is very little knowledge about the treatment of biowaste and the activities are very low. The laws and regulations governing the management of waste, in general, are weak and inefficient. An important factor for the inefficiency of the laws is the lack of possibilities to obey them.

Some insight into organic waste accumulation and attempts of composting are given. The paper highlights the special aspects of biowaste treatment that should be considered in countries like Nigeria. It emphasizes that sociological factors may be of major consideration to achieve efficient and sustainable waste management in the country.

Introduction

The population of Nigeria is about 120 million. The annual rate of population increase in the country is proportionally 3 percent. The rapid increase in the urban population in the country rapidly increases the generation of wastes and puts more pressure on urban land use. Lagos is populated with about 10 million people while Ibadan has a population of about 4 million. Because of its cosmopolitan nature and because it is the most industrial of the cities in Nigeria, Lagos generates more plastic, paper and other non-organic wastes in the country.

In many developing countries, especially in Africa, roughly half of the waste is organic (Yhdego 1994). It is, however, mixed with all sorts of waste, including infectious medical wastes. Wastes are carelessly dumped around the cities. The regulations regarding the collection, disposal and treatment of wastes are very loose and usually not enforced. This has been the situation since the colonial rule that lasted till 1960.

The oil boom of the 70s encouraged the government in co-operation with the World Bank to focus on investing large amount of funds on expensive incinerators, expensive trucks and dumping of organic waste at the edge of the cities. The

[1]University of Jyväskylä, Department of Biological and Environmental Sciences,
[2]Friends of the Environment, 106/110 Lewis Street, P.O.Box 10627, Lagos, Nigeria

H. Insam, N. Riddech, S. Klammer (Eds.)
Microbiology of Composting

available "easy revenue" from crude oil somehow "stifled" the minds of the law-makers. All wastes were seen by the Nigerian policy-makers as things that should be discarded of and disposed in the dumpsites, according to the tradition of collecting and disposal of municipal wastes.

In reality there was gradual and sometimes rapid increase in the funds allocated for waste management but the policies and funds were poorly managed. Now the expensive incinerators that were built have become white elephants in the city of Lagos, and the truck-loading has become inefficient. Some Non Governmental Organizations (NGOs) have also started to be involved in municipal solid waste management.

The sociological aspects of bio-waste treatment are not usually considered in studies concerning waste treatment in the developing countries. The importance of sociological consideration of waste management was suggested in the study of Werlin (1995). He referred to it in his study dealing with comparative solid waste management in Nigeria, as "political hardware" and " political software".

However, the importance of considering sociological aspects in treating bio-waste in a developing country like Nigeria cannot be overlooked. Firstly, it is important to avoid the problems, or negative effects on people, that could result from some methods or technology used in the treatment of bio-waste. Secondly, as part of social policy, official environmental policies are tailored to suit the mind-set of the elite class. Even though people participate in unorganised recycling (scavenging), the traditional and general attitude towards waste is that it is something to be discarded of, and not reused. Finally, the economic, political and health incentives are equally important to consider.

There is no in-depth study of the waste problems in Nigeria. We attempt at giving a general picture of biowaste accumulation and what should be considered for its treatment in Nigeria.

Generation of Organic Waste in the Municipalities

The total waste generated in Lagos is on the average about 0.5 kg/capita/day (LAWMA, 1997). In Ibadan it is estimated to be around 0.6 kg/capita/day (2,000 tons per day) (Olushola & Habila 1999). Ibadan is known for its high generation of organic waste in form of leaves, vegetables and fruits. The organic waste in the city has been attributed to be the cause of floods. In Nigeria most of the wastes are from households. There is however, no valid statistics to estimate the amount of waste generated by the industries or according to municipal areas in Lagos or Ibadan. All wastes are dumped together without clear records of their sources. Only hazardous and chemical waste can be traced to its source if necessary.

The colonial pattern of development still persists in Nigerian cities. Housing policies are organised in a way that the majority of the low income earners live in high-density and/or informal housing, while high-income groups live in better planned and low density areas. This kind of arrangement is a factor that militates against proper management of wastes in the low-income areas. The markets are also located in areas where the low income earners live, thus making it possible for the government officials to overlook dirt in such area. Such areas are often given illegal status and the inaccessibility of these settlements makes local authorities reluctant to provide them with urban services. The urban poor are left on their own to contend with waste disposal. The result of such official neglect is dumping waste into canals, open sewerages and other illegal sites.

Organic, biodegradable waste is generated mainly by the households (including restaurants and food vendors), markets and big hotels in the municipalities. The restaurants and food vendors may or may not be in the market places. Many of the restaurants are located in the same buildings where the owners and managers live. Others are in form of a make shift kiosk. Usually it is a family business that may not even be registered officially.

The markets in cities generate a considerable amount of organic waste. For instance, Bodija market which is one of the numerous markets in Ibadan, generates about 40 tons of solid waste per day. The type of wastes generated are leaves, vegetables, fruits, wastes from poultry, eggs, fish, meat and bio-degradable materials used in making baskets or containers. Many other markets generate a lot of organic wastes from cocoa, rice husks, vegetable peelings, wheat, leaves for wrapping foods, bananas, guinea corn. There is also substantial daily generation of slaughterhouse wastes in and around the cities. Ibadan and Lagos have a peculiar problem of organic wastes blocking drains, especially in the rainy seasons.

In the two cities breweries and other industries are also constantly generating organic waste. These industries usually discharge such waste into the river or sewerage. Sometimes animal waste like dead animals and garden waste are common in the cities. However, disposal of such waste, is done by the Lagos State Waste Management Authority (LAWMA) and the Waste Management Board in Ibadan.

Small-scale wood processing factories and markets where wooden planks are sold generate a lot of sawdust and wood chips as waste. These wastes could be used as bulking agent in compost but at the moment are dumped into the sea or burnt. It is an everyday activity of open burning of sawdust in popular wood processing and selling markets in Lagos. This kind of scenario is also common in Ibadan and other cities and towns in Nigeria. Garden wastes, especially generated by city farmers and city elites are also a substantial portion of solid organic waste that can be composted. Usually all the wastes end up now in the dumpsites, or in the case of city farmers, are often openly burnt.

The wastes generated in the households vary from one household to the other. Waste generation is also affected by climate, vegetation, culture and type of food crops available in the regions (Odele 1999). For example, the waste from households in the high-income earners areas may not contain much organic waste as those in the low income areas or slums. This is because the Nigerian elites and even those in the middle-income group tend to use more products that come in plastic bags or packages. However, they also are likely to produce more garden waste that is compostable.

In the low-income area or slums, the garbage is mostly organic but very much mixed with remnants of plastics bags and other wastes. Usually leaves, rotten fruits, human excreta, metals, papers, plastics, and broken glasses are the common contents of the heap of refuse littered around in these areas. There is usually very little left-over food as waste in the slums or low income earning areas.

Composting of organic waste in Nigeria

Unlike for Asian and Latin American cities, very few studies have been done and documented on the potential for waste re-use in African cities. However, some studies have revealed that treatment of bio-waste in form of limited composting had been practiced in Nigeria 50 years ago and that various States in Nigeria still practice it according to traditional methods (Sridhar 1993). The study made by Sridhar indicates that some composting had been done in Kaduna, Kano and Maiduguri, all towns in the northern part of Nigeria. There also used to be, according to the study, a compost plant in Kano from 1936 to 1942. The plant was used to compost city refuse mixed with night soil and slaughterhouse waste.

Farmers in eastern part of Nigeria compost by digging pits that are 2 -5 m deep with a circumference of 10 m. Wastes are dumped into the pits and periodically turned until the compost is ready in about three months. Another method used in northern Nigeria is mixing human and animal wastes with crop residues, in a 1:3 ratio. The mixed wastes are then sprayed with water and turned regularly. The compost is said to be ready in about 30 days (ibid). These methods are labour intensive, may not destroy the pathogens and *Ascaris* eggs (depending the temperature obtained) and may not be that attractive to farmers anymore, even though seen as suitable for rural farmers.

A practical study on small-scale composting started in Lagos in spring 2000. It is the approach of Finnish Swedish Martha Association (NGO) and the University of Jyväskylä with some Nigerian NGOs (African Hazard Research and Study Center, AHRSC) on composting of market wastes in the market place and also organic waste generated by the schools in the school premises. It is a small-scale project that is geared towards examining the possibility of an in-depth research for a possible use of local materials and sustainable methods of composting organic municipal wastes and proper collection and disposal of medical wastes. For the experiment, wooden boxes (1.80m x 0.85m x 0.85m) were built after consultations with the market women and a school. The boxes are expected to be stationed in the market area for dumping of organic wastes generated from the market. Building of the boxes was funded by NGOs but handed over to selected market women. Their task is to supervise the dumping of refuse into the box and also do the periodical turning of the waste until maturity stage. The humus from the compost is expected for use as fertiliser on the small farms or garden.

The project will be linked up with research institutions that can make scientific studies of the type of wastes, the regularity of supply, volume of garbage, attitudes of people and the composting process. It will be part of the attempts at coming up with suitable ways of recycling municipal organic wastes in Nigeria.

Large-scale composting plants have proven unsuccessful in other African countries because of their too complicated technology and too big expectations for logistics of waste stream. The siting of the plants and the inability of the people to manage them are also factors for their failure. In some countries like Ghana, a composting plant has become a "dumping plant".

The advantages of biowaste treatment and technology in Nigeria have so far been minimal. There are attempts by UNICEF to experiment composting of organic waste in some of the local government areas in Lagos. The Lagos State Government has also indicated its interest in bio treatment of organic waste. This year (2000) the Federal Government of Nigeria invited bidders for the construction of compost plants in different parts of the country. It is, however, important for the authorities to know that modern techniques and technology of treating biowaste need to be improved according to the new information from waste management studies.

Sociological consideration

The usefulness of organic waste composting for use as fertiliser in African countries has been stressed by Asomani-Boateng and Haight (1999). Economically, composting of organic wastes in countries like Nigeria could be profitable. However, care should be taken in laying too much emphasis on the cash incentive of end product form composting. Many of the government officials in Nigeria take for granted that composting of municipal organic waste will generate a lot of funds for the government or the private sector. This notion is probably born out of lack of in depth assessment of the cost, and proper understanding of compost plants. Cointreau-Levine (1998) pointed out in her study that composting could be 2 -3 times more costly than open dumping in developing countries. It is essential to secure the constant availability of the organic waste, short hauling distance and quality humus for the market. Badly managed compost may not produce sufficient nutrients for crops and the temperature should be controlled to kill pathogens.

The advantages of proper treatment of biowaste are many. Composting of biowaste, for example, will reduce the breeding of rodents and other disease-carrying vectors that could pose a threat to public health. The treatment of biowaste also helps in decreasing the refuse load on landfill or dump sites.

Even though governments can embark on public projects without the prior and full awareness of the people, treatment of municipal organic wastes should involve the proper participation of the people.

In developed countries waste management is highly organised, and also organic waste is in some countries, like Finland, Germany, Sweden, Austria, source separated and composted in large highly automatised composting plants that can treat large masses of organic waste. The quality of the end product is carefully controlled. The legislation entitles the operative authorities of waste management to charge the residents for the costs. This is an indirect means to support household composting because residents are then saving the collection cost of biowaste fraction. Adequate compost bins and advisory help are easily accessible in the markets at reasonable price. Studies (Hänninen etal, 2000) have indicated that about 70% of the Finns are sorting their kitchen biowaste. In the city of Jyväskylä about 50%

of the detached houses select small-scale composting, suggesting that the importance of waste recycling is understood by the people in Finland.

Research activities in the field of biowaste treatment have intensified in the Western countries and results can be gainfully used in developing countries. For example studies in Finland have shown that cardboard-based liquid containers can be successfully composed (Hänninen et al. 1997). Such compostable materials are becoming common in Africa also. However, the intensity of studies and reports also show that there are issues that should be taken care of. Even researches in composting of organic waste from the farm or household are yielding results that indicate the need for more studies (Heimonen and Hänninen, 1997).

Legislations formulated to manage with sanitary situation in many of the developing countries are very poor. In many cases ordinances or legislations affecting public services in countries like Nigeria can also look like “oppressive laws”. This is because they do not consider the financial burden they add to the problems of impoverished people. In Nigeria the waste management legislation in general is very inadequate. Even the inadequate one is not enforced. There are also problems with constitutional and statutory division of environmental and refuse disposal powers between the three tiers of government.

The awareness level of biowaste treatment in Nigeria is very low. Even though there are traits of using humus from composted organic waste as fertiliser in the "city farms", organic waste composting is not organised. City farmers go to dump sites to collect untreated organic waste that they use as fertiliser on their plots. There is no source separation of waste in Nigeria.

Few researches on composting in Nigeria have been carried out, they have not, however, been able to promote the acceptability of waste composting in Nigeria as a method for dealing with the waste management problems in the country (Sridhar et al. 1993). The knowledge, so to say, has remained within the narrow scientific circle of University researchers.

Some State governments like Lagos have shown interest in bio-waste treatment, but such interest has so far remained mere talks. The attitudes of the government officials and private sector dealing with waste management have not changed from the "collect, dump and burn" approach. Such attitudes seem to be encouraged by international donors. Many people also see such approach as the easiest way of disposing wastes.

Economic hardship forces a growing number of migrants to urban areas in search of employment increases the already high number of unemployed and the urban poor. More wastes are generated and greater strain is placed on urban pressure points like solid waste management. However, majority of the urban dwellers cannot pay for an efficient municipal waste management system.

The urban areas are also badly planned. Roads in many residential areas in Lagos and Ibadan are narrow and in accessible for trucks to collect refuse. Houses, shops/kiosks and markets are built in a way that makes it difficult for collection of waste especially in the low-income areas. Many of the areas that generate organic waste that could be biotreated are inaccessible for collection because of the lack of accessible roads and general neglect of such places by the government authorities.

It is observed that trucks collecting wastes cannot get into 40% of the inner part of the municipalities in Lagos State because the roads are too narrow or blocked by illegal structure (Falomo, 1995). In 1999 this observation was still confirmed (Odele 1999). So in Lagos for example, it can be estimated that one-third of the population received no refuse collection service (Werlin 1995).

So the town plans and housing lay-outs are determinants affecting the collection and disposal of waste. The ways the town planning authorities implement official policies and the attitudes of people to the rules are factors that can affect waste management. Here we have an important factor for the inefficiency of the waste laws: the lack of possibilities to obey it.

In case of treatment of biowaste in Nigeria another important factor for the inefficiency of the waste laws is the low public knowledge. Even environmental agencies and research institutions are not involved in biowaste treatment research. The few attempts made by some NGOs towards scientific treatment of bio-wastes have been haphazard. So the poor environmental laws in the country are inefficient, and are not even respected by the public (Raheem, etal, 1999).

Health and environmental related issues of inefficient waste management should not be overlooked. Even though composting of organic waste has not been identified as a cause of disease, there are possibilities that pathogenic agents, toxic emissions or odours from the compost plant or bin could affect people's health.

Source separation of waste has to be carefully checked so that it does not contain hazardous materials. Compost that have heavy metal and refractory organic concentrations may be injurious to soil structure, toxic to plants and potentially carcinogenic if bio-accumulated through the food chain (ibid).

Conclusions

We have attempted at giving a general picture of biowaste accumulation and composting in Nigeria. The two cities selected for this paper are where organic waste has been a serious problem for the government and people since the early 1970s. There are of course other cities in Nigeria that have a high density of population and waste management problems, however, Lagos and Ibadan have the most prominently reported by the media. And even the World Bank assistance in form of loans to address the problems of waste management in Nigeria has been directed mostly at the two cities.

Introducing new techniques and technology for bio-waste treatment in countries like Nigeria needs careful consideration. New methods or enhancing old ones may need a community-based approach. The living environment should also be considered before setting up medium or large plants for bio-waste treatment.

Another option that may be examined is composting at the household level, which could encourage individuals to get more interested in the proper management of waste. However, it is important that people have the basic understanding of the microbiology of household composting. Educational and awareness programmes are important to achieve success.

Government policy concerning waste management in Nigeria does not encourage research institutions to carry out studies on the possibility of bio-waste treatment in the country. Even though there have been political discussions about treatment of bio-waste in Nigeria, enough efforts have not been made by the authorities to praticalise the ideas. Private sectors have been involved in the issues of waste management in some cities but all have been on how to efficiently collect and dump wastes in the dumping sites. More information on the various techniques and technology in bio-waste treatment is needed for the authorities concerned with waste management.

References

Asomani-Boateng R, Haight M (1999) Reusing organic solid waste in urban farming in African cities: A challenge for urban planners. In: Smith OB (Ed), Urban agriculture in West Africa, contributing to food security and urban sanitation. CRDI/CTA, ISBN 0-88936-890-2.

Cointreau-Levine S (1998) Occupational and environmental health issues of solid waste management, with special emphasis on developing countries. Draft unpublished paper.

Falomo AA (1995) City Waste as a Public Nuisance. Paper presented at the Annual Conference of the Nigerian Environmental Society.

Heimonen R, Hänninen K (1997) Straw composting as a source of CO_2 for greenhouse. In: Stentiford EI (Ed) Organic recovery & biological treatment into the next millenium International Conference, harrogate, United Kingdom. 3 - 5 September, 1997.

Hänninen K, Kovanen T, Alen R (1997) Degradation of cardboard-based liquid containers in composting. In: Drozd J, Gonet SS, Senesi N, Weber J (Eds) The role of humic substances in the ecosystems and in environmental protection. IHSS - Polish Society of Humic Substances, Wroclaw, Poland.

Hänninen K, Lappi S, Tolvanen O (2000) Technique and microbiology of household composting. Dept. Biological & Environmental Sciences, Univ. of Jyväskylä, Finland.

Odele M (1999) Solid waste management in Lagos State. Unpubl. Report, Lagos, Nigeria.

Olushola I, Habila O (1999) Urban Peri-urban Water and Sanitation Project: UNICEF Experience. Unpublished report, UNICEF, Lagos, Nigeria.

Raheem K, Hänninen K, Huagie T (1999) developing a sustainable system for solid waste management in West African countries. In: Bidlingmaier W, de Bertoldi M, Diaz L, Papadimitriou E (Eds): Orbit 1999, Organic Recovery & Biological Treatment, Part iii. Organic Waste Management in developing Countries, pp. 85 - 856.

Sridhar MKC, Adeoye GO, Omueti JA I, Yinda G, Reece ZD (1993) Waste recycling through composting in Nigeria. - Compost Science and Utilization 1: 69 - 74.

Werlin HH (1995) Comparative Solid Waste Management. The Technical assistance Implications. Journal of Asian and Afican Studies - JAAS, Vol. XXX, No. 3- 4.

Yhdego M (1994) Institutional organic wastes as soil conditioner in Tanzania. - Resources, Conservation and Recycling 12: 185-194.

Plant and Human Pathogens

Use of Actinobacteria in Composting of Sheep Litter

S. Baccella[1,2], A.L. Botta[2], S. Manfroni, A. Trinchera[3], P. Imperiale[1], A. Benedeti[3], M. Del Gallo*[2] and A. Lepidi[2]

Abstract. Selected Actinobacteria have been utilised in composting sheep litter/manure for use as organic fertiliser and to promote litter sanitation. For this purpose, 14 strains of Actinobacteria were isolated from composted sheep litter and checked for their antibiotic capability against *E.coli* and *Clostridium* spp. All strains were grown separately in the laboratory and used to inoculate sheep litter. Chemical and microbiological changes were studied during the composting process in lab-scale fermentors. After 2 months of treatment, the inoculated sheep litter resembled stable compost: chemical analyses confirmed humic-fulvic carbon increase, a C/N ratio reduction and an improvement of humification indexes. Isoelectric focusing of extracted humic-like matter showed an increase during the composting process of more stabilised organic compounds, directly correlated with the humification parameters. Microbiological analyses showed that in the inoculated litter, the number of pathogenic *Clostridia* rapidly decreased until total removal at the end of the composting process.

Introduction

Sheep raising has been for centuries a leading activity of the peoples inhabiting the Apennines. Today, the economic relevance of sheep flocks wandering the grazing lands depends on the capability of joining ancient values with new ones. In this framework, zootechnical practices are changing with increasing sedentarization and settlement of animals. This leads to two main problems: (1) hazards deriving from exposure of the animals to pathogenic agents (mainly microorganisms) which proliferate in manure and litter and (2) accumulation of increasing amounts of organic wastes. The present research aims at developing a biotechnological process toward an integrated solution to some of the problems. It is based on the possibility of exploiting selected microbial agents as composting bioactivator of sheep litter, and as antagonists of the pathogens. Actinobacteria are good candidates for this purpose as possible producers of humus-like materials. These microbes, *Streptomyces* in particular, are mostly soil-resident – their number ranging from 10^6 to 10^7 colony-forming units (CFU) g^{-1} of dry soil, increasing

[1] Science and Technology Park of Abruzzo L'Aquila, Italy
[2] Department of Basic and Applied Biology, University of L' Aquila, Italy
[3] Inst. of Advanced Research in Plant Nutrition, MIPAF, Rome, Italy;
e mail: delgallo@aquila.infn.it

H. Insam, N. Riddech, S. Klammer (Eds.)
Microbiology of Composting

with soil fertility (Quaroni et al. 1987). The capability of many members of Actinobacteria to degrade lignocellulosic compounds, resulting in increased carboxylic acids and phenolic hydroxyl content of the substrate has been reported (Godden and Penninckx 1984; McCarthy 1987; Ball et al. 1990;). Another characteristic of Actinomycetes is their ability to produce melanin dark-brown pigments, contributing to increase the humic acid in the substrate (Andreyuk and Gordienko 1981; Gomes et al. 1996). Moreover, Actinobacteria – *Streptomyces* spp. in particular – are known to produce antibiotics useful for sheep litter sanitation eliminating both human and animal pathogens and bacterial plant pathogens. From the agronomic point of view, the distribution of well composted manure having a large actinomycete population can exert further benefit by reducing plant phytopathogens from soil (Quaroni et al. 1987). In the present work, *Clostridia*, the most relevant cattle pathogens, are considered as contamination indexes. They produce a large variety of exotoxins that can seriously damage cattle tissue and organs (intestine, kidney, liver, muscles etc.). *Clostridia* are naturally present in humus-rich soil, but they can also be found in the intestines of healthy animals, where they may cause disease in combination with aggravating factors.

Materials and Methods

Microbial Strains

Actinobacteria

The microbial strains utilised throughout the work were isolated from soil. They were then inoculated into sheep litter, and the prevailing strains were reisolated again. For the isolation, 1 g of dry soil was mixed with 99 ml of sterile water supplemented with two drops of Tween-20. One ml of this suspension was diluted 1:10 in sterile water and heated for 16 h at 45 °C. Then 0.1 ml of suspension were plated onto sodium caseinate agar (SCA): Na-caseinate 0.2 g l^{-1}, K_2HPO_4 0.5 g l^{-1}, $MgSO_4$ 0.2 g l^{-1}, $FeCl_3$ 0.01 g l^{-1}, agar 15 g l^{-1}, pH 6.5. Plates were incubated at room temperature for 5 to 7 days. Strains were kept as spore suspensions in 20% glycerol in sterile water (v/v) at –20 °C and recovered by culturing on R_2Ye solid medium (Hopwood et al. 1985).

Clostridia

The pathogenic strains of *Clostridia* utilised in the present work were isolated from sheep manure on Perfringens Agar Base SFP medium (PAB; Oxoid Manual 1998) and on reinforced *Clostridium* agar (RCA; Oxoid Manual 1998). The isolates were incubated in anaerobic and aerobic conditions and strictly anaerobic colonies were identified with API System for anaerobic microorganisms (Rapid ID 32 A, bioMérieux Italia S.p.A.).

E. coli

A laboratory strain, pRK2, was routinely grown at 37 °C in LB medium (Hopwood et al. 1985) and was kept an LB agar at 4 °C or in LB broth at –20 °C.

Preparation of Actinobacterial Inocula

Inocula of Actinobacteria were prepared as liquid cultures of each strain by growing them in a medium consisting of (g l^{-1}) K_2HPO_4 0.50, $FeCl_3$ 0.01, $MgSO_4$ 0.2, asparagine 2, sodium caseinate 0.5. After 24 h incubation in an orbital shaker (200 rpm) at 37 °C, the cultures were mixed just before their inoculation on the litter. Average spore numbers in the inoculum were around $2 \cdot 10^8$ SFUml^{-1}).

Composting Procedure

Composting of litter was performed in two lab-scale reactors. The reactors were commercial composters (ALKO-400) containing 400 l of sheep litter. Each reactor was homogeneously inoculated with 20 l, the first with the liquid inoculum described above and the second (control) with sterile medium. The composting mass was aerated by compressed air. Temperature and pH were measured weekly and water was routinely supplied to keep the moisture above 40%.

Microbiological Analyses

The SFU (spore-forming units) of *Actinomycetes* and *Clostridia* were numbered by serial dilution by plating on SCA and on PAB (Oxoid), respectively. For counting *Actinomycetes*, litter suspensions and incubation were done as described for isolation. For *Clostridia* counting, litter suspension (10 g in 90 ml of physiological solution) was heated at 80 °C for 10 min before serial dilution. Plates were incubated at 37 °C for 24 h in a Gaspak device.

Chemical Analyses

Moisture and pH were measured according to the official protocol (S.I.S.S. 1985). Total N was analyzed by a LECO FP 228 (Bremner and Tabatabai 1971). Total organic C was measured according to Springer and Klee (1954), reported by Sequi and De Nobili (2000). Extractable C (C_{ext}) and humic and fulvic C (C_{HA+FA}) were separated and purified according to Ciavatta et al. (1990). The isoelectric focusing technique (IEF) separation was carried out in a Multiphore II, LKB electrophoretic cell, according to Govi et al. (1994); 25 mL of $NaOH/Na_4P_2O_7$ extracts were acidified with HCl 5% until reaching pH<2. After centrifugation at 3500 rpm for 20 min, precipitated humic acids were resolubilised with NaOH 0.5 N; 10 ml of this solution were dialyzed in 6000–8000-Da dialysis tubes and then lyophilised. Samples were electrofocused in a pH range 3.5–8.0, on a 5.06% T and 3.33% C polyacrylamide slab gel, using a mixture of carrier ampholytes (Pharmacia Biotech) constituted by 50% ampholine pH 3.5-5, 25% ampholine pH 5.0-7.0 and 25% ampholine pH 6.0-8.0. A prerun (2 h; 1200 V; 1 °C) was performed and the pH gradient formed in the slab was checked by a specific surface electrode. The electrophoretic run (2 h 30; 1200 V; 1 °C) was carried out loading the lyophilized humic acids, resolubilised with bidistilled water in order to obtain the same concentration for all the composts samples (100 µg C $sample^{-1}$). The bands obtained were stained with an aqueous solution of Basic Blue 3 (30%) and scanned by an Ultrascan-XL Densitometer.

Inhibition of Pathogens

Escherichia coli

200 µl of liquid culture of *E. coli* was plated into LB plates, together with a spot seeding of each actinobacterial strain. Incubation was done at 37 °C and the development of inhibition haloes was daily examined.

Clostridia

RCA plates (Oxoid) were inoculated with a spot seeding of each actinobacterial strain. After 10 days of incubation at 37 °C, plates were covered with 1 ml of each *Clostridium* culture grown on RCM (Oxoid). Inhibition haloes around the actinobacterial colonies were observed after 24 h of incubation at 37 °C in a Gaspak device.

Results and Discussions

Composting of Litter

Microbial Parameters

Fourteen strains of Actinobacteria, labelled LT1 to LT5, LT6-1, LT6-2, LT7 to LT13, were isolated from composted sheep litter. All actinobacterial strains were morphologically characterised as *Streptomyces* spp. Sheep litter transformation in the presence of massive inoculation of the selected strains has been studied. The purpose was to verify both agronomic and hygienic quality of the product and to collect the kinetic data needed for process development. The composting was carried out for about two months and the sampling every week. Figure 1 shows that in both cases *Clostridia* decreased quickly during the first 2 weeks and remained stable for the remaining incubation period. In the inoculated litter the sanitation appeared to be completed within 15 days. The control instead showed constant residual presence of clostridia of about 500 SFP g^{-1} residue which could grow again if favourable conditions occurred.

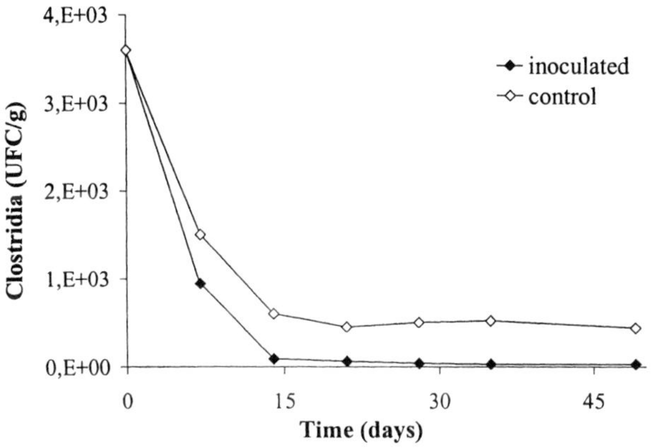

Fig. 1. Kinetics of *Clostridia* during sheep litter composting, with (*inoculated*) and without (*control*) actinobacterial starter

In Figs. 2 and 3, actinobacterial kinetics and temperature profiles are shown. From the temperature profiles it is evident that the biooxidative phase was faster in the inoculated litter. This might affect pathogen inactivation in terms of quantitative efficiency (Stentiford 1986). Differently from the control, all Actinobacteria in the inoculated trial grew quickly during the first 7 days. This can be explained by native Actinobacteria being less resistant to high temperatures. In comparison to the control, which showed a slow but consistent decrease in the actinomycetal strain population, inoculation resulted in a considerable number (10^9) after 7 days of incubation. The presence of the inoculum, however, did not affect the temperature of the process, indicating that the lowering of the clostridial population is due to an antagonistic and/or antibiotic effect by the inoculated strains.

Humification Parameters and Isoelectric Focusing

The main chemical parameters analysed for the inoculated and control composts, for the different sampling events, are reported in Table 1.

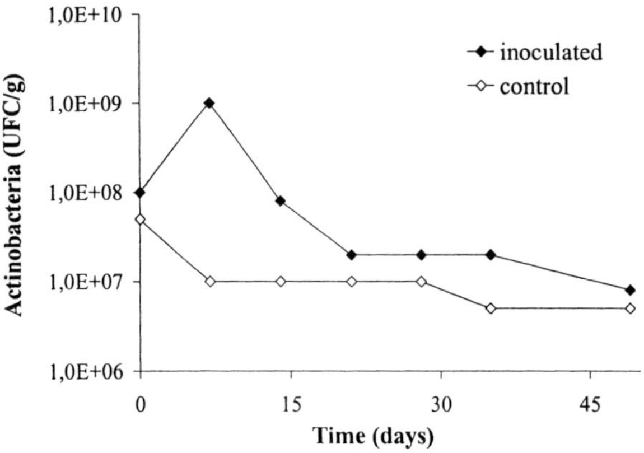

Fig. 2. Kinetics of total Actinobacteria during sheep litter composting, with (*inoculated*) and without (*control*) actinobacterial starter

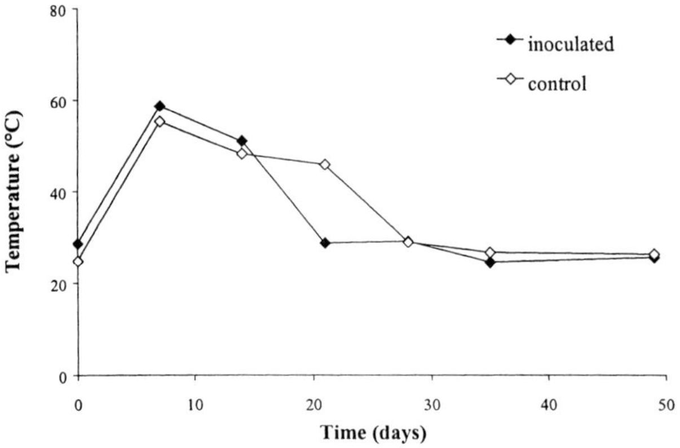

Fig. 3. Evolution of temperature during sheep litter composting, with (*inoculated*) and without (*control*) actinobacterial starter

Table 1. Chemical changes and humification parameters in sheep litter during composting

Inoculated Time (days)	N%	C/N	TOC%	TEC%	$C_{(ha+fa)}$%	HR%	DH%	HI
0	2.2	17.6	38.8	17.3	11.8	30.4	68.2	0.47
21	2.8	13.3	37.5	17.1	12.8	35.1	74.9	0.33
35	2.9	12.6	36.9	18.4	14.0	35.1	76.1	0.31
49	2.9	12.5	36.5	21.9	18.3	49.3	74.9	0.33

Table 1. (cont.)

Control Time (days)	N%	C/N	TOC%	TEC%	$C_{(ha+fa)}$%	HR%	DH%	HI
0	2.2	17.6	38.8	17.3	11.8	30.4	68.2	0.47
21	3.0	11.9	36.3	17.5	12.8	35.3	74.9	0.37
35	3.2	10.9	35.8	18.9	14.6	38.0	76.1	0.29
49	3.0	12.1	35.7	22.0	18.3	51.3	74.9	0.35

The results obtained for both inoculated and control composts showed that total organic carbon decreased (from 17.6 to 12–12.5) and total nitrogen increased (from 2.2 to 3.0%), resulting in a decrease in C/N ratio from 17.6% to 12.1–12.5%. In parallel, fractionation of organic matter showed humic and fulvic carbon increased from 11.8 to 18.3 % for both composts and, consequently, the humification rates (HR) and the degrees of humification (DH) became higher during composting, reaching values of ~50% and ~75%, respectively. Humification indexes, which give information on the ratio between the not-humified and the humified carbon, decreased significantly, confirming that in both the composting processes, the organic matter sustained a stabilisation, corresponding to the formation of humic and fulvic-like substances.

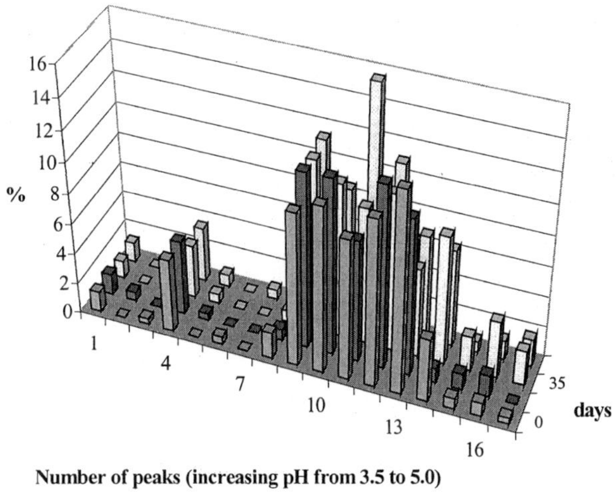

Fig. 4. Histogram of peak areas obtained from isoelectric focusing profiles of the humic-like matter purified from control compost. The compost was sampled at 0, 21, 35 and 49 days

Histograms of peak areas obtained from isoelectric focusing profiles of the humic-like matter purified from both control and inoculate composts are shown in Figs. 4 and 5. There were no relevant differences between the two series of histograms, even though the percentages of the single peak area varied for control and inoculated compost as absolute values.

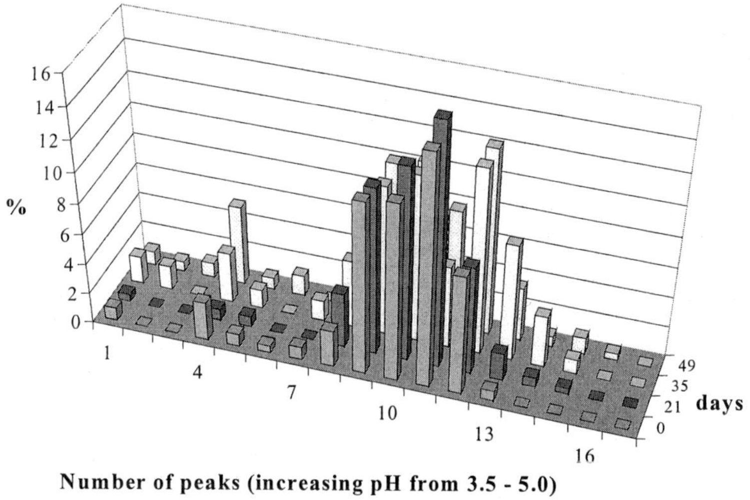

Fig. 5. Histogram of peak areas obtained from isoelectric focusing profiles of the humic-like matter purified from inoculate compost. The compost was sampled at 0, 21, 35 and 49 days

It is possible to note that during the composting process, peaks at higher pH values (peaks 14–17 for CNI and peaks 13–17 for CI) increased their relative areas. Since generally more chemically stable compounds focus at higher values of pH, this could be correlated to the level of stability of the humic-like fraction during the composting processes, (Govi et al. 1994). The study of organic matter evolution during composting showed that a transformation in the organic compounds took place in both composts, particularly because of the high values of humification parameters reached at the end of the processes.

On the other hand, at least from the chemical point of view, no significant differences were present between the inoculated and control composts. Actinobacteria inoculation does not appear to directly influence the humification process, since all quantitative parameters were similar to the control, even if the qualitative characterisation of humic-like matter revealed some differences.

Inhibition of Pathogens

An *E.coli* laboratory strain was utilised as a marker to evaluate actinobacterial antagonist activity (Fig. 6). The results show that in most cases inhibition was highest after 10 days (Fig. 7). However, selected strains behaved in two different ways: strain LT1, LT4, LT6, LT12 showed antagonistic effects, constantly increasing during the monitored period; all other strains showed an inhibition peak at 10 d. This is possibly due to the production of a antibiotic substance inactivated or degraded after such a period. Successively, actinobacterial antagonism was tested against the pathogenic clostridial strains previously isolated from sheep litter (Table 2).

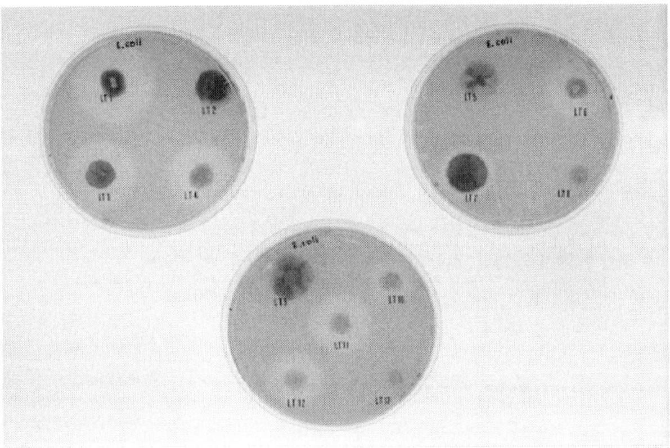

Fig. 6. Antagonist activity of actinobacterial strains against *E.coli*: inhibition haloes are visible around the spot of actinobacterial colonies

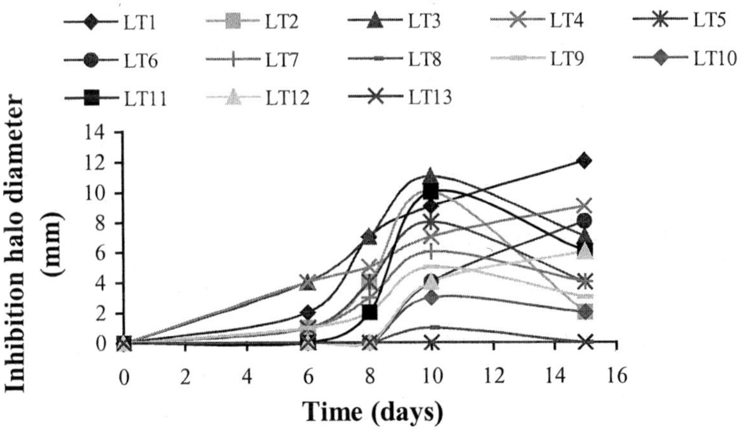

Fig. 7. Antagonist activity of actinobacterial strains against *E.coli*

Table 2. Pathogenic clostridial strains isolated from sheep litter

Strain profile	Species	API System %ID
C4	*C. perfringens*	99.9%
C1	*C. fallax*	99.9%
C13	*C. difficile*	63%
C22	*C. tetani*	Uncertain profile
K1	*C. histoliticum*	85.7%
K4	*C. sordelli*	84.6%

All tested strains, except LT2, LT5, LT9 and LT13, were able to inhibit pathogens (Table 3, Fig. 8). No inhibition was observed against C13 (C. difficile) and K1 (C. histoliticum).

Table 3. *Clostridium* growth inhibition: diameters (cm) of haloes are reported

Actinobacterium strain	C4 (*C. perfringens*)	C1 (*C. fallax*)	C22 (*C. tetani*)	K4 (*C. sordelli*)
LT1	1.0	1.0	-	0.2
LT2	-	-	-	-
LT3	0.5	1.2	0.3	0.3
LT4	0.5	0.8	0.4	0.3
LT5	-	-	-	-
LT6-1	0.7	1.1	0.4	0.4
LT6-2	0.4	1.1	1.0	0.2
LT7	1.5	0.3	0.5	0.4
LT8	1.0	0.5	0.7	0.3
LT9	-	-	-	-
LT10	0.7	0.6	0.4	0.2
LT11	0.8	1.3	-	0.3
LT12	0.7	1.6	0.6	0.2
LT13	-	-	-	-

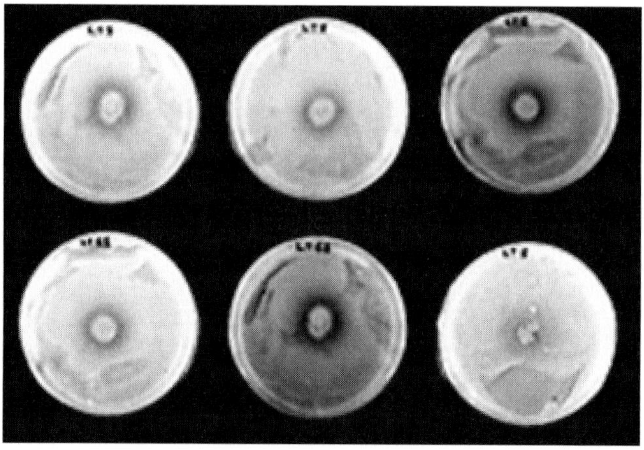

Fig. 8. Antagonist activity of actinobacterial strains against *C. sordelli*: inhibition haloes are visible around antagonist actinobacterial colonies

Concluding Remarks

Recovery of organic matter via compost production – an ever-interesting topic of research and innovation – is hampered by hygienic and/or sanitary constraints. Solid and liquid effluents of large aggregates of humans and animals are often contaminated with pathogenic and/or parasitic entities whose elimination is a precondition for their reuse. In the present chapter, a process was set up to eliminate the main obstacle to the reutilisation of the areas where flocks reside during the grazing season (stazzi) and of the litters accumulated during the winter time when animals are housed for a long time.

The results obtained indicate the possibility to obtain good sheep litter sanitation, by adding actinobacterial strains at the begining of the composting process. Laboratory tests show that some selected strains are able to inhibit *E.coli* and several pathogen clostridial strains native to the litter. Treatment of sheep litter with single actinobacterial strains will allow us to obtain other information about the hygienic and quality of this compost as a fertilizer. A further selection of the microbial starters is envisaged to optimize the biotechnological process.

Acknowledgements. This study is a part of the project Innovative methods for extensive sheep breeding in protected areas c/o S. Marco Sheep Breeding Research Centre (Castel del Monte, AQ, Italy), Regione Abruzzo Pom 94/96, Misura 3.1, Ricerca. The researchers A.L. Botta and S. Manfroni were supported by a grant from the Province of L'Aquila.

References

Andreyuk EI, Gordienko SA (1981) Biosynthesis of humus-like substances by *Actinomycetes* of the *Chromogenes* and *Niger* genus. Mikrobiol Zh (Kiev) 43/2: 146-151

Ball AS, Godden B, Helvenstein P, Penninckx MJ, Mc Carthy (1990) Lignocarbohydrate solubilization from straw by *Actinomycetes*. Appl Environ Microbiol 56, 10: 3017-3022

Benedetti A (1992) Cinque anni di sperimentazione sui casting da lombrico. Agric Ric 132: 21-32

Bremner JM, Tabatabai MA (1971) Use of automated combustion techniques for total carbon, total nitrogen, and total sulfur analysis of soils. In: (Walsh ed) Instrumental methods for analysis of soils and plant tissue. SSSA, Madison, WI pp 1-15

Ciavatta C, Govi M, Vittori Antisari L, Sequi P (1990) Characterization of humified compounds by extraction and fractionation on soild polyvynilpyrrolidone. J Chromatogr 509: 141-146

Godden B, Penninckx MJ (1984) Identification and evolution of the cellulolytic microflora present during composting of cattle manure: on the role of *Actinomycetes* sp. Ann Microbiol (Inst Pasteur) 135 B: 69-78

Godden B, Legon T, Helvenstein P, Penninckx M (1989) Regolation of the production of Hemicellulolytic and cellulolytic enzymes by *Streptomyces* sp. growing on lignocellulose. J Gen Microbiol 135: 285-292

Gomes RC, Mangrich AS, Coelho RRR, Linhares LF (1996) Elemental, functional group and infrared spectroscopic analysis of actinomycete melanins from brazilian soils. Biol Fertil Soil 21: 84-88

Govi M, Ciavatta C, Gessa C (1994) Evaluation of the stability of the organic matter in slurries, sludge and composts using humification parameters and isoelectric focusing.

In: Senesi S, Miano TM (eds) Humic substances in the global environment and implications on human health. Elsevier Science, Amsterdam, pp 1311-1316

Hopwood DA, Bibb MJ, Chater KF, Kieser T, Bruton CJ, Kieser HM, Lydiate DJ, Smith CP, Ward JM, Schrempf H (1985) Genetic manipulation of *Streptomyces*. A laboratory manual. The John Innes Foundation, Norwich

McCarthy AJ(1987) Lignocellulose-degrading *Actinomycetes*. FEMS Microbiol Rev 46: 145-163

Ministero delle Risorse Agricole, Alimentari e Forestali (1994) Metodi ufficiali di analisi chimica del suolo. Roma

Quaroni S, Petrolini B, Saracchi M, Sardi P (1987) Indagini preliminari sulla utilizzazione degli attinomiceti nel miglioramento della produzione vegetale. Nat Ric 75/76: 49-54

Senesi N, Miano TM (1991) Riciclo di biomasse di rifiuto e di scarto e fertilizzazione organica del suolo. Pàtron, Bologna

Sequi P, De Nobili M (2000) Cap.VII Carbonio Organico. In: Violante P (ed) Metodi di analisi chimica del suolo, Osservatorio Nazionale Pedologico, Ministero delle Politiche Agricole. Franco Angeli, Bologna

Società Italiana della Scienza del Suolo (1985) Metodi Normalizzati di Analisi del Suolo. Ed agricole, Bologna

Springer U, Klee J (1954) Prüfung der Leistungsfähigkeit von einigen wichtigeren Verfahren zur Bestimmung des Kohlenstoffs mittels Chromschwefelsäure sowie Vorschlag einer neuen Schnellmethode. Z Pflanzenernaehr Dueng Bodenkd 64: 1

Stentiford EI (1986) Recent developments in composting. Compost: production, quality and use. Udine, Italy, 17-19 April 1986 Proc Symp, pp 52-60

Methods for Health Risk Assessment by *Clostridium botulinum* in Biocompost

H. Böhnel, B.-H. Briese and F.Gessler [1]

Abstract. In recent investigations in 66 out of 143 tested samples of biocompost or substrates containing biocompost *Clostridium botulinum* was detetcted.

In commercial potting soils, containing 50% biocompost, even at the purchase day, botulinum toxin was found in the plastic bags.

Obviously,

- the number of cases of botulism in animals has been constantly increasing in the past years;
- demands for quality control of human food make it necessary to reduce any contamination of the soil;
- physicians need to become aware of the importance of human botulism.

The pathogen *Clostridium botulinum* endangers health and life of man and animals by production of a very potent metabolite, the Botulinum Neurotoxin (BoNT). It is an anaerobic bacterium, ie., it multiplies and forms toxin under exclusion of oxygen. Under certain conditions it is even able to create its own anaerobic micro-environment in aerobic atmosphere, supporting multiplication. Clostridia are sporeformers, ie., they may survive adverse conditions very well. By the so-called hygienisation during composting possibly not all spores are destroyed. As a result surviving spores will multiply during subsequent curing and storage in compost. The nutrients, warmer temperatures, humidity, and exclusion from air form ideal growing conditions.

The different influence factors will be discussed. The laboratory proof of bacteria and toxins is complicated due to the fact that internationally only the mouse bio-assay for food and pathological samples is accepted which takes at least 5-10 days for completion. New field and laboratory tests for compost and soils were established. First results are presented.

Problem

The disease botulism is caused by neurotoxins (BoNTs) produced by *Clostridium (C. botulinum)*. The classical form of this normally fatal disease is the inhibition of neurotransmitter release at the neuromuscular synapses, which lead to paralysis of muscles. If the thoracical muscles are affected, death will follow by suffocation;

[1]Institute for Tropical Animal Health, University of Goettingen, Kellnerweg 6, 37077 Goettingen, Germany, e-mail: hboehne@gwdg.de

H. Insam, N. Riddech, S. Klammer (Eds.)
Microbiology of Composting

the sensorium remains undisturbed (Smith and Sugyiama 1988; Seifert and Böhnel 1995). BoNTs are the most potent biological metabolites.

In standard human and veterinary medicine, botulism is almost completely neglected. There are two reasons for this: evidence of its occurence is rare, and these cases are very often misdiagnosed. The experience of our laboratory shows that in reality there are numerous botulism cases in man and animals. Estimates suspect about 500–1000 dead infants per year in Germany from sudden infant botulism, although officially only 1–2 cases are reported anually (Behrens et al. 1998; Böhnel et al. 2001b). In cattle we would estimate certainly more than 1000 cases per year (Böhnel 1999). However, there is another recently detected form of the disease, mainly in cattle, the so-called visceral botulism. Here, the BoNTs act on visceral nerves, which leads to chronic forms of wasting and reduced milk production in cows, and increased calf mortality. There are no figures available. We are convinced that thousands of animals are affected (Böhnel et al. 2001a).

C. botulinum exists in many types (officially A–G, and other BoNT-producing *clostridia*). They are found "all over the world", as written in many text books. This is not true. The different types are of distinct geographic distribution, they are site-specific, (topotypes), depending on soil, (sea) water, agricultural use, prevailing animals and other unknown factors. All types may be pathogenic for man and animals, although the types A, B and E are frequently found in man, whereas B is common in horses and C and D in cattle. Birds are rather resistant, but there are sometimes outbreaks with thousands of deaths (Baird-Parker and Freame 1967; Smith and Sugyiama 1988; Westphal 1991). Besides BoNT, several other metabolites contribute to pathogenicity (Böhnel 1988).

C. botulinum is a soil bacterium, i.e. it may proliferate directly in soil (without toxin production), and may survive for decades by spore formation (Mitscherlich and Marth 1984). In many animals and humans it multiplies in the intestine and is spread by faecal excretion without any toxin production. These latter facts are retained by many bacteriologists or medical people, leading to their conviction that *C. botulinum* is "everywhere", "no pathogen", and thus "no problem".

The problem is that many physicians or veterinarians are not aware of the disease botulism. Since *C. botulinum* is known for its highly lethal toxin, special laboratory equipment for culture and handling is needed. Due to these difficulties only few scientific groups are dealing with these bacteria. This probably leads to the fact that botulism is underestimated in most European countries (de Groot and Steenhof 1997).

Introduction

In recent investigations we detected toxigenic *C. botulinum* in more than 50% of biocompost samples, and in some cases even free BoNTs (Böhnel 1999; Böhnel and Lube 2000). Additionally,

- the number of cases of botulism in animals has been constantly increasing in the past years;

- demands for quality control of human food make it necessary to reduce any contamination of the soil;
- physicians are becoming aware of the importance of human botulism.

Hence, it is important to assess the health risk from *C. botulinum* and biocompost (see Table 1).

Table 1. Overview on *C. botulinum* and possible health hazards

Clostridium botulinum	produces	botulino-neurotoxins (BoNTs)
BoNTs	cause	botulism in man and animals
Botulism	is	an intoxication (uptake of toxin) or an infection (uptake of bacteria)
	may be	lethal
May biocompost influence the cycle *Clostridium botulinum*–botulino-neurotoxins (BoNTs)–botulism?		

In an aerobic environment, the pathogen *C. botulinum* will create its own anaerobic microenvironment to start multiplication. *Clostridia* are spore-formers, i.e. they may survive adverse conditions well. Spores may remain alive in the environment for many decades (Mitscherlich and Marth 1984). However they may exist as dormant spores.

There might be a mechanism whereby different types of *C. botulinum* could be intermittently replenished in cultivated soil in spite of the diluting, dispersing and destructive effects of rainfall, drainage and exposure. Areas of spore abundance may be created (Dolman 1964). By distributing poultry litter containing *C. botulinum* on agricultural land, the risk of botulism in cattle and probably in man may be enhanced (Popoff and Argente 1996).

By the so-called hygienisation during composting, not all spores which could be brought in with the feedstock are destroyed. Even when blowing air through the composting material, there remain unaffected anaerobic compartments (Grabbe 1996). The survivors will multiply during subsequent curing and storage in the final product compost. Nutrients, warmer temperatures, humidity and seclusion from air form ideal growing conditions.

Thus, the commercial products biocompost or substrate containing biocompost may carry spores, vegetative forms and even free toxins. Hence, they would be a source for conveying the pathogen, or contaminating nature or house gardens.

Transmission of the pathogen and its toxins to man and animals may be via

- soil/food, feed
- soil/water
- soil, dust/air.

Hence, soil is a crucial epidemiological factor. Any health risk may remain so for decades through surviving spores (Mitscherlich and Marth 1984).

Quality assessment of biocompost follows national regulations (Bundesumweltministerium 1998). Here, mainly chemical analyses are important. The final product must be almost free from viable plant seeds. The production systems are bacteriologically evaluated by freedom of *Salmonella senftenbergii W775*, a non-spore forming Gram-negative bacterium. Pathogenic spore-formers, such as *Clostridia* spp. are not listed, and their presence would not interfere with the actual legally accepted quality (Bundesumweltministerium 1998; Deutsche Bundesstiftung Umwelt 1998). In Germany, microbial toxins are not included in reference lists for environmental toxins (Ewers et al. 1999).

The European population is becoming increasingly aware of questions concerning health, food and environment. The acceptance by the population of the recycling product biocompost may subsequently be reduced or even completely endangered. Obviously, not each production system and/or facility is contaminated, as we found samples of compost of good quality without any traces of *C. botulinum*.

Methods

Sampling

The official methods for collecting statistically adequate samples during production or storage of the final product (Bundesgütegemeinschaft Kompost 1994) are equally suited for biological tests. However cross-contamination by using the same equipment must be avoided: the plastic sheet used for mixing has to be covered each time with a new thin plastic foil as it is used by painters (0.3 mm thickness). Shovels must be sterilized by immersion in 10% formalin for 60 min at room temperature, and rinsed with sterile water afterwards. Five-L plastic buckets with firmly closing lids must be sterile as well.

Transport of the samples must be at temperatures below 15 °C and should reach the laboratory within 24 h. Storage at the laboratory is at 6 °C.

Laboratory Preparatory Work

All laboratory work has to be done according to standard bacteriological methods to avoid contamination. The samples are thoroughly mixed to homogenize the material which might have settled during transport.

- Mixing on a plastic sheet according to the sampling method (dry sample/room temperature)
- Shaking with glass beads (addition of water or buffer/overnight, 6 °C)
- Mixing in a stomacher (5 min, room temperature)

The last method is recommended for wet samples which do not contain large and sharp-edged items. The use of a strainer may be recommended, but may negatively influence the results.

Bacteriological Testing

Each bacteriological procedure for the presence of *C. botulinum* is terminated by the mouse bioassay for BoNT (Smith and Sugyiama 1988; CDC 1998).
Three methods were used:

- Elution of natural free toxin by buffer solution.
- Artificial stimulation of toxin production in bacteriological medium under anaerobic conditions.
- Environmental simulation by addition of water to the substrate and anaerobic incubation (Böhnel and Lube 2000).

Toxin elution is done with phosphate buffered saline (PBS) or gelatine-phosphate buffer (GPB) at pH 7.2. Spores are activated by heat (60 °C 30 min^{-1}) and treated afterwards like bacteria.

Multiplication and toxin production is tested in RCM (reinforced *Clostridium* medium). The presence of toxin is detected according to the international accepted mouse bioassay (Smith and Sugyiama 1988; CDC 1998). As all types of BoNT are pathogenic, it seemed not to be necessary to differentiate between the different types. To simplify handling and especially to save laboratory animals, for convenience two groups of BoNT were formed. The monospecific polyclonal antitoxins of the types A, B, E and C, D, respectively, were pooled and tested together. A laboratory non-animal test is actually evaluated in order to be used for soil and compost samples.

Results and Interpretation

Initial results have been published concerning biocompost (Böhnel and Lube 2000). Additionally, in Table 2 some results of on-going research work show trends of comparing those results, including some additional tests, with other recycling products and soil. As shown, about almost half of the compost samples contained toxinogenic *C. botulinum* and about one third of the potting soil blended with biocompost was proven to be positive as well. Free toxin was demonstrated in some samples of compost and sewage from biogas facilities and in agricultural soil, manured with animal faeces of different origin. Those samples of compost containing free toxin are relatively small in number compared with sewage of biogas production.

Table 2. Results of toxicity tests in different types of specimens

Specimen	Positive tests		
Type	Free toxin	Bacterial forms	Bacterial forms / toxigenic factors
	Elution	Enrichment	Environmental simulation
Biocompost	5/125	54/125	18/73
Potting soil	0/20	7/20	0/16
Sewage, biogas	5/13	5/13	5/12
Sewage	0/3	1/3	0/2
Sewage sludge	0/5	3/5	0/2
Agricultural soil	0/31	2/31	0/6
Agricultural soil, manured	3/20	3/20	0/20

About 15% of those tests lethal for mice could not be neutralized with those antitoxins available at our laboratory. This may be due to divergent field strains of *C. botulinum*, or other BoNT producing clostridia. Especially in compost, other mouse-lethal metabolites, e.g. biogenic amines, may be present.

The mouse bioassay is the only internationally accepted test and is the most important. It has the advantage of being able to prove the presence of toxins (and active toxigenic bacteria). A big disadvantage is that only positive results can be evaluated.

A negative toxin test may be due to the following reasons:

- *C. botulinum* not present
- Spores not activated or dormant
- Bacteria present, but non-toxic strain
- Toxigenic bacteria present but genetically not activated
- Genetically active bacteria present, but
  - competitive bacteria suppress or reduce toxin formation, or
  - medium components are not adequate for toxin production, or
  - water activity, temperature, oxygen tension render any toxin production impossible
- Toxin formed, but detoxified by other metabolites or substances present
- Toxin formed, but quantity not sufficient to kill a mouse
- Mouse too resistant for minor quantities of toxin

For toxin production, three components are necessary: genetically determined toxigenic bacteria, supportive substances, and substrate for multiplication of bacteria. For the development of the clinical picture of the disease botulism-receptible animals (or man) must be affected (Table 3).

PCR was not done. This method may detect the presence of known toxin genes. It is actually not possible to detect all genes, especially in locality-specific strains. It is not possible to distinguish between live and dead bacteria, and whether the genes are active expriming toxins. Mixed infections, new strains or other bacteria-producing BoNTs may further complicate the evaluation of results.

In nature, the number of spores, vegetative bacteria and/or toxins may be diminished through dilution and dispersed by rain, snow or wind. However, they may be concentrated in lower wet areas (concentration areas). They may be taken up and excreted somewhere else by healthy and unaffected animals (distributors).

Each rural area is part of the agricultural production system which mainly aims to produce food. The pathogen *C. botulinum* may find its way into human food, leading possibly to disease and death. In this way, farm animals, wildlife, and companion animals may be affected as well.

Table 3. Epidemiology of botulism

Microbial factor				Toxin production	Susceptible animal	
Toxigenic bacteria	Supportive substances	Substrate	Sum		Intoxication	Infection
+	+	+	+++	External	+	
+	+	+	+++	Internal		+
+(–)	+(–)	+(–)	++–/+––/–––	Nil	–	–

It is common knowledge that

- without bacteria there is no toxin,
- the presence of toxin means the presence of bacteria,
- there are bacteria without toxin,
- there are spores surviving for longer periods,
- there are dormant spores which cannot be activated.

Any risk assessment (Haas et al. 1999) for human and animal health in connection with biocompost should be done according to the fact that *C. botulinum* is a soil bacterium which may become a pathogen by toxin production.

- Negative results: practically no evidence of either acute or future health risk.
- Positive results: the presence of toxinogenic bacteria (enrichment) is alarming, additional toxinogenic factors show an actual high risk (simulation).

Free toxin seems to be no health hazard in biocompost, other substrates, or soil. In liquid samples, the number of tested specimens is too low to evaluate the obtained results.

Almost nothing is known for risk assessment, when either bacteria or toxinogenic factors are spread to areas where the complementary partner for toxinogenesis is already invisibly prevalent. Especially in animal and human cases of enterotoxaemia, where toxin is formed in the intestines, hitherto unknown courses of diseases may take place. If spores are abundant, they will eventually enter the food chain in modern agriculture (Böhnel et al. 2000, 2001a).

The process of biocompost production, from collecting biobins and other feed stock to composting, curing, stockage and sale of the final substrate, may be prone to contamination by and multiplication of *C. botulinum,* and BoNT formation (Fig. 1).

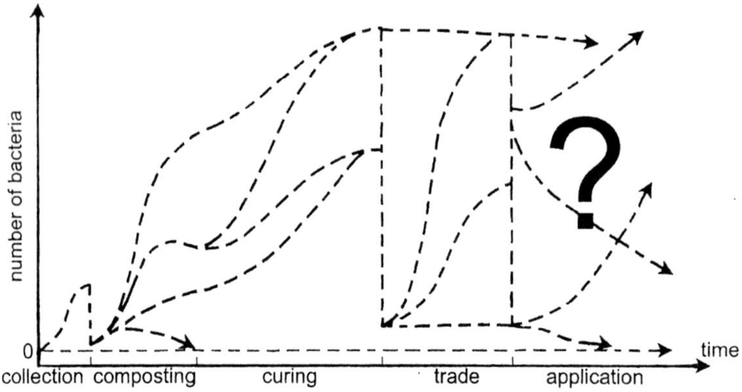

Fig. 1. Estimates of possible numerical developments of *C. botulinum* during compost production and commercialisation

As it is shown in Fig. 1 and Table 4 the long-term fate of *C. botulinum* and BoNT, distributed in the environment, are not yet established. Apparently, there is no lower limit defining a number representing a cut-off value as no danger. Scientific evidence shows that small numbers of spores/bacteria may germinate/multiply within some days up to a maximum number. This quantity is determined only by growing conditions of *C. botulinum* and competing/enhancing microorganisms (Baird-Parker and Freame 1967).

Table 4. Influence of man-made distribution on the natural occurrence of *C. botulinum* in soil (+ = increase, – = decrease, o = no effect)

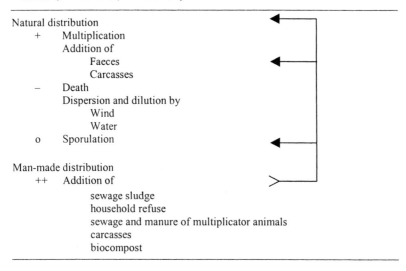

Natural distribution		
+	Multiplication	
	Addition of	
		Faeces
		Carcasses
–	Death	
	Dispersion and dilution by	
		Wind
		Water
o	Sporulation	
Man-made distribution		
++	Addition of	
		sewage sludge
		household refuse
		sewage and manure of multiplicator animals
		carcasses
		biocompost

The use of biocompost as fertilizer or soil improver may initiate a reaction in the environment (Böhnel et al. 2000). The final presence of bacteria, toxigenic factors and nutrients in the environment, wherever they may originate from, may pose a health hazard.

The European population is ever increasingly aware of the environment and health hazards caused by food. Thus the future acceptance of biocompost should be taken into account by political and technological decisions (Table 5).

Table 5. Quality of biocompost, presence of C. botulinum, and acceptance by the population

Product	Certified *C. botulinum*	for Acceptance	Rejection
	Not present	Unlimited	
	Present	Limited recultivation(?)	Partial
	Not certified	Nil	Complete

As we have also found some biocompost samples with negative results, it would be worthwhile to look into details of the whole composting process and to find a Hazard Analysis Critical Control Points (HACCP) concept for these systems. Further research in production and use of biocompost is urgently needed. A possible interaction in the human food chain has to be considered as well. As botulism is a factorial disease, interdisciplinary cooperation is a conditio sine qua non. What we need is not just an accumulation of specialized results but a holistic approach to the whole complex.

Acknowledgements. Part of our actual research work was financially supported by The German Federal Foundation for the Protection of the Environment, The German Federal Environmental Agency, The Institute for Applied Biotechnology in the Tropics, and The Farm Animal Disease Compensation Agency of Lower Saxony.

References

Baird-Parker AC., Freame B (1967) Combined effect of water activity, pH and temperature on the growth of *Clostridium botulinum* from spore and vegetative cell inocula. J Appl Bacteriol 30: 420-429

Behrens S., Sukop U, Saternus K-S, Böhnel H (1998) SID and botulism: can a correlation be proved? Res Leg Med 18: 121-126

Böhnel H (1988) The toxins of *Clostridia* (in German). J Vet Med B 35: 29-47

Böhnel H (1999) Botulism – a forgotten diasease? (in German) Berl Münch Tierärztl Wochenschr 112: 139-145

Böhnel H, Lube K (2000) *Clostridium botulinum* and bio-compost. A contribution to the analysis of potential health hazards caused by bio-waste recycling. J Vet Med B 47: 785-795

Böhnel H, Briese B-H, Gessler F (2000) Bio-compost and *Clostridium botulinum* – a health hazard for man and animals? Microbiology and Composting, Oct 2000, Innsbruck

Böhnel H, Schwagerick B, Gessler F (2001a) Visceral botulism – a new form of bovine *Clostridium botulinum* toxication. J Vet Med A 48: 373-383

Böhnel H, Behrens S, Loch P, Lube K, Saternus K-S, Gessler F (2001b) Is there a link between infant botulism and sudden infant death? Bacteriological results obtained from central Germany. Eur J Pediatr 160: 623-628

Bundesgütegemeinschaft Kompost (BGK) (1994) Handbook of methods for compost analyses (in German). Abfall Now, Stuttgart

Bundesumweltministerium (1998) Bioabfallverordnung–BioAbfV BGBl I: 2955-2971, Bundesdruckerei, Berlin

CDC (1998) Botulism in the United States, 1899-1996. Handbook for epidemiologists, clinicians, and laboratory workers. Center for Disease Control and Prevention, Atlanta

de Groot M, Steenhof V (1997) Composting in the European Union. DHV Environment and Infrastructure, Final report. European Commission, DG XI, Environment, nuclear safety and civil protection. Report No K 1089-61-001/AT-973090, Brussels

Deutsche Bundesstiftung Umwelt (1998) Hygiene in bio-waste composting (in German). Zeller, Osnabrück

Dolman CE (1964) Botulism as a world health problem. Environ Health Serv Food Protect 1: 5-32

Ewers U, Krause C, Schulz C, Wilhelm M (1999) Reference values and human biological monitoring values for environmental toxins. Int Arch Occup Environ Health 72: 255-260

Grabbe M (1996) Fundamentals of bioprocess management in composting biogenic residuals and their relevance to produce compost of reproducible quality (in German). In: Wiemer K, Kern M (eds) Biological waste treatment III (in German). M.I.C. Baeza, Witzenhausen, pp 171-214

Haas CN, Rose JB, Gerba CP (1999) Quantitative microbial risk assessment. Wiley, New York

Mitscherlich E, Marth EH (1984) Microbial survival in the environment. Springer, Berlin Heidelberg New York

Popoff MR, Argente G (1996) Animal botulism, is it a menace for man? (in French) Bull Acad Vét Fr 69: 373-382

Seifert HSH, Böhnel H (1995) Clostridioses (in German). In: Blobel H, Schliesser T (eds) Handbook for bacterial infections in animals, vol II/4 (in German). Fischer, Jena

Smith LD, Sugyiama H (1988) Botulism. The organism, its toxins, the disease, 2nd edn. Charles C Thomas, Springfield

Westphal U (1991) Avian botulism (in German). Aula, Wiesbaden

The Fate of Plant Pathogens and Seeds During Backyard Composting of Vegetable, Fruit and Garden Wastes

J. Ryckeboer, S. Cops and J. Coosemans[1]

Abstract. Backyard composting is widely used to process vegetable, fruit and garden wastes (biowastes) in gardens. Unfortunately, little is known about the hygienisation efficiency during this small-scale composting process. In a 6-month experiment, the eradication of tobacco mosaic virus (TMV), *Plasmodiophora brassicae*, *Heterodera schachtii* and of tomato seeds was followed during composting in four 200-l vessels. In conformity with the German BioAbfV norms which were used as guideline for this research, the pathogens were incorporated as TMV-infected tobacco leaves, as a mixture of clubroot-infected cauliflower roots and infested soil, as cysts or as seeds packed in nylon fibre nets. Two vessels were totally filled; two others were half-filled. At the end of the 6-month composting period, only TMV was completely eliminated, while *H. schachtii* and the tomato seeds were almost completely eliminated. On the other hand, *P. brassicae* survived the composting process. Destruction of the tomato seeds, of TMV and of *H. schachtii* could partly be explained on the basis of heat treatment. Other factors may have contributed to the sanitising effect observed during this rather low-temperature composting process.

Introduction

Composting is an alternative way to convert vegetable, fruit and garden wastes (biowastes) to a useful product. In cities, source-separated household wastes are mostly composted in commercial composting plants, while in the countryside composting of these biowastes often occurs in gardens in small compost heaps or in a composting vessel (bin). In spite of the fact that home composting is a general practice in several countries, little is known about the hygienic aspects after composting of this waste. Nevertheless, a lot of pathogens including viruses, bacteria, fungi and nematodes, insects and mites but also weeds and weed seeds can be present in the waste (Bollen 1993; Ryckeboer et al. 1997a,d). Eradication of pathogens from organic wastes during composting may be due to (1) heat generated during the thermophilic phase(s) of the composting process, (2) toxic compounds (i.e. organic acids, ammonia, etc.) released during or after the self-heating process, (3) enzymatic degradation, (4) microbial antagonism (i.e. parasitism, antibiotics, etc.) and (5) competition for nutrients in the sublethal outer temperature zones of

[1]Laboratory of Phytopathology and Plant Protection, Katholieke Universiteit Leuven, W. de Croylaan 42, 3001 Leuven, Belgium

H. Insam, N. Riddech, S. Klammer (Eds.)
Microbiology of Composting

the pile or later during maturation (Golueke 1982; Hoitink and Fahy 1986; Bollen 1993; Bollen and Volker 1996; Ryckeboer et al. 1997a). In contrast with composting in commercial composting plants, small-scale compost piles may not achieve adequate temperature-time exposure necessary for a complete elimination of harmful organisms. Bollen (1985) posed that the formation of toxic compounds would be too low to achieve sufficient inactivation of the pathogens. He also claimed that such compost could be a source of pathogens rather than a product to suppress plant diseases. In contrast, Wijnen et al. (1983) found eradication of artificially incorporated *Botrytis allii, Fusarium oxysporum, Phomopsis sclerotioides, Phytophthora cryptogea, Plasmodiophora brassicae, Rhizoctonia solani, Sclerotinia sclerotiorum, Sclerotium cepivorum* and *Stromatinia gladioli* after 4 months of composting in a vessel containing vegetable, fruit and garden wastes, horse manure and basic slag. During peak heating, the temperatures increased to 47 to 56 °C, depending on the position in the vessel. Wijnen et al. (1983) warned that these conclusions refer only to optimal set up compost piles and not to compost piles which were regularly fed with fresh waste. Yuen and Raabe (1984) investigated the elimination of plant pathogens in compost bins with a volume of 0.9 m^3 filled with herbaceous plants and sawdust. These compost piles were turned every 2 to 3 days. In the centre of the pile, where the temperatures rose above 50 °C during 10 to 14 days, all incorporated pathogens (*Armillaria mellea, Rhizoctonia solani, Sclerotium rolfsii* and *Verticillium dahliae*) were completely eliminated. There were also organisms incorporated in the corner of the compost bin, where the temperature did not achieve more than 30 °C. When the pile was turned, they were incorporated again in the corner. All the organisms survived, with exception of the sclerotia of *Sclerotium rolfsii* which were possibly inactivated by microbial antagonism and toxic compounds. Yuen and Raabe (1984) concluded from this experiment that home composting in small-scale piles, which were turned regularly to achieve thorough heating of all parts of the waste, is efficient enough to eradicate plant pathogens. The results of the organisms which were placed in the corner of the pile suggest that compost produced without turning the piles should not be utilised.

The aim of this study was to investigate the fate of plant pathogens and seeds during composting of vegetable, fruit and garden wastes. Therefore, a 6-month experiment was performed with four compost vessels from autumn until spring. To simulate a less optimal composting process, two compost vessels were filled to half capacity.

The prescribed test organisms of the German BioAbfV norms were used in the experiment (LAGA Merkblatt M10 1994; BioAbfV 1998). These test agents are the tobacco mosaic virus (TMV), the biotrophic fungus *Plasmodiophora brassicae* and tomato seeds. They were selected based on their particular resistance against biological influences and high temperatures which prevail during composting. Based on the elimination results of these organisms, the phytohygienic safety of the compost can be predicted. According to these norms, a complete elimination is not required. The BioAbfV protocols, originally written for composting under practical conditions, were modified for these small-scale experiments. Besides these test organisms the elimination of the beet cyst nematode *Heterodera schachtii* was also investigated.

Materials and Methods

Tobacco leaves infected with TMV, *P. brassicae*-infected cauliflower roots and infested soil, tomato seeds and *H. schachtii* cysts were incorporated on several depths in four compost vessels, filled with vegetable, fruit and garden wastes (Ryckeboer et al. 1997b).

Preparation of the Test Organisms

The preparation of the pathogens and of the samples was executed as described in the BioAbfV norms (LAGA Merkblatt M10 1994; BioAbfV 1998). Ten grams TMV-infected tobacco leaves and 100 g vegetable, fruit and garden wastes were mixed together and brought into a nylon fibre net (20 cm x 20 cm, mesh 1 mm). These leaves were harvested from *Nicotiana tabacum* cv. Samsun. An infection of this cultivar with TMV has a systemic character and is expressed as mosaic spots on the leaves. A fine-mesh nylon net (3 cm x 3 cm, mesh 200 μm) filled with 100 cysts of *H. schachtii* and an identical net filled with 400 tomato (*Lycopersicon esculentum* L. cv. Moneymaker) seeds (germination rate: 94%) were also added to this nylon fibre net.

A second nylon fibre net (30 cm x 20 cm, mesh 1 mm) was filled with a mixture of 30 g *P. brassicae*-infected cauliflower roots, 430 g *P. brassicae*-infested sandy loam soil and 200 g vegetable, fruit and garden wastes. This gave a composition of about 5% (w/w) tuberous roots, 65% (w/w) soil and 30% (w/w) compost. Contrary to the BioAbfV norms, the same mixtures were made as controls and stored at -21°C.

Each treatment was replicated three times, i.e. for each organism three replicates were foreseen for each of the five sample data.

Experimental Design

Four cylindrical compost vessels (brands Milko, VAM, Fusion and Danaïd) of the same size (200 l) and design were filled with vegetable, fruit and garden wastes, grass clippings, conifer wood clippings, weeds and leaves. Previously, a 10-cm layer of hardwood chips was brought on the bottom of the vessels. To simulate a backyard composting at low process temperatures, two vessels were half-filled. To achieve "normal" backyard composting, the two other vessels were totally filled.
The nylon fibre nets containing the samples were divided over the four compost vessels in which they were placed randomly. Until day 132 the vessels were regularly fed with fresh vegetable, fruit and garden wastes to keep filling levels con-

stant. The vessels were turned after 29, 79 and 121 days of composting. During each turning time, a 10-cm layer of hardwood clippings was brought on the bottom of the vessels. Then, the vessels were refilled with the composting waste and the nets were buried again. The nets which were located before at the bottom of the vessel were reincorporated at the top of the vessel and vice versa. After day 132 the compost vessels received no further treatment.

At regular moments, temperatures in the centre of the vessels and the environment were measured with a metal probe. The compost temperatures of the half-filled vessels were measured at a depth of approx. 30 cm, while temperatures of the completely filled vessels were measured at two depths, i.e. approx. 30 and 50 cm. Normally, on a regular basis, a poker is used to stir the material and to improve aeration during backyard composting, thus stimulating the composting process. This could not be done in this experiment, due to the presence of the nylon fibre nets on several depths in the vessels.

To evaluate eradication of the pathogens and seeds, samples were removed after 11, 29, 79, 121 and 184 days of composting.

Pathogen Survival Bioassays

Tobacco Mosaic Virus

The half-leaf method was used to detect infectious virus particles (Walkey 1991). Compared with other virus assay methods, which depend on physical, chemical or serological procedures, it has the advantage of quantifying the relative amount of infectious virus, rather than the total amount of nucleoprotein, all of which may not be infectious (Walkey 1991). Contrary to the BioAbfV norms (LAGA Merkblatt M10 1994; BioAbfV 1998) which prescribe the use of entire plants, for practical reasons we used detached leaves in this work. The detached leaves with a size of approx. 90 cm² were the second and third leaves from plants in the 6–8 leaf stage grown under constant conditions (Walkey 1991). The leaves were harvested from an adult plant of *Nicotiana tabacum* cv. Samsun NN. This cultivar expresses TMV infection as local lesions (small, round spots with a necrotic centre) visible within a few days.

The entire content of the metal fibre net was cut with scissors into a 1-l beaker. The material was then ground with a pestle in 30 ml of phosphate buffer (0.05 mol l^{-1}, pH 7.0) and squeezed through a nylon net (mesh 1 mm) above the two right halves of two tobacco leaves powdered with silicium carbide (carborundum, 600 mesh). The extract was spread on the leaf with rotating movements using a glass spatula. Previously, on the left halves of the same silicium carbide-powdered leaves a control extract was applied in the same way. Both extracts were allowed to interact for 30 s with the carborundum-damaged leaves. The leaves were then washed with running tap water during 10 s. The control samples consisted of a mixture of 10 g TMV-infected tobacco leaves and 100 g vegetable, fruit and garden wastes stored at the beginning of the experiment at -21 °C. The inoculated leaves were then placed on 2 moist filter papers in a Petri dish and incubated in a growth chamber (22.5–24 °C; 16 h 210 $\mu E\ m^{-2}\ s^{-1}$). After approximately 6 days of incubation, the number of lesions was enumerated.

The BioAbfV norms prescribe that the sum of TMV lesions of the two leaf halves treated with extracts of the composted sample have to be smaller than or equal to eight (LAGA Merkblatt M10 1994; BioAbfV 1998).

Plasmodiophora brassicae

The survival of *Plasmodiophora brassicae* was evaluated with a *Brassica juncea* bioassay. As recommended in the BioAbfV norms, a potting mix with $pH(CaCl_2)$ 6.0 was made with the composted sample, dry white sand and white peat. For this, 325 ml of the composted sample was mixed with 82.5 ml sand and 192.5 ml peat. This potting mix was brought into a 600-ml pot and four *Brassica juncea* plants were planted. The plant material was previously grown during 15 days on perforated polystyrene trays filled with a potting mix [$pH(H_2O)$: 6.0–6.5] based on white and dark peat. After planting into the pot, the plants were grown in a heated greenhouse with photosynthetic illumination (SON-T) at 21 to 22 °C during 5 weeks. After this growth period, a disease severity rating of each plant was made according to the following scale: *0* no visible symptoms; *1* slight gall development on lateral roots; *2* moderate gall development on lateral and main roots; and *3* severe gall development on lateral and main roots (Buczacki et al. 1975; LAGA Merkblatt M10 1994; BioAbfV 1998). Finally, the disease index of each replication was calculated with the following equation:

$$\text{Disease index} = \frac{\sum(\text{Number of infected plants x disease severity rate})}{\text{Total number of plants}} .$$

According to the BioAbfV norms, the mean clubroot disease index of the test plants of each replication may not be higher than 0.5.

Tomato Seeds

After each sampling, the fine-mesh nylon nets were rinsed with tap water. From each net 50 tomato seeds in 4 replicates were removed and brought on 2 moist filter papers into 4 Petri dishes. These Petri dishes were incubated in a growth chamber (22.5–24 °C; 16 h 210 $\mu E\ m^{-2}\ s^{-1}$). Every 7 days, the number of germinated seeds was counted until no more germination was observed. If after 3 weeks not a single seed germinated, the germination capacity is supposed to be lost for 100%. The germination index was calculated as follows:

$$\text{Germination index} = \frac{\text{Germination (\%) of composted seeds}}{\text{Germination (\%) of untreated seeds}} \text{ x } 100 .$$

The remaining 200 seeds were dried and stored at room temperature for a possible counter assessment (LAGA Merkblatt M10 1994; BioAbfV 1998). According to the BioAbfV norms, the germination index of the tomato seeds of each replication should be less than 2%.

Heterodera schachtii

After rinsing with tap water, the fine-mesh nylon nets filled with the 100 cysts of *H. schachtii* were soaked in a funnel filled with a $ZnCl_2$ solution (620 µg ml^{-1}) (Clarke and Perry 1977). After 3 weeks, the $ZnCl_2$ solution was tapped and the number of juveniles hatched from the cysts was counted under a light microscope.

Statistical Analysis

Each treatment was replicated three times. To determine the significance of differences among means, analysis of variance was performed with the LSD test ($P \leq 0.05$) for multiple comparisons of the SPSS program (version 6.1, SPSS Inc., Chicago, IL). If necessary, data were log-transformed to improve homogeneity of variance. Because the clubroot disease index is a categorical variable, analysis of this data was performed with the CATMOD procedure ($P \leq 0.05$) of the SAS program (version 6.12, SAS Institute Inc., Cary, NC).

Results and Discussion

Due to low ambient temperatures during the course of the experiment, the temperatures in the vessels were very low (Fig. 1). Each time after the vessels were turned, the temperatures rose.

Figure 2 presents the decrease in the mean number of lesions caused by the frozen controls and the TMV-infected composted samples respectively on two leaf halves for the half-filled (Fig. 2A) and the totally filled (Fig. 2B) vessel treatments, respectively. In the half-filled and the totally filled vessels, the BioAbfV norms were achieved for TMV within, respectively, 79 and 121 days of composting. Total inactivation of TMV was achieved in all vessels within 184 days of composting. A comparison of the trends in the mean number of TMV lesions on the test leaves and the temperature profiles is remarkable. Although temperatures in the half-filled vessels were much lower, TMV inactivation was comparable in both vessel treatments. Because the temperature of the half-filled vessels never exceeded 31 °C, eradication of TMV in this type of composting process probably was related to enzymatic degradation or affected by toxic conversion products during organic matter decomposition.

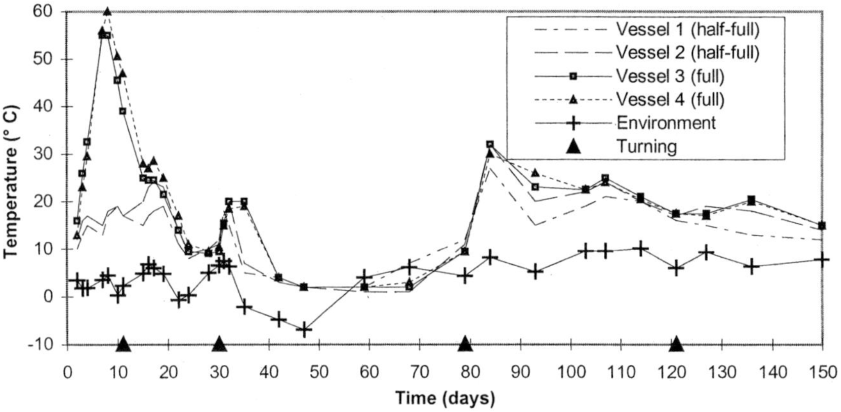

Fig. 1. Ambient temperatures and mean temperatures measured in the centre of the compost vessels. Compost temperatures remained low after 150 days of composting (data not shown)

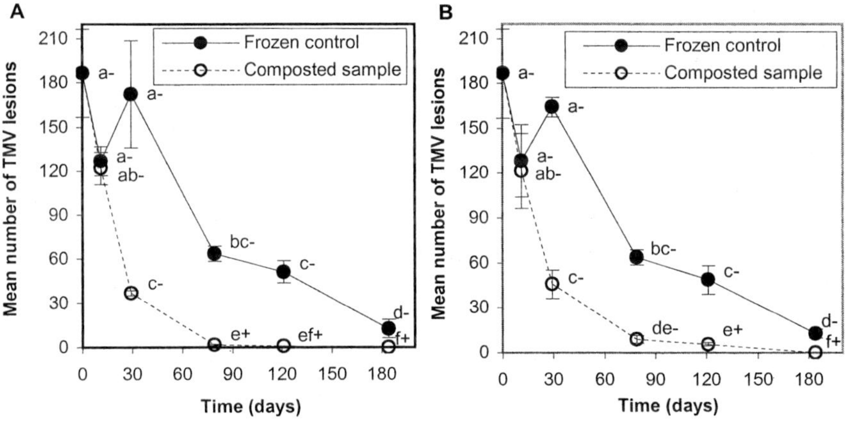

Fig. 2A,B. Effect of composting time on the mean number of TMV lesions caused by TMV-infested samples per two leaf halves after composting in half-filled (**A**) and totally filled vessels (**B**). The controls illustrate the mean number of TMV lesions caused by TMV-infested samples stored at -21 °C. *Means indicated with a same letter* do not differ significantly [$P \leq 0.05$; LSD; data log(x+1)-transformed]; while + and – mean that the BioAbfV norms are fulfilled or not fulfilled, respectively. *Error bars* represent standard deviation

P. brassicae was not eliminated in either the half-filled or the totally filled vessels after 184 days of composting. Figure 3 illustrates clearly that the disease index is higher than the tolerated 0.5 for all treatments.

H. schachtii has not been described as an indicator-organism in the BioAbfV norms. The composting process obviously reduced the number of nematodes which hatched from the cysts (Fig. 4). The viability of the cysts declined faster in the totally filled vessels, which can be explained by the higher temperatures that developed in these vessels, especially during the first 15 days of the experiment.

However, total eradication of *H. schachtii* was not achieved in any of the vessels. At day 185, in the half-filled and the totally filled vessels mean numbers of juveniles (hatched from 100 cysts) were 60 and 13, respectively.

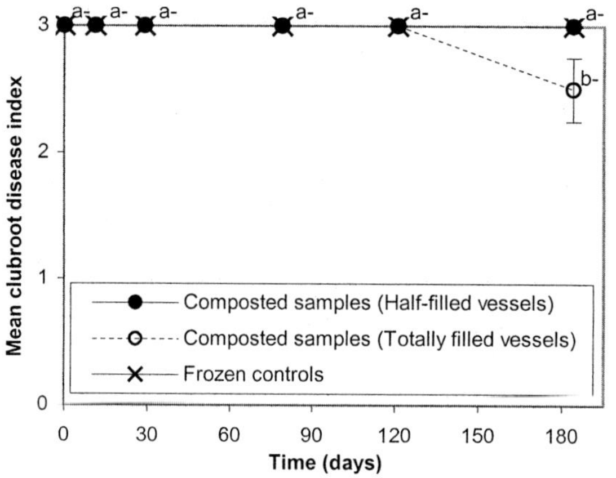

Fig. 3. Mean clubroot disease index (*Brassica juncea*) after composting in half-filled and totally filled vessels. *Means indicated with the same letter* do not differ significantly (CATMOD; $P \leq 0.05$), while + and – mean that the BioAbfV norms are fulfilled or not fulfilled, respectively. *Error bars* represent standard deviation. Disease index: *0* no visible symptoms; *1* slight gall development on lateral roots; *2* moderate gall development on lateral and main roots; and *3* severe gall development on lateral and main roots

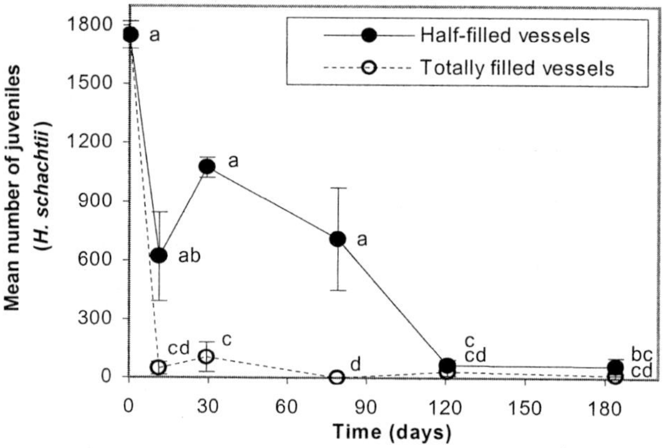

Fig. 4. Mean number of juveniles of *Heterodera schachtii* which hatched from 100 cysts, after composting in half-filled and totally filled vessels. *Means indicated with the same letter* do not differ significantly [LSD; $P \leq 0.05$; data log(x+1)-transformed]. *Error bars* represent standard deviation

Taking into account the BioAbfV norms, the germination index for all replicates of the composted tomato seeds should not be higher than 2%. The germination index of the composted tomato seeds of the half-filled vessel treatment did not comply with these norms for one replicate after 184 days of composting.
In the half-filled and the totally filled vessels, the mean germination index was 1.8 and 0.2, respectively. The germination index for the totally filled vessel treatment met the BioAbfV norms. The variable decline of the germination index of the composted tomato seeds in the totally filled vessels is remarkable (Fig. 5).

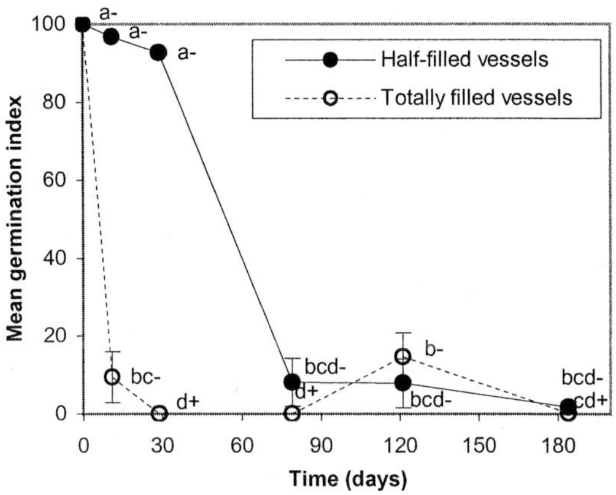

Fig. 5. Mean germination index of tomato (*Lycopersicon esculentum*) seeds after composting in half-filled and totally filled vessels. Means indicated by the same letter do not differ significantly ($P \leq 0.05$; LSD; data Log(x+1) transformed), while + and – mean that the BioAbfV norms are fulfilled or not fulfilled, respectively. Error bars represent standard deviation

Conclusions

Temperatures measured in the centre of the vessels were not as high as those reported in other studies on small or larger scales (Yuen and Raabe 1984). Temperatures in the completely filled vessels obviously were higher than half-filled vessels, probably as result of the higher mass and the lower rate of heat loss in the completely filled vessel (Ryckeboer et al. 1997c,d).

Based on the results and the literature, we can conclude that death of pathogens and seeds during composting depends on the type of organism. Of the four test organisms (TMV, *P. brassicae*, *H. schachtii* and tomato seeds) used in this study, only TMV was eliminated within the test period. Although viruses can be very heat-resistant, they are sensitive to enzymatic degradation (Avgelis and Manios 1989; Kegler et al. 1995; Bollen and Volker 1996; Idelmann et al. 1997). If composting takes enough time, problems with this virus should not occur. In the experiment described above, a 6-month composting period was required to eradicate TMV.

H. schachtii and tomato seeds were almost entirely eliminated during 184 days of composting. The BioAbfV norms for tomato seeds were achieved. Even though nematodes are very heat-sensitive (Bollen 1993; Bollen and Volker 1996), they were not completely eliminated, possibly due to lower temperatures caused by edge effects (Yuen and Raabe 1984). Temperature seemed to play an important role in the elimination of tomato seeds (Cops 1997). This temperature effect was clearly illustrated by the differences in eradication of tomato seeds among half-filled and totally filled compost vessels. Lethal temperatures are severely limited spatially in small compost piles (Yuen and Raabe 1984); they appear mostly in the centre of the pile (Golueke 1982). Tomato seeds that survived composting in the completely filled vessels at day 121 possibly did so as result of localised cool spots, e.g. the border of compost piles or vessels (Herrmann et al. 1994; Grundy et al. 1998). Possibly the presence of cool spots also explains the high variability within the replicates of some treatments.

Biotrophic fungi, such as *Plasmodiophora brassicae*, are highly resistant to conditions that prevail during composting (Bollen and Volker 1996). The heavily walled resting spores survive dehydration and heat. This experiment illustrates clearly that backyard composting in small vessels is not sufficient for the destruction of *P. brassicae*. The disease index of the test plants was only slightly reduced after composting of infested material in the totally filled vessels. Consequently, dissemination of clubroot with compost originating from small vessels is possible. According to Bollen (1985), heat is the most important factor in the elimination of *P. brassicae*. According to Bollen and Volker (1996) the resting spores of these fungi need to be composted for at least several hours at 60 °C under humid circumstances to be eliminated.

Addition of plants or plant parts to compost vessels till the end of the composting period is a common practice during backyard composting. The results presented here clearly illustrated slow elimination of most of the test pathogens. We may thus conclude that this practice is not safe from a plant-hygienic point of view, especially when detrimental organisms are present (Wijnen et al. 1983).

Based on this experiment, we can conclude that a longer composting period and a 'minimal' heat peak is required to achieve sufficient hygienisation. Small-scale compost piles may not achieve adequate temperature-time exposure, but industrial-scale operations should achieve effective pathogen kill, if adequate precautions are taken (Bollen 1993; Ryckeboer et al. 1997c,d).

Acknowledgements. J. R. was financed with a scholarship from the Flemish Institute for the stimulation of the scientific-technologic research in industry (IWT). The authors want to thank the Comité Jean Pain (G. Van Dale) for assistance and B. Lievens and K. Van den Bergh for reading of the text.

References

Avgelis AD, Manios VI (1989) Elimination of tomato mosaic virus by composting tomato residues. Neth J Plant Pathol 95 (3):167-170

BioAbfV (1998) Ordinance on the utilisation of bio-wastes on land used for agricultural, silvicultural and horticultural purposes. Bundesministerium für Umwelt, Naturschutz und Reaktorsicherheit, Bonn, Germany, 49 pp

Bollen GJ (1985) The fate of plant pathogens during composting of crop residues. In: Gasser JKR (ed) Composting of agricultural and other wastes. Elsevier, London, pp 282-290

Bollen GJ (1993) Factors involved in inactivation of plant pathogens during composting of crop residues. In: Hoitink HAJ, Keener KM (eds) Science and engineering of composting. The Ohio State University, Ohio, pp 301-318

Bollen GJ, Volker D (1996) Phytohygienic aspects of composting. In: De Bertoldi M, Sequi P, Lemmes B, Papi T (eds) The science of composting - Part 1. Chapman and Hall, London, pp 247-254

Buczacki ST, Tocopeus H, Mattusch P, Johnston TD, Dixon GR, Hobolth LA (1975) Study of physiologic specialization in *Plasmodiophora brassicae*: proposals for rationalisation through an international approach. Trans Br Mycol Soc 65:295-303

Clarke AJ, Perry RN (1977) Hatching of cyst-nematodes. Nematologica 23:350-368

Cops S (1997) Fytohygiënisatie van GFT– en groenafval door aërobe en anaërobe compostering. M Sc Dissertation. Katholieke Universiteit Leuven, Leuven, 115 pp

Golueke CG (1982) When is compost "safe"? A review of criteria for assessing the destruction of pathogens in composting. Biocycle March/April:28-38

Grundy AC, Green JM, Lennartsson M. (1998) The effect of temperature on the viability of weed seeds in compost. Compost Sci Util 6(3):26-33

Herrmann I, Meissner S, Bächle E, Rupp E, Menke G, Grossmann F (1994) Einfluβ des Rotteprozesses von Bioabfall auf das Überleben von Phytopathogenen Organismen und von Tomatensamen. Z Pflanzenkr Pflanzenschutz 101(1):48-65

Hoitink HAJ, Fahy PC (1986) Basis for the control of soilborne plant pathogens with composts. Annu Rev Phytopathol 24:93-114

Idelmann M, Marciniszyn E, Waldow F (1997) F&E Vorhaben Phytohygiene der Bioabfallkompostierung. Abschlußbericht Februar 1997. PlanCoTec, Witzenhausen, Germany, 8pp

Kegler H, Fuchs E, Spaar D, Kegler J (1995) Viren im Boden und Grundwasser. Arch Phytopathol Pflanz 29:349-371

LAGA Merkblatt M10 (1994) Qualitätskriterien und Anwendungsempfehlungen für Kompost. Endfassung zur Vorlage für die 44. ATA-Sitzung am 17./18. Januar 1995. Vorgelegt vor der LAGA-AG Biokompost. Berlin, 12 Dezember 1994, 88 pp

Ryckeboer J, Coosemans J, Cops S (1997a) Hygiënisatie tijdens het thuiscomposteren: Worden plantenziekten afgedood? Deel 1: Literatuurstudie. Humus News 13(1):11-12

Ryckeboer J, Coosemans J, Cops S (1997b) Hygiënisatie tijdens het thuiscomposteren: Worden plantenziekten afgedood? Deel 2: Proefopzet. Humus News 13(2):9-11

Ryckeboer J, Coosemans J, Cops S (1997c) Hygiënisatie tijdens het thuiscomposteren: Worden plantenziekten afgedood? Deel 3: Resultaten. Humus News 13(3):5-8

Ryckeboer J, Coosemans J, Cops S (1997d) De hygiënisatie van groente-, fruit-, en tuinafval tijdens het composteren. Vlacovaria 3:6-10

Walkey DGA (1991) Applied plant virology, 2nd edn. Chapman and Hall, London, 217 pp

Wijnen AP, Volker D, Bollen GJ (1983) De lotgevallen van pathogene schimmels in een composthoop. Gewasbescherming 14:5

Yuen GY, Raabe RD (1984) Effects of small-scale aerobic composting on survival of some fungal plant pathogens. Plant Dis 68:134-136

Survival of Phytopathogen Viruses During Semipilot-Scale Composting

F. Suárez-Estrella, M.J. López, M.A. Elorrieta, M.C. Vargas-García and J. Moreno[1]

Abstract. Underplastic protected cultures of horticultural plants constitute nowadays the primary income for the people in Almería (southeast of Spain). This intensive agriculture yields annually a large amount of plant wastes, which represent a serious environmental problem. These vegetable residues constitute an important inoculum source of phytopathogen microorganisms such as bacteria, fungi and viruses.

Since composting is proposed as an important means of plant disease control when applied to plant wastes, we have investigated the suppression of several phytopathogenic viruses during different trials of horticultural waste composting.

Vegetable residues infected with viruses PMMV (pepper mild mottle virus), TSWV (tomato spotted wilt virus) and MNSV (melon necrotic spot virus) were included into compost piles. Studies were conducted in several compost windrows subjected to different treatments. Results showed an effective suppression of viral infective activity of MNSV, during the initial 14 days of composting. PMMV was the most resistant pathogenic virus, remaining viable up to 8–9 weeks. Short-time sampling demonstrated suppression of TSWV after 60 h of composting.

Composting process was effective for biocontrol of phytopathogenic microorganisms in the investigated plant wastes.

Introduction

Underplastic culture of horticultural plants constitutes an important economical source in the province of Almería (southeast of Spain) that yields a large amount of plant wastes annually (around 10^6 t), representing a serious environmental problem. These vegetable residues constitute an important inoculum source of pathogens (Fletcher 1984). In fact, there are many reports showing that soil-borne phytopathogenic microorganisms may survive on residues (Conway 1996). Thus, management of crop residues in the field after harvest is necessary to minimize its adverse effects, particularly if the involved pathogens can infect subsequent crops (Conway 1996).

There are different sanitation possibilities that are not applicable in areas where underplastic cultures are developed on soil covered with sand (Conway 1996). In

[1]Unidad de Microbiologia, Departamento de Biología Aplicada, Escuela Politécnica Superior, Universidad de Almería, La Cañada de San Urbano, 04120 Almeria, Spain
e-mail: fsuarez@ual.es; Tel: 950015891; Fax: 950015476

H. Insam, N. Riddech, S. Klammer (Eds.)
Microbiology of Composting

Almería, plant wastes are often piled and either left to decay near the greenhouses or burned. These practices imply several undesirable agronomic and environmental effects.

Our research group is mainly involved in composting of horticultural crop residues generated in the greenhouses located at Almeria. One of the benefits ascribed to composting is the suppression of vegetable pathogens placed in these residues. Furthermore, the usefulness of compost for the control of phytopathogens was already known and practiced in the early stages of agriculture (Cook and Baker 1983). However, only recently, after the discovery of some soil-borne plant pathogenic microorganisms, interest has grown to take advantage of the plant--protecting properties of compost (Lumsden et al. 1983; Hoitink and Fahy 1986; Kostov et al. 1996).

Composting is proposed as an important means of plant disease control when applied to plant wastes (Hoitink and Fahy 1986). Three main factors seem to be implied in suppression activity of composting process: (1) The high temperatures generated during termophile phase (above 40–50 °C) (Yuen and Raabe 1984; Bollen 1985); (2) The production of antimicrobial compounds such as phenolics generated during lignocellulosic material decay (Sáez 1989); (3) The colonization of compost with many different organisms that either compete with pathogens for nutrients and/or produce general antibiotics that reduce pathogen survival and growth (Hoitink et al. 1997; Hoitink and Boehm 1999). Thus, the success of disease suppression with composting depends on a number of factors such as temperature, feedstock source, moisture, location, as well as the nature of the pathogen. The proper knowledge of those factors is crucial to ensure a high-quality compost.

In our province, several important plant diseases caused by typical phytopathogens are known. Damping-off by *Pythium aphanidermatum* and/or *Rhizoctonia solani* or *stem rot* caused by *R. solani* are diseases hard to control (Gómez 1993). Furthermore, *Fusarium oxysporum* f.sp. *melonis* (Tello et al. 1987) together with MNSV (melon necrotic spot virus) were the most critical and determinant diseases (Gómez 1993). The last is a long-term persistent virus in soil due to its association to fungal spores of *Olpidium radicale* (Campbell 1996). MNSV is a worldwide-distributed virus in underplastic-cultivated melon and cucumber plants. The great stability of this virus and the fungal spores explains the existence of long-term infected soils.

There are some other harmful viruses affecting pepper cultures in Almería such as PMMV (pepper mild mottle virus) (Cuadrado et al. 1992) and TSWV (tomato spotted wilt virus) (Cuadrado et al. 1991; Cuadrado 1994; Gea 1996). Both have been responsible for pepper crop losses in the past years. The hosts are mainly Solanaceae, Compositeae and Leguminosae and the transmission is mediated either by several thrips species or seeds.

Since pepper (*Capsicum annuum* L.) constitutes the most important crop in this area, it can be considered the first source of wastes for composting. The application of pepper compost to farmland depends on product quality. The absence of phytopathogens is a quality parameter. Pepper mild mottle tobamovirus, so named by Wetter et al. (1984), systematically infects all *Capsicum* spp., and is easily transmitted by mechanical inoculation, handling during cultivation and through contaminated seeds (Wetter and Conti 1988). Furthermore, its transmission through roots of plants cultivated in experimental pots with infected substrate has been shown (Pares and Gunn 1989). Environmental conditions in underplastic- or glass-protected cultures are particularly favourable for the dissemination of tobamoviruses (Conti and Lovisolo 1982). Thus, infection may rapidly reach 100% of the greenhouse surface, drastically diminishing the yield of marketable fruits (Conti and Marte 1983; Marte and Wetter 1986).

Bearing in mind that composting could be the best alternative for this kind of waste management and owing to the absence of data about inactivation of PMMV, TSWV and MNSV by composting, the aim of this work was to investigate survival of these viruses during different semipilot-scale trials.

Material and Methods

Collection of PMMV-, TSWV- and MNSV- Infected Plants and Experimental Conditions

To study the survival of these viruses in infected plants exposed to composting, pepper plants (*Capsicum annuum* L. cv. California) infected with PMMV or TSWV were collected from several plastic greenhouses located in Almería province (Spain). In the case of MNSV, infected melon plants were collected. Once harvested, vegetable wastes were placed in muslin bags and exposed to the composting process.

Composting Trials

To test the effect of composting on viral survival, different trials were carried out at semipilot-scale.

A composting process at semipilot-scale was performed in three piles (A1, B1 and C1) 1.5 m long, 1.2 m wide and 1.2 m high, made of horticultural wastes. Piles were set up with a mixture of pepper, cucumber and bean wastes. Wood chips were added as bulking agent. C/N value of the mixture was 20–25. Piles were turned over weekly and aerated at 2-day intervals. Air-forced composting of each pile was achieved through perforated PVC tubing (5 cm diameter, 120 cm long) placed below a fine mesh screen near the bottom of the piles. Air was supplied by a blower S&P CBB-60. Temperature was measured daily until the maturation phase. At the beginning of the process, PMMV- and TSWV-infected pepper leaves and MNSV-infected melon stems were distributed into samples of approximately 20 g and individually introduced into muslin bags, which were put 60 cm deep inside the composting piles. One muslin bag from each pile was taken at 0, 14, 28, 42, 56 and 70 days. No samples were taken once the maturation phase of composting had started. At sampling times, the content of each muslin bag was subdivided into five subsamples and PMMV, TSWV and MNSV survival and infective capacity were determined.

A second composting trial was set up in four piles (A2, B2, C2 and D2) 1.5 m long, 1.2 m wide and 1.2 m high, containing pepper wastes only. Conditions similar to those employed in the first trial were used. However, samples were taken at shorter intervals: 0, 12, 36 and 60 h. In this case, the content of each muslin bags was subdivided into three subsamples. Furthermore, TSWV was the only virus analyzed in this process, since in the first trial it was not detected.

Determination of Viral Survival

Viral survival was determined in all cases by both coat protein antigen detection and infectivity assays.

Viral Antigen Detection

Viral antigen detection was performed by double-antibody sandwich (DAS) ELISA (Clark and Adams 1977). Commercial ELISA kits for PMMV, TSWV and MNSV (Loewe Biochemica GmbH, Sauerlach, Germany) were used following the manufacturer's recommendations. Each plate also contained two known infected samples, two healthy and two buffer controls (Hill and Moran 1996). Extracts of test samples were obtained from ground leaves or stems and homogenized in commercial extraction buffer at 1:15 (w/v) (TSWV) or 1:20 (PMMV and MNSV). Absorbance was measured after 1 h (TSWV) or 1 and 3 h (PMMV and MNSV) at 405 nm in a Bio-Rad 450 microplate reader (Bio-Rad Laboratories, Hercules, CA). A sample was considered positive if the optical density was greater than twice the mean of the healthy controls (Chamberlin et al. 1992; Culbreath et al. 1993; Rodony et al. 1994). All analyses were carried out in duplication.

Viral Infectivity

Infective capacity of samples was determined by mechanical inoculation of healthy pepper (PMMV and TSWV) and melon (MNSV) plants. Viral saps were extracted from pepper and melon tissues (1.5 to 2 g) in cold 0.2 M sodium phosphate buffer pH 7.2 (PMMV), 0.01 M sodium phosphate pH 7 (TSWV) and 0.03 M sodium phosphate pH 8 (MNSV) (Pategas et al. 1989). To 1 ml of sap, 0.075 g charcoal and 0.075 g of carborundum powder were added prior to inoculation. Between two and four seedlings were rubbed with each cold extract.

Inoculated plants were kept in the greenhouse and observed for 2 months. Viral antigen detection was performed by ELISA on all test plants. Young leaves (or necrotic areas in the case of MNSV) not used in the mechanical inoculation were tested in this assay. Negative (inoculated with buffer) and positive control plants were included and handled identically.

Statistical Analysis

Data obtained throughout these studies were subjected to statistical analysis, using Statgraphics Plus 4.0 software (Manugistics, Inc. Rockville, MD).

One-way analysis of variance was performed to compare either the persistence or infectivity mean values for the different levels of sampling time and to test whether there were any significant differences amongst the means at the 95% confidence level. In order to determine which means were significantly ($P<0.05$) different from which others, multiple comparison tests (Fisher's least significant difference), were used.

Results

All experiments were started with infected plant material as shown by serological and infectivity assays (Figs. 1, 2 and 3). Positive and negative controls inoculated in all infectivity assays carried out were serologically and symptomatically positive and negative, respectively, as expected.

First Trial

At sampling times, a muslin bag containing infected material from each composting pile, was taken, divided into five subsamples and analyzed. Percentage average of positive and infective subsamples from the piles are represented in Figs. 1 and 2, for PMMV and MNSV respectively, as well as the evolution of mean temperature in composting piles. Results referred to TSWV are not shown for this experiment since this virus was not isolated.

PMMV was the most resistant pathogenic virus remaining in composted wastes up to 70 days. Viral antigen percentages decreased during the time of this assay, and only 20% of samples were ELISA positive at the end of composting (Fig. 1). Virus infective capacity inactivation was achieved before the maturation phase of composting. Thus, at 56 days composting only 13–14% of samples were positive and at the end of the assay (70 days), all samples were non-infective (Fig. 1).

Although PMMV viral antigen were found in 20% of samples at the end of the process, a significant reduction of PMMV persistence at 56 and 70 days from the beginning, could be observed. The infective capacity of PMMV infected samples decreased more rapidly than persistence. In this case, a significant infectivity decrease was observed at 14, 56 and 70 days of composting (see table of means in Fig.1).

Results for MNSV showed a significant decrease in both persistence and infective capacity after the initial 14 days of composting (Fig. 2). In respect to viral infectivity, the statistical analysis supported total elimination at 28 days. However, viral antigen disappeared significantly after 42 days of composting (see table of means in Fig.2).

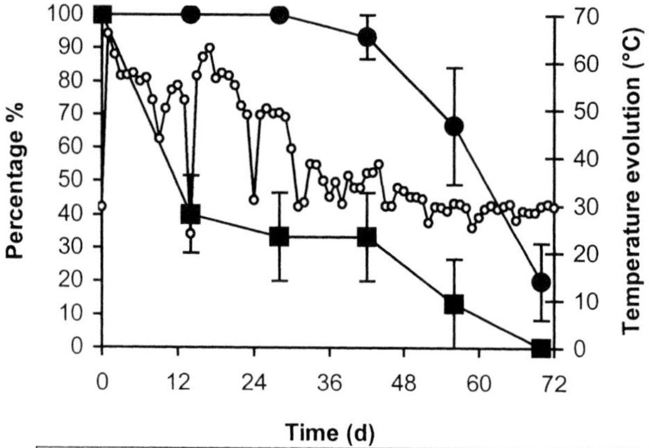

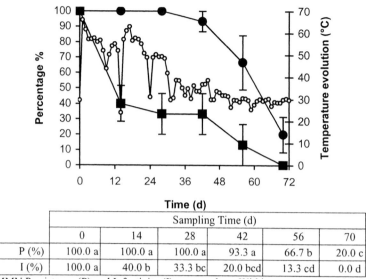

	Sampling Time (d)					
	0	14	28	42	56	70
P (%)	100.0 a	100.0 a	100.0 a	93.3 a	66.7 b	20.0 c
I (%)	100.0 a	40.0 b	33.3 bc	20.0 bcd	13.3 cd	0.0 d

PMMV Persistence (P) and Infectivity (I) mean values. Within a row, means with the same letter are not significantly different.

Fig. 1. Effect of a semipilot-scale composting (first trial) on pepper mild mottle virus antigen detection (●) and infectivity (■). Mean temperature of the three piles is included (O)

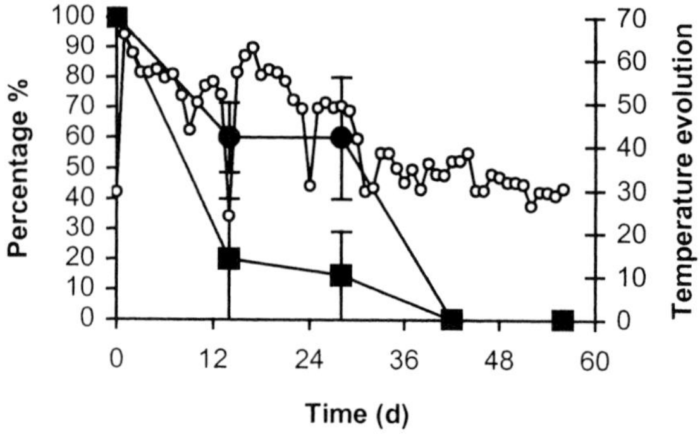

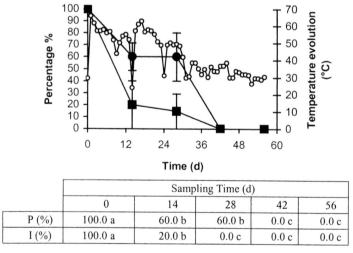

	Sampling Time (d)				
	0	14	28	42	56
P (%)	100.0 a	60.0 b	60.0 b	0.0 c	0.0 c
I (%)	100.0 a	20.0 b	0.0 c	0.0 c	0.0 c

MNSV Persistence (P) and Infectivity (I) mean values. Legend see Fig. 1.

Fig. 2. Effect of a semipilot-scale composting (first trial) on melon necrotic spotted virus antigen detection (●) and infectivity (■). Mean temperature of the three piles is included (O)

Second Trial

TSWV persistence was investigated in a second composting trial (Fig. 3). This process demonstrated a significant decrease of TSWV at 36 h of composting. After 60 h viral antigens were not detected anymore. Viral infectivity diminution was significantly reached at 12 and 60 h after the beginning of the process TSWV infectivity was totally eliminated (see table of means in Fig. 3).

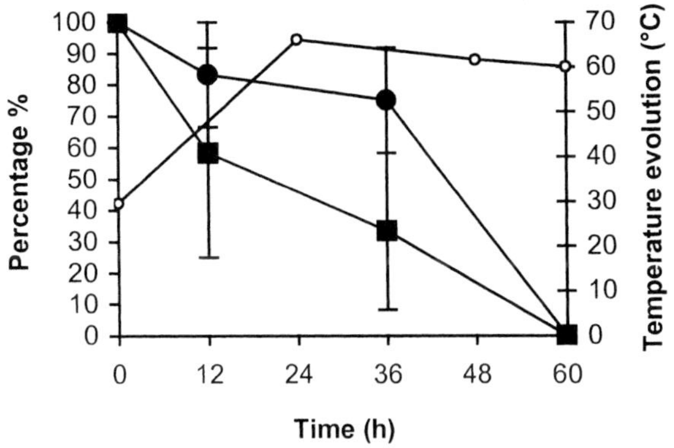

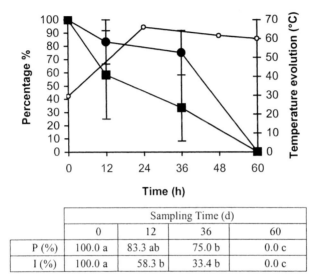

	Sampling Time (d)			
	0	12	36	60
P (%)	100.0 a	83.3 ab	75.0 b	0.0 c
I (%)	100.0 a	58.3 b	33.4 b	0.0 c

TSWV Persistence (P) and Infectivity (I) mean values. Legend see Fig. 1

Fig. 3. Effect of a semipilot-scale composting (second trial) on tomato spotted wilt virus antigen detection (●) and infectivity (■). Mean temperature of the four piles is included (O)

Results obtained in both trials suggest that there is not a perfect correlation between the presence of viral antigen and the infective capacity. In fact, the viral infective capacity is a determinant factor for the contamination potential of horticultural wastes. For this reason it is imperative to analyze both antigen persistence and viral activity (Figs. 1, 2 and 3).

Discussion

Results obtained throughout this work clearly showed a higher persistence of PMMV during the composting process (Fig. 1). On the contrary TSWV was easily eliminated during the initial three days of composting (Fig. 3). Both results presented here, pointed out the sanitation potential of composting.

In all cases, viral inactivation was achieved during the active phase of composting (before the maturation phase), suggesting the need to perform waste composting processes in which high temperatures are maintained over several weeks.

The US EPA (Environmental Protection Agency of United States of America) recommends to maintain temperatures above 40 °C or higher for 5 days. For 4 h during the 5-day period, the temperature of the compost pile must exceed 55 °C (US EPA 1979).

Our results support those of Hoitink and Fahy (1986), who pointed out the disease control possibility by horticultural waste composting. Similar results were obtained in other processes reaching temperatures up to 60–73 °C (Hoitink et al. 1976; Wijnen et al. 1983) and eliminating phytopathogens such as *Phytophthora cinnamomi, Pythium irregulare, Rhizoctonia solani, Botrytis cinerea or Sclerotinia sclerotiorum.* An investigation developed on *Pseudomonas phaseolicola* by López-Real and Foster (1985) showed bacterial inactivation at 35 °C in 4 days of composting.

Temperatures reached at the semipilot-scale fluctuated between 60 and 65 °C. Such temperatures are considered the first cause for phytopathogen inactivation in plant wastes (Ylimäki et al. 1983; Yuen and Raabe 1984). More rapid viral elimination would be possible if temperatures exceed 70 °C. However, temperatures higher than 70 °C are not recommended because they impair microbial recolonization (Golueke 1992). Therefore, results obtained at semipilot-scale are promising due to the inactivation of viruses which was significantly achieved (although in a longer time in the case of PMMV).

Our results about longevity of PMMV support those of Jarvis (1992) respect to stability of the related tobamovirus TMV (tobacco mosaic virus) in plant wastes. Wild strains of this virus infect species and cultivars of the genus *Capsicum.* These strains are economically important because they may be the most deleterious viruses in this crop (Demski 1981). This virus, like PMMV, is considered very persistent on residues left on the soil surface (Jarvis 1992). This high persistence capacity of PMMV has been verified in vitro by application of temperatures between 90 and 95 °C (Wetter et al. 1984).

Viral destruction during composting is of great importance due to the use of compost as humic amendment in the farmland or as substrate for culture in containers (Bollen 1985). Several investigations including viral phytopathogen elimination by composting support our results. Such studies have shown that tomato mosaic virus, tobacco necrosis virus and cucumber mottle mosaic virus disappeared from plant material during composting (Ylimäki et al. 1983; Hoitink and Fahy 1986; Avgelis and Manios 1992).

Beside the temperature factor, the effect of the composting upon phytopathogenic microorganisms had been explained on the basis of the combined action of several environmental factors such as toxicity of several by-products from the plant wastes decomposition, microbial activity etc. (Ylimäki et al. 1983). These factors could also explain the lower persistence of antigenic viral particles exposed to composting effect.

Therefore, this procedure is able to recycle horticultural wastes from the environment, and concomitantly suppress infective capacity of several harmful pathogens of pepper and melon crops in this country.

Acknowledgements. This study was supported by the Comisión Interministerial de Ciencia y Tecnología (CICYT), Ministerio de Educación y Ciencia, Project AMB96-1171.

References

Avgelis AD, Manios VI (1992) Elimination of cucumber mottle mosaic tobamovirus by composting infected cucumber residues. Acta Hortic 302: 311-314

Bollen GJ (1985) The fate of plant pathogens during composting of crop residues. In: Gasser JKR (ed) Composting of agricultural and other wastes. Elsevier, London, pp 282-290

Campbell RN (1996) Fungal transmission of plant viruses. Annu Rev Phytopathol 34: 87-108

Chamberlin JR, Tood JW, Beshear RJ, Culbreath AK, Demski JW (1992) Overwintering hosts and windform of thrips, Frankinella spp., in Georgia (Thysanoptera: Thripidae): implications for management of spotted wilt disease. Entomol Soc Am 21: 121-128

Clark MF, Adams AN (1977) Characteristics of the microplate method of enzyme-linked immunosorbent assay for the detection of plant viruses. J Gen Virol 34: 475-483

Conti M, Lovisolo O (1982) Virus problems in protected vegetable crops. Acta Hortic 127: 83-100

Conti M, Marte M (1983) Virus, virosi e micoplasmosi del peperone. Ital Agric 120: 132-152

Conway KE (1996) An overview of the influence of sustainable agricultural systems on plant diseases. Crop Prot 15: 223-228

Cook RJ, Baker KF (1983) The nature and practice of biological control of plant pathogens. Am Phytopathol Soc, St Paul

Cuadrado IM (1994) Las virosis de las hortalizas en los cultivos de invernadero de Almería. In: Consejería de Agricultura y Pesca (ed) Direción General de Investigación, Tecnología y Formación Agroalimentaria y Pesquera. Junta de Andalucía, Sevilla

Cuadrado IM, Alonso ES, de Juan E (1992) Diagnóstico específico de razas del virus del mosaico del tabaco (TMV) en pimiento. In: Consejería de Agricultura y Pesca (ed) Identificación de las razas del TMV que están afectando los cultivos de pimiento en Almería. Información técnica 20/92. Junta de Andalucía, Sevilla

Culbreath AK, Bertrand PF, Csinos AS, McPherson RM (1993) Effect of seedling source on incidence of tomato spotted wilt in flue-cured tobacco. Tobacco Sci 37: 9-10

Demski JW (1981) Tobacco mosaic virus is seedborne in pimiento peppers. Plant Dis 65: 723-724

Fletcher JT (1984) Diseases of greenhouse plants. Longman, London

Gea M (1996) Comparación de las técnicas serológicas, E.L.I.S.A. e inmunosupresión, para la detección del virus del bronceado o TSWV (Tomato Spotted Wilt Virus) en plantas de pimiento y tomate. Proyecto monográfico Fin de Carrera. E.P.S. Universidad de Almería, Almería

Golueke CG (1992) Bacteriology of composting. Biocycle Jan 55-57

Gómez J (1993) Enfermedades de melón en los cultivos "sin suelo" de la provincia de Almería. In: Junta de Andalucía (ed) Comunicación I+D Agroalimentaria 3/93, Sevilla

Hill MF, Moran JR (1996) The incidence of tomato spotted wilt tospovirus (TSWV) in Australian nursery plants. Aust Plant Pathol 25: 114-119

Hoitink HAJ, Boehm MJ (1999) Biocontrol within the context of soil microbial communites: a substrate-dependent phenomenon. Annu Rev Phytopathol 37: 427-446

Hoitink HAJ, Fahy PC (1986) Basis for the control of soilborne plant pathogens with composts. Annu Rev Phytopathol 24: 93-114

Hoitink HAJ, Herr LJ, Schmitthenner AF (1976) Survival of some plant pathogens during composting of hardwood tree bark. Phytopathology 66: 1369-1372

Hoitink HAJ, Stone AG, Han DY (1997) Suppression of plant diseases by compost. HortScience 32: 184-87

Jarvis WR (1992) Managing diseases in greenhouse crops. Am Phytopathol Soc, St. Paul

Kostov O, Tzvtkov Y, Petkova G, Lynch JM (1996) Aerobic composting of plant wastes and their effect on the yield of ryegrass and tomatoes. Biol Fertil Soils 23: 20-25

López-Real J, Foster M (1985) Plant pathogen survival during composting of agricultural wastes. In: Gasser JKR (ed) Composting of agricultural and other wastes. Elsevier, London, pp 291-300

Lumsden RD, Lewis EA, Papavizas GC (1983) Effect of organic amedments on soilborne plant diseases and pathogen antagonist. In: Lockeretz W (ed) Environmentally sound agriculture. Praeger, New York, pp 51-70

Marte M, Wetter C (1986) Occurrence of pepper mild mottle virus in pepper cultivars from Italy and Spain. J Plant Dis Prot 93: 37-43

Pares RD, Gunn LV (1989) The role of non-vectored soil transmission as a primary source of infection by pepper mild mottle and cucumber mosaic viruses in glasshouse-grown capsicum in Australia. J Phytopathol 126: 353-360

Pategas KG, Schuerger AC, Wetter C (1989) Management of tomato mosaic virus in hydroponically grown pepper (Capsicum annuum). Plant Dis 73: 570-573

Rodony C, Hepworth G, Richardson C, Moran JR (1994) The use of sequential bath testing procedure and ELISA to determine the incidence of five viruses in Victorian cut–flower Sim carnations. Aust J Agric Res 45: 223-230

Sáez LE (1989) Contenido fenólico del alpechín y actividad antibacteriana. Memoria de Licenciatura. Universidad de Granada, Granada

Tello JC, Gómez J, Salinas J, Lacasa A (1987) La fusariosis vascular del melón en los cultivos de Almería. Cuad Fitopatol 10: 31-37

US EPA (1979) Preproposal draft regulation on distribution and marketing of sewage sludge products. US EPA 40 CFR part 257 Washington, D.C. Sept. 1979

Wetter C, Conti M (1988) Pepper mild mottle virus. AAB descriptions of plant viruses. Association of Applied Biologists, Wellesbourne, Warwick,No 330

Wetter C, Conti M, Altschuh D, Tabillion R, Van Regenmortel MHV (1984) Pepper mild mottle virus, a tobamovirus infecting pepper cultivars in Sicily. Phytopathology 74: 405-410

Wijnen AP, Volker D, Bollen GJ (1983) De lotgevallen van pathogene schimmels in een composthoop. Gewasbescherming 14:5

Ylimäki A, Toiviainen A, Kallio H, Tikanmäki E (1983) Survival of some plant pathogens during industrial-scale composting of wastes from a food processing plant. Ann Agric Fenn 22: 77-85

Yuen GY, Raabe RD (1984) Effects of small-scale aerobic composting on survival of some fungal plant pathogens. Plant Dis 68: 134-136

Air-Borne Emissions and Their Control

Composting Conditions Preventing the Development of Odorous Compounds

E. Binner, D. Grassinger and M. Humer[1]

Abstract. Laboratory tests allow to compost biowaste under defined boundary conditions. By varying these conditions we found out parameters responsible for the production of odorous compounds during the composting process. Parameters describing the composting conditions were pH value, content of fatty acids in the rotting material and CO_2 production. The most important influences to prevent the production of odorous compounds are fast and careful mechanical pretreatment of the biowastes and sufficient well oxygen supply.

Introduction

Organic wastes are the most important component of secondary materials in municipal waste. Since 1995 it has been mandatory to collect organic wastes separately in Austria (Verordnung über die getrennte Sammlung biogener Abfälle 1992). Therefore Austria has had to develop a system for separate collection. Organic household wastes (or biowaste) such as kitchen waste as well as grass and leaves from gardens are collected in so-called biobins, a secondary waste-collecting system in Austria's municipalities. Large-sized garden wastes must be brought to so-called collecting centres. In 1995 the quantity of organic wastes collected separately in Austria amounted to approximately 1.4 million Mg year^{-1}. The potential is approximately 2.2 million Mg year^{-1}.

Most of the organic waste is composted in open windrow systems. Odour emissions and/or quality failures of the final compost are common due to processing demands. For this reason, Department 22 (MA22) of the City of Vienna has financed a series of laboratory and field tests in a composting facility to explain the influences on the development of odorous compounds during the composting process (Binner et al. 1997). This chapter presents the results on the influence of oxygen supply.

A number of laboratory tests were performed with biowaste under defined conditions. By varying these conditions (pretreatment of the feed mixture, pH value, temperature and oxygen supply) during the intensive degradation phase, we tried to identify factors responsible for producing odorous compounds. We analyzed the following physical and chemical parameters: the pH value (Ö-NORM S2023), content of organic acids (fatty acids, method Mostbauer and Balderas 1991) and ammonia nitrogen (Ö-NORM S2023) in the composting material.

---

[1]Department of Waste Management (ABF-BOKU), Universität für Bodenkultur- Wien, Nußdorfer Lände 29-31, 1190 Vienna, Austria

H. Insam, N. Riddech, S. Klammer (Eds.)
Microbiology of Composting

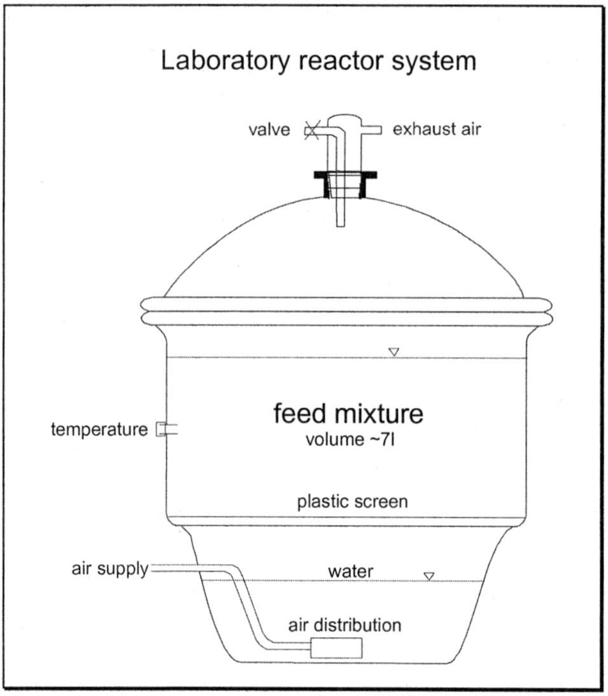

Fig. 1. Laboratory reactor system (Grassinger 1998)

In varying the methods of Ö-NORM S2023, we used a solid-to-liquid (deionized water) ratio of 1:10 for all parameters in order to avoid unnecessary wastage of the sample amount. In the exhaust air (= waste air), we measured CO_2, odour emissions (by olfactometry) and organic acids.

The following boundary conditions were varied:

- Feed mixtures (biowaste with and without bulking agents, and/or with and without the addition of lime). The effects on addition of lime are not shown in this chapter.
- Temperature (30 and 45 °C, as our climate room did not allow for a higher temperature), not shown in this chapter; and
- Oxygen supply.

As test reactors (Fig. 1), we used six adapted desiccators (volume ~ 7 l, Ø = 30 cm) located in a climate room (0 to 45 °C). The air supply was measured by flow meters and blown in through a bath of water at the bottom of the test reactor to prevent the compost from drying. An IR gas analyzer measured the CO_2 content of the exhaust air, which is released from the top of the test reactor, every 2 h (Fig. 2). The amount of air taken into the test reactor was regulated by the CO_2 content of the exhaust air. For sufficient oxygen supply, CO_2 was regulated at < 2 vol%, while it was regulated at > 15 vol% for an insufficient oxygen supply.

The feedstock in the laboratory tests was a mixture of separate collection from biobins in Vienna and shredded garden waste (weight 80:20). For the field test, the content of garden waste was raised to > 45 % (weight).

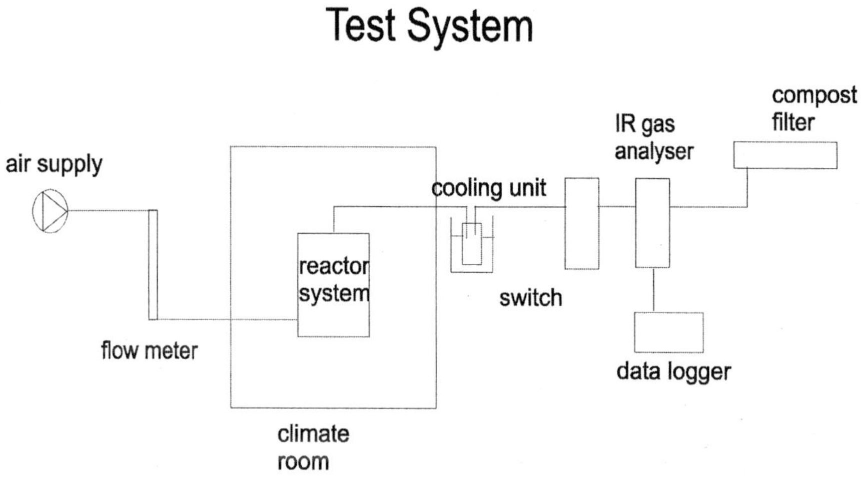

Fig. 2. Laboratory test system (Grassinger 1998)

Due to the limited capacity of the climate room, it was necessary to stabilize the feedstock through drying. When interpreting the results, one has to consider that drying and rewetting the feedstock reduces the biological processes as well as odour emissions. One also has to realize that laboratory tests proceed much faster than naturally aerated windrows. Therefore, the test results permit only a qualitative comparison of the different boundary conditions.

Fundamental Perspectives

Before the start of the decomposition process, there is fairly long lag phase. The duration of this lag phase depends upon the treatment and storage of the feedstock, the oxygen supply, temperature and other factors. During this phase, a low level of CO_2 is produced in combination with a low pH value (≤ 5.5) and a high organic acids (OA) content (Fig. 3).

As the organic acid level decreases, the pH value increases. The temperature rises (as a result of self-heating), and the degradation of organic matter starts exponentially (the CO_2 content in the waste air rises). If we know the concentrations of the organic acids – used as "guide" substances – it is possible to calculate an odour potential (O_{pot}) of the rotting material. Therefore the odour index for each organic acid (Verschueren 1983) has been multiplied by its concentration.

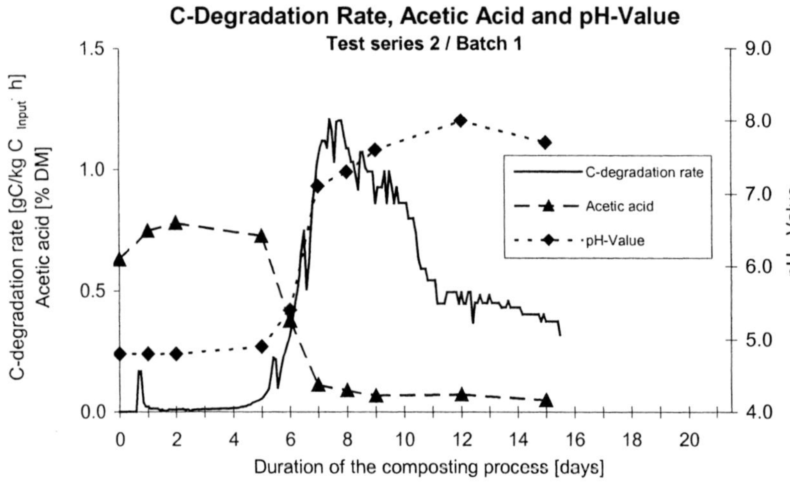

Fig. 3. Relationship between degradation rate (CO_2 content in the waste air), pH value and organic acids (for example acetic acid) in the solid substance during intensive degradation phase

All of these odour values are added to the odour potential, an absolute measure characterizing the potential of possible odour emissions from composting material (Binner and Nöhbauer 1994) which could be transferred to the gaseous phase under unfavourable conditions (e.gturning the material). Thanks to the relationships described above (Fig. 3), it seems possible to conduct an adequately accurate assessment of the odour potential of biowaste by calculating the characteristic courses of the pH value and CO_2 production rates.

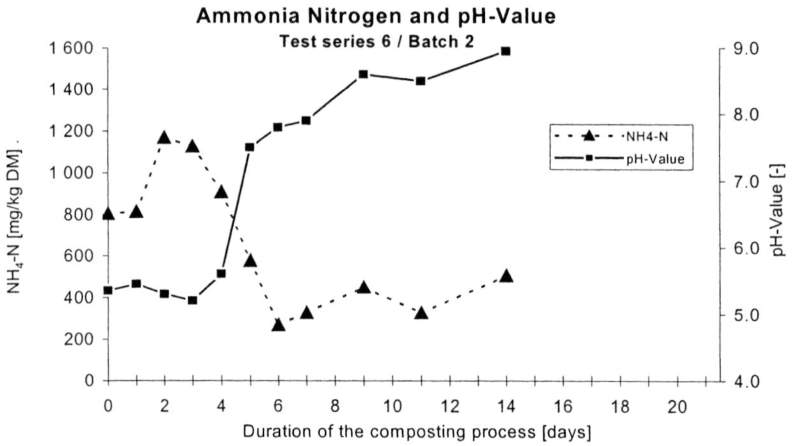

Fig. 4. Relationship between pH value and concentration of ammonia nitrogen during the intensive degradation phase

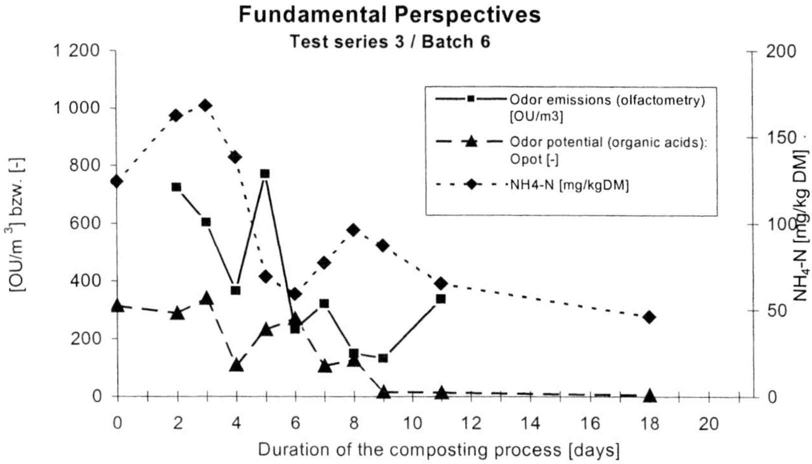

Fig. 5. Characteristic course of odour emissions (olfactometry), odour potential (organic acids in solid substance) and concentration of ammonia nitrogen during the intensive degradation phase

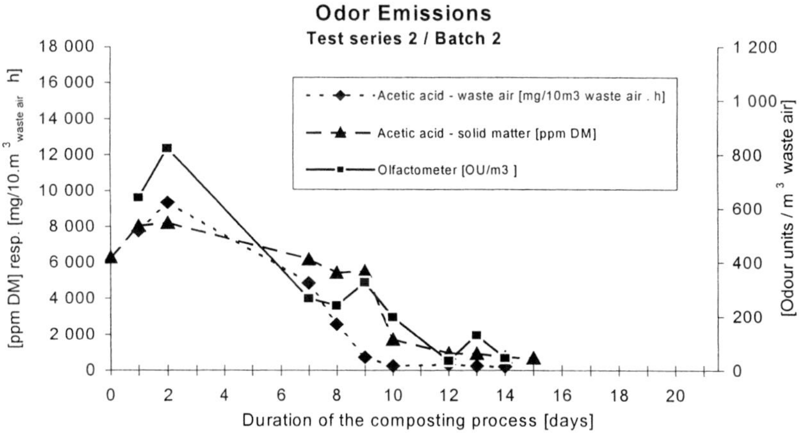

Fig. 6. Relationship between odour emissions (olfactometry) and organic acids (acetic acid) in waste air and solid substance during the intensive degradation phase

The course of ammonia nitrogen is similar to that of organic acids (Fig. 4). The decrease in NH_4-N depends mainly on emission of ammonia (NH_3). The transfer of ammonia to the gaseous phase starts at pH values of about 7 and increases rapidly at pH values greater than 8. NH_3 leads to high odour emissions, which cannot be described by calculating the odour potential of organic acids. This is one reason why odour potential (measured by organic acids) and odour emissions (odour units = OU, measured by olfactometry) sometimes do not correlate. Decreasing levels of organic acids (= decreasing odour potential) raise the pH value and the emission of NH_3, which in turn increases the odour emissions (Fig. 5).

Tandem to analyzing the solid substance, we also measured on the exhaust air. Odour emissions can be analyzed by olfactometry. Figure 6 shows the (parallel) relationship between olfactometric method and the content of organic acids (for example, acetic acid) in solid substance and waste air. Such a parallel relationship occurs when the air is evenly spread over the material. If there are channels in the material or between the material and the wall of the reactor, the odour potential is not transferred to the exhaust air. In this case we find a high odour potential, but low odour emissions (as well as a low content of organic acids and CO_2 in the exhaust air). This is why there is often no parallel relationship between odour potential and olfactometry. During our field tests, we tried to take waste air samples (using lances) as well as solid samples from the same location in the windrow. In this instance the correlation coefficient between the odour units (olfactometry) and odour potential (organic acids) was $r = 0.781$ (Blochberger 1998).

Results

It is possible to characterize the quality of the oxygen supply by interpreting the (carbon) degradation rates (duration of the lag phases) and the course of the pH value. A sufficient and even oxygen supply (= high O_2 in Fig. 7) leads to short lag phases and hastens the degradation of organic substances and the increase in pH. The concentrations of organic acids and ammonia decrease faster, which shortens the phase of high odour emissions.

While concentrations of acetic acid and odour emissions (olfactometry), which are both measured in the exhaust air, often show large differences (independent of the quality of the oxygen supply), the calculation of the odour potential (by analyzing organic acids in the solid substance) does a good job of characterizing the milieu conditions. The results of odour potential during the laboratory tests correspond to the conditions found at the composting facilities (field tests): a high odour potential during the lag phase and the first part of the intensive degradation phase, a decrease during the intensive degradation phase and low odour potential following the intensive degradation phase.

A sufficient and even oxygen supply shows a strong effect on the duration of the high odour emissions, but no effect on the amount of the emissions (Fig. 8). Independent of the oxygen supply, a similar trend can be seen in the odour potential (OA), as well as the odour emissions in the waste air (olfactometry). When the oxygen supply is insufficient, acetic acid in the solid substance, odour emissions (Fig. 9) and odour potential (Fig. 8) remain at a high level for a much longer duration.

As field tests show (Blochberger 1998), it is possible to assess the oxygen supply of naturally aerated windrows by taking air samples from the windrow with measuring lances. Oxygen (Fig. 10) and carbon dioxide can be detected using portable gas analyzers. Some analyzers also allow methane detection. Under aerobic conditions the sum of O_2 + CO_2 is about 20.8 vol% (Fig. 11). When the amount exceeds 20.8 vol%, the composting conditions worsen. This insufficient oxygen supply leads to anaerobic processes in parts of the composting material, and the odour potential rises. By selecting a large number of measuring points, it is possible to assess the milieu conditions for the entire windrow.

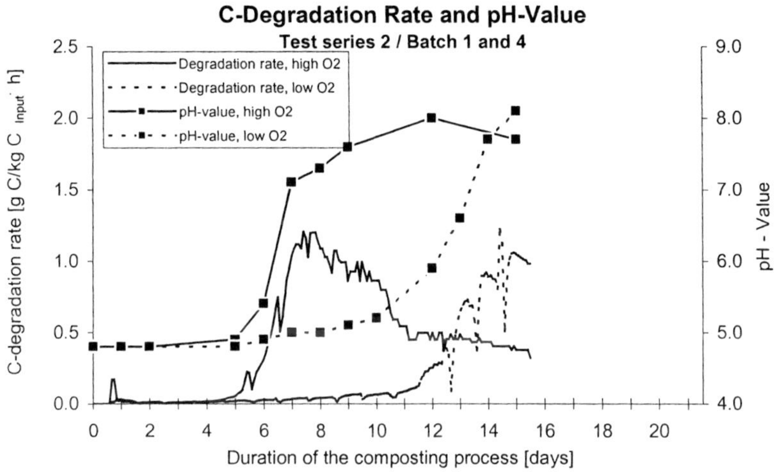

Fig. 7. Influence of oxygen supply on degradation rate and pH value during the intensive degradation phase (temperature 45 °C). *High* O_2 is used as a synonym for sufficient and even oxygen supply; *low* O_2 for insufficient and uneven oxygen supply

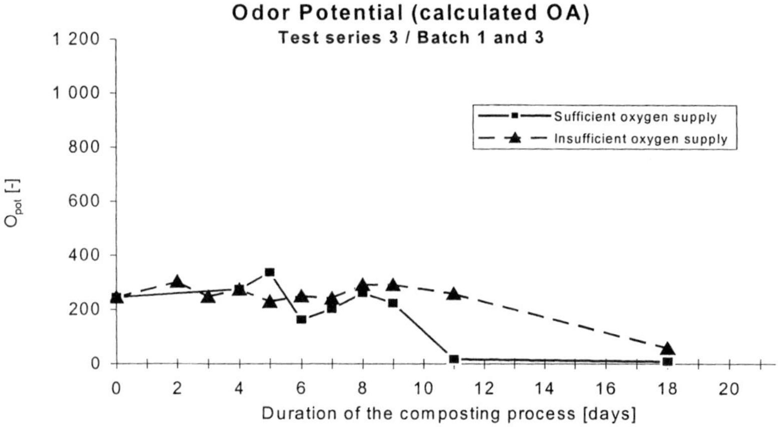

Fig. 8. Influence of oxygen supply on odour potential during the intensive degradation phase (temperature 45 °C)

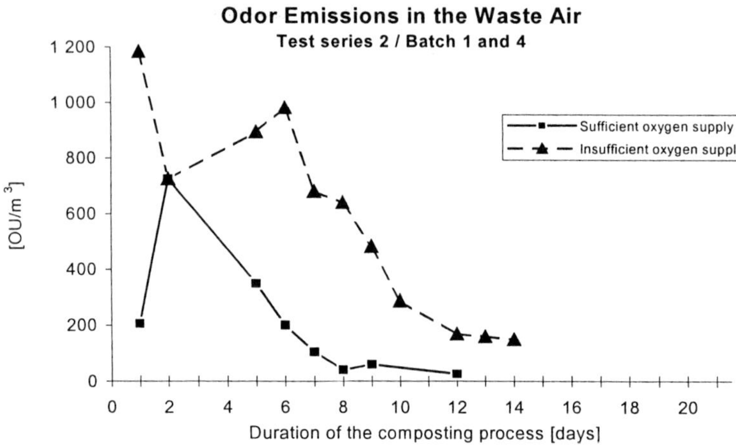

Fig. 9. Influence of oxygen supply on odour emissions in the waste air (olfactometry) during the intensive degradation phase (temperature 45 °C)

Blochberger (1998) found that pretreatment of the feedstock influences the duration of high odour potential (Fig. 12) just as well as the oxygen supply. Careful pretreatment leads to a higher level of FAS (=free air space) in the feedstock, which in turn allows a more sufficient oxygen supply.

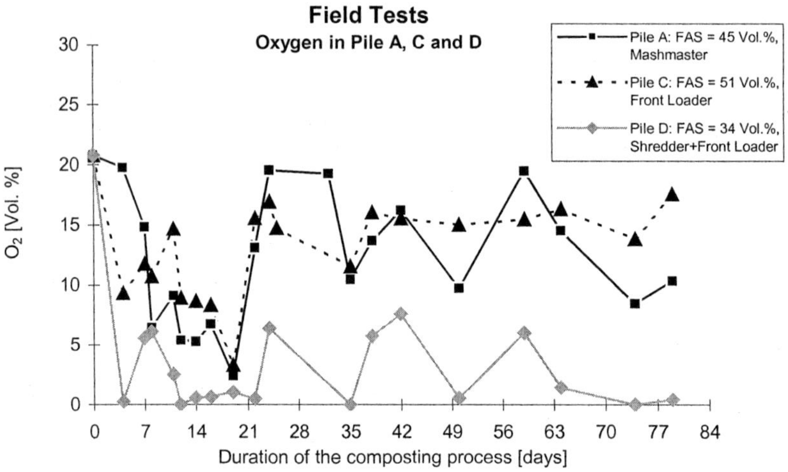

Fig. 10. Influence of pretreatment on *FAS* (= free air space) and concentration of oxygen in naturally aerated windrows. Pretreatment: *Pile A* grinding and mixing by Mashmaster; *Pile C* mixing by front loader; *Pile D* grinding by shredder and mixing by front loader

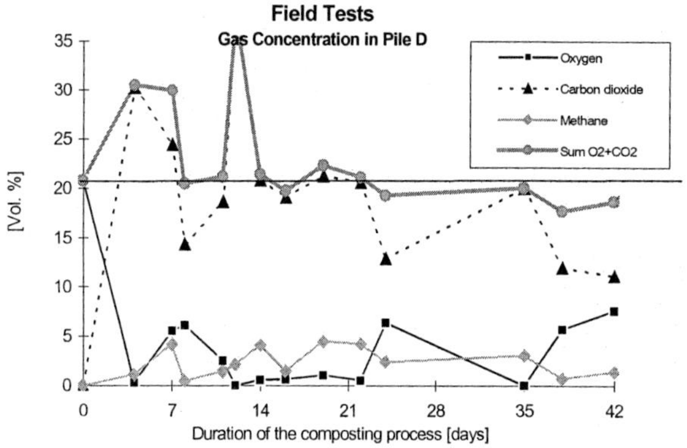

Fig. 11. Concentrations of oxygen, carbon dioxide and methane in pile D (grinding by shredder and mixing by front loader) and calculated sum of O_2+CO_2

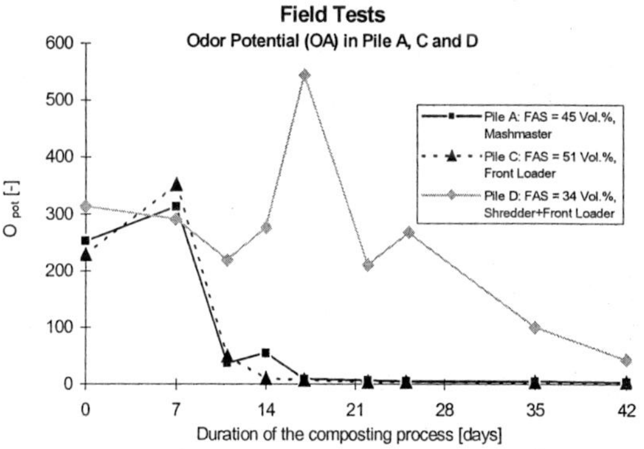

Fig. 12. Influence of pretreatment on *FAS* (= free air space) and odour potential in naturally aerated windrows. Pretreatment: *Pile A* grinding and mixing by Mashmaster; *Pile C* mixing by front loader; *Pile D* grinding by shredder and mixing by front loader

Conclusions

The physical and chemical analysis of rotting material and exhaust (waste) air (46 tests / 7 test series + 4 windrows in field tests) showed a very good correlation between the duration of the lag phases, degradation rates, pH, concentrations of ammonia nitrogen and organic acids (fatty acids), odour potentials (calculated by the amounts of organic acids) and odour emissions in the waste air (olfactometry).

Short lag phases lead to high degradation rates in combination with a rapid increase in the pH value as well as a rapid decrease in ammonia and organic acids. These represent important factors for lower odour potential.

To minimize odour emissions,

- an efficient and careful pretreatment of feedstock is important;
- a sufficient and even oxygen supply is necessary (preventing channels in the composting material);
- high aeration rates do not always guarantee positive effects (channels, dry and cold zones);
- during windrow composting the quality of the oxygen supply can be assessed by measuring O_2, CO_2, CH_4 in the windrow (samples taken by measuring lance). If the sum of $O_2 + CO_2$ exceeds 20.8 vol%, the composting conditions worsen. This insufficient oxygen supply leads to anaerobic processes in parts of the composting material.

The results (influence of O_2 supply and pH) of laboratory tests are verified by practical tests in composting facilities (Binner et al. 1997; Blochberger 1998). Laboratory tests (due to forced acration and optimized conditions) proceed much faster than practical tests. High odour potential lasts only 3 to 4 d under optimized laboratory conditions, whereas in naturally aerated windrows the critical phase lasts 1 to 2 weeks or even longer.

References

Anonymus (1992) Verordnung des Bundesministeriums f. Umwelt, Jugend und Familie über die getrennte Sammlung biogener Abfälle. BGBl. 68/1992, Vienna

Binner E, Nöhbauer F (1994) Bestimmung des Geruchspotentials von Rottegut – ein neuer Ansatz zur Bewertung der Geruchssituation bei Kompostanlagen. Österreichische Wasser- und Abfallwirtschaft, Jahrgang 46, Heft 9/10, Springer, Wien, pp239-244

Binner E, Humer M, Grassinger D, Blochberger F, Lechner P (1997) Einflüsse auf die Geruchsfreisetzung während der Kompostierung. Research program ordered by City of Vienna, Department MA22, Vienna

Blochberger F (1998) Einfluß der Materialaufbereitung auf Geruchspotential und Rotteverlauf bei der Mietenkompostierung von Bioabfällen. Diplomarbeit am ABF-BOKU, Universität für Bodenkultur Wien (Vienna)

Grassinger D (1998) Einfluß von Temperatur und Sauerstoffgehalt auf die Humifizierung und die Mineralisierung bei der Verrottung von Bioabfall. Dissertation am ABF-BOKU, Universität für Bodenkultur Wien (Vienna)

Mostbauer P, Balderas G (1991) Gaschromatographische Bestimmung Niederer Carbonsäuren in Kompost und Sickerwässern. GIT Spezial Chromatographie 1/91: pp 4-9

Ö-NORM S2022 (1.11.1993) Quality requirements for biowaste compost – marking of conformity. Austrian Standards Institute, Vienna

Ö-NORM S2023 (1.11.1993) Analytic methods and quality control of compost. Austrian Standards Institute, Vienna

Verschueren K (1983) Handbook of environmental data on organic chemicals. 2nd edn., Van Nostrand Reinhold, New York

Odour Emissions During Yard Waste Composting: Effect of Turning Frequency

N. Defoer and H. Van Langenhove[1]

Abstract. In Flanders, yard waste (270 000 tons year^{-1}) is collected separately and treated in 15 open-air composting plants. Although the method is widely used, few systematic scientific studies have been done on factors controllable by the compost operator who processes yard trimmings. These factors include turning frequency and its effect on odour generation. To obtain a clearer view on the point of turning frequency and odour production, a monitoring campaign was set up at three yard waste composting plants with different turning frequencies (plant A: every 6 weeks, plant B: every week and plant C: every 8 weeks).

Two different odour measurement strategies were applied. First of all, in order to evaluate the total impact of composting plants, sniffing team measurements were carried out. Secondly, on each plant one windrow was followed up by olfactometric measurements from the beginning of the composting process during 3 months.

The variation in the sniffing team measurements was high and due to this fact it was difficult to draw conclusions concerning the effect of the turning frequency on the odour emission. For the different activities on a composting plant (rest, size reduction, sieving and turning), however, we could say that the highest odour emission occurs during size reduction and turning.

The results of the olfactometric measurements were more obvious. Here, we saw that the composting plant with the highest turning frequency had the lowest odour emission expressed in ou_E ton compost^{-1}h^{-1}. Also the lower height of the compost windrows of plant B can be an additional explanation for the lower odour emission of plant B. Furthermore, the olfactometric results showed that the highest odour emissions occurred during the first 10 days of the composting process and the moments immediately after turning.

Introduction

In Flanders, yard waste (270 000 tons year^{-1}) is collected separately and treated in 15 open-air composting plants. Although the method is widely used, few systematic scientific studies have been done on factors controllable by the compost operator who processes yard trimmings. These factors include turning frequency and its effect on odour generation. Turning frequency is commonly used to improve the

[1]Department of Organic Chemistry, Faculty of Agricultural and Applied Biological Sciences, University of Gent, Coupure Links 653, 9000 Gent, Belgium

H. Insam, N. Riddech, S. Klammer (Eds.)
Microbiology of Composting

its effect on odour generation. Turning frequency is commonly used to improve the rate of composting. However, composting costs are strongly affected by the number of times the material is turned, and odours are released during turning (Michel and Reddy 1996).

The main question for the compost operators still remains: To minimise odour emissions during composting, is it better to turn frequently or infrequently? Unfortunately, there is no simple answer. On the one hand, turning releases odorous compounds trapped within the windrow. One the other hand, lack of turning allows anaerobic conditions, and the associated odorous compounds can proliferate within the compost windrow. This increases the severity of odours that are released when the windrow is eventually disturbed.

To obtain a clearer view on the point of turning frequency and odour production, a monitoring campaign was set up at three Flemish yard waste-composting plants with different turning frequencies.

Material and Methods

The Monitoring Campaign

In the monitoring campaign, three different yard waste-composting plants with different turning frequencies were followed up. Some general information about the turning frequencies, the duration of the composting process, the height of the windrows and the composting capacity for the different examined composting plants is given in Table 1.

Table 1. Turning frequency, duration of composting process, height of windrows and composting capacity of the three examined composting plants

Composting Plant	Turning frequency	Duration of process	Height of Windrows (m)	Composting Capacity (ton $year^{-1}$)
A	Every 6 weeks	6 to 8 months	4	15000
B	Every week	3 to 4 months	2	6000
C	Every 8 weeks	5 to 6 months	3.5	25000

Two different odour measurements were applied. First of all, in order to evaluate the total impact of the composting plants, sniffing team measurements were carried out. Secondly, on each plant one windrow was followed up from the beginning to the end of the composting process by olfactometric measurements.

Sniffing Team Measurements

Sniffing team measurements allow the evaluation of the total (including diffusive sources) odorous impact on the neighbourhood, but do not allow discriminating the relative importance of different emission points of a source. This approach has been used by the research group for several years and, although it is not as standardised as olfactometric measurements, it gives reasonable results. Over the years, we have developed our own standard operational procedures, which results in intrapanel standard deviations of 20–30%.

During a sniffing team measurement, one walks around the composting plant starting downwind and afterwards going to the composting plant perpendicular to the wind direction. Every time odour is noticed, it is noted on a topographical map. Afterwards, the central axis of the odour spot can be drawn. The maximum distance of odour observation along this axis, is called the maximum distance of odour perception (MDOP) or the sniffing border. With this distance and the meteorological conditions during the sniffing team measurements, one can calculate the overall odorous emission, which is expressed in sniffing units per second (su s^{-1}), using a short-term atmospheric dispersion model (Van Broeck and Van Langenhove 1999).

Olfactometric Measurements

Sampling Method

The aim of this study was to obtain odour emissions for the different composting plants expressed in ou_E ton compost^{-1} h^{-1}. Expressing the emission per weight unit of compost allows a comparison between the different composting plants. To achieve this unit, not only the odour concentration was needed but also the flow rate of the air emitted by the compost windrows. Compost windrows are naturally ventilated surface sources, which makes it difficult to measure the flowrate of the outcoming air. Furthermore flow rates are very low. In order to convert the surface source to a guided source, a part of the compost windrow was covered with a measuring tent. In Fig. 1 a picture of a measurement with the tent is given.

Fig. 1. Picture of the measuring tent

The velocity of the air in the PVC pipe was measured with a vane anemometer. The tent covered a surface of 9 m^2. By measuring also the height of the windrows and the density of the compost, the odour emission expressed in ou_E ton $compost^{-1}$ h^{-1} could be calculated. This measurement procedure was based on the work of Derickx et al. (1990) and Sterckx (1998).

Olfactometric Measurements

For the olfactometric measurements a method was used according to CEN (1998). Hereby, the odour concentration of a gaseous sample is determined by presenting a panel of selected and screened human persons with that sample. Varying concentrations were presented by diluting samples with neutral gas, in order to determine the dilution factor at the 50% detection threshold. At that dilution factor the odour concentration is 1 ou_E m^{-3} by definition. The odour concentration of the examined sample is then expressed as a multiple (equal to the dilution factor) of one European Odour Unit per cubic metre (ou_E m^{-3}) at standard conditions for olfactometry. In CEN (1998) the threshold for the reference compound is set at 40 ppbv.

Results

Sniffing Team Measurements

On a yard waste-composting plant different activities take place and they all are considered to cause different odour emissions. When the yard waste arrives at the composting plant, the size of the material is reduced with a shipper or a grinder. There are different reasons for the size reduction; including volume reduction and increasing surface area for more rapid decomposition (Norstedt et al. 1993). From time to time the composting material is turned to speed up the composting process. At the end of the composting process the composted material is sieved in two or three fractions. During the different activities, sniffing team measurements were carried out around three composting plants to see if there was a difference between them. The results of these sniffing team measurements are given in Tables 2 - 4.

Table 2. Results of sniffing team measurements for composting plant A

Date	MDOP (m)	Odour emission (su s^{-1})	Activity	Date	MDOP (m)	Odour emission (su s^{-1})	Activity
Dec. 1	640	59 193	Rest	Mar. 16	560	35 100	Rest
Dec. 29	580	76 388	Rest	Jan. 11	200	12 068	Sieving
Jan. 28	400	25 856	Rest	Dec. 14	750	42 461	Size reduction
Jan. 31	430	28 814	Rest	Feb. 24	760	43 314	Size reduction
Feb. 17	480	44 829	Rest	Jan. 26	600	24 137	Turning

MDOP = maximum distance of odour perception

MDOP = maximum distance of odour perception

Table 3. Results of sniffing team measurements for composting plant B

Date	MDOP (m)	Odour emission ($su\ s^{-1}$)	Activity	Date	MDOP	Odour emission ($su\ s^{-1}$)	Activity
Dec. 2	400	27 193	Rest	Feb. 17	290	12 630	Sieving
Dec. 13	340	16 039	Rest	March 2	450	40 078	Sieving
Dec. 29	320	26 596	Rest	March 16	360	19 771	Sieving
Jan. 20	370	14 929	Rest	March 20	120	5 008	Sieving
Dec. 9	400	74 343	Turning	March 30	180	3 139	Sieving
Jan. 20	390	19 714	Turning	Dec. 22	530	25 633	Size reduction
Feb. 22	160	7 612	Turning	Jan. 25	350	40 332	Size reduction

Table 4. Results of sniffing team measurements for composting plant C

Date	MDOP (m)	Odour emission ($su\ s^{-1}$)	Activity	Date	MDOP	Odour emission ($su\ s^{-1}$)	Activity
Oct. 25	910	117 344	S & R[a]	Feb. 4	660	61 993	Sieving
Feb. 11	510	48 737	S & R	March 17	460	20 382	Sieving
April 20	390	12 651	S & R	Dec. 13	650	90 669	S & T[b]
May 9	300	36 864	S & R	Jan. 6	370	23 008	S & T
Nov. 30	650	78 149	Sieving	April 11	430	36 693	S & T
Jan. 14	420	8 664	Sieving	March 3	350	71 288	Size reduction

[a] S & R: sieving and size reduction. [b] S & T: sieving and turning.

In Table 5 an overview is given of the average, the median, the P25 (25 percentile) and P75 (75 percentile) for all the sniffing team measurements at the three examined composting plants.

Table 5. Overview of average, median, P25 and P75 of MDOP (m) and odour emission ($su\ s^{-1}$) for the three composting plants

Activity	Plant A		Plant B		Plant C	
	MDOP (m)	Emission ($su\ s^{-1}$)	MDOP (m)	Emission ($su\ s^{-1}$)	MDOP (m)	Emission ($su\ s^{-1}$)
Average	540	39 216	323	23 787	508	50 537
Median	570	38 781	345	19 743	445	42 801
P25	442	26 596	257	13 205	385	22 352
P75	630	44 450	392	27 044	650	73 003
P25/median	0.78	0.69	0.75	0.67	0.87	0.52
P75/P25	1.11	1.67	1.14	2.05	1.69	3.27

Olfactometric Measurements

For each composting plant one compost windrow was followed up from the start during 3 months. Every week, an olfactometric sample was taken and the measurements were done more frequently during periods of turning of the compost windrows. The results from the olfactometric measurements on the three composting plants are given in Figs. 2 and 3. The arrows in the figures refer to the moments of turning at the respective companies. Composting plant A was followed up for 120 d, composting plant B for 103 d and composting plant C for 98 d.

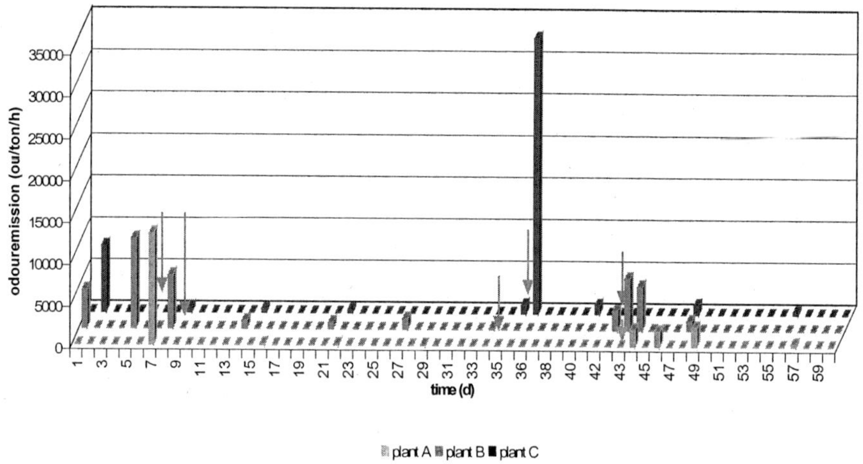

Fig. 2. Olfactometric results for the three different yard waste-composting plants (day 1–day 60)

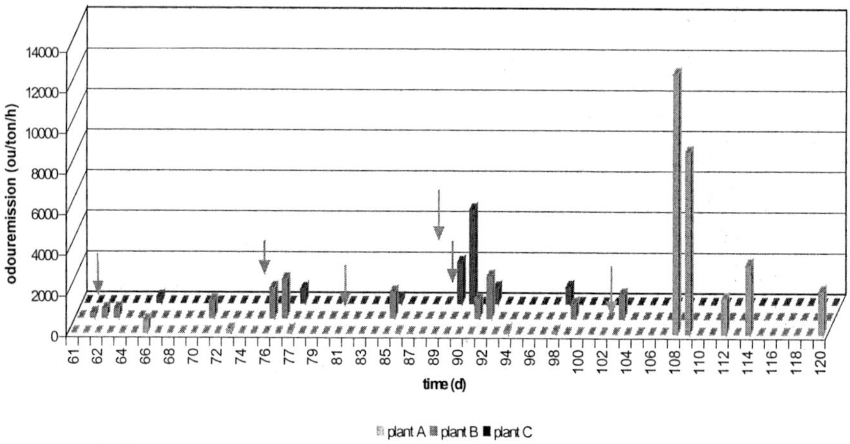

Fig. 3. Olfactometric results for the three different yard waste-composting plants (day 61–day 120)

Discussion

Sniffing Team Measurements

When the results of the sniffing team measurements in Tables 2 - 4 are considered, there is a great variation in the results for the same activity, both for the MDOP as well as for the odour emissions. So with the present limited number of results per activity, it is not possible to distinguish activities which result in significantly larger odour emissions than others. Further, the recalculation of the MDOP to an odour emission with a short-term atmospheric model introduces extra variability since the difference between the rate P25/median and P75/P25 is larger for the calculated odour emissions than for the MDOP. Basic reasons for this phenomenon are firstly that the model uses stability classes as input instead of a continuous stability parameter, secondly it is sometimes difficult to obtain the appropriate meteorological data (especially stability parameters) for the site under investigation.

When the odour emissions of the different composting plants during different activities are taken into consideration, the conclusion can be made that, in general, size reduction and turning causes the highest odour emissions. For all three composting plants the results are the lowest for sieving, although we expected that during rest the lowest values would be measured.

When the average results in Table 5 are considered, it can be concluded that composting plant B has the lowest average odour emission and MDOP. This is logical because composting plant B has the lowest composting capacity (Table 1). For composting plants A and C, the conclusion is not that clear because composting plant C has the highest composting capacity and also the highest mean odour emission but when we look at the MDOP, composting plant A has the highest average MDOP although it does not have the highest composting capacity.

Olfactometric Measurements

As can be seen in Fig. 2, the odour emission is high for the three composting plants during the first 10 days of the composting process (about 5000–10000 ou_E ton^{-1} h^{-1}). These emissions were probably, at least partially, the result of the size reduction of the yard waste before it was set on compost windrows. During size reduction, odours from the yard waste material itself are emitted. Also Kryzmien et al. (1999) and a Dutch study carried out by TNO (1994) mention that the odour emission is high during the first 2 weeks of the composting process. Afterwards, there is a fast decline of the odour emission to the same range for all three plants (100–700 ou_E ton^{-1} h^{-1}).

Iacoboni et al. (1980) evaluated odour emissions during the composting of biosolids in windrows. They stated that the highest odour emissions occurred immediately after turning. After 15 min the odours reached the levels prior to turning.

From Figs. 2 and 3 it is clear that there is an increase in odour emission every time the compost windrows are turned. These peak emissions are also the lowest for composting plant B with the highest turning frequency. The peak emissions of plant B were lower by a factor 2 than these of plant C and a by a factor 6 than those of plant A. This corresponds to the statement of Epstein (1997), that windrows that are turned more frequently have lower peak emissions.

The decline in odour emissions after these peak emissions is slower than mentioned by Iacoboni et al. (1980). For example, at composting plant B, on day 43 the compost windrow was turned. Just before turning, an odour emission of 2318 ou_E ton^{-1} h^{-1} was measured. Immediately after turning the odour emission rose to 6146 ou_E ton^{-1} h^{-1}. Twenty four h later (day 44), emissions decreased to 5220 ou_E ton^{-1} h^{-1}, which is still higher than before turning. So, according to the present experiments, it took 3 to 4 d before the odour emission declined to the level of odour emission before turning.

Not all the composting plants in our project were followed up until the end of the composting process; only at composting plant B was the process finished. However, from the conclusions above (high and nearly the same odour emissions for all three plants in the beginning, afterwards decline to the same odour emission with peak emissions after every turning that were the lowest for composting plant B), we could say that composting plant B with the highest turning frequency had the lowest overall odour emission. Also the lower height of the compost windrows of plant B can be an additional reason for the lower odour emission of plant B.

In general, we can also mention that the influence of the turning frequency on the odour emission was smaller than expected. This fact agrees with the findings of Michel et al. (1996), that windrow turning frequency affected compost bulk density but did not significantly affect temperature or oxygen concentration, the time to produce stable compost, or the characteristics of finished compost. So the effect of the turning frequency on different parameters was also rather low. However, based on the empirical evaluation of facility operators, the time required to produce a compost ready for curing was 60 to 80 days in well-turned windrows (seven times every 4 weeks) and 120 to 150 days in infrequently turned windrows.

Conclusion

For the sniffing team measurements, median values of 570 m were found for composting plant A, 345 m for plant B and 445 m for plant C. The variation in the sniffing team measurements was high and due to this fact it was difficult to draw conclusions concerning the effect of the turning frequency on the odour emission. For the different activities, however, we could say that the highest odour emission occurs during size reduction and turning.

Results of the olfactometric measurements allowed to compare odour emissions from composting windrows. When the odour emissions of the three composting plants were compared, it became clear that all three composting plants had high and nearly the same odour emissions during the first 10 d (about 5,000–10,000 ou_E ton^{-1} h^{-1}). Every time the compost was turned, there was a peak emission of odour that declined to the level of odour emission before turning within 3 to 4 days, which is more than mentioned in literature. These peak emissions were the lowest for composting plant B. Afterwards, odour emission level was of the same range for all three plants (100–700 ou_E ton^{-1} h^{-1}).

Only composting plant B was followed up until the end, but although it can be said that composting plant B, with the highest turning frequency, had the lowest overall odour emission. Also the lower height of the compost windrows of plant B can be an additional reason for the lower odour emission of plant B.

Acknowledgement. The authors acknowledge financial support by Vlaco (The Flemish Compost Organisation).

REFERENCES

CEN (1998) Draft prEN XXXXX Air quality – determination of odour concentration by dynamic olfactometry, European Committee for Standardisation, Brussels, May 1998

Derickx PJL, Op den Camp HJM, Van der Drift C, Van Griesven LJLD, Vogels GD (1990) Odorous sulfur compounds emitted during production of compost used as a substrate in mushroom cultivation. Appl and Environ Microbiol, 56(1): 176-180

Epstein E (1997) The science of composting. Technomic Publishing, Lancaster, 487 pp

Iacoboni MD, Lebrun TJ, Livinston J (1980) Deep windrow composting of dewatered sewage sludge. In: Proc Natl Conf Municipal & Industrial Sludge Composting, Philadelphia, PA

Kryzmien M, Day M, Shaw K, Zaremba L (1999) An investigation of odors and volatile organic compounds released during composting. J Air Waste Manage Assoc, 49: 804-813

Michel FC, Reddy CA (1996) Analyzing key factors in yard trimmings composting. Biocycle 1: 77-82

Michel FC, Forney LJ, Huang AJ-F, Drew S, Czuprenski M, Lindeberg JD, Reddy CA (1996) Effects of turning frequency, leaves to grass mix ratio and windrow vs. pile configuration on the composting of yard trimmings. Compost Sci Util, 4(1): 26-43

Norstedt RA, Barkdoll AW, Schroeder RM (1993) Composting of yard wastes. In: Hoitink HAJ, Keener HM (eds) Science and engineering of composting, The Ohio State University, Wooster, Ohio, pp 154-168

Sterckx B (1998) Geur-en ammoniaktechnologie bij de industriële bereiding van substraat voor agaricus bisporus. Thesis, Faculty of Agricultural and Applied Biological Sciences, Ghent University, Gent

TNO-report 94-202 (1994) Compostering van groenafval (geen GFT-afval). Branchegeuronderzoek in opdracht van de BVOR, Apeldoorn, The Netherlands, 141 pp

Van Broeck G, Van Langenhove H (1999) Development of a methodology to set up odour quality objectives for sewage treatment plants in Flanders. Technical papers from the CIWEM and IAWQ joint international conference, control and prevention of odours in the water industry, Sept 1999, London

Immission of Microorganisms from Composting Facilities

P. Kämpfer, C. Jureit, A. Albrecht and A. Neef[1]

Abstract. The dispersion and composition of microbial aerosols in the surrounding of three composting plants in central Germany were examined. Immission measurements were carried out at specifically chosen sampling points downwind to the plants including reference points upwind. Only sampling days (nights) were chosen for which the meteorological situation allowed a minimum dilution of the bioaerosol emissions downwind to the plants. The samples were analysed for the presence of six groups of organisms by selective cultivation, including mesophilic bacteria and moulds, xerophilic moulds as well as thermophilic moulds and thermophilic actinomycetes.

Determined concentrations of CFU showed marked differences in the immission situations around the three plants ranging from 0 to more than 106 cfu m^{-3}. Concentrations of microorganisms downwind to the composting plants correlated with distances of the sampling points from the plant. The specific meteorological conditions (especially drainage flow situations) during sampling and local topography obviously have a strong impact on the results of bioaerosol sampling. On the basis of these results a more general sampling strategy was developed.

Introduction

Composting of organic material is an integral part of modern waste management based on a recycling approach. The aim is a mass and volume reduction of the biodegradable part of mainly municipal wastes. Compost as the product of the process can be used as organic fertiliser. In Germany around 550 full-scale composting plants process about 8 million tons of biological wastes (Kämpfer and Weißenfels 2001).

The compost process comprises the degradation of organic materials containing high quantities of easily biodegradable carbohydrates, proteins and lipids. Within this complex process, spontaneous self-heating to 55–60 °C is an inherent part of the process (Lacey and Crook 1988). Especially this part is responsible for the development of thermophilic and thermotolerant microorganisms, among them mucoraceous fungi and aspergilli, especially *Aspergillus fumigatus.* As a consequence, spores of these organisms can be released into the surrounding air during the various processing steps of biological waste treatment (Reiß 1995). Often, the biodegradable materials are initially shredded, repeatedly moved during the rotting

[1]Institut für Angewandte Mikrobiologie, IFZ für Umweltsicherung, Justus-Liebig-Universität Giessen, Heinrich-Buff-Ring 26-32, 35392 Giessen, Germany

H. Insam, N. Riddech, S. Klammer (Eds.)
Microbiology of Composting

phase, and when the composting process is finished, the compost is sieved to remove large particles including materials which are not biodegradable. All these operation processes can lead to emissions of microorganisms into the air.

Bioaerosols which originate from compost can contain various microorganisms, among them vegetative cells and spores from bacteria especially actinomycetes and fungi (Millner et al. 1994). Often they are attached to each other in aggregates or bound to organic particles (Henningson and Ahlberg 1994). It is well known that the vast majority of airborne microorganisms (with the exception of some Gram-positive bacteria and fungal spores) are rapidly inactivated as a result of environmental stress, e.g. desiccation, temperature and UV radiation (Mohr 1997). For this reason, the viable and hence the culturable fraction of bioaerosols represents only a small part of the total microorganisms.

During the past 10–15 years the majority of composting facilities in Germany were built and it was often very difficult to find suitable locations. Certain topographical and especially meteorological facts were not carefully considered.

Because of the close distances of composting plants to settlements, the question whether or not biological emissions orginating from composting plants might have a negative impact on the health of people is open to controverse discussion. It has long been known that bioaerosols can be hazardous to humans in a number of ways. Infection, for example, can occur as a result of direct contact with inhalation or ingestion of viable pathogenic or potentially pathogenic microorganisms. Although infection events due to presence of pathogens in organic dust is rarely reported, even among highly exposed workers, it is a fact that exposure to airborne microorganisms can inititate an allergic response. For this process, microorganisms do not need to be viable because dead cells and cell debris may provoke the same reaction as viable cells. As a consequence, the total number of airborne microorganisms is as important as CFU (colony-forming unit) numbers (Griffiths and DeCosemo 1994; Millner et al. 1994). Also fragments of microbial cells (e.g. cell walls, flagella and genetic material) and microbial metabolites (e.g. volatile organic compounds, endotoxins and mycotoxins) could be of interest (Stetzenbach 1997).

In addition to the problems of odours occurring in connection with the composting processes, the aerial transport of vegetative cells and spores from composting plants to settlements close to these facilities can also lead to acceptance problems for the composting technology in general. Despite these problems, up to now data on immission concentrations and transport phenomena of microorganisms are relatively scarce.

In the light of this complex situation in 1997 extensive investigations were carried out at three German composting plants, to find out if and to what extent microorganisms can be transported from composting material into the surroundings by aerial dispersion. Site-specific sampling strategies were developed considering the topographic and meteorological conditions which enabled aerial transport from the plant to the individually selected sampling points (worst-case strategy).

On the basis of these investigations and the results, it was obvious that a more generalized sampling strategy should be developed. The aim of this contribution is the presentation of a more generalized sampling strategy for measuring the emissions of microorganisms from composting facilities and detection of immissions in their surroundings.

Materials and Methods

Composting Plants

Three full-scale composting plants in Hesse (Germany) were investigated. The technologies of composting, distances to residential areas and turnover of biological wastes are summarised in Table 1.

Sampling Strategy of the Initial Study

The immission concentrations of microorganisms and endotoxins were studied in the surroundings of three open compost plants. At each plant six sampling points were selected considering the specific topographic and meteorological situations (Table 2). One or two of the sampling points were located upwind to the plant. They were used as reference points for evaluating background levels of aerosol concentrations. Sampling was performed in summer 1997 at each site and only during meteorological situations when almost undisturbed transport of emissions from the plants in direction to the sampling points downwind could be expected. To generate high emission concentrations of bioaerosols, the compost material was intensively moved and shredded before and during the sampling periods (worst-case situations).

Table 1. Process of composting, distances to residential areas and turnover of the compost facilities investigated

Compost plant	Process of composting	Distance to residential areas	Turnover (t a^{-1})
A (GS)	Plant residues In house composting	800–1200 m	13,650
B (HE)	Plant residues Open compost stack	800 m	13,500
C (KS)	Plant and domestic bioresidues Open compost stack	150 m	5400 plant residues and 12 600 domestic bioresidues

Bioaerosol sampling was performed simultaneously at all six points. The main meteorological parameters (direction and speed of wind, relative humidity, ambient air temperature) were recorded some hours before and during each sampling period to assess conditions for aerial transport during the sampling events.

Table 2. Description of sampling points. Type (upwind, U, or downwind, D), distance to the plant and local situation (in m) are given

Point	Plant A		Plant B		Plant C	
1	U 500	Beside a road, between forest and farmland	U 300	Meadow	U 500	Road in between farmland
2	U 220	Field path in-between farmland	D 100	Meadow	D 200	Built-up area
3	D 750	Open area	U 800	Road in between farmland and built-up area	D 250	Built-up area
4	D 1350	Open area	D 1800	Built-up area	D 320	Built-up area
5	D 1400	Open area	D 1600	Built-up area	D 300	Built-up area
6	D 1650	Built-up area	D 3400	Built-up area	D 550	Built-up area

Collecting System and Procedure

Sampling of airborne microorganisms was done at a height of 1.5 m using filtration samplers MD8 (Sartorius, Göttingen, Germany). An air volume of 1.3 m^3 was collected within 10 min (using a flow rate of 8 m^3 h^{-1}) through gelantin membrane filters (pore size 3.0 µm; Sartorius, Göttingen, Germany). The bioaerosols were sampled repeatedly with each five filters at each point on 2 different sampling days with comparable meteorological conditions. The filters were processed in the laboratory after a maximum of 24 h following standard procedures (Anonymous 1996; TRBA 430). For direct cultivation two filters were divided in six equal parts and these transferred on solid agars (Table 3). For indirect cultivation, each of the three filters were dissolved in 10 ml of 0.9% NaCl solution, containing 0.01% Tween-80 at pH 7. For homogenisation the solutions were shaken for 15 min at 35 °C until they became clear. After serial dilutions, selective media were inoculated in three replicates with 0.1 ml. Temperature of incubation is also listed in Table 3. After an incubation lasting between 2 and 14 days, numbers of grown colonies were counted for the plates with appropriate numbers (10–100 cfu). Concentrations of microorganisms were calculated as a mean of the three replicates of every culture media used for each collecting point. Airborne microbial concentrations were expressed as cfu m^{-3} air. Thermophilic groups were incubated in closed plastic bags in a humid atmosphere. The limit of detection under the used conditions (1.3 m^3 tested volume of air) was about 5 cfu m^{-3} air for direct cultivation and for indirect cultivation about 180 cfu m^{-3} air. The used culture media and conditions used for differentiation of airborne microorganisms are listed in Table 3.

Table 3. Media and conditions used for the cultivation of airborne microorganisms

Medium	Selectivity	(°C)	Specific supplements
CASO-Agar (Oxoid)	No specific selectivity	36	
R2A-Agar (Oxoid)	Bacteria	25	Cycloheximide (Serva) 200 µg ml^{-1}
Malt-Extract (Merck)	Moulds	30	Chloramphenicol (Serva) 100 µg ml^{-1}, Oxytetracycline (Serva) 40 µg ml^{-1}
Malt-Extract (Merck)	Thermophilic moulds [b)]	45	Chloramphenicol (Serva) 100 µg ml^{-1}, Oxytetracycline (Serva) 40 µg ml^{-1}
Dichloran-Glycerol (DG 18)-Agar (Oxoid)	Xerophilic moulds	25	Chloramphenicol (Serva) 100 µg ml^{-1}, Oxytetracycline (Serva) 40 µg ml^{-1}
Glycerol-Arginine (GA)-Agar [a)]	Thermophilic and thermo-tolerant actinomycetes [b)]	50	Cycloheximide (Serva) 200 µg ml^{-1}

[a)] Modified of El-Nakeeb and Lechevalier 1963 [g l^{-1}]: Glycerol 12.5; L-Arginine 1.0; NaCl 1.0; KH_2PO_4 0.3; K_2HPO_4 0.7; $MgSO_4 * 7H_2O$ 0.5; Agar 15.0; pH 7.2. 2 ml trace elements solution (Drews 1983). [b)] Were incubated in closed plastic bags.

Fungal and actinomycete colonies were distinguished macroscopically on the basis of colony-morphological characters, e.g. structure of mycelia and colour. For differentiation of moulds, a binocular microscope was used. Detailed differentiation of exemplary colonies was done light-microscopially with prepared slides.

Endotoxins

Endotoxins were analysed using the *Limulus* assay (Anonymous 1996).

Results

Quantification of Microorganisms

In Table 4 an overview of the concentrations of microorganisms on all six agar media is given. Minimum and maximum values for pooled data of all sampling points located upwind and downwind, respectively, are presented. Concentrations between minima and maxima spanned two to three decadic orders of magnitude in some cases. Highest concentrations were normally in the range of 10^2–10^4 cfu m^{-3} upwind to the plants and between 10^3 and 10^6 cfu m^{-3} downwind. In some cases, minimum values were below the detection limit of the dilution assay (approximately 40 cfu m^{-3}), especially at the reference sampling points. Presence of culturable organisms could then be shown only by direct transfer of gelatine filters to agar. In some cases not any members of the thermophilic groups could be detected.

In general, downwind concentrations were markedly higher than those at the upwind sampling points, and the extent of the increase of concentration was clearly dependent on the distance from the plant. The highest increase occurred at plant B, where concentrations at the downwind distance of 100 m were approximately 100 times higher than the background values. On the other hand, at the sampling points with the highest distance to all three plants (distances of 550 m, 1900 m and 3400 m, respectively), concentrations were in the same range as the background levels.

Table 4. Comparison of ranges of upwind and downwind concentrations for all three plants. Pooled data form two sampling days with each three raw samples analysed by the dilution method and two raw samples analysed directly

	Upwind		Downwind	
	Minimum	Maximum	Minimum	Maximum
Plant A				
CASO 36 °C	$5 \cdot 10^2$	$5 \cdot 10^3$	$4 \cdot 10^1$	$2 \cdot 10^4$
R2A, 25 °C	$< 4 \cdot 10^1$	$3 \cdot 10^3$	< 5	$2 \cdot 10^4$
Malt, 30 °C	$8 \cdot 10^2$	$8 \cdot 10^3$	$2 \cdot 10^2$	$6 \cdot 10^3$
Malt, 45 °C	$<4 \cdot 10^1$	$7 \cdot 10^1$	< 5	$8 \cdot 10^2$
DG18, 25 °C	$3 \cdot 10^3$	$1 \cdot 10^4$	$2 \cdot 10^3$	$1 \cdot 10^4$
GA, 50 °C	n.d.	$7 \cdot 10^2$	n.d.	$5 \cdot 10^3$
Plant B				
CASO 36 °C	$< 5 / < 4 \cdot 10^1$	$5 \cdot 10^2$	$< 4 \cdot 10^1$	$6 \cdot 10^4$
R2A, 25 °C	$8 \cdot 10^1$	$7 \cdot 10^2$	$< 4 \cdot 10^1$	$4 \cdot 10^4$
Malt, 30 °C	$< 4 \cdot 10^1$	$3 \cdot 10^2$	$3 \cdot 10^1$	$3 \cdot 10^4$
Malt, 45 °C	n.d.	$8 \cdot 10^1$	$< 4 \cdot 10^1$	$4 \cdot 10^4$
DG18, 25 °C	$3 \cdot 10^3$	$5 \cdot 10^4$	$3 \cdot 10^3$	$5 \cdot 10^4$
GA, 50 °C	n.d.	n.d.	$< 4 \cdot 10^1$	$4 \cdot 10^4$
Plant C				
CASO 36 °C	$4 \cdot 10^2$	$2 \cdot 10^5$	$1 \cdot 10^3$	$1 \cdot 10^6$
R2A, 25 °C	$8 \cdot 10^2$	$1 \cdot 10^5$	$8 \cdot 10^2$	$1 \cdot 10^6$
Malt, 30 °C	$3 \cdot 10^2$	$4 \cdot 10^3$	$6 \cdot 10^2$	$2 \cdot 10^5$
Malt, 45 °C	$< 4 \cdot 10^1$	$2 \cdot 10^4$	$< 5 / < 4 \cdot 10^1$	$1 \cdot 10^5$
DG18, 25 °C	$2 \cdot 10^3$	$8 \cdot 10^3$	$1 \cdot 10^3$	$1 \cdot 10^5$
GA, 50 °C	n.d.	$4 \cdot 10^5$	$< 4 \cdot 10^1$	$3 \cdot 10^6$

On DG18 agar all counts were rather similar and in the range of about 10^4 cfu m^{-3} independently of the sampling point. High differences between upwind and downwind values were found for the two groups of thermophilic/thermotolerant microorganisms: the thermotolerant fungi and thermophilic actinomycetes. Up-wind concentrations were generally quite low. At some sampling points these organisms could not be detected at all. For the other three groups upwind/downwind differences were lower but still significant (Table 4).

Thermophilic and Thermotolerant Actinomycetes

In Fig. 1 the airborne concentrations of thermophilic and thermotolerant actinomycetes are shown. Upwind to plant A no thermophilic actinomycetes or only low numbers of them (<100 cfu m^{-3} air) were found. At sampling day 1, concentrations at the four sampling sites downwind were in the same range. At sampling day 2 higher concentrations were detected. In distances of 750 and 1350 m (sampling points 3 & 4), concentrations between $1.2 \cdot 10^3$ and $5.3 \cdot 10^3$ cfu m^{-3} were found, respectively. Also at points 5 & 6 concentrations up to $1.7 \cdot 10^3$ cfu m^{-3} were found.

In the surroundings of compost facility B, the concentrations of thermophilic actinomycetes were similar at both sampling days. At the reference sampling sites, these organisms could not be detected. However, downwind they were found at all sampling sites. The highest concentrations of approximately $2–3 \cdot 10^4$ cfu m^{-3} were found closest to the plant (point no. 2). The concentrations in larger distances (between 1600 and 3400 m) were about 10^2 cfu m^{-3} air.

At composting plant C high concentrations between 10^4 and 10^6 cfu m^{-3} were detected at all points downwind at both sampling days. At day 1 the mean concentrations decreased from $9.6 \cdot 10^5$ in a distance of 200 m to $4.9 \cdot 10^4$ cfu m^{-3} in a distance of 550 m. Thermophilic actinomycetes were not detectable upwind at sampling day 2. Although the concentrations at the five sampling points downwind were lower, similar results to those of sampling day 1 were obtained. Highest concentrations of about $2.0 \cdot 10^5$ cfu m^{-3} were found at a distance of 200 m and decreased continuously to less than 10^3 cfu m^{-3} at a distance of 550 m.

Thermophilic and Thermotolerant Fungi

In Fig. 2 the concentrations of fungi cultivated on malt extract agar (MEA) at 45 °C are given. In the surrounding of composting plant A there were no significant differences between the upwind and downwind concentrations. In general, the concentrations were very low (around 10^2 cfu m^{-3}). Also for composting facility B nearly all concentrations were very low with only one exception (sampling point no. 2 in a distance of 100 m downwind).

High concentrations between 10^3 and 10^5 cfu m^{-3} were measured downwind to composting facility C at both sampling days. Concentrations measured during the first day's sampling were higher than those of the second day. This corresponds with the observations made for the concentrations of thermophilic actinomycetes.

With respect to the type of thermophilic fungi, *Aspergillus fumigatus* was dominant in the direct surroundings of plants. *A. flavus* was also present.

The results for the other groups of organisms are summarized in Table 4 and presented in detail elsewhere (HMFUFL 1999). Significant upwind/downwind differences were found for xerophilic fungi on DG18 incubated at 25 °C at plants B and C. The genera *Cladosporium* and *Alternaria* were most abundant upwind to the plants and downwind in larger distances. However, downwind close to the plants, aspergilli and penicillia were dominant.

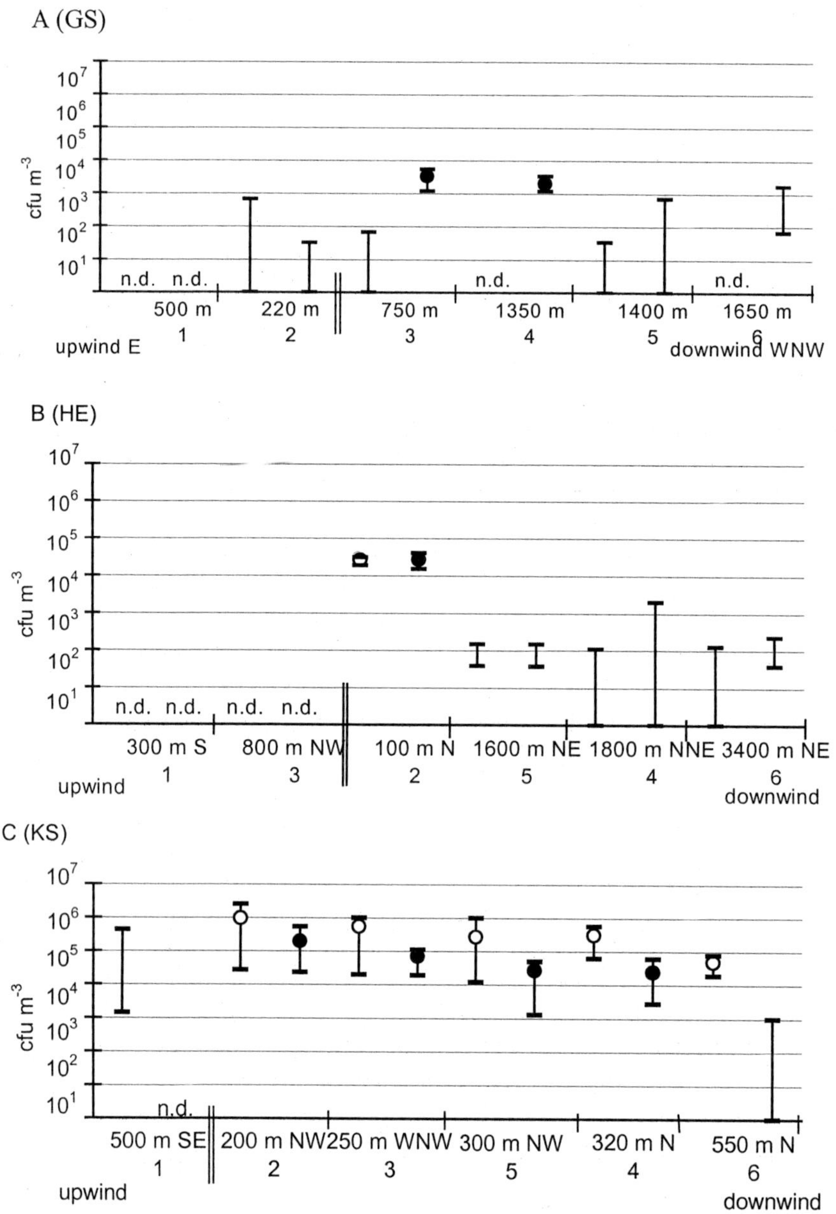

Fig. 1. Concentrations of thermophilic actinomycetes with glycerine-arginine agar at 50 °C at plants A, B and C. Values are given for both sampling days at the six sampling points (*1–6*). *Left column* Day 1; *right column* day 2. Distance in m and compass direction relative to the plant for reference points and downwind sampling points are displayed *from left to right* regarding their location. Indicated are the span between highest and lowest individual concentration datum and the mean in all cases it could be validly calculated; *n.d.* not detectable

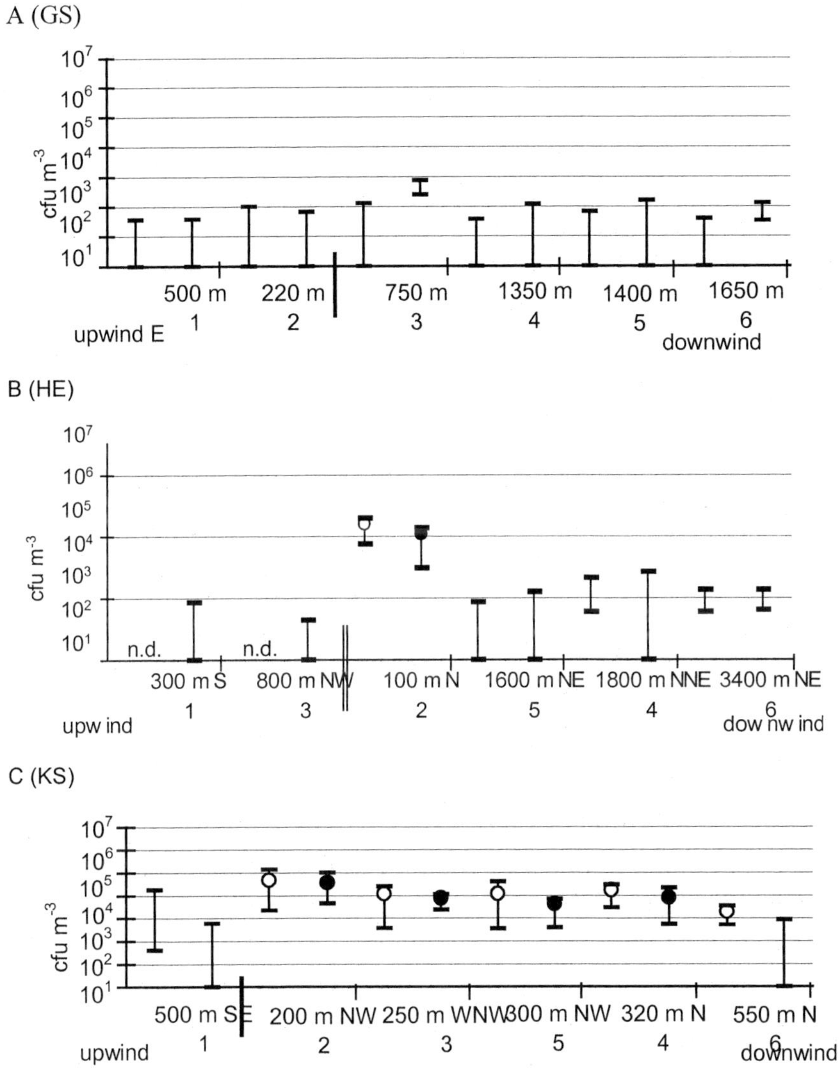

Fig. 2. Concentrations of thermophilic moulds with MEA at 45 °C at plants A, B and C. Data presentation, abbreviations and symbols are as in Fig. 1

Endotoxins

Endotoxin concentrations in the surroundings of all composting plants were in a range between 0.4 and 20.6 EU m^{-3} (Table 5). Clearly increased concentrations relative to the downwind locations on both days could only be detected in the close vicinity at plant B. At the other two plants, only partly elevated concentrations were found at the points closest to the facilities. Low concentrations of endotoxins were also found at the reference sampling points which were not influenced by emissions from the composting facilities.

Table 5. Endotoxin concentrations determined in the vicinity of the three composting facilities

Compost plant	Distance (m)	Endotoxin (EU m^{-3}) Sampling day 1	Sampling day 2
A (GS)	220 (downwind)	5.9	0.4
	750 (upwind)	1.3	4.8
	1650 (upwind)	1.3	1.2
B (HE)	300 (downwind)	0.8	0.6
	100 (upwind)	20.6	14.9
	1800 (upwind)	n.m.	0.7
C (KS)	500 (downwind)	15.3	1.9
	250 (upwind)	10.6	10.6
	320 (upwind)	20.5	1.6

Values are averages from two successive sampling events. n.m. not measured.

Discussion

Thermophilic and thermotolerant microorganisms (actinomycetes and fungi) are important microbial groups responsible for transformation processes especially during the thermophilic phase of composting. These organisms are found in higher concentrations in air samples in the surroundings of composting plants, in contrast to reference samples. For this reason, they can be regarded as indicative for aerial transport processes originating from compost material. Sampling results at the three examined plants support this.

Quantification of Microorganisms

Correspondingly for all three plants, higher concentrations of airborne thermophilic and thermotolerant microorganisms (actinomycetes and fungi) were found downwind to the composting plant in comparison with the upwind concentrations. Generally, Thermophilic species are found both in self-heating situations in the natural environment as well as at composting facilities (Millner et al. 1994). Therefore the total number of the process-emitted microorganisms will have to be estimated against a background of naturally occurring microorganisms like done in this study. Large numbers of microorganisms could be dispersed into the air because the compost material was intensively moved and mixed during the measurements.

In addition, because of the selected worst-case strategy, microorganisms could be transported even over longer distances. The data show the possibility of aerial transport of organisms over more than 500 m, and also indicate that there is no generally recommendable minimum distance between compost plants and residential areas. In general, the specification of appropriate buffer distances depends on the specific site location, design of the plant, local micrometeorological conditions and emission controls (Millner et al. 1994).

The classical cultivation-based detection of microorganisms on solid media can, however, significantly underestimate the total numbers of microorganisms (Albrecht and Kämpfer 2000; Kämpfer and Neef, this Vol.). It has been estimated that culturable counts record only about 10% or less of the total population within environmental samples (Griffiths and DeCosemo 1994). Furthermore, the counts of all culturable microorganisms give no specific informations about the microorganisms' ability to colonise and/or infect a host (Griffiths and DeCosemo 1994). Individual spores are small and can be carried for some distance by very light wind currents; if inhaled, the spores can enter the lungs (Millner et al. 1994). Airborne thermophilic actinomycetes (e.g. *Thermoactinomyces vulgaris* and *Saccharopolyspora rectivirgula*) have been implicated in several cases of hypersensitivity pneumonitis and other allergic reactions. These organisms grow naturally in organic materials that have self-heated to temperatures up to 65–70 °C, e.g. during composting (Lacey and Crook 1988; Stetzenbach 1997). The only known mesophilic actinomycetes implicated in allergic alveolitis are *Streptomyces olivaceus* and *S. albus* (Lacey and Crook 1988).

Airborne thermophilic and thermotolerant fungi are also very important disease agents. Especially in the surrounding of plant C, high concentrations between 10^4 and 10^5 cfu m^{-3} were found. High concentrations were also found in a distance of 100 m of plant B. The dominance of *Aspergillus* (e.g. *A. niger, A. fumigatus* and *A. flavus*) and *Penicillium* downwind to the plants in contrast to the dominance of *Cladosporium* and *Alternaria* upwind, is a clear indication for downwind transport of emissions from the compost plants. *Aspergillus* and *Penicillium* species are some of the most abundant fungi during composting. When compost material is mixed, these microorganisms are dispersed into the air in large numbers (Lacey and Crook 1988). On the other hand, the dominance of *Cladosporium* and *Alternaria* in natural environments is well known (Millner et al. 1994). The thermotolerant *A. fumigatus*, an allergenic and opportunistic pathogenic mould, has been isolated worldwide and is a common organism in the environment, but its concentration in the natural environment is usually low in comparison to other fungi, such as *Cladosporium* and *Alternaria* (Millner et al. 1994). When there are sources of self-heating material in the environment, e.g. a compost plant, the content of *A. fumigatus* increases (Lacey and Crook 1988). The dominance of *A. fumigatus* in the downwind vicinity in the group of thermophilic and thermotolerant moulds is therefore another indication for the release of emissions from the compost plants.

Endotoxins

The endotoxin concentrations found in the surroundings of the three composting plant were very low. Only at a distance of 100 m and 250 m downwind of the plants was a low increase of endotoxins detected. Endotoxins are released into the environment after lysis or during active cell growth of Gram-negative bacteria. Low concentrations at the control locations upwind to plants were also found in other studies (Danneberg et al. 1997). Because of the ubiquitous nature of Gram^{-} bacteria, endotoxins are widely omnipresent. Airborne endotoxin concentrations between 50 and 100 ng m^{-3} air are discussed as being a health risk at work places (Danneberg et al. 1997). In the presented and other studies such concentrations were not found. For this reason, endotoxin concentrations seem not to be suitable as indicators for emissions from composting plants.

Conclusion

It can be concluded that in any sampling strategy, the plant-specific emission/immission situation must be considered.

Especially the specific topography and meteorological situations are very important. The results of the study of the three composting facilities have shown the suitability of thermophilic fungi and thermophilic actinomycetes as indicator organisms. Dependent on local meteorological situation (especially in drainage flow situations) a transport of microorganisms even to distances of 500 m downwind to the plants is possible.

In order to evaluate the impact of bioaerosols on populations living in areas near composting facilities, more data are needed regarding the natural concentrations of compost process-associated microbes in ambient air (Millner et al. 1994).

For this reason a more generalized strategy was developed, in which the following points are considered:

1. The type of composting plant must be considered and the sources of emissions. Open systems can be totally different to an in house composting technology.
2. The topography in connection with the local meteorological situation is of essential importance. Especially drainage flow situations (often in the evening or night hours) can lead to a more or less undiluted dispersion of microorganisms.
3. The typical operation situations on the plant must be considered. Mechanical movement of the biological wastes, disruption of larger particles and sieving of compost material can produce highly concentrated bioaerosols.
4. Considering points 1 to 3, emission and immission measurements should be performed at points in increasing distance to the source (propagation behaviour).
5. Sampling must be performed under standard conditions (most frequent meteorological situation) and/or in worst-case situations to assess the highest possible concentrations.

6. It is desirable to measure simultaneously the meteorological data (wind speed, wind direction, humidity etc.)
7. It is also important to assess the odours originating from composting plants to find out correlations to microorganism dispersion.
8. A quantification and qualification of different microorganisms (in particular: thermophilic fungi and thermophilic actinomycetes) is recommended.

In a subsequent research project this strategy will be applied to nine different composting facilities of different types. The following sampling parameters (filtration as major sampling technique) will be chosen: Quantification of the following microorganisms: thermophilic actinomycetes, on glycerol-arginine agar (50 °C); fungi, DG-18 agar (25 °C); thermophilic fungi, malt-extract agar (45 °C); total bacterial/fungal counts, CASO-agar (36 °C). At least four replicates at each sampling point with 5-7 sampling points downwind to the plant, one to two sampling points upwind to the plant (reference samples), and 2-5 sampling points directly in the composting plant area. In addition, impingement will be used for comparison at three sampling points for comparison.

Acknowledgements. The results reported here were part of a research project supported by the Hessisches Ministerium für Umwelt, Energie, Jugend, Familie und Gesundheit (Wiesbaden, Germany). Thanks are due to the Hessische Landesanstalt für Umwelt, Wiesbaden. The subsequent project is supported by the German Ministry of Education and Research (Bonn, Germany).

References

Albrecht A, Kämpfer P (2000) Wachstum und koloniemorphologisches Erscheinungsbild thermotoleranter und thermophiler Actinomyceten. Gefahrstoffe - Reinhaltung der Luft 60, 4:139-145

Anonymous (1996) Methodik zur lufthygienischen Untersuchung in Kompostierungsanlagen in Niedersachsen. Niedersächsisches Landesamt für Ökologie, Gewerbeärztlicher Dienst. Hannover, Germany

Danneberg G, Grüneklee E, Seitz M, Hartung J, Driesel A (1997) Microbial and endotoxin immissions in the neighbourhood of a compost plant. Ann Agric Environ Med 4: 169-174

Drews G (1983) Mikrobiologisches Praktikum. Springer, Berlin Heidelberg New York

El-Nakeeb MA, Lechevalier DHA (1963) Selective isolation of aerobic actinomycetes. Appl Microbiol 11:75-77

Griffths WD, DeCosemo GAL (1994) The assessment of bioaerosols: critical review. J Aerosol Sci 25, 8:1425-1458

Henningson EW, Ahlberg MS (1994) Evaluation of microbiological aerosol samplers: review. J Aerosol Sci 25/8:1459-1492

Hessisches Ministerium für Umwelt, Landwirtschaft und Forsten (HMFUFL) (ed) (1999) Umweltmedizinische Relevanz von Emissionen aus Kompostierungsanlagen für die

Anwohner. Immissionsmessungen und epidemiologische Daten und Befunde im Umfeld von drei Kompostierungsanlagen. Wiesbaden. ISBN 3-89274-172-7, 1-235

Kämpfer P, Weißenfels DW (2001) Biologische Behandlung organischer Abfälle. Springer, Berlin Heidelberg New York

Lacey J, Crook B (1988) Fungal and actinomycete as pollutants of the workplace and occupational allergens. Ann Occup Hyg 32:515-533

Millner PD, Olenchock SA, Epstein E, Rylander R, Haines J (1994) Bioaerosols associated with composting facilities. Compost Sci Util 2:6-57

Mohr AJ (1997) Fate and Transport of Microorganisms in Air. Manual of Environmental Microbiology. Washington, pp 641-650

Reiß J (1995) Moulds in containers with biological wastes. American Society for Micobiology. Washington, DC. Microbiol Res 150:93-98

Stetzenbach L (1997) Introduction to aerobiology. Manual of Environmental Microbiology. American Society for Microbiology. Washington, DC. pp 617-628

TRBA 430 (1997) Technische Regeln für Biologische Arbeitsstoffe (TRBA) 430. Verfahren zur Bestimmung der Schimmelpilzkonzentration in der Luft am Arbeitsplatz. Bonn. Bundesarbeitsblatt 10:74-77

Molecular Identification of Airborne Microorganisms from Composting Facilities

A. Neef[1] and P. Kämpfer

Abstract. Exposure to bioaerosols from composting facilities can have severe effects on the occupational health of workers and can also influence residents living in the close vicinity of these plants. New methods for the specific detection of airborne microorganisms originating from these plants need to be developed as standards. Present techniques for analyses of microbial bioaerosols are based on cultivation using selective agar media, whereby the non-culturable cells and cell fragments are not detected. In contrast to this methodology, cultivation-independent approaches are adapted here for a more direct assessment of bioaerosol concentrations. For this purpose different sampling methods, such as filtration, impingement and impaction, are evaluated. Direct detection following sampling is performed by epifluorescence microscopy. This means in particular the determination of whole cell numbers and identification of key organisms with taxon-specific ribosomal RNA (rRNA)-targeted oligonucleotide probes by fluorescence in situ hybridization (FISH). In a comparative study, suitable combinations of sampling and detection methods will be developed. Due to their significance in composting processes, thermotolerant and thermophilic actinomycetes, e.g. *Saccharopolyspora rectivirgula,* are selected as target organisms for rRNA probing. Using such molecular techniques, highly specific detection of the cellular fraction of bioaerosols within 1 to 2 working days instead of the present 1 to 3 weeks should be feasible.

Introduction

Due to its low humidity and lack of nutrients, the air cannot be considered as a primary habitat for microorganisms. Despite this fact, microorganisms are often present in the air. Because of their low masses and small dimensions they can be released from various sources, including soil, the phyllosphere, water, animals etc., and persist in air for longer periods, depending on (often taxon-specific) resistance mechanisms (Lindemann et al. 1982; Pillai et al. 1996).
Biotechnical processes like composting allow microorganisms extensive enumeration and these microorganisms can be released into the air as a consequence of handling, mainly in the course of the mixing, mechanical disruption of the material or the sieving of the compost. In addition, biofilters can contribute to microorganism emissions.

[1] Gesellschaft für Biotechnologische Forschung, Division Mikrobiologie, Mascheroder Weg 1, D-38124 Braunschweig, Germany, e-mail: aln@gbf.de

H. Insam, N. Riddech, S. Klammer (Eds.)
Microbiology of Composting

The resulting bioaerosols are most often composed of organic particles or small water droplets to which the microorganisms are attached. These particles can persist for up to several minutes in the air (Zentner 1966; Thomson et al. 1992) and, depending on speed and direction of current winds, they can be transported over distances up to more than 1 km (Lighthart and Kim 1989). Different microorganisms can be detected in bioaerosols, among them vegetative cells of bacteria and fungi, hyphae and spores. In addition, cell fragments, viruses and biomolecules of different size and origin can be found as constituents of bioaerosols. Their concentrations in the air are mainly influenced by their emission rate and the extent of air movement (Lighthart and Kim 1989). The composition of a bioaerosol depends on the survival capacity of their members under the specific conditions (Marshall et al. 1988; Handley and Webster 1993).

Some of the airborne microorganisms are of medical relevance. They can cause infections in the upper respiratory tract and for some other organisms an allergic and toxicological potential has been described (Strom 1985; Albrecht and Kämpfer 2000).

Large-scale composting of organic wastes in specialized facilities resulted in an increased public interest in the question of microbial emissions from respective plants over the past few years. However, the methodology to examine airborne organisms has not yet been adapted to the requirements of modern analytical practice (Griffiths and DeCosemo 1994; Kämpfer and Neef 1999).

In general, suitable sampling and detection methods have to be combined for the enumeration and/or identification of bioaerosol organisms. The method of choice for bioaerosol collection depends on the source and composition of the bioaerosols. Important factors are the bioaerosol concentration and the ability of the airborne organisms to withstand air passage. Different methods for bioaerosol sampling are based on the physical principles of filtration, impaction and impingement (Henningson and Ahlberg 1994). For bioaerosols containing only a few organisms, the impaction or even simple sedimentation methods are suitable. Aerosols of higher concentrations are often sampled with impingers. The filtration techniques can be used for different applications, ranging from low to very high concentrated bioaerosols. With respect to the "sampling stress", the impingement is the most preserving method. Impaction causes mechanical forces which can result in cell disruption. Filtration can lead to cell damage through drying effects (Jensen et al. 1994; Burge 1995).

Standard techniques to measure the abundance of distinct microorganisms in aerosols still depend on cultivation (Griffiths and DeCosemo 1994). Differentiation of certain organisms with selective media is rather time-consuming and often has low specificity. To overcome these drawbacks, molecular-biological methods are being adapted to the needs for detection of airborne cells (Alvarez et al. 1994; Neef et al. 1995; Lange et al. 1997).

Probing techniques, i.e. hybridization assays, have the potential for fast monitoring and access to quantitative results (Schleifer et al. 1993). Nucleic acid probes targeting certain groups of organisms allow clear identification in samples of mixed composition. rRNA molecules present in all living organisms are popular markers for the phylogenetically based identification of microorganisms (Head et al. 1998). Presence or absence of signature information specific for distinct taxa can be easily and rapidly tested in oligonucleotide hybridization assays. A positive detection is achieved when the probe complements perfectly the target sequence.

Different formats are attractive for use to detect airborne organisms. Here we describe two approaches to detect microbial emissions from compost plants:
Colony hybridization combines the conventional cultivation step with specific detection of filter-transferred colonies with probes (Grunstein and Hogness 1975).

The basic principle of a FISH experiment is simple as well (Amann et al. 1995): fixed cells are incubated with dye-labelled probes which pass the cell wall barrier and bind to the intracellular ribosomal targets. Reliable detection, performable with any epifluorescence microscope, is always possible when the cells contain enough ribosomes. FISH enables the identification of cells present in natural samples independent of cultivation.

As stated above, aerosol sampling can be performed by different principles. Here we started with a widely used filtration sampler to combine it with probe-based detection.

Material and Methods

Aerosol Sampling

Sample collection on a compost plant treating ca. 20 000 t a^{-1} organic household waste and decaying plant material (Niddatal, Central Hesse, FRG) was performed with MD8 air samplers (Sartorius, Göttingen, FRG) and the accompanying 3.0-µm gelatine filters. In the laboratory the filters were directly transferred onto uncharged nylon membranes (Pall, Dreieich, FRG) atop of CASO agar and incubated at 30 °C for 24 to 48 h before cell lysis.

Colony Hybridization

The cell lysis procedure for filter-grown colonies was a modification of the original protocol described by Tatzel et al. (1994). Fig. 1 presents a process scheme. Hybridization with POD-labelled probes was done as described previously (Neef et al. 1995).

FISH Test of Reference Strains

Thermophilic actinomycetes type strains of different genera as well as strains isolated from the surroundings of different composting facilities in FRG (Table 1) were probed with the group-specific probe HGC69a (Roller et al. 1994) using a standard protocol. Organisms were grown in suspended cultures and harvested for ethanol fixation (Roller et al. 1994) in time series between day 1 and 16. Prior to the hybridization, the fixed cells were pretreated in a combined acetic acid/lysozyme incubation according to Zarda et al. (1997).

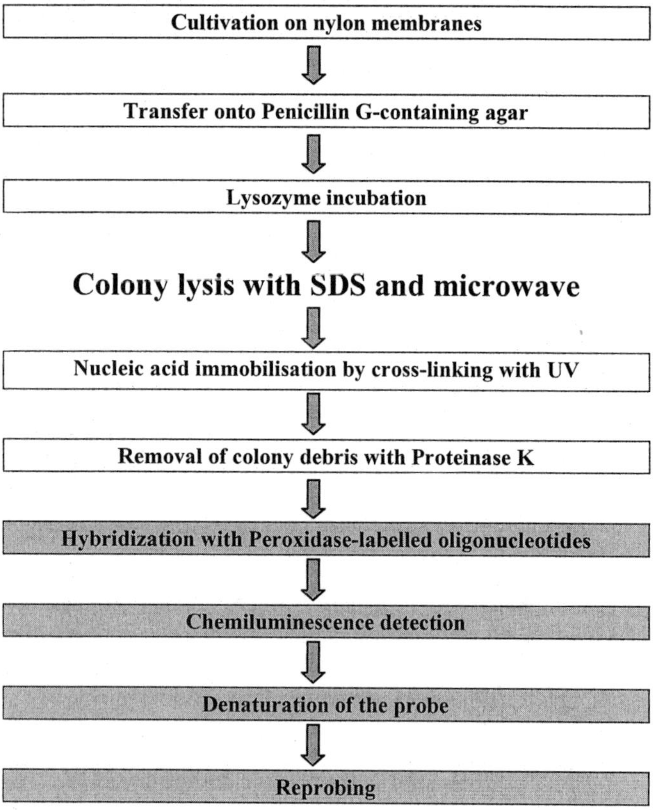

Fig. 1. Process scheme of cell lysis and colony hybridization

Results

Differentiation of Aerosols from a Compost Plant via Colony Hybridization

Sampling points at distances of approximately 50 m from rotting material gave counts ranging from 10^2 and $2.5 \cdot 10^3$ cfu m^{-3} ($n = 16$). Hybridization with the probes UNIV and EUB confirmed that nearly all colonies were members of the domain *Bacteria*. Tests were carried out using four membranes. On all of them more than 50% of the colonies hybridized with probe HGC whereas approximately 25% of the cfu gave signals with the LGC probe mix. Gram--negative bacteria could not be identified with probes for α, β and γ-subclasses of the *Proteobacteria* or the *Cytophaga-Flexibacter* cluster. An example is given in Fig. 2.

FISH of Thermophilic Actinomycetes

Cells grown in liquid batch cultures showed widely different results in the probing assay (Table 1). Some strains showed strong hybridization results for a significant fraction of the cells (e.g. *Micromonospora aurantiaca* and *Streptomyces thermoviolaceus* subsp. *apingens*).

Due to the partly negative results the pretreatment protocol was varied. In fact, using a shortened acetic acid step the detectability of *Micromonospora aurantiaca* was improved (Table 1, Fig. 3B), with the fraction of hybridized cells increasing significantly.

However, other organisms were hardly or not detectable (e.g. *Thermoactinomyces vulgaris*; Fig. 3 *A*). Often detectability of the strains decreased when cells became older.

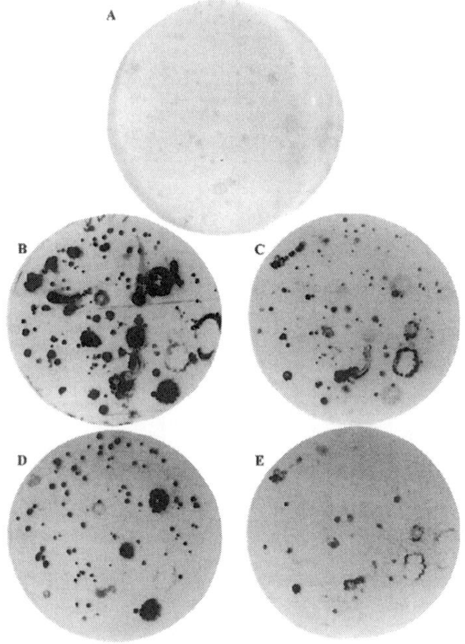

Fig. 2 A–E. Probe-based identification of airborne organisms. **A** Growth on CASO agar, **B–E** Chemiluminescent detection after hybridization with POD-labelled probes UNIV1390 (**B** Zheng et al. 1996), EUB338 (**C** Amann et al. 1990), HGC69a (**D**) and LGC354 (**E** Meier et al. 1999)

Table 1. Detectability of thermophilic actinomycetes by FISH

Origin	Species	Strain designation	Cultivation conditions	Hybridization after cultivation for # days: 1	2	3	4	7	10	16
DSMZ	*Micromonospora aurantiaca*	DSM 43813[T]	TS-B / 36 °C	+	+/o		+	o/-	+/o	-/--
				++/+[a]				+[a]		
	Saccharomonospora caesia	DSM 43044	TS-B / 28 °C	+			--	-	o	
	Saccharopolyspora rectivirgula	DSM 43747[T]	TS-B / 55 °C	-	o	-				
	Streptomyces thermoviolaceus subsp. *apingens*	DSM 41392[T]	TS-B / 50 °C	++	+	o/-	-/--		--	
	Thermoactinomyces vulgaris	DSM 43016[T]	TS-B / 55 °C	--						
Isolates originating from emissions of compost plants	*Nocardiopsis* sp.[b]	KS 1 R 12	TS-B / 25 °C	+/o		o / -		o		
	Streptomyces sp.[b]	KS 1 R 21	TS-B / 25 °C							
	Streptomyces sp.[b]	KS 2 C 7	TS-B / 36 °C					o / -		
	Nocardiopsis sp.[b]	KS 2 C 9	TS-B / 36 °C							
	Streptomyces sp.[b]	HE 2 C 5	TS-B / 36 °C					- / --		
	[c]	DS 2 R 1a	TS-B / 25 °C							
	[c]	LB I E	TS-B / 55 °C	- / --						
	[c]	LB III C	TS-B / 55 °C	+	o	--				
	[c]	LB II 5	TS-B / 55 °C	+/o		o / -		- / --		

The standard protocol included an incubation with acetic acid for 1 h and subsequent lysozyme treatment for 15 min at 37 °C. Fraction of hybridized cells: ++, all cells show signals; +, almost all cells stained; *o*, approximately 50% stained; -, less than 50% stained; --, hardly any cell stained. [a] Improved result after acetic acid preincubation for only 15 min instead of 1 h. [b] Genus affiliation on the basis of fatty acid methylester analysis. [c] Strains could not be affiliated to known genera yet.

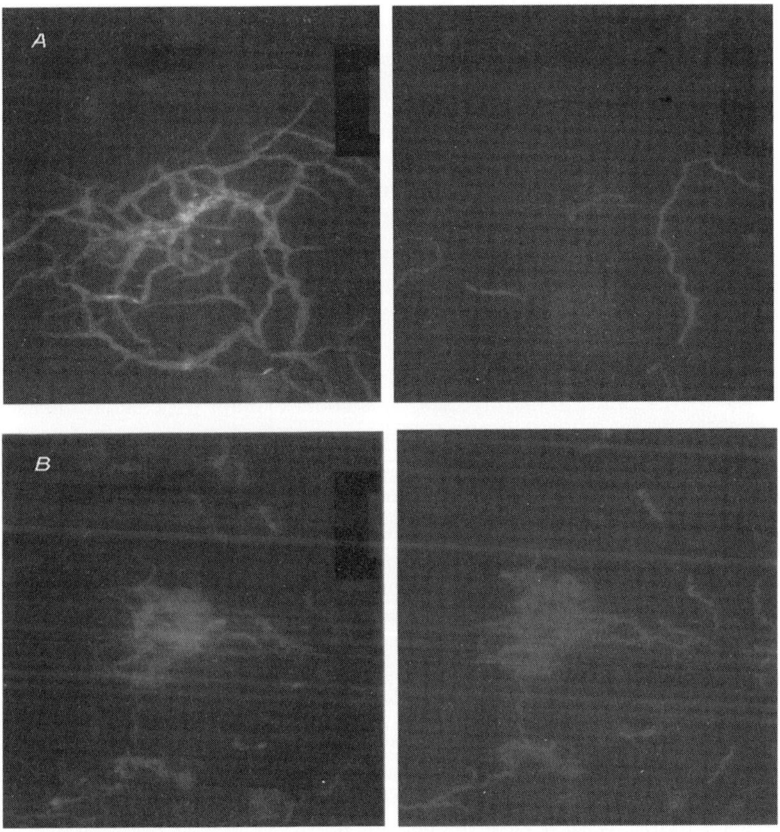

Fig. 3. A Detection of cells of the airborne isolate LB I E with DAPI stain (*left panel*) and via FISH (*right panel*) using the *Actinobacteria*-specific probe HGC69a. **B** Detection of *Micromonospora aurantiaca* DSM 43016^{T} with DAPI stain (*left*) and via FISH (HGC69a, *right*) after improved cell wall lysis. A high fraction of cells was readily hybridized. *Bar* 20 μm

Discussion

Conventionally, the detection of organisms present in bioaerosols is cultivation-based, i.e. the use of selective agar media allow the growth of a specific fraction of organisms and the numbers of grown organisms are counted as colony-forming units (cfu). Therefore, colony hybridization also involving a cultivation step on solid media (Grunstein and Hogness 1975) was chosen as a first step in the application of probing assays to environmental aerosols. rRNA-targeted probes could here be used for the differentiation of the culturable fraction of airborne microorganisms using media with minor selectivity.

Routine application of this technique would help to minimize material and personnel for cultivation-based analyses.

Regarding cultivation-independent detection by FISH (Amann et al. 1995), successful attempts have already been made with artificial bioaerosols (Neef et al. 1995). In combination with a universal stain (e.g. DAPI), in order to visualize all filter-sampled organisms, relative cell numbers can be obtained. However, this method has not yet been tested for natural aerosols.

In the case of emissions from compost plants, distinct difficulties have to be overcome for FISH application. Particular groups of microorganisms, e.g. thermophilic actinomycetes, cannot be easily detected. Generally, signal intensities are dependent on the cellular amount of ribosomes, thus growth rates, and the cell wall permeability. FISH of bacteria can be hampered by low growth rates as well as Gram-positive wall types (Amann et al. 1995). Both characteristics are encountered in the thermophilic actinomycetes. This could explain why signal intensities were frequently rather low. The observed decrease of fluorescence signals with increasing cell age is probably caused by a decline of target molecules (rRNAs, ribosomes), but changes in cell wall composition may also occur. However, as shown, by variation of permeabilization protocols the detectability could be clearly improved.

Use of hybridization probes offers significant advantages in comparison with other molecular-biological detection assays like PCR or immunological methods. FISH enables the identification of organisms which are resistant to selective cultivation conditions (Amann et al. 1995; Head et al. 1998). It is superior to PCR in that it allows the quantification of cell numbers. In contrast to serological techniques, the prior isolation of target strains for production of specific antibodies is not necessary with the use of taxon-specific probes directed against rRNA.

As a basis for the design of specific probes for certain indicator species (e.g. *Saccharopolyspora rectivirgula*), sequencing of 16S rDNA of thermophilic actinomycetes is under way. Such probes can then be used in combination with the technical development of appropriate sampling methods and hybridization assays.

Finally, the ultimate goal of specific and rapid quantification of airborne microorganisms which are emitted from compost plants and often discussed as a major threat for the health of workers as well as nearby living residents, can be achieved.

Conclusions

Application of colony hybridization enables specific quantification of cfu numbers of bacteria in ambient air and their accelerated differentiation with taxon-specific probes.

In aerosol samples from a composting facility the majority of colonies grown on CASO agar were identified as members of the class *Actinobacteria*.

Detection of thermophilic actinomycetes, an important indicator group often emitted from composts, by FISH is possible. However, specific protocols have to be developed to meet the special requirements of tracking recalcitrant members from the airborne state.

Acknowledgements. Our studies are financially supported by the German Federal Environmental Fund (Deutsche Bundesstiftung Umwelt, DBU; Osnabrück, FRG). We gratefully acknowledge the technical assistance of Gundula Will and Oliver Waitze, and are indebted to Roger Cresswell (Lincoln University, Canterbury, NZ) for valuable suggestions on the text and to Claudia Jureit for help during preparation of the final MS.

References

Albrecht A, Kämpfer P (2000) Wachstum und koloniemorphologisches Erscheinungsbild thermotoleranter und thermophiler Actinomyceten. Gefahrstoffe – Reinhaltung der Luft 60:139–145

Alvarez AJ, Buttner MP, Toranzos GA, Dvorsky EA, Toro A, Heikes TB, Mertikas-Pifer LA, Stetzenbach LD (1994) Use of solid-phase PCR for enhanced detection of airborne microorganisms. Appl Environ Microbiol 60:374–376

Amann RI, Binder BJ, Olson RJ, Chisholm SW, Devereux R, Stahl DA (1990) Combination of 16S rRNA-targeted oligonucleotide probes with flow cytometry for analyzing mixed microbial populations. Appl Environ Microbiol 56:1919–1925

Amann RI, Ludwig W, Schleifer K-H (1995) Phylogenetic identification and in situ detection of individual microbial cells without cultivation. Microbiol Rev 59:143–169

Burge HA (1995) Bioaerosol investigations. In: Burge HA (ed) Bioaerosols. CRC Press, Boca Raton, pp 1-23

Griffiths WD, DeCosemo GAL (1994) The assessment of bioaerosols: critical review. J Aerosol Sci 25:1425–1458

Grunstein M, Hogness DS (1975) Colony hybridization: method for the isolation of cloned DNAs that contain a specific gene. Proc Natl Acad Sci USA 72:3961–3965

Handley BA, Webster AJF (1993) Some factors affecting airborne survival of *Pseudomonas fluorescens* indoors. J Appl Bacteriol 75:35-42

Head IM, Saunders JR, Pickup RW (1998) Microbial evolution, diversity and ecology: decade of ribosomal RNA analysis of uncultivated microorganisms. Microb Ecol 35:1–21

Henningson EW, Ahlberg MS (1994) Evaluation of microbiological aerosol samplers: review. J Aerosol Sci 25:1459-1492

Jensen PA, Lighthart B, Mohr AJ, Shaffer BT (1994) Instrumentation used with microbial bioaerosol. In: Lighthart B, Mohr AJ (eds) Atmospheric microbial aerosols. Chapman & Hall, New York, pp 226-284

Kämpfer P, Neef A (1999) Klassische und molekularbiologische Methoden zur Detektion luftgetragener Organismen. VDI Berichte 1443. Neuere Entwicklungen bei der Messung und Beurteilung der Luftqualität. VDI-Verlag, Düsseldorf, pp 411–424

Lange JL, Thorne PS, Lynch N (1997) Application of flow cytometry and fluorescent in situ hybridization for assessment of exposures to airborne bacteria. Appl Environ Microbiol 63:1557–1563

Lighthart B, Kim J (1989) Simulation of airborne microbial droplet transport. Appl Environ Microbiol 55:2349-2355

Lindemann J, Constantinidou HA, Barchet WR, Upper CD (1982) Plants as sources of airborne bacteria, including ice nucleation-active bacteria. Appl Environ Microbiol 44:1059-1063

Marshall B, Flynn P, Kamely D, Levy SB (1988) Survival of *Escherichia coli* with and without ColE1:Tn5 after aerosol dispersal in a laboratory and a farm environment. Appl Environ Microbiol 54:1776-1783

Meier H, Amann R, Ludwig W, Schleifer K-H (1999) Specific oligonucleotide probes for in situ detection of a major group of Gram-positive bacteria with low DNA G+C content. Syst Appl Microbiol 22:186–196

Neef A, Amann R, Schleifer K-H (1995) Detection of microbial cells in aerosols using nucleic acid probes. Syst Appl Microbiol 18:113–122

Pillai SD, Widmer KD, Dowd SE, Ricke SC (1996) Occurrence of airborne bacteria and pathogen indicators during land application of sewage sludge. Appl Environ Microbiol 62:296-299

Rollcr C, Wagner M, Amann R, Ludwig W, Schleifer K-H (1994) In situ probing of gram-positive bacteria with high DNA G+C content by using 23S rRNA-targeted oligonucleotides. Microbiology 140:2849–2858

Schleifer K-H, Ludwig W, Amann R (1993) Nucleic acid probes. In: Goodfellow M, O'Donnell AG (eds) Handbook of new bacterial systematics. Academic Press, London, pp 463–510

Strom PF (1985) Identification of thermophilic bacteria in solid-waste composting. Appl Environ Microbiol 50:906-913

Tatzel R, Ludwig W, Schleifer K-H, Wallnöfer PR (1994) Identification of *Bacillus licheniformis* by colony hybridization with 23S rRNA-targeted oligonucleotide probes. Syst Appl Microbiol 17:99–103

Thomson CMA, Chanter N, Wathes CM (1992) Survival of toxigenic *Pasteurella multocida* in aerosols and aqueous liquids. Appl Environ Microbiol 58:932-936

Zarda B, Hahn D, Chatzinotas A, Schönhuber W, Neef A, Amann RI, Zeyer J (1997) Analysis of bacterial community structure in bulk soil by in situ hybridization. Arch Microbiol 168:185–192

Zentner RJ (1966) Physical and chemical stresses of aerosolization. Bacteriol Rev 30:551-557

Zheng D, Alm EW, Stahl DA, Raskin L (1996) Characterization of universal small-subunit rRNA hybridization probes for quantitative molecular microbial ecology studies. Appl Environ Microbiol 62:4504–4513

Bioaerosols and Public Health

S.D. Pillai[1]

Abstract. There is an increasing emphasis on the recycling of municipal and agricultural wastes around the world using processes such as composting. Theoretically, composting should result in the reduction or total destruction of microbial pathogens. However, variabilities associated with the process can result in improper temperatures and moisture conditions with the result that the final product can harbor pathogenic microorganisms. Though there is a significant amount of information on the occurrence of fungal spores around composting facilities, information regarding other human pathogens such as viruses and bacteria are almost nonexistent. This is particularly significant considering that many communities around the world are considering using municipal biosolids as feed material in large-scale composting operations. Biosolids contain large numbers of potentially pathogenic organisms and the re-growth of *Salmonella* spp. in composted biosolids is a possibility. Mechanical agitation of biosolids material in open windrows will result in the aerosolization of microbial pathogens, and bioaerosols around composting operations that involve municipal biosolids need to be appropriately monitored to determine the possible health risks. There are, however, some significant impediments to determining these risks accurately. These include availability of efficient portable samplers, analytical methodologies for specific pathogens, and mathematical models to quantitate the risks.

Introduction

Composting is a microbiologically mediated process which is employed for the decomposition of organic matter. The term, composting is rather encompassing in its definition since it includes the simplistic backyard composting to large commercial composting facilities. The primary driving force behind composting operations is the need to reduce the solid organic wastes. The disposal of domestic sewage sludge (biosolids) has become a pressing problem. Today, in the United States more than 33% of all biosolids are disposed of on lands and or in landfills (McLamarra and Pruit 1995; Scanlan et al. 1989; Semenza 1995). Since landfill operations are expensive, alternate approaches such as composting tend to be favored. The US EPA has challenged a number of communities around the United States to reduce and recycle their municipal solid wastes (USEPA 1989). Many states consider composting as a form or recycling and have therefore implemented

[1]Poultry Science Department and Institute of Food Science & Engineering
Texas A&M University, College Station, Texas 77843-2472, USA
Tel: (979) 845-2994; Fax: (979) 845-1921; E-mail: spillai@poultry.tamu.edu.

H. Insam, N. Riddech, S. Klammer (Eds.)
Microbiology of Composting

composting plans as part of their recycling goals (Millner et al. 1994). Composting operations are currently being employed to degrade a wide variety of substrates (Table 1).

Table 1. Substrates used as feedstock in composting operations (Herman and Maier 2000)

Compost substrate	Sources
Biosolids	Sewage treatment plants
Municipal solid wastes	Yard wastes, paper
Manure	Poultry wastes, animal feedlots
Special wastes	Explosives, petroleum sludge
Industrial wastes	Food and vegetable processing industry, agricultural residues

Since composting is a microbiologically driven process, the success of a particular composting process relies very heavily on the microbial populations that are involved which, in turn, rely very heavily on the abiotic factors that control their proliferation. The primary abiotic factors controlling the diverse microbial populations are temperature, moisture content, nutrients, and pH. The degradation of the organic substrate is dependent on the elevated temperature that is normally encountered within compost piles. Temperature initially rises and then levels off. In-vessel composting operations provide the best level of temperature control, while open windrow operations offer the least (Epstein 1997). Temperature control is critical not only in terms of the efficiency of organic matter degradation, but also for the destruction of human and plant pathogens. Since human and plant pathogens are present in a variety of substrates such as biosolids and yard wastes, precise temperature control is critical for controlling their survival and proliferation within compost piles. This chapter addresses the issues related to bioaerosols from municipal biosolid-based composting operations.

Microbial Pathogens in Biosolids

Biosolids previously known as sewage sludge, is a by product of municipal sewage sludge treatment plants. More precisely, it is a by-product of physical (primary treatment), biological (activated sludge, trickling filters, or rotating biological contractors) and physicochemical (precipitation with lime, ferric chloride, or alum) treatment of wastewaters. Pathogens that are present in raw wastewater are present in biosolids as well, albeit in reduced concentrations (Straub et al. 1993). The concentrations and types of pathogens in biosolids depend significantly on the incidence of infection within a community and the type of treatment process (Straub et al. 1993). Even though aerobic and anaerobic treatment processes do reduce pathogen numbers, significant numbers remain in the final biosolid product (Tables 2,3).

Table 2. Selected viral and bacterial pathogens associated with biosolids

Organism	Symptom
Adenovirus	Conjunctivitis
Astrovirus	Acute gastroenteritis
Calicivirus	Acute gastroenteritis
Enteroviruses	Febrile illness, respiratory illness
Hepatitis A virus	Jaundice
Norwalk and Norwalk-like viruses	Acute gastroenteritis, vomiting
Aeromonas	Diarrhea
Campylobacter	Acute gastroenteritis
Enterohemorrhagic *E. coli*	Bloody diarrhea
Enteropathogenic *E. coli*	Diarrhea
Enteroinvasive *E. coli*	Diarrhea
Enterotoxigenic *E. coli*	Diarrhea
Salmonella enterica serovars	Diarrhea, typhoid fever
Shigella	Dysentery
Yersinia enterocolitica	Bloody diarrhea

Table 3. Selected fungal, protozoan, and helminth pathogens associated with biosolids

Organism	Symptom
Aspergillus fumigatus	Respiratory infections
Crypotococcus neoformans	Subacute chronic meningitis
Entamoeba histolytica	Abdominal pain, bloody diarrhea
Giardia lamblia	Abdominal pain, diarrhea
Microsporidium	Gastroenteritis
Ascaris lumbricoides	Ascariasis
Anclyostoma duodenale	Anemia

Aerobic and anaerobic treatment units are 90 - 97% efficient in removing fecal coliforms (Morozzi et al. 1988). Stelzer et al. (1988) have also reported significant reduction of campylobacters when the sewage undergoes activated sludge treatment. The reduction of pathogens during treatment processes can vary depending on how precisely the process is controlled. Even with a 1 – 2 order of magnitude decrease in bacterial and viral numbers, the actual concentrations of pathogens in the treated biosolids can be still significantly high. (Pepper and Gerba 1989; Soares 1990). Thus it has to be assumed that even secondary treated biosolids (which are often used as feedstock for composting) still harbors significant numbers of microbial pathogens.

Bioaerosols-Theory and Principles

Bioaerosols are defined as a collection of aerosolized biological particles. The composition, size and concentration of the microbial populations comprising the bioaerosol vary with the source, dispersal mechanisms in the air, and, more importantly, the environmental conditions prevailing at the particular site. Bioaerosols generated from water sources (such as during splashing and wave action) are different from those generated from soil or nonaqueous surfaces in that they are usually formed with a thin layer of moisture surrounding the microorganisms. They often consist of aggregates of several microorganisms (Wickman 1994). Bioaerosols released into the air from soil surfaces, such as those surrounding biosolid and composting facilities, are often single units or associated with particles. In many instances, the presence is particulate matter serves as "raft" for microorganisms (Lighthart and Stetzenbach 1994). The extent of dispersal (transport) and the settling of a bioaerosol are affected by its physical properties and the environmental parameters that it encounters while airborne. The size, density, and shape of the droplets/particles comprise the most important physical characteristics, while the magnitude of air currents, relative humidity, and temperature are the significant environmental parameters. (Lighthart and Mohr 1987; Pedgley 1991). It must be emphasized that bioaerosols vary greatly in size ranging from 0.02 to 100 µm in diameter and are classified based on their size (Dowd and Maier 2000) (Fig. 1).

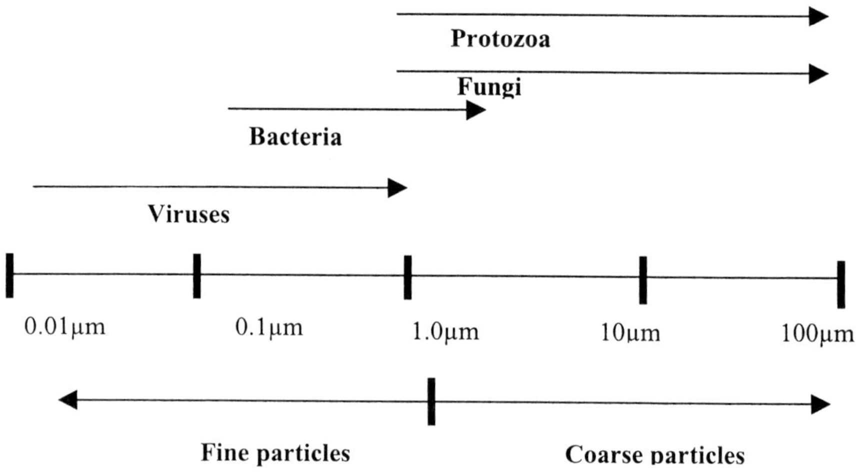

Fig. 1 Schematic representation of the relative sizes of bioaerosols (Dowd and Maier 2000)

Aerosols can be launched from either point sources, linear or area sources. A biosolid pile is an example of a point source, while an agricultural field, which has been spread with biosolids, is an example of an area source (Dowd et al. 2000). Point sources are well-defined sources and the dispersion from these sites displays conical shape dispersion. Linear sources exhibit a particulate wave shaped dispersion. An example of a linear source is the contrail behind a jet aircraft flying at high altitudes. Point sources can be further categorized into instantaneous point sources (e.g., sneezes) or continuous sources (for example, the release of bioaerosols from a biosolids pile). The transport of bioaerosols can be defined in terms of distance and time. Submicroscale transport involves very short periods of time under 10 min, as well as relatively short distances under 100 m. This type of transport is common within indoor environments. Microscale transport ranges from 10 min to 1 hour and from 100 m to 1 km and is the most common and significant type of bioaerosol transport from a human health standpoint. Mesoscale and macroscale transport refers to longer duration bioaerosol transport patterns (Hugh-Jones and Wright 1970). The diffusion of bioaerosols during their transport is one of the primary means by which their concentration decreases. Atmospheric turbulence significantly influences the diffusion of bioaerosols. Thus, with biosolid-based composting, the storage conditions of the feedstock, the length and size of the open windrows, the atmospheric conditions during storage and windrow turning, and the distance to the closest population center have to be taken into careful consideration when evaluating whether composting associated bioaerosols can have public health implications.

Aerosolization of Pathogens

Inhalation, ingestion, and dermal contact are routes of human exposure for aerosolized microorganisms. Since the average human being inhales about 10 m^3 of air per day, inhalation is the predominant route resulting in adverse health effects. Large aerosolized particles are lodged in the upper respiratory tract (nose and nasopharynx). Particles < 6 μm in diameter are transported to the lung with the greatest retention of 2-1'- μm-sized particles in the alveoli (Randall and Ledbetter 1966; Sattar and Ijaz 1987; Salem and Gardner 1994). Microbial pathogens such as *Legionella pneumophila*, *Mycobacterium tuberculosis,* and *Hantavirus* infections are known to be aerosol-transmitted and are capable of causing severe infections. Asthma, hypersensitivity pneumonitis, and other respiratory illnesses are also associated with exposure to bioaerosols containing respiratory pathogens. The typical route of exposure for organisms that are primarily associated with intestinal infections such as *Salmonella* spp., *Campylobacter* spp, and enteric viruses is based upon the inhalation of bioaerosols (containing these pathogens), which are then deposited in the throat and upper airway and swallowed (Wathes et al. 1988). Additionally, the inhaled enteric pathogens may establish throat and respiratory infections that can, in turn, increase the risk of swallowing an infectious dose (Clemmer et al. 1960). This could possibly explain why the infectious dose of enteric organisms is lower when these organisms are inhaled as opposed to ingested (Darlow et al. 1961).

A number of publications over the past few years have documented that aerosolization of microbial pathogens is strongly linked to waste application practices, biosolids handling, wind patterns, and micrometeorological fluctuations (Brenner et al. 1988; Lighthart and Shaffer 1995; Pillai et al. 1996; Dowd et al. 1997). Pathogens originally present in the feedstock materials may remain in the finished composted material if the composting process is not adequately controlled and monitored. Additionally, the pathogens present in the original biosolids material may undergo proliferation during the end of the composting if the conditions are not adequately controlled. The surficial layers of a compost pile will not reach elevated temperatures unless the material is constantly turned. Under such circumstances, improper temperature regimes within the compost pile can lead to pathogen regrowth due to the sudden increase in available carbon and nitrogen nutrients. Furthermore, the very process of turning over or mechanical agitation of biosolids material at the initial stages of the composting process or during the process itself can generate large amounts of microbial pathogens. Studies conducted around biosolids land application processes have shown that when the biosolids material is physically agitated *Salmonella* and fecal indicator viruses can be released into the surroundings (Dowd et al. 1997). At an arid location in the United States, Dowd and coworkers detected bioaerosols averaging 300 most probable number (MPN) of *Salmonella* cells m^{-3} of air at biosolids loading and application sites. The levels of fecal indicator viruses averaged around 1000 virus particles (PFU)m^{-3}. On occasions, *Salmonella* at levels up to 3000 MPNm^{-3} were detected 4 miles downwind. The detection of microbial pathogens at distances away from the point source is indicative of how wind gusts and wind patterns can transport bioaerosols over distances. The amount of pathogens bioaerosolized and transported are dictated by the source material, wind patterns,and mechanical agitation of the biosolids material. It is the aerosolization of pathogens from the mechanical agitation of biosolids under the various stages of composting (especially open windrow systems) that is environmentally significant, and not necessarily pathogen aerosolization from the finished compost. Millner et al. (1980) have reported that mechanical agitation of compost material was a major source of airborne emissions. Neef et al. (1999) monitored airborne microorganisms for 12 months around two different composting facilities (open and closed) in Germany. Surprisingly, the emission levels of aerosolized molds were higher around the closed composting facility as compared to the open facility. They suggested that additional fungal indicators such as *Saccharopolyspora* spp. and *Thermoactinomyces* spp. should be included in monitoring programs.

Though there are studies documenting the presence of *Aspergillus fumigatus* spores around composting facilities, very few studies have been published regarding the presence of specific pathogens such as *Salmonella* spp. Millner et al. (1977) have reported on the presence of *A. fumigatus* spores during composting of sewage sludges in close proximity to the plant. Nersting et al. (1991) from a study in Denmark have reported that total numbers of microorganisms in composting plants can range between a 500 CFUm^{-3} and 105 CFUm^{-3} while Gram-negative bacteria can range between 200 CFUm^{-3} and 50000 CFUm^{-3}. The increased occupational risk to compost workers in such settings has been reported (Bunger et al. 2000; Douwes et al. 2000). Douwes et al. (2000) recently reported on the upper airway based occupational exposure of compost workers to microbial agents (endotoxin and beta 1,3 glucan). Bunger et al. (2000) have also reported that high exposure to bioaerosols in compost workers is significantly associated with higher frequency of health complaints and diseases as well as higher concentrations of specific antibodies against molds and actinomycetes.

Quantitative Microbial Risk Assessment

When microorganisms are aerosolized, one of the primary questions is how far and in what concentrations they will travel (Dowd et al. 2000). Mathematical models have been designed to predict the transport of microorganism associated aerosols. Pasquill (1961) described a classic model of particulate airborne transport of aerosols launched from a continual point source, as might be represented by a biosolids pile at a windrow composting facility. Lighthart and Frisch (1976) modified Pasquill's equation to include a microbial inactivation constant. They rationalized that UV radiation and desiccation will lead to microbial inactivation during transport. Bioaerosol sampling used in conjunction with aerosol transport models can be used to estimate exposure during inhalation. This in turn could be used in dose-response models (for a given microorganism) to determine the risks of infection (Haas et al. 1999). Based on actual field sampling data, Dowd et al. (2000) have developed projections of concentrations of organisms per m^3 as predicted by point source and area source models for distances from 100 -10,000 m (Table 4).

Table 4. Concentrations of organisms as predicted by point source and area source models (Dowd et al. 2000)

Windspeed ms^{-1}	100 m	500 m	1000 m	10,000 m
		Virus area source		
2	$5.3 \cdot 10^{-4}$	$2 \cdot 10^{-6}$	$1.8 \cdot 10^{-9}$	$7 \cdot 10^{-63}$
5	$2.9 \cdot 10^{-3}$	$2.6 \cdot 10^{-4}$	$1.4 \cdot 10^{-5}$	$2.3 \cdot 10^{-27}$
10	$7.8 \cdot 10^{-3}$	$2 \cdot 10^{-3}$	$4.1 \cdot 10^{-4}$	$2.3 \cdot 10^{-15}$
20	$2 \cdot 10^{-2}$	$8 \cdot 10^{-3}$	$3.2 \cdot 10^{-3}$	$3.3 \cdot 10^{-9}$
		Bacteria area source		
2	13.8	10.5	8.1	0.6
5	34.5	26.1	19.8	2.7
10	69.1	52.8	40.5	6.5
20	138.4	106.2	82	14.6
		Virus point source		
2	$2.2 \cdot 10^{-4}$	$4 \cdot 10^{-8}$	$1.2 \cdot 10^{-11}$	$2 \cdot 10^{-66}$
5	$1.3 \cdot 10^{-3}$	$5.8 \cdot 10^{-6}$	$9.7 \cdot 10^{-9}$	$7.6 \cdot 10^{-31}$
10	$3.3 \cdot 10^{-3}$	$4.5 \cdot 10^{-5}$	$2.9 \cdot 10^{-6}$	$8.1 \cdot 10^{-19}$
20	$7.5 \cdot 10^{-3}$	$1.8 \cdot 10^{-4}$	$2.2 \cdot 10^{-5}$	$1.2 \cdot 10^{-12}$
		Bacteria point source		
2	62.7	2.4	0.6	$2 \cdot 10^{-3}$
5	157.8	6.2	1.5	0.01
20	633.2	25.4	6.3	0.05

These models were based on dose-response curves developed previously from human feeding studies and normal human lung ventilation patterns. The predicted rate of release of viruses was about 4-5 fold less than that of bacteria, probably due to the smaller number of viral than bacterial numbers within the biosolids.

Table 5. Predicted risk from bacteria originating from point sources (Dowd et al. 2000)

Exposure (h)	100 m	500 m	1000 m	10, 000 m
		Wind velocity 2 ms^{-1}		
1	0.02	$7.8 \cdot 10^{-4}$	$2 \cdot 10^{-4}$	$5.5 \cdot 10^{-7}$
8	0.12	$6.2 \cdot 10^{-3}$	$1.6 \cdot 10^{-3}$	$5.2 \cdot 10^{-6}$
24	0.25	0.02	4.6 · 10-3	$.6 \cdot 10^{-5}$
		Wind velocity 5 ms^{-1}		
1	0.05	$2 \cdot 10^{-3}$	$4.9 \cdot 10^{-4}$	$3.3 \cdot 10^{-6}$
8	0.23	0.02	$3.9 \cdot 10^{-3}$	$2.6 \cdot 10^{-5}$
24	0.40	0.04	0.01	$7.8 \cdot 10^{-5}$
		Wind velocity 10 ms^{-1}		
1	0.09	$4.1 \cdot 10^{-3}$	$1 \cdot 10^{-3}$	$9.8 \cdot 10^{-6}$
8	0.33	0.03	$7.9 \cdot 10^{-3}$	$7.8 \cdot 10^{-5}$
24	0.50	0.08	0.02	$2.3 \cdot 10^{-4}$
		Wind velocity 20 ms^{-1}		
1	0.15	$8.1 \cdot 10^{-3}$	$2.4 \cdot 10^{-3}$	$1.6 \cdot 10^{-5}$
8	0.44	0.06	0.02	$1.3 \cdot 10^{-4}$
24	0.60	0.14	0.05	$3.9 \cdot 10^{-4}$

Based on the above projections, the investigators modeled the probability of infection to an exposed individual (Table 5). The risk of bacterial infections to workers at the biosolids "application site" was predicted to be higher for bioaerosol originating from the point sources rather than the area sources. At 100 m, after 1 h of exposure to bioaerosol downwind from the site, a 2% infection risk was predicted. Higher wind velocities were found to increase the infection risks. These studies thus suggest that the possible health risks from biosolids-generated bioaerosol should not be overlooked. However, it is imperative that epidemiological studies be conducted alongside microbial monitoring studies so that the accuracy of such predictions can be verified using epidemiological data. Thus, when open windrow based biosolid-based composting plants are being designed and operated, acomprehensive microbiological and epidemiological monitoring plans have to be carefully designed and implemented.

Current Limitations

There is a paucity of published datasets on the occurrence of specific bioaerosolized bacterial and viral pathogens around composting facilities. The absence of data can be attributed to the challenges associated with developing and implementing an effective bioaerosol monitoring program. This is due to the challenges imposed on implementing a rigorous sampling scheme, the need to integrate sampling with micrometerological fluctuations, as well as the lack of efficient and portable bioaerosol samplers. Other than the ASTM standard sampling protocol for evaluating the microbiological quality around facilities handling municipal solid wastes (ASTM 1993), there are no standardized sampling schemes for determining the bacteriological and viral quality around composting operations. Even though there have been improvements in bioaerosol sampler design, recent studies have shown that there are significant differences in their efficacies (Juozaitis et al. 1994). In addition to the lack of adequate samplers and sampling methods, the issue of viable but nonculturable microorganisms is still a major limitation of culture-based analytical methodologies. Microbial cells during their transport, deposition and sampling are exposed to inactivation/desiccation which could injure the bacterial cells (Terzieva et al. 1996; Lange et al. 1997). These "injured" cells may be incapable of being cultured on routine microbiological media. Thus, assays which rely on culture-based enumerations can be underestimating the actual number of viable cells within bioaerosols. Though molecular biology-based assays such as gene probe hybridization and gene amplifications have the promise to detect and characterize specific microbial groups within bioaerosols, the methods still suffer from technical shortcomings such as inhibitory sample effects, sample processing deficiencies and laborious protocols and possible laboratory-based contamination (Alvarez et al. 1995; Pena et al. 1999). Droffner and Brinton (1995) have reported on the detection of *Salmonella* – specific nucleic acids within thermophilic compost piles, suggesting that microbial nucleic acids can be resistant to degradation even at the elevated temperatures found within compost piles. The detection of stable nucleic acid sequences does not imply viable organsisms and care should be taken when interpreting molecular analysis such as gene probe hybridizations and gene amplifications.

Future Research Areas

Significantly more research is needed to accurately determine the conditions that permit the aerosolization of microbial pathogens from compost piles. The precise composition of composted materials and bioaerosols from composting facilities needs to be identified using a variety of conventional and contemporary molecular tools such as qualitative and quantitative PCR assays (Alvarez et al. 1995; McGilloway et al. 2001), denaturing gradient gel electrophoresis (DGGE), in situ hybridizations (Neef et al. 1995), and flow cytometry (Lange et al. 1997) so that the process can be optimized and standardized using microbial populations as end-point controls rather than relying solely on temperature end-points. Such approaches would also obviate the need to rely solely on culture-based methods to characterize bioaerosol. If monitoring is performed at sites surrounding commercial composting facilities, bacterial isolates should be archived so the presence of specific pathogens can be confirmed with molecular techniques. Also, the availability of archived isolates permits the use of DNA fingerprinting methodologies for source-tracking purposes to identify whether the isolates are originating from the composting operations (Dowd and Pillai 1999).

Monitoring data covering a variety of composting operations are urgently needed so that appropriate management strategies could be instituted if needed. It is dangerous to assume that there are no public health risks from bioaerosol from compost operations based on the lack of published information. In addition to the monitoring studies, epidemiological studies involving compost workers as well as individuals from the surrounding population centers have to be undertaken to improve our understanding of composting and associated public health risks. Significant technological improvements have to be made in bioaerosol sampler design to facilitate efficient sampling over extended periods of time. Additionally, rugged and portable samplers have to be developed so that they are capable of being used repeatedly at remote field sites that have limited electrical and other utilities. In the meantime, however, biosolid and compost management strategies such as reducing mechanical agitation of open feed stocks and composted materials, improved facility design and sitting to control off-site emissions, bioaerosol dispersion control and maximizing buffer distances between compost facilities and population centers should be implemented to reduce human and animal exposure to pathogen-laden bioaerosols.

Acknowledgements. This work was supported by funds from the Texas Higher Education Coordinating Board's Advanced Technology Program Projects # 000517-0361-1999 and 000517-0165-1997 and Hatch grant H 8708. The assistance provided by Ms. Patricia Horsman in the preparation of this manuscript is greatly appreciated.

References

Alvarez AJ, Buttner MP, Stetzenbach LD (1995) PCR for bioaerosol monitoring: sensitivity and environmental interference Appl Envir Microbiol 61: 3639-3644

American Society for Testing and Materials (1993) Standard practice for sampling microorganisms at municipal solid-waste processing facilities E 884-82, pp 42-46

Brenner KP, Scarpino PV, Clark SC (1988) Animal viruses, coliphages and bacteria in aerosols and wastewater at a spray irrigation site. Appl Environ Microbiol 54: 409-415

Bunger J, Antlauf-Lammers M, Schulz TG, Westphal GA, Muller MM, Ruhnau P, Hallier E (2000) Health complaints and immunological markers of exposure to bioaerosol among biowaste collectors and compost workers. Occup Environ Med 57(7): 458-464

Clemmer DI, Hickey JLS, Gridges JF, Schliessmann DJ, Shaffer MF (1960) Bacteriologic studies of experimental air-borne salmonellosis in chicks. J Infect Dis 106: 197-210

Darlow HM, Bale WR, Carter GB (1961) Infection of mice by the respiratory route with *Salmonella typhimurium*. J Hyg 59: 303-308

Douwes J, Wouters I, Dubbeld H, van Zwieten L, Steerenberg P, Doekes G, Heedrik D (2000) Upper airway inflammation assessed by nasal lavage in compost workers: a relation with bio-aerosol exposure. Am J Ind Med 37(5): 459-468

Dowd SE, Maier RM (2000) Aeromicrobiology. In: Maier RM, Pepper IL,Gerba CP (eds) Environmental microbiology. Academic Press, San Diego, pp 91-122

Dowd SE, Pillai SD (1999) identifying the sources of biosolids derived pathogen indicator organisms in aerosols by ribosomal DNA fingerprinting. J Env sci Health -A (34): 1061-1074

Dowd SE, Widmer KW, Pillai SD (1997) Thermotolerant clostridia as an airborne pathogen indicator during land application of biosolids. J Environ Qual 26: 194-199

Dowd SE, Gerba CP, Pepper IL, Pillai SD (2000) Bioaerosol transport modeling and risk assessment in relation to biosolids placement. J Environ Qual 29: 343-348

Droffner ML, Brinton WF (1995) Survival of *E.coli* and *Salmonella* populations in aerobic thermophilic composts as measured with DNA gene probes. Zentralbl Hyg Umweltmed 197: 387-397

Epstein E (1997) The science of composting. Technomic Press, PA.

Haas CN, Rose JB, Gerba CP (1999) Quantified microbial risk assessment. Wiley, New York

Herman D, Maier R (2000) Consequences of biogeochemical cycles gone wild. In: Maier RM, Pepper IL,Gerba CP (eds) Environmental microbiology. Academic Press, San Diego, pp 347-361

Hugh-Jones, ME, Wright PB (1970) Studies on the 1967–8 foot and mouth disease epidemics: the relation of weather to the spread of disease. J Hyg 91: 33

Juozaitis A, Willeke K, Grinshpun SA, Donelly JA (1994) Impaction onto a glass slide or agar versus impingement into a liquid for the collection and recovery of airborne microorganisms. Appl Env Microbiol 66: 861-870

Lange, JL, Thorne PS, Lynch N (1997) Application of flow cytometry and fluorescent in situ hybridization for assessment of exposures to airborne bacteria. Appl Env Microbiol 63: 1557-1563

Lighthart B, Frisch AS (1976) Estimation of viable airborne microbes downwind from a point source. Appl Environ Microbiol 60: 861-870

Lighthart B, Mohr AJ (1987) estimating downwind concentrations of viable airborne microorganisms in dynamic atmospheric conditions. Appl Env Microbiol 53: 1580-1583

Lighthart B, Shaffer BT (1995) Airborne bacteria in the atmospheric surface layer: temporal distribution above a grass seed field. Appl Environ Microbiol 61: 1492-1496

Lighthart B, Stetzenbach LD (1994) Distribution of microbial bioaerosol. In: Lighthart B, Mohr AJ (eds) Atmospheric microbial aerosols, theory and applications. Chapman and Hall, New York, pp 68-98

McGilloway R, Weaver R, Pillai SD (2001) TaqMan-based approached for estimating Nitrobacter spp in clinoptilolite compounds. Abstr Ann Meeting Am Soc Microbiol Orlando, Fl. May

McLamarra J, Pruitt J (1995) Beneficial reuse in the southeast. Ind Wastewater 3(2): 22-24

Millner PD, Olenchock SA, Epstein E, Rylander R, Haines J, Walker J, Ooi BL, Maritato M (1994) Bioaerosols associated with composting facilities. Compost Sci Util 2: 6-57

Millner PD, Bassett D, Marsh PB (1980) Dispersal of Aspergillus fumigatus from sewage sludge compost piles subjected to mechanical agitation in open air. Appl Environ Microbiol 39: 1000-1009

Millner PD, Marsh PB, Snowden RB, Parr JF (1977) Occurrence of *Aspergillus fumigatus* during composting of sewage sludge. Appl Environ Microbiol 34: 765-772

Morozzi G, Sportolari R, Caldini G, Cenci G, Morosi A (1988) The effect of anaerobic and aerobic wastewater treatment on faecal coliforms and antibiotic resistant fecal coliforms. Zentralbe BaktMikrobiol Hyg 185: 340-349

Neef A Albrecht A, Tilkes F, Harpel S, Herr C, Liebl K, Eikmann T, Kampfer P (1999) Measuring the spread of airborne microorganisms in the area of composting sites. Schriftenr Ver Wasser Boden Lufthyg 104:655-664

Neef A, Amann R, Schleifer K-H (1995) Detection of microbial cells in aerosols using nucleic acid probes. Syst Appl Microbiol 18: 113-122

Nersting L, Malmros P, Sigsgaard T, Petersen C (1991) Biological health risks associated with resource recovery, sorting of recycled waste and composting. Grana 30: 454-457

Pasquill F (1961) The estimation of dispersion of windborne material. Meteorol Mag 90: 33-49

Pedgley DE (1991) Aerobiology: the atmosphere as a source and sink for microbes. In: Andrews JH, Hirano SS (eds) Microbial ecology of leaves. Springer, Berlin, pp 43-59

Pena J, Ricke SC, Shermer CL, Gibbs T, Pillai SD (1999) A gene amplification-hybridization sensor based methodology to rapidly screen aerosol samples for specific bacterial gene sequences. J Env Sci Health 34(A): 529-556

Pepper IL, Gerba CP (1989) Pathogens. In: Agricultural sludge reclamation study, part 6. Pima County Wastewater Management Dept. Tucson, AZ, pp 94-146

Pillai SD, Widmer KW, Dowd SE (1996) Occurrence of airborne bacteria and pathogen indicators during land application of sewage sludge. Appl Env Microbiol 62: 296-299

Randall CW, Ledbetter JO (1966) Bacterial air pollution from activated sludge units. Am Ind Hyg Assoc J Nov-Dec: 506-519

Salem H, Gardner De (1994).Health aspects of bioaerosol. In: Lighthart B, Mohr AJ (eds) Atmospheric microbial aerosols, theory and applications. Chapman and Hall, New York, pp 304-330

Sattar SA, Ijaz MK (1987) Spread of viral infections by aerosols. Crit Rev Environ Control 17: 89-131

Scanlan JW, Psaris PJ, Kuchewither RD, Nelson M, Metcalf M, Reinero RS, Akero TG, Flynn BP, Schafer PL, Kelly JM (1989) Review of EPA sewage sludge technical regulations. J Water Pollut Control Fed 61: 1206-1213

Semenza R (1995) California food processors turn to land application. Ind Wastewater 3(2): 25-26

Soares AC (1990) Occurrence of enteroviruses and Giardia cysts in sludge before and after-anaerobic digestion. MS Thesis, Dept of Microbiology and Immunology, Univ of Arizona, Tucson, AZ

Stelzer W, Mochmann H, Richter U, Dobberkan HJ (1988) Characterization of *Campylobacter jejuni* and *Campylobacter coli* isolated from wastewater. Zent Bakt Mikrobiol Hyg 269: 188-196

Straub TM, Pepper IL, Gerba CP (1993) Hazards from pathogenic microorganisms in land-disposed sewage sludge. Rev Environ Contam Toxicol 132: 55-91

Terzieva S, Donnelly J, Ulevicius V, Grinshpun SA, Willeke K, Stelma GN, Brenner KP (1996) Comparison of methods for detection and enumeration of airborne microorganisms collected by liquid impingement Appl Environ Microbiol 62: 2264-2272

USEPA (1989) The solid waste dilemma: an agenda for action. Office of Solid Waste. EPA/503-SW-89-019 Washington D.C.

Wathes CM, Zaidan WAR, Pearson GR, Hinton M, Todd N (1988) Aerosol infection of calves and mice with *Salmonella typhimurium*. Vet Rec 123: 590-594

Wickman HH (1994) Deposition, adhesion and release of bioaerosol. In: Lighthart B, Mohr AJ (eds) Atmospheric microbial aerosols, theory and applications. Chapman and Hall, New York, pp 99-165

Passively Aerated Composting of Straw-Rich Organic Pig Manure

A. Veeken, V. de Wilde, G. Szanto and B. Hamelers[1]

Abstract. Composting of animal manures using forced aeration generates high ammonia emissions. Although passively aerated composting results in lower ammonia emissions, there are concerns it may lead to high methane emissions and poor organic matter degradation. This study examined these issues using pig manure originating from organic farming systems, which could be directly composted by passive aeration because it contains high amounts of straw, which served as a bulking agent. Experiments were performed in 2-m^3 reactors (2 m in height and closed walls) with a bottom grid providing vertical aeration of the compost pile. A series of experiments evaluated the effects of the compost bed structure and monthly turning of the piles on the composting process and on the emissions of ammonia, nitrous oxide and methane. Effectiveness of the composting process strongly depended on the density of the compost. Above a critical maximum density natural convection would not be initiated, aerobic degradation would fail, and anaerobic conditions would lead to emissions of methane and odorous compounds. Below a critical lower density the high rate of natural convection would keep the temperature low, thereby preventing the destruction of pathogens and weeds. Best results were observed at a density of 700 kg m^{-3}, where both aerobic degradation and drying were adequate and temperatures were high enough to kill pathogens and weeds. Monthly turning shortened the process from 8 to 3 months, as it provided a better compost bed structure. Moreover, turning gave a more homogeneous end product. Composting did not result in significant ammonia emissions (1–3% of total nitrogen initially present) and nitrous oxide emissions were low (2%) for turned piles but significant (7%) for undisturbed piles. Anaerobic regions were present inside the pile but methane emissions were not observed as methane was oxidised in the top layer of the bed. Mass balance analysis indicated that the major part of nitrogen (50%) was emitted as dinitrogen, presumably a result of simultaneous nitrification-denitrification. Surprisingly, nitrification also took place at temperatures higher than 50 °C, where thermophilic methanotrophs may have been responsible for oxidation of ammonia.

[1] Department of Agrotechnology and Food Sciences, Environmental Technology, Wageningen University, P.O. Box 8129, 6700 EV Wageningen, The Netherlands
E-mail: adrie.veeken@algemeen.mt.wag-ur.nl
Telephone: +31 317 483344; Fax: +31 317 482108

H. Insam, N. Riddech, S. Klammer (Eds.)
Microbiology of Composting

Introduction

Modern livestock industries produce high emissions of ammonia, odours and greenhouse gases (methane and nitrous oxide) from buildings and during storage and application of manure. These emissions lead to a large number of environmental problems (Burton 1997). Animal manures can also be seen as a highly valuable resource through the recycling of nutrients and organic matter to soil systems, leading to biological control, reduced use of artificial fertilisers and pesticides and peat (Hoitink and Grebus 1994; Richard and Choi 1999). Manure treatment increasingly is viewed as a way to reduce negative environmental impacts associated with livestock production, and a wide range of technologies are being developed (Burton 1997). Composting is among the most promising techniques for treatment of the solid fraction of animal manures, as it is inexpensive and produces a valuable endproduct (Lopez-Real and Baptista 1996). The composting process results in breakdown and stabilisation of organic matter, mass reduction (Haug 1993) and destruction of pathogens and plant seeds (Bollen 1993). The resulting compost can be used as an organic fertiliser in agriculture without negative impacts on plant growth and the environment. Commonly applied composting strategies of forced aeration and windrow turning, so-called intensive composting systems, result in short composting times but will also lead to high emissions of ammonia (Witter and Lopez-Real 1988). Composting of animal manures in passively aerated static piles, so-called extensive composting, has been shown to result in a significant reduction in ammonia emissions (Lopez-Real and Baptista 1996).

Extensive composting can be used on-farm because it approximates normal handling procedures by farmers. As passively aerated piles are not turned frequently, and forced aeration is not applied, capital and operating costs are low (Mattur 1997). Oxygen supply and heat and water removal in passively aerated composting are regulated by natural convection (Lynch and Cherry 1995). Natural convection is driven by the difference in air density between the air inside the compost and the density of the ambient air, resulting in the so-called chimney effect (Kubler 1982). The velocity of natural convection increases with increasing permeability of the compost bed (Kubler 1982). In time, compaction of the compost bed results in low air permeability and reduces the velocity of natural convection. Compaction thus reduces aerobic degradation and may lead to anaerobic regions and emissions of methane and nitrous oxide, both harmful greenhouse gases (Czepiel et al. 1996).

This chapter presents the effect of the porosity of the compost bed and the effect of periodic turning (once a month) on passively aerated composting of straw-rich pig manure, generated by an organic farming operation where high amounts of straw are applied. The composting process was evaluated with respect to drying, organic matter degradation and emissions of ammonia, nitrous oxide and methane. The passively aerated, static pile composting process was physically simulated in 2-m^3 reactors with closed sidewalls and a perforated bottom grid of 1-m^2.

Materials and Methods

Composting Experiment

The composting reactor had a volume of 2-m^3 with a height of 2 m and a square bottom of 1 · 1 m^2. At 0.2 m from the bottom a metal grid with 2 · 2 cm^2 holes was fixed which supported the composting bed and facilitated the inflow of ambient air from below. The side walls of the reactor were made of insulating sandwich panels to minimise heat loss. At 0.4 m above the compost bed a roof was constructed which prevented rainfall from moistening the compost bed. The chosen experimental setup gave a realistic physical simulation of the composting process, but the one-dimensional profile facilitated the interpretation of the composting process while simplifying sampling and analysis.

Temperature sensors (PT-100) were fitted in 16 · 1.8 mm PE tubes which were distributed at regular intervals of approx. 0.3 m over the height of the compost bed. The temperature sensors were connected to a multichannel automatic data acquisition system (Data logger DT100, Data Electronics Ltd., Australia) and temperatures were recorded at 1-h intervals. Gas samples of 2–5 l were collected periodically from the same PE tubes in 5-l gas bags (Teseraux, Tecobag, PETP/AL/PE 12/12/75) and analysed for oxygen, carbon dioxide, nitrous oxide and methane. Ammonia emissions were measured periodically by determining the ammonia concentration in the air next to the top of the compost bed. The velocity of natural convection was measured by an anemometer device measuring the velocity of the air leaving the top of the compost bed. Further details of the experimental setup were described by Veeken et al. (2001a).

Samples were taken from the compost bed at three different heights, with samples collected at each turning of the piles as well as at the end of the composting process. The samples were cut and homogenised in a 5-l stainless steel commercial food chopper and used for determination of total solids (TS), volatile solids (VS), Kjeldahl nitrogen (Nkj), total ammonia (NH_3) and nitrate (NO_3^-). The total weight of the compost bed was determined before composting, after each turning and after composting.

The straw-rich manure was collected at an organic pig farm (Uden, The Netherlands) and two sets of experiments were performed. In the first experiment, three composting runs were made at significantly different porosities. Reactor 1 was filled with the straw-rich manure, while 3% extra wheat straw (on total mass basis) was blended into the mixture for reactor 2 and 5% extra wheat straw added to the mixture in reactor 3 (see Table 1).

In the second experiment, one reactor was turned monthly (A) and one reactor was left undisturbed (B) for 4 months, with the same medium density mixture in each reactor (see Table 1).

Table 1. Characteristics of compost bed at the start and the end of composting

Effect of porosity	Reactor 1		Reactor 2		Reactor 3	
	Start	End	Start	End	Start	End
Time (d)	46		137		98	
Weight (kg)	1380	1320	900	550	400	280
Height (m)	1.7	1.5	1.6	1.1	1.1	0.8
Density (kg m^{-3})	1100	1290	700	560	560	540
TS (kg .kg^{-1})	0.23	0.27	0.25	0.28	0.30	0.32
VS (kg Kg^{-1} TS)	0.71	0.71	0.74	0.73	0.79	0.76
PH (-)	8.6	8.2	8.6	7.4	8.6	7.3
VFA (g kg^{-1} TS)	3.9	0.2	4.2	0.1	2.5	0.2
N-org (g kg^{-1} TS)	24.4	22.7	20.4	20.0	21.6	16.1
NH_3 (g kg^{-1} TS)	15.0	17.3	13.1	3.3	9.9	4.3
NO_3 (g kg^{-1} TS)	0.3	0.2	0.2	5.9	0.2	9.3

Effect of turning	Turned		Undisturbed	
	Start	End	Start	End
Time (days)	118		118	
Weight (kg)	1132	449	1045	611
Height (m)	1.30	0.72	1.30	0.95
Density (kg m^{-3})	871	623	804	643
TS (kg kg^{-1})	28.2	43.5	28.4	34.3
VS (kg kg^{-1} TS)	71.1	49.3	70.7	62.8
PH (-)	8.4	8.3	8.5	8.4
VFA (g kg^{-1} TS)	69.9	0.2	51.8	2.9
N-org (g kg^{-1} TS)	15.8	19.8	12.7	4.1
NH_3 (g kg^{-1} TS)	18.4	5.7	17.6	12.1
NO_3 (g kg^{-1} TS)	0.1	0.4	0.1	3.3

Analyses

Total solids (TS), volatile solids (VS), Kjeldahl nitrogen (Nkj) and total ammonia (NH_3) of the compost samples were determined according to standard methods (APHA 1992). The amount of organic nitrogen (N-org) was calculated as N-org = Nkj - NH_3 - NO_3^-. The composition of the gas phase (O_2, N_2, CO_2 and CH_4) was analysed using a Hewlett Packard 5890 gas chromatograph equipped with a molecular sieve as described by Ten Brummeler and Koster (1990). N_2O was measured according to Weiss (1981) with a Hewlett Packard 5890A gas chromatograph. The ammonia concentration in the air adjacent to the top of the compost bed was measured manually with Draeger gas detection tubes and a bellow pump according to the manufacturer's directions (Draeger 1996).

Results

Effect of Porosity on the Composting Process

Temperature

Fig. 1. shows average temperatures of the three composting runs, which were monitored online.

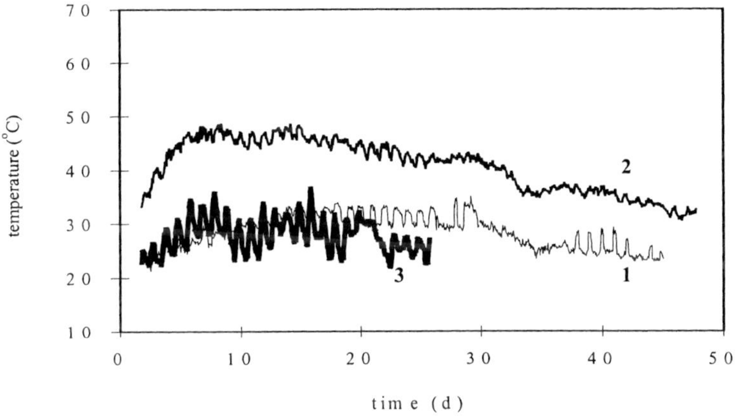

Fig. 1. Temperature course during composting for different compost bed densities (1: 1100 kg m^{-3}; 2: 700 kg m^{-3}; 3: 560 kg m^{-3})

Both reactors 1 and 3 showed only a small temperature increase with respect to ambient temperature between days 10 and 30 and only reactor 2 showed temperature development typical of a well-operating composting process. After several weeks of online monitoring, the temperatures of the reactors gradually dropped to ambient levels.

Organic Matter Degradation and Drying

The absolute loss of water and organic matter can be calculated accurately on the basis of the total weight, TS and VS before and after composting. The water loss amounted to 1 kg (0.4%) for reactor 1, 75 kg (33%) for reactor 2 and 31 kg (26%) for reactor 3. The degradation of organic matter amounted to 4.8 kg (2 %) for reactor 1, 31.3 kg (33%) for reactor 2, and 27.7 kg (29%) for reactor 3, which correspond to linear VS degradation rates of 0.05%, 0.24% and 0.30% VS per day, respectively. Despite the small changes in TS and VS for all reactors as measured on a concentration basis (Table 1), water removal and organic matter degradation were significant for reactors 2 and 3.

Nitrogen Balance

The concentration of nitrogenous compounds reported on a TS basis (Table 1) does not provide any information about the absolute changes of the various nitrogen compounds during composting because organic matter (part of the TS) is degraded. In these experiments, changes in absolute amounts of nitrogenous compounds (in kg) could be calculated because the absolute changes in total weight and composition of the compost bed were measured (Table 1).

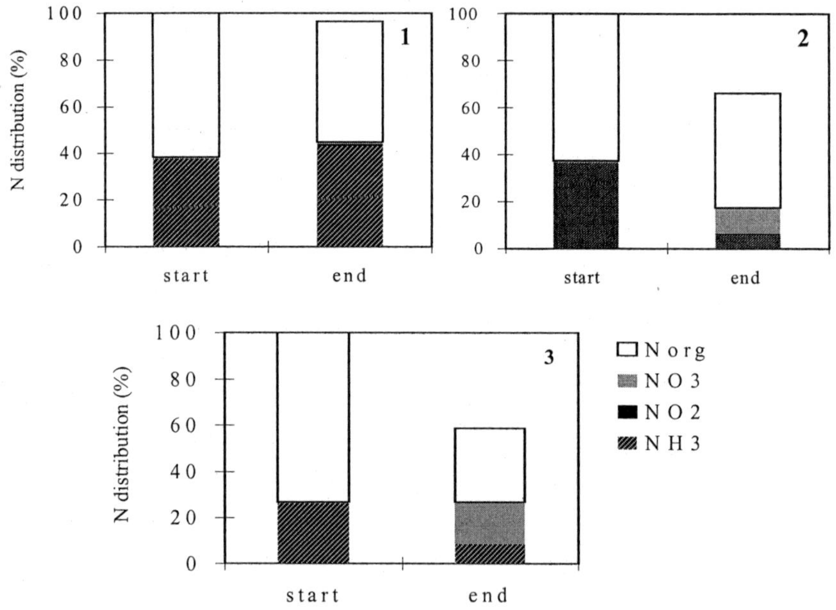

Fig. 2. Nitrogen distribution before and after composting for different compost bed densities (1: 1100 kg m^{-3}, 2: 700 kg m^{-3}, 3: 560 kg m^{-3})

The changes in the absolute quantities of these components (in % of the total N initially present) for the three reactors are given in Fig. 2. In reactor 1 no significant changes in the N distribution took place. The 3% loss of N was within the sampling and analytical error due to the heterogeneity of the compost bed. The manure bed of reactor 2 showed a significant shift in N distribution, with a large part of the ammonia converted to nitrate and some to nitrite. The decrease in N_{org} is due to mineralisation of N-containing VS (e.g. proteins). Reactor 3 also showed a high reduction in ammonia and even a greater accumulation of nitrate. There was no leachate loss from these reactors, so the N which disappeared from reactors 2 and 3 had to have been emitted as ammonia, nitrous oxide or N_2 gas (see later).

Effect of Turning on the Composting Process

Temperature

Fig. 3 shows the temperature profiles over the compost bed of the turned (Fig. 3A) and undisturbed (Fig. 3B) reactors along with the ambient temperature.

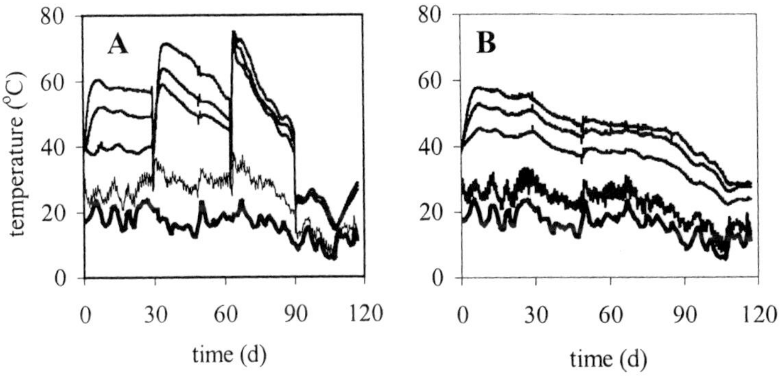

Fig. 3. Temperature profiles during composting of turned (**A**) and undisturbed (**B**) reactors

Both reactors showed a strong increase in temperature directly after the start of the experiment and both systems reached a maximum temperature of 60 °C after 7 d in the top of the compost bed. In the undisturbed reactor, the maximum temperature remained around 55 °C for 30 d. It dropped down to 45 °C thereafter and this temperature was maintained for 60 d. It then levelled off to 27 °C. This temperature profile is typical for composting, with the temperature decreasing as the rate of aerobic degradation decreases. The turned reactor also maintained its maximum temperature near 55 °C for the first 30 d, but within 5 d after turning the temperature increased to 70 °C. Thereafter, the temperature dropped gradually. After the second turning, the temperature rose again to 70 °C within 4 d, but the subsequent drop in temperature was faster. After the third turning, the temperature did not rise again, indicating that rapid aerobic degradation has completed.

Degradation and Drying

As for the density trials, absolute removal of water (in kg) was calculated on the basis of the total mass, TS and VS before and after composting (Table 1). At the end of the experiment, 520 kg of water had been removed from the turned system and only 360 kg from the undisturbed system. For the turned system, 57% of VS was degraded within 4 months, compared to 39% for the undisturbed system. This corresponded to linear degradation rates of 0.48% and 0.33% VS d^{-1}, respectively.

Due to mixing during monthly turning, TS and VS values of the turned reactor showed no gradient over the height of the reactor. However, large differences in TS and VS were observed for the undisturbed reactor (Table 1).

Table 1. Composition of the compost bed over the height of the reactors during composting (d)

Parameter	Turned			Undisturbed		
Day	Top	Middle	Bottom	Top	Middle	Bottom
TS 0	28	28	29	33	30	27
30	40	30	30	-	-	-
63	40	37	41	-	-	-
91	46	43	39	-	-	-
118	39	43	42	58	35	29
VS 0	75	70	73	72	72	71
30	63	62	61	-	-	-
63	56	55	53	-	-	-
91	55	50	52	-	-	-
118	48	52	50	51	55	68
N-org 0	13.8	20.1	10.9	12.3	17.1	16.7
30	15.0	16.3	17.2	-	-	-
63	10.4	7.0	3.7	-	-	-
91	17.3	11.5	20.8	-	-	-
118	22.3	15.8	15.0	3.1	7.9	11.0
NH_3 0	15.8	17.0	16.6	14.9	16.2	17.6
30	14.6	16.0	13.4	-	-	-
63	7.9	10.4	8.8	-	-	-
91	7.1	6.7	6.5	-	-	-
118	4.9	5.6	5.8	7.1	12.1	11.4
NO_3 0	0.05	0.02	0.02	0.01	0.07	0.02
30	0.03	0.05	0.08	-	-	-
63	0.27	0.09	0.08	-	-	-
91	0.39	0.08	0.46	-	-	-
118	0.57	0.78	0.99	1.53	0.66	0.50

Natural Convection

Fig. 4 shows the velocity of natural convection as function of the temperature difference observed over time (data points are the averaged values per day).

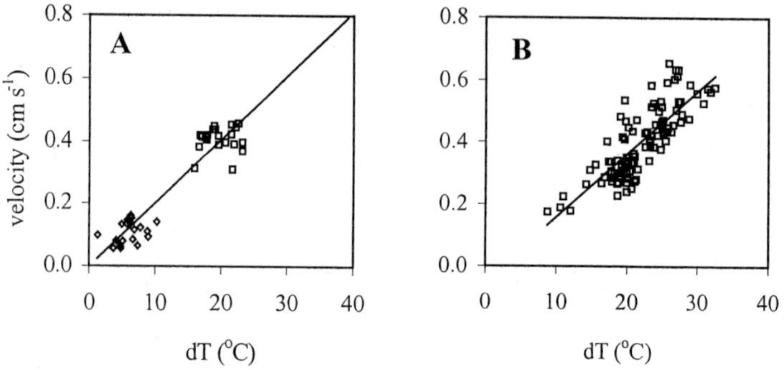

Fig. 4. Velocity of natural convection versus temperature difference for turned (**A**) and undisturbed (**B**) reactors

Emission Measurements

The gas phase composition in the top, middle and bottom of the reactors is shown in Fig. 5.

The oxygen and carbon dioxide profiles were as expected for static pile composting. Oxygen-saturated air entered the bottom of the reactor by natural convection. As the air rises in the composting pile, oxygen is consumed by aerobic degradation. As a result, the concentration of oxygen drops. The reduction in oxygen levels inside the bed is accompanied by an increase in carbon dioxide. However, the drop in oxygen concentration is not a direct measure of aerobic activity, as the rate of aeration is also important, i.e. a low oxygen content can be an indication of either a high aerobic degradation rate or a low air replacement rate, or sometimes a combination of the two. Both systems showed a large drop in oxygen in the first month (as low as 5%). The oxygen level remained low in the undisturbed reactors (10%) but for the turned reactor oxygen approximated levels in ambient air (15–20%).

In the turned reactor, the methane content in the interior of the pile was high in the first month but very low for the rest of the composting period.

In the undisturbed reactor, the methane content in the interior remained high throughout the composting process. The high concentration of methane in the gas phase indicates the extent of anaerobic metabolism in the compost bed. The methane content was low again in the top of the reactors, as methane is readily oxidised by methanotrophic bacteria in oxygen-rich regions (Figueroa 1993). High methane and low oxygen concentrations in the compost bed showed that anaerobic zones were present. The anaerobic zones in the interior of the piles were due to agglomeration of particles so that oxygen diffusion was limited. The anaerobic zones were more pronounced in the undisturbed system, as compaction of the compost bed resulted in more agglomeration of particles, while turning reduced compaction because it broke apart these agglomerations. Nitrous oxide was produced as a product of denitrification inside the compost pile (concentration in ambient air 0.3 ppmv) in the anaerobic zones (Fig. 5.). The concentration in the turned system generally remained low during the course of the experiment. In the undisturbed system, the concentration dramatically increased after day 60, which corresponded to the time that significant regions of the pile began to cool below the 40 °C threshold for nitrification (Schlegel 1993).

Ammonia concentrations in the air adjacent to the top of the compost bed were very low for both systems (0–30 ppmv; Fig 6). In comparison, intensive composting systems (forced aeration or daily turning) can reach levels of 2000–8000 ppmv of ammonia in the gas phase (Burton 1997).

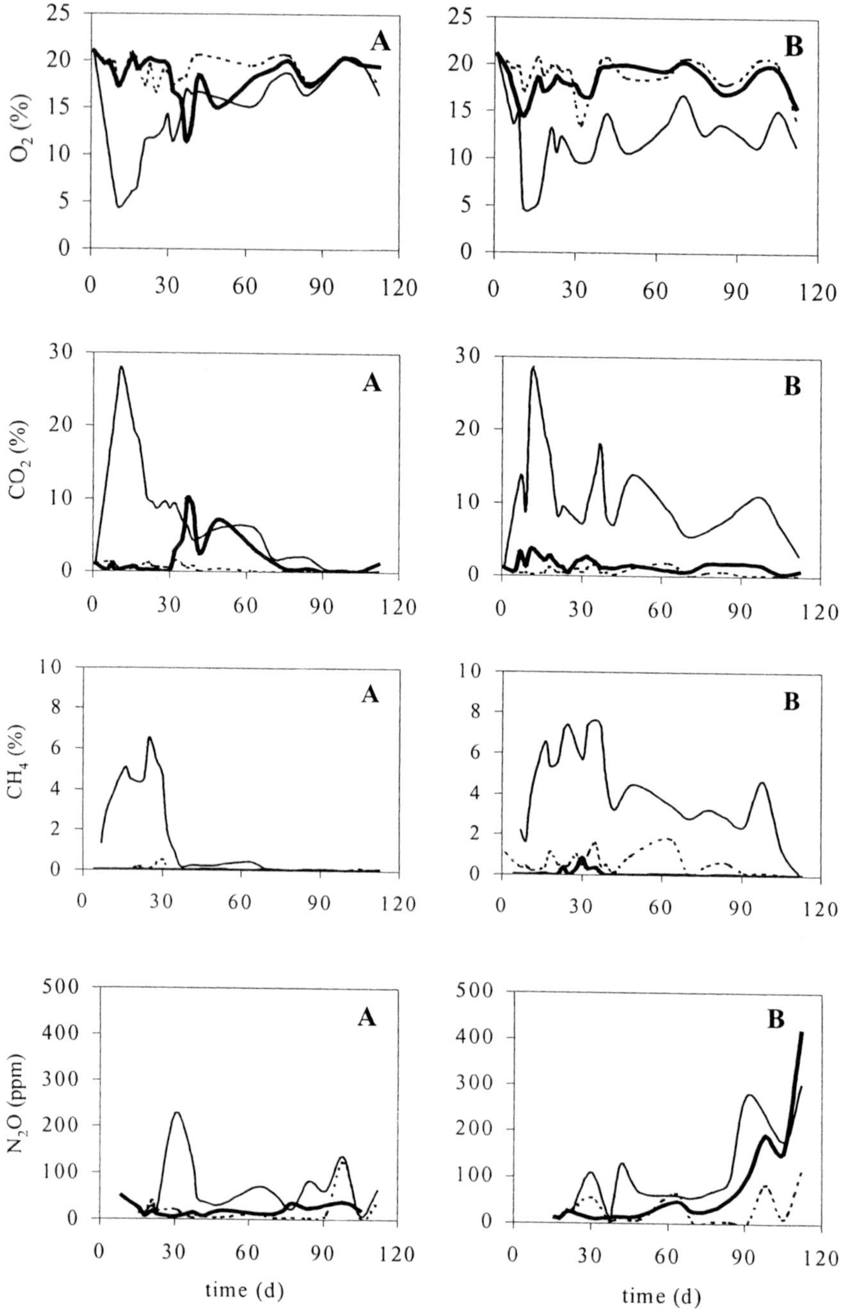

Fig. 5. Gas phase composition (O_2, CO_2, CH_4 and N_2O) in the compost bed for turned (**A**) and undisturbed (**B**) reactors (—— top; —— middle; ----bottom)

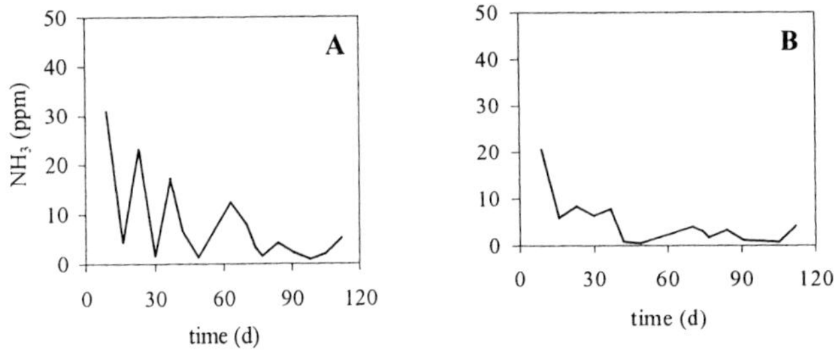

Fig. 6. Ammonia concentrations in the air adjacent to the top of the compost bed for turned (**A**) and undisturbed (**B**) reactors

Nitrogen balance

Table 1 shows the nitrogen distribution (N-org, NH_3 and NO_3^-) in the top, middle and bottom of the turned and undisturbed systems, while the average composition of the complete compost bed before and after composting are given in Table 1. In the turned system, ammonia was removed within 3 months but it was still present after 4 months in the undisturbed system. As almost no ammonia was lost due to emissions (discussed later), the decrease in ammonia can largely be attributed to nitrification. For the turned system, the decrease in ammonia was highest in the bottom part of the compost bed and lowest in the top part (Table 1). Part of the ammonia was present in the solid fraction (Veeken et al. 2001b) which is less readily available for nitrifying bacteria. As nitrate accumulated in only small amounts during composting, nitrate had to have been converted to nitrogen gas or nitrous oxide by denitrifying bacteria.

The emissions of ammonia and nitrous oxide were calculated on the basis of the velocity of natural convection (or airflow rate) and the concentration of ammonia and nitrous oxide in the air leaving the compost bed (details see Veeken et al. 2001b). It was assumed that ammonia in the air adjacent to the top of the compost bed and the concentration of nitrous oxide in the top of the compost bed were representative for the concentrations in the air emitted from the compost bed. Based on the total emissions and calculations of the absolute changes in nitrogenous compounds (on basis of the total weight and the compost bed composition), the total N balance was set up (Fig. 7).

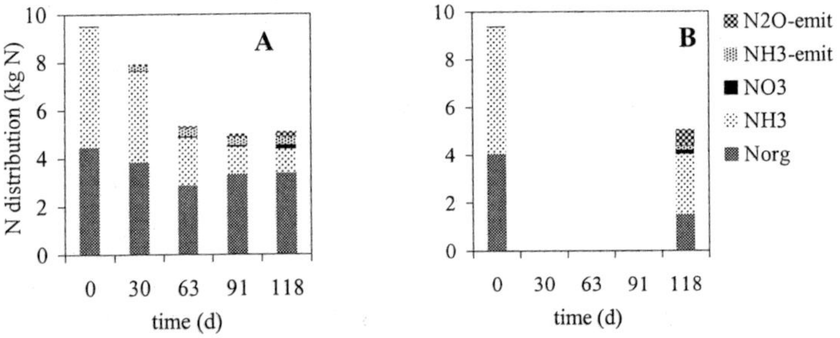

Fig. 7. Nitrogen distribution during composting for turned (**A**) and undisturbed (**B**) reactors

For the turned system, 47% of the nitrogen was lost from the compost bed, 3.5% as ammonia and 2.2% as nitrous oxide. Since there was no leachate loss from these piles, the rest must have been emitted as dinitrogen gas. For the undisturbed system, 46% of the nitrogen was lost, 1.8% as ammonia and 7.3% as nitrous oxide.

Discussion

Composting Process

The temperature rise during composting largely is due to the heat production by microorganisms degrading organic matter (Haug 1993). In passively aerated piles, heat is predominantly removed from the piles through water evaporation and subsequent removal of the water-saturated air by natural convection. However, heat losses by conduction through the surfaces of the pile can be significant also (Van Ginkel 1996). As the compost bed is heated largely by aerobic degradation, air heats up and evaporates water from the compost bed. Because hot and water-saturated air has a lower density, the air rises and leaves the reactor from the top of the bed. Initially, natural convection has to be triggered by aerobic degradation consuming the small amount of oxygen initially present in the pores of the compost bed, plus whatever oxygen can enter the pile through diffusion and wind action. Our results show that natural convection cannot be triggered above a critical maximum density. Only a small temperature rise and a low degradation rate were observed in dense samples (Fig. **1**). However, when the density was too low, degradation did take place but the temperature remains low due to high natural aeration rates. The low temperature of the compost bed in this case would prevent the destruction of pathogens and weeds.

The temperature profiles of both the turned and disturbed system were typical for a well-performing passively aerated composting process (Fernandes et al. 1994; Joshua et al. 1998). The maximum temperature of the composting process reached 60–70 °C, a temperature range where many microorganisms become less active (Haug 1993). Over time, the temperature gradually declined as the degradation rate of organic matter became smaller. The linear relation between the velocity of natural convection versus the temperature difference between the compost bed and the open air (Fig 4.) confirmed that natural convection was controlling the composting process. The process of natural convection was thus driven by the heat generated by predominantly aerobic degradation, and the velocity was determined by the aerobic degradation rate and the air permeability of the compost bed.

The rates of biodegradation and water removal were improved by monthly turning of the compost bed. This also resulted in a homogeneous product and it would also destroy pathogens more efficiently as mixing allowed the complete compost mass to experience the high temperatures in the middle and top part of the compost bed.

A complete water balance for the compost bed involves the production of water by microbial degradation and water removal by the air flowing through the compost bed. Water production during composting amounts to about 0.6 kg kg^{-1} VS degraded (Haug 1993). As natural convection is the dominating process for water removal, the water removal can be calculated from the velocity of natural convection and the water content of the warm air leaving the compost bed. Assuming typical but somewhat conservative values from our results, including a degradation rate of 0.4% per day, air leaving the compost bed at 40 °C and 100% relative humidity, and a velocity of natural convection of 0.003 m s^{-1}, the compost bed of the turned system should have reached a TS content of 70% within 75 d. The most probable reason why the compost bed did not reach the final TS of 70% within 118 d is the fact that a high amount of heat was lost by conduction through the sidewalls, despite the insulating panels.

Nitrogen Conversions

Both the turned and undisturbed systems showed a drop in organic nitrogen (N-org) due to the hydrolysis of nitrogen-containing organic compounds in pig manure, mainly proteins (Fig. 7). The small increase in N-org for the turned system might suggest that nitrogen was fixed in the system. This can be due to incorporation of nitrogen into biomass and chemical reactions of ammonia with humic substances. However, sample heterogeneity makes us believe that this small change might also be within the sampling error. The ammonia concentration decreased during composting due to nitrification. However, oxidation of ammonia during the first 2 months in the high-temperature top and middle parts of the compost bed is very surprising (Table 1), as it is common knowledge that nitrifying bacteria are unable to thrive at thermophilic temperatures (Schlegel 1993). Pel et al. (1996) demonstrated that methanotrophs are capable of ammonia oxidation under thermophilic conditions. The environmental conditions for ammonia oxidation by methanotrophs were ideal in the compost bed as both ammonia and methane were present at the interface of aerobic and anaerobic zones. The oxidation of ammonia did not consume all the ammonia but instead reached a platform, probably because residual ammonia was present in the solid phase and secluded from bacteria. Oxidised ammonia did not accumulate as nitrate and nitrite (Table 1), but instead appeared to have been converted by denitrifying bacteria to nitrogen gas and/or nitrous oxide. Nitrate probably would have been reduced in the anaerobic zones present in the compost bed, because nitrate is used as a final electron acceptor during the degradation of organic matter under anoxic conditions. Under optimum conditions, nitrate is completely reduced to N_2, but under stressed conditions N_2O becomes an important end product (Robertson and Kuenen 1991). The turned system showed a low level of production of nitrous oxide (2.2%) but in the undisturbed system the evolution of nitrous oxide was significant (7.7%). This was probably due to the combined effects of anaerobic zones in the undisturbed system and less ideal conditions such as high concentrations of ammonia and volatile fatty acids.

The evidence presented in this chapter of simultaneous methane production-oxidation and nitrification-denitrification show that both aerobic and anaerobic regions were present in the compost bed. The anaerobic regions were present as regions where oxygen was not able to diffuse, due to low porosity and large regions of aggregation (Hamelers 1993). Methane, which is produced in the anaerobic zones, is oxidised in aerobic zones, while nitrate, which is produced in aerobic zones, is reduced in anaerobic zones. Both aerobic and anaerobic products of the C and N cycles can even meet at the interface to permit ammonia oxidation by methane-oxidising bacteria under thermophilic temperatures. These aerobic and anaerobic zones can be present on both macro- and microscales. Compounds can diffuse to the other zones on a macroscale basis but also on microscale in the biofilms around solid particles or aggregates (Pel et al. 1996; De Beer and Schramm 1999).

Acknowledgements. This research was funded by the Economy, Environment and Technology Program of the Dutch Ministry of Economic Affairs (Hercules project) and by the Ministry of Agriculture, Nature Management and Fisheries. We thank Tom Richard (on sabbatical leave from Iowa State University) for valuable discussions on the paper.

References

APHA (1992) Standard methods for the examination of water and wastewater. 18th ed., part 2540D. American Public Health Association, Washington, DC

Bollen GJ (1993) Factors involved in inactivation of plant pathogens during composting of crop residues. In: Hoitink AJ (ed) Science and engineering of composting: design, environmental, microbiological and utilization aspects. Renaissance Publications, Ohio, pp 301-318

Burton CH (1997) Manure management: treatment strategies for sustainable agriculture. Silsoe Research Institute, Bedford

Czepiel P, Douglas E, Harris R, Crill P (1996) Measurement of N_2O from composting organic wastes. Environ Sci Technol 30:2519-2525

De Beer D, Schramm A (1999) Micro-environments and mass transfer phenomena in biofilms studied by microsensors. Water Sci Technol 39:173-178

Draeger 1996. Draeger gas detection tubes catalogue. Draeger, Lübeck, Germany

Fernandes L, Zhan W, Patni NK, Jui PY (1994) Temperature distribution and variation in passively aerated static compost piles. Biores Technol 48:257-263

Figueroa RA (1993) Methane oxidation in landfill top soils. In: Stegmann R (ed) Proc 4th Int Landfill Symp, Cagliari, pp 701-716

Hamelers HVM (1993) A theoretical model of composting kinetics. In: Hoitink AJ (ed) Science and engineering of composting: design, environmental, microbiological and utilization aspects. Renaissance Publications, Ohio, pp.36-58

Haug RT (1993) The practical handbook of compost engineering. Lewis, Boca Raton, Florida, USA

Hoitink HAJ, Grebus ME (1994) Status of biological control of plant diseases with compost. Compost Sci Util 2:6-12

Joshua RS, Macauley BJ, Mitchell HJ (1998) Characterization of temperature and oxygen profiles in windrow processing systems. Compost Sci Util 6:15-28

Kirchmann H, Lundvall A (1998) Treatment of solid animal manures: identification of low NH_3 emission practices. Nutr Cycl Agroecosyst 51:65-71

Kubler H (1982) Air convection in self-heating piles of wood chips. Tappi 65:79-83

Lopez-Real J, Baptista M (1996) A preliminary comparative study of three manure composting systems and their influence on process parameters and methane emissions. Compost Sci Util 4:71-82

Lynch NJ, Cherry RS (1995) Design of passively aerated compost piles: vertical air velocities between the pipes. Biotechnol Prog 12:624-629

Mattur SP (1997) Composting processes. In: Martin AM (ed) Bioconversion of waste materials to industrial products. Chapmann & Hall, London, pp 154-196

Pel R., Oldenhuis R, Brand W, Vos A, Gottschal JC, Zwart KB (1996) Stable-isotope analysis of a combined nitrification-denitrification sustained by thermophilic methanotrophs under low-oxygen conditions. Appl Environ Microbiol 63:474-481

Richards TL, Choi HL (1999) Eliminating waste: strategies for sustainable manure management. Asian-Aust J Anim Sci 12:1162-1169

Robertson LA, Kuenen JG (1991) Physiology of nitrifying and denitrifying bacteria. In: Rogers JE, Whitman WB (eds) Greenhouse gases: methane, nitrogen oxides, and halomethanes. American Society of Microbiology, Washington, DC, pp 189-200

Schlegel HG (1993) General microbiology, 7th edn. Cambridge University Press, New York

Ten Brummeler E., Koster IW (1990) Enhancement of dry anaerobic batch digestion of municipal solid waste by an aerobic treatment step. Biol Wastes 31:199-210

Van Ginkel JT (1996) Physical and biochemical processes in composting material. PhD Thesis Wageningen Agricultural University, Wageningen, The Netherlands

Veeken A, De Wilde V, Hamelers A (2001a) Passively aerated composting of straw-rich pig manure: effect of compost bed porosity. Compost Sci Util (accepted)

Veeken A, De Wilde V, Szanto G, Hamelers B (2001b) Gaseous emissions during passively aerated composting of straw-rich pig manure. J Environ Qual (submitted)

Weiss RF (1981) Determinations of carbondioxide and methane by dual catalyst flame ionization chromatography and nitrous oxide by electron capture chromatography. J Chromatogr Sci 19: 611-616

Witter E, Lopez-Real J (1988) Nitrogen losses during composting of sewage sludge, and the effectiveness of clay soil , zeolite, and compost in adsorbing the volatilized ammonia. Biol Wastes 23:279-294

Index

A

acetic acid 177–181, 185–187, 189, 197–200
Acetobacter 99, 104–107, 186
Achromobacter 145
Actinobacillus
  capsulatus 33f., 36
  equuli 33–36
  hominis 33–36
  lignieresii 34f.
  pleuropneum 34
  suis 33–36
Actinobacteria 505-510, 512, 591f.
Actinomadura sp. 273, 280
  rubrobrunea 280
Actinomycetes 12, 29, 39, 53, 55, 59, 61f., 104, 108, 279f., 487, 571f., 575–578, 580–583, 585, 587, 590, 592, 600
Adenovirus 597
aeration 119, 123f., 129, 132, 134, 143, 145, 147, 151, 153, 166, 186, 188–191, 194–200, 204, 218, 224, 229, 231, 234, 237f., 242, 245, 253, 255, 259, 283, 300, 345, 376, 376-378, 408, 446, 452, 459, 487, 560, 607f., 615, 618
Aeromonas sp. 99f., 104f.
  hydrophila 100
Afipia felis 33, 37f.
Agaricus bisporus 17, 23
agricultural wastes 324
Agromyces mediolanus 35
Alcaligenes 145
Alicyclobacillus acidoterrestris 34, 36
alpechin 324, 326f., 329-331
alpeorujo 324, 326, 328
Alternaria 578, 581
AluI 87, 89, 91f.
amino acids 120, 144–146, 221, 309f.
ammonia 140, 143f., 146–150, 152f., 155, 161–164, 188, 203f., 214–216, 224, 238, 242, 247, 249, 255, 257, 259–261, 335, 338, 375, 428, 445f., 527, 551, 554–556, 559, 607–610, 612, 615–621
  -accumulation 449
  -concentration 447
  -emission 247, 259, 261, 446, 449, 452-455, 607f.
  -formation 143, 150
  -volatilisation 445f., 452, 455
ammonification 49, 143f., 149–153, 238, 241
ammonium 70, 143–146, 148–150, 152, 205, 221, 237, 239, 241, 243f., 336, 341, 345, 347, 349f., 374, 376
amplified ribosomal DNA restriction analysis (ARDRA) 66, 81, 83f., 87f., 91f., 94f., 97f.
anaerobic activity 224
anaerobic digestion 155, 161f., 164
Anaerobranca horikoshii 36
Anclyostoma duodenale 597
Aphanomyces euteiches 222
Aporothielavia leptoderma 20f.
apples 147
Archaea 54, 58, 60, 62
Armillaria mellea 528
Arthrobacter globiformis 68, 77
Ascaris lumbricoides 597
Aspergillus 330
  flavus 577, 581
  fumigatus 487, 489-492, 494, 571, 577, 581, 597, 600, 605
  niger 581
ATP
  -concentration 378
  -content 165–167, 171f., 174
average well color development (AWCD) 45, 387f., 393
Azotobacter 83, 86, 88, 90–93, 95

B

Bacillus sp. 25, 32, 34–40, 56, 58, 60f., 78f., 99, 104f., 107, 257, 330f., 334
  alcalophilus 35
  alvei 34
  anthracis 34, 36
  brevis 37
  cereus 34, 36
  circulans 56
  fusiformis 34, 36

macquariensis 34
medusa 34, 36
mycoides 34, 36–38
pabuli 34
polymyxa 34
stearothermophilus 56
subtilis 330
thuringiensis 34, 36
Bacteria 575, 589, 601

Bacteroides 38, 99, 104–107
bark 3–7, 11f. 14, 117, 132, 147, 190, 201, 222f., 259, 261
Bdellovibrio 99, 106f.
Beijerinckia indica 33
bioaerosols 572–574, 582, 584–586, 591–593, 595f., 598–600, 603f.
biochemical oxygen demand 3–6, 8–10, 14f., 155, 159, 162
biodegradability 265–268, 277f., 283, 285f., 308, 457, 459f., 470, 480
biodegradation 112, 115, 120, 216, 271–274, 277, 283, 285f., 288, 296, 300, 304, 308, 313, 315, 317f., 334, 350, 373, 375f., 379, 381, 618
biogas 156, 473
bioindicators 166, 168, 174
Biolog 43–45, 49–51, 100, 383f., 386, 395
biostabiliser 111–117
biowaste 457f., 473, 476-478, 495f., 499-501, 527, 551
Blumeria (Erysiphe) graminis f.sp. hordei 436, 441
Botrytis
allii 528
cinerea 222, 492, 546
Botulinum Neurotoxin 517
Brassica juncea 531, 534
Brochothrix thermosphacta 36
Buchnera aphidicola 33–38
bunker systems 217
butyric acid 179, 189, 197–200

C

C/N ratio 119–121, 124f., 129, 131, 135, 139, 156, 162, 166, 169, 173, 195, 221, 237, 242, 244, 247, 259, 336, 342, 345, 349f., 353, 358f., 374,376, 384f., 398, 400-402, 425-429, 448, 505, 511, 541
C/P ratio 428
cadmium 425, 457, 459, 488, 494
calcium 305
calorimetry 117f.
Campylobacter 597, 599, 606
Capsicum annuum 121, 540f., 548
carbohydrates 46, 49f.
carbon dioxide 119, 189, 205, 208, 289
-evolution 376, 379-381
carbon source utilization tests 100
carboxylic acids 46, 49f.
Carnobacterium divergens 35
funditum 35
piscicola 35
cation exchange capacity 121f., 132, 374
Caulobacter 104
cellulose 17, 54, 273, 275–277, 279–286, 332, 360, 362f., 366, 370f., 408, 410-412, 419, 427, 430
-acetate 273–284f., 295, 297
-degradation 410
Chaetomium elatum 20f.
globosum 20f.
chemical oxygen demand 3f., 6, 8f., 155, 159, 162
chlorsulfuron 309–311, 313–315, 318f.
chromium 425, 457, 488
Citrobacter freundii 33f., 36f.
Cladosporium 578, 581
Clostridium 25, 34–38, 40, 78f., 99, 104–107, 177, 186, 505-509, 514, 517, 519-521, 525f.
botulinum 517-525
difficile 38
herbivorans 34–36
oroticum 38
sticklandii 37
CO_2
-concentration 55, 195
-content 146
-emission 182
-evolution 137, 277, 284, 287, 289, 291, 295, 338, 342
-production 53, 55–57, 60, 111–113, 137, 287, 289, 291, 551, 554
-respiration 335, 337
cobalt 488

community level physiological profiles (CLPPs) 43, 45, 50 , 371, 283f., 394
compost maturity 142, 157, 164, 175, 189, 201, 324, 327, 335, 341-343, 346, 350, 352-355, 382-384, 394f., 406, 438
composting
backyard 133–135, 473, 527
large-scale 133f., 237
small-scale 117, 133–135
conductivity 121, 136, 224, 238, 266, 289, 337f., 345f., 349f., 376
container media 335-337, 339
cooling phase 148f., 152f., 189, 191, 194–196, 199
copper 487-494
Coriolus versicolor 120
corn 275
Corynebacterium 104
cotton 222f., 324, 327
Crabtree-effect 177, 187
cress 338, 341f., 347, 430, 435-437
Crypotococcus neoformans 597
Cryptosporidium 225
cucumber *(Cucumis sativus)* 119, 121, 223, 435f., 438f., 441f., 444, 540f., 546, 547f.
Cyanophora paradoxa 33, 37
Cytophaga 589, 330

D

Dano-System 111
denaturing gradient gel electrophoresis (DGGE) 66, 84
denitrification 143, 145, 152f., 237f., 241f., 247f., 252–257, 259f., 607, 615, 619, 621,
detoxification 443, 492
Dewar test 335f.
diarrhea 597
diether lipids 54, 58, 63
differential scanning calorimetry (DSC) 357, 359-364, 366-368, 370f.
diseases resistance 443
DNA extraction 312
drum-composters 217
dry matter 291, 302–304, 337, 348, 369, 447, 450, 460-467, 469f.
-content 203, 211f., 259
dry solids 179
dry weight 11, 45, 56, 122, 167, 250, 275, 290

E

E. coli 19, 25, 32–40, 87, 173, 187f., 278, 505, 512f., 515
EcoPlates 49, 383, 386
Eisenia
foetida 99f., 105, 108
lucens 100, 108
ELISA 542f., 548
endotoxins 575, 580, 582
Entamoeba histolytica 597
environmental impacts 445, 457f.
enzymatic
-activities 165–167, 171–173, 345f., 348, 350-353
-degradation 527, 532, 535
epifluorescence microscopy 585
Erisyphe
polygoni 222
graminis 222
Erwinia carotovora 36f.
Erythrobacter longus 33
Escherichia coli see *E.coli*
Eubacterium 78f.
EU Landfill Directive 473, 479
European Odour Unit 564
evaporation 116, 129, 136
Excellospora viridilutea 280
external heating 53, 55–57, 60

F

fatty acids 101f., 108, 165f.,190, 205, 551, 559, 619
fertiliser 165f., 215, 220, 505
flame atomic absorption spectrometry 101
FLASH bioluminescent test 375-379
Flavobacterium sp. 35–38, 78f., 330f.
balustinum 35f.
gleum 35f.
indologenes 35f.
indolthetic 35f.
lutescens 37
odoratum 36

Flexibacter 78f., 589
flow cytometry 593, 603, 605
fluorescence *in situ* hybridization (FISH) 585, 587, 589–592
fluorescence spectrometry 459
Fourier Transform Infrared Spectrophotometry (FTIR) 407, 409, 416, 418-421
friability 7, 11, 16
fulvic acids 408, 415, 426, 428f.
fungal inoculation 119, 287
Fusarium
conglutinans 222
culmorum 222
oxysporum 222, 528, 540

G

garden wastes 527-530
gas chromatography (GC) 7, 55, 62f., 98–102, 104, 107f., 266, 289, 448, 460
gas chromatography-mass spectrometry (GC-MS) 55, 62, 99–102, 104, 107f., 179 , 289–291, 293
gas emissions 247, 257, 260, 452
gastroenteritis 597
gel electrophoresis 26f., 41, 66, 81, 84, 89, 97f., 489
germination
-index 156, 161–163, 189, 325, 349f., 352f., 430, 531f., 535
-rate 529
-test 345, 347
Giardia lamblia 597
Gini coefficient 387, 389f., 394f.
Globicatella sanguis 35f., 78
Gloeophyllum trabeum 332
Gluconobacter sp. 186
grass 147
green bean 121
green residue 384, 390, 393
green wastes 147, 151, 222f., 457f.
gypsum 232

H

Haemophilus
paracuniculus 35–37
parasuis 35–37
Hafnia alvei 33f., 37
Hantavirus 599
HapII 87, 89, 91
Hazard Analysis Critical Control Points 525
health 175, 215, 225, 232, 260, 516f., 519f., 523, 525f., 572, 582, 585, 592, 595, 599f., 602, 604, 606
heat generation 49, 376f., 381
heat production 111–113, 117
heavy metals 99, 101, 103, 108, 305, 425-427, 430, 432, 457-461, 463f., 467f., 470-472, 488, 490, 492, 494, 501
hemicellulose 360, 363, 427, 430
herbicides 309–313, 315319
Heterodera schachtii 527, 529, 532, 537
HhaI 25, 27–37, 39, 65, 68f., 73–78
Hirschia baltica 36
horticultural wastes 119–121, 130f.
HPLC 191, 289, 291, 294, 312, 315, 459
humic acids 355, 407-409, 412, 414-420, 426, 428f., 508
humic extract 122, 127
Humicola
insolens 17, 19, 21
grisea 17, 19, 21
humidity 167, 426, 438, 517, 519
humification 346, 358, 366, 370f., 395, 397f., 401-403, 405f., 408, 414, 426, 428-431, 433f., 505, 510-512, 515
-degree 425, 430, 511
-index 397f., 401, 426, 430, 505
-parameters 374, 400, 405, 426, 428,
-rate 397, 400f., 430, 511
hybridization 586–589, 591–594, 603, 605f.
hydrogen peroxide 144, 227, 323-334
hydrogen sulfide 195, 338, 340, 342
hygienisation 517, 519, 527, 536
Hyphochytrium catenoides 20f.

I

immobilization 143, 145, 150, 152, 238
incineration 457, 474, 481-485
infectivity assays 542f.

in-vessel 65, 67, 70–72, 74, 78, 80, 217–219, 225, 300, 336
isoelectric focusing (IEF) 397f., 400-403, 405, 505, 507

K

Kinetoplastibacterium critidum 38
kitchen waste 147
Klebsiella 35–37

L

laboratory-scale 323f.
lactic acid 189, 197–200
Lactobacillus sp. 257
  maltaromicus 35
  thermophilus 34, 37
Lactococcus
  garvieae 36, 78
  lactis 36
landfill 373, 408-412, 414, 417-420
landfill simulation reactor 408
leachate 3–6, 8–11, 14f., 146, 154, 446, 450f., 454, 612, 618
  -loss 445, 452, 612, 618
leaves 147, 223
Legionella pneumophila 35, 37f., 599
Lepidium sativum 140f., 338, 345, 347, 349, 430
lignin 54, 120, 130, 145f., 151f., 190, 327, 332, 334, 357, 360, 410
ligninolytic enzymes 287f., 292, 295f.
lignocellulose 143
lime 147
Lindstrom's coefficients 393
Listeria sp. 225
  grayi 38
  ivanovii 34, 36
  murrayi 35f., 38
  seeligeri 34, 36
Lulworthia fucicola 20f.
Lycopersicon esculentum 529, 535

M

Maduromycetes 99, 105, 107
magnesium 305
Magnetospirillum 38
manure 4, 11 , 16, 18, 39, 49, 51, 63, 65–68, 70–74, 76f., 80–82, 101, 103, 105, 108, 436, 445-458, 460, 462-464, 471, 473, 505f., 515, 524, 528
  -cattle 99–101, 103, 106, 222f., 445
  -farm yard (FYM) 446-451, 453f., 457f., 460-464, 467-469
  -horse 528
  -sheep 505-510, 513, 515
Marinomonas vaga 37
mass
  -balance 189, 213, 607
  -losses 450, 454
  -retention time 155–157, 159–163
  -spectrometry 459
mechanical-biological pretreatment 407, 419
melon 539-542, 544, 546
melon necrotic spot virus (MNSV) 539-543
mesophilic phase 123, 147, 189
methane 557, 559, 607–609, 615, 619, 621
  -emissions 179, 257, 607, 621
Methylobacillus flagellatus 33f., 38
Methylococcus capsulatus 37, 102
Methylosinus 37f.
metsulfuron-methyl 309f., 313–319
microbial
  -activity 120, 136f., 152, 177f., 185f., 188, 210, 224f., 247, 252f., 255, 257, 324, 326f., 329, 345f., 357, 359, 374, 376-379, 381, 385, 393, 395, 446, 452, 471
  -aerosols 571, 593, 605f.
  -biomass 16, 43–45, 49–51, 53f., 56, 60–62, 108
  -respiration 136
Micromonospora aurantiaca 589–591
Microsporidium 597
mineralisation 49, 129f., 137, 143f., 221, 229, 237, 359, 406f., 426, 428, 446, 457, 461
  -rates 54
mixing 117, 121, 133f., 136–141, 190, 207, 219f., 259, 451, 487, 498, 558f., 585, 613, 618
moisture 6, 8, 11, 26, 44f., 49, 55, 62, 66, 82, 101, 112, 115–117, 121f., 129, 133, 135f., 147, 165, 169, 188, 212, 218, 220, 232–234, 247f., 252f., 255,

257, 323f., 330, 337, 345, 347-349, 358-361, 364, 373, 376, 379-381, 385, 399f., 507, 540, 595f., 598
-content 7, 46, 119, 121, 359, 373, 379, 381
moulds 438, 571, 575, 579, 582
MspI 25, 27–37, 39, 65, 68f., 73–78
mucorales 487
mushroom 17–19, 21–24, 32, 62, 188, 222, 224, 227–229, 231f., 235f., 325, 328f., 371
Mycobacterium tuberculosis 599
Mycoplasma sp. 34f., 37f., 78f.
capricolum 37f.
genitalium 34, 38
gallisepticum 34
pneumoniae 34, 38
putrefaciens 38
species strain PG 50 35, 37

N

N see nitrogen
NH_3 see ammonia
N_2O 248f., 259–261, 607–609, 612, 617–619, 621
-concentrations 250
-emission 445, 448f.
-production 252, 254f.
natural convection 614
Nicotiana tabacum 529f.
nitrate 144–146, 149, 152f., 221, 237, 273, 336, 338, 374f.
nitrification 6, 70, 139, 143–145, 151f., 238, 241, 247f., 250, 252,–257, 259f., 607, 615, 617, 619, 621
nitrite 139, 144–146, 149, 152, 237, 239, 241–244,
Nitrobacter winogradskyi 144
nitrogen 4, 85, 99, 101, 103, 105f., 122, 130f., 136f., 139, 142, 155, 157, 161–164, 169, 205, 207, 211, 213–215, 221, 224, 227f., 230, 237f., 245, 247f., 251, 256f., 259f., 273f., 280, 282, 289, 305f., 345, 347, 349f., 357, 359, 374f., 399f., 406-408, 417, 419f., 425, 427-429, 443, 445f., 451, 453, 455-459, 471, 600, 607, 609f., 612, 617–619, 621
-content 428, 443
-dissolved 136f.
-emission 257
-immobilisation 461, 470
-losses 445, 449f., 452-454
-mineralisation 461
-transformation 241
Nitrosolobus multiformis 35, 37, 79
Nitrosomonas europeana 144
Nocardia 99, 104f., 107
Nocardiopsis 590

O

O_2 see oxygen
Oceanospirillum sp. 79
commune 34–36, 38
odorous compounds 551, 562, 607
odour emission 553,–556, 558–564, 567–569
olfactometry 552, 555f., 558f., 564, 569
olive 165f., 168f., 173, 175, 324, 327, 334, 343, 382
-husks 165, 168f.
-pulp 324
Olpidium radicale 540
optical density 45, 48f., 267
organic acids 140, 155, 157–159, 163, 177–179, 189, 197, 200f., 527, 551–556, 559
organic carbon 85, 359, 400f., 404
-mineralisation 400, 404
organic matter 26, 28, 39, 49, 51, 54, 61, 65, 67, 70f., 85, 99, 101, 103, 105–107, 118, 121f., 125, 129–134, 136, 140, 157, 160, 177–179, 187, 189, 201, 203, 205–207, 211, 213, 215, 231, 238, 324, 326f., 345, 347, 349, 353, 355, 357-360, 363f., 366, 369-371, 373f., 386, 397-407, 410, 412, 419f., 511f., 514f., 532, 553, 595f., 607f., 611f., 618f.
-content 101, 463
-degradation 596, 607f., 611
-stability 397f., 405
orujo 324-332
Oscillatoria williamsii 38
oxidase activity 267
oxygen
-concentration 54, 195, 568, 615

-consumption 112, 179, 324
-content 339, 341
-depletion 335
-limitation 140
-supply 55, 62, 137, 139f., 145, 203, 207, 435, 444, 446, 551–553, 556–558, 560
-uptake 26, 28, 162f., 182f., 185, 206f., 210–212, 215, 224

P

packaging 265, 285, 299, 302, 304f., 308
Pantoea sp. 25, 27, 32–34, 37–39
agglomerens 33f.
ananas 33f., 37f.
herbicola pathovar gypso 33f., 37f.
pastazzo 358, 360, 362f., 366, 370, 384f., 388, 391, 393, 397-399
Pasteurella
haemolytica 33–37
species strain.Bisga 33–36
pasteurisation 224f.
pathogens 175, 225, 227f., 436-439, 441, 445f., 451, 454, 489, 498f., 505, 514, 527-530, 535-537, 539f., 546-548, 572, 595–597, 599f., 602f., 607f., 618, 620
-removal 210
PCR 17f., 21–28, 31, 33, 39–42, 60, 63, 65–67, 72, 81f., 84, 87–89, 95, 97f., 168, 174f., 278, 309, 317, 592f., 603f.
peat 112, 117, 220–222, 430, 438f.
Penicillium 330, 333, 581
pepper 124, 539-542, 544, 546, 548
pepper mild mottle virus (PMMV) 539-543, 545f.
permeability 265f., 268
perturbation 3, 15
Pestalotiopsis westerdijkii 274
pH 3f., 6–11, 16, 18, 26, 28f., 43–46, 49f., 54, 60, 67, 70, 85–90, 92, 103, 123, 136, 143–145, 147, 153, 167, 169, 177–180, 185, 189, 191, 194, 196, 205, 207, 212, 216, 219, 229, 237f., 241–244, 267f., 274, 278, 280–282, 289f., 294, 310, 325, 329, 337f., 345-349, 374, 376, 379, 383, 386, 399f., 402f., 405, 506f., 512, 521, 525, 530f., 542, 551, 553–557, 559f., 574f., 596
Phanerochaete chrysosporium 287f., 291, 295–287, 332
Phaseolus vulgaris 121
phenolic compounds 388, 394
phenols 130, 165f., 540
Phoma medicaginis 222
Phomopsis sclerotioides 528
phosphate 267, 305
phospholipid fatty acids (PLFAs) 7, 8, 11–16, 39, 50, 53–63, 100
-analysis 3, 16, 41, 51, 63
phosphorous 5, 99, 101, 457f., 461
Photobacterium leiognathi 38
phytopathogens 506, 540, 546
Phytophtora sp. 223
capsici 223
cinnamomi 223, 546
cryptogea 528
fragariae 223
infestans 223
nicotianae 223
phytotoxicity 137, 166, 190, 201, 324, 338, 342f., 345-347, 349f., 353, 371, 376, 382, 427, 430, 433, 438, 440
plant
-disease 222, 226, 435f., 441, 443, 528, 540, 547f.
-disease suppression 226, 435
-growth 341f.
-health 435, 444
-pathogens 435, 438, 506, 527f.
-residues 358, 360, 362f., 366, 370
Plasmodiophora brassicae 223, 527f., 531, 536f.
plastics 408, 410
Plesiomonas shigelloides 36f.
polychlorinated biphenyls (PCB) 287–295, 311f., 314, 457, 459f., 462, 464, 466, 468, 470f.
polycyclic aromatic hydrocarbons (PAH) 457, 459f., 462-465, 468, 470, 472
polyethylene 272, 299–302, 308
polyhydroxyalkanoates (PHA) 83, 85f., 88, 92, 94–96
polymers 388f., 394
polyolefins 268, 300
polyphenols 165
polypropylene 265f., 272, 299

porosity 173, 220, 227, 327, 335, 337, 345, 358, 375, 425, 427, 432, 452
Posidonia oceanica 425f., 432-434
potassium 86f., 99, 101, 305
potatoes 147
process dynamics 177
propionic acid 189, 197–200
protein degradation 144
Proteus
  mirabilis 33f., 36f.
  vulgaris 33, 35, 37f.
Pseudomonas sp. 25, 27, 32, 35–40, 99, 104–108, 145
  cepacia 36
  chlororaphis 38
  corrugata 35
  flavescens 35f.
  fluorescens 36, 309, 313
  paucimobilis 274
  phaseolicola 546
  putida 313, 319
  saccharophila 37
  testosteroni 37
Pythium
  aphanidermatum 223, 540
  graminicola 223
  irregulare 223, 546
  myriotylum 223
  ultimum 223, 435-439, 441f.

R

Raphanus sativus (radish) 156, 164
rDNA 17–22, 66, 68, 87, 95, 278–280, 592
  -16S 280, 592
  -18S 17, 21f., 24, 29
recirculation 147, 205, 208f., 219
recycling 120, 133, 189, 217, 237, 383, 384, 425, 473f., 478-485, 496, 498, 500, 502, 571, 595, 608
redox-potential 376
respiration tests 273
respirometry 323f., 333, 343
Rhizoctonia solani 223, 435f., 441f., 444, 528, 540, 546
Rhizopus sp. 330
  stolonifer 492, 494
Rhodobacter sphaeroides 33
Rhodococcus 104
Rhodopseudomonas sp. 38
  marina 38
Rhodospirillum centenum 35f., 38
Riemann's sum 387, 390f., 393
risk assessment 601
root
  -development 339
  -growth 430
Rottegrad 374, 379, 381
rRNA 24–33, 40–42, 60f., 63, 65–69, 73, 76–78, 81f., 97f.
  -16S 25–33, 40–42, 60f., 65–69, 73, 76–78, 81–84, 87, 95, 97f.
  -probing 585
RsaI 25, 27–36, 39, 65, 68f., 73–78
R*uminobacter amylophilus* 36

S

Saccharomonospora caesia 590
Saccharopolyspora sp. 581, 585, 590, 592, 600
  rectivirgula 581, 585, 590, 592
safety 225
Salmonella sp. 166, 168, 173f., 225, 245, 427, 430, 595, 597, 599f., 603–606
  lili 168, 173
  enteritidis 33f., 36f.
  senftenbergii 520
salt content 335
sanitation 120, 129, 134f., 141, 203, 505f., 509, 515, 539, 545
sawdust 275, 528
scanning electron microscopy (SEM) 267–272
Sclerotinia
  minor 223
  sclerotiorum 528, 546
Sclerotium
  cepivorum 528
  rolfsii 528
Scytalidium thermophilum 21, 24, 224
seaweed 425, 434
self-heating 147, 177f., 553, 571, 580f., 621
  -test 323f., 327, 329, 333, 347, 374
Serratia marcescens 36f.
Shannon diversity index 65, 67, 69, 71, 76

sheep litter see manure
Shigella 597
single-strand-conformation polymorphism (SSCP) 23, 42, 61, 63, 66, 81, 84, 98
sludge 51, 81, 99–101, 103, 106, 324, 334, 343, 354f., 357f., 360-363, 366, 370, 383-385, 388, 393-395, 397-399, 405, 408, 414, 425f., 430, 432, 457f., 461-463, 465, 467f., 470-472, 480, 487f., 493f., 569, 594–597, 605f., 621
-sewage 408, 457f., 487, 569f., 594–597, 600, 605f., 621
soil
-amender 155f., 161f.
-respiration 400
-steaming 438
Solvita tests 335, 337
spectrophotometry 427
Sphaerotheca fuliginea 223
Sphingobacterium 99, 104f., 107
Sphingomonas
paucimobilis 33
capsulata 33
Spirillum sp. 38, 99, 106f.
volutans 38
Spiroplasma
species strain DU-1 35, 37f.
species strain TG- 1 34f., 37f.
sporeformers 517
Staphylococcus
aerophilus 35
epidermidis 35
starch 54, 272f., 276, 282
straw 132, 147, 190, 203f., 224, 255
Streptococcus
anginosus 34f.
bovis 36, 78
salivarius 36
sanguis 36, 78
Streptomyces sp. 39, 99, 104f., 107, 330-332, 334, 505, 508, 516f., 581, 589f.
albus 581
coelicolor 330
olivaceus 581
thermoviolaceus subsp. apingens 589f.
Stromatinia gladioli 528
structure material 136–141
sulfonylurea 309, 318
Synechococcus 38
Synechocystis 35, 37
synthetic polymers 265, 272

T

tannins 130, 165f.
temperature gradient gel electrophoresis (TGGE) 66, 81, 84
tensile tests 266
terminal restriction fragment length polymorphisms (T-RFLP) 25–30, 31-37, 39f., 65–69, 71–78, 80
terpene 265f., 272
thallium 425
thermal stability 323, 329, 358, 360f., 363f., 370
-index 323, 329, 357, 364, 370
Thermoactinomyces sp. 104f., 581, 589f., 600
vulgaris 581, 589f.
thermogravimetry (TG) 357, 359f., 362-364, 366, 368, 370
Thermomonospora curvata 274, 286
thermophilic phase 123f., 147, 149, 152, 195
Thermus 41, 60, 62f.
tobacco mosaic virus (TMV) 527f., 546-548
tomato 222f., 527-529, 531f., 535f., 539f., 545-548
-seeds 527-529, 531f., 535f.
tomato spotted wilt virus (TSWV) 539-543, 545, 547f.
Torula thermophila 17, 19, 21
total carbon 122
total carbon content 305
total mass loss 445
total nitrogen 4, 85, 99, 136, 349, 459, 607
total organic carbon 136, 359, 363, 400, 427-429, 459, 474, 511
total oxygen uptake 155
toxicity 334f., 343, 355, 373-377, 379, 382, 430, 490, 494
toxin 517-519, 521-523
Trichoderma sp. 119–122, 129, 131
Tricholoma
albobrunneum 231
crassum 231, 236

equestre 231
flavovirens 231
georgii 231
matsutake 231
paneolum 231
terreum 231
Triticum aestivum 338
trypsin 272, 489
turning 119, 122–124, 129, 142, 146f., 190, 218, 220, 224, 245, 249f., 253, 257, 301, 428, 436, 447, 452f., 498, 561f., 565–569, 599f., 607–610, 613, 615, 618
-frequency 147

U

urea 275–277

V

Vagococcus salmoninarum 35
ventilation 5, 85
Venturia inaequalis 223
vermiculite 273, 275, 277, 282, 285
Verticillium dahliae 528
vessels 5, 67, 527-530, 532-536
Vibrio sp. 32–39, 78, 99f., 104f., 107
aestuarianus 33f.
anguillarum 33f., 36, 38
damsela 33f., 36, 38.
diazotrophicus 33, 35f., 38
fisheri 373, 379
gazogenes 33f., 36f.
marinus 34
metschnikovi 33
navarrensis 33f., 36
ordalii 33f., 36f.
splendidus 38
vulnificus 33f., 36
virus 530, 535f., 539f., 542-548
melon necrotic spot virus (MNSV) 539-543
pepper mild mottle virus (PMMV) 539-543, 545f.
tobacco mosaic virus (TMV) 527f., 546-548
tomato spotted wilt virus (TSWV) 539-543, 545, 547f.
volatile fatty acids 335
volatile organic acids 155, 159, 161f., 189f.
volatile solids 28, 157f., 474, 609f.

W

wastegas 204
wastewater 15, 99, 106f., 324, 343
water content 67, 70, 85, 134–136, 138, 143, 146, 152, 165, 179, 190, 200, 237f., 242, 244, 248, 259, 330, 364, 367, 370, 376, 379, 381
water-holding capacity 370, 376, 379f.
watering 119, 233, 301
Weeksella virosa 36
wheat 132, 147, 203f., 222, 338, 342f., 371, 382
windrow 16, 25f., 28–30, 41, 44, 62, 65, 67, 70–72, 74, 78, 80, 121, 217f., 224, 229, 245, 301, 436, 459, 553, 556, 558–564, 566–569, 595, 599
Wolbachia persica 38
Wolinella 99, 104–107
wood chips 121

X

Xanthomonas sp. 25
campestris 38, 68, 73, 77
oryzae 38
phasedi 38
Xylella fastidiosa 33f., 38

Y

yard trimmings 25f., 28–30, 39, 41
Yersinia sp. 597
enterocolitica 36f.
Yucca sp. 247f., 250, 253, 255, 259f.
schidigera 248

Druck:	Strauss Offsetdruck, Mörlenbach
Verarbeitung:	Schäffer, Grünstadt